SECOND EDITION

ESSENTIAL MATHEMATICS

Mary Kay Beavers
City College of San Francisco

HarperCollinsCollegePublishers

To my nieces: Stephanie, Courtney, Nellie, and Caitlin

Sponsoring Editor: Karin E. Wagner
Development Editor: PSI/Lori Toscano
Project Editor: Lisa A. De Mol
Design Administrator: Jess Schaal
Text Design: Lesiak/Crampton Design Inc.: Cynthia Crampton
Cover Design: Lesiak/Crampton Design Inc.: Cynthia Crampton
Cover Illustration: Tom James
Production Administrator: Randee Wire
Compositor: Interactive Composition Corporation
Printer and Binder: R.R. Donnelley & Sons Company
Cover Printer: R.R. Donnelley & Sons Company

Photos on page 7 are courtesy of Casio, Inc.

Essential Mathematics, Second Edition

Copyright © 1995 by HarperCollins College Publishers

All rights reserved. Printed in the United States of America. No part of this book may be used or reproduced in any manner whatsoever without written permission, except in the case of brief quotations embodied in critical articles and reviews. For information address HarperCollins College Publishers, 10 East 53rd Street, New York, NY 10022.

Library of Congress Cataloging-in-Publication Data
Beavers, Mary Kay, 1942-
    Essential mathematics / Mary Kay Beavers. -- 2nd ed.
      p.   cm.
    Includes index.
    ISBN 0-06-040602-X—0-06-501954-7 (annotated instructor's ed.)
    1. Arithmetic.   I. Title.
QA107.B4 1994
513--dc20
                                                                                         94-12269
                                                                                          CIP

3 4 5 6 7 8 9 10 11 - DOW - 03 02 01 00 99 98

CONTENTS

Preface   xi
Acknowledgments   xviii

1   *Whole Numbers: Place Value, Addition, and Subtraction*   1

    1.1   Place Value, Comparing, and Rounding   1
         Using the Calculator #1 *Types of Calculators*   7
    1.2   Addition and Subtraction   8
         Developing Number Sense #1 *Approximating by Rounding*   13
         Using the Calculator #2 *Addition and Subtraction*   14
    1.3   Language   14
    1.4   Applications   17
    Chapter 1 Summary   24
    Chapter 1 Review Exercises   25
    In Your Own Words   27
    Chapter 1 Practice Test   28

2   *Whole Numbers: Multiplication and Division*   29

    2.1   Multiplication   29
    2.2   Division   36
         Developing Number Sense #2 *Approximating by Rounding*   45
         Using the Calculator #3 *Multiplication and Division*   46
    2.3   The Missing Number in a Multiplication Statement, Multiples, Divisibility Rules, and Prime Factoring   47
         Using the Calculator #4 *Introduction to the Memory Keys*   57
    2.4   Language   58
    2.5   Applications   62
         Using the Calculator #5 *The M+ and M− Keys*   78
    Chapter 2 Summary   79
    Chapter 2 Review Exercises   80
    In Your Own Words   82
    Chapter 2 Practice Test   82
    Cumulative Review Exercises: Chapters 1–2   84

3   *An Introduction to Fractions*   85

    3.1   Picturing Fractions by Using Shaded Regions   85
    3.2   Reading and Writing Fractions   90
    3.3   Raising Fractions to Higher Terms and Converting Whole Numerals to Fractions   93
    3.4   Converting Improper Fractions to Mixed Numerals and Vice Versa   98
    3.5   Reducing Fractions to Lowest Terms   102
         Using the Calculator #6 *An Introduction to the $a\frac{b}{c}$ Key*   109
    3.6   A Fraction as Division, Ratio, and What Fraction of One Quantity Is Another Quantity   110
    Chapter 3 Summary   121
    Chapter 3 Review Exercises   122
    In Your Own Words   124
    Chapter 3 Practice Test   125

4   *Fractions: Multiplication and Division*   127
    4.1   Multiplication   127
          Developing Number Sense #3 *Multiplying by a Number Less Than, Equal to, or More Than 1*   134
    4.2   Division   135
    4.3   The Missing Number in a Multiplication Statement   139
    4.4   Language   142
    4.5   Applications   150
    Chapter 4 Summary   170
    Chapter 4 Review Exercises   170
    In Your Own Words   172
    Chapter 4 Practice Test   172

5   *Fractions: Addition and Subtraction*   174
    5.1   Adding with Like Denominators   174
    5.2   Finding the Least Common Multiple by Prime Factoring   178
    5.3   Adding with Unlike Denominators   183
    5.4   Comparing Fractions   190
          Developing Number Sense #4 *Comparing a Fraction to $\frac{1}{2}$*   194
    5.5   Subtracting   194
          Developing Number Sense #5 *Approximating by Rounding*   200
    5.6   Language   201
    5.7   Applications   205
          Using the Calculator #7 *More on the $a\frac{b}{c}$ Key*   217
    Chapter 5 Summary   218
    Chapter 5 Review Exercises   218
    In Your Own Words   220
    Chapter 5 Practice Test   221
    Cumulative Review Exercises: Chapters 1–5   222

6   *Decimals: Place Value, Addition, and Subtraction*   224
    6.1   An Introduction to Decimals   224
    6.2   Money and Check Writing   229
    6.3   Comparing Decimals   234
    6.4   Rounding Off, Rounding Up, and Truncating   237
    6.5   Addition and Subtraction   240
    6.6   Language   246
    6.7   Applications   249
    Chapter 6 Summary   256
    Chapter 6 Review Exercises   256
    In Your Own Words   258
    Chapter 6 Practice Test   259

7   *Decimals: Multiplication and Division*   260
    7.1   Multiplication   260
    7.2   Division   268
          Developing Number Sense #6 *Multiplying by 0.1, 0.01, and 0.001 Mentally*   279

	7.3 The Missing Number in a Multiplication Statement	280
	Developing Number Sense #7 Multiplying and Dividing by a Decimal Less Than, Equal to, or More Than 1	282
	7.4 Language	283
	7.5 Applications	287
	Chapter 7 Summary	304
	Chapter 7 Review Exercises	305
	In Your Own Words	307
	Chapter 7 Practice Test	307
	Cumulative Review Exercises: Chapters 1–7	308
8	***Problem Solving: Whole Numbers, Fractions, and Decimals***	310
	8.1 The Number Line	310
	8.2 Comparing Numbers	321
	8.3 Language	330
	Using the Calculator #8 Multiplying and Dividing Fractions as Decimals	334
	8.4 Applications: Choosing the Correct Operation	335
	Using the Calculator #9 Adding and Subtracting Fractions as Decimals	346
	8.5 Applications: More Than One Step	347
	8.6 Solving Proportions (Optional)	357
	Chapter 8 Summary	365
	Chapter 8 Review Exercises	366
	In Your Own Words	368
	Chapter 8 Practice Test	369
9	***Percent***	370
	9.1 Converting Percents to Decimals and Fractions	370
	9.2 Finding a Percent of a Number	379
	9.3 Converting Decimals and Fractions to Percent	386
	9.4 The Missing Number in a Percent Statement	395
	9.5 Solving Percent Applications by Translating	404
	9.6 Applications: Sales Tax, Commission, Interest, and Tips	411
	Developing Number Sense #8 Figuring the Restaurant Tip Mentally	422
	9.7 Applications: Percent Increase, Percent Decrease, and Markup	423
	Using the Calculator #10 The % Key	433
	9.8 Using Proportions to Solve Percent Problems (Optional)	434
	Chapter 9 Summary	449
	Chapter 9 Review Exercises	450
	In Your Own Words	452
	Chapter 9 Practice Test	452
	Cumulative Review Exercises: Chapters 1–9	453
10	***Measurement: English and Metric Systems***	455
	10.1 Length	455
	Developing Number Sense #9 Converting Metric Units by Using a Horizontal Display of the Units	466
	10.2 Area and Volume	468
	10.3 Weight, Mass, and Time	481
	10.4 Rates	488
	10.5 Measurements Involving More Than One Unit	493

Chapter 10 Summary ... 504
Chapter 10 Review Exercises ... 504
In Your Own Words ... 507
Chapter 10 Practice Test ... 507

11 *Introduction to Statistics* ... 509

11.1 Tables, Bar Graphs, and Line Graphs ... 509
11.2 Pictographs and Circle Graphs ... 517
11.3 Measures of Central Tendency: Mean, Median, and Mode ... 529
Chapter 11 Summary ... 537
Chapter 11 Review Exercises ... 538
In Your Own Words ... 542
Chapter 11 Practice Test ... 543
Cumulative Review Exercises: Chapters 1–11 ... 544

12 *Order of Operation, Signed Numbers, and Variables* ... 547

12.1 Order of Operation ... 547
    Using the Calculator #11 *Calculators and the Order of Operation* ... 558
    Using the Calculator #12 *The Parentheses Keys* ... 559
12.2 An Introduction to Signed Numbers ... 560
12.3 Addition of Signed Numbers ... 565
12.4 Subtraction of Signed Numbers ... 571
12.5 Multiplication and Division of Signed Numbers ... 582
12.6 Signed Fractions and Decimals ... 592
    Using the Calculator #13 *The +/− Key* ... 601
12.7 Variables, Simple Algebraic Expressions, and Formulas ... 602
Chapter 12 Summary ... 607
Chapter 12 Review Exercises ... 609
In Your Own Words ... 610
Chapter 12 Practice Test ... 611

13 *Exponents and Square Roots* ... 612

13.1 Positive Integer Exponents ... 612
    Using the Calculator #14 *The X^y Key* ... 616
13.2 Zero and Negative Exponents ... 617
13.3 Laws of Exponents ... 621
13.4 Powers of Ten and Scientific Notation ... 627
    Using the Calculator #15 *Handling Very Large and Very Small Numbers* ... 634
13.5 Square Roots ... 635
    Developing Number Sense #10 *Roughly Approximating a Square Root without Using a Calculator* ... 638
    Using the Calculator #16 *The $\sqrt{\ }$ Key* ... 640
Chapter 13 Summary ... 642
Chapter 13 Review Exercises ... 642
In Your Own Words ... 644
Chapter 13 Practice Test ... 644
Cumulative Review Exercises: Chapters 1–13 ... 645

14 *Introduction to Algebra* — 647
  14.1 Simplifying Basic Algebraic Expressions — 647
  14.2 Combining Like Terms — 652
  14.3 Solving Equations with the Variable Appearing Only on One Side — 664
  14.4 Applications — 675
  14.5 Solving Equations with the Variable Appearing on Both Sides — 691
  14.6 Basic Algebraic Fractions — 697
  14.7 Solving Proportions and Other Equations with Fractions — 712
  Chapter 14 Summary — 729
  Chapter 14 Review Exercises — 730
  In Your Own Words — 732
  Chapter 14 Practice Test — 733

15 *Geometry* — 734
  15.1 Points, Lines, and Angles — 734
  15.2 Triangles and Quadrilaterals — 748
  15.3 Perimeter and Area — 757
  15.4 Circles — 772
     Using the Calculator #17 *Working with the Number π* — 776
  Chapter 15 Summary — 784
  Chapter 15 Review Exercises — 785
  In Your Own Words — 787
  Chapter 15 Practice Test — 788
  Cumulative Review Exercises: Chapters 1–15 — 789

Appendix Tables
  1 Multiplication Facts — 791
  2 Primes Less Than 100 — 792
  3 Common Conversions: Fraction-Decimal-Percent — 792
  4 Perfect Squares — 792
  5 Choosing the Correct Operation — 793
  6 Percent Formulas — 794
  7 Units of Time — 794
  8 Units of Measurement: English System — 795
  9 Metric Prefixes — 795
  10 Units of Measurement: Metric System — 796
  11 English-Metric Conversions — 796
  12 Geometric Formulas — 797

Answers to Selected Exercises — 799

Index — 827

PREFACE

Essential Mathematics, second edition, is designed for a one- or two-semester basic mathematics course. The book provides thorough coverage of whole numbers, fractions, decimals, percent, and measurement, as well as an introduction to statistics and some elementary algebra and geometry topics. The second edition is not just a new edition; it has been completely rewritten to enhance the strengths of the first edition.

The text is designed so an average student can learn from it with minimal help from an instructor, tutor, or media such as videotapes and computer-assisted instruction. The following features demonstrate this.

- Numerous examples, detailed explanations, diagrams that promote visualizing concepts, a discovery writing style, and margin exercises all keep students actively involved in the learning process as they read the text.
- The order of topics is application driven, thus giving students the chance to use what they learn while progressing through the text. The numerous real-world application exercises are accompanied by a well-designed plan for teaching the student how to think about and solve applications.
- To support applications that require translation from words to symbols, great attention is paid throughout the book to language development.
- For motivation and enrichment, special topics called **Developing Number Sense** and **Using the Calculator** are dispersed strategically throughout the text.
- At the end of each chapter review exercise set there are discussion questions titled **In Your Own Words**. The content of these writing exercises is directly related to the material discussed in the chapter.

KEY CONTENT CHANGES
FOR THIS EDITION

Changes in this edition include the following.

- In Chapter 1 the "missing number" in an addition statement has been added, and the applications have been grouped in categories. The same alterations are found in Chapters 5 and 6 on addition and subtraction of fractions and decimals, and in Chapter 8 on problem solving.
- The multiplication and division of whole numbers have been combined into one chapter, Chapter 2. This allows students to learn whether to multiply or divide in an application involving whole numbers before facing the same issue with fractions. This chapter also now includes divisibility rules, prime numbers, and prime factoring. Rates are given more attention, without putting them in fractional form. Application problems with more than one step include problems that involve addition and subtraction as well as multiplication and division.
- Chapter 3, introducing fractions, has been completely reorganized. The English name for fractions has been placed earlier in the chapter, but all other language development and applications have been moved to the end of the chapter. The concept of ratio has been added.

- The applications of multiplying and dividing fractions in Section 4.5 have been categorized, and more examples and exercises have been provided. The language of ratio is discussed, and a discussion of rates as multipliers and ratios is included. This theme holds elsewhere in the book, as in Chapter 7 on multiplication and division of decimals and in Chapter 8 on problem solving.
- Chapter 5 on addition and subtraction of fractions has been altered slightly. Now Section 5.2 is devoted to prime factoring and finding the least common multiple by prime factoring. Two former separate sections—covering the addition of fractions with unlike denominators, common denominator given, and finding the least common multiple by listing multiples—have been omitted. To ready students for algebra, addition and subtraction problems are solved in a horizontal format as well as a vertical format.
- Two chapters on decimals have been combined into one Chapter 6, covering place value, addition, and subtraction.
- In Chapter 8 a new Section 8.3 called "Language" reviews all the language introduced so far in the text. This should aid students in solving the greater variety of applications presented by this point in the course. A new optional section on solving proportions has been added to the end of the chapter for those who wish to solve equivalent-rate problems by using proportions.
- Chapter 9 on percent has been rewritten extensively. The introductory sections on conversion have been compressed to help students see the connection between the different types of conversions. The section on applications, however, has been expanded from one to three sections. New material on markup based on the cost and on the selling price provides additional real-world applications. A new optional section on using proportions to solve percent problems concludes the chapter.
- Bar graphs and circle graphs are placed strategically throughout the text for the students to read and interpret. Before Chapter 11, which introduces statistics, the graphs are meant for motivation and enrichment rather than material to be mastered. This new statistics chapter is designed to help students learn not only how to read these graphs but also how to construct them. Measures of central tendency are also covered.
- The introduction to algebra has been expanded to two chapters, 12 and 14. More detailed coverage of signed numbers, including signed fractions and decimals, is provided. Further, the topics of combining like terms, solving equations with variables on both sides, and applications requiring equations with variables on both sides have been added.
- Chapter 13, exponents and square roots, now includes negative integer exponents and zero exponents as well as positive integer exponents.
- While basic geometry topics such as perimeter and area of rectangular regions, including composites of rectangular regions, are covered throughout the text, a new Chapter 15 offers a more extensive introduction to geometry.
- The feature called Try These Problems that provide quick feedback about their understanding of the concepts are now placed in the margin and are dispersed more frequently throughout the text than in the previous edition.

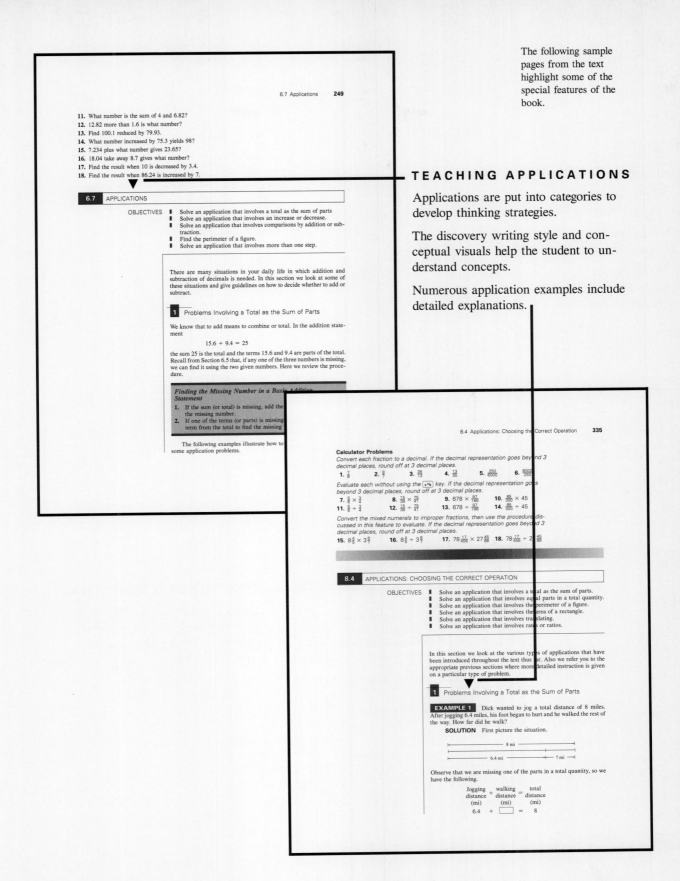

TEACHING APPLICATIONS

Applications are put into categories to develop thinking strategies.

The discovery writing style and conceptual visuals help the student to understand concepts.

Numerous application examples include detailed explanations.

LANGUAGE DEVELOPMENT

Separate sections concentrate on translating English to math symbols.

Language charts appear in application sections as well as language sections.

Numerous translation examples appear in both pure language settings and application settings.

TRY THESE PROBLEMS

Located in the margins, the *Try These Problems* keep the student actively involved in the learning process.

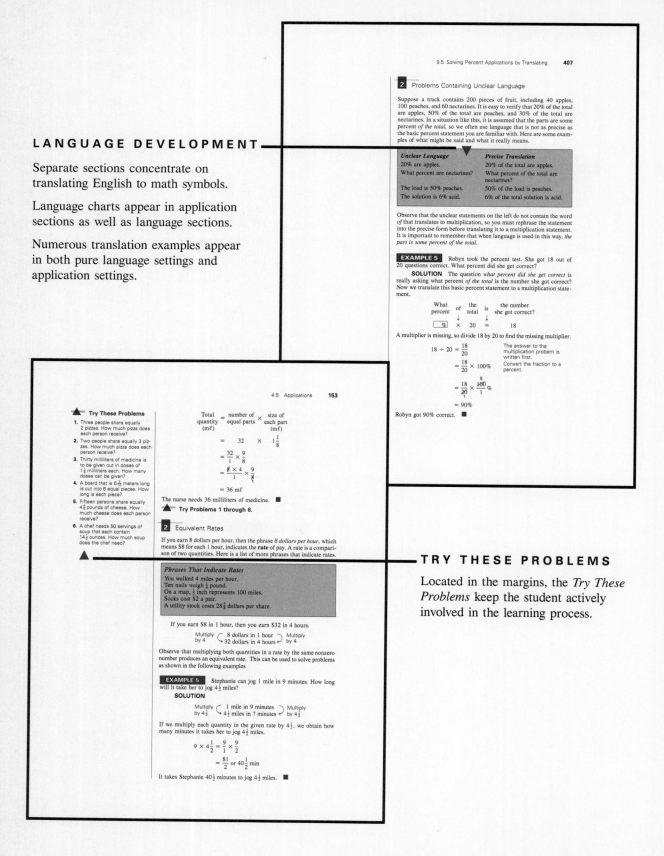

134 Chapter 4 Fractions: Multiplication and Division

25. $\frac{40}{340} \times 1700$ **26.** $12\frac{3}{4} \times \frac{4}{51}$ **27.** $\frac{30}{63} \times 9\frac{1}{4}$ **28.** $\frac{21}{50} \times \frac{5}{105}$
29. $6\frac{7}{8} \times \frac{40}{55}$ **30.** $\frac{74}{87} \times 13\frac{1}{2}$ **31.** $\frac{47}{52} \times \frac{21}{13}$ **32.** $\frac{28}{33} \times \frac{33}{93}$
33. $\frac{31}{97} \times \frac{104}{2}$ **34.** $\frac{7}{2000} \times \frac{3}{3000}$ **35.** $\frac{300}{569} \times 3\frac{1}{2}$ **36.** $16\frac{3}{9} \times 11\frac{7}{9}$
37. $\frac{2}{3} \times \frac{6}{9} \times \frac{12}{5}$ **38.** $\frac{1}{10} \times \frac{3}{8} \times \frac{27}{4}$ **39.** $2\frac{1}{2} \times \frac{3}{3} \times 3\frac{1}{4}$ **40.** $6\frac{2}{3} \times \frac{1}{2} \times 1\frac{1}{3}$
41. $\frac{3}{4} \times \frac{6}{9} \times \frac{2}{3} \times \frac{12}{5}$ **42.** $\frac{2}{3} \times \frac{3}{4} \times 8\frac{1}{2} \times 5\frac{1}{3}$

Solve.

43. Jeffrey bought $3\frac{3}{4}$ pounds of round steak at $4 a pound. How much did he pay for the steak?

44. Suppose you buy 60 shares of a stock selling for $16\frac{3}{4}$ dollars per share. What is the total cost?

45. Susann bought $4\frac{2}{3}$ yards of fabric that sells for $9 a yard. How much did she pay for the fabric?

46. A desk top measures $4\frac{1}{2}$ meters long and $2\frac{1}{4}$ meters wide. Find the area of the desk top.

▼

DEVELOPING NUMBER SENSE #3

MULTIPLYING BY A NUMBER LESS THAN, EQUAL TO, OR MORE THAN 1

Here we show 200 multiplied by 3, 2, and $1\frac{1}{2}$. How do the products compare with 200 and why?

$3 \times 200 = 600 \quad 2 \times 200 = 400$
$1\frac{1}{2} \times 200 = 300$

Observe that the results—600, 400, and 300—are each more than 200 because the numbers 3, 2, and $1\frac{1}{2}$ are each more than 1.
If we multiply 200 by 1, we obtain 200.

$1 \times 200 = 200$

In general, any number multiplied by 1 is that number.
What happens if we multiply 200 by a fraction that is less than 1? Here we look at a few examples.

$\frac{3}{4} \times 200 = 150 \quad \frac{1}{2} \times 200 = 100$
$\frac{1}{4} \times 200 = 50$

Observe that the results 150, 100, and 50 are each less than 200 because the numbers $\frac{3}{4}, \frac{1}{2}$, and $\frac{1}{4}$ are each less than 1.

Number Sense Problems
Without computing, decide whether the result is less than, equal to, or more than 84.

1. 5×84 **2.** $2\frac{1}{4} \times 84$ **3.** $\frac{2}{3} \times 84$ **4.** 1×84 **5.** $\frac{4}{10} \times 84$ **6.** $\frac{3}{2} \times 84$

Without computing, decide whether the result is less than, equal to, or more than 682.

7. $\frac{1}{3} \times 682$ **8.** $\frac{7}{3} \times 682$ **9.** 12×682 **10.** $13\frac{3}{4} \times 682$
11. $\frac{8}{8} \times 682$ **12.** $\frac{13}{14} \times 682$

Without computing, decide whether the result is less than, equal to, or more than $25\frac{1}{2}$.

13. $\frac{1}{2} \times 25\frac{1}{2}$ **14.** $5\frac{3}{8} \times 25\frac{1}{2}$ **15.** $\frac{9}{8} \times 25\frac{1}{2}$ **16.** $1 \times 25\frac{1}{2}$

NUMBER SENSE FEATURES

The *Developing Number Sense* features help the student gain approximating skills and mental computational skills.

Number Sense Problems within the *Developing Number Sense* feature allow students to practice the skills introduced in that feature.

9.7 Applications: Percent Increase, Percent Decrease, and Markup   **433**

USING THE CALCULATOR #10

THE % KEY

Most calculators have a [%] key. On a scientific calculator this key may be a second function key so that you must enter the [2ndF], [INV], or [Shift] key before entering the [%] key.
The percent key can be used to find a percent of a number without first changing the percent to a fraction or decimal. Here we show an example.

To Compute   50% of 60
Enter   60 [×] 50 [%]
Result   30.

On most calculators you must enter the percent after the number you are taking the percent of. Also, if the percent key is a second function key on your calculator, you must enter the [2ndF], [INV], or [Shift] key before entering the [%] key.
The percent key can also be used to find the final amount after a quantity has been increased by a certain percent. The procedure is not the same on all calculators. Here we show how to compute 40 increased by 25% on two different calculators.
ON SOME CALCULATORS

To Compute   40 + (25% of 40)
Enter   40 [×] 25 [%] [+]
Result   50

ON OTHER CALCULATORS

To Compute   40 + (25% of 40)
Enter   40 [+] 25 [%]
Result   50

If the percent key is a second function key on your calculator, you must enter the [2ndF], [INV], or [Shift] key before entering the [%] key.
You can also use the percent key to find the final amount after a quantity has been decreased by a certain percent. The procedure is exactly like the above except enter the [−] key instead of the [+] key. Here we show how to compute 40 decreased by 25% on two different calculators.
ON SOME CALCULATORS

To Compute   40 − (25% of 40)
Enter   40 [×] 25 [%] [−]
Result   30

ON SOME CALCULATORS

To Compute   40 − (25% of 40)
Enter   40 [−] 25 [%]
Result   30

If the percent key is a second function key on your calculator, you must enter the [2ndF], [INV], or [Shift] key before entering the [%] key.
The percent key applies to only very special situations. You will not be able to use it to solve all percent problems.

Calculator Problems
Evaluate each by using the [%] key.

1. 900 × 40%   **2.** 75 × 3.6%   **3.** 85% of 560   **4.** 125% of $9.80
5. 600, increased by 10%   **6.** 600, decreased by 10%
7. $45.76, increased by 75%   **8.** $45.76, decreased by 75%
9. 850, increased by 150%   **10.** 850, decreased by 0.06%

Evaluate each on the calculator, but without using the [%] key. If the decimal representation of the answer goes beyond 3 decimal places, then round off at 3 decimal places.

11. Find $83\frac{1}{3}$% of 600.   **12.** 81 is 15% of what number?
13. Find 90, increased by $3\frac{5}{9}$%.   **14.** Find 120, decreased by $66\frac{2}{3}$%.
15. What percent of 15.2 is 4.75?   **16.** $\frac{1}{4}$% of what number is 16?

CALCULATOR FEATURES

The *Using the Calculator* features, dispersed strategically throughout the text, actually show how to enter the keys on both basic and scientific calculators.

Calculator Problems within the calculator feature allow the student to practice the key-punching skills introduced in that feature.

xiii

EXERCISES Approximately 5400 exercises are included throughout the text to provide students ample practice with concepts. The following types of exercise sets are provided.

- Margin exercises called **Try These Problems**
- Section **Exercises** at the end of each section
- **Review Exercises** at the end of each chapter
- Writing exercises called **In Your Own Words** located at the end of the Review Exercises in each chapter
- A **Practice Test** at the end of each chapter
- **Cumulative Review Exercises** at the end of Chapters 2, 5, 7, 9, 11, 13, and 15
- **Number Sense Problems** at the end of each *Developing Number Sense* feature
- **Calculator Problems** in all *Using the Calculator* features except the first one

Answers to section exercises are placed at the end of the text, rather than immediately following each exercise set. Only answers to the odd-numbered section exercises are given in the text. Answers to even-numbered section exercises are given in the *Instructor's Resource Manual*. This manual also includes a set of supplementary exercises for each section of the text, so instructors may provide extra practice for students or create quizzes and tests.

TO THE STUDENT I have written this book especially for you. I know that you probably have never before been able to read a math book, but studying from this book will be a different experience! You can read and learn from this book on your own with minimal help from an instructor, tutor, or other educational aids such as videotapes and computer-assisted instruction.

Of course, some topics are difficult to present in a written text or are difficult for you to learn. I know that among these topics are solving applications and working with fractions. For additional help you can use the videotapes made by the author or the computer-assisted instruction developed by the publisher. Ask your instructor about the availability of these materials on your campus.

TO THE INSTRUCTOR **Using This Book in a Self-paced Format** The readability of the first edition of this text enabled it to be very successful in programs where students are not given a lecture. Because of improvements in teaching how to think about applications, the second edition should be even more readable. Since this text has been designed to work for mastery learning, two chapters on whole numbers, three chapters on fractions, and two chapters on decimals are included to ensure that students master the beginning material. More chapters on this material also allow early and frequent attention to a wide variety of applications.

Using This Book in a Lecture Format A well-designed lecture is a more powerful medium than a written text. Naturally, if you design your lectures to suit your goals and style rather than to be exactly like the text, you will progress at a much faster pace. You also may avoid reviewing some of the material that is repeated throughout the book. (The repetition is necessary for students in an individualized program who may have placed out of earlier chapters.) Even if you do not lecture with the same level of detail as the text, having such detail present in the book is helpful for students if they miss a lecture, fall behind, or need an alternative presentation.

SUPPLEMENT PACKAGE TO ACCOMPANY THE TEXT

FOR THE INSTRUCTOR

The *Annotated Instructor's Edition* includes answers to all of the problems in the text in place on the page. The only exception is that answers to In Your Own Words problems are not given, since answers are discussions and will vary.

The *Instructor's Resource Manual,* written by the author, contains a supplementary set of exercises for each section of the text and an extensive test bank. There are six forms of chapter tests for each chapter, including four open-response and two multiple-choice forms. Two additional open-response forms for Chapters 1 through 9 are also available. In addition, there are some cumulative chapter tests covering the whole number chapters, the fraction chapters, and the decimal chapters. Two forms of a multiple-choice chapter placement test and several possible final examinations are provided. The answers to the supplementary exercises, the tests, and the even-numbered section exercises in the text are also included.

Transparency masters, prepared by the author, include language charts, graphs, rule lists, and other material that will aid the lecturer in presenting key information efficiently.

The *HarperCollins Test Generator/Editor for Mathematics with QuizMaster* is available in IBM and Macintosh versions. Fully networkable, the Test Generator enables instructors to select questions by objective, section, or chapter, or to use a ready-made test for each chapter. The Editor enables instructors to edit any preexisting data or to easily create their own questions. The software is algorithm driven, allowing the instructor to regenerate constants while maintaining problem type, providing a nearly unlimited number of available test or quiz items in multiple-choice and/or open-response formats for one or more text forms. The system features printed graphics and accurate mathematics symbols. QuizMaster enables instructors to create tests and quizzes using the Test Generator/Editor and save them to disk so students can take the test or quiz on a stand-alone computer or network. QuizMaster then grades the test or quiz and allows the instructor to create reports on individual students or entire classes. CLAST and TASP versions of this package are also available for IBM and Mac machines.

FOR THE STUDENT

A *Student's Solutions Manual* includes complete worked-out solutions to the odd-numbered problems in the text except for In Your Own Words problems since answers will vary. Use ISBN 0-06-501955-5 to order.

A *videotape series* has been developed by the author to accompany *Essential Mathematics,* second edition. These video lessons offer a detailed presentation of solving some applications and working with other trouble areas. The examples on the tapes are different from those in the text, offering a real extension of the text instruction. A cross-reference sheet explains which video lessons correspond to which topics within a section in the text. Also, at the end of each video lesson, the viewer is given the text location of additional relevant examples and an appropriate assignment in the text to provide exercises for practice.

The *Interactive Tutorial Software with Management System* is available in IBM—both DOS and Windows applications—and Macintosh versions and is fully networkable. As with the Test Generator/Editor, this software is algorithm driven, which automatically regenerates constants so a student will not see the numbers repeat in a problem type if he or she revisits any particular section. The tutorial is objective-based, self-paced, and provides unlimited opportunities to review lessons and to practice problem solving. If students give a wrong answer, they can request to see the problem worked out and get a textbook page reference. The program is menu driven for ease of use, and on-screen help can be obtained at any time with a single keystroke. Students' scores are automatically recorded and can be printed for a permanent record. The optional Management System lets instructors record student scores on disk and print diagnostic reports for individual students or classes. TASP and CLAST versions of this tutorial are also available for both IBM and Macintosh machines. This software may also be purchased by students for home use. IBM DOS version ISBN 0–06–502474–5; Windows version ISBN 0–06–502498–7; Macintosh version ISBN 0–06–502473–7.

ACKNOWLEDGMENTS

I am greatly indebted to the following instructors who offered many helpful criticisms and suggestions in their review of the manuscript.

Jacquelyn P. Briley, Guilford Technical Community College

Carol Curtis, Fresno City College

Diane Daniels, Mississippi State University

Mike Farrell, Carl Sandburg College

John Harp, Yakima Valley Community College

Linda L. Hermann, Parkland College

Clista H. LeGrand, Greenville Technical College

Debra Madrid-Doyle, Santa Fe Community College

Carl J. Mancuso, William Paterson College

Gael Mericle, Mankato State University

Philip J. Metz, Passaic County Community College

Cindy Moody, City College of San Francisco

Robert P. Peters, Cuyahoga Community College

Debbie Polito, North Harris College

John Rossi, Daytona Beach Community College

Marcine E. Smith, Lansing Community College

Patricia Stanley, Ball State University

Lenore A. Vest, Lower Columbia College

Joseph Williams, Essex County College

It is a pleasure for me to acknowledge the many people who have contributed to the development of this book. The staff at HarperCollins has given me excellent support: Karin Wagner, acquisitions editor; Linda Youngman and Lori Toscano, developmental editors; and Lisa De Mol, project editor. I would like to thank them for their expertise, patience, and cooperation during the development and production processes.

The following colleagues of mine at City College of San Francisco have used the first edition of this book in their basic mathematics classes over the past ten years: Glenn Aguiar, Mary Allen, Diedre Baker, Charles Burke, Gonzalo Castro-Gonzalez, Frank Cerrato, Jim Cribbs, Mark Davis, Guy De Primo, Lydia Gans, Greg Gregory, Tom Haggerty, Lee Kaiser, Jimmy Kan, Bill King, Ted Lee, Gary Ling, Leon Luey, Keith McAllister, Cindy Moody, Jerry Morell, Dennis Piontkowski, David Ross, Fred Safier, Gus Srouji, Tom Swartz, Bie Han Tan, Paul Tang, Tom Walsh, and Ulf Wostner. Their support and suggestions have been invaluable in rewriting this book. Special thanks go to Cindy Moody and Dennis Piontkowski for helping me with the *Instructor's Resource Manual*.

I wish to thank my mother, Nell Beavers, who independently worked every problem in the manuscript to check my answers. I also appreciate the contributions of Jan Jesensky, Dudley Brooks, Nell Beavers, and Mary Allen in checking the answers further during the production process.

Mary Kay Beavers

CHAPTER 1

Whole Numbers: Place Value, Addition, and Subtraction

1.1 PLACE VALUE, COMPARING, AND ROUNDING

OBJECTIVES
- Specify the digit in a numeral that has a certain place value.
- Specify the place value of a certain digit in a numeral.
- Compare whole numbers.
- Write the English name for a whole numeral.
- Write the numeral when given the English name of a whole number.
- Round off a whole number to a given place value.

1 Numbers Less Than 10,000

Betty has 26 dollars and Susan has 206 dollars. You know that Susan has more money than Betty because 206 is more than 26.

$$206 = 2 \text{ hundreds} + 0 \text{ tens} + 6 \text{ ones}$$
$$= 200 + 0 + 6$$

but

$$26 = 2 \text{ tens} + 6 \text{ ones}$$
$$= 20 + 6$$

In the above display, the symbol = (read *equals*) is used to indicate that symbols are representing quantities with the same value. The symbol + (read *plus*) for addition means to combine or total. The numeral 206 is read "two hundred six." The numeral 26 is read "twenty-six." When we write out the numeral in words, we are giving the English name of the numeral.

Each whole number is represented by a symbol called a **whole numeral.** Each numeral consists of one or more of the symbols 0, 1, 2, 3, 4, 5, 6, 7, 8, and 9. Each of these symbols is called a **digit.**

2 Chapter 1 Whole Numbers: Place Value, Addition, and Subtraction

 Try These Problems

1. Use the numeral 8034 to answer each question.
   a. What digit is in the hundreds place?
   b. What digit is in the tens place?
   c. What is the place value of the digit 4?
   d. What is the place value of the digit 8?

Write the numeral for each English name.
2. Four hundred thirteen
3. Nine thousand eighty-two

Write the English name for each numeral.
4. 635
5. 3407

Each digit in a whole numeral has a **place value.** For example, in the illustration that follows, the place value for each digit in the numeral 7136 is indicated.

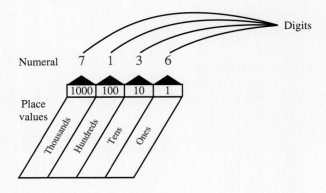

Thus, the numeral 7136 means

$$7 \text{ thousands} + 1 \text{ hundred} + 3 \text{ tens} + 6 \text{ ones}$$
$$= 7000 + 100 + 30 + 6$$
$$= 7136$$

The numeral 7136 is read "seven thousand one hundred thirty-six," which is the English name of the numeral.

The place values one (1), ten (10), a hundred (100), and a thousand (1000) are related in the following way.

$$10 = 10 \text{ ones}$$
$$100 = 10 \text{ tens}$$
$$1000 = 10 \text{ hundreds}$$

 Try Problems 1 through 5.

2 Larger Numbers

The following diagram gives the place values for the numeral 312,036,070,408, which is read "three hundred twelve billion, thirty-six million, seventy thousand, four hundred eight." The commas, placed every third digit starting from the right, serve as an aid in reading the numeral.

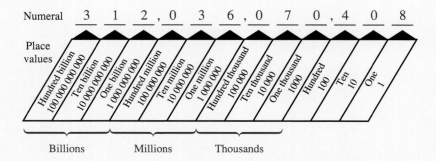

1.1 Place Value, Comparing, and Rounding

 Try These Problems

6. Use the numeral 17,904,083,265 to answer each question.
   a. What digit is in the ten millions place?
   b. What digit is in the billions place?
   c. What is the place value of the digit 9?
   d. What is the place value of the digit 8?

Write the numeral for each English name.

7. Seventy-three thousand, nine hundred two

8. Fourteen billion, two hundred four million, eight hundred

In Problems 9 and 10, write the English name for each numeral.

9. 107,034

10. 230,018,276

11. Which is less, 54 or 36?

12. List from smallest to largest: 50; 45; 400; 396; 2100

The place values are related in the following way.

$10 = 10$ ones
$100 = 10$ tens
$1000 = 10$ hundreds
$10,000 = 10$ thousands
$100,000 = 10$ ten thousands
$1,000,000 = 10$ hundred thousands

Observe that as you move to the left from digit to digit in a whole numeral, each place value is 10 of the previous place values. This pattern continues indefinitely.

The following chart gives the English names for several whole numerals.

Whole Numeral	English Name
3200	Three thousand two hundred *or* thirty-two hundred
26,367	Twenty-six thousand, three hundred sixty-seven
704,900	Seven hundred four thousand, nine hundred
85,040,001	Eighty-five million, forty thousand, one
3,052,000,000	Three billion, fifty-two million

Observe that the word *and* should never be used when reading a whole numeral.

Sometimes we use a combination of a numeral and the English language in referring to a whole number. This is especially true for very large whole numbers. For example,

$$3 \text{ million} = 3,000,000$$
$$56 \text{ million} = 56,000,000$$
$$728 \text{ billion} = 728,000,000,000$$

 Try Problems 6 through 10.

3 Comparing Numbers

The first few **whole numbers** are listed here.

$$0, 1, 2, 3, 4, 5, 6, 7, 8, 9, 10, 11, 12, 13, 14$$

This list continues indefinitely. In this list, a number to the right is considered more than (or larger than) a number to the left. Also, a number to the left is considered less than (or smaller than) a number to the right. For example,

$$13 \text{ is more than } 4$$

and

$$16 \text{ is less than } 25.$$

 Try Problems 11 and 12.

Try These Problems

13. Is the number 587 closer to 500 or 600?
14. Is the number 12,368 closer to 12,000 or 13,000?
15. Round off the number 5489 to the thousands place.
16. Round off the number 67,683 to the nearest hundred.

Round off the given number to the given place value.

17. 7653; hundreds
18. 385; tens
19. 146,518; thousands
20. 81,932; hundreds

4 Rounding Off Numbers

Suppose you purchase a car for $12,348. In speaking with your friends you might say the car costs $12,000. The price $12,000 is an approximation of the actual cost $12,348. We use the symbol $\approx$ or $\doteq$ to mean *approximately equal to*. More precisely, the number 12,000 is the result of **rounding off** the number 12,348 to the nearest thousand. Observe that 12,348 is closer to 12,000 than it is to 13,000 because the digit after the thousands place is 3, which is less than 5.

$$12{,}348$$
— The digit after the thousands place is less than 5.

$$\approx 12{,}000$$

Here we have rounded to the nearest thousand.

We can round off numbers to any specified place value. Here are more examples.

EXAMPLE 1 Round off 34,792 to the nearest hundred.

SOLUTION First locate the hundreds place in the number 34,792.

$$34{,}792$$
— Hundreds place

We want to know whether this number is closer to 34,700 or 34,800. To see this, look at the digit that is to the right of the hundreds digit. It is a 9. Because it is more than 5, we conclude that the number is closer to 34,800 than 34,700.

$$34{,}792$$
— The digit after the hundreds place is more than 5.

$$\approx 34{,}800 \quad 34{,}792 \text{ is closer to } 34{,}800 \text{ than } 34{,}700.$$ ■

 Try Problems 13 through 16.

EXAMPLE 2 Round off 235 to the tens place.

SOLUTION First locate the tens place in the number 235.

$$235$$
— Tens place

We want to know whether this number is closer to 230 or 240. The number 235 is exactly halfway between 230 and 240. In this case, we agree to use the higher number for the approximation.

$$235$$
— The digit after the tens place is equal to 5.

$$\approx 240 \quad \text{We agree to increase the tens digit from 3 to 4 because the next digit is equal to 5.}$$ ■

 Try Problems 17 through 20.

Try These Problems

Round off the given number to the given place value.
21. 392; tens
22. 65,983; hundreds
23. 345,713; thousands
24. 345,713; ten thousands
25. 4,671,324; ten thousands
26. 4,671,324; thousands
27. 108,972; hundreds
28. 8,993,504; hundred thousands

EXAMPLE 3 Round off 129,657 to the nearest thousand.

SOLUTION First locate the thousands place in the number 129,657.

129,657
   └── Thousands place

We want to know if this number is closer to 129,000 or 130,000. To see this, look at the digit to the right of the thousands digit 9. Because it is a 6, which is 5 or more, the number 129,657 is closer to 130,000 than 129,000.

129,657
   └── The digit after the thousands place is more than 5.
≈ 130,000   129,657 is closer to 130,000 than 129,000.

 Try Problems 21 and 22.

Finally we display a chart that gives several numbers rounded off to specified place values.

ROUNDING OFF NUMBERS TO GIVEN PLACE VALUES

Number	Round Off to the Tens Place	Round Off to the Hundreds Place	Round Off to the Thousands Place	Round Off to the Ten-thousands Place
28	30	0	0	0
569	570	600	1000	0
5325	5330	5300	5000	10,000
13,973	13,970	14,000	14,000	10,000
760,542	760,540	760,500	761,000	760,000
3,800,400	3,800,400	3,800,400	3,800,000	3,800,000

Study the chart carefully enough to make the following observations.

- The digit in the round-off place is unchanged if the first digit to the right of the round-off place is less than 5.
- The digit in the round-off place is increased by 1 if the first digit to the right of the round-off place is 5 or more.
- The digits to the right of the round-off place are replaced by zeros.
- The digits to the left of the round-off place change only when the digit in the round-off place is a 9 and the first digit to its right is 5 or more.

 Try Problems 23 through 28.

Answers to Try These Problems
1. a. 0 b. 3 c. 1 (one) d. 1000 (thousand)
2. 413   3. 9082   4. Six hundred thirty-five
5. Three thousand four hundred seven
6. a. 0 b. 7 c. 100,000,000 (hundred million)
   d. 10,000 (ten thousand)   7. 73,902   8. 14,204,000,800
9. One hundred seven thousand, thirty-four
10. Two hundred thirty million, eighteen thousand, two hundred seventy-six
11. 36   12. 45; 50; 396; 400; 2100   13. 600   14. 12,000
15. 5000   16. 67,700   17. 7700   18. 390   19. 147,000
20. 81,900   21. 390   22. 66,000   23. 346,000   24. 350,000
25. 4,670,000   26. 4,671,000   27. 109,000   28. 9,000,000

EXERCISES 1.1

Use the numeral 78,534 to answer Exercises 1 through 4.
1. What digit is in the hundreds place?
2. What digit is in the ten thousands place?
3. What is the place value of the digit 3?
4. What is the place value of the digit 8?

Use the numeral 52,730,149,869 to answer Exercises 5 through 8.
5. What digit is in the ten billions place?
6. What digit is in the hundred thousands place?
7. What is the place value of the digit 7?
8. What is the place value of the digit 4?
9. Which is larger, 7983 or 10,000?
10. List from smallest to largest: 700; 698; 1000
11. Aron is 63 inches tall. Her sister Ann is 36 inches tall. Who is taller?
12. Last week a head of lettuce was selling for 79 cents. This week a head of lettuce sells for 92 cents. Did the price increase or decrease?
13. Candy has $500 in her checking account. She writes a check for $465. Is there enough money in her account to cover the check?
14. A real estate company purchased an apartment building in Chicago for $1,700,000. A similar building in Houston cost $1,098,000. Which building was more expensive?

Write a whole numeral for each English name.
15. Fourteen
16. Ninety-eight
17. Two hundred eighty
18. Three thousand three hundred eleven
19. Twenty-six hundred
20. One hundred seven thousand, eight
21. Fifteen million, five hundred thousand
22. 75 billion

23. The population of San Francisco one year ago was seven hundred thousand people. Write the population as a whole numeral.
24. The profit last year for an architect's firm was ten thousand dollars. Write the profit as a whole numeral.

Write the English name for each numeral.

25. 47
26. 450
27. 806
28. 3542
29. 5500
30. 78,360
31. 103,085
32. 4,008,701
33. 5,070,000,000

34. A car salesperson sold 52 cars last year. Write the English name for the number of cars sold.
35. As an accountant, Sylvia will earn $35,000 this year. Write the English name for Sylvia's yearly salary.

Round off the given number to the given place value.

36. 74; tens
37. 863; hundreds
38. 86,317; thousands
39. 6829; tens
40. 8654; hundreds
41. 4532; thousands
42. 764,938; ten thousands
43. 5,087,063; ten thousands
44. 6,357,876; hundred thousands
45. 45,490,785; millions

USING THE CALCULATOR #1

TYPES OF CALCULATORS

There are many types of calculators. Four of the commonly used ones are basic, scientific, business, and graphing calculators. In this text, we discuss only the basic and scientific calculators that use an algebraic entry logic. We do not discuss calculators that use reverse Polish notation, nor do we discuss calculators that are programmable. Here we picture an example of the types that will be discussed in this text.

Scientific calculators are not much more expensive than basic calculators and can do a lot more. If you have not already purchased a calculator, and you plan to take more math or science, you should consider buying a scientific calculator. Your mathematics instructor can help you decide which kind of calculator is best for you.

Basic Calculator

Scientific Calculator

1.2 ADDITION AND SUBTRACTION

OBJECTIVES
- Add two or more whole numbers.
- Solve an application involving addition of whole numbers.
- Subtract two whole numbers.
- Solve an application involving subtraction of whole numbers.
- Find the missing number in an addition statement.

1 Addition

The Jackson family spent $20 on movie tickets and $8 on popcorn and drinks. To find the total amount spent, we **add** $20 and $8. The symbol + (read *plus*) is used to indicate addition.

$$\$20 + \$8 = \$28$$

The total amount spent was $28.

To add means to combine or total. It does not matter in what order the numbers are written. For example,

$$3 + 8 = 11 \quad \text{and} \quad 8 + 3 = 11$$

When adding more than two numbers, it does not matter which two numbers are added first. For example,

$$4 + 3 + 5 = 7 + 5 = 12$$

and

$$4 + 3 + 5 = 4 + 8 = 12$$

and

$$4 + 3 + 5 = 9 + 3 = 12$$

In an addition problem, the numbers being added are called **terms** and the answer to the addition problem is called the **sum**. For example, in the statement

$$\underbrace{8 + 7 + 5}_{\text{Terms}} = \underbrace{20}_{\text{Sum}}$$

the numbers 8, 7, and 5 are terms and the number 20 is the sum.

When adding larger numbers, care must be taken to add digits with like place values. The procedure is illustrated in the following examples.

EXAMPLE 1 Add: $314 + 2 + 53$

SOLUTION Arrange the numbers vertically so that digits with like place values form a column.

Try These Problems

Add.
1. $7 + 6 + 4 + 3$
2. $13 + 241 + 2105$
3. $2564 + 368$
4. $\;7382$
   $\;\;675$
   $+ \;3982$

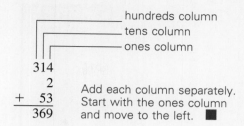

Add each column separately. Start with the ones column and move to the left. ■

EXAMPLE 2   Add: $285 + 372$

SOLUTION   Arrange the numbers vertically so that digits with like place value form a column.

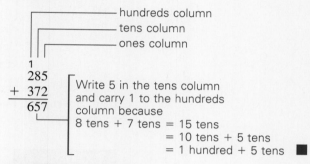

Write 5 in the tens column and carry 1 to the hundreds column because
8 tens + 7 tens = 15 tens
$= 10$ tens + 5 tens
$= 1$ hundred + 5 tens ■

 Try Problems 1 through 4.

 Subtraction

If you had \$12 and spent \$7, to find the amount you have left you would **subtract** 7 from 12. The symbol $-$ (read *minus*) is used to indicate subtraction.

$$\$12 - \$7 = \$5$$

The amount you have left is \$5.

To subtract means to take away. Subtraction is the reverse of addition. For example,

$$12 - 7 = 5 \quad \text{because} \quad 5 + 7 = 12$$

It does matter in which order you write down a subtraction problem. For example,

$$12 - 7 = 5$$

but

$$7 - 12 = -5$$

The symbol -5 represents the number "negative five." This number, -5, does not equal 5. We study negative numbers in Chapter 12 of this book.

When subtracting larger numbers, be careful to subtract digits with like place values.

Try These Problems

Subtract.
5. 62 − 19
6. 5837 − 215
7. 865 − 280
8. 13,425 − 7817

EXAMPLE 3 Subtract: 7284 − 31

SOLUTION Arrange the numbers vertically so that the larger number is on the top and digits with like place values form a column.

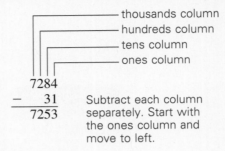

```
  7284
−   31
  7253
```
Subtract each column separately. Start with the ones column and move to left.

CHECK
```
  7253
+   31
  7284
```
∎

In any one column, when the bottom digit is larger than the top digit, **borrowing** must be used. The following examples illustrate how to use borrowing.

EXAMPLE 4 Subtract: 3475 − 752

SOLUTION
```
  ²3̸¹475
−   752
   2723
```
In the hundreds column, 7 is larger than 4, so we must borrow.
Borrow 1 thousand from 3 thousands. Mark out 3 and write 2.
The 1 thousand we borrow is put in the hundreds column to make 14 hundreds because
 1 thousand + 4 hundreds
= 10 hundreds + 4 hundreds
= 14 hundreds

CHECK
```
  ¹752
+ 2723
  3475
```
∎

 Try Problems 5 through 8.

Sometimes there is a 0 digit in the column where you want to borrow. The following examples illustrate how to handle this situation.

EXAMPLE 5 Subtract: 503 − 37

SOLUTION
```
   ⁴ ⁹
   5̸ 0̸ ¹3
 −   3 7
     4 6 6
```
In the ones column, 7 is larger than 3, so we must borrow. We cannot borrow 1 ten from 0 tens, so we borrow 1 ten from 50 tens. Mark out 50 and write 49. The 1 ten we borrow is put in the ones column to make 13 ones.

1.2 Addition and Subtraction

▲ **Try These Problems**

Subtract.
9. 604 − 78
10. 700 − 216
11. 50,302 − 7485
12. 91,000 − 6040

Find the missing number.
13. 7 + ☐ = 15
14. ☐ = 25 + 16
15. 136 = ☐ + 57
16. 807 + ☐ = 2004

CHECK

```
   11
  466
+  37
  503  ■
```

EXAMPLE 6 Subtract: 600,274 − 3521

SOLUTION

```
    5 9 9
    6 0 0¹2 7 4
  −     3 5 2 1
    5 9 6 7 5 3
```

In the hundreds column 5 is larger than 2, so we must borrow. We borrow 1 thousand from 600 thousands. Mark out 600 and write 599. The 1 thousand we borrow is put in the hundreds column to make 12 hundreds.

CHECK

```
     111
    3521
+ 596753
  600274  ■
```

▲ **Try Problems 9 through 12.**

3 Finding the Missing Number in an Addition Statement

The statement 6 + 3 = 9 is an addition statement. If any one of the three numbers is omitted, you want to know how to find the missing number. For example, three problems could arise:

1. 6 + 3 = ☐   Here the answer to the addition problem is missing, so we *add* to find the missing number.

    The missing number is 6 + 3 = 9.

2. ☐ + 3 = 9   Here one of the terms is missing, so we *subtract* to find the missing number.

    The missing number is 9 − 3 = 6.

3. 6 + ☐ = 9   Here one of the terms is missing, so we *subtract* to find the missing number.

    The missing number is 9 − 6 = 3.

▲ **Try Problems 13 through 16.**

▲ **Answers to Try These Problems**

1. 20   2. 2359   3. 2932   4. 12,039   5. 43   6. 5622
7. 585   8. 5608   9. 526   10. 484   11. 42,817   12. 84,960
13. 8   14. 41   15. 79   16. 1197

EXERCISES 1.2

Add.

1. $8 + 36$
2. $46 + 5$
3. $5 + 7 + 8$
4. $3 + 9 + 7$
5. $6 + 7 + 5 + 3$
6. $6 + 5 + 8 + 9$

7. 792
   36
   + 19

8. 3099
   8764
   + 278

9. $9674 + 20{,}911 + 450$
10. $10{,}346 + 9807 + 123$
11. $8 + 317 + 86$
12. $370{,}816 + 2115 + 92 + 348$

13. 124,577
    80,675
    986,501
    + 3,845

14. 66,177
    9,323
    50,689
    + 7,135

15. Mr. Maguire purchases these items at the supermarket.
    Beef $18
    Potatoes $ 3
    Dog food $ 5
    Beer $ 6
    What is the total cost?

16. What is the total length of this shaft?

    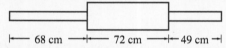

    |— 68 cm —|— 72 cm —|— 49 cm —|

17. Ms. Nichols had $915 in her checking account. She deposited $97 in the account. What is her balance now?

18. Ross weighed 176 pounds before going on his vacation. He gained 14 pounds on his trip. What did Ross weigh after his vacation?

Subtract. Check your answers by adding.

19. $43 - 5$
20. $759 - 42$
21. $638 - 46$
22. 2367
    - 124

23. 4238
    - 1705

24. 72,378
    - 6,084

25. 3107
    - 783

26. 6324
    - 583

27. 64,013
    - 4,208

28. 6004
    - 36

29. 50,026
    - 1,451

30. $700 - 8$

31. $300 - 67$
32. $5000 - 407$
33. $6000 - 1038$
34. $8000 - 350$

35. $70{,}000 - 300$
36. 200,300
    - 7,502

37. 520,036
    - 71,578

38. 308,010
    - 73,524

39. 400,000
    - 70,293

40. 1030
    - 987

41. 41,020
    - 9,382

42. 310,015
    - 73,046

43. 610,003
    - 685

44. 100,000
    - 73,600

45. Carlos had $50 in his pocket. He spent $12 on lunch for himself and his friend Dorothy. How much money does he have left?

46. Don's goal is to make $3500 during the summer. By the end of July he had earned $2793. How much does he have to earn to reach his goal?

47. At the start of a trip the odometer of a car read 52,037. At the end of the trip the odometer read 54,100. If the odometer measures distance in miles, how far has the car traveled on this trip?

48. Tony weighs 239 pounds. Mike weighs 251 pounds. Who weighs more and by how much?

Find the missing number.

49. ▢ = 9 + 7
50. 8 + ▢ = 11
51. ▢ + 13 = 32
52. 47 = 29 + ▢
53. 135 + ▢ = 300
54. 269 + 87 = ▢
55. 1305 + ▢ = 2002
56. 30,001 = ▢ + 7983

DEVELOPING NUMBER SENSE #1

APPROXIMATING BY ROUNDING

Whether you are computing by using a calculator or paper and pencil, it is always possible to make a careless mistake that causes your answer to be not only wrong, but ridiculously wrong. It is a good idea to approximate the answer to check whether your answer is reasonable. In some cases, an approximation may be all that you are interested in. One method for approximating is to round the numbers being computed to make them easier to work with, then compute the rounded numbers.

For example, suppose that we want to find an approximation for the following addition problem.

$$1280 + 788 + 69$$

We could round off each number to the nearest hundred, then add.

$1280 \approx 1300$
$788 \approx 800$
$69 \approx 100$

Each number is rounded off to the hundreds place.

Adding 1300, 800, and 100, we obtain 2200 for the approximation.

There are no set rules on how to round the numbers to approximate. We could have done the rounding in many ways. Here we show the approximation obtained in another way. This time we round off so that each rounded number has only one nonzero digit. In this way the numbers are extremely easy to work with.

$1280 \approx 1000$
$788 \approx 800$
$69 \approx 70$

Each number is rounded so that the result has only one nonzero digit.

Adding 1000, 800, and 70, we obtain 1870 for the approximation. The actual sum is 2137.

Number Sense Problems

Approximate each answer by rounding off each number to the nearest hundred.

1. 794 + 315
2. 643 + 86 + 128
3. 2370 − 1496
4. 1682 − 379

Approximate each answer by rounding off each number to the nearest thousand.

5. 26,321 − 9718
6. 1379 + 899 + 18,956 + 7917

Approximate each answer by rounding off each number so that it contains only one nonzero digit.

7. 534 + 79 + 3598
8. 768,000 + 5743 + 52,113
9. 214,987 − 5634

USING THE CALCULATOR #2

ADDITION AND SUBTRACTION

Suppose you want to add 24 and 48 on the calculator. Whether you have a basic calculator or scientific calculator, here are the steps you take.

To Compute 24 + 48
Enter 24 [+] 48 [=]
Result 72

Suppose you want to add several numbers. For example, you want to compute 20 + 75 + 35 + 20.

To Compute 20 + 75 + 35 + 20
Enter 20 [+] 75 [+] 35 [+] 20 [=]
Result 150

If you enter [+] at the end instead of [=], the same result is obtained. Also, because addition can be done in any order, it does not matter in what order you enter the terms.

When computing a subtraction problem, you must be careful that the numbers are entered in the correct order. Here is an example.

To Compute 120 − 80
Enter 120 [−] 80 [=]
Result 40

Observe that the numbers are entered in the same order as they are written down when using the subtraction symbol −. If you enter the numbers 120 and 80 in the reverse order, you obtain the negative number −40, because −40 is the result of subtracting 120 from 80. Negative numbers are discussed in Chapter 12.

Calculator Problems

Add or subtract as indicated.

1. 7 + 5
2. 28 + 57
3. 479 + 8796
4. 23,986 + 567,982
5. 8 + 2 + 12
6. 17 + 9 + 47 + 185
7. 807 − 469
8. 6003 − 747
9. Subtract 18 from 72.
10. Subtract 537 from 7000.

1.3 LANGUAGE

OBJECTIVES
- Translate an English statement to an addition or subtraction statement using math symbols.
- Solve problems by translating.

1 Translating

The math symbols that we use often have many translations to the English language. Knowing these translations can help in solving application problems. The following chart gives the many English translations for the symbol =.

Math Symbol	English
=	equals
	is equal to
	is the same as
	is
	was
	represents
	gives
	makes
	yields
	will be
	were
	are

The addition statement, $7 + 5 = 12$, is written using math symbols. Some of the ways to read this in English are given in the following chart.

Math Symbols	English
$7 + 5 = 12$	7 plus 5 equals 12.
	The *sum* of 7 and 5 is 12.
	5 more than 7 is equal to 12.
	7 *increased by* 5 yields 12.
	12 represents 7 added to 5.
	7 and 5 is 12.

Because $7 + 5$ equals $5 + 7$, the order that you read the terms 5 and 7 makes no difference; however, when reading the subtraction statement $12 - 5 = 7$, be sure to read 12 and 5 in the correct order, because $12 - 5$ does not equal $5 - 12$.

$$12 - 5 = 7 \quad \text{but} \quad 5 - 12 = -7$$

The number 7 does not equal -7. The number -7 is a negative number. This book covers negative numbers in Chapter 12. The following chart gives several correct ways to read $12 - 5 = 7$.

Math Symbols	English
$12 - 5 = 7$	12 minus 5 equals 7.
	12 take away 5 is 7.
	12 diminished by 5 gives 7.
	12 reduced by 5 yields 7.
	12 subtract 5 is 7.
	5 subtracted from 12 equals 7.
	5 less than 12 is 7.
	The *difference* between 12 and 5 is 7.
	7 represents 12 *decreased by* 5.

Try These Problems

Translate to an addition or subtraction statement using math symbols.

1. The sum of 25 and 17 is 42.
2. The difference between 300 and 112 equals 188.
3. 73 represents 61 increased by 12.
4. 60 minus 18 yields 42.
5. 28 less than 75 is equal to 47.
6. 15 subtracted from 120 gives 105.

Solve.

7. The sum of 79 and 895 yields what number?
8. What number plus 74 equals 3001?
9. 295 represents 81 more than what number?
10. 487 less than 700 is what number?

Note that when using the phrases *subtracted from* or *less than* for the symbol −, we read the numbers 12 and 5 in reverse order. That is,

$$5 \text{ subtracted from } 12 \quad \text{means} \quad 12 - 5$$

and

$$5 \text{ less than } 12 \quad \text{means} \quad 12 - 5$$

Try Problems 1 through 6.

2 Solving Problems by Translating

The following examples illustrate how we can use translating to solve problems.

EXAMPLE 1 What number increased by 18 yields 50?

SOLUTION Translate the question to math symbols.

What number increased by 18 yields 50?
$$\square \quad + \quad 18 \quad = \quad 50$$

The missing number is one of the terms in an addition statement, so we subtract the given term 18 from the sum 50 to find the missing term.

$$50 - 18 = 32$$

The answer is 32. ∎

EXAMPLE 2 What number represents 205 decreased by 38?

SOLUTION The question translates to math symbols as follows.

What number represents 205 decreased by 38?
$$\square \quad = \quad 205 \quad - \quad 38$$

The missing number is the answer to the subtraction problem, $205 - 38$.

$$\begin{array}{r} \overset{1}{2}\overset{9}{\cancel{0}}{}^1 5 \\ -3\,8 \\ \hline 1\,6\,7 \end{array}$$

The answer is 167. ∎

Try Problems 7 through 10.

Answers to Try These Problems

1. $25 + 17 = 42$ 2. $300 - 112 = 188$ 3. $73 = 61 + 12$
4. $60 - 18 = 42$ 5. $75 - 28 = 47$ 6. $120 - 15 = 105$
7. 974 8. 2927 9. 214 10. 213

EXERCISES 1.3

Translate each English statement to an addition or subtraction statement using math symbols.

1. The sum of 18 and 17 is 35.
2. 63 increased by 9 equals 72.
3. 54 represents 63 decreased by 9.
4. The difference between 202 and 78 is 124.
5. 50 more than 360 is equal to 410.
6. 910 take away 14 yields 896.
7. 1023 plus 978 is 2101.
8. 308 subtracted from 4200 equals 3892.
9. 3 is 4 less than 7.
10. 28 represents the sum of 20 and 8.

Solve.

11. 407 minus 82 equals what number?
12. Find the difference between 566 and 463.
13. Find the result when 2000 is increased by 1802.
14. Find the result when 2000 is decreased by 1802.
15. What number increased by 956 yields 3050?
16. 619 subtracted from 1023 gives what number?
17. Find the sum of 5090 and 936.
18. What number is 13,825 plus 86?
19. What number represents 7000 decreased by 64?
20. 30,000 minus 7800 yields what number?

1.4 APPLICATIONS

OBJECTIVES
- Solve applications that involve a total as the sum of parts.
- Find the perimeter of a figure.
- Solve applications that involve increases and decreases.
- Solve applications that involve comparing quantities by addition or subtraction.

There are many situations in your daily life in which addition and subtraction of numbers are needed. In this section we look at some of these situations and give guidelines on how to decide whether to add or subtract.

1 Problems Involving a Total as the Sum of Parts

You know that to add means to combine or total. In the addition statement

$$7 + 9 = 16$$

the sum 16 is the total and the terms 7 and 9 are parts of the total. Recall from Section 1.2 that if any one of the three numbers is missing, you can find the missing number by using the two given numbers. Here we review the procedure.

> *Finding the Missing Number in a Basic Addition Statement*
> 1. If the sum (or total) is missing, add the terms (or parts) to find the missing number.
> 2. If one of the terms (or parts) is missing, subtract the given term from the total to find the missing term.

The following examples illustrate how to use this concept to solve some application problems.

EXAMPLE 1 The marked price of a car is $9350. After adding the tax, the total price is $9911. How much is the tax?

SOLUTION The problem indicates the following.

$$\begin{array}{c}\text{marked} \\ \text{price}\end{array} + \text{tax} = \text{total price}$$

$$\$9350 + \boxed{} = \$9911$$

The tax is a missing term in an addition statement so we subtract.

$$\begin{array}{r} 9\,\overset{8}{\cancel{9}}{}^{1}1\,1 \\ -\ 9\,3\,5\,0 \\ \hline 5\,6\,1 \end{array} \begin{array}{l} \text{— total price} \\ \text{— marked price} \\ \text{— tax}\end{array}$$

The tax is $561. ∎

EXAMPLE 2 Find the perimeter of this triangle.

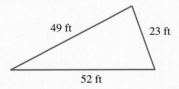

SOLUTION The **perimeter** of a figure means the total distance around it, so we have the following.

$$\text{Perimeter} = 49 + 23 + 52$$

$$\boxed{} = 49 + 23 + 52$$

Try These Problems

1. Barbara bought some chicken breasts and a ham at the supermarket. The total bill was $23. The ham cost $14. What was the cost of the chicken?

2. What is the perimeter of this rectangle?

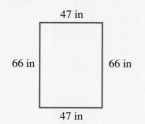

3. How long is the bottom portion of this pineapple?

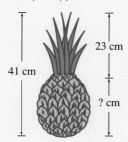

4. A swimming pool holds 5000 gallons of water. The swimming pool contains 1870 gallons of water now. How many more gallons are needed to fill the pool?

We add to find the total.

$$\begin{array}{r} \overset{1}{}49 \\ 23 \\ +\ 52 \\ \hline 124 \end{array}$$

The perimeter is 124 feet. ∎

EXAMPLE 3 The seat of a chair is 49 centimeters from the floor. The total height of the chair is 108 centimeters. How long is the back of the chair?

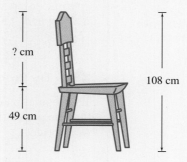

SOLUTION The total height of the chair is found by adding the height of the seat and the length of the back.

$$\begin{array}{c} \text{Height of} \\ \text{seat} \end{array} + \begin{array}{c} \text{length of} \\ \text{back} \end{array} = \begin{array}{c} \text{total} \\ \text{height} \end{array}$$

$$49 \quad + \quad \boxed{} \quad = \quad 108$$

We are missing a part of the total so we subtract.

$$\begin{array}{r} {}^{0}\cancel{1}\,{}^{9}\cancel{0}\,{}^{1}8 \\ -\ \ 4\ 9 \\ \hline 5\ 9 \end{array}$$ — total height
— parts

The length of the back is 59 centimeters. ∎

Try Problems 1 through 4.

2 Problems Involving Increases and Decreases

The price of a videocassette recorder decreased from $450 to $300. The amount that the price went down is called the **decrease.** To find the decrease we subtract.

$$\begin{aligned} \text{Decrease} &= \begin{array}{c}\text{higher}\\\text{price}\end{array} - \begin{array}{c}\text{lower}\\\text{price}\end{array} \\ &= \$450 - \$300 \\ &= \$150 \end{aligned}$$

The decrease in price is $150.

Try These Problems

5. The price of apples decreased from 75 cents a pound to 59 cents a pound. What is the decrease in price?
6. An elephant's weight went from 2000 pounds to 1872 pounds. Find the loss in weight.
7. Last year the profit for a small company was $186,000. This year the profit was $200,000. What was the increase in profit?
8. You receive a salary raise. Your original salary was $2460 monthly and now you make $2700 monthly. What was your raise for one month?

Other words can be used to mean decrease. Here we list some situations where a value has gone down and give the corresponding words that can be used to indicate the amount that the value went down.

Situation	Words Indicating the Amount the Value Went Down
There is a sale on sheets. An $80 value is now selling for $60.	The *discount* is $20. The *savings* is $20. The *reduction* in price is $20. The *decrease* in price is $20.
A man lost weight. His weight dropped from 280 pounds to 220 pounds.	The *loss* in weight is 60 pounds. The *reduction* in weight is 60 pounds. The *decrease* is 60 pounds.

 Try Problems 5 and 6.

The population of a town increased from 13,200 to 15,500. The amount that the population went up is called the **increase.** To find the increase we subtract.

$$\text{Increase} = \frac{\text{larger}}{\text{population}} - \frac{\text{smaller}}{\text{population}}$$
$$= 15{,}500 - 13{,}200$$
$$= 2300$$

The increase in population is 2300.

Other words can be used to mean increase. Here we list some situations where a value has gone up and give the corresponding words that can be used to indicate the amount that the value went up.

Situation	Words Indicating the Amount the Value Went Up
Your annual salary increased. It went from $40,000 to $42,000.	Your *raise* is $2000. The *gain* in salary is $2000. The *increase* in salary is $2000.
A woman gained weight. Her weight went from 120 pounds to 130 pounds.	Her *gain* in weight is 10 pounds. The *increase* is 10 pounds.

 Try Problems 7 and 8.

 3 Problems Involving Comparisons

Two unequal numbers, such as 80 and 95, can be compared using addition and subtraction. Here is a list of several addition and subtraction statements involving 80 and 95 and the corresponding comparison statements in English.

Try These Problems

9. Tom drove 407 miles on Monday and 600 miles on Tuesday. How much farther did he drive on Tuesday?

10. It took Mike 16 hours to paint the exterior of a house. It took him 8 hours more to paint a larger house. How long did it take Mike to paint the larger house?

11. Cathy's annual salary is less than Steve's annual salary by $2900. If Cathy's salary is $38,000, find Steve's salary.

12. The populations of two towns in Florida differ by 38,000 people. If the larger population is 156,000, what is the smaller population?

Comparing Two Numbers

Math Symbols	English
80 + 15 = 95	15 more than 80 is 95.
95 − 15 = 80	15 less than 95 is 80.
95 − 80 = 15	95 and 80 differ by 15.
	The difference between 95 and 80 is 15.
	80 less than 95 is 15.

Now we look at examples of problems involving comparisons.

EXAMPLE 4 The height of a eucalyptus tree is 102 feet and the height of a pine tree is only 75 feet. How much taller is the eucalyptus tree?

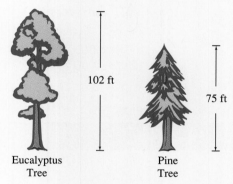

SOLUTION To find how much taller the eucalyptus is, we subtract.

```
  0 9
  1 0¹2      height of the eucalyptus
−   7 5      height of the pine
─────────
    2 7      the difference
```

The eucalyptus tree is 27 feet taller than the pine tree. ■

EXAMPLE 5 Bob jogged 49 miles last week. Susan jogged 13 miles more than Bob. How many miles did Susan jog?

SOLUTION Because Susan jogged 13 miles more than Bob, we have

$$\text{Susan's distance} = \text{Bob's distance} + 13$$
$$= 49 + 13$$
$$= 62$$

Susan jogged 62 miles. ■

Try Problems 9 through 12.

Now we summarize the material presented in this section by giving guidelines that will help you decide whether to add or subtract.

Situations Requiring Addition or Subtraction

Operation	Situation
+	1. You are looking for the total or whole.
	2. You are looking for the result when a quantity has been increased.
−	1. You are looking for one of the parts in a total or whole.
	2. You are looking for the result when a quantity has been decreased.
	3. You are looking for how much larger one quantity is than another.
	4. You are looking for how much smaller one quantity is than another.

Answers to Try These Problems

1. $9   2. 226 in   3. 18 cm   4. 3130 gal   5. 16¢   6. 128 lb
7. $14,000   8. $240   9. 193 mi   10. 24 hr   11. $40,900
12. 118,000

EXERCISES 1.4

Solve.

1. Ms. Ridens purchased these items at the department store.

    Sweater $45
    Skirt $76
    Shoes $89

    What was the total cost of the three items?

2. Irene had $792 in a savings account. She withdrew $165. How much is left in the account?

3. The population of a town in Florida increased from 18,003 to 25,000 over a five-year period. What was the increase in population?

4. How far is it from A to B?

   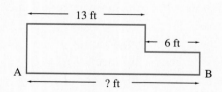

5. Frank operates a printing machine. He printed 40,000 pages on Wednesday. On Thursday he printed 8000 fewer pages. How many pages did he print on Thursday?

6. A certain calculator used to cost $25. Over the past two years the price has decreased by $7. What is the cost of the calculator now?

7. The profit for a company was $1,200,000 last year. This year the profit is $1,098,000. Did the profit increase or decrease, and by how much?

8. An airplane travels 580 miles per hour when the air is still. A 32 mile-per-hour wind is blowing in the direction that the plane is flying. How fast is the plane traveling?

9. Loula is selling her house. Yesterday the Allen family offered her $70,860. Today the Burton family offered her $71,200. Who offered more money and how much more?

10. Judy lost 15 pounds on her vacation in Hawaii. If she weighed 112 pounds after returning, how much did she weigh before the vacation?

11. How tall is the base of this lamp?

12. During the night the temperature dropped from 40°F to 26°F. What was the decrease in temperature?

13. Find the perimeter of this rectangle.

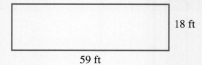

14. Find the perimeter of this floor.

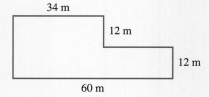

15. Antonio's college tuition for one year was $1400 less than Mohammed's college tuition. If Antonio's tuition was $5100, what was Mohammed's tuition?

16. Rita received a raise. Her salary was $589 a week before the raise. Now she makes $600 a week. What was her raise?

Mr. Yano owns a service station and several apartment buildings. The bar graph gives Mr. Yano's annual salary for each of five years. Use the graph to answer Exercises 17 through 22.

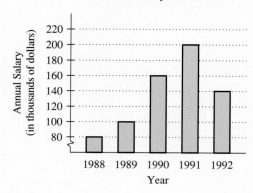

17. What was Mr. Yano's salary in 1991?
18. What was Mr. Yano's salary in 1992?
19. How much more did he earn in 1990 than in 1988?
20. How much more did he earn in 1991 than in 1989?
21. From 1990 to 1991 did his salary increase or decrease, and by how much?
22. From 1991 to 1992 did his salary increase or decrease, and by how much?

CHAPTER 1 SUMMARY

KEY WORDS AND PHRASES

whole numeral [1.1]
digit [1.1]
place value [1.1]
whole number [1.1]
round off [1.1]
addition [1.2]

term [1.2]
sum [1.2]
subtraction [1.2]
borrowing [1.2]
translating [1.3]
increased by [1.3]

difference [1.3]
decreased by [1.3]
perimeter [1.4]
decrease [1.4]
increase [1.4]

SYMBOLS

= means *is equal to* [1.1]
≈ means *is approximately equal to* [1.1]
≐ means *is approximately equal to* [1.1]
+ means *to add* [1.2]
− means *to subtract* [1.2]

IMPORTANT RULES

The Meaning of Addition [1.2]
To add means to total or combine.

The Meaning of Subtraction [1.2]
To subtract means to take away. Subtraction is the reverse of addition, that is, $9 - 5 = 4$ because $4 + 5 = 9$.

How to Find the Missing Number in an Addition Statement [1.2]
- If the sum (or total) is missing, then add the terms (or parts) to find the sum.
- If one of the terms (or parts) is missing, then subtract the known term from the sum (or total) to obtain the missing term.

The Perimeter of a Figure [1.4]
The perimeter of a figure means the total distance around the figure.

CHAPTER 1 REVIEW EXERCISES

Use the numeral 40,356 to answer Exercises 1 through 4.
1. What digit is in the hundreds place?
2. What digit is in the tens place?
3. What is the place value of the digit 6?
4. What is the place value of the digit 4?
5. In the numeral 57,036,894, what digit is in the hundred thousands place?
6. In the numeral 1,238,049,000, what is the place value of the digit 3?
7. List from smallest to largest: 3000; 2099; 2200

In Exercises 8 through 11, round off to the given place value.
8. 723; tens
9. 1086; hundreds
10. 17,982; hundreds
11. 802,541; thousands

In Exercises 12 through 16, write a whole numeral for each English name.
12. Twelve
13. Three hundred eight
14. Fifty-seven thousand, twenty-four
15. Two million, eighty thousand, nine hundred eleven
16. 65 billion

In Exercises 17 through 20, write the English name for each numeral.
17. 18
18. 489
19. 206,801
20. 3,000,800,005
21. The distance between Atlanta, Georgia, and Miami, Florida, is six hundred sixty-five miles. Write the whole numeral for this distance.

Add.
22. 135 + 92
23. 9 + 705 + 36
24. 5604
     738
  + 2897
25.   5,395
    6,027
 + 98,730
26. 72,634
  50,987
 + 53,668
27. 2059
   500
 +   18
28. 12,938 + 7069 + 258
29. 12,000,813,700 + 20,813,000

Subtract.

30.	915 − 207	31.	8231 − 923	32.	7034 − 237	33.	78,096 − 8,028
34.	213,750 − 4,387	35.	402 − 83	36.	3004 − 98	37.	5007 − 72

38. 6000 − 703
39. 40,000 − 7036
40. 200,000 − 51,000
41. 2,003,005 − 76,324

42. 9,040,080
    − 24,567

43. 211,320
    − 7,857

44. 510
    − 82

45. 32,100
    − 7,986

46. 1000
    − 317

47. 611,000
    − 17,903

48. 100,000
    − 7,200

49. 210,003,100
    − 7,034,826

Find the missing number.

50. 218 + ☐ = 392
51. 1046 + 18,739 = ☐
52. 810 = ☐ + 482
53. 6000 = 910 + ☐
54. ☐ = 29,038 + 156,400
55. ☐ + 299 = 10,802

In Exercises 56 through 59, translate each English statement to an addition or subtraction statement using math symbols.

56. The difference between 1000 and 72 is 928.
57. 4300 increased by 486 is 4786.
58. The sum of 26 and 193 is 219.
59. 800 reduced by 108 is 692.

Solve.

60. How much larger is 51 than 38?
61. Find the sum of 732 and 970.
62. Subtract 34 from 2082.
63. What number is 68 more than 3736?
64. Lisa jogged along the highway for 8 miles, then along the beach for 6 miles. How far did Lisa jog?
65. From 1950 to 1960 the population of Dallas increased from 434,462 to 679,684. What was the increase in population?
66. Chuck has $651 in his checking account. He deposits $1369 in the account. What is the balance after the deposit?
67. The area of Memphis is 141 square miles and the area of San Francisco is 45 square miles. How much larger in area is Memphis than San Francisco?
68. Find the perimeter of this triangle.

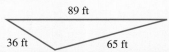

69. Last week Margaret paid the following bills.
   Gas and electric $ 36
   Mortgage $387
   Mastercard $172
   Auto insurance $230
   What was the total amount paid?

70. During the day the temperature in Phoenix went from 52°F to 91°F. What was the increase in temperature?

71. What is the total height of this sailboat, from the bottom of the keel to the top of the sail?

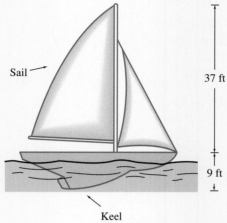

72. The Gateway Arch in St. Louis is 75 feet taller than the Washington Monument. The Gateway Arch is 630 feet tall. What is the height of the Washington Monument?

73. Last year the brake service for Harvey's car cost $175. This year the same service cost $17 more. What did Harvey pay this year?

74. Ashley has a charge account with Sears. His bill this month shows a balance of $223. If he pays $50, what is the new balance?

75. Mr. Wong is driving his car on a business trip. The total distance he must travel is 600 miles. He has already gone 237 miles. How much farther must he go?

76. As a computer programmer, Nancy earns $22,100 annually. As an auto mechanic, Richard earns $17,980. Who earns more money and how much more?

✍ In Your Own Words

Write complete sentences to discuss each of the following. Support your comments with examples or pictures, if appropriate.

77. A student incorrectly writes the numeral 600407 for six hundred forty seven. Discuss how you would help this student to understand the error made.

78. Discuss how the following two problems differ.
   a. ☐ + 82 = 116   **b.** 116 + 82 = ☐

79. Discuss a real-life situation where you need to calculate the perimeter of a figure.

80. Discuss a real-life situation where you need to subtract two numbers.

81. Discuss a real-life situation where you need to add more than two numbers.

CHAPTER 1 PRACTICE TEST

1. In the numeral 36,178, what digit is in the thousands place?
2. In the numeral 2,315,406, what digit is in the hundred thousands place?
3. In the numeral 4863, what is the place value of the digit 8?
4. In the numeral 715,320,400, what is the place value of the digit 1?
5. Write the numeral for seven hundred forty-six.
6. Write the English name for 90,000.
7. List from smallest to largest: 1002; 799; 801; 98

In Exercises 8 and 9, round off to the given place value.

8. 674; hundreds
9. 98,456; thousands

Add.

10. 7346 + 682 + 95
11. 27 + 314 + 82,376
12. 72,894
    5,843
    + 96,732
13. 536
    4728
    966
    + 857

Subtract.

14. 514
    − 24
15. 703
    − 288
16. 2136 − 540
17. 8000 − 2174
18. 50,040 − 386
19. 100,000 − 60,820

Find the missing number.

20. 2103 = ☐ + 932
21. 87 + 95 = ☐

Solve.

22. What number increased by 719 yields 2050?
23. What number is 365 more than 87?
24. Find the difference between 5000 and 92.
25. Ginny has $702 in her checking account. She writes a check for $83. How much is in the account now?
26. What is the total height of this house?

27. George's annual salary is less than Marilyn's annual salary by $1300. If George's salary is $26,500, find Marilyn's salary.
28. A company needs to hire 1000 new workers. They have already hired 681. How many more workers do they need?

CHAPTER 2

Whole Numbers: Multiplication and Division

2.1 MULTIPLICATION

OBJECTIVES
- Multiply two whole numbers.
- Use a shortcut to multiply whole numbers involving zeros.
- Multiply more than two numbers.
- Solve an application using multiplication of whole numbers.

1 Introduction to Multiplication

One shirt costs $7. What is the cost of 4 shirts? To answer this question, we multiply $7 by 4. The symbol $\times$ (read *times*) is used to indicate **multiplication.**

$$4 \times \$7 = \$7 + \$7 + \$7 + \$7$$
$$= \$28$$

The 4 shirts cost $28.

In the multiplication statement

$$4 \times 7 = 28$$

the numbers 4 and 7 are called **factors** or **multipliers,** and the result 28 is called the **product.** Observe that

$$4 \times 7 = 7 + 7 + 7 + 7 = 28$$

and also

$$7 \times 4 = 4 + 4 + 4 + 4 + 4 + 4 + 4 = 28$$

That is, it does not matter in what order the factors are written.

Although the operation of multiplication can be interpreted as repeated addition, you want to be able to compute multiplication without adding repeatedly. That is, you should know that 4 times 7 is 28 without adding.

Here is a completed table of the basic multiplication facts that you must memorize.

X	0	1	2	3	4	5	6	7	8	9
0	0	0	0	0	0	0	0	0	0	0
1	0	1	2	3	4	5	6	7	8	9
2	0	2	4	6	8	10	12	14	16	18
3	0	3	6	9	12	15	18	21	24	27
4	0	4	8	12	16	20	24	28	32	36
5	0	5	10	15	20	25	30	35	40	45
6	0	6	12	18	24	30	36	42	48	54
7	0	7	14	21	28	35	42	49	56	63
8	0	8	16	24	32	40	48	56	64	72
9	0	9	18	27	36	45	54	63	72	81

There are 100 entries in the above table. However, if you make some observations, you see that there are a lot fewer than 100 facts to memorize. Here we list the observations you should make.

- Any number multiplied by 0 is 0.
- Any number multiplied by 1 is that number.
- It does not matter in what order the multipliers are written.

Now you see that you only need to memorize the facts that are outlined in color.

Here are some examples of multiplying numbers with more than one digit by numbers with only one digit.

EXAMPLE 1   Multiply: 3024 × 2

SOLUTION

```
    3024
  ×    2
    6048
```

EXAMPLE 2   Multiply: 23 × 4

SOLUTION

```
     1
    23
  ×  4
    92
```

Step 1: 4 × 3 = 12.
Write 2 in the ones column.
Carry 1 to the tens column,
because 12 = 1 ten + 2 ones

Step 2: 4 × 2 = 8, then 8 + 1 = 9.
Write 9 in the tens column.

 **Try These Problems**

Multiply.

1. 78 × 6
2. 203 × 4
3. 2070 × 5
4. 5117 × 8
5. 70,846 × 9

EXAMPLE 3 Multiply: 5307 × 8

SOLUTION

```
  2 5
 5307
×   8
42456
```

Step 1: 8 × 7 = 56. Write 6, carry 5.
Step 2: 8 × 0 = 0, then 0 + 5 = 5. Write 5.
Step 3: 8 × 3 = 24. Write 4, carry 2.
Step 4: 8 × 5 = 40, then 40 + 2 = 42. Write 42.

The answer is 42,456. ∎

 Try Problems 1 through 5.

Now we look at multiplying two numbers where both numbers have more than one digit.

EXAMPLE 4 Multiply: 412 × 23

SOLUTION Arrange the numbers vertically so that like place values form a column. It doesn't matter which number is on top, but the work is easier if the numeral with more digits is on top.

```
  412
×  23
 1236
  824
 9476
```

Step 1: 3 × 412 = 1236. Since 3 is in the ones column, write 6 in the ones column.
Step 2: 2 × 412 = 824. Since 2 is in the tens column, write 4 in the tens column. 2 tens × 412 = 824 tens.
Step 3: Add.

The answer is 9476. ∎

EXAMPLE 5 Multiply: 513 × 274

SOLUTION

```
    513
 ×  274
   2052
   3591
   1026
 140562
```

— Write 2 in the ones column.
— Write 1 in the tens column.
— Write 6 in the hundreds column.

The answer is 140,562. ∎

Try These Problems

Multiply.

6. 94
    × 23

7. 2135
    × 61

8. 5064
    × 402

9. 5021 × 8316

10. 657 × 30,984

EXAMPLE 6   Multiply: 3108 × 430,165

SOLUTION

```
      430165
   ×    3108
     3441320
     000000      Be neat so
     430165      that the digits
    1290495      line up properly.
   1336952820
```

The answer is 1,336,952,820. ■

 Try Problems 6 through 10.

2 Shortcuts Involving Zeros

It will be helpful for you to learn a couple of shortcuts for multiplication problems involving many zeros. First we look at a shortcut for multiplying numbers when one or both of them end in zeros.

EXAMPLE 7   Multiply: 5 × 300

SOLUTION
LONG METHOD

```
    300
   ×  5
   1500
```

SHORTCUT

5 × 300 = 1500   5 × 3 = 15, then attach the 2 zeros

The answer is 1500. ■

EXAMPLE 8   Multiply: 24 × 6000

SOLUTION
LONG METHOD

```
     6000
   ×   24
    24000
    12000
   144000     Observe that these three zeros were
              caused by the three zeros in 6000.
```

SHORTCUT

```
      24
   × 6000
   144000
```
Multiply 6 times 24, then attach three zeros.

The answer is 144,000. ■

Try These Problems

Multiply.
11. 7 × 6000
12. 326 × 900
13. 70 × 11,200
14. 8500 × 365,000
15. 780 × 15,000
16. 2360 × 135,000

EXAMPLE 9 Multiply: 40,000 × 700

SOLUTION
LONG METHOD

```
      40000
   ×    700
      00000
     00000
    280000
   28000000
```
Observe that these six zeros were caused by the six zeros in 40000 and 700.

SHORTCUT

40000 × 700 = 28000000

six zeros

4 × 7 = 28, then attach six zeros

The answer is 28,000,000. ■

EXAMPLE 10 Multiply: 3250 × 1800

SOLUTION Multiply 325 × 18, then attach three zeros.

```
      3250
   ×  1800   } three zeros
      2600
      325
    5850000
```
18 × 325
Attach three zeros.

The answer is 5,850,000. ■

 Try Problems 11 through 16.

Next we look at a shortcut that can be used when the number placed on bottom has zero digits elsewhere than at the end.

EXAMPLE 11 Multiply: 512 × 306

SOLUTION
LONG METHOD

```
     512
  ×  306
    3072
     000     This row of zeros
    1536     contributes nothing
   156672    to the sum.
```

SHORTCUT

```
     512
  ×  306
    3072
    1536     Since 3 is in the hundreds
   156672    column, you must write
             6 in the hundreds column.
```

The answer is 156,672. ■

Try These Problems

Multiply.

17. 732
 × 905

18. 4082
 × 7008

19. 98,736
 × 50,008

20. 500,302 × 710,275

EXAMPLE 12 Multiply: 7308 × 6004

SOLUTION

LONG METHOD

```
        7308
      × 6004
       29232
        0000
       0000
      43848
    43877232
```
In the shortcut we omit these zeros.

SHORTCUT

```
        7308
      × 6004
       29232
      43848
    43877232
```
Write 8 in the thousands column, since 6 is in the thousands column.

The answer is 43,877,232. ■

 Try Problems 17 through 20.

3 Multiplying More Than Two Numbers

To multiply more than two numbers, multiply any two of them first, then multiply that result by another one, and continue until all of the numbers have been multiplied. The following examples illustrate the procedure.

EXAMPLE 13 Multiply: 5 × 2 × 3

SOLUTION

METHOD 1

$$5 \times 2 \times 3 = 10 \times 3 = 30$$

METHOD 2

$$5 \times 2 \times 3 = 5 \times 6 = 30$$

METHOD 3

$$5 \times 2 \times 3 = 15 \times 2 = 30 \quad \blacksquare$$

No matter in what order the numbers are multiplied, the result is 30.

EXAMPLE 14 Multiply: 7 × 5 × 8 × 4

SOLUTION

METHOD 1

$$7 \times 5 \times 8 \times 4 = 40 \times 28 = 1120$$

METHOD 2

$$7 \times 5 \times 8 \times 4 = 56 \times 20 = 1120$$

Try These Problems

Multiply.
21. $4 \times 5 \times 8$
22. $3 \times 3 \times 5 \times 2$
23. $7 \times 6 \times 8 \times 4$
24. $26 \times 5 \times 13$
25. $11 \times 17 \times 5 \times 9$
26. $23 \times 41 \times 3 \times 7$

METHOD 3

$$7 \times 5 \times 8 \times 4 = 35 \times 8 \times 4$$
$$= 280 \times 4$$
$$= 1120 \ \blacksquare$$

No matter in what order the numbers are multiplied, the result is 1120.

 Try Problems 21 through 26.

 Answers to Try These Problems

1. 468 2. 812 3. 10,350 4. 40,936 5. 637,614 6. 2162
7. 130,235 8. 2,035,728 9. 41,754,636 10. 20,356,488
11. 42,000 12. 293,400 13. 784,000 14. 3,102,500,000
15. 11,700,000 16. 318,600,000 17. 662,460 18. 28,606,656
19. 4,937,589,888 20. 355,352,003,050 21. 160 22. 90
23. 1344 24. 1690 25. 8415 26. 19,803

EXERCISES 2.1

Multiply.

1. $\begin{array}{r} 73 \\ \times\ 6 \\ \hline \end{array}$
2. $\begin{array}{r} 84 \\ \times\ 9 \\ \hline \end{array}$
3. $\begin{array}{r} 132 \\ \times\ 7 \\ \hline \end{array}$
4. $\begin{array}{r} 208 \\ \times\ 4 \\ \hline \end{array}$
5. $\begin{array}{r} 900 \\ \times\ 6 \\ \hline \end{array}$
6. $\begin{array}{r} 5026 \\ \times\ 3 \\ \hline \end{array}$
7. $\begin{array}{r} 7763 \\ \times\ 5 \\ \hline \end{array}$
8. $\begin{array}{r} 8184 \\ \times\ 8 \\ \hline \end{array}$
9. $\begin{array}{r} 86 \\ \times\ 32 \\ \hline \end{array}$
10. $\begin{array}{r} 80 \\ \times\ 46 \\ \hline \end{array}$
11. $\begin{array}{r} 341 \\ \times\ 25 \\ \hline \end{array}$
12. $\begin{array}{r} 237 \\ \times\ 74 \\ \hline \end{array}$
13. $\begin{array}{r} 704 \\ \times\ 286 \\ \hline \end{array}$
14. $\begin{array}{r} 800 \\ \times\ 314 \\ \hline \end{array}$
15. $\begin{array}{r} 7236 \\ \times\ 52 \\ \hline \end{array}$
16. $\begin{array}{r} 5056 \\ \times\ 304 \\ \hline \end{array}$
17. 3046×92
18. 75×3379
19. 2114×3756
20. $824 \times 93,406$

Multiply. Use a shortcut.

21. 400×3
22. 7000×20
23. $90,000 \times 800$
24. 600×320
25. $8 \times 24,000$
26. $31,200 \times 8000$
27. $2500 \times 68,000$
28. $126,000 \times 18,000$
29. $\begin{array}{r} 215 \\ \times\ 108 \\ \hline \end{array}$
30. $\begin{array}{r} 392 \\ \times\ 205 \\ \hline \end{array}$
31. $\begin{array}{r} 3146 \\ \times\ 4001 \\ \hline \end{array}$
32. $\begin{array}{r} 5104 \\ \times\ 6007 \\ \hline \end{array}$
33. $2030 \times 91,732$
34. $50,070 \times 2906$
35. $39,000 \times 5004$
36. $20,004 \times 13,682$

Multiply.

37. $5 \times 4 \times 3$
38. $2 \times 3 \times 3$
39. $9 \times 8 \times 6$
40. $7 \times 2 \times 2 \times 5$
41. $9 \times 6 \times 3 \times 4$
42. $14 \times 6 \times 36 \times 8$
43. $25 \times 30 \times 19$
44. $7 \times 16 \times 200 \times 38$

Solve.

45. One window costs $158. What is the cost of 18 windows?

46. A bicyclist goes 17 miles in one hour. How far can she go in 6 hours?

47. Anita's weekly salary is $570. Charles makes 7 times that much. Find Charles' weekly salary.

48. An old machine prints 682 pages in an hour. A new machine prints 15 times that much. How many pages can the new machine print in an hour?

2.2 DIVISION

OBJECTIVES
- Understand how division relates to multiplication.
- Perform long division by using a one, two or three-digit divisor.
- Perform long division when there are zeros in the quotient.
- Solve an application using division of whole numbers.

1 The Meaning of Division

Grandma Hendren has $20 that she wants to distribute equally among her 4 grandchildren. How much money does each child receive? To solve this problem, we divide $20 by 4. The symbols $\overline{)}$ and $\div$ are used to indicate division.

$$4\overline{)20}^{\,5} \quad \text{or} \quad 20 \div 4 = 5$$

Each child receives $5.

Division is the reverse of multiplication. For example,

$$4\overline{)20}^{\,5} \quad \text{because} \quad 4 \times 5 = 20$$

and

$$20 \div 4 = 5 \quad \text{because} \quad 4 \times 5 = 20$$

It does matter in what order you write down a division problem. For example,

$$20 \div 4 = 5$$

but

$$4 \div 20 = \frac{4}{20} = \frac{1}{5} \text{ or } 0.2$$

The number $\frac{1}{5}$ (in fraction form) or 0.2 (in decimal form) is a number less than 1. Fractions are introduced in Chapter 3 and decimals are introduced in Chapter 6.

The division statement $4\overline{)20}^{\,5}$ can be read "4 divided into 20 equals 5" or "20 divided by 4 is 5." Either of these statements can also be used to read the statement $20 \div 4 = 5$. Notice that when we use the symbol

÷ to indicate division, the numbers are written in the reverse order than when using the symbol $\overline{)}$. That is,

$$20 \div 4 \text{ means } 4\overline{)20}$$

The 4 is written second here.   The 4 is written first here.

Each number in a division statement has a special name.

$$8\overline{)48} \overset{6 \longleftarrow \text{quotient}}{} \qquad 48 \div 8 = 6 \longleftarrow \text{quotient}$$

divisor   dividend   dividend   divisor

In the above example, the **divisor** multiplied by the **quotient** is exactly equal to the **dividend**. When this happens, we say *the division comes out evenly.*

Now we look at a situation where the division does not come out evenly. How many groups of 5 are in 14?

(XXX XX) (XXX XX) XXX X

There are 2 groups of 5 with 4 left over. This concept is symbolized using a division statement with **remainder.**

$$\begin{array}{r} 2 \longleftarrow \text{quotient} \\ 5\overline{)14} \\ \underline{10} \\ 4 \longleftarrow \text{remainder} \end{array}$$

The answer is written 2 R4.

Observe that a division problem with remainder is related to multiplication and addition.

$$\begin{array}{r} 2 \\ 5\overline{)14} \\ \underline{10} \\ 4 \end{array} \qquad \begin{array}{r} 2 \\ \times\ 5 \\ \hline 10 \\ +\ 4 \\ \hline 14 \end{array} \begin{array}{l} \text{Quotient times divisor.} \\ \\ \text{Add the remainder} \\ \text{You get back the dividend.} \end{array}$$

Use this relationship to check your work when doing division problems.

EXAMPLE 1   Divide: $51 \div 6$

SOLUTION

$$6\overline{)51}^{\ ?}$$

How many groups of 6 are in 51?

$$\begin{array}{ccccc} 6 & 6 & 6 & 6 & 6 \\ \times\ 5 & \times\ 6 & \times\ 7 & \times\ 8 & \times\ 9 \\ \hline 30 & 36 & 42 & 48 & 54 \end{array}$$

There are 8 groups of 6 in 51.

$$\begin{array}{r} 8 \\ 6\overline{)51} \\ \underline{48} \longleftarrow 6 \times 8 = 48 \\ 3 \longleftarrow \text{Subtract to get the remainder.} \end{array}$$

Try These Problems

Divide.
1. 63 ÷ 7
2. 42 ÷ 6
3. 36 ÷ 6
4. 51 ÷ 8
5. 89 ÷ 9
6. 37 ÷ 5

CHECK

```
    8
  × 6
  ---
   48
  + 3
  ---
   51
```

The answer is 8 R3. ■

Sometimes you will choose the wrong quotient. The following example gives some pointers on how to recognize this so that you can back up and start over.

EXAMPLE 2 Divide: 44 ÷ 9

SOLUTION

UNSUCCESSFUL ATTEMPT

```
      3 ← ?
    _____
  9 ) 44
      27
      --
      17
```
The remainder must be smaller than the divisor. 17 is larger than 9, so the 3 must be changed.

UNSUCCESSFUL ATTEMPT

```
      5 ← ?
    _____
  9 ) 44
      45
```
This number must be smaller than or equal to 44 so we can subtract. 45 is larger than 44, so the 5 must be changed.

SUCCESSFUL ATTEMPT

```
      4
    _____
  9 ) 44
      36     — 36 is smaller than 44.
      --
       8     — 8 is smaller than 9.
```

The answer is 4 R8. ■

Try Problems 1 through 6.

2 Long Division by A One-Digit Divisor

Now we look at a process called **long division.** This procedure enables you to find the quotient and remainder when the numbers are larger. First we look at examples that involve a one-digit divisor.

EXAMPLE 3 Divide: 159 ÷ 5

SOLUTION

```
        3 1
      _____
    5 ) 1 5 9
        1 5 ↓
        -----
        0 | 9
            5
            -
            4
```

Step 1: 5 divided into 1? It goes 0 times, but we do not have to write the 0 at the beginning.
Step 2: 5 divided into 15? 3
Step 3: 3 times 5? 15
Step 4: 15 subtract 15? 0
Step 5: Bring down 9.
Step 6: 5 divided into 9? 1
Step 7: 1 times 5? 5
Step 8: 9 subtract 5? 4

There are 31 groups of 5 in 159 with 4 left over.
CHECK

$$\begin{array}{r} 31 \\ \times5 \\ \hline 155 \\ +4 \\ \hline 159 \end{array}$$

The answer is 31 R4. ∎

EXAMPLE 4 Divide: $7\overline{)4256}$

SOLUTION

$$7\overline{)4256}$$ quotient 608

Step 1: 7 divided into 42? 6
Step 2: 6 times 7? 42
Step 3: 42 subtract 42? 0
Step 4: Bring down 5.
Step 5: 7 divided into 5? It goes 0 times. You must write a 0 in the quotient!
Step 6: 0 times 7? 0
Step 7: 5 subtract 0? 5
Step 8: Bring down 6.
Step 9: 7 divided into 56? 8
Step 10: 8 times 7? 56
Step 11: 56 subtract 56? 0

CHECK

$$\begin{array}{r} 608 \\ \times7 \\ \hline 4256 \end{array}$$

The answer is 608. ∎

EXAMPLE 5 Divide: $20{,}334 \div 4$

SOLUTION

$$4\overline{)20334}$$ quotient 5083

Bring down only one digit at a time. Each time you bring down a digit, you must divide.

After you subtract, be sure to bring down the next digit before you divide.

CHECK

$$\begin{array}{r} 5083 \\ \times4 \\ \hline 20332 \\ +2 \\ \hline 20334 \end{array}$$

The answer is 5083 R2. ∎

Chapter 2 Whole Numbers: Multiplication and Division

 Try These Problems

Divide.

7. $7\overline{)616}$
8. $3\overline{)970}$
9. $8\overline{)63{,}216}$
10. $1863 \div 6$
11. $21{,}049 \div 7$
12. $75{,}000 \div 5$

EXAMPLE 6 Divide: $8\overline{)2885}$

SOLUTION

```
        3 6 0
     _____
   8 ) 2 8 8 5
       2 4 ↓
       _____
         4 8
         4 8 ↓
         _____
           0 5
             0
           ___
             5
```

— After bringing down the last digit, you must divide, multiply, a subtract one more time.

CHECK

```
    360
  ×   8
  _____
   2880
  +   5
  _____
   2885
```

The answer is 360 R5.

Now we summarize the procedure for long division.

Long Division Process

The long division process involves repeating these four steps.

1. Divide.
2. Multiply.
3. Subtract. (This result must be less than the divisor.)
4. Bring down the next digit.

Bring down only one digit at a time. Each time you bring down a digit, you must divide. After you bring down the last digit, you must divide, multiply, and subtract one more time.

 Try Problems 7 through 12.

 Division by Two- and Three-Digit Numbers

To divide by a two-digit number or a three-digit number, use the same long division process that is used to divide by a one-digit number. The following examples illustrate the procedure.

EXAMPLE 7 Divide: $25\overline{)824}$

SOLUTION

```
      3 2
25)8 2 4
   7 5 ↓
     7 4
     5 0
     2 4
```

Step 1: 25 into 8? It goes 0 times. We do not have to write a 0 at the beginning.

Step 2: 25 into 82?

$$\begin{array}{ccc} 25 & 25 & 25 \\ \times\ 2 & \times\ 3 & \times\ 4 \\ \hline 50 & 75 & 100 \end{array}$$

3 R7
Write 3 above the last digit in 82.

Step 3: Bring down 4.
Step 4: 25 into 74?
2 R24

CHECK

```
     32
  ×  25
    160
     64
    800
  +  24
    824
```

The answer is 32 R24. ■

The following example illustrates how to recognize when you have chosen the wrong digit in the quotient.

EXAMPLE 8 Divide: 4552 ÷ 92

SOLUTION

UNSUCCESSFUL ATTEMPT

```
       3 ←?
  92)4552
     276
     179
```

— The subtraction result must be smaller than the divisor, 92. 179 is larger than 92, so the 3 must be changed.

UNSUCCESSFUL ATTEMPT

```
       5 ←?
  92)4552
     460
```

— This number must be smaller than 455 so we can subtract. 460 is larger than 455, so the 5 must be changed.

SUCCESSFUL ATTEMPT

```
       4 9
  92)4 5 5 2
     3 6 8 ↓
         8 7 2
         8 2 8
             4 4
```

The subtraction results, 87 and 44, are each smaller than 92.

The answer is 49 R44. ■

Chapter 2 Whole Numbers: Multiplication and Division

EXAMPLE 9 Divide: $615\overline{)44{,}286}$

SOLUTION

$$\begin{array}{r} 72 \\ 615\overline{)4\,4\,2\,8\,6} \\ \underline{4\,3\,0\,5}\downarrow \\ 1\,2\,3\,\vert 6 \\ \underline{1\,2\,3\,0} \\ 6 \end{array}$$

Step 1: 615 into 442? It goes 0 times. We do not have to write a 0 digit at the beginning.
Step 2: 615 into 4428?

$$\begin{array}{r} 615 \\ \times6 \\ \hline 3690 \end{array} \quad \begin{array}{r} 615 \\ \times7 \\ \hline 4305 \end{array} \quad \begin{array}{r} 615 \\ \times8 \\ \hline 4920 \end{array}$$

7 R123
Write 7 above the last digit in 4428.
Step 3: Bring down 6.
Step 4: 615 into 1236?

$$\begin{array}{r} 615 \\ \times1 \\ \hline 615 \end{array} \quad \begin{array}{r} 615 \\ \times2 \\ \hline 1230 \end{array}$$

2 R6

CHECK

$$\begin{array}{r} 615 \\ \times72 \\ \hline 1230 \\ 4305 \\ \hline 44280 \\ +6 \\ \hline 44286 \end{array}$$

The answer is 72 R6. ∎

EXAMPLE 10 Divide: $26\overline{)9626}$

SOLUTION

$$\begin{array}{r} 370 \\ 26\overline{)9\,6\,2\,6} \\ \underline{7\,8}\downarrow \\ 1\,8\,\vert 2 \\ \underline{1\,8\,2}\downarrow \\ 0\,\vert 6 \\ \underline{0} \\ 6 \end{array}$$

After you bring down the last digit, you must divide, multiply, and subtract one more time.

The answer is 370 R6. ∎

Try These Problems

Divide.
13. 750 ÷ 60
14. 9706 ÷ 23
15. 16,281 ÷ 23
16. 40)256,120
17. 15)92,250
18. 36)151,225
19. 9450 ÷ 210
20. 706,854 ÷ 662

EXAMPLE 11   Divide: 930,512 ÷ 186

SOLUTION

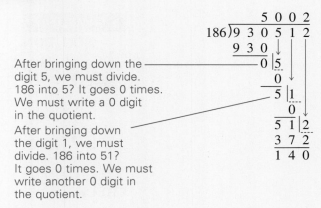

After bringing down the digit 5, we must divide. 186 into 5? It goes 0 times. We must write a 0 digit in the quotient.

After bringing down the digit 1, we must divide. 186 into 51? It goes 0 times. We must write another 0 digit in the quotient.

CHECK

$$
\begin{array}{r}
5002 \\
\times\ 186 \\
\hline
30012 \\
40016 \\
5002 \\
\hline
930372 \\
+140 \\
\hline
930512
\end{array}
$$

The answer is 5002 R140. ■

 Try Problems 13 through 20.

4   Paying Special Attention to Zeros in the Quotient

Suppose you are one of the 6 persons who are to share equally an inheritance of $42,540. What is your share? To solve this problem we divide 42,540 by 6.

```
      7 0 9 0
  6)4 2 5 4 0       Your share
    4 2                is $7090.
    ───
      0 5
      0
      ───
        5 4
        5 4
        ───
          0 0
          0
          ───
            0
```

The zero digits in the quotient are very important. You want to be careful not to leave them out, since $7090 is a lot more money than $709 or $79. Also be careful that you do not insert extra zeros. The other five persons would not appreciate your claiming a share of $70,090 when there is only $42,540 to begin with.

The following examples give guidelines to help you with division problems involving zeros in the quotient.

EXAMPLE 12 Divide: 45,032 ÷ 75

SOLUTION

```
         6 0 0
    75)4 5 0 3 2
       4 5 0 ↓
         0 3
             0 ↓
             3 2
                 0
             3 2
```

Slow down! Don't try to take unnecessary shortcuts.

CHECK

```
       75
    ×  600
    45000
  +    32
    45032
```

The answer is 600 R32. ■

EXAMPLE 13 Divide: 840,480 ÷ 408

SOLUTION

```
          2 0 6 0
   408)8 4 0 4 8 0
       8 1 6 ↓
         2 4 4
             0 ↓
         2 4 4 8
         2 4 4 8
                 0 0
                   0
                   0
```

After the division gets started, there is a digit in the quotient to correspond with each of the remaining digits in the dividend.

```
          2 0 6 0
   408)8 4 0 4 8 0
```

The answer is 2060. ■

EXAMPLE 14 Divide: 252,352 ÷ 63

SOLUTION

```
          4 0 0 5
    63)2 5 2 3 5 2
       2 5 2 ↓
           0 3
             0 ↓
             3 5
                 0 ↓
             3 5 2
             3 1 5
                 3 7
```

After you subtract, bring down *only one* digit, then divide. Place the digit in the quotient directly above the digit you brought down.

The answer is 4005 R37. ■

Try These Problems

Divide.
21. 5)35,010
22. 55)15,400
23. 28)560,858
24. 7,645,677 ÷ 152
25. 81,479,320 ÷ 783

Try Problems 21 through 25.

Answers to Try These Problems

1. 9 2. 7 3. 6 4. 6 R3 5. 9 R8 6. 7 R2 7. 88
8. 323 R1 9. 7902 10. 310 R3 11. 3007 12. 15,000
13. 12 R30 14. 422 15. 707 R20 16. 6403 17. 6150
18. 4200 R25 19. 45 20. 1067 R500 21. 7002 22. 280
23. 20,030 R18 24. 50,300 R77 25. 104,060 R340

EXERCISES 2.2

Divide.

1. 8)75
2. 3)97
3. 8)526
4. 6)624
5. 5)4374
6. 7)5322
7. 7)7218
8. 4)29,296
9. 380 ÷ 40
10. 2280 ÷ 76
11. 53,020 ÷ 80
12. 4680 ÷ 42
13. 7645 ÷ 35
14. 16,482 ÷ 15
15. 204,555 ÷ 23
16. 5,328,200 ÷ 75
17. 800)26,000
18. 700)9368
19. 280)17,460
20. 170)137,360
21. 428)100,580
22. 564)4,386,278
23. 372)15,713,280
24. 685)3,122,230
25. 4804 ÷ 8
26. 35,058 ÷ 5
27. 456,600 ÷ 76
28. 141,363 ÷ 67
29. 1,401,120 ÷ 28
30. 56,630 ÷ 700
31. 3,424,876 ÷ 428
32. 3,964,590 ÷ 651

Solve.

33. Mr. Yen earned $1512 in a 6-week period. How much did he earn per week?
34. Folding tables are packed 5 to a box. How many boxes are needed to pack 510 tables?
35. A chandelier holds 16 bulbs. How many chandeliers can be outfitted with 650 bulbs? How many bulbs, if any, are left over?
36. A box of candy holds 24 pieces. How many boxes of candy can be filled with 12,125 pieces of candy? How many pieces, if any, are left over?
37. A machine cuts 800 washers from a sheet of metal in one hour. How long does it take the machine to cut out 32,000 washers?
38. Ms. Marcott ordered 7200 ounces of punch for a reception. She expects to have approximately 225 guests at the reception. How much punch does this allow for each guest?

DEVELOPING NUMBER SENSE #2

APPROXIMATING BY ROUNDING

In this feature we look at approximating multiplication and division by rounding. We will round each number so that the resulting multiplication or division is as easy as possible.

For example, suppose you want to approximate 38 × 714. First we round off each number so that the result has only one nonzero digit.

$$38 \approx 40$$
$$714 \approx 700$$

Multiplying 40 by 700, we obtain 28,000 for the approximation. The actual product is 27,132.

Next we look at approximating a division problem. Suppose you want an approximation for

Cont. page 46

$7704 \div 36$. First we round off each number so that the result has only one nonzero digit.

$$7704 \approx 8000$$
$$36 \approx 40$$

Next we divide 8000 by 40 to obtain the approximation.

$$\begin{array}{r} 200 \\ 40\overline{)8000} \\ \underline{80} \end{array}$$

The approximation is 200. The actual quotient is 214.

When approximating a division problem, do not expect to be able to divide the rounded numbers mentally. However, dividing the rounded numbers will be a lot easier than dividing the original numbers if you are approximating without the help of a calculator.

Number Sense Problems

Approximate each by rounding off each number so the result has only one nonzero digit.

1. 8×71
2. 12×58
3. 6×319
4. 63×782
5. 478×311
6. 7×3289
7. 76×2374
8. $197 \times 67{,}392$
9. $215 \div 5$
10. $644 \div 28$
11. $2958 \div 58$
12. $93{,}294 \div 438$

USING THE CALCULATOR #3

MULTIPLICATION AND DIVISION

Multiplying numbers on a calculator is similar to adding numbers on the calculator. Simply enter $\boxed{\times}$ instead of $\boxed{+}$. Here is an example.

To Compute	25×5
Enter	$25\ \boxed{\times}\ 5\ \boxed{=}$
Result	125

Here we show an example of multiplying several numbers.

To Compute	$8 \times 2 \times 5 \times 6$
Enter	$8\ \boxed{\times}\ 2\ \boxed{\times}\ 5\ \boxed{\times}\ 6\ \boxed{=}$
Result	480

If you enter another $\boxed{\times}$ instead of $\boxed{=}$, the result is also 480. Also, since multiplication can be done in any order, it does not matter in what order you enter the factors.

When computing a division problem, you must be careful that the numbers are entered in the correct order. Here is an example.

To Compute	$8 \div 2$ or $2\overline{)8}$
Enter	$8\ \boxed{\div}\ 2\ \boxed{=}$
Result	4

Observe that the numbers are entered in the order that they are written down when using the division symbol $\div$, but the numbers are entered in the reverse order than they are written down when using the division symbol $\overline{)}\ $. In either case, remember that the number being divided into is entered first and the divisor is entered second.

If you enter the numbers 8 and 2 in the reverse order, you would obtain the decimal 0.25, since 0.25 is the result of dividing 2 by 8. Decimals are discussed in Chapters 6 and 7.

Cont. page 47

Calculator Problems

Multiply or divide as indicated.

1. 8×9
2. 500×7800
3. $2 \times 13 \times 7 \times 29$
4. 769×876
5. $12 \div 6$
6. $32{,}616 \div 36$
7. $9\overline{)7983}$
8. $675\overline{)178{,}200}$
9. Divide 300 by 75.
10. Divide 82 into 498,560.

2.3 THE MISSING NUMBER IN A MULTIPLICATION STATEMENT, MULTIPLES, DIVISIBILITY RULES, AND PRIME FACTORING

OBJECTIVES
- Find the missing number in a multiplication statement.
- Determine whether a number is divisible by a given number.
- Determine whether a number is a multiple of a given number.
- Determine whether a number is divisible by 2, 3, 5, 10, 100, or 1000 by using divisibility rules.
- Write a number as the product of primes.

1 Finding the Missing Number in a Multiplication Statement

In a multiplication statement such as

$$4 \times 7 = 28 \quad \text{or} \quad 28 = 4 \times 7$$

the numbers 4 and 7 are called **multipliers** (or **factors**) and the answer 28 is called the **product.** We can ask three questions by omitting any one of the three numbers.

1. $4 \times 7 = \boxed{}$ — missing number is 28

   We are missing the answer to a multiplication problem, so we *multiply* to find the missing number.

2. $4 \times \boxed{} = 28$

   $4\overline{)28}^{7}$

   We are missing one of the multipliers, so we *divide* to find the missing number.

   Divide the answer 28 by the known multiplier 4. The missing number is 7.

3. $\boxed{} \times 7 = 28$

   $7\overline{)28}^{4}$

   We are missing one of the multipliers, so we *divide* to find the missing number.

   Divide the answer 28 by the known multiplier 7. The missing number is 4.

Try These Problems

Find the missing number.
1. 5 × ☐ = 500
2. 448 = ☐ × 16
3. ☐ = 97 × 3600
4. ☐ × 27 = 21,654

EXAMPLE 1   ☐ × 7 = 119

SOLUTION   We are missing a multiplier, so we divide to find the missing number.

$$\text{multipliers} \longrightarrow 7\overline{)119}\longleftarrow \text{answer to the multiplication problem}$$

$$\begin{array}{r} 17 \\ 7\overline{)119} \\ \underline{7} \\ 49 \\ \underline{49} \end{array}$$

The answer is 17. ■

EXAMPLE 2   ☐ = 82 × 794

SOLUTION   We are missing the answer to a multiplication problem, so we multiply to find the missing number.

$$\begin{array}{r} 794 \\ \times82 \\ \hline 1588 \\ 6352 \\ \hline 65108 \end{array}$$

The answer is 65,108. ■

EXAMPLE 3   750 = 25 × ☐

SOLUTION   We are missing a multiplier, so we divide to find the missing number.

$$\text{multipliers} \longrightarrow 25\overline{)750}\longleftarrow \text{answer to the multiplication problem}$$

$$\begin{array}{r} 30 \\ 25\overline{)750} \\ \underline{75} \\ 00 \\ \underline{0} \end{array}$$

The answer is 30. ■

Try Problems 1 through 4.

2 Multiples and Divisibility Rules

If we multiply 6 by a whole number, we create a **multiple** of 6. Here are some multiples of 6.

$$\begin{array}{r} 6 \\ \times\,1 \\ \hline 6 \end{array} \quad \begin{array}{r} 6 \\ \times\,2 \\ \hline 12 \end{array} \quad \begin{array}{r} 6 \\ \times\,3 \\ \hline 18 \end{array} \quad \begin{array}{r} 6 \\ \times\,4 \\ \hline 24 \end{array}$$

$$\begin{array}{r} 6 \\ \times\,5 \\ \hline 30 \end{array} \quad \begin{array}{r} 6 \\ \times\,6 \\ \hline 36 \end{array} \quad \begin{array}{r} 6 \\ \times\,7 \\ \hline 42 \end{array} \quad \begin{array}{r} 13 \\ \times\,6 \\ \hline 78 \end{array} \quad \begin{array}{r} 28 \\ \times\,6 \\ \hline 168 \end{array}$$

2.3 The Missing Number in a Multiplication Statement, Multiples, Divisibility Rules, and Prime Factoring **49**

A multiple of 6 such as 78 is special with relation to 6 because 6 divides evenly into 78. We say, 78 is **divisible by** 6.

$$\begin{array}{r} 13 \\ 6\overline{)78} \\ \underline{6} \\ 18 \\ \underline{18} \end{array}$$

EXAMPLE 4 Which of these numbers are divisible by 7?

a. 27 **b.** 91 **c.** 406

SOLUTION

a.

$$\begin{array}{r} 3 \\ 7\overline{)27} \\ \underline{21} \\ 6 \end{array}$$

No, 27 is not divisible by 7 because the remainder is 6. The division does not come out evenly.

b.

$$\begin{array}{r} 13 \\ 7\overline{)91} \\ \underline{7} \\ 21 \\ \underline{21} \end{array}$$

Yes, 91 is divisible by 7 because the remainder is 0. The division comes out evenly.

c.

$$\begin{array}{r} 58 \\ 7\overline{)406} \\ \underline{35} \\ 56 \\ \underline{56} \end{array}$$

Yes, 406 is divisible by 7 because the remainder is 0. The division comes out evenly. ■

EXAMPLE 5 Which of these numbers are multiples of 13?

a. 65 **b.** 377 **c.** 397

SOLUTION

a. 65 is a multiple of 13 if 65 is divisible by 13.

$$\begin{array}{r} 5 \\ 13\overline{)65} \\ \underline{65} \end{array}$$

Yes, 65 is a multiple of 13.

Try These Problems

5. Which of these numbers are divisible by 3?
   a. 42 **b.** 51 **c.** 117 **d.** 317

6. Which of these numbers are multiples of 11?
   a. 56 **b.** 88 **c.** 187 **d.** 462

Answer yes or no.
7. Indicate whether each number is divisible by 2.
   a. 2461 **b.** 688 **c.** 356 **d.** 4003

b. 377 is a multiple of 13 if 377 is divisible by 13.

$$\begin{array}{r} 29 \\ 13\overline{)377} \\ \underline{26} \\ 117 \\ \underline{117} \end{array}$$

Yes, 377 is a multiple of 13.

c. Does 13 divide evenly into 397?

$$\begin{array}{r} 30 \\ 13\overline{)397} \\ \underline{39} \\ 7 \\ \underline{0} \\ 7 \end{array}$$

No, 397 is not a multiple of 13. ∎

 Try Problems 5 and 6.

The multiples of 2 are 2, 4, 6, 8, 10, 12, 14, 16, 18, . . . , and so on. These numbers are called **even numbers.** Each of them is divisible by 2. It is easy to recognize that a number is divisible by 2 because it always ends in 0, 2, 4, 6, or 8. For example, the number 1336 is divisible by 2 because the last digit is 6 and 6 is an even number.

$$\begin{array}{r} 668 \\ 2\overline{)1336} \\ \underline{12} \\ 13 \\ \underline{12} \\ 16 \\ \underline{16} \end{array}$$

Divisibility by 2

Any whole number ending in a 0, 2, 4, 6, or 8 is divisible by 2.

EXAMPLE 6   Which of these numbers is divisible by 2?

a. 572 **b.** 31,580 **c.** 3907 **d.** 90,144

SOLUTION

a. Yes, 572 is divisible by 2 because it ends in a 2.

b. Yes, the number 31,580 is divisible by 2 because it ends in a 0.

c. No, 3907 is not divisible by 2 because it does not end in a 0, 2, 4, 6 or 8. It ends in a 7.

d. Yes, the number 90,144 is divisible by 2 because it ends in a 4. ∎

 Try Problem 7.

Try These Problems

Answer yes or no.
8. Indicate whether each number is divisible by 3.
   a. 72 **b.** 813 **c.** 244 **d.** 8052

The multiples of 3 are 3, 6, 9, 12, 15, 18, 21, 24, 27, 30, 33, 36, 39, 42, . . . and so on. Note that in each case the sum of the digits is a multiple of 3. In general, all multiples of 3 have digits whose sum is a multiple of 3. We can use this to help us recognize numbers that are divisible by 3. For example, the number 117 is divisible by 3 because the sum of the digits, $1 + 1 + 7 = 9$, is divisible by 3. Here we check by dividing.

$$\begin{array}{r} 39 \\ 3\overline{)117} \\ \underline{9} \\ 27 \\ \underline{27} \end{array}$$

Divisibility by 3
Any whole number is divisible by 3 if the sum of its digits is divisible by 3.

EXAMPLE 7 Which of these numbers are divisible by 3?

a. 51   **b.** 852   **c.** 923

SOLUTION

a. Yes, 51 is divisible by 3 because $5 + 1 = 6$ and 6 is divisible by 3.

b. Yes, 852 is divisible by 3 because $8 + 5 + 2 = 15$ and 15 is divisible by 3.

c. No, 923 is not divisible by 3 because $9 + 2 + 3 = 14$ and 14 is not divisible by 3. ■

 Try Problem 8.

The multiples of 5 are 5, 10, 15, 20, 25, 30, 35, 40, 45, 50, 55, 60, . . . and so on. Note that each multiple of 5 ends in a 0 or a 5. We can use this to help us recognize when a number is divisible by 5. For example, 3015 is divisible by 5 because it ends in 5. Here we check by dividing.

$$\begin{array}{r} 603 \\ 5\overline{)3015} \\ \underline{30} \\ 1 \\ \underline{0} \\ 15 \\ \underline{15} \end{array}$$

Divisibility by 5
Any whole number ending in 0 or 5 is divisible by 5.

Try These Problems

Answer yes or no.
9. Indicate whether each number is divisible by 5.
   a. 85 b. 3008 c. 1045
   d. 70,000
10. Indicate whether each number is divisible by 10.
    a. 955 b. 1048 c. 830
    d. 8000

EXAMPLE 8 Which of these numbers are divisible by 5?

a. 70 b. 135 c. 8004

SOLUTION

a. Yes, 70 is divisible by 5 because it ends in a 0.

b. Yes, 135 is divisible by 5 because it ends in 5.

c. No, 8004 is not divisible by 5 because it does not end in a 0 or a 5. It ends in 4. ∎

 Try Problem 9.

The multiples of 10 are 10, 20, 30, 40, 50, 60, 70, 80, 90, 100, 110, 120, 130, and so on. Note that multiples of 10 end in a 0. This can help us to recognize when a number is divisible by 10. For example, the number 4130 is divisible by 10 because it ends in a 0. Here we check by dividing.

$$
\begin{array}{r}
413 \\
10\overline{)4130} \\
\underline{40} \\
13 \\
\underline{10} \\
30 \\
\underline{30}
\end{array}
$$

Divisibility by 10
Any whole number ending in a 0 is divisible by 10.

EXAMPLE 9 Which of these numbers are divisible by 10?

a. 156 b. 300 c. 14,780

SOLUTION

a. No, 156 is not divisible by 10 because it does not end in 0. It ends in 6.

b. Yes, 300 is divisible by 10 because it ends in 0.

c. Yes, 14,780 is divisible by 10 because it ends in 0. ∎

 Try Problem 10.

The multiples of 100 are 100, 200, 300, 400, 500, 600, 700, 800, 900, 1000, 1100, 1200, and so on. Observe that each multiple of 100 ends in two 0s. This gives us a way to recognize when a number is divisible by 100. For example, the number 11,600 is divisible by 100 because it ends in two 0s. Here we check by dividing.

$$
\begin{array}{r}
116 \\
100\overline{)11600} \\
\underline{100} \\
160 \\
\underline{100} \\
600 \\
\underline{600}
\end{array}
$$

Try These Problems

Answer yes or no.
11. Indicate whether each number is divisible by 100.
    a. 6002 b. 9300 c. 88,000
    d. 7050

Divisibility by 100
Any whole number ending in two 0s is divisible by 100.

EXAMPLE 10   Which of these numbers are divisible by 100?

a. 1030 b. 3500 c. 80,000

SOLUTION

a. No, 1030 is not divisible by 100 because it does not end in two 0s. It ends in only one 0.

b. Yes, 3500 is divisible by 100 because it ends in two 0s.

c. Yes, the number 80,000 is divisible by 100 because it ends in two 0s. ∎

 Try Problem 11.

We have seen that the multiples of 10 end in one 0, and the multiples of 100 end in two 0s. This pattern continues; that is, the multiples of 1000 end in three 0s, the multiples of 10,000 end in four 0s, and so on.

Now we give a summary of the divisibility rules that you should be familiar with.

Divisibility Rules
1. Any whole number ending in 0, 2, 4, 6, or 8 is divisible by 2.
2. Any whole number is divisible by 3 if the sum of its digits is divisible by 3.
3. Any whole number ending in 0 or 5 is divisible by 5.
4. Any whole number
   is divisible by 10 if it ends in one 0,
   is divisible by 100 if it ends in two 0s,
   is divisible by 1000 if it ends in three 0s,
   and so on.

3 Prime Numbers and Prime Factoring

A whole number, other than 1, that is divisible only by itself and 1 is called a **prime number.** For example, the number 13 is a prime number because it is divisible only by 13 and 1. Here is a list of the first ten prime numbers.

First Ten Primes									
2	3	5	7	11	13	17	19	23	29

The number 15 is *not* prime because 15 is divisible by 5 and 3. We can write 15 as the product of primes.

$$15 = 3 \times 5$$

Try These Problems

Write each number as the product of primes.
12. 21
13. 49
14. 10
15. 39
16. 115

In fact, any whole number that is not prime, except for 0 and 1, can be written as the product of primes. The process of writing numbers as the product of primes is called **prime factoring.** The following examples illustrate how to write numbers as the product of primes.

EXAMPLE 11   Write each number as the product of primes.

a. 49   **b.** 35   **c.** 22   **d.** 51

SOLUTION

a. $49 = 7 \times 7$

b. $35 = 5 \times 7$

c. Because 22 ends in 2, it is divisible by 2. Divide 22 by 2 to find the other factor.

$$2\overline{)22} \quad \begin{array}{c} 11 \\ \underline{22} \end{array}$$

$22 = 2 \times 11$

d. 51 is divisible by 3 because $5 + 1 = 6$ and 6 is divisible by 3. Divide 51 by 3 to find the other factor.

$$3\overline{)51} \quad \begin{array}{c} 17 \\ \underline{3} \\ 21 \\ \underline{21} \end{array}$$

$51 = 3 \times 17$. ■

In Example 11, observe that numbers like 49 and 35 are easy to factor because you recognize them as multiplication facts. To factor other numbers, such as 22 and 51, you need to use the divisibility rules and division.

 Try Problems 12 through 16.

Sometimes it takes several steps to get a number written as the product of primes. The following examples illustrate the technique.

EXAMPLE 12   Write each number as the product of primes.

a. 40   **b.** 54   **c.** 2900   **d.** 2205

SOLUTION

a. Because 40 ends in a 0, 40 is divisible by 10.

$$40 = 4 \times 10$$
$$= 2 \times 2 \times 2 \times 5$$

You can get started with any two numbers that multiply together to give 40. Do not stop until all the factors are prime numbers.

2.3 The Missing Number in a Multiplication Statement, Multiples, Divisibility Rules, and Prime Factoring   55

 Try These Problems

Write each number as the product of primes.
17. 60
18. 56
19. 8100
20. 147
21. 425

b. Because 54 ends in 4, 54 is divisible by 2. Divide 54 by 2 to find the other factor.

$$2\overline{)54} \atop \underline{4} \atop 14 \atop \underline{14}$$ gives 27

$54 = 2 \times 27$ Begin with 2×27.
$ = 2 \times 3 \times 9$ Write 27 as 3×9.
$ = 2 \times 3 \times 3 \times 3$ Finally, $9 = 3 \times 3$.

c. Because 2900 ends in two 0s, it is divisible by 100.

$2900 = 29 \times 100$ Now 29 is prime, but 100 is not prime.
$ = 29 \times 10 \times 10$
$ = 29 \times 2 \times 5 \times 2 \times 5$ Now all the factors are prime.

d. Because 2205 ends in 5, it is divisible by 5. Divide 2205 by 5 to find the other factor.

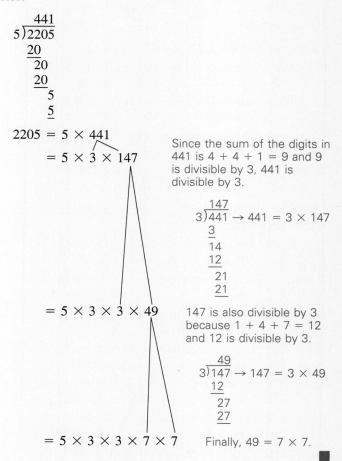

$2205 = 5 \times 441$
$ = 5 \times 3 \times 147$ Since the sum of the digits in 441 is $4 + 4 + 1 = 9$ and 9 is divisible by 3, 441 is divisible by 3.

$3\overline{)441} \rightarrow 441 = 3 \times 147$

$ = 5 \times 3 \times 3 \times 49$ 147 is also divisible by 3 because $1 + 4 + 7 = 12$ and 12 is divisible by 3.

$3\overline{)147} \rightarrow 147 = 3 \times 49$

$ = 5 \times 3 \times 3 \times 7 \times 7$ Finally, $49 = 7 \times 7$.

 Try Problems 17 through 21.

Answers to Try These Problems

1. 100   2. 28   3. 349,200   4. 802
5. a. yes b. yes c. yes d. no
6. a. no b. yes c. yes d. yes
7. a. no b. yes c. yes d. no
8. a. yes b. yes c. no d. yes
9. a. yes b. no c. yes d. yes
10. a. no b. no c. yes d. yes
11. a. no b. yes c. yes d. no
12. 3×7 13. 7×7 14. 2×5 15. 3×13
16. 5×23 17. $2 \times 3 \times 2 \times 5$ 18. $2 \times 2 \times 2 \times 7$
19. $3 \times 3 \times 3 \times 3 \times 2 \times 5 \times 2 \times 5$ 20. $3 \times 7 \times 7$
21. $5 \times 5 \times 17$

EXERCISES 2.3

Find the missing number.

1. $9 \times \boxed{} = 108$
2. $8 \times \boxed{} = 136$
3. $572 = 52 \times \boxed{}$
4. $\boxed{} = 19 \times 83$
5. $135{,}000 = \boxed{} \times 450$
6. $28{,}800 = \boxed{} \times 360$
7. $506 \times 8000 = \boxed{}$
8. $\boxed{} \times 85 = 25{,}840$

In Exercises 9 through 12, answer yes or no.

9. Indicate whether each number is divisible by 7.
   a. 42   b. 105   c. 177   d. 322
10. Indicate whether each number is divisible by 11.
    a. 99   b. 255   c. 666   d. 495
11. Indicate whether each number is a multiple of 25.
    a. 75   b. 85   c. 400   d. 3025
12. Indicate whether each number is a multiple of 36.
    a. 110   b. 540   c. 28,800   d. 6120

In Exercises 13 through 22, answer yes or no. Use the divisibility rules for 2, 3, 5, 10, 100, and 1000.

13. Indicate whether each number is divisible by 2.
    a. 30   b. 112   c. 248   d. 35,714
14. Indicate whether each number is divisible by 2.
    a. 81   b. 94   c. 356   d. 1078
15. Indicate whether each number is divisible by 3.
    a. 81   b. 93   c. 255   d. 533
16. Indicate whether each number is divisible by 3.
    a. 73   b. 84   c. 306   d. 1509
17. Indicate whether each number is divisible by 5.
    a. 35   b. 570   c. 3545   d. 5052
18. Indicate whether each number is divisible by 5.
    a. 60   b. 315   c. 7003   d. 9000
19. Indicate whether each number is divisible by 10.
    a. 203   b. 200   c. 2130   d. 7000

2.3 The Missing Number in a Multiplication Statement, Multiples, Divisibility Rules, and Prime Factoring

20. Indicate whether each number is divisible by 10.
    a. 40   b. 3200   c. 1905   d. 12,460
21. Indicate whether each number is divisible by 100.
    a. 120   b. 800   c. 13,000   d. 40,500
22. Indicate whether each number is divisible by 1000.
    a. 1300   b. 2060   c. 15,000   d. 40,500,000

Write each number as the product of primes.

23. 9   24. 14   25. 12   26. 20   27. 45
28. 48   29. 55   30. 77   31. 26   32. 85
33. 57   34. 92   35. 114   36. 88   37. 245
38. 600   39. 150   40. 360   41. 1200   42. 4500
43. 297   44. 117   45. 87   46. 69   47. 504
48. 1125

USING THE CALCULATOR #4

INTRODUCTION TO THE MEMORY KEYS

Most basic and scientific calculators have a memory. This means the calculator has the capability of storing a number that can be recalled later. A number put in the memory remains there while other calculations are done. The memory is usually unaffected by turning off the calculator.

The memory keys on a basic calculator may include all or some of the following.

[M+] Entering a number, then this key, will add the number to the existing memory.

[MR] This is the memory recall key. Enter this key to find out what number is currently in the memory and to ready it for use.

[M−] Entering a number, then this key, will subtract the number from the existing memory. This key can be used to clear the memory on a basic calculator.

[MC] This key clears the memory. Not all basic calculators have this key.

Most scientific calculators have the [M+], [MR], and [M−] keys, but usually do not have the [MC] key. In addition to these memory keys, scientific calculators have a memory-in key that can be used to replace the existing number in the memory with another number.

[Min] or [x→M]   This is the memory-in key. Entering a number, then this key, will replace the existing memory with the number. This key can be used to clear the memory on a scientific calculator by putting the number 0 in the memory.

Observe that entering the [C] or [AC] key does not clear the memory. Also, turning off the calculator does not clear the memory.

Calculator Problems

Try each of the following activities on your calculator.

1. Find out what number is currently in the memory of your calculator. If it is not 0, then clear the memory, that is, put a 0 in the memory.

Cont. page 58

2. Put the number 8 in the memory of your calculator. Check to make sure that 8 is really in the memory.
3. Clear the number 8 from the memory, that is, put a 0 in the memory.
4. Put the number 15 in the memory. Now enter 5, then enter the key [M+]. What number is in the memory now?

Solve the following problems by putting the number 7500 in the memory, then recalling it as needed.

5. $7500 - 657$ **6.** 7500×18 **7.** $7500 \div 12$ **8.** $172{,}500 \div 7500$

Write each of these numbers as the product of primes. Store the number in memory before beginning to check for prime divisors.

9. 221 **10.** 161 **11.** 2109 **12.** 4199

2.4 LANGUAGE

OBJECTIVES
- Translate English statements involving multiplication and division to math symbols.
- Translate a division statement written with math symbols to English.
- Solve problems by using translations.

1 Translating

The math symbols that we use often have many translations to the English language. Knowing these translations can help in solving application problems. Here we review the many English translations for the symbol $=$.

English	Math Symbol
equals	$=$
is equal to	
is the same as	
is the result of	
is	
was	
represents	
gives	
makes	
yields	
will be	
are	
were	

2.4 Language

 Try These Problems

Write a division statement using the symbol ⟌.
1. 7 divided into 63 equals 9.
2. 100 divided by 25 is 4.

Write a division statement using the symbol ÷.
3. 5 divided into 35 equals 7.
4. 60 divided by 12 is 5.

Fill in the blank with the appropriate word, by *or* into.

5. $4\overline{)24}^{\,6}$ is read "24 divided _____ 4 equals 6."
6. $7\overline{)70}^{\,10}$ is read "7 divided _____ 70 equals 10."
7. 84 ÷ 12 = 7 is read "84 divided _____ 12 is 7."
8. 500 ÷ 25 = 20 is read "25 divided _____ 500 is 20."

The multiplication statement 3 × 7 = 21 is written with math symbols. Some of the ways to read this in English are given in the following chart.

English	Math Symbols
Three times seven equals twenty-one.	3 × 7 = 21
The *product* of 3 and 7 is 21.	21 = 3 × 7
21 is 3 times as large as 7.	
3 multiplied by 7 gives 21.	
21 represents 3 multiplied times 7.	

Because 3 × 7 equals 7 × 3, the order that you say the factors 3 and 7 makes no difference. However, when reading the division statement 21 ÷ 3 = 7, take care to say 21 and 3 in the correct order, because 21 ÷ 3 does not equal 3 ÷ 21.

$$21 \div 3 = 7, \quad \text{but} \quad 3 \div 21 = \frac{3}{21} = \frac{1}{7}$$

The numbers 7 and $\frac{1}{7}$ are *not* equal. The number $\frac{1}{7}$ is a fraction less than 1. We study fractions in Chapter 3. The following chart gives correct ways to read 21 ÷ 3 = 7 or $3\overline{)21}^{\,7}$.

Math Symbols	English
21 ÷ 3 = 7	21 *divided by* 3 is 7.
$3\overline{)21}^{\,7}$	3 *divided into* 21 is 7.

 Try Problems 1 through 8.

Special language is used when the number 2 is a factor in a multiplication statement or 2 is the divisor in a division statement. The following chart illustrates this.

Math Symbols	English
2 × 8 = 16	Two times 8 equals 16.
	Twice 8 equals 16.
	Eight *doubled* is 16.
	16 represents the product of 2 and 8.
16 ÷ 2 = 8	16 divided by 2 equals 8.
$2\overline{)16}^{\,8}$	2 divided into 16 equals 8.
	Half of 16 is 8.

 2 Solving Problems by Using Translating

The following examples illustrate how we can use translating to solve problems.

Try These Problems

Solve.

9. The product of a number and 17 is 1360. Find the number.
10. What number is twice 268?
11. Find 1659 divided by 7.
12. 408 represents twice what number?
13. Divide 30 into 2070.
14. What number is 450 multiplied by itself?

EXAMPLE 1 Twice a number yields 134. Find the number.

SOLUTION The sentence, "twice a number yields 134," translates to a multiplication statement.

$$\underbrace{\text{Twice}}_{2\,\times}\ \underbrace{\text{a number}}_{\Box}\ \text{yields 134.} = 134$$

Because a multiplier is missing, we divide to find the missing multiplier.

$$\begin{array}{r}67\\2\overline{)134}\\\underline{12}\\14\\\underline{14}\end{array}$$

The number is 67.

EXAMPLE 2 Find the product of 50 and 75,000.

SOLUTION To find the product of two numbers means to multiply.

$$\underbrace{\text{Product of 50 and 75,000}}_{50\,\times\,75{,}000} = \Box$$

$$\begin{array}{r}75000\\\times\ \ 50\\\hline 3750000\end{array}$$

Multiply 75 by 5 and attach 4 zeros.

The answer is 3,750,000.

EXAMPLE 3 6 divided into 30,048 is what number?

SOLUTION The question translates to a division statement.

$$\underbrace{\text{6 divided into 30,048}}_{30{,}048\,\div\,6}\ \text{is what number?} = \Box$$

We divide 30,048 by 6 to find the missing number.

$$\begin{array}{r}5008\\6\overline{)30048}\\\underline{30}\\0\\\underline{0}\\4\\\underline{0}\\48\\\underline{48}\end{array}$$

The number is 5008.

Try Problems 9 through 14.

Answers to Try These Problems

1. $7\overline{)63}$ with 9 on top 2. $25\overline{)100}$ with 4 on top 3. $35 \div 5 = 7$ 4. $60 \div 12 = 5$
5. by 6. into 7. by 8. into 9. 80 10. 536 11. 237
12. 204 13. 69 14. 202,500

EXERCISES 2.4

Write a division statement using the symbol $\overline{)}$.

1. 9 divided into 54 equals 6.
2. 4 divided into 52 is 13.
3. 30 divided into 1350 is 45.
4. 15 divided by 3 equals 5.
5. 75 divided by 15 is 5.
6. 400 divided by 40 is 10.

Write a division statement using the symbol $\div$.

7. 8 divided into 16 equals 2.
8. 7 divided into 119 equals 17.
9. 50 divided into 600 equals 12.
10. 84 divided by 28 is 3.
11. 621 divided by 3 is 207.
12. 598 divided by 26 equals 23.

Fill in the blank with the appropriate word, by *or* into.

13. $4\overline{)28}$ with 7 on top is read "4 divided _____ 28 equals 7."
14. $4\overline{)28}$ with 7 on top is read "28 divided _____ 4 equals 7."
15. $9\overline{)117}$ with 13 on top is read "117 divided _____ 9 is 13."
16. $9\overline{)117}$ with 13 on top is read "9 divided _____ 117 is 13."
17. $12 \div 4 = 3$ is read "12 divided _____ 4 equals 3."
18. $12 \div 4 = 3$ is read "4 divided _____ 12 equals 3."
19. $144 \div 9 = 16$ is read "9 divided _____ 144 is 16."
20. $144 \div 9 = 16$ is read "144 divided _____ 9 is 16."

Translate each to a multiplication or division statement using math symbols.

21. The product of 17 and 38 is 646.
22. Twenty-five divided into 350 yields 14.
23. Thirty-three represents 3 times 11.
24. 5 equals 490 divided by 98.
25. Twice 230 gives 460.
26. 900 represents twice 450.

Solve.

27. Eighty-two multiplied by what number is 984?
28. 214 represents twice what number?
29. Divide 810 by 5.
30. Divide 20 into 1040.
31. What number is the product of 1300 and 60?
32. The product of 135 and what number is 8775?
33. What number equals twice 9500?
34. Find half of 712.

2.5 APPLICATIONS

OBJECTIVES
- Solve an application involving equal parts in a total quantity.
- Solve an application involving equivalent rates.
- Solve an application that involves a rate as a multiplier.
- Solve an application that involves the area of a rectangle.
- Solve an application by using translations.
- Find the average (or mean) of a collection of numbers.
- Solve an application involving more than one step.

1 Equal Parts in a Total Quantity

Suppose you have a string that is 20 inches long. If the string is cut into 4 equal pieces, then each piece is 5 inches long. Here we picture the situation.

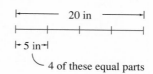

Observe that the numbers are related as follows.

$$\begin{array}{ccc} \text{Number of} & & \text{Size of} & & \text{Total} \\ \text{equal parts} & \times & \text{each part} & = & \text{quantity} \\ 4 & \times & 5 & = & 20 \end{array}$$

In general, if a situation involves a number of equal parts making up a total quantity, the following formula always applies.

$$\begin{array}{ccc} \textbf{Number of} & & \textbf{Size of} & & \textbf{Total} \\ \textbf{equal parts} & \times & \textbf{each part} & = & \textbf{quantity} \end{array}$$

A problem can be presented by giving any two of the three quantities and asking for the third quantity. The following examples illustrate this.

2.5 Applications

EXAMPLE 1 A nurse is to give 15 doses of medicine. Each dose contains 2 milliliters of medicine. How much medicine is needed?

SOLUTION This problem is about 15 doses that are 2 milliliters each, making up a total quantity. We use the formula,

$$\begin{array}{c}\text{Number of}\\\text{equal parts}\end{array} \times \begin{array}{c}\text{Size of}\\\text{each part}\\(\text{m}\ell)\end{array} = \begin{array}{c}\text{Total}\\\text{quantity}\\(\text{m}\ell)\end{array}$$

$$15 \quad \times \quad 2 \quad = \quad \boxed{}$$

The 15 doses is the number of equal parts and 2 milliliters is the size of each part. We are missing the total quantity, so we multiply.

$$15 \times 2 = 30$$

30 milliliters of medicine is needed. ■

EXAMPLE 2 Juan earns $13,000 over a 52-week period. If he earns the same amount each week, what is his weekly salary?

SOLUTION This problem is about 52 equal payments amounting to a total of $13,000. We use the formula,

$$\begin{array}{c}\text{Number of}\\\text{equal parts}\end{array} \times \begin{array}{c}\text{Size of}\\\text{each part}\\(\$)\end{array} = \begin{array}{c}\text{Total}\\\text{quantity}\\(\$)\end{array}$$

$$52 \quad \times \quad \boxed{} \quad = \quad 13{,}000$$

The number of equal parts is 52 and $13,000 is the total quantity. We are missing one of the multipliers, so we divide to find the missing number.

$$\begin{array}{r}250\\52\overline{)13000}\\\underline{104}\\260\\\underline{260}\\00\\\underline{0}\end{array}$$

The weekly salary is $250. ■

EXAMPLE 3 Stuart sells corn from his garden. He puts 1290 ears of corn in packages with 6 ears in each package. How many packages are made?

SOLUTION This problem is about a total quantity, 1290 ears of corn, being split apart with each part containing 6 ears of corn. We use the formula,

$$\begin{array}{c}\text{Number of}\\\text{equal parts}\end{array} \times \begin{array}{c}\text{Size of}\\\text{each part}\\(\text{ears of corn})\end{array} = \begin{array}{c}\text{Total}\\\text{quantity}\\(\text{ears of corn})\end{array}$$

$$\boxed{} \quad \times \quad 6 \quad = \quad 1290$$

Try These Problems

Solve.
1. A wire that is 92 inches long is cut into 4 equal pieces. How long is each piece?
2. A truck carries a load of 15 cases. The total load weighs 1230 pounds. What is the weight of each case?
3. One hundred fifty pieces of track that are each 36 feet long are to be laid one right after the other. How far will the track extend?
4. Cindy is transporting 3300 pounds of sand in truckloads of 220 pounds each. How many truckloads will it take to transport all the sand?
5. How many 25-minute speeches fit in a 320-minute time period? How much time, if any, is left over?
6. Kenny and his 5 sisters share equally an inheritance. If each person's share is $2060, what was the total amount of the inheritance?

We are missing a multiplier, the number of equal parts, so we divide.

$$\begin{array}{r} 215 \\ 6\overline{)1290} \\ \underline{12} \\ 09 \\ \underline{6} \\ 30 \\ \underline{30} \end{array}$$

215 packages can be made. ■

EXAMPLE 4 Stephanie has 1260 centimeters of string. She wants to cut it into pieces that are each 16 centimeters long. How many pieces can she cut? How much string, if any, is left over?

SOLUTION We divide the total quantity, 1260 centimeters, by the size of each piece, 16 centimeters, to find out how many pieces she can cut. The remainder will be the amount of string left over.

$$\begin{array}{r} 78 \text{— number of pieces}\\ 16\overline{)1260} \\ \underline{112} \\ 140 \\ \underline{128} \\ 12 \text{— amount of string left over} \end{array}$$

She can cut 78 pieces with 12 centimeters of string left over. ■

 Try Problems 1 through 6.

2 Equivalent Rates

If you drive 50 miles in 1 hour, then the phrase *50 miles in 1 hour,* or *50 miles per hour,* is specifying the **speed** that you are traveling. Speed is one of the many examples of **rate**. A rate is a comparison of two quantities. Here we list some more phrases that indicate rates.

> *Phrases that Indicate Rates*
> You type 65 words per minute.
> Chuck paid $30 for each shirt.
> Cynthia jogs 5 miles in 30 minutes.
> Four pounds of coffee cost $20.
> On a map, 1 inch represents 25 miles.

If you drive 50 miles in 1 hour, then you can go 150 miles in 3 hours.

Multiply by 3. 50 miles in 1 hour   Multiply by 3.
              150 miles in 3 hours

Observe that multiplying both quantities in a rate by the same nonzero number produces an equivalent rate. We can use this to solve problems involving equivalent rates, as shown in the following examples.

EXAMPLE 5 You type 65 words per minute. How many words can you type in 20 minutes?

SOLUTION The phrase *65 words per minute,* means 65 words every 1 minute.

Multiply by 20. ⎧ 65 words every 1 minute ⎫ Multiply by 20.
⎩ ? words every 20 minutes ⎭

If we multiply each quantity in the given rate by 20, we can see how many words can be typed in 20 minutes.

$$65 \times 20 = 1300$$

You can type 1300 words in 20 minutes. ■

EXAMPLE 6 The faucet leaks 1 ounce every 5 hours. How long will it take the faucet to leak 12 ounces?

SOLUTION

Multiply by 12. ⎧ 1 ounce every 5 hours ⎫ Multiply by 12.
⎩ 12 ounces in ? hours ⎭

If we multiply each quantity in the given rate by 12, we obtain the number of hours it takes the faucet to leak 12 ounces.

$$5 \times 12 = 60$$

It takes 60 hours for the faucet to leak 12 ounces. ■

In the previous examples, you saw that multiplying both quantities of a rate by the same nonzero number produces an equivalent rate. Now we illustrate that you can also divide both quantities of a rate by the same nonzero number and obtain an equivalent rate.

If 4 pounds of coffee cost $20, then 1 pound of this coffee costs $5.

Divide by 4. ⎧ 4 pounds cost $20 ⎫ Divide by 4.
⎩ 1 pound costs $5 ⎭

Observe that dividing both quantities in a rate by the same nonzero number produces an equivalent rate. We can use this to solve problems involving equivalent rates, as shown in the following examples.

EXAMPLE 7 It takes Cynthia 40 minutes to jog 5 miles. How long will it take her to jog 1 mile?

SOLUTION

Divide by 5. ⎧ 40 minutes for 5 miles ⎫ Divide by 5.
⎩ ? minutes for 1 mile ⎭

Dividing each quantity in the given rate by 5 gives the number of minutes it takes Cynthia to jog 1 mile.

$$40 \div 5 = 8$$

It takes 8 minutes for Cynthia to jog 1 mile. ■

Try These Problems

7. One ounce of gold is worth $545. What is 3 ounces of gold worth?

8. Steve drives 45 miles in 1 hour. How far can he go in 6 hours?

9. Ellen runs 6 miles in 48 minutes. How long does it take her to run 1 mile?

10. Henry paid $30 for 15 pounds of mushrooms. What is the cost of 1 pound?

11. A cable 17 feet long weighs 68 pounds. What is the weight per foot?

12. On a map, 1 inch represents 200 miles. If two cities are 5 inches apart on the map, what is the actual distance between them?

EXAMPLE 8   Bruce types 2400 words in 30 minutes. How many words can he type in 1 minute?

SOLUTION

Divide by 30.   2400 words in 30 minutes   Divide by 30.
? words in 1 minute

Dividing both quantities in the given rate by 30 gives the number of words Bruce can type in 1 minute.

$$\begin{array}{r} 80 \\ 30\overline{)2400} \\ \underline{240} \\ 00 \\ \underline{0} \end{array}$$

Bruce types 80 words per minute. ■

 Try Problems 7 through 12.

3 More About Rates

Because a rate is a comparison of two quantities, the units associated with a rate are like.

miles per hour

cost per foot

ounces for each person

dollars per pound

If you buy 8 pounds of coffee at $6 per pound, then you pay a total of $48 for this coffee.

Dollars per pound × Number of pounds = Total dollars

6 × 8 = 48

Note that the rate, $6 per pound, is multiplied by the number of pounds to get the total cost. Notice how the units agree.

Dollars per pound × pounds = dollars
(units agree)

If you pay close attention to the units associated with a rate, it can help you to solve problems involving rates. Here we show more examples.

miles per hour × number of hours = total miles

cost per foot × number of feet = total cost

ounces for each person × number of persons = total ounces

Try These Problems

13. A rope weighs 28 ounces per foot. How much does 350 feet of this rope weigh?

14. Steve types 65 words per minute. How long will it take him to type 2470 words?

15. The owner of a small business purchases 12 computers that cost $6000 each. What is the total cost?

16. A tank is leaking 8 gallons per minute. How long will it take the tank to leak 472 gallons?

EXAMPLE 9 The Miller family bought 5 opera tickets at $38 each. What was the total cost?

SOLUTION $38 is a rate because it is the cost per ticket.

$$\begin{array}{c} \text{cost per ticket} \\ (\$) \end{array} \times \begin{array}{c} \text{number of} \\ \text{tickets} \end{array} = \begin{array}{c} \text{total cost} \\ (\$) \end{array}$$

$$38 \quad \times \quad 5 \quad = \quad \boxed{}$$

We multiply to find the total cost.

$$38 \times 5 = 190$$

The total cost was $190. ∎

EXAMPLE 10 A car travels 816 miles averaging 48 miles per hour. How long did the trip take?

SOLUTION The quantity 48 miles per hour is a rate.

$$\text{miles per hour} \times \begin{array}{c} \text{number of} \\ \text{hours} \end{array} = \begin{array}{c} \text{total} \\ \text{miles} \end{array}$$

$$48 \quad \times \quad \boxed{} \quad = \quad 816$$

A multiplier is missing, so we divide.

$$\begin{array}{r} 17 \\ 48\overline{)816} \\ \underline{48} \\ 336 \\ \underline{336} \end{array}$$

The trip took 17 hours. ∎

 Try Problems 13 through 16.

4 Area of a Rectangle

Area is a measure of the extent of a region. A square that measures 1 unit on each side is said to have an area of 1 square unit.

1 unit ▢ Area = 1 square unit
1 unit

A rectangle that is 3 units wide and 5 units long contains 15 of the 1-square-unit squares. Therefore, the area is 15 square units.

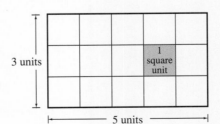

Area = 15 square units

Try These Problems

17. Find the area of this rectangle.

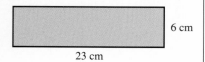
6 cm
23 cm

18. A window is 9 feet wide and 5 feet high. How many square feet of glass are needed for this window?

19. The floor of a room is in the shape of a rectangle with area 360 square feet. If the width is 15 feet, find the length.

20. Anthony bought 910 square inches of glass in the shape of a rectangle. If the glass is 26 inches wide, how long is it?

Observe that the area of this rectangle can be obtained by multiplying the length by the width.

$$\begin{aligned}\text{Area} &= \text{length} \times \text{width} \\ &= 5 \times 3 \\ &= 15 \text{ square units}\end{aligned}$$

In general, the area of any rectangle can be computed by multiplying the length by the width.

Area of a Rectangle
The area of a rectangle is the length times the width.
Area = length × width

EXAMPLE 11 Find the area of a rectangle whose width is 18 feet and length is 25 feet.

SOLUTION

$$\begin{array}{cccc}\text{Area} & = & \text{length} & \times & \text{width} \\ (\text{sq ft}) & & (\text{ft}) & & (\text{ft}) \\ \square & = & 25 & \times & 18\end{array}$$

Multiply 25 by 18 to find the area.

```
     25
   × 18
   ----
    200
     25
   ----
    450
```

The area is 450 square feet. ■

EXAMPLE 12 The area of a piece of fabric is 3900 square inches. If the fabric is 52 inches wide, how long is it?

SOLUTION We assume the piece of fabric is in the shape of a rectangle.

$$\begin{array}{cccc}\text{Area} & = & \text{length} & \times & \text{width} \\ (\text{sq in}) & & (\text{in}) & & (\text{in}) \\ 3900 & = & \square & \times & 52\end{array}$$

The area 3900 square inches is given, and the width 52 inches is given. We are missing the length, one of the multipliers, so we divide.

```
       75
   52)3900
      364
      ---
       260
       260
```

The fabric is 75 inches long. ■

 Try Problems 17 through 20.

▲ Try These Problems

21. There are seven times as many men as women attending the convention. Forty-nine women are at the convention. How many men are there?

22. Ralph bought a condominium for $47,000. The value has doubled in five years. What is the condominium worth after the five years?

23. The cost of a car is 4 times what it was ten years ago. If the car now costs $21,000, what did it cost ten years ago?

24. The length of a rectangle is 9 times as long as the width. If the length is 963 centimeters, what is the width?

5 Problems Involving Translations

In Section 2.4 we saw how translating English phrases to math symbols can be used to solve problems. Now we look at examples involving real-life situations.

EXAMPLE 13 It takes Barbara 3 times as long to paint her deck as it does a professional painter. It took a professional painter 5 hours to do the job. How long would it take Barbara to paint the deck?

SOLUTION The first sentence, "It takes Barbara 3 times as long to paint her deck as it does a professional painter," can be translated to a multiplication statement. The sentence is saying,

Barbara's time (hr)	is	3	times as long as	the professional's time (hr)
☐	=	3	×	5

We are missing Barbara's time which is the answer to the multiplication statement, so we multiply.

$$3 \times 5 = 15$$

It takes Barbara 15 hours to paint the deck. ■

EXAMPLE 14 With a lot of practice, Phil now types 76 words per minute, which is twice his speed a month ago. How fast did he type a month ago?

SOLUTION The sentence, "Phil now types 76 words per minute, which is twice his speed a month ago," translates to a multiplication statement.

Phil's speed now (wpm)	is	twice	his speed a month ago (wpm)
76	=	2 ×	☐

We are missing one of the multipliers, so we divide.

$$\begin{array}{r} 38 \\ 2\overline{)76} \\ \underline{6} \\ 16 \\ \underline{16} \end{array}$$

Phil's typing speed a month ago was 38 words per minute. ■

▲ **Try Problems 21 through 24.**

6 Averaging

Fred took three exams. His scores were 75, 88, and 77. He has a total of 240 points.

$$\begin{array}{r} 75 \\ 88 \\ +\ 77 \\ \hline 240 \end{array}$$

Try These Problems

25. Find the average of 121 and 233.

26. In preparing for her vacation, Teresita bought six dresses at the following prices:

    $23 $19 $32 $26 $15 $41

    What was the average price of these dresses?

What same score could Fred have made on each of the three exams and still have a total of 240?

$$3\overline{)240}^{\,80} \qquad \left.\begin{array}{r}80\\80\\+\,80\\\hline 240\end{array}\right\} \text{three scores of 80 give the same total}$$

The score 80 is called the **average** (or **mean**) of the scores 75, 88, and 77. The average of a set of data is a measure of the middle or center of the data. The average of a collection of numbers can be found by adding the numbers, then dividing by how many numbers there are.

To Find the Average of a Collection of Numbers
1. Add all of the numbers in the collection.
2. Divide the sum by how many numbers are in the collection.

EXAMPLE 15 Find the average of 17, 133, 82, and 36.

SOLUTION First, add the four numbers.

$$17 + 133 + 82 + 36 = 268$$

Second, divide the sum 268 by 4.

$$4\overline{)268}^{\,67}$$
$$\underline{24}$$
$$28$$
$$\underline{28}$$

The average of the four numbers is 67. ■

EXAMPLE 16 A salesperson earned the following commissions in the last five weeks:

$168 $194 $216 $186 $136

What is the average weekly commission?

SOLUTION First, add the five commissions. Second, divide the total 900 by 5.

```
    168              180  — average
    194          5)900
    216              5
    186              —
  + 136              40
  ——                 40
    900 — total      —
                      0
                      0
                      —
```

The average weekly commission is $180. ■

 Try Problems 25 and 26.

7 Applications Involving More Than One Step

Now we look at examples of applications that require more than one step to solve.

EXAMPLE 17 You plan to put weather stripping around a rectangular window that is 4 feet by 5 feet. If the stripping costs $2 per foot, what is the total cost?

SOLUTION Since the weather stripping costs $2 per foot, the total cost can be found as follows.

Cost per foot × number of feet = total cost
2 × ☐ = ☐

Before finding the total cost, we need the number of feet around the rectangular window; that is, we need the *perimeter* of the window.

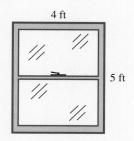

Perimeter = 5 + 5 + 4 + 4
= 10 + 8
= 18 ft

Total Cost = cost per foot × number of feet
= 2 × 18
= $36 ■

EXAMPLE 18 If carpeting costs $19 per square yard, what is the total cost of carpeting this floor?

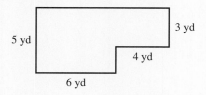

SOLUTION Since the carpeting costs $19 per square yard, the total cost can be found as follows.

Cost per square yard × number of square yards = total cost
19 × ☐ = ☐

Try These Problems

27. A rectangular window is 3 times as wide as it is high. If the height is 6 feet and glass costs $5 per square foot, what is the cost of enough glass for this window?

28. A farmer wishes to fence in a region as shown. If fencing costs $25 per foot, what is the total cost of fencing this region?

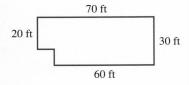

29. You purchase 8 shirts at $15 each, 5 pairs of socks at $3 each, and 6 pairs of slacks at $34 each. After paying for these items, what will you have left out of $500?

Before finding the total cost, we need the number of square yards that cover the floor space; that is, we need the *area* of the floor. To find the area of the floor, view it as two rectangles.

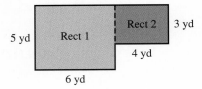

Area of Rect 1
$= 5 \times 6$
$= 30$ sq yd

Area of Rect 2
$= 3 \times 4$
$= 12$ sq yd

Total Area of the floor $= 30 + 12 = 42$ sq yd

$$\text{Total Cost} = \frac{\text{cost per}}{\text{square yard}} \times \frac{\text{number of}}{\text{square yards}}$$
$$= 19 \times 42$$
$$= \$798 \quad \blacksquare$$

 Try Problems 27 and 28.

EXAMPLE 19 A tank contained its total capacity of 2000 liters of water, and then began to leak at a rate of 3 liters per hour. How much water was in the tank after 9 hours?

SOLUTION First we need to find the amount of water that the tank has lost.

$$\text{Amount lost} = \frac{\text{amount lost}}{\text{per hour}} \times \frac{\text{number}}{\text{of hours}}$$
$$= 3 \times 9$$
$$= 27 \text{ liters}$$

To find the amount of water remaining in the tank, subtract 27 from 2000.

$$\text{Amount remaining} = 2000 - 27 = 1973 \text{ liters} \quad \blacksquare$$

 Try Problem 29.

Try These Problems

30. Mr. Harrison, the owner of a small flower market, purchased 3612 flowers. After arranging them in bunches of 18, he took home the left-over flowers and split them equally among four of his neighbors. How many flowers did each neighbor receive?

31. Harriet's annual salary has increased from $37,020 to $50,100 over a 12-year period. On the average, how much did her salary increase each year?

EXAMPLE 20 A machine fills and caps 300 bottles of soda each hour. After 18 hours how many 6-packs of soda have been filled and capped?

SOLUTION First we find the total number of bottles filled and capped in 18 hours.

$$\text{Total number of bottles} = \text{bottles each hour} \times \text{number of hours}$$
$$= 300 \times 18$$
$$= 5400$$

Now we find how many 6-packs can be made from 5400 bottles. Divide 5400 by 6.

$$6\overline{)5400} = 900$$

After 18 hours, 900 6-packs have been filled and capped. ■

 Try Problem 30.

EXAMPLE 21 After having a massive heart attack, a patient's weight dropped from 270 pounds to 218 pounds in 4 weeks. On the average, how much weight did the patient lose per week?

SOLUTION First find the total weight lost by subtracting 218 from 270.

$$\text{Weight lost} = 270 - 218$$
$$= 52 \text{ pounds}$$

The weight lost per week is a rate. We know that

$$\text{weight lost per week} \times \text{number of weeks} = \text{total weight lost}$$
$$\square \times 4 = 52$$

We are missing a multiplier, so we divide 52 by 4 to find the missing multiplier.

$$4\overline{)52} \quad \begin{array}{r} 13 \\ \underline{4} \\ 12 \\ \underline{12} \end{array}$$

The patient lost 13 pounds per week. ■

 Try Problem 31.

Now we summarize the material in this section by giving guidelines that will help you to decide whether to multiply or divide.

Situations that Require Multiplication

× 1. You are looking for a total quantity.

$$\text{Total quantity} = \text{size of each part} \times \text{number of equal parts}$$

2. You are looking for a total, as in the following examples.

$$\text{Total miles} = \text{miles per hour} \times \text{number of hours}$$

$$\text{Total cost} = \text{cost per pound} \times \text{number of pounds}$$

3. You are looking for a number that is a certain amount times as large as another number.
4. You are looking for the area of a rectangle.

$$\text{Area} = \text{length} \times \text{width}$$

Situations that Require Division

÷ 1. A total quantity is being separated into a number of equal parts.

$$\text{Size of each part} = \text{total quantity} \div \text{number of equal parts}$$

$$\text{Number of equal parts} = \text{total quantity} \div \text{size of each part}$$

2. You are looking for a missing multiplier in a multiplication statement.

$$\text{Missing multiplier} = \text{answer to the multiplication problem} \div \text{given multiplier}$$

3. You are looking for an average.

$$\text{Average of a collection of numbers} = \text{sum of the numbers} \div \text{how many numbers}$$

$$\text{Average cost per item} = \text{total cost} \div \text{number of items}$$

$$\text{Average miles per hour} = \text{total miles} \div \text{number of hours}$$

▲ Answers to Try These Problems

1. 23 in   2. 82 lb   3. 5400 ft   4. 15 truckloads
5. 12 speeches, 20 min left over   6. $12,360   7. $1635
8. 270 mi   9. 8 min   10. $2   11. 4 lb   12. 1000 mi
13. 9800 oz   14. 38 min   15. $72,000   16. 59 min
17. 138 sq cm   18. 45 sq ft   19. 24 ft   20. 35 in
21. 343 men   22. $94,000   23. $5250   24. 107 cm
25. 177   26. $26   27. $540   28. $5000   29. $161   30. 3
31. $1090

EXERCISES 2.5

Solve.

1. A program consists of 8 speeches that are each 15 minutes long. If the speeches are given one right after the other, how long is the program?
2. A cord that is 2232 centimeters long is cut into 36 equal pieces. How long is each piece?
3. Carlos has 3145 pounds of fertilizer dust that he wants to spray equally over 185 acres of land. How much should he spray over each acre?
4. Mr. Benitez purchased 18,300 square yards of land in Texas. He wants to divide it into lots of 500 square yards each. How many lots will he have? How much land, if any, is left over?
5. A chandelier holds 18 bulbs. How many chandeliers can be outfitted with 3700 bulbs? How many bulbs, if any, are left over?
6. Twelve partners in a company share equally a profit. If each person's share is $39,000, what is the total profit?
7. Susan walks 4 miles in 1 hour. How far can she walk in 8 hours?
8. One share of a utility stock costs $98. What is the cost of 50 shares of this stock?
9. Five ounces of gold is worth $2650. What is 1 ounce of gold worth?
10. Smoked salmon costs $36 for 3 pounds. What is the cost of 1 pound?
11. On an architect's drawing, 5 centimeters represents an actual distance of 200 yards. What distance is represented by 1 centimeter?
12. A cable weighs 38 ounces every foot. What is the weight of 760 feet of this cable?
13. An Illinois farmer sold 8400 bushels of corn at $6 per bushel. What was the total income?
14. A machine seals potato-chip sacks at the rate of 360 sacks per hour. How many sacks are sealed at the end of 48 hours?
15. A car traveled 715 miles in 13 hours. What is the average number of miles traveled in 1 hour?
16. During the 21 school days last March, the total attendance in a school was 8526. What was the average daily attendance?
17. A car is traveling at an average speed of 50 miles per hour. How long does it take to go 1600 miles?
18. A 15-ounce can of tomatoes costs 75 cents. What is the cost per ounce?
19. A tank leaks 3 ounces per hour. How long will it take the tank to leak 315 ounces?
20. Carpeting costs $15 per square yard. How many square yards of carpeting can you buy for $840?
21. Find the area of this rectangle.

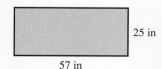

22. A rectangular floor is 15 feet wide and 26 feet long. How many square feet of carpeting are needed to cover this floor?

23. A rectangular piece of fabric has an area of 1620 square inches. If it is 36 inches wide, how long is it?
24. A piece of paper has an area of 88 square inches. If the paper is 11 inches long, how wide is it?
25. The width of a rectangle is 18 inches. The length is 5 times the width. Find the area of the rectangle.
26. The width of a rectangle is 52 inches. The length is 9 inches more than the width. Find the perimeter of the rectangle.
27. A carpet measures 5 yards by 8 yards. It costs $19 per square yard. What is the total cost of the carpet?
28. Mr. Bergeron bought a plot of land that is 300 feet wide and 350 feet long. After keeping 35,000 square feet for himself, he wants to divide the remaining land into lots of equal size for his 14 grandchildren. How much land will each grandchild receive?
29. Find the total area of this floor.

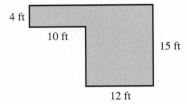

30. Find the total area of this region.

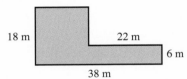

31. Carpeting costs $16 per square yard. What is the cost of carpeting this room?

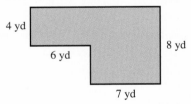

32. Warren's annual salary is six times what it was ten years ago. Ten years ago his salary was $9200. What is his salary now?
33. One nail is twice as long as another nail. If the shorter nail is 55 millimeters long, what is the length of the longer nail?
34. The current value of a house is 3 times what it was fifteen years ago. If the value now is $240,000, what was the value fifteen years ago?
35. Brenda earns twice as much as Ricor. If Brenda's salary is $75,000, how much does Ricor earn?
36. Curtis used to have an automobile that went only 9 miles per gallon of gasoline. His new compact car goes 4 times farther on a gallon of gas. How far does his new car go on a gallon of gasoline?
37. Janet is a receptionist and bookkeeper in an office. She makes $850 per month. Her friend Ann is an engineer making three times as much as Janet. What is Ann's monthly salary?

38. Kathy's bowling scores for 3 games were the following:
    120   141   135
    What was her average score?

39. The rainfall for a city during a 4-year period was as follows:

    1989   37 inches
    1990   24 inches
    1991   28 inches
    1992   35 inches

    What was the average yearly rainfall?

40. A car traveled a total of 315 miles in 5 days. What was the average number of miles traveled each day?

41. A student's quiz scores are five 80s, three 70s, and two 60s. What is the average score?

42. Tony has a piece of string that is 850 centimeters long. After cutting off 200 centimeters, he wants to divide the remaining string into 26 equal pieces. How long will each piece be?

43. Elaine walks 4 miles each hour. She jogs 6 miles each hour. If Elaine walks for 3 hours and jogs for 2 hours, how far does she go?

44. George types 62 words per minute and Gertrude types 55 words per minute. How many words can they type if they both type for 25 minutes?

45. Steven and his two sisters inherited $153,000. After paying taxes of $15,600, they divided the remaining money equally. How much did each person receive?

46. A rectangular plot of land is twice as long as it is wide. If the width is 80 feet, and you plan to put fencing around the plot that costs $19 per foot, what is the cost of the fencing?

47. Dede purchases 17 refrigerators at $715 each and 29 stoves at $568 each. How much did she spend?

48. You purchase 6 pounds of coffee at $7 per pound and 15 coffee cups at $4 each. How much money do you have left out of $150?

49. A man weighed 300 pounds. He then began to lose 5 pounds per week. After 8 weeks, how much does he weigh?

50. Over a 12-hour period, the temperature dropped from 110° Fahrenheit to 62° Fahrenheit. On the average, how much did the temperature drop each hour?

The bar graph gives the weight of 6 persons. Use the graph to answer Exercises 51 through 54.

51. How much does Edna weigh?

52. How much does Art weigh?

53. What is the average weight of the 6 persons?

54. What is the average weight of the 3 women: Edna, Linda, and Sue?

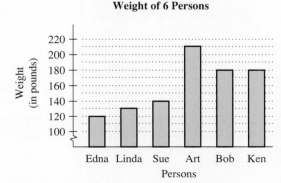

Weight of 6 Persons

USING THE CALCULATOR #5

THE M+ AND M− KEYS

Suppose you purchase 4 books at $38 each and 5 binders at $3 each. Without tax, what is the total cost of these items? Here we show the problem solved without a calculator.

$$4 \text{ books at } \$38 \text{ each} \rightarrow 4 \times \$38 = \$ 152$$
$$5 \text{ binders at } \$3 \text{ each} \rightarrow 5 \times \$3 = \underline{\$\ \ 15}$$
$$\text{Total cost} = \$ 167$$

To solve this problem we compute two multiplications, then we add the results of the two multiplications.

One way to compute this on a calculator is to store the sum of the multiplication results in the memory by using the [M+] key twice. Here we show the procedure.

Before beginning, be sure that the memory is clear, that is, that it contains the number 0. If your calculator does not have a clear-the-memory key, you can enter [MR] then enter [M−] to clear the memory. On a scientific calculator, you can clear the memory by entering the number 0, then enter [Min] or [x→M]

To Compute	(4 × 38)
	+ (5 × 3)
Enter	4 [×] 38 [M+]
Then Enter	5 [×] 3 [M+]
Finally Enter	[MR]
Result	167

If you want to subtract a number from the existing memory, use the [M−] key instead of the [M+] key. Here we show an example.

To Compute	(4 × 50)
	− (5 × 25)
Enter	4 [×] 50 [M+] 5 [×] 25 [M−] [MR]
Result	75

Observe that you enter the [M−] key after the number you are subtracting. You still enter the [M+] key after the number you are subtracting from.

To compute the above problems on a scientific calculator, you do not need to use the memory. Here we show how to do the addition problem. (A subtraction problem would be done similarly by simply entering the [−] key instead of the [+] key.)

ON A SCIENTIFIC CALCULATOR

To Compute	(4 × 38)
	+ (5 × 3)
Enter	4 [×] 38 [+] 5 [×] 3 [=]
Result	167

The scientific calculator knows to perform the two multiplications before the addition or subtraction, because it has been programmed to use the rules for the order to operate. These rules are discussed in Chapter 12. Basic calculators are not programmed to use these rules.

Calculator Problems
Evaluate each.

1. Find the sum of 5 × 9 and 6 × 5.
2. Find the sum of 256 × 48 and 786 × 39.
3. Find the difference between 6 × 8 and 4 × 7.
4. Subtract 67 × 75 from 125 × 82.
5. Ms. Vasquez, the manager of a store, ordered 8 dresses at $40 each, 12 pairs of slacks at $28 each, and 15 skirts at $35 each. What is the total cost of these items?
6. Mr. Jackson purchased two vacant rectangular lots. The larger lot is 85 feet long and 46 feet wide, and the smaller lot is 69 feet long and 54 feet wide. What is the difference in the areas of these two lots?

CHAPTER 2 SUMMARY

KEY WORDS AND PHRASES

multiplication [2.1]
factor [2.1, 2.3]
multiplier [2.1, 2.3]
product [2.1, 2.3, 2.4]
division [2.2]
divisor [2.2]
dividend [2.2]
quotient [2.2]
remainder [2.2]
long division [2.2]
multiple [2.3]
divisible by [2.3]
even number [2.3]
prime number [2.3]
prime factoring [2.3]
divided by [2.4]
divided into [2.4]
speed [2.5]
rate [2.5]
area [2.5]
average (mean) [2.5]

SYMBOLS

$\times$ means to multiply [2.1]
) means to divide [2.2]
$\div$ means to divide [2.2]

IMPORTANT RULES

The Meaning of Multiplication [2.1]

Multiplication indicates repeated addition, for example, 3×5 means $5 + 5 + 5$.

The Meaning of Division [2.2]

Division is the reverse of multiplication, that is, $30 \div 5 = 6$ because $5 \times 6 = 30$.

The Long Division Process [2.2]

The long division process involves repeating these four steps

- Divide
- Multiply
- Subtract (This result must be less than the divisor.)
- Bring down the next digit.

Bring down only one digit at a time. Each time you bring down a digit, you must divide. After you bring down the last digit, you must divide, multiply, and subtract one more time.

How to Find the Missing Number in a Multiplication Statement [2.3]

- If the product is missing, multiply to find the missing product.
- If a multiplier (or factor) is missing, divide the product by the given multiplier to find the missing multiplier.

Divisibility Rules [2.3]

- Any whole number ending in 0, 2, 4, 6, or 8 is divisible by 2.
- Any whole number is divisible by 3 if the sum of its digits is divisible by 3.
- Any whole number ending in a 0 or 5 is divisible by 5.
- Any whole number
  is divisible by 10 if it ends in one 0,
  is divisible by 100 if it ends in two 0s,
  is divisible by 1000 if it ends in three 0s,
  and so on.

The Meaning of a Prime Number [2.3]

A prime number is a whole number, other than 1, that is divisible only by itself and 1. Examples of prime numbers are 2, 3, 5, 7, and 11.

The Area of a Rectangle [2.5]
Area measures the extent of the surface of a region. The area of a rectangle is found by multiplying the length by the width.

$$\text{Area of a rectangle} = \text{length} \times \text{width}$$

How to Find the Average (or Mean) of a Collection of Numbers [2.5]
- Add all of the numbers in the collection.
- Divide the sum by how many numbers are in the collection.

CHAPTER 2 REVIEW EXERCISES

Multiply.

1. 28×41
2. 971×13
3. 207×54
4. 8062×27
5. 16307×458
6. 3056×859
7. 535×288
8. 9382×716
9. 250×1700
10. $192 \times 30,000$
11. 208×307
12. 5138×3009
13. $4 \times 6 \times 2$
14. $9 \times 2 \times 7$
15. $2 \times 2 \times 3 \times 5$
16. $7 \times 7 \times 4 \times 5$

Divide.

17. $5\overline{)38}$
18. $4\overline{)120}$
19. $8\overline{)6712}$
20. $6\overline{)32,494}$
21. $70\overline{)160}$
22. $30\overline{)18,022}$
23. $15\overline{)8325}$
24. $96,160 \div 48$
25. $98,216 \div 93$
26. $500\overline{)8346}$
27. $800\overline{)65,820}$
28. $910\overline{)783,000}$
29. $403\overline{)706,000}$
30. $615\overline{)3,111,900}$

Find the missing number.

31. $9 \times \boxed{} = 225$
32. $\boxed{} \times 18 = 126$
33. $8 \times 48 = \boxed{}$
34. $975 = \boxed{} \times 65$
35. $4796 = 11 \times \boxed{}$
36. $\boxed{} = 40 \times 900$

Answer yes or no.

37. Indicate whether each number is divisible by 12.
    a. 86   b. 336   c. 1308   d. 11,650
38. Indicate whether each number is divisible by 35.
    a. 320   b. 565   c. 2100   d. 10,710
39. Indicate whether each number is a multiple of 25.
    a. 675   b. 1050   c. 2300   d. 6085
40. Indicate whether each number is a multiple of 19.
    a. 57   b. 171   c. 199   d. 5738
41. Indicate whether each number is divisible by 5.
    a. 300   b. 715   c. 8326   d. 2010
42. Indicate whether each number is divisible by 3.
    a. 51   b. 1212   c. 792   d. 3456

43. Indicate whether each number is divisible by 100.
 a. 9090 **b.** 5005 **c.** 3000 **d.** 21,500

44. Indicate whether each number is divisible by 1000.
 a. 4008 **b.** 30,500 **c.** 9000 **d.** 70,000

Write each number as the product of primes.

45. 18 **46.** 54 **47.** 168 **48.** 441 **49.** 750
50. 140 **51.** 2700 **52.** 24,000 **53.** 351 **54.** 255

Write a division statement using the symbol $\overline{)}$.

55. 4 divided into 12 equals 3.

56. 100 divided by 5 is 20.

Write a division statement using the symbol ÷.

57. 8 divided into 72 is 9.

58. 325 divided by 25 equals 13.

Fill in the blank with the appropriate word, by *or* into.

59. $5\overline{)35}^{\,7}$ is read "5 divided _____ 35 is 7."

60. $5\overline{)35}^{\,7}$ is read "35 divided _____ 5 is 7."

61. 130 ÷ 10 = 13 is read "130 divided _____ 10 equals 13."

62. 130 ÷ 10 = 10 is read "10 divided _____ 130 equals 13."

Translate each English statement to a multiplication or division statement using math symbols.

63. The product of 3 and 16 is 48.

64. 800 is a number twice as large as 400.

65. 29 equals 174 divided by 6.

66. 13 divided into 585 is 45.

67. 68 equals 4 times 17.

68. 60 multiplied times 400 yields 24,000.

69. Eighty-one is nine multiplied times itself.

70. Sixteen divided into eight thousand eighty yields five hundred five.

Solve.

71. What number multiplied by 15 equals 3015?

72. Find the product of 750 and 8000.

73. Find the difference between 51 and 17.

74. What number is 57 divided by 19?

75. Find twice 950.

76. Find the average of 36, 42, and 54.

77. List the first five multiples of 8.

78. Is 68 a multiple of 17? If so, write 68 as the product of 17 and another whole number.

79. Eight partners share equally a profit of $18,000. How much money does each partner receive?

80. Crabmeat sells for $7 a pound. What is the cost of 15 pounds of crabmeat?
81. Pam's bowling scores were as follows: 119   127   125   117
    What was her average score?
82. A city allocates $810,000 for salaries for park recreation assistants during the summer. How many assistants can be hired with a salary of $750 for each?
83. A machine cuts the threads in a screw at the rate of 75 screws per hour. How long does it take the machine to cut the threads in 3150 screws?
84. A bookstore purchased 900 textbooks from a publisher at $15 a book. What was the total cost to the bookstore?
85. What is the total area of this floor?

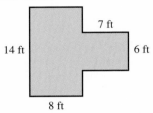

86. The length of a rectangle is 3 times its width. If the width measures 49 feet, find the area of the rectangle.
87. If Virna can walk 5 miles in 75 minutes, how long will it take her to walk 1 mile?
88. The area of a rectangle is 1360 square meters. If the width is 17 meters, find the perimeter.
89. Last year Chris earned $65,000. This year he earned 3 times that much. What is the increase in his earnings?
90. A parking garage charges $3 for the first hour, then $2 per hour for each additional hour. You enter the garage at 9 AM and leave at 2 PM. How much do you pay?

In Your Own Words

Write complete sentences to discuss each of the following. Support your comments with examples or pictures, if appropriate.

91. Discuss two ways of deciding whether a number is divisible by 5.
92. Discuss two ways of deciding whether a number is divisible by 3.
93. Discuss how the following two problems differ.
    a. ☐ × 28 = 420   b. 28 × 420 = ☐
94. Discuss the difference between finding the perimeter of a rectangle and finding the area of a rectangle.
95. Give examples of at least three different rates that you experience in your life.

CHAPTER 2 PRACTICE TEST

Multiply.

1. 9237 × 8
2. 8037 × 67
3. 532 × 417
4. 600 × 8000
5. 73,000 × 13
6. 208 × 307
7. 5482 × 3009
8. 7 × 2 × 3
9. 3 × 5 × 5 × 7

Divide.
10. 8)6712
11. 6)36,494
12. 15)8325
13. 96,160 ÷ 48
14. 783,000 ÷ 910
15. 615)3,111,900

Find the missing number.
16. 25 × ☐ = 30,750
17. ☐ = 450 × 900

Answer yes or no.
18. Is 184 a multiple of 23?
19. Is 351 divisible by 3?

Write each number as the product of primes.
20. 42
21. 850

Fill in the blank with the appropriate word by *or* into.
22. 8)40 (with 5 above) is read "40 divided _____ 8 equals 5."
23. 666 ÷ 6 = 111 is read "6 divided _____ 666 is 111."

Solve.
24. Twice what number yields 908?
25. Find the average of 682, 1016, and 801.
26. Seven friends share equally the cost of renting a ski cabin for the winter season. The total rent is $2275. What is each person's share?
27. Karin sells corn from her garden. She puts 2000 ears of corn in packages with 8 ears in each package. How many packages can she make? How many, if any, ears of corn are left over?
28. The area of a desk top is 1050 square inches. The desk is 25 inches wide. How long is it?
29. Mr. Martinez works in the food and beverage department at a hotel. He ordered 4350 ounces of punch for a reception. He is expecting approximately 150 guests. How much punch does this allow for each guest?
30. George has a piece of string that is 850 centimeters long. After cutting off 200 centimeters, he wants to divide the remaining string into 26 equal pieces. How long will each of these pieces be?
31. What is the cost of carpeting this floor space if the carpeting costs $28 per square yard?

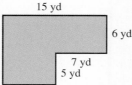

32. A jet travels for 8 hours at 450 miles per hour, then travels for 4 hours at 580 miles per hour. How much farther did the jet travel during the 8-hour period than during the 4-hour period?
33. A tank leaks 20 ounces of water every 300 minutes. How long does it take for the tank to leak 1 ounce?

CUMULATIVE REVIEW EXERCISES: CHAPTERS 1–2

1. Write the English name for 64,098.
2. Write the numeral for two million, thirteen thousand, five hundred.
3. Round off 148,106 to the nearest ten thousand.

Perform the indicated operations.

4. $8796 + 652 + 376{,}552$
5. $8000 - 752$
6. $70{,}000 \times 3400$
7. $253{,}680 \div 28$

Find the quotient and remainder.

8. $8986 \div 35$
9. $600{,}700 \div 150$
10. $50{,}796 \div 516$

Write each number as the product of primes.

11. 36
12. 275
13. 9000
14. 637

Find the missing number.

15. $\Box \times 75 = 450$
16. $701 = \Box + 189$

Solve.

17. Find the difference between 5010 and 349.
18. Find the product of 4580 and 308.
19. Find the sum of 240, 673, and 548.
20. Find the average of 68, 74, and 98.
21. What number increased by 65 is 200?
22. What number times 65 is 1170?
23. A shopping center has a rectangular parking lot that is 350 feet by 280 feet.
    a. Find the area of the lot.   b. Find the perimeter of the lot.
24. Ms. Sanchez has saved $850 to buy a sofa. The sofa she wants costs only $685. How much money will she have left after buying the sofa?
25. Frank's checking account balance is $405. He writes checks for $82, $78, and $136. What is his balance now?
26. Melissa bought a car for $9900. She plans to pay for it by making 36 equal monthly payments. How much is her monthly payment?
27. A race car travels 450 miles in 3 hours. At this rate, how far does the car go in 1 hour?
28. Joshua's watch loses 1 minute every 15 hours. At this rate, how long does it take his watch to lose 5 minutes?
29. Irene had grades of 92, 86, 75, 88, and 94 on five exams.
    a. What is the total number of points on all five exams?
    b. What is the average of the five scores?
    c. How much larger is the highest score than the lowest score?
30. A restaurant contains 8 booths that each seat 6 persons, 5 booths that seat 2 persons, 12 tables that seat 4 persons, and 6 tables that seat 6 persons. How many customers can eat in the restaurant at one time?
31. The Hamilton family budgeted $1300 for the month of November. Their expenses for that month amounted to $287 for food, $435 for rent, $138 for clothing, $68 for utilities, $65 for entertainment, $148 for medical bills, and $125 for other miscellaneous items. Are their expenses over or under their budget, and by how much?

CHAPTER 3

An Introduction to Fractions

3.1 PICTURING FRACTIONS BY USING SHADED REGIONS

OBJECTIVES
- Write a fraction or a mixed numeral represented by a shaded region.
- Recognize whether a fraction is less than, equal to, or more than the number 1.

1 Fractions Less Than or Equal to One Whole

Ian bought a pizza and cut it into 6 *equal* parts as shown in the figure. Each slice is $\frac{1}{6}$ (one-sixth) of the pizza.

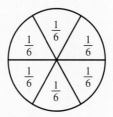

If he eats 2 slices, then he eats $\frac{2}{6}$ (two-sixths) of the pizza.

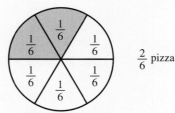

$\frac{2}{6}$ pizza

Try These Problems

Write a fraction represented by the shaded region. Assume each figure represents one whole.

1.

2.

3.

4.

If he eats 5 slices, then he eats $\frac{5}{6}$ (five-sixths) of the pizza.

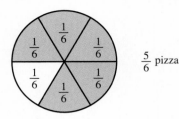

If he eats 6 slices, then he eats $\frac{6}{6}$ (six-sixths) of the pizza or one whole pizza.

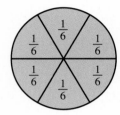

The numbers $\frac{1}{6}$, $\frac{2}{6}$, $\frac{5}{6}$, and $\frac{6}{6}$ are called **fractions.** Fractions help us to talk about part or all of a whole. The horizontal bar is called the **fraction bar,** and the numbers above and below the fraction bar are called the **numerator** and **denominator,** respectively.

$$\text{Fraction bar} \longrightarrow \frac{5 \longleftarrow \text{Numerator}}{6 \longleftarrow \text{Denominator}}$$

Here are some more examples where fractions represent part of one whole.

EXAMPLE 1 Write a fraction represented by the shaded region. Assume each figure represents one whole.

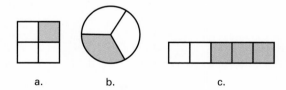

a. b. c.

SOLUTION

a. Because 1 out of 4 equal parts is shaded, the fraction is $\frac{1}{4}$ (one-fourth).

b. Because 1 out of 3 equal parts is shaded, the fraction is $\frac{1}{3}$ (one-third).

c. Because 3 out of 5 equal parts are shaded, the fraction is $\frac{3}{5}$ (three-fifths). ■

Try Problems 1 through 4.

2 Fractions More Than or Equal to One Whole

Antoinette bought 3 pizzas and cut each one into 8 *equal* parts as shown in the figure. Each slice is $\frac{1}{8}$ (one-eighth) of a pizza.

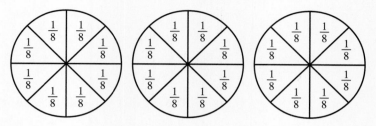

If she and her friends eat 8 slices, then they eat $\frac{8}{8}$ (eight-eighths) pizza or 1 pizza.

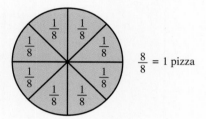

If she and her friends eat 11 slices, then they eat $\frac{11}{8}$ (eleven-eighths) pizzas or $1\frac{3}{8}$ (one and three-eighths) pizzas.

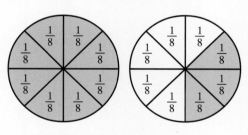

$\frac{11}{8}$ or $1\frac{3}{8}$ pizzas

If she and her friends eat 21 slices, then they eat $\frac{21}{8}$ (twenty-one eighths) pizzas or $2\frac{5}{8}$ (two and five-eighths) pizzas.

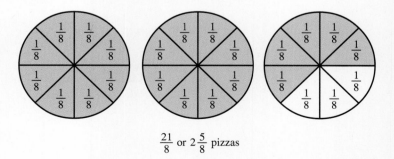

$\frac{21}{8}$ or $2\frac{5}{8}$ pizzas

Fractions like $\frac{11}{8}$ and $\frac{21}{8}$, that have a numerator more than the denominator, are numbers more than 1 whole. These numbers can be written in two ways.

$$\frac{11}{8} = \frac{8}{8} + \frac{3}{8} = 1 + \frac{3}{8} = 1\frac{3}{8} \rightarrow \text{Therefore, } \frac{11}{8} = 1\frac{3}{8}$$

$$\frac{21}{8} = \frac{8}{8} + \frac{8}{8} + \frac{5}{8} = 1 + 1 + \frac{5}{8} = 2\frac{5}{8} \rightarrow$$

$$\text{Therefore, } \frac{21}{8} = 2\frac{5}{8}$$

The fraction $\frac{11}{8}$ is called an **improper fraction,** and when it is written in the form $1\frac{3}{8}$, it is called a **mixed numeral.** In general, an improper fraction is a fraction with the numerator more than the denominator, and its value is more than 1.

EXAMPLE 2 Write an improper fraction and a mixed numeral or whole numeral that represent the shaded regions. Assume each separate figure represents 1 whole.

SOLUTION Because each separate figure is divided into 3 equal parts, each part is $\frac{1}{3}$.

$$\frac{3}{3} \quad + \quad \frac{2}{3} \quad = \quad \frac{5}{3} \quad \text{five-thirds}$$

$$1 \quad + \quad \frac{2}{3} \quad = \quad 1\frac{2}{3} \quad \text{one and two-thirds}$$

The improper fraction is $\frac{5}{3}$ and the mixed numeral is $1\frac{2}{3}$. ■

EXAMPLE 3 Write an improper fraction and a mixed numeral or whole numeral that represent the shaded regions. Assume each separate figure represents 1 whole.

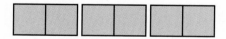

SOLUTION Because each separate figure is divided into 2 equal parts, each part is $\frac{1}{2}$.

Try These Problems

Write an improper fraction and a mixed numeral or a whole numeral that represent the shaded regions. Assume each separate figure represents one whole.

5.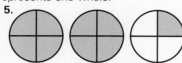

6.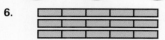

7.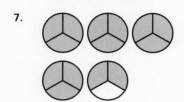

8.

9. Which of these fractions are equal to one whole?
$\frac{12}{12}, \frac{7}{12}, \frac{19}{12}, \frac{8}{4}, \frac{8}{8}, \frac{4}{4}$

10. Which of these fractions are more than one whole?
$\frac{7}{30}, \frac{30}{7}, \frac{75}{25}, \frac{25}{75}, \frac{2}{3}, \frac{3}{2}$

11. Which of these fractions are less than one whole?
$\frac{9}{10}, \frac{10}{9}, \frac{4}{12}, \frac{12}{4}, \frac{33}{80}, \frac{80}{33}$

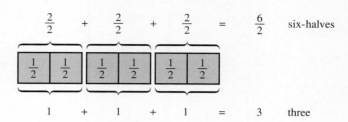

$$\frac{2}{2} + \frac{2}{2} + \frac{2}{2} = \frac{6}{2} \quad \text{six-halves}$$

$$1 + 1 + 1 = 3 \quad \text{three}$$

The improper fraction is $\frac{6}{2}$ and the whole numeral is 3. ∎

Try Problems 5 through 8.

After picturing fractions represented by shaded regions, you should have discovered the following basic properties of fractions.

> **Some Basic Properties of Fractions**
> 1. A fraction with a numerator less than the denominator is a number less than 1 whole. (For example, the fractions $\frac{1}{4}, \frac{2}{7}$, and $\frac{19}{20}$ are each less than 1.)
> 2. A fraction with a numerator more than the denominator is a number more than 1 whole. (For example, the fractions $\frac{5}{4}$, $\frac{12}{7}$, and $\frac{41}{20}$ are each more than 1.)
> 3. A fraction with a numerator that equals the denominator is equal to 1 whole. (For example, the fractions $\frac{4}{4}, \frac{7}{7}$, and $\frac{20}{20}$ are each equal to 1.)

Try Problems 9 through 11.

Answers to Try These Problems

1. $\frac{1}{5}$ 2. $\frac{1}{6}$ 3. $\frac{2}{3}$ 4. $\frac{7}{7} = 1$ 5. $\frac{9}{4}, 2\frac{1}{4}$ 6. $\frac{15}{5}, 3$ 7. $\frac{14}{3}, 4\frac{2}{3}$
8. $\frac{20}{4}, 5$ 9. $\frac{12}{12}, \frac{8}{8}, \frac{4}{4}$ 10. $\frac{30}{7}, \frac{75}{25}, \frac{3}{2}$ 11. $\frac{9}{10}, \frac{4}{12}, \frac{33}{80}$

EXERCISES 3.1

Write a fraction represented by the shaded region. Assume each figure represents one whole.

1.   2.   3.   4.

5.   6.   7.   8.

9.   10.

Write an improper fraction and a mixed numeral or a whole numeral that represent the shaded regions. Assume each separate figure represents one whole.

11. 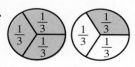   12.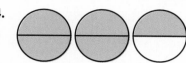

13. 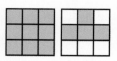   14.

15.    16.

17.   18.

Choose the fractions in each group that are more than one whole.

19. $\frac{5}{7}, \frac{7}{5}, \frac{8}{3}, \frac{3}{8}, \frac{30}{13}, \frac{13}{30}$ 20. $\frac{4}{4}, \frac{7}{7}, \frac{9}{8}, \frac{8}{9}, \frac{23}{50}, \frac{50}{23}$

Choose the fractions in each group that equal one whole.

21. $\frac{4}{8}, \frac{8}{4}, \frac{8}{8}, \frac{4}{4}, \frac{12}{13}, \frac{13}{13}$ 22. $\frac{2}{2}, \frac{6}{2}, \frac{2}{6}, \frac{9}{9}, \frac{17}{17}, \frac{18}{17}$

Choose the fractions in each group that are less than one whole.

23. $\frac{2}{5}, \frac{5}{2}, \frac{5}{5}, \frac{7}{12}, \frac{12}{7}, \frac{12}{12}$ 24. $\frac{1}{8}, \frac{9}{8}, \frac{8}{8}, \frac{21}{25}, \frac{28}{25}, \frac{3}{4}$

3.2 READING AND WRITING FRACTIONS

OBJECTIVES
- Write a fraction for the English name of a fraction.
- Write the English name of a fraction.
- Write a mixed numeral for the English name of a mixed numeral.
- Write the English name of a mixed numeral.

1 The English Name of a Fraction

You need to become familiar with the English names for fractions. Some examples follow.

3.2 Reading and Writing Fractions 91

 Try These Problems

Write a fraction for each English name.
1. Two-halves
2. Two-thirds
3. Three-fifths
4. Eleven-ninths

Write the English name for each fraction.
5. $\frac{1}{6}$
6. $\frac{5}{2}$

Write a fraction for each English name.
7. Nine-tenths
8. Ten-seventeenths
9. Eighteen forty-seconds
10. Sixty-five sixty-thirds

Write the English name for each fraction.
11. $\frac{12}{13}$
12. $\frac{8}{50}$

Fraction	English Name
$\frac{1}{2}$	one-half
$\frac{3}{2}$	three-halves
$\frac{1}{3}$	one-third
$\frac{12}{4}$	twelve-fourths
$\frac{4}{5}$	four-fifths
$\frac{7}{6}$	seven-sixths
$\frac{5}{7}$	five-sevenths
$\frac{33}{8}$	thirty-three eighths

Observe that when the denominator is 4, 5, 6, 7, . . . , the English name ends in *-ths*.

 Try Problems 1 through 6.

Now let's look at fractions with larger denominators. Study the following examples.

Fraction	English Name
$\frac{7}{10}$	seven-tenths
$\frac{11}{18}$	eleven-eighteenths
$\frac{1}{20}$	one-twentieth
$\frac{3}{72}$	three seventy-seconds
$\frac{77}{83}$	seventy-seven eighty-thirds
$\frac{9}{95}$	nine ninety-fifths

 Try Problems 7 through 12.

Later, when you study decimals, it will be especially important for you to be familiar with the English names for fractions with denomina-

Try These Problems

Write a fraction for each English name.

13. Eight-tenths
14. Thirty hundredths
15. Seven hundredths
16. Sixty-two tenths
17. Thirteen thousandths
18. Three hundred-thousandths
19. Fifty ten-thousandths
20. Ninety-six thousandths

Write the English name for each fraction.

21. $\frac{18}{100}$
22. $\frac{33}{10}$
23. $\frac{4}{1000}$
24. $\frac{27}{10,000}$

Write a mixed numeral for each English name.

25. Ten and three-fourths
26. Eight and nine-thirteenths
27. Twelve and seventeen-fortieths
28. Sixty-seven and twenty-three hundredths

Write the English name for each mixed numeral.

29. $4\frac{5}{6}$
30. $13\frac{2}{11}$
31. $50\frac{17}{100}$

tors like 10, 100, 1000, 10,000 and so on. At this time let's take a closer look at some of these.

Fraction	English Name
$\frac{23}{10}$	twenty-three tenths
$\frac{1}{100}$	one hundredth
$\frac{15}{100}$	fifteen hundredths
$\frac{3}{1000}$	three thousandths
$\frac{90}{10,000}$	ninety ten-thousandths
$\frac{82}{100,000}$	eighty-two hundred-thousandths

 Try Problems 13 through 24.

2 The English Name of a Mixed Numeral

A mixed numeral like $4\frac{2}{3}$ has a whole part and a fraction part. The word *and* is used in the English name to separate the two parts clearly.

$4\frac{2}{3}$ is read "four and two-thirds."

Here are some additional examples.

Mixed Numeral	English Name
$5\frac{2}{7}$	five and two-sevenths
$13\frac{9}{10}$	thirteen and nine-tenths
$25\frac{12}{25}$	twenty-five and twelve twenty-fifths
$9\frac{3}{100}$	nine and three hundredths

 Try Problems 25 through 31.

 Answers to Try These Problems

1. $\frac{2}{2}$ 2. $\frac{2}{3}$ 3. $\frac{3}{5}$ 4. $\frac{11}{9}$ 5. one-sixth   6. five-halves   7. $\frac{9}{10}$
8. $\frac{10}{17}$ 9. $\frac{18}{42}$ 10. $\frac{65}{63}$ 11. twelve-thirteenths   12. eight-fiftieths
13. $\frac{8}{10}$ 14. $\frac{30}{100}$ 15. $\frac{7}{100}$ 16. $\frac{62}{10}$ 17. $\frac{13}{1000}$ 18. $\frac{3}{100,000}$ 19. $\frac{50}{10,000}$
20. $\frac{96}{1000}$ 21. eighteen hundredths   22. thirty-three tenths
23. four thousandths   24. twenty-seven ten-thousandths

25. $10\frac{3}{4}$ 26. $8\frac{9}{13}$ 27. $12\frac{17}{40}$ 28. $67\frac{23}{100}$
29. four and five-sixths 30. thirteen and two-elevenths
31. fifty and seventeen hundredths

EXERCISES 3.2

Write a fraction for each.

1. One-half
2. Nine-fourths
3. Twelve-fifths
4. Eighty-five thirds
5. Two-thirteenths
6. Ninety-three two-hundredths
7. Eleven-thirtieths
8. One thousand twenty-thirds
9. Five-tenths
10. Forty-five hundredths
11. Thirty-six thousandths
12. Eighteen hundred-thousandths

Write the English name for each.

13. $\frac{7}{2}$
14. $\frac{3}{7}$
15. $\frac{6}{11}$
16. $\frac{86}{90}$
17. $\frac{1}{100}$
18. $\frac{5}{10,000}$

Write a mixed numeral for each.

19. Six and two-fifths
20. Fifteen and seven-tenths
21. Sixty and three hundredths
22. Five and thirteen thousandths

Write the English name for each.

23. $9\frac{3}{10}$
24. $80\frac{1}{2}$
25. $14\frac{17}{100}$
26. $27\frac{8}{13}$

3.3 RAISING FRACTIONS TO HIGHER TERMS AND CONVERTING WHOLE NUMERALS TO FRACTIONS

OBJECTIVES
- Write two equal fractions represented by a shaded region.
- Find a missing numerator so that two fractions are equal.
- Find a missing numerator so that a whole numeral is equal to a fraction.
- Raise a fraction to higher terms to have a specified denominator.
- Convert a whole numeral to a fraction having a specified denominator.

1 Raising Fractions to Higher Terms

If you bought a pizza, there are many ways you could eat $\frac{1}{2}$ of the pizza. If you cut the pizza into 2 equal parts and ate 1 part, you would eat $\frac{1}{2}$ of the pizza.

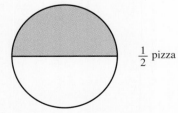

$\frac{1}{2}$ pizza

If you cut the pizza into 4 equal parts and ate 2 parts, you would eat $\frac{2}{4}$ of the pizza, which is the same as $\frac{1}{2}$ of the pizza.

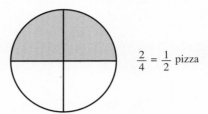

$\frac{2}{4} = \frac{1}{2}$ pizza

If you cut the pizza into 6 equal parts and ate 3 parts, you would eat $\frac{3}{6}$ of the pizza, which is the same as $\frac{1}{2}$ of the pizza.

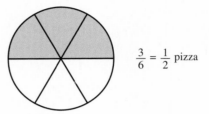

$\frac{3}{6} = \frac{1}{2}$ pizza

If you cut the pizza into 8 equal parts and ate 4 parts, you would eat $\frac{4}{8}$ of the pizza, which is the same as $\frac{1}{2}$ of the pizza.

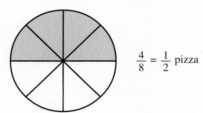

$\frac{4}{8} = \frac{1}{2}$ pizza

The fractions $\frac{1}{2}$, $\frac{2}{4}$, $\frac{3}{6}$, and $\frac{4}{8}$ are equal numbers.

$$\frac{1}{2} = \frac{2}{4} = \frac{3}{6} = \frac{4}{8}$$

Observe that the fraction $\frac{2}{4}$ can be obtained from $\frac{1}{2}$ by multiplying the numerator and denominator by 2.

$$\frac{1}{2} = \frac{1 \times 2}{2 \times 2} = \frac{2}{4}$$

The fraction $\frac{3}{6}$ can be obtained from $\frac{1}{2}$ by multiplying the numerator and denominator by 3.

$$\frac{1}{2} = \frac{1 \times 3}{2 \times 3} = \frac{3}{6}$$

The fraction $\frac{4}{8}$ can be obtained from $\frac{1}{2}$ by multiplying the numerator and denominator by 4.

$$\frac{1}{2} = \frac{1 \times 4}{2 \times 4} = \frac{4}{8}$$

Try These Problems

Write two equal fractions represented by the shaded region. Assume each figure represents one whole.

1.
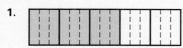

2.

3.

Find the missing numerator so that the two numbers are equal.

4. $\frac{3}{5} = \frac{?}{10}$
5. $\frac{5}{6} = \frac{?}{24}$
6. $\frac{1}{12} = \frac{?}{60}$
7. $\frac{5}{9} = \frac{?}{153}$

In general, we can multiply the numerator and denominator of a fraction by the *same* nonzero number without changing the value of the fraction. The process of multiplying the numerator and denominator by the same number is called **raising the fraction to higher terms.**

Raising a Fraction to Higher Terms
Multiplying the numerator and denominator of a fraction by the same nonzero number does not change the value of the fraction.

EXAMPLE 1 Write two equal fractions represented by the shaded region. Assume the figure represents one whole.

SOLUTION First, view the figure as divided into 12 equal parts. Nine of these parts are shaded, so the fraction is $\frac{9}{12}$.

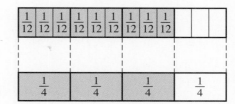

Second, view the figure as divided into 4 equal parts. Three of these parts are shaded, so the fraction is $\frac{3}{4}$. The fractions are $\frac{9}{12}$ and $\frac{3}{4}$. ■

▲ Try Problems 1 through 3.

EXAMPLE 2 Find the missing numerator so that the two fractions are equal.

a. $\frac{2}{5} = \frac{?}{45}$ **b.** $\frac{3}{7} = \frac{?}{161}$

SOLUTION

a.
$$\frac{2}{5} = \frac{2 \times 9}{5 \times 9}$$
$$= \frac{18}{45}$$

Divide 45 by 5 to obtain 9. $5\overline{)45}^{\,9}$
Multiply numerator and denominator of $\frac{2}{5}$ by 9 to obtain an equal fraction with denominator 45.

The missing number is 18.

b.
$$\frac{3}{7} = \frac{3 \times 23}{7 \times 23}$$
$$= \frac{69}{161}$$

Divide 161 by 7 to obtain 23. $7\overline{)161}^{\,23}$
Multiply numerator and denominator of $\frac{3}{7}$ by 23 to obtain an equal fraction with denominator 161.

The missing numerator is 69. ■

▲ Try Problems 4 through 7.

Try These Problems

Find the missing numerator so that the two numbers are equal.

8. $1 = \frac{?}{9}$

9. $1 = \frac{?}{12}$

2 Writing Whole Numerals as Fractions

Recall from Section 3.1 that the number 1 can be written as a fraction in many ways. The pictures that follow illustrate this.

$1 = \frac{1}{1}$ one

$1 = \frac{2}{2}$ one equals **two-halves**

$1 = \frac{3}{3}$ one equals **three-thirds**

$1 = \frac{4}{4}$ one equals **four-fourths**

We can write 1 in fractional form simply by making the numerator equal the denominator.

 Try Problems 8 and 9.

The number 2 can also be written as a fraction in many ways. The pictures that follow illustrate this.

$2 = \frac{2}{1}$ two

$2 = \frac{4}{2}$ two equals **four-halves**

$2 = \frac{6}{3}$ two equals **six-thirds**

$2 = \frac{8}{4}$ two equals **eight-fourths**

We can write 2 as a fraction with any nonzero denominator. The following example illustrates how to do this.

EXAMPLE 3   Find the missing numerator so that the two numbers are equal.

a. $2 = \frac{?}{5}$ **b.** $2 = \frac{?}{12}$

3.3 Raising Fractions to Higher Terms and Converting Whole Numerals to Fractions

 Try These Problems

Find the missing numerator so that the two numbers are equal.

10. $2 = \frac{?}{8}$
11. $2 = \frac{?}{15}$
12. $4 = \frac{?}{3}$
13. $6 = \frac{?}{2}$
14. $7 = \frac{?}{40}$
15. $18 = \frac{?}{6}$

SOLUTION

a. $\quad 2 = \frac{2}{1} \quad$ Write 2 as $\frac{2}{1}$.

$\quad = \frac{2 \times 5}{1 \times 5} \quad$ Multiply numerator and denominator by 5 to obtain an equal fraction with denominator 5.

$\quad = \frac{10}{5}$

The missing numerator is 10.

b. $\quad 2 = \frac{2}{1} \quad$ Write 2 as $\frac{2}{1}$.

$\quad = \frac{2 \times 12}{1 \times 12} \quad$ Multiply numerator and denominator by 12 to obtain an equal fraction with denominator 12.

$\quad = \frac{24}{12}$

The missing numerator is 24. ∎

 Try Problems 10 and 11.

Any whole numeral can be written as a fraction with any nonzero denominator. The following example illustrates how to do this.

EXAMPLE 4 Find the missing numerator so that the two numbers are equal.

a. $3 = \frac{?}{9}$ **b.** $12 = \frac{?}{4}$

SOLUTION

a. $\quad 3 = \frac{3}{1} \quad$ Write 3 as $\frac{3}{1}$.

$\quad = \frac{3 \times 9}{1 \times 9} \quad$ Multiply numerator and denominator by 9 to obtain an equal fraction with denominator 9.

$\quad = \frac{27}{9}$

The missing numerator is 27.

b. $\quad 12 = \frac{12}{1} \quad$ Write 12 as $\frac{12}{1}$.

$\quad = \frac{12 \times 4}{1 \times 4} \quad$ Multiply numerator and denominator by 4 to obtain an equal fraction with denominator 4.

$\quad = \frac{48}{4}$

The missing numerator is 48. ∎

 Try Problems 12 through 15.

Answers to Try These Problems

1. $\frac{9}{15}, \frac{3}{5}$ 2. $\frac{6}{8}, \frac{3}{4}$ 3. $\frac{2}{3}, \frac{6}{9}$ 4. 6   5. 20   6. 5   7. 85   8. 9
9. 12   10. 16   11. 30   12. 12   13. 12   14. 280   15. 108

EXERCISES 3.3

Write two equal fractions represented by the shaded region. Assume each figure represents one whole.

1.   2.    3.

4.    5.    6.

7.   8.

Find the missing numerator so that the two numbers are equal.

9. $\frac{3}{4} = \frac{?}{20}$ 10. $\frac{5}{9} = \frac{?}{18}$ 11. $\frac{10}{3} = \frac{?}{12}$ 12. $\frac{7}{6} = \frac{?}{54}$

13. $\frac{11}{5} = \frac{?}{65}$ 14. $\frac{7}{4} = \frac{?}{96}$ 15. $\frac{2}{25} = \frac{?}{350}$ 16. $\frac{5}{36} = \frac{?}{288}$

17. $1 = \frac{?}{8}$ 18. $1 = \frac{?}{37}$ 19. $3 = \frac{?}{2}$ 20. $3 = \frac{?}{5}$

21. $5 = \frac{?}{7}$ 22. $8 = \frac{?}{9}$ 23. $6 = \frac{?}{21}$ 24. $25 = \frac{?}{5}$

25. Convert $\frac{1}{3}$ to a fraction with denominator 24.
26. Write a fraction with denominator 30 that has the same value as $\frac{7}{6}$.
27. Write a fraction with denominator 42 that has the same value as $\frac{5}{7}$.
28. Write a fraction with denominator 8 that is equal to the whole number 3.
29. Write a fraction with denominator 15 that is equal to the whole number 4.
30. Convert 16 to a fraction with denominator 6.

3.4 CONVERTING IMPROPER FRACTIONS TO MIXED NUMERALS AND VICE VERSA

OBJECTIVES
- Convert an improper fraction to a mixed or whole numeral.
- Convert a mixed numeral to an improper fraction.
- Recognize whether or not a fraction can be converted to a mixed numeral.

1 Converting Improper Fractions to Mixed Numerals or Whole Numerals

In Section 3.1 you learned that an improper fraction (a fraction more than one whole) can be written as a whole numeral or mixed numeral. The picture that follows illustrates that $\frac{12}{3}$ is equal to 4.

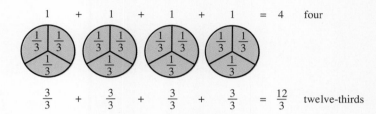

$$1 + 1 + 1 + 1 = 4 \quad \text{four}$$

$$\frac{3}{3} + \frac{3}{3} + \frac{3}{3} + \frac{3}{3} = \frac{12}{3} \quad \text{twelve-thirds}$$

Observe that we can obtain the whole numeral 4 from the improper fraction $\frac{12}{3}$ without viewing the picture. The numerator 12 divided by the denominator 3 yields 4.

$$\frac{12}{3} = 12 \div 3 = 4$$

or

$$\frac{12}{3} \rightarrow 3\overline{)12}^{\,4} \rightarrow \frac{12}{3} = 4$$

In some cases an improper fraction does not equal a whole numeral but can be written as a mixed numeral. The following picture illustrates that $\frac{11}{4}$ is equal to $2\frac{3}{4}$.

$$1 + 1 + \frac{3}{4} = 2 + \frac{3}{4} = 2\frac{3}{4} \quad \text{two and three-fourths}$$

$$\frac{4}{4} + \frac{4}{4} + \frac{3}{4} = \frac{11}{4} \quad \text{eleven-fourths}$$

Observe that we can obtain the mixed numeral $2\frac{3}{4}$ from the improper fraction $\frac{11}{4}$ without viewing the picture. Divide the numerator 11 by the denominator 4.

$$\frac{11}{4} \rightarrow 4\overline{)11}^{\,2}\underline{8}3 \rightarrow \frac{11}{4} = 2\frac{3}{4}$$

After performing the division, the quotient 2 tells you how many wholes are in $\frac{11}{4}$ and the remainder 3 tells you how many $\frac{1}{4}$s are left over. Note that the denominator 4 in the mixed numeral is the same as the original denominator in the improper fraction. Now we look at more examples.

Try These Problems

Convert each of these improper fractions to a mixed or whole numeral.

1. $\frac{42}{7}$
2. $\frac{75}{25}$
3. $\frac{38}{5}$
4. $\frac{74}{9}$
5. $\frac{3240}{30}$
6. $\frac{252}{25}$

EXAMPLE 1 Convert each improper fraction to a mixed or whole numeral.

a. $\frac{91}{13}$ b. $\frac{565}{8}$

SOLUTION

a. $\frac{91}{13}$ means 91 divided by 13.

$$13\overline{)91} \atop \underline{91}^{7}$$

$$\frac{91}{13} = 7$$

b. $\frac{565}{8}$ means 565 divided by 8.

$$8\overline{)565} \atop \underline{56}^{70} \atop 5 \atop \underline{0} \atop 5$$

$$\frac{565}{8} = 70\frac{5}{8} \quad \blacksquare$$

 Try Problems 1 through 6.

2 Converting Mixed Numerals to Improper Fractions

You have learned to convert improper fractions to mixed numerals by dividing the numerator by the denominator. For example,

$$\frac{22}{5} \rightarrow 5\overline{)22}^{4} \atop \underline{20} \atop 2 \rightarrow \frac{22}{5} = 4\frac{2}{5}$$

We can reverse this process and convert a mixed numeral to an improper fraction. Let's begin here with $4\frac{2}{5}$ and show how to obtain the improper fraction $\frac{22}{5}$.

$4\frac{2}{5} = 4 + \frac{2}{5}$ The mixed numeral $4\frac{2}{5}$ is a shortcut notation for $4 + \frac{2}{5}$.

$\phantom{4\frac{2}{5}} = \frac{4 \times 5}{1 \times 5} + \frac{2}{5}$ Write 4 as $\frac{4}{1}$, then multiply numerator and denominator by 5 so that 4 is written as a fraction with the same denominator as $\frac{2}{5}$.

$\phantom{4\frac{2}{5}} = \frac{20}{5} + \frac{2}{5}$ Observe that $\frac{20}{5}$ is the same as 4.

$\phantom{4\frac{2}{5}} = \frac{22}{5}$ Add the numerators and place this over the common denominator.

By using the above procedure, we convert the whole numeral 4 to a fraction with denominator 5, then add this fraction to the fraction $\frac{2}{5}$. The result is $\frac{22}{5}$. There is a shortcut for obtaining the result $\frac{22}{5}$. Observe that

Try These Problems

Convert each of these mixed numerals to an improper fraction.

7. $2\frac{1}{3}$
8. $7\frac{3}{4}$
9. $1\frac{7}{20}$
10. $24\frac{9}{13}$

the denominator 5 is the same denominator that was in the mixed numeral $4\frac{2}{5}$. Also the numerator 22 can be obtained by multiplying the whole numeral 4 by the denominator 5, then adding the original numerator 2. For example,

$$4\frac{2}{5} = \frac{22}{5}$$ To obtain 22, multiply 4 by 5, then add 2. Keep the same denominator 5.

EXAMPLE 2 Convert $6\frac{7}{12}$ to an improper fraction.

SOLUTION
LONG METHOD

$$6\frac{7}{12} = 6 + \frac{7}{12}$$

$$= \frac{6 \times 12}{1 \times 12} + \frac{7}{12}$$

$$= \frac{72}{12} + \frac{7}{12} \quad \text{Convert 6 to } \frac{72}{12}.$$

$$= \frac{79}{12} \quad \text{Add the fractions.}$$

SHORTCUT

$$6\frac{7}{12} = \frac{79}{12}$$ To obtain 79, multiply 6 by 12, then add 7. Keep the same denominator 12.

CHECK

$$\frac{79}{12} \rightarrow 12\overline{)79} \rightarrow \frac{79}{12} = 6\frac{7}{12} \quad \blacksquare$$

Now we state a general rule for converting a mixed numeral to an improper fraction. In this rule we state both the long method and the shortcut.

Converting a Mixed Numeral to an Improper Fraction

Long Method

1. Write the whole numeral part as an improper fraction with a denominator that is the same as the denominator in the fractional part.
2. Add the two fractions by adding the numerators and placing this over the common denominator.

Shortcut

Multiply the whole numeral by the denominator, then add the numerator to obtain the numerator of the improper fraction. Keep the same denominator.

 Try Problems 7 through 10.

Answers to Try These Problems

1. 6   2. 3   3. $7\frac{3}{5}$ 4. $8\frac{2}{9}$ 5. 108   6. $10\frac{2}{25}$ 7. $\frac{7}{3}$ 8. $\frac{31}{4}$
9. $\frac{27}{20}$ 10. $\frac{321}{13}$

EXERCISES 3.4

Convert each of these improper fractions to a mixed or whole numeral.

1. $\frac{20}{4}$ 2. $\frac{54}{6}$ 3. $\frac{36}{5}$ 4. $\frac{63}{8}$ 5. $\frac{245}{6}$
6. $\frac{191}{9}$ 7. $\frac{200}{8}$ 8. $\frac{480}{15}$ 9. $\frac{211}{20}$ 10. $\frac{5115}{17}$
11. $\frac{6622}{22}$ 12. $\frac{5200}{13}$

Convert each of these mixed numerals to an improper fraction.

13. $3\frac{1}{4}$ 14. $2\frac{8}{9}$ 15. $7\frac{2}{5}$ 16. $1\frac{12}{17}$ 17. $13\frac{3}{4}$
18. $25\frac{1}{6}$ 19. $34\frac{4}{5}$ 20. $18\frac{5}{7}$ 21. $3\frac{8}{11}$ 22. $8\frac{1}{20}$
23. $15\frac{9}{10}$ 24. $36\frac{3}{41}$

A fraction less than one whole cannot *be written as a whole or mixed numeral. Write each of these numbers as a whole or mixed numeral if possible. If not possible, say* not possible.

25. $\frac{1}{3}$ 26. $\frac{3}{3}$ 27. $\frac{3}{1}$ 28. $\frac{5}{3}$ 29. $\frac{4}{2}$
30. $\frac{3}{4}$ 31. $\frac{4}{4}$ 32. $\frac{13}{4}$ 33. $\frac{13}{27}$ 34. $\frac{13}{13}$
35. $\frac{27}{13}$ 36. $\frac{49}{7}$

3.5 REDUCING FRACTIONS TO LOWEST TERMS

OBJECTIVES
- Reduce a fraction to lowest terms (or simplify a fraction).
- Decide whether or not a fraction can be reduced to lowest terms.

Numerators and Denominators that Factor Easily

Recall from Section 3.3 that you can multiply the numerator and denominator of a fraction by the *same* nonzero number without changing the value of the fraction. For example,

$$\frac{3}{4} = \frac{3 \times 5}{4 \times 5} = \frac{15}{20}$$

This process is called **raising the fraction to higher terms.** We can reverse this process and divide the numerator and denominator by the *same* nonzero number without changing the value of the fraction. For example,

$$\frac{15}{20} = \frac{15 \div 5}{20 \div 5} = \frac{3}{4}$$

Try These Problems

Reduce to lowest terms.
1. $\frac{10}{25}$
2. $\frac{9}{15}$
3. $\frac{18}{24}$
4. $\frac{14}{49}$
5. $\frac{56}{72}$

This process is called reducing the fraction to lower terms. In fact, we say that $\frac{3}{4}$ is **reduced to lowest terms** (or **simplified**) because the only common divisor of 3 and 4 is 1. The fraction $\frac{15}{20}$ is *not* reduced to lowest terms (or *not* simplified) because a common divisor of 15 and 20 is 5, a number other than 1.

To reduce a fraction to lowest terms you must be able to find common divisors of the numerator and denominator. Viewing the numerator and denominator in factored form can be helpful. For example,

$$\frac{15}{21} = \frac{\cancel{3} \times 5}{\cancel{3} \times 7} = \frac{5}{7} \qquad \text{Write 15 in factored form.} \\ \text{Write 21 in factored form.}$$

Cancelling the common factor 3 from the numerator and denominator is equivalent to dividing the numerator and denominator by 3.

EXAMPLE 1   Reduce to lowest terms: $\frac{42}{63}$

SOLUTION   There is more than one way to write the steps when simplifying fractions. We show this problem done in two different ways.

METHOD 1

$$\frac{42}{63} = \frac{6 \times \cancel{7}}{\cancel{7} \times 9} \qquad \text{Cancel the common factor 7. This is equivalent to dividing the numerator and denominator by 7.}$$

$$= \frac{6}{9} \qquad \text{Do not stop until all common factors have been cancelled.}$$

$$= \frac{2 \times \cancel{3}}{3 \times \cancel{3}} \qquad \text{Cancel the common factor 3. This is equivalent to dividing the numerator and denominator by 3.}$$

$$= \frac{2}{3}$$

METHOD 2

$$\frac{42}{63} = \frac{42 \div 3}{63 \div 3} \qquad \text{Divide numerator and denominator by 3.}$$

$$= \frac{14}{21}$$

$$= \frac{14 \div 7}{21 \div 7} \qquad \text{Divide numerator and denominator by 7.}$$

$$= \frac{2}{3} \blacksquare$$

 Try Problems 1 through 5.

If you factor the numerator and denominator as the product of primes, and there are no common prime factors, then the fraction is reduced to lowest terms. For example, consider the fraction $\frac{25}{36}$.

$$\frac{25}{36} = \frac{5 \times 5}{6 \times 6} = \frac{5 \times 5}{2 \times 3 \times 2 \times 3}$$

Try These Problems

Reduce to lowest terms, if possible. If not possible, say not possible.

6. $\frac{64}{72}$
7. $\frac{12}{25}$
8. $\frac{21}{81}$
9. $\frac{49}{54}$

Because there are no common prime factors in the numerator and denominator, the fraction is reduced to lowest terms. Recall that the first few prime numbers are as follows.

Primes Less Than 20

| 2 | 3 | 5 | 7 | 11 | 13 | 17 | 19 |

A prime number has no divisors, other than itself and one.

EXAMPLE 2 Reduce to lowest terms, if possible. If not possible, say *not possible*.

a. $\frac{28}{48}$ b. $\frac{35}{54}$

SOLUTION

a.
$$\frac{28}{48} = \frac{4 \times 7}{6 \times 8}$$
$$= \frac{\cancel{2} \times \cancel{2} \times 7}{\cancel{2} \times 3 \times \cancel{2} \times 2 \times 2}$$

Cancel two factors of 2. This is equivalent to dividing numerator and denominator by 4.

$$= \frac{7}{12}$$

b.
$$\frac{35}{54} = \frac{5 \times 7}{6 \times 9}$$
$$= \frac{5 \times 7}{2 \times 3 \times 3 \times 3}$$

There are no common prime factors to cancel.

$$= \frac{35}{54}$$

It is not possible to reduce $\frac{35}{54}$. ∎

Try Problems 6 through 9.

In case the fraction can be written as a mixed numeral, then either the improper fraction form or the mixed numeral form is considered simplified (reduced to lowest terms) as long as the fraction that appears has no common factors in the numerator and denominator, other than 1, and the fractional part of the mixed numeral is less than 1 whole.

EXAMPLE 3 Simplify: $\frac{30}{18}$.

SOLUTION

$$\frac{30}{18} = \frac{6 \times 5}{2 \times 9}$$
$$= \frac{\cancel{2} \times \cancel{3} \times 5}{\cancel{2} \times \cancel{3} \times 3}$$

Cancel the common factors 2 and 3. This is equivalent to dividing the numerator and denominator by 6.

$$= \frac{5}{3} \text{ or } 1\frac{2}{3} \quad ∎$$

Try These Problems

Simplify.
10. $\frac{16}{12}$
11. $\frac{36}{12}$
12. $\frac{63}{42}$
13. $\frac{63}{45}$

In Example 3, both $\frac{5}{3}$ and $1\frac{2}{3}$ are considered simplified.

In case the fraction can be written as a whole numeral, then only the whole numeral form is considered simplified. For example, the fractions $\frac{12}{3}$ and $\frac{8}{1}$ are *not* simplified. The simplified forms are 4 and 8, respectively.

$$\frac{12}{3} = \frac{\cancel{3} \times 4}{\underset{1}{\cancel{3}}} = 4 \qquad \frac{8}{1} = 8$$

Try Problems 10 through 13.

We summarize the previous discussion by writing a rule for reducing fractions to lowest terms.

Reducing Fractions to Lowest Terms (Simplifying Fractions)

1. Write the numerator and denominator as the product of primes.
2. Cancel all common factors from the numerator and denominator. (This is equivalent to dividing the numerator and denominator by the same nonzero number.)
3. If the fraction can be written as a mixed numeral, either the improper fraction form or the mixed numeral form is accepted as simplified.
4. If the fraction can be written as a whole numeral, then only the whole numeral form is considered simplified.

---

2 Larger Numerators and Denominators

Now we look at strategies for reducing more difficult fractions to lowest terms. You want to be able to recognize more prime numbers. Here we list the first 16 primes.

First Sixteen Primes

2	3	5	7	11	13	17	19
23	29	31	37	41	43	47	53

Also, you want to recall the divisibility rules that were covered in Chapter 2. Here we list them for review.

> *Divisibility Rules*
>
> 1. Any whole number ending in 0, 2, 4, 6, or 8 is divisible by 2. (For example, 20, 32, 74, 126, and 708 are all divisible by 2.)
> 2. Any whole number with digits that add up to a number divisible by 3 is itself divisible by 3. (For example, 42, 51, 72 and 462 are all divisible by 3.)
> 3. Any whole number ending in 5 or 0 is divisible by 5. (For example, 785 and 810 are both divisible by 5.)
> 4. Any whole number ending in
>   one 0 is divisible by 10
>   two 0s is divisible by 100
>   three 0s is divisible by 1000
>   and so on. (For example, 3050 is divisible by 10, and 17,000 is divisible by 1000.)

The following examples illustrate how to simplify more difficult fractions.

EXAMPLE 4 Simplify: $\frac{28}{91}$

SOLUTION Because the numerator 28 is easy to factor, we write 28 as the product of primes, then check to see if any of these prime factors are divisors of 91.

$$28 = 4 \times 7$$
$$= 2 \times 2 \times 7$$

The prime factors of 28 are 2 and 7. Is 91 divisible by 2? No, because 91 does not end in 0, 2, 4, 6, or 8. (91 is not even.) Is 91 divisible by 7? Divide 91 by 7 to find out.

$$\begin{array}{r} 13 \\ 7\overline{)91} \\ \underline{7} \\ 21 \\ \underline{21} \end{array}$$

Yes, $91 = 7 \times 13$. Now we can reduce the fraction to lowest terms.

$$\frac{28}{91} = \frac{4 \times \cancel{7}}{\cancel{7} \times 13} \qquad \text{Cancel the common factor 7.}$$
$$= \frac{4}{13}$$

The fraction $\frac{4}{13}$ is reduced to lowest terms because there are no common prime factors in the numerator and denominator.

$$\frac{4}{13} = \frac{2 \times 2}{13} \qquad \text{This fraction cannot be reduced to lowest terms.}$$

Therefore, $\frac{28}{91} = \frac{4}{13}$. ∎

3.5 Reducing Fractions to Lowest Terms

EXAMPLE 5 Simplify: $\frac{117}{360}$

SOLUTION Write either the numerator or denominator as the product of primes, whichever is easier.

$$360 = 36 \times 10$$
$$= 6 \times 6 \times 5 \times 2$$
$$= 2 \times 3 \times 2 \times 3 \times 5 \times 2$$

The prime factors of 360 are 2, 3, and 5. Check to see if 2, 3, or 5 are divisors of 117. Is 117 divisible by 2? No, because 117 does not end in 0, 2, 4, 6, or 8. Is 117 divisible by 3? Yes, because the sum of the digits is 9 and 9 is divisible by 3. Divide 117 by 3 to find the other factor.

$$\begin{array}{r} 39 \\ 3\overline{)117} \\ \underline{9} \\ 27 \\ \underline{27} \end{array}$$

Therefore, $117 = 3 \times 39$. Note that 3 is also a divisor of 39.

$$117 = 3 \times 39$$
$$= 3 \times 3 \times 13$$

Now we can see how to reduce the original fraction to lowest terms.

$$\frac{117}{360} = \frac{\cancel{3} \times \cancel{3} \times 13}{2 \times \cancel{3} \times 2 \times \cancel{3} \times 5 \times 2} \quad \text{Cancel two factors of 3.}$$

$$= \frac{13}{2 \times 2 \times 5 \times 2} \quad \text{There are no more common prime factors in the numerator and denominator.}$$

$$= \frac{13}{40} \ \blacksquare$$

EXAMPLE 6 Simplify: $\frac{204}{85}$

SOLUTION Begin by factoring either the numerator or the denominator as the product of primes. We choose 85 because it seems easier. Because 85 ends in 5, it is divisible by 5.

$$\begin{array}{r} 17 \\ 5\overline{)85} \\ \underline{5} \\ 35 \\ \underline{35} \end{array}$$

Therefore,

$$85 = 5 \times 17$$

and the only prime factors of 85 are 5 and 17. Now check to see if the numerator 204 is divisible by 5 or 17. Is 204 divisible by 5? No, because it does not end in 5 or 0. Is 204 divisible by 17?

Try These Problems

Simplify.

14. $\frac{36}{150}$
15. $\frac{24}{68}$
16. $\frac{147}{180}$
17. $\frac{133}{76}$
18. $\frac{104}{195}$
19. $\frac{253}{207}$
20. $\frac{40}{120}$
21. $\frac{130}{2500}$
22. $\frac{25{,}000}{1500}$
23. $\frac{1200}{90}$

$$17\overline{)204}$$
$$\underline{17}$$
$$34$$
$$\underline{34}$$

Yes, $204 = 12 \times 17$. Now we can reduce the fraction to lowest terms.

$$\frac{204}{85} = \frac{12 \times \cancel{17}}{5 \times \cancel{17}}$$ Cancel the common factor 17.

$$= \frac{2 \times 2 \times 3}{5}$$ Make sure there are no common prime factors left to cancel.

$$= \frac{12}{5} \text{ or } 2\frac{2}{5} \ \blacksquare$$

 Try Problems 14 through 19.

EXAMPLE 7 Simplify: $\frac{3600}{42{,}000}$

SOLUTION

$$\frac{3600}{42{,}000} = \frac{36 \times 100}{42 \times 1000}$$

$$= \frac{36 \times \cancel{10} \times \cancel{10}}{42 \times \cancel{10} \times \cancel{10} \times 10}$$ Cancel the two factors of 10.

$$= \frac{\cancel{6} \times 6}{\cancel{6} \times 7 \times 10}$$ Cancel the common factor 6.

$$= \frac{\cancel{2} \times 3}{7 \times \cancel{2} \times 5}$$ Cancel the common factor 2.

$$= \frac{3}{35}$$

 Try Problems 20 through 23.

 Answers to Try These Problems

1. $\frac{2}{5}$ 2. $\frac{3}{5}$ 3. $\frac{3}{4}$ 4. $\frac{2}{7}$ 5. $\frac{7}{9}$ 6. $\frac{8}{9}$ 7. not possible 8. $\frac{7}{27}$
9. not possible 10. $\frac{4}{3}$ or $1\frac{1}{3}$ 11. 3 12. $\frac{3}{2}$ or $1\frac{1}{2}$ 13. $1\frac{2}{5}$
14. $\frac{6}{25}$ 15. $\frac{6}{17}$ 16. $\frac{49}{60}$ 17. $\frac{7}{4}$ or $1\frac{3}{4}$ 18. $\frac{8}{15}$ 19. $\frac{11}{9}$ or $1\frac{2}{9}$ 20. $\frac{1}{3}$
21. $\frac{13}{250}$ 22. $\frac{50}{3}$ or $16\frac{2}{3}$ 23. $\frac{40}{3}$ or $13\frac{1}{3}$

EXERCISES 3.5

Reduce to lowest terms, if possible. If not possible, say not possible.

1. $\frac{6}{9}$
2. $\frac{10}{15}$
3. $\frac{12}{8}$
4. $\frac{30}{18}$
5. $\frac{24}{40}$
6. $\frac{27}{54}$
7. $\frac{54}{72}$
8. $\frac{45}{56}$
9. $\frac{64}{81}$
10. $\frac{60}{48}$
11. $\frac{21}{70}$
12. $\frac{35}{20}$
13. $\frac{36}{8}$
14. $\frac{28}{49}$
15. $\frac{48}{42}$
16. $\frac{13}{20}$
17. $\frac{17}{21}$
18. $\frac{72}{80}$
19. $\frac{54}{90}$
20. $\frac{120}{75}$

Simplify.

21. $\frac{42}{66}$ 22. $\frac{72}{117}$ 23. $\frac{15}{85}$ 24. $\frac{16}{84}$ 25. $\frac{91}{14}$

26. $\frac{80}{15}$ 27. $\frac{77}{140}$ 28. $\frac{60}{204}$ 29. $\frac{165}{45}$ 30. $\frac{126}{147}$

31. $\frac{90}{315}$ 32. $\frac{147}{196}$ 33. $\frac{55}{132}$ 34. $\frac{77}{143}$ 35. $\frac{39}{104}$

36. $\frac{26}{65}$ 37. $\frac{78}{52}$ 38. $\frac{156}{117}$ 39. $\frac{85}{68}$ 40. $\frac{153}{34}$

41. $\frac{68}{136}$ 42. $\frac{102}{255}$ 43. $\frac{133}{171}$ 44. $\frac{152}{57}$ 45. $\frac{228}{285}$

46. $\frac{380}{570}$ 47. $\frac{525}{390}$ 48. $\frac{380}{360}$ 49. $\frac{117}{135}$ 50. $\frac{288}{135}$

51. $\frac{441}{504}$ 52. $\frac{560}{693}$ 53. $\frac{120}{42}$ 54. $\frac{63}{1800}$ 55. $\frac{810}{4500}$

56. $\frac{70}{3600}$ 57. $\frac{2100}{14,000}$ 58. $\frac{11,200}{8400}$ 59. $\frac{25,300}{781,000}$ 60. $\frac{5200}{31,200}$

USING THE CALCULATOR #6

AN INTRODUCTION TO THE $a\frac{b}{c}$ KEY

Fractions are usually entered into a calculator as decimals by dividing the numerator by the denominator. We discuss this in more detail after decimals are introduced in Chapter 6.

Some scientific calculators are actually capable of handling fractions in their fractional form. These calculators have an [a b/c] key. In this feature, we discuss how to use this key to enter a fraction and change the form of a fraction.

First we look at how to enter a fraction into the calculator and what it looks like on the display screen.

To Display $5\frac{3}{4}$

Enter   5 [a b/c] 3 [a b/c] 4

Result   5 ⌐ 3 ⌐ 4.

Observe how the calculator displays a mixed numeral. The whole number is written first, the numerator of the fraction part second, and last the denominator. The mark ⌐ is used to separate the parts. Some calculators use a mark that is facing in the reverse direction.

The [a b/c] key can also be used to enter a single fraction. Here is an example.

To Display $\frac{7}{8}$

Enter   7 [a b/c] 8

Result   7 ⌐ 8.

Scientific calculators have an [INV], [2ndF] or [Shift] key which is usually located near the upper left-hand corner of the calculator. This key is often entered just before a function key to access the alternative function of the key.

After entering a fraction or mixed numeral, if you enter [INV] and [a b/c], the form of the fraction will change. If the fraction entered is not reduced, this action will reduce the fraction. If the fraction is reduced, then entering [INV] and [a b/c] repeatedly will switch the form of the fraction from mixed numeral to improper fraction, and vice versa.

Another way to simplify a fraction or mixed numeral is to enter the fraction, then enter [=].

A scientific calculator that has the [a b/c] key is usually limited to handling fractions with numerators and denominators that have three or fewer digits.

Calculator Problems

1. Using the [a b/c] key, try to enter each of these fractions into your calculator to learn the limitations of your calculator. $\frac{13}{20}$ $\frac{123}{500}$ $\frac{1233}{7000}$ $\frac{12,333}{80,000}$

Use the [a b/c] key to convert each mixed numeral to an improper fraction.

2. $2\frac{1}{3}$ 3. $18\frac{10}{12}$ 4. $12\frac{23}{79}$ 5. $20\frac{27}{36}$

Cont. page 110

Use the [a b/c] *key to convert each improper fraction to a mixed numeral.*

6. $\frac{28}{9}$ **7.** $\frac{347}{15}$ **8.** $\frac{678}{24}$ **9.** $\frac{161}{42}$

Use the [a b/c] *key to reduce to lowest terms.*

10. $\frac{6}{9}$ **11.** $\frac{150}{51}$ **12.** $\frac{112}{280}$ **13.** $8\frac{75}{120}$

14. Without using the [a b/c] key, can you discover a procedure for using the calculator to help you convert a mixed numeral to an improper fraction? Using the example $6\frac{2}{3}$, clearly explain your procedure.

15. Using your procedure from Problem 14, convert $890\frac{126}{6255}$ to an improper fraction.

16. Without using the [a b/c] key, can you discover a procedure for using the calculator to help you reduce a fraction to lowest terms? Using the example $\frac{70}{504}$, clearly explain your procedure.

17. Using your procedure from Problem 16, reduce the fraction $\frac{1771}{5083}$ to lowest terms.

3.6 A FRACTION AS DIVISION, RATIO, AND WHAT FRACTION OF ONE QUANTITY IS ANOTHER QUANTITY

OBJECTIVES
- Solve an application involving a total quantity split into a number of equal parts.
- Write a fraction that represents a division problem.
- Write the division problem represented by a fraction.
- Find the ratio of one quantity to another quantity.
- Find what fraction of one quantity is another quantity.
- Solve an application involving ratio.
- Solve an application involving finding what fraction of one quantity is another quantity.

1 A Fraction as Division

In Section 3.4 you learned that an improper fraction can be converted to a whole or mixed numeral by dividing the numerator by the denominator. For example,

$$\frac{12}{3} = 12 \div 3 = 4$$

and

$$\frac{14}{3} \rightarrow 3\overline{)14} \rightarrow \frac{14}{3} = 4\frac{2}{3}$$

Here we see that the fraction bar indicates division. That is,

$$\frac{\text{Numerator}}{\text{Denominator}} = \text{Numerator} \div \text{Denominator}$$

In this section, we look at several ways to interpret a fraction as division.

Suppose you have 6 gallons of paint to be divided equally among 3 workers. How much paint does each worker receive? To solve this problem, we divide 6 by 3.

$$3\overline{)6}^{\,2} \quad \text{or} \quad 6 \div 3 = 2 \quad \text{or} \quad \frac{6}{3} = 2$$

In this problem a total quantity is being split into a number of equal parts and the following general relationship applies.

$$\begin{aligned}\text{Size of each part} &= \text{total quantity} \div \text{number of equal parts} \\ &= 6 \div 3 \\ &= 2\end{aligned}$$

or

$$\begin{aligned}\text{Size of each part} &= \frac{\text{total quantity}}{\text{number of equal parts}} \\ &= \frac{6}{3} \\ &= 2\end{aligned}$$

Each worker receives 2 gallons of paint.

Now suppose that you have only 2 gallons of paint to split equally among 3 workers. How much paint does each worker receive? Here the total amount of paint to be split apart is 2 gallons, and the number of equal parts is 3. If we use the above relationship, we have

$$\begin{aligned}\text{Size of each part} &= \frac{\text{total quantity}}{\text{number of equal parts}} \\ &= \frac{2 \text{ gallons}}{3} \\ &= \frac{2}{3} \text{ gallon}\end{aligned}$$

Therefore, each worker receives $\frac{2}{3}$ gallon paint. We can check the answer. If 3 workers each have $\frac{2}{3}$ gallon paint, is the total 2 gallons?

$$\frac{2}{3} + \frac{2}{3} + \frac{2}{3} = \frac{6}{3} = 2$$

Yes, the total is 2 gallons. Therefore, $\frac{2}{3}$ gallon is the correct amount for each worker.

Separating a Quantity Into a Number of Equal Parts

$$\text{Size of each part} = \text{total quantity} \div \text{number of equal parts}$$

$$= \frac{\text{total quantity}}{\text{number of equal parts}}$$

EXAMPLE 1 Seven persons share equally a pizza. How much pizza does each person receive?

SOLUTION We want to separate 1 pizza into 7 equal parts.

$$\text{Size of each part} = \frac{\text{total quantity}}{\text{number of equal parts}}$$

$$= \frac{1 \text{ pizza}}{7}$$

$$= \frac{1}{7} \text{ pizza}$$

Each person receives $\frac{1}{7}$ pizza. ∎

EXAMPLE 2 A rope that is 3 meters long is cut into 5 equal pieces. How long is each piece?

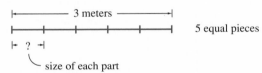

SOLUTION We want to separate 3 meters of rope into 5 equal pieces.

$$\text{Size of each part} = \frac{\text{total quantity}}{\text{number of equal parts}}$$

$$= \frac{3 \text{ meters}}{5}$$

$$= \frac{3}{5} \text{ meter}$$

CHECK $\quad \frac{3}{5} + \frac{3}{5} + \frac{3}{5} + \frac{3}{5} + \frac{3}{5} = \frac{15}{5} = 3$

Each piece measures $\frac{3}{5}$ meter. ∎

EXAMPLE 3 Twelve persons share equally 20 pounds of cheese. How much cheese does each person receive? Simplify the answer.

SOLUTION We want to separate 20 pounds of cheese into 12 equal parts.

3.6 A Fraction as Division, Ratio, and What Fraction of One Quantity Is Another Quantity

Try These Problems

Solve.

1. Four families share equally 12 quarts of milk. How much milk does each family receive? Simplify the answer.

2. Twelve families share equally 4 quarts of milk. How much milk does each family receive? Simplify the answer.

3. Mary Thomas split a pecan pie equally among 8 persons. How much pie did each person receive?

4. A board that is 7 feet long is cut into 10 equal pieces. How long is each piece?

5. A board that is 10 feet long is cut into 7 equal pieces. How long is each piece?

6. Write as a fraction.
   a. $4 \div 5$
   b. $17 \div 5$
   c. $7 \overline{)20}$
   d. $9 \overline{)4}$

7. Write as a division problem using the symbol $\div$.
   a. $\frac{1}{6}$
   b. $\frac{3}{5}$
   c. $\frac{8}{3}$
   d. $\frac{10}{7}$

8. Write as a division problem using the symbol $\overline{)}$.
   a. $\frac{1}{9}$
   b. $\frac{2}{3}$
   c. $\frac{9}{4}$
   d. $\frac{11}{5}$

Write a fraction for each.

9. Nine-halves
10. Thirty-sevenths
11. Eight divided by three
12. Eight divided into three

$$\frac{\text{Size of}}{\text{each part}} = \frac{\text{total quantity}}{\text{number of equal parts}}$$

$$= \frac{20 \text{ pounds}}{12}$$

$$= \frac{20}{12} \text{ pounds}$$

$$= \frac{5 \times \cancel{4}}{3 \times \cancel{4}} \text{ pounds}$$

$$= \frac{5}{3} \text{ or } 1\frac{2}{3} \text{ pounds}$$

Each person receives $\frac{5}{3}$ or $1\frac{2}{3}$ pounds of cheese. ■

Try Problems 1 through 5.

The fraction bar indicates division as shown here.

$$\frac{\text{Numerator}}{\text{Denominator}} = \text{Numerator} \div \text{Denominator}$$

The denominator is *always* the divisor. This gives us three ways to write a division problem as illustrated by the following chart.

Fraction	Division Using $\div$	Division Using $\overline{)}$	English
$\frac{1}{8}$	$1 \div 8$	$8\overline{)1}$	One-eighth One divided by eight Eight divided into one
$\frac{2}{5}$	$2 \div 5$	$5\overline{)2}$	Two-fifths Two divided by five Five divided into two
$\frac{5}{2}$	$5 \div 2$	$2\overline{)5}$	Five-halves Five divided by two Two divided into five
$\frac{14}{3}$	$14 \div 3$	$3\overline{)14}$	Fourteen-thirds Fourteen divided by three Three divided into fourteen

It is important that you know how to write the two numbers in the correct order and that you use the correct language to read the fraction or division problem.

Try Problems 6 through 12.

2 Ratio—A Comparison By Division

Suppose a committee consists of 5 persons: 2 women and 3 men. We can compare the number of women to the number of men by division.

▲ **Try These Problems**

Write a fraction for each. Simplify, if possible.
13. The ratio of 2 to 11.
14. The ratio of 11 to 2.
15. The ratio of 36 to 27.
16. The ratio of 27 to 36.

We say, "the **ratio** of women to men is 2 to 3." We write,

$$\text{Ratio of women to men} = \frac{2}{3} \begin{array}{l} \text{— number of women} \\ \text{— number of men} \end{array}$$

We can compare the quantities in the reverse order.

$$\text{Ratio of men to women} = \frac{3}{2} \begin{array}{l} \text{— number of men} \\ \text{— number of women} \end{array}$$

We can also compare either one of the quantities to the total number of persons on the committee.

$$\text{Ratio of men to the total} = \frac{3}{5} \begin{array}{l} \text{— number of men} \\ \text{— total} \end{array}$$

To recognize a comparison by division pay close attention to the language. The word *ratio* indicates to form a fraction, and the quantity that follows the word *to* is put in the denominator.

EXAMPLE 4   Write a fraction for each. Simplify.

a. The ratio of 18 to 30.   **b.** The ratio of 49 to 35.

SOLUTION

a. The ratio of 18 to 30 $= \dfrac{18}{30}$

$= \dfrac{\cancel{3} \times 6}{\cancel{3} \times 10}$

$= \dfrac{\cancel{2} \times 3}{\cancel{2} \times 5}$

$= \dfrac{3}{5}$

The quantity following the word *to* is put in the denominator.

b. The ratio of 49 to 35 $= \dfrac{49}{35}$

$= \dfrac{7 \times \cancel{7}}{5 \times \cancel{7}}$

$= \dfrac{7}{5}$

The quantity following the word *to* is put in the denominator.

Because a ratio compares two quantities by division, we prefer to leave the answer $\frac{7}{5}$ as an improper fraction rather than a mixed numeral. ■

 **Try Problems 13 through 16**.

EXAMPLE 5   Ed, a mechanic, earned $28,000 last year while Alice, his girlfriend, earned $70,000 as a financial planner.

a. Find the ratio of Ed's earnings to Alice's earnings.

b. Find the ratio of Alice's earnings to their total earnings.

Simplify both results.

3.6 A Fraction as Division, Ratio, and What Fraction of One Quantity Is Another Quantity   115

 Try These Problems

17. A board 15 feet long is cut into two pieces. One piece is 9 feet long, the other is 6 feet long.
a. What is the ratio of the shorter piece to the longer piece?
b. What is the ratio of the longer piece to the shorter piece?
c. What is the ratio of the longer piece to the total?

18. A football team played a total of 52 games. They won 13 of them and lost the rest.
a. What is the ratio of wins to losses?
b. What is the ratio of wins to total games?

SOLUTION

a.
$$\text{Ratio of Ed's earnings to Alice's earnings} = \frac{\text{Ed's earnings}}{\text{Alice's earnings}}$$
$$= \frac{28{,}000}{70{,}000}$$
$$= \frac{28 \times \cancel{1000}}{70 \times \cancel{1000}}$$
$$= \frac{4 \times \cancel{7}}{\cancel{7} \times 10}$$
$$= \frac{\cancel{2} \times 2}{\cancel{2} \times 5}$$
$$= \frac{2}{5}$$

b.
$$\text{Ratio of Alice's earnings to their total earnings} = \frac{\text{Alice's earnings}}{\text{Total earnings}}$$
$$= \frac{70{,}000}{28{,}000 + 70{,}000}$$
$$= \frac{70{,}000}{98{,}000}$$
$$= \frac{70 \times \cancel{1000}}{98 \times \cancel{1000}}$$
$$= \frac{7 \times 10}{2 \times 49}$$
$$= \frac{\cancel{7} \times \cancel{2} \times 5}{\cancel{2} \times \cancel{7} \times 7}$$
$$= \frac{5}{7} \ \blacksquare$$

 Try Problems 17 and 18.

3 What Fraction of One Quantity Is Another Quantity?

There is another way to verbalize a comparison by division other than using the word *ratio*. If there are 5 committee members including 2 women and 3 men, we say

$$\text{The fraction of the committee that is women} = \frac{2}{5} \begin{array}{l} \text{— number of women} \\ \text{— total} \end{array}$$

Here we are comparing the number of women to the total number of persons on the committee by division. Note that the phrase *fraction of* indicates to form a fraction. The quantity that follows the phrase *fraction of* is put in the denominator and the other quantity mentioned is put in the numerator. Here are some examples.

 **Try These Problems**

19. What fraction of 54 is 49?
20. 132 is what fraction of 88?

What fraction of 10 is 3? $\dfrac{3}{10}$

What fraction of 3 is 10? $\dfrac{10}{3}$ or $3\dfrac{1}{3}$

140 is what fraction of 210? $\dfrac{140}{210} = \dfrac{2 \times \cancel{7} \times \cancel{10}}{3 \times \cancel{7} \times \cancel{10}} = \dfrac{2}{3}$

210 is what fraction of 140? $\dfrac{210}{140} = \dfrac{3 \times \cancel{7} \times \cancel{10}}{2 \times \cancel{7} \times \cancel{10}} = \dfrac{3}{2} = 1\dfrac{1}{2}$

 Try Problems 19 and 20.

EXAMPLE 6   What fraction of these figures are circles?

SOLUTION   What fraction of these figures are circles? $\dfrac{3}{5}$

The quantity mentioned after the phrase *fraction of* is put in the denominator. The other quantity is put in the numerator. $\dfrac{3}{5}$ of these figures are circles. ■

EXAMPLE 7   The legs of this bar stool are what fraction of the total height?

SOLUTION   The legs of this bar stool are what fraction of the total height?

$$\dfrac{37}{12 + 37} \begin{array}{l}\text{—— height of the legs}\\ \text{—— total height}\end{array}$$

$$= \dfrac{37}{49}$$

The legs are $\dfrac{37}{49}$ of the total height. ■

EXAMPLE 8   Dorothy owed a friend $30. She paid the friend $14. What fraction of the debt does she still owe? Simplify.

Try These Problems

21. What fraction of the figures are triangles?

22. What fraction of these circles is *not* shaded?

23. The Miami Dolphins won 12 of the 15 games they played and lost the rest. What fraction of the games did they lose? Simplify.

24. Susanne purchased these items at the grocery store.

Ground beef $5
Chicken breasts $8
Apples $2
Bananas $3
Lettuce $1
Bread $1

The amount spent on fruit is what fraction of the total bill? Simplify.

SOLUTION What fraction of the debt does she still owe?

$$\frac{36 - 14}{30} \begin{array}{l} \text{— amount she still owes} \\ \text{— original debt} \end{array}$$

$$= \frac{16}{30}$$

$$= \frac{\cancel{2} \times 8}{\cancel{2} \times 15}$$

$$= \frac{8}{15}$$

Dorothy still owes $\frac{8}{15}$ of the debt. ■

 Try Problems 21 through 24.

Now we summarize three ways that a fraction can represent division.

A Fraction as Division

1. A total quantity is separated into a number of equal parts and you are looking for the size of each part.

$$\text{Size of each part} = \text{total quantity} \div \text{number of equal parts}$$

$$= \frac{\text{total quantity}}{\text{number of equal parts}}$$

2. You want to know the ratio of one quantity to another quantity.

$$\text{Ratio of 4 to 7} = \frac{4}{7}$$

The quantity following the word *to* is put in the denominator.

3. You want to know what fraction of one quantity is another quantity.

$$\text{What fraction of 7 is 4?} \quad \frac{4}{7}$$

The quantity following the phrase *fraction of* is put in the denominator.

Answers to Try These Problems

1. 3 qt   2. $\frac{1}{3}$ qt   3. $\frac{1}{8}$ pie   4. $\frac{7}{10}$ ft   5. $\frac{10}{7}$ or $1\frac{3}{7}$ ft
6. a. $\frac{4}{5}$ b. $\frac{17}{5}$ c. $\frac{20}{7}$ d. $\frac{4}{9}$ 7. a. $1 \div 6$
7. b. $3 \div 5$ c. $8 \div 3$ d. $10 \div 7$ 8. a. $9\overline{)1}$ b. $3\overline{)2}$
8. c. $4\overline{)9}$ d. $5\overline{)11}$ 9. $\frac{9}{2}$ 10. $\frac{30}{7}$ 11. $\frac{8}{3}$
12. $\frac{3}{8}$ 13. $\frac{2}{11}$ 14. $\frac{11}{2}$ 15. $\frac{4}{3}$ 16. $\frac{3}{4}$
17. a. $\frac{2}{3}$ b. $\frac{3}{2}$ c. $\frac{3}{5}$ 18. a. $\frac{1}{3}$ b. $\frac{1}{4}$
19. $\frac{49}{54}$ 20. $\frac{3}{2}$ 21. $\frac{4}{7}$ 22. $\frac{8}{15}$ 23. $\frac{1}{5}$ 24. $\frac{1}{4}$

EXERCISES 3.6

Answer with a fraction. Simplify, if possible.

1. Twenty gallons of paint are distributed equally among 5 workers. How much paint does each worker receive?
2. Five gallons of paint are distributed equally among 20 workers. How much paint does each worker receive?
3. Six persons share equally 9 pounds of cheese. How much cheese does each person receive?
4. Nine persons share equally 6 pounds of cheese. How much cheese does each person receive?
5. A string that is 1 foot long is cut into 5 equal pieces. How long is each piece?
6. Six children share equally 1 gallon of ice cream. How much ice cream does each child receive?
7. Bob, Steve, and Marge share equally a quart of milk. How much milk does each person receive?
8. A pound of butter is distributed equally among Carol, Sue, Jedd, and John. How much butter does each person receive?
9. Ed and Dick go trekking in the Himalayan mountains. The total distance of the trip is 100 kilometers. They want to spread this equally over a 3-day period. How far will they trek each day?
10. A wire is cut into 150 equal pieces. If the wire is 45 inches long, how long is each piece?

Write as a fraction. Simplify, if possible.

11. $2 \div 7$
12. $7 \div 2$
13. $20\overline{)8}$
14. $8\overline{)20}$
15. $35 \div 15$
16. $15 \div 35$
17. $5\overline{)3}$
18. $3\overline{)5}$

Write as a division problem using the symbol $\div$

19. $\frac{1}{8}$
20. $\frac{4}{9}$
21. $\frac{19}{5}$
22. $\frac{23}{8}$
23. $4\overline{)3}$
24. $2\overline{)17}$
25. 4 divided by 15
26. 3 divided into 5

Write as a division problem using the symbol $\overline{)}$.

27. $\frac{1}{12}$
28. $\frac{5}{8}$
29. $\frac{12}{5}$
30. $\frac{32}{9}$
31. $8 \div 13$
32. $26 \div 3$
33. 6 divided into 1
34. 25 divided by 4

Write a fraction for each. Simplify, if possible.

35. Two-thirds
36. Eleven-halves
37. Eight-tenths
38. Twenty-four ninths
39. One divided by five
40. Twelve divided by eight
41. Thirty-six divided into nine
42. hree divided into seven
43. The ratio of 3 to 11.
44. The ratio of 15 to 18.
45. The ratio of 17 to 5.
46. The ratio of 48 to 90.
47. What fraction of 50 is 35?
48. What fraction of 12 is 52?
49. 2 is what fraction of 400?
50. 28 is what fraction of 21?

51. What fraction of these figures are arrows?

52. What fraction of these figures are *not* circles?

53. What is the ratio of the number of triangles to the number of circles? Simplify.

54. What is the ratio of the number of circles to the total number of figures? Simplify.

55. Find the ratio of the height of the sail to the total height of the sailboat. Simplify.

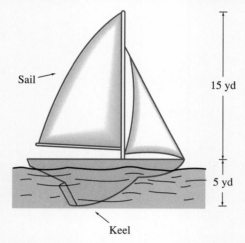

56. What is the ratio of the length of the cab to the total length of the truck? Simplify.

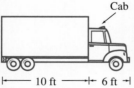

57. The woman's torso is what fraction of her total height?

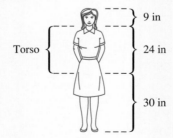

58. The thread portion of the screw is what fraction of its total length?

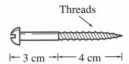

59. If you had $50 and you spent $23, what fraction of your money did you spend?

60. There are 7 students absent in a class of 32. What fraction of the class is present?

61. A shipment contains 350 pink grapefruit and 400 white grapefruit. What fraction of the shipment is pink grapefruit?

62. A grocery bill is as follows:
Bananas $1
Apples $2
Ham $8
Chicken $5
Frozen vegetables $3
Tax $1
What fraction of the bill is spent on fruit?

63. A baseball player got 35 hits out of 100 times at bat. Find the ratio of the number of hits to the number of times at bat? Simplify.

64. Out of 2000 workers, 1600 are nonsmokers. Find the ratio of smokers to nonsmokers. Simplify.

65. A rectangle has length 68 feet and width 17 feet. What is the ratio of the area to the perimeter? Simplify.

66. The height of a rectangular window is 42 inches. The width is 3 times the height. What is the ratio of the height to the perimeter?

67. The price of a microwave oven decreased from $450 to $400. The decrease in price is what fraction of the original price?

68. The price of a stock increased from $50 to $65. The increase in price is what fraction of the original price?

The total population of a town in southern California is 60,000 persons. The bar graph gives the distribution of the population by race. Use the graph to answer Exercises 69 through 72.

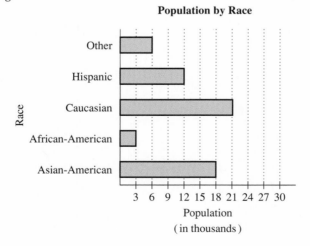

69. Find the ratio of the number of African-Americans to the number of Asian-Americans.
70. Find the ratio of the number of Caucasians to the number of Asian-Americans.
71. What fraction of the total population is Hispanic?
72. What fraction of the total population is African-American?

Lisa, Liz, and Kent bought an apartment building together for $325,000. The circle graph gives the portion of the building that each person owns. Use the graph to answer Exercises 73 through 76.

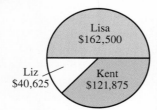

Portion of Building Owned

73. What fraction of the building does Kent own?
74. What fraction of the building does Liz own?
75. What is the ratio of Liz's share to Lisa's share?
76. What is the ratio of Lisa's share to Kent's share?

CHAPTER 3 SUMMARY

KEY WORDS AND PHRASES
fraction [3.1]
fraction bar [3.1]
numerator [3.1]
denominator [3.1]
improper fraction [3.1, 3.4]
mixed numeral [3.1, 3.4]
raising a fraction to higher terms [3.3]
reduce to lowest terms [3.5]
simplify [3.5]
ratio [3.6]

SYMBOLS
The fraction bar indicates division. For example, $\frac{3}{4}$ means $3 \div 4$

IMPORTANT RULES
When a Fraction Is Less Than, Equal to, or More Than the Number 1 [3.1]
- A fraction that has a numerator less than the denominator is less than the number 1. For example, the fractions $\frac{7}{8}$ and $\frac{12}{25}$ are each less than 1.
- A fraction that has a numerator more than the denominator is more than the number 1. For example, the fractions $\frac{7}{4}$ and $\frac{30}{9}$ are each more than 1.
- A fraction that has a numerator equal to the denominator is equal to the number 1. For example, the fractions $\frac{6}{6}$ and $\frac{50}{50}$ are each equal to 1.

Raising a Fraction to Higher Terms [3.3]
Multiplying the numerator and denominator by the same nonzero number does not change the value of a fraction.

Converting an Improper Fraction to a Mixed or Whole Numeral [3.4]
Divide the numerator by the denominator. The quotient is the whole part. If there is a remainder, other than 0, the remainder over the divisor (the original denominator) is the fraction part.

Converting a Mixed Numeral to an Improper Fraction [3.4]
LONG METHOD
- Write the whole numeral part as an improper fraction with a denominator that is the same as the denominator in the fractional part.
- Add the two fractions by adding the numerators and placing this over the common denominator.

SHORTCUT
Multiply the whole numeral by the denominator, then add the numerator to obtain the numerator of the improper fraction. Keep the same denominator.

Reducing a Fraction to Lowest Terms (Simplifying Fractions) [3.5]
- Dividing the numerator and denominator of a fraction by the same nonzero number does not change the value of the fraction. This is equivalent to canceling common factors from the numerator and denominator.
- A fraction is reduced to lowest terms (or simplified) when there are no common divisors of the numerator and denominator, other than 1.
- When a fraction is more than 1, and does not equal a whole numeral, either the improper fraction form or the mixed numeral form is considered simplified.
- When a fraction equals a whole numeral, only the whole numeral form is considered simplified.

Divisibility Rules [3.5]
These rules are listed on page 106.

A Fraction as Division [3.6]
A summary of how a fraction represents division is on page 117.

CHAPTER 3 REVIEW EXERCISES

Write the English name for each.

1. $\frac{1}{2}$
2. $\frac{2}{3}$
3. $\frac{6}{13}$
4. $3\frac{1}{4}$
5. $14\frac{2}{5}$
6. $20\frac{13}{100}$

Write a fraction represented by the shaded region. Assume each figure represents one whole.

7.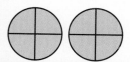

8.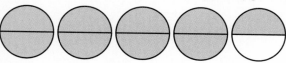

Write an improper fraction and a mixed numeral or a whole numeral that represents the shaded regions. Assume each separate figure represents one whole.

9.

10.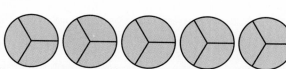

11.

12.

Choose the fractions in each group that are more than one whole.

13. $\frac{7}{9}, \frac{21}{5}, \frac{13}{17}, \frac{13}{13}, \frac{11}{2}$ **14.** $\frac{130}{21}, \frac{19}{4}, \frac{3}{4}, \frac{15}{43}, \frac{9}{9}$

Choose the fractions in each group that are equal to one whole.

15. $\frac{1}{3}, \frac{4}{1}, \frac{4}{4}, \frac{6}{2}, \frac{30}{30}$ **16.** $\frac{15}{15}, \frac{8}{1}, \frac{1}{1}, \frac{7}{7}, \frac{1}{12}$

Name two equal fractions represented by the shaded region. Assume each figure represents one whole.

17. **18.** **19.** **20.**

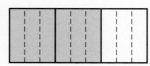

Find the missing numerator so that the two numbers are equal.

21. $\frac{4}{5} = \frac{?}{35}$ **22.** $\frac{13}{9} = \frac{?}{108}$ **23.** $1 = \frac{?}{12}$ **24.** $6 = \frac{?}{30}$

25. Convert $\frac{2}{7}$ to a fraction with denominator 56.

26. Write a fraction with denominator 4 that is equal to the whole number 8.

Convert each of these improper fractions to a mixed or whole numeral.

27. $\frac{47}{8}$ **28.** $\frac{300}{15}$ **29.** $\frac{475}{25}$ **30.** $\frac{83}{4}$

Convert each of these mixed numerals to an improper fraction.

31. $5\frac{2}{7}$ **32.** $35\frac{1}{4}$ **33.** $12\frac{5}{6}$ **34.** $7\frac{6}{23}$

A fraction less than one whole cannot *be written as a whole or mixed numeral. Write each of these numbers as a whole or mixed numeral, if possible. If not possible, say* not possible.

35. $\frac{16}{3}$ **36.** $\frac{3}{16}$ **37.** $\frac{16}{16}$ **38.** $\frac{1}{4}$ **39.** $\frac{4}{1}$ **40.** $\frac{125}{6}$

Reduce to lowest terms, if possible. If not possible, say not possible.

41. $\frac{15}{50}$ **42.** $\frac{42}{28}$ **43.** $\frac{36}{49}$ **44.** $\frac{32}{56}$ **45.** $\frac{54}{42}$ **46.** $\frac{27}{70}$

Simplify.

47. $\frac{24}{45}$ **48.** $\frac{50}{135}$ **49.** $\frac{126}{140}$ **50.** $\frac{294}{36}$ **51.** $\frac{165}{75}$

52. $\frac{198}{385}$ **53.** $\frac{132}{65}$ **54.** $\frac{68}{102}$ **55.** $\frac{7000}{1400}$ **56.** $\frac{36{,}000}{240{,}000}$

Write a fraction for each. Simplify, if possible.

57. $13 \div 3$ **58.** $3 \div 13$ **59.** $54\overline{)45}$ **60.** $45\overline{)54}$

61. 12 divided by 20 **62.** 12 divided into 20

63. Five-thirds **64.** Thirteen-halves **65.** Seven-fifths

66. Twenty-one sixty-thirds

67. The ratio of 4 to 12.

68. The ratio of 180 to 3600.

69. What fraction of 800 is 160?

70. 620 is what fraction of 155?

71. What fraction of these triangles is shaded?

72. What fraction of these figures are circles?

73. What is the ratio of the number of shaded squares to the number of unshaded squares?

74. A wine glass has measurements as shown in the diagram. What is the ratio of the height of the stem to the total height.

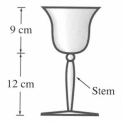

75. Courtney and her two sisters share equally 5 pounds of nuts. How many pounds of nuts does each person receive?

76. A string is 3 meters long. Mr. Lau cuts the string into 12 equal pieces. How long is each piece?

77. A church is giving away food to the homeless. They started with 50 gallons of milk. On Tuesday, they gave away 15 gallons. On Wednesday, the remaining milk was distributed equally to 60 persons. How much milk did each person receive on Wednesday?

78. Josette has 18 gallons of white paint and 7 gallons of blue paint. If she mixes it all together and then distributes it equally among 8 workers, how much paint does each worker receive?

79. Mr. Wilson borrowed $18,000 to buy a car. After the first year, he had paid $3600 on his loan. What fraction of the loan does he still owe?

80. The length of a rectangle is 6 feet more than its width. If the width is 72 feet, find the ratio of the length to the perimeter.

81. The population of a city increased from 175,000 to 250,000 in 15 years. What is the ratio of the increase in population to the original population?

82. The price of a software program decreased from $650 to $520. The decrease in price is what fraction of the original price?

In Your Own Words

Write complete sentences to discuss each of the following. Support your comments with examples or pictures, if appropriate.

83. Discuss how you can decide whether a fraction is less than, more than or equal to the number 1.

84. Add the same number to the numerator and denominator of a fraction and discuss whether this procedure generates fractions equal to the original fraction.

85. Discuss what can be done to a fraction without changing its value.

86. Multiples of 10 (numbers like 10, 100, 1000, etc.) have only two different types of prime factors. What are the two primes? Discuss how you can use this fact to quickly conclude that a fraction like $\frac{321}{1000}$ is reduced to lowest terms.

87. Discuss the three ways to indicate division using math symbols; in each case, give three ways to read the statement using English.

CHAPTER 3 PRACTICE TEST

1. What fraction of the figure is shaded? Assume the figure represents one whole.

2. What fraction of these figures are squares?

3. Write an improper fraction and a mixed numeral represented by the shaded region. Assume each separate figure represents one whole.

4. Write *two* equal fractions represented by the shaded region. Assume the figure represents one whole.

5. Choose the fractions that are less than one whole.
   $\frac{5}{8} \quad \frac{7}{6} \quad \frac{6}{7} \quad \frac{8}{8} \quad \frac{1}{8} \quad \frac{8}{1}$

Find the missing numerator so that the two numbers are equal.

6. $\frac{5}{8} = \frac{?}{48}$
7. $\frac{3}{14} = \frac{?}{84}$
8. $6 = \frac{?}{4}$

Reduce to lowest terms, if possible. If not possible, say not possible.

9. $\frac{16}{56}$
10. $\frac{45}{36}$
11. $\frac{200}{250}$
12. $\frac{78}{117}$

13. Write $\frac{2}{5}$ as a division problem using the symbol $\overline{)}$.
14. Write the English name for $\frac{7}{11}$.
15. Write the English name for $4\frac{3}{5}$.

Convert each to a whole or mixed numeral.

16. $\frac{13}{5}$
17. $\frac{449}{18}$
18. $\frac{5740}{82}$

Convert each to an improper fraction.

19. $12\frac{3}{8}$
20. $9\frac{17}{20}$

Write a fraction for each. Simplify, if possible.

21. 5 ÷ 7
22. 49 divided into 140
23. The ratio of 150 to 725.
24. What fraction of 42 is 12?
25. A basket contains 150 pieces of fruit, consisting of apples and pears. If there are 67 apples, what fraction of the total pieces of fruit are pears?
26. The lower trunk of the tree is what fraction of the total height of the tree?

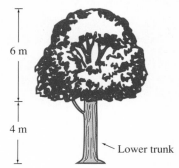

27. A calculus class consists of 24 men and 6 women. What is the ratio of the number of women to the total number of persons in the class?
28. Allan wants to cut his 27-centimeter piece of wire into 20 equal pieces. How long will each piece be?
29. A nurse takes care of 40 patients. If he has 15 centiliters of medicine to distribute among them equally, how much medicine does each patient receive?
30. Mr. Pakey received a $12,000 bonus. He bought 4 suits at $350 each, a computer for $4200, and new ski boots for $250. What is the ratio of the money spent to the total bonus?

CHAPTER 4

Fractions: Multiplication and Division

4.1 MULTIPLICATION

OBJECTIVES
- Multiply two fractions.
- Cancel correctly when multiplying fractions.
- Multiply more than two fractions.
- Solve an application involving multiplication of fractions.

1 Multiplying Fractions

The recipe for a lemon cheesecake calls for $\frac{2}{3}$ cup of evaporated milk. The chef of a small cafe wants to make 4 of these cheesecakes. How much evaporated milk will the chef need? To solve this problem we need to multiply $\frac{2}{3}$ by 4.

$$4 \times \frac{2}{3} = ?$$

Because one of the factors, 4, is a whole number, we can interpret this multiplication problem as repeated addition.

$$4 \times \frac{2}{3} = \frac{2}{3} + \frac{2}{3} + \frac{2}{3} + \frac{2}{3}$$
$$= \frac{2 + 2 + 2 + 2}{3}$$
$$= \frac{8}{3} \quad \text{or} \quad 2\frac{2}{3}$$

Observe that we can obtain the correct answer without using repeated addition.

Try These Problems

Multiply and simplify.
1. $3 \times \frac{2}{5}$
2. $16 \times \frac{9}{7}$
3. $\frac{1}{8} \times \frac{13}{2}$
4. $2\frac{1}{2} \times \frac{3}{4}$
5. $8 \times 4\frac{2}{3}$
6. $7\frac{3}{5} \times 2\frac{1}{3}$

$$4 \times \frac{2}{3} = \frac{4}{1} \times \frac{2}{3}$$ Convert 4 to a fraction. $4 = \frac{4}{1}$

$$= \frac{4 \times 2}{1 \times 3}$$ Multiply the numerators to obtain the numerator of the product. Multiply the denominators to obtain the denominator of the product.

$$= \frac{8}{3} \text{ or } 2\frac{2}{3}$$ The answer may be left as an improper fraction or as a mixed numeral.

This illustrates that the procedure for multiplying fractions is to multiply the numerators to obtain the numerator of the answer, and multiply the denominators to obtain the denominator of the answer. Here are more examples.

EXAMPLE 1 Multiply: $\frac{4}{5} \times \frac{9}{11}$

SOLUTION

$$\frac{4}{5} \times \frac{9}{11} = \frac{4 \times 9}{5 \times 11}$$ Multiply the numerators. Multiply the denominators.

$$= \frac{36}{55} \blacksquare$$

EXAMPLE 2 Multiply: $5\frac{3}{8} \times 7$

SOLUTION

$$5\frac{3}{8} \times 7 = \frac{43}{8} \times \frac{7}{1}$$ Convert $5\frac{3}{8}$ to a fraction. $5\frac{3}{8} = \frac{43}{8}$

Convert 7 to a fraction. $7 = \frac{7}{1}$

$$= \frac{43 \times 7}{8 \times 1}$$ Multiply the numerators. Multiply the denominators.

$$= \frac{301}{8} \text{ or } 37\frac{5}{8}$$ The answer may be left as an improper fraction or as a mixed numeral. $\blacksquare$

 Try Problems 1 through 6.

2 Multiplication Involving Canceling

Consider the multiplication problem

$$\frac{10}{9} \times \frac{4}{15}$$

If we multiply the numerators and the denominators, we obtain an answer that is not reduced to lowest terms.

$$\frac{10}{9} \times \frac{4}{15} = \frac{10 \times 4}{9 \times 15}$$

$$= \frac{40}{135} \quad \text{The numbers 40 and 135 are both divisible by 5.}$$

$$= \frac{\cancel{5} \times 8}{\cancel{5} \times 27}$$

$$= \frac{8}{27}$$

There is an easier way to do this problem. Observe that one of the numerators, 10, and one of the denominators, 15, have a common factor 5. The common factor 5 can be canceled *before* multiplying. Here we show the problem done in this way.

$$\frac{10}{9} \times \frac{4}{15} = \frac{2 \times \cancel{5}}{9} \times \frac{4}{3 \times \cancel{5}} \quad \text{Cancel the common factor 5 from the numerator and denominator before multiplying.}$$

$$= \frac{2 \times 4}{9 \times 3}$$

$$= \frac{8}{27}$$

Here are more examples.

EXAMPLE 3 Multiply: $12\frac{1}{2} \times \frac{8}{15}$

SOLUTION

$$12\frac{1}{2} \times \frac{8}{15}$$

$$= \frac{25}{2} \times \frac{8}{15} \quad \text{Convert } 12\frac{1}{2} \text{ to } \frac{25}{2} \text{ before canceling common factors.}$$

$$= \frac{5 \times \cancel{5}}{\underset{1}{\cancel{2}}} \times \frac{4 \times \cancel{2}}{3 \times \cancel{5}}$$

$$= \frac{20}{3} \quad \text{or} \quad 6\frac{2}{3} \quad \blacksquare$$

EXAMPLE 4 Multiply: $\frac{3}{800} \times \frac{5200}{27}$

SOLUTION

$$\frac{3}{800} \times \frac{5200}{27}$$

$$= \frac{\overset{1}{\cancel{3}}}{8 \times \cancel{100}} \times \frac{52 \times \cancel{100}}{\cancel{3} \times 9} \quad \text{Both 800 and 5200 are divisible by 100. Both 3 and 27 are divisible by 3.}$$

$$= \frac{1}{8} \times \frac{52}{9} \quad \text{Cancel common factors from 8 and 52 before multiplying.}$$

$$\begin{aligned} 8 &= 2 \times 4 = 2 \times 2 \times 2 \\ 52 &= 2 \times 26 \\ &= 2 \times 2 \times 13 \end{aligned}$$

Try These Problems

Multiply and simplify.

7. $\frac{11}{5} \times \frac{15}{16}$
8. $\frac{21}{23} \times \frac{3}{14}$
9. $\frac{25}{8} \times \frac{18}{65}$
10. $3\frac{1}{3} \times \frac{9}{400}$
11. $4\frac{2}{3} \times \frac{27}{35}$
12. $180 \times 2\frac{5}{6}$

$$= \frac{1}{2 \times \cancel{4}} \times \frac{\cancel{4} \times 13}{9}$$

$$= \frac{1 \times 13}{2 \times 9}$$

$$= \frac{13}{18} \ \blacksquare$$

Therefore, 8 and 52 have a common factor 4.

Try Problems 7 through 12.

Now consider the problem

$$\frac{36}{42} \times \frac{25}{7}$$

Observe that the numerator and denominator of the first fraction have a common factor 6.

$$\frac{36}{42} \times \frac{25}{7} = \frac{6 \times 6}{6 \times 7} \times \frac{25}{7}$$

It is easier to cancel this common factor 6 from the numerator and denominator before multiplying. This is all right because dividing the numerator and denominator of a fraction by the same nonzero number does not change the value of the fraction.

$$\frac{36}{42} \times \frac{25}{7} = \frac{\cancel{6} \times 6}{\cancel{6} \times 7} \times \frac{25}{7}$$

$$= \frac{6}{7} \times \frac{25}{7}$$ No more canceling can be done.

$$= \frac{6 \times 25}{7 \times 7}$$

$$= \frac{150}{49} \quad \text{or} \quad 3\frac{3}{49}$$

Here are more examples.

EXAMPLE 5   Multiply: $\frac{33}{55} \times \frac{39}{52}$

SOLUTION

$$\frac{33}{55} \times \frac{39}{52}$$

$$= \frac{3 \times 11}{5 \times 11} \times \frac{3 \times 13}{2 \times 2 \times 13}$$

Write each numerator and denominator as the product of primes.
$33 = 3 \times 11$
$55 = 5 \times 11$
$39 = 3 \times 13$
$52 = 4 \times 13$
$\quad = 2 \times 2 \times 13$

$$= \frac{3 \times \cancel{11}}{5 \times \cancel{11}} \times \frac{3 \times \cancel{13}}{2 \times 2 \times \cancel{13}}$$

$$= \frac{3 \times 3}{5 \times 2 \times 2}$$

$$= \frac{9}{20} \ \blacksquare$$

Cancel the common factor 11 from the numerator and denominator of the first fraction. Cancel the common factor 13 from the numerator and denominator of the second fraction.

Try These Problems

Multiply and simplify.

13. $\frac{36}{45} \times \frac{8}{25}$
14. $\frac{3}{7} \times \frac{12}{16}$
15. $\frac{30}{150} \times \frac{26}{65}$
16. $\frac{21}{63} \times \frac{60}{25}$
17. $9000 \times \frac{50}{75}$
18. $\frac{55}{77} \times 4\frac{3}{8}$

EXAMPLE 6 Multiply: $\frac{40}{100} \times 235$

SOLUTION

$$\frac{40}{100} \times 235$$

$$= \frac{4 \times \cancel{10}}{10 \times \cancel{10}} \times \frac{235}{1} \quad \text{Both 40 and 100 are divisible by 10.}$$

$$= \frac{4}{10} \times \frac{235}{1}$$

$$= \frac{\cancel{2} \times 2}{\cancel{2} \times \cancel{5}} \times \frac{\cancel{5} \times 47}{1} \quad \text{More canceling can be done before multiplying.}$$

$$= \frac{2 \times 47}{1 \times 1}$$

$$= \frac{94}{1} = 94 \quad \blacksquare$$

EXAMPLE 7 Multiply: $33\frac{1}{3} \times \frac{63}{7000}$

SOLUTION

$$33\frac{1}{3} \times \frac{63}{7000}$$

$$= \frac{100}{3} \times \frac{63}{7000} \quad \text{Convert } 33\frac{1}{3} \text{ to } \frac{100}{3} \text{ before beginning the canceling process.}$$

$$= \frac{\overset{1}{\cancel{100}}}{3} \times \frac{63}{\cancel{100} \times 70} \quad \text{Both 100 and 7000 are divisible by 100.}$$

$$= \frac{1}{3} \times \frac{63}{70}$$

$$= \frac{1}{\underset{1}{\cancel{3}}} \times \frac{\cancel{3} \times 3 \times \cancel{7}}{\cancel{7} \times 10} \quad \text{Remember to cancel the common factor 7 in the numerator and denominator of the second fraction.}$$

$$= \frac{1 \times 3}{1 \times 10}$$

$$= \frac{3}{10} \quad \blacksquare$$

 Try Problems 13 through 18.

Now we summarize the procedure for multiplying fractions.

> ### *Multiplying Fractions*
> 1. Convert the mixed numerals to improper fractions and convert the whole numerals to fractions.
> 2. Cancel as much as possible. Each time you cancel a factor from a numerator, you must cancel the same factor from a denominator.
> 3. Multiply the numerators to obtain the numerator of the answer. Multiply the denominators to obtain the denominator of the answer.
> 4. Check to make sure the answer is simplified. When the answer is an improper fraction, you may leave it as an improper fraction or you may convert it to a mixed numeral.

3 Multiplying More Than Two Fractions

Recall from Chapter 2 that to multiply more than two numbers, we must first multiply any two of them, then multiply that result by another one, and continue until all of the numbers have been multiplied. Here we show a problem done in several ways.

$$6 \times 5 \times 2 \times 8 = 30 \times 2 \times 8$$
$$= 60 \times 8$$
$$= 480$$

$$6 \times 5 \times 2 \times 8 = 6 \times 5 \times 16$$
$$= 6 \times 80$$
$$= 480$$

$$6 \times 5 \times 2 \times 8 = 48 \times 10$$
$$= 480$$

Now we look at a problem involving fractions done in several ways.

$$\frac{1}{2} \times \frac{3}{5} \times 8 = \frac{3}{\overset{5}{\cancel{10}}} \times \frac{\overset{4}{\cancel{8}}}{1} = \frac{12}{5}$$ Here we multiply $\frac{1}{2}$ by $\frac{3}{5}$ first.

$$\frac{1}{2} \times \frac{3}{5} \times 8 = \frac{1}{\underset{1}{\cancel{2}}} \times \frac{\overset{12}{\cancel{24}}}{5} = \frac{12}{5}$$ Here we multiply $\frac{3}{5}$ by 8 first.

$$\frac{1}{2} \times \frac{3}{5} \times 8 = \frac{1}{\underset{1}{\cancel{2}}} \times \frac{3}{5} \times \frac{\overset{4}{\cancel{8}}}{1}$$ Observe that it is all right to cancel the common factor 2 before multiplying.

$$= \frac{1 \times 3 \times 4}{1 \times 5 \times 1}$$

$$= \frac{12}{5}$$

4.1 Multiplication

▲ Try These Problems

Multiply and simplify.

19. $\frac{15}{16} \times \frac{3}{5} \times \frac{24}{7}$
20. $\frac{11}{36} \times \frac{5}{3} \times \frac{2}{22}$
21. $\frac{12}{15} \times \frac{3}{27} \times 15 \times \frac{3}{4}$
22. $20 \times \frac{4}{35} \times \frac{3}{10} \times \frac{7}{8}$

EXAMPLE 8   Multiply: $1\frac{1}{2} \times 15 \times \frac{5}{9}$

SOLUTION

$$1\frac{1}{2} \times 15 \times \frac{5}{9}$$

$$= \frac{\overset{1}{\cancel{3}}}{2} \times \frac{15}{1} \times \frac{5}{\underset{3}{\cancel{9}}}$$ Cancel the common factor 3 from the numerator and denominator.

$$= \frac{1 \times \overset{5}{\cancel{15}} \times 5}{2 \times 1 \times \cancel{3}}$$ Cancel another factor of 3.

$$= \frac{25}{2} \quad \text{or} \quad 12\frac{1}{2} \;\blacksquare$$

EXAMPLE 9   Multiply: $\frac{49}{14} \times \frac{5}{9} \times 50 \times \frac{3}{11}$

SOLUTION

$$\frac{49}{14} \times \frac{5}{9} \times \frac{50}{1} \times \frac{3}{11}$$

$$= \frac{7 \times \cancel{7}}{\cancel{2} \times \cancel{7}} \times \frac{5}{\cancel{3} \times 3} \times \frac{\cancel{2} \times 25}{1} \times \frac{\overset{1}{\cancel{3}}}{11}$$ Cancel common factors 7, 2, and 3.

$$= \frac{7 \times 5 \times 25}{3 \times 11}$$

$$= \frac{875}{33} \quad \text{or} \quad 26\frac{17}{33} \;\blacksquare$$

▲ Try Problems 19 through 22.

▲ Answers to Try These Problems

1. $\frac{6}{5}$ or $1\frac{1}{5}$ 2. $\frac{144}{7}$ or $20\frac{4}{7}$ 3. $\frac{13}{16}$ 4. $\frac{15}{8}$ or $1\frac{7}{8}$ 5. $\frac{112}{3}$ or $37\frac{1}{3}$
6. $\frac{266}{15}$ or $17\frac{11}{15}$ 7. $\frac{33}{16}$ or $2\frac{1}{16}$ 8. $\frac{9}{46}$ 9. $\frac{45}{52}$ 10. $\frac{3}{40}$
11. $\frac{18}{5}$ or $3\frac{3}{5}$ 12. 510 13. $\frac{32}{125}$ 14. $\frac{9}{28}$ 15. $\frac{2}{25}$ 16. $\frac{4}{5}$
17. 6000 18. $\frac{25}{8}$ or $3\frac{1}{8}$ 19. $\frac{27}{14}$ or $1\frac{13}{14}$ 20. $\frac{5}{108}$ 21. 1 22. $\frac{3}{5}$

EXERCISES 4.1

Multiply and simplify.

1. $5 \times \frac{2}{3}$
2. $9 \times \frac{3}{14}$
3. $\frac{2}{11} \times \frac{3}{7}$
4. $\frac{5}{8} \times \frac{13}{9}$
5. $6\frac{8}{9} \times \frac{5}{27}$
6. $12\frac{1}{8} \times \frac{9}{8}$
7. $130 \times \frac{1}{9}$
8. $2000 \times \frac{2}{7}$
9. $\frac{7}{8} \times \frac{8}{19}$
10. $\frac{6}{7} \times \frac{5}{9}$
11. $44 \times \frac{8}{11}$
12. $48 \times \frac{7}{30}$
13. $5\frac{2}{5} \times \frac{20}{63}$
14. $250 \times 66\frac{2}{3}$
15. $\frac{26}{35} \times \frac{28}{39}$
16. $\frac{48}{135} \times 2\frac{13}{36}$
17. $\frac{15}{20} \times 1\frac{1}{2}$
18. $12 \times \frac{5000}{700}$
19. $\frac{60}{100} \times 153$
20. $\frac{11}{3} \times \frac{32}{56}$
21. $\frac{24}{40} \times \frac{15}{27}$
22. $\frac{18}{27} \times \frac{22}{550}$
23. $6\frac{1}{4} \times \frac{27}{75}$
24. $\frac{6}{100} \times 3500$

25. $\frac{40}{340} \times 1700$ 26. $12\frac{3}{4} \times \frac{28}{21}$ 27. $\frac{30}{36} \times \frac{91}{35}$ 28. $\frac{98}{27} \times \frac{36}{105}$
29. $6\frac{7}{8} \times \frac{40}{900}$ 30. $\frac{75}{77} \times 13\frac{1}{5}$ 31. $\frac{57}{24} \times \frac{21}{38}$ 32. $\frac{68}{85} \times \frac{65}{22}$
33. $\frac{33}{91} \times \frac{104}{42}$ 34. $\frac{13}{2000} \times \frac{8}{3900}$ 35. $\frac{360}{500} \times 3\frac{1}{6}$ 36. $16\frac{2}{7} \times 11\frac{2}{3}$
37. $\frac{2}{3} \times \frac{6}{7} \times \frac{15}{4}$ 38. $\frac{1}{10} \times \frac{5}{8} \times \frac{25}{3}$ 39. $2\frac{1}{2} \times \frac{3}{5} \times 3\frac{1}{4}$ 40. $6\frac{2}{3} \times \frac{15}{24} \times 1\frac{1}{2}$
41. $\frac{3}{2} \times \frac{6}{7} \times \frac{5}{9} \times \frac{14}{25}$ 42. $\frac{2}{5} \times \frac{3}{4} \times 8\frac{1}{2} \times 5\frac{1}{3}$

Solve.

43. Jeffrey bought $3\frac{3}{4}$ pounds of round steak at $4 a pound. How much did he pay for the steak?

44. Suppose you buy 60 shares of a stock selling for $16\frac{3}{4}$ dollars per share. What is the total cost?

45. Susann bought $4\frac{2}{3}$ yards of fabric that sells for $9 a yard. How much did she pay for the fabric?

46. A desk top measures $4\frac{1}{2}$ meters long and $2\frac{1}{4}$ meters wide. Find the area of the desk top.

DEVELOPING NUMBER SENSE #3

MULTIPLYING BY A NUMBER LESS THAN, EQUAL TO, OR MORE THAN 1

Here we show 200 multiplied by 3, 2, and $1\frac{1}{2}$. How do the products compare with 200 and why?

$$3 \times 200 = 600 \qquad 2 \times 200 = 400$$
$$1\frac{1}{2} \times 200 = 300$$

Observe that the results—600, 400, and 300—are each more than 200 because the numbers 3, 2, and $1\frac{1}{2}$ are each more than 1.

If we multiply 200 by 1, we obtain 200.

$$1 \times 200 = 200$$

In general, any number multiplied by 1 is that number.

What happens if we multiply 200 by a fraction that is less than 1? Let's look at a few examples.

$$\frac{3}{4} \times 200 = 150 \qquad \frac{1}{2} \times 200 = 100$$
$$\frac{1}{4} \times 200 = 50$$

Observe that the results 150, 100, and 50 are each less than 200 because the numbers $\frac{3}{4}$, $\frac{1}{2}$, and $\frac{1}{4}$ are each less than 1.

Number Sense Problems

Without computing, decide whether the result is less than, equal to, or more than 84.

1. 5×84 2. $2\frac{1}{4} \times 84$ 3. $\frac{2}{3} \times 84$ 4. 1×84 5. $\frac{4}{10} \times 84$ 6. $\frac{3}{2} \times 84$

Without computing, decide whether the result is less than, equal to, or more than 682.

7. $\frac{1}{6} \times 682$ 8. $\frac{7}{5} \times 682$ 9. 12×682 10. $13\frac{3}{4} \times 682$
11. $\frac{8}{8} \times 682$ 12. $\frac{12}{13} \times 682$

Without computing, decide whether the result is less than, equal to, or more than $25\frac{1}{2}$.

13. $\frac{1}{2} \times 25\frac{1}{2}$ 14. $5\frac{3}{4} \times 25\frac{1}{2}$ 15. $\frac{5}{2} \times 25\frac{1}{2}$ 16. $1 \times 25\frac{1}{2}$

4.2 DIVISION

OBJECTIVES
- Divide two fractions.
- Perform division of fractions indicated by a fraction bar.
- Solve an application problem involving division.

1 Dividing Fractions

A piece of wire $4\frac{1}{2}$ feet long is to be cut into pieces that are each $\frac{3}{4}$ foot. How many pieces can be cut? We want to know how many lengths of $\frac{3}{4}$ foot there are in a total length of $4\frac{1}{2}$ feet. In Section 2.5, we learned that in a situation like this we must divide the total quantity by the size of each part to obtain the number of equal parts. The only difference here is that fractions are involved. Here we picture the situation.

$$\underbrace{\frac{3}{4}\text{ft} + \frac{3}{4}\text{ft} + \frac{3}{4}\text{ft} + \frac{3}{4}\text{ft} + \frac{3}{4}\text{ft} + \frac{3}{4}\text{ft}}_{4\frac{1}{2}\text{ ft}}$$

How many $\frac{3}{4}$s are in $4\frac{1}{2}$?

$$\begin{array}{ccc}
\text{Number of equal parts} \times \text{size of each part} & = & \text{total quantity} \\
? \times \frac{3}{4} & = & 4\frac{1}{2}
\end{array}$$

To solve this problem we need to divide $4\frac{1}{2}$ by $\frac{3}{4}$. From the picture, we see that 6 of the $\frac{3}{4}$-foot pieces make up the total length of $4\frac{1}{2}$ feet. We can check this by multiplying $\frac{3}{4}$ by 6.

$$6 \times \frac{3}{4} = \frac{6}{1} \times \frac{3}{4}$$
$$= \frac{\cancel{2} \times 3}{1} \times \frac{3}{\cancel{2} \times 2}$$
$$= \frac{3 \times 3}{1 \times 2}$$
$$= \frac{9}{2} \quad \text{or} \quad 4\frac{1}{2}$$

We have shown that $4\frac{1}{2} \div \frac{3}{4} = 6$, but we want to be able to divide fractions without viewing a picture. Observe that the following procedure gives the correct answer 6 for the problem $4\frac{1}{2} \div \frac{3}{4}$.

$$4\frac{1}{2} \div \frac{3}{4} = \frac{9}{2} \div \frac{3}{4}$$ Convert $4\frac{1}{2}$ to $\frac{9}{2}$.

$$= \frac{9}{2} \times \frac{4}{3}$$ Change the division to multiplication and invert the second fraction.

$$= \frac{\cancel{9}^{\,3} \times 3}{\cancel{2}_{\,1}} \times \frac{\cancel{4}^{\,2} \times 2}{\cancel{3}_{\,1}}$$ Cancel the common factor 2. Cancel the common factor 3.

$$= \frac{3 \times 2}{1 \times 1}$$ Multiply the fractions.

$$= 6$$

Observe that dividing by $\frac{3}{4}$ is the same as multiplying by $\frac{4}{3}$. The fraction $\frac{4}{3}$ is called the **reciprocal** of $\frac{3}{4}$. Therefore, *dividing by a fraction is the same as multiplying by its reciprocal.* Here are more examples.

EXAMPLE 1   Divide: $\frac{5}{2} \div \frac{1}{4}$

SOLUTION

$$\frac{5}{2} \div \frac{1}{4}$$

$$= \frac{5}{2} \times \frac{4}{1}$$ Change $\div$ to $\times$ and invert the second fraction.

$$= \frac{5}{\cancel{2}_{\,1}} \times \frac{\cancel{4}^{\,2} \times 2}{1}$$ Cancel the common factor 2.

$$= \frac{10}{1}$$

$$= 10 \quad \blacksquare$$

EXAMPLE 2   Divide: $\frac{15}{19} \div 3$

SOLUTION

$$\frac{15}{19} \div 3$$

$$= \frac{15}{19} \div \frac{3}{1}$$ Convert 3 to a fraction. $3 = \frac{3}{1}$

$$= \frac{15}{19} \times \frac{1}{3}$$ Change $\div$ to $\times$ and invert the second fraction.

$$= \frac{5 \times \cancel{3}}{19} \times \frac{1}{\cancel{3}_{\,1}}$$ Cancel the common factor 3.

$$= \frac{5 \times 1}{19 \times 1}$$ Multiply the fractions.

$$= \frac{5}{19} \quad \blacksquare$$

Try These Problems

Divide and simplify.
1. $\frac{7}{3} \div \frac{1}{9}$
2. $\frac{2}{3} \div \frac{12}{5}$
3. $8 \div 10\frac{2}{3}$
4. $\frac{25}{125} \div 20$
5. $3\frac{2}{3} \div \frac{2}{27}$
6. $12\frac{3}{5} \div 1\frac{4}{5}$
7. $140 \div 42$
8. $48 \div 200$

EXAMPLE 3 Divide: $5 \div 12\frac{3}{5}$

SOLUTION

$$5 \div 12\frac{3}{5}$$
$$= \frac{5}{1} \div \frac{63}{5} \quad \text{Convert 5 to a fraction. } 5 = \frac{5}{1}$$
$$\quad\quad\quad\quad\quad \text{Convert } 12\frac{3}{5} \text{ to an improper fraction.}$$
$$\quad\quad\quad\quad\quad 12\frac{3}{5} = \frac{63}{5}$$
$$= \frac{5}{1} \times \frac{5}{63} \quad \text{Change } \div \text{ to } \times \text{ and invert the second fraction.}$$
$$= \frac{5 \times 5}{1 \times 63} \quad \text{Multiply the fractions.}$$
$$= \frac{25}{63} \quad \blacksquare$$

Here is a summary of the procedure for dividing fractions.

> ### Dividing Fractions
> 1. Convert each mixed numeral to an improper fraction. Convert the whole numerals to fractions.
> 2. Change division ($\div$) to multiplication ($\times$) and invert the divisor. The divisor is the second number.
> 3. Cancel as much as possible, then multiply the fractions.
> 4. Check to make sure the answer is simplified. When the answer is an improper fraction, you may leave it as an improper fraction or you may write it as a mixed numeral.

Try Problems 1 through 8.

2 Indicating Division with the Fraction Bar

Recall from Chapter 3 that the fraction bar used to name fractions indicates division. For example,

$\frac{12}{3}$ means $12 \div 3$

$\frac{3}{12}$ means $3 \div 12$

$\frac{1\frac{1}{2}}{\frac{1}{3}}$ means $1\frac{1}{2} \div \frac{1}{3}$

$\frac{\frac{1}{3}}{1\frac{1}{2}}$ means $\frac{1}{3} \div 1\frac{1}{2}$

Chapter 4 Fractions: Multiplication and Division

 Try These Problems

Divide and simplify.

9. $\dfrac{\frac{5}{3}}{4}$

10. $\dfrac{\frac{3}{7}}{\frac{1}{9}}$

11. $\dfrac{8}{2\frac{1}{2}}$

12. $\dfrac{4\frac{3}{5}}{\frac{4}{5}}$

13. $\dfrac{\frac{7}{21}}{\frac{8}{27}}$

14. $\dfrac{\frac{54}{12}}{8\frac{1}{4}}$

EXAMPLE 4 Divide: $\dfrac{\frac{8}{9}}{5}$

SOLUTION

$\dfrac{\frac{8}{9}}{5} = \dfrac{8}{9} \div 5$ The numerator $\frac{8}{9}$ is written first. The denominator 5 is written second.

$= \dfrac{8}{9} \times \dfrac{1}{5}$

$= \dfrac{8}{45}$ ∎

EXAMPLE 5 Divide: $\dfrac{2\frac{1}{2}}{5\frac{3}{4}}$

SOLUTION

$\dfrac{2\frac{1}{2}}{5\frac{3}{4}} = 2\dfrac{1}{2} \div 5\dfrac{3}{4}$ The numerator $2\frac{1}{2}$ is written first. The denominator $5\frac{3}{4}$ is written second.

$= \dfrac{5}{2} \div \dfrac{23}{4}$

$= \dfrac{5}{2} \times \dfrac{4}{23}$

$= \dfrac{5}{\underset{1}{\cancel{2}}} \times \dfrac{\cancel{2} \times 2}{23}$

$= \dfrac{10}{23}$ ∎

 Try Problems 9 through 14.

 Answers to Try These Problems

1. 21   2. $\frac{5}{18}$ 3. $\frac{3}{4}$ 4. $\frac{1}{100}$ 5. $\frac{99}{2}$ or $49\frac{1}{2}$ 6. 7   7. $\frac{10}{3}$ or $3\frac{1}{3}$
8. $\frac{6}{25}$ 9. $\frac{5}{12}$ 10. $\frac{27}{7}$ or $3\frac{6}{7}$ 11. $\frac{16}{5}$ or $3\frac{1}{5}$ 12. $\frac{23}{4}$ or $5\frac{3}{4}$
13. $\frac{9}{8}$ or $1\frac{1}{8}$ 14. $\frac{6}{11}$

EXERCISES 4.2

Divide and simplify.

1. $\frac{3}{5} \div \frac{1}{2}$
2. $\frac{5}{9} \div \frac{1}{5}$
3. $\frac{9}{10} \div \frac{2}{3}$
4. $\frac{3}{4} \div \frac{7}{8}$
5. $\frac{7}{29} \div 1\frac{5}{9}$
6. $\frac{3}{200} \div 2\frac{7}{16}$
7. $5 \div \frac{1}{3}$
8. $27 \div \frac{3}{5}$
9. $1050 \div \frac{2}{5}$
10. $5400 \div 1\frac{1}{5}$
11. $1\frac{2}{3} \div 40$
12. $\frac{27}{81} \div 3$
13. $13\frac{1}{2} \div 15$
14. $8\frac{3}{4} \div 10$
15. $3 \div 7$
16. $7 \div 3$
17. $6 \div 10$
18. $10 \div 6$
19. $560 \div 40$
20. $40 \div 560$
21. $2\frac{1}{3} \div \frac{2}{15}$
22. $3\frac{4}{7} \div \frac{30}{28}$
23. $9\frac{3}{8} \div \frac{45}{24}$
24. $1\frac{1}{5} \div 12\frac{1}{3}$
25. $16\frac{2}{3} \div 5\frac{1}{5}$
26. $4\frac{4}{11} \div 1\frac{23}{33}$
27. $9\frac{3}{4} \div 9\frac{3}{4}$
28. $\frac{8}{19} \div \frac{8}{19}$
29. $\dfrac{\frac{3}{10}}{4}$
30. $\dfrac{\frac{14}{21}}{8}$
31. $\dfrac{36}{2\frac{2}{3}}$
32. $\dfrac{150}{4\frac{1}{6}}$

33. $\dfrac{7\frac{1}{2}}{3\frac{4}{5}}$ 34. $\dfrac{3\frac{9}{10}}{6\frac{1}{2}}$ 35. $\dfrac{\frac{15}{250}}{\frac{12}{8}}$ 36. $\dfrac{6\frac{3}{4}}{\frac{14}{63}}$

37. $\dfrac{2800}{\frac{49}{50}}$ 38. $\dfrac{15\frac{3}{5}}{\frac{35}{91}}$

Solve and simplify.

39. A board 5 feet long is separated into 3 equal pieces. How long is each piece?

40. A board 3 feet long is separated into 5 equal pieces. How long is each piece?

41. The contents of a can is to be divided into 5 equal parts. If the can holds $\frac{7}{8}$ of a liter, how much will be in each part?

42. A wire is to be cut into 20 equal pieces. If the wire has a total length of $2\frac{1}{2}$ meters, how long is each piece?

43. Warren wants to make bows for his Christmas tree. Each bow requires $1\frac{1}{2}$ feet of ribbon. He has purchased 75 feet of ribbon. How many bows can he make?

44. A wire that is $49\frac{1}{2}$ centimeters long is to be cut into $2\frac{1}{4}$-centimeter pieces. How many of the smaller pieces can be cut?

4.3 THE MISSING NUMBER IN A MULTIPLICATION STATEMENT

OBJECTIVE ■ Find the missing number in a multiplication statement.

In the multiplication statement

$$4 \times 6 = 24 \quad \text{or} \quad 24 = 4 \times 6$$

the numbers 4 and 6 are called the **factors** or **multipliers**. The answer to the multiplication problem, 24, is called the **product** of 4 and 6. Note that the statement can be written with the product located after the equality symbol or before the equality symbol.

Three questions can be asked by omitting any one of the three numbers.

1. $4 \times 6 = \boxed{}$ The answer to the multiplication problem is missing, so multiply.
   ↑
   24 is the missing number.

2. $\boxed{} \times 6 = 24$ A multiplier is missing, so divide. The answer to the multiplication problem is divided by the given multiplier.
   $24 \div 6 = \boxed{}$
   ↑
   4 is the missing number.

3. $4 \times \boxed{} = 24$ A multiplier is missing, so divide. The answer to the multiplication problem is divided by the given multiplier.
   $24 \div 4 = \boxed{}$
   ↑
   6 is the missing number.

When both multipliers are *more* than 1, the product is the largest of the three numbers. However, do not expect this to be true when one of the multipliers is *less* than 1. Study these examples:

140 Chapter 4 Fractions: Multiplication and Division

 Try These Problems

Find the missing number.

1. $\frac{1}{4} \times 120 = \boxed{}$
2. $\frac{3}{5} \times 63 = \boxed{}$
3. $\boxed{} = 2\frac{5}{6} \times 27$
4. $\boxed{} = 3\frac{1}{5} \times \frac{15}{24}$

1. $8 \times 9 = 72$
2. $2 \times 3\frac{1}{2} = 7$ } Here both multipliers are more than 1, so the product is the largest of the three numbers.
3. $\frac{3}{4} \times 200 = 150$ Because $\frac{3}{4}$ is less than 1, the product 150 is less than 200.

You can use this observation to help you decide if your answer is reasonable. Now we look at some examples of finding the missing number in a multiplication statement.

EXAMPLE 1 $\boxed{} = \frac{4}{5} \times 200$

SOLUTION   The answer to the multiplication problem is missing, so multiply $\frac{4}{5}$ by 200.

$$\frac{4}{5} \times 200 = \frac{4}{5} \times \frac{200}{1} = \frac{4}{\cancel{5}_1} \times \frac{\cancel{5} \times 40}{1} = \frac{160}{1} = 160 \;\blacksquare$$

In Example 1 we expect to get an answer less than 200 because $\frac{4}{5}$ is less than 1.

 Try Problems 1 through 4.

EXAMPLE 2 $5 \times \boxed{} = \frac{3}{4}$

SOLUTION   The number $\frac{3}{4}$ is the answer to the multiplication problem, and the number 5 is the known multiplier. The other multiplier is missing, so divide $\frac{3}{4}$ by 5.

$$\text{Missing multiplier} = \text{answer to the multiplication problem} \div \text{known multiplier}$$

$$= \frac{3}{4} \div 5$$

$$= \frac{3}{4} \div \frac{5}{1}$$

$$= \frac{3}{4} \times \frac{1}{5}$$

$$= \frac{3}{20}$$

CHECK

$$5 \times \frac{3}{20} \stackrel{?}{=} \frac{3}{4}$$

$$\frac{\cancel{5}^1}{1} \times \frac{3}{\cancel{20}_4} \stackrel{?}{=} \frac{3}{4}$$

$$\frac{3}{4} = \frac{3}{4}$$

The missing multiplier is $\frac{3}{20}$. $\blacksquare$

Try These Problems

Find the missing number.

5. $4 = \frac{1}{4} \times \boxed{}$
6. $\frac{9}{200} \times \boxed{} = 45$
7. $3\frac{1}{2} = \boxed{} \times 1\frac{1}{2}$
8. $\frac{2}{3} = 9 \times \boxed{}$

EXAMPLE 3 $\quad 8\frac{1}{3} = 2\frac{1}{2} \times \boxed{}$

SOLUTION The number $8\frac{1}{3}$ is the answer to the multiplication problem, and the number $2\frac{1}{2}$ is the known multiplier. The other multiplier is missing, so divide $8\frac{1}{3}$ by $2\frac{1}{2}$.

$$\text{Missing multiplier} = \text{answer to the multiplication problem} \div \text{known multiplier}$$

$$= 8\frac{1}{3} \div 2\frac{1}{2}$$

$$= \frac{25}{3} \div \frac{5}{2}$$

$$= \frac{\overset{5}{\cancel{25}}}{3} \times \frac{2}{\underset{1}{\cancel{5}}}$$

$$= \frac{10}{3} \text{ or } 3\frac{1}{3} \quad \blacksquare$$

 Try Problems 5 through 8.

Now we summarize the procedure for finding the missing number in a multiplication statement.

> *Finding the Missing Number in a Multiplication Statement*
> 1. If you are missing the answer to the multiplication problem, then multiply.
> 2. If you are missing one of the multipliers, then divide. Be careful not to divide backward. Here is how it works.
>
> $$\text{Missing multiplier} = \text{answer to the multiplication problem} \div \text{known multiplier}$$

 Answers to Try These Problems

1. 30   2. $\frac{189}{5}$ or $37\frac{4}{5}$ 3. $\frac{153}{2}$ or $76\frac{1}{2}$ 4. 2   5. 16   6. 1000
7. $\frac{7}{3}$ or $2\frac{1}{3}$ 8. $\frac{2}{27}$

EXERCISES 4.3

Find the missing number and simplify.

1. $\frac{1}{6} \times 18 = \boxed{}$
2. $\frac{1}{8} \times 32 = \boxed{}$
3. $\frac{1}{5} \times 42 = \boxed{}$
4. $16 \times \boxed{} = 2$
5. $12 \times \boxed{} = 3$
6. $35 \times \boxed{} = 7$
7. $\boxed{} \times 63 = 9$
8. $\boxed{} \times 48 = 24$
9. $\boxed{} \times 30 = 5$
10. $90 = \frac{1}{4} \times \boxed{}$
11. $8 = \frac{1}{9} \times \boxed{}$
12. $17 = \frac{1}{3} \times \boxed{}$

13. $\boxed{} = \frac{3}{5} \times 20$
14. $\boxed{} = \frac{2}{3} \times 36$
15. $\boxed{} = \frac{4}{7} \times 200$
16. $45 = \boxed{} \times 54$
17. $14 = \boxed{} \times 49$
18. $21 = \boxed{} \times 56$
19. $40 = 50 \times \boxed{}$
20. $28 = 40 \times \boxed{}$
21. $18 = 120 \times \boxed{}$
22. $\frac{3}{400} \times \boxed{} = 24$
23. $\frac{5}{120} \times \boxed{} = 30$
24. $\frac{5}{80} \times \boxed{} = 150$
25. $50{,}000 = 4\frac{3}{10} \times \boxed{}$
26. $\boxed{} \times 5\frac{1}{5} = 1300$

4.4 LANGUAGE

OBJECTIVES
- Find a fraction of a number.
- Translate an English statement to a multiplication or division statement using math symbols.
- Solve problems using translations.

1 Finding a Fraction of a Number

The picture shows 12 small squares.

To shade $\frac{1}{3}$ of the 12 squares means to shade 1 out of 3 equal parts.

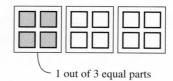

1 out of 3 equal parts

From the picture you see that $\frac{1}{3}$ of the 12 squares is 4 squares. Observe that we can find $\frac{1}{3}$ of 12 without viewing the picture because the correct result 4 can be obtained by multiplying 12 by $\frac{1}{3}$.

$$\frac{1}{3} \text{ of } 12$$
$$= \frac{1}{3} \times 12 \quad \text{The word } of \text{ used in this way translates to multiplication.}$$
$$= \frac{1}{3} \times \frac{12}{1}$$
$$= \frac{12}{3}$$
$$= 4$$

Try These Problems

Solve.
1. $\frac{1}{4}$ of 60
2. $\frac{3}{4}$ of 60
3. $\frac{2}{7}$ of $5\frac{3}{5}$
4. What number is $\frac{5}{12}$ of $\frac{2}{25}$?
5. Find $7\frac{1}{2}$ of $9\frac{3}{5}$.
6. $\frac{36}{240}$ of 900 is what number?

To shade $\frac{2}{3}$ of the 12 squares means to shade 2 out of 3 equal parts.

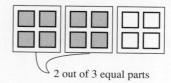

2 out of 3 equal parts

From the picture you see that $\frac{2}{3}$ of the 12 squares is 8 squares. Observe that we can find $\frac{2}{3}$ of 12 without viewing the picture because the correct result 8 can be obtained by multiplying 12 by $\frac{2}{3}$.

$$\frac{2}{3} \text{ of } 12$$
$$= \frac{2}{3} \times 12 \qquad \text{The word } of \text{ used in this way translates to multiplication.}$$
$$= \frac{2}{3} \times \frac{12}{1}$$
$$= \frac{2}{\cancel{3}} \times \frac{4 \times \cancel{3}}{1}$$
$$= 8$$

In general, to find a fraction of a number, translate the word *of* to multiplication. Here are some more examples.

1. $\frac{1}{4}$ of $20 = \frac{1}{4} \times \frac{20}{1} = \frac{20}{4} = 5$
2. $\frac{3}{4}$ of $20 = \frac{3}{4} \times \frac{20}{1} = \frac{3}{\cancel{4}} \times \frac{\cancel{4} \times 5}{1} = 15$
3. $3\frac{1}{2}$ of $12 = 3\frac{1}{2} \times 12 = \frac{7}{2} \times \frac{12}{1} = \frac{7}{\cancel{2}} \times \frac{\cancel{2} \times 6}{1} = 42$
4. $\frac{3}{10}$ of $2\frac{1}{2} = \frac{3}{10} \times \frac{5}{2} = \frac{3}{2 \times \cancel{5}} \times \frac{\cancel{5}}{2} = \frac{3}{4}$

Finding a Fraction of a Number
To find a fraction of a number, multiply the fraction by the number, that is, the word *of* used in this way translates to multiplication.

 Try Problems 1 through 6.

2 Language That Translates to Multiplication

Here is a review of the many English translations for the symbol =.

English	Math Symbols
equals	=
is equal to	
is the same as	
is the result of	
is	
was	
represents	
gives	
makes	
yields	
will be	
are	
were	

The multiplication statement $\frac{2}{3} \times 12 = 8$ or $8 = \frac{2}{3} \times 12$ is written with math symbols. Some of the ways to read this statement in English are given in the following chart.

English	Math Symbols
Two-thirds times twelve equals eight.	$\frac{2}{3} \times 12 = 8$
$\frac{2}{3}$ of 12 is 8.	$8 = \frac{2}{3} \times 12$
8 is $\frac{2}{3}$ times as large as 12.	
8 is $\frac{2}{3}$ as large as 12.	
8 is the *product* of $\frac{2}{3}$ and 12.	
$\frac{2}{3}$ multiplied by 12 yields 8.	

Because $\frac{2}{3} \times 12$ equals $12 \times \frac{2}{3}$, the order in which we say the multipliers makes no difference.

The following examples illustrate how to use translating to solve problems.

EXAMPLE 1 20 equals $\frac{3}{4}$ multiplied by what number?

SOLUTION

20 equals $\frac{3}{4}$ multiplied by what number

20 = $\frac{3}{4}$ × ☐

A multiplier is missing, so divide the answer 20 by the given multiplier $\frac{3}{4}$.

$$\begin{aligned}
\text{Missing multiplier} &= \text{answer to the multiplication problem} \div \text{known multiplier} \\
&= 20 \div \frac{3}{4} \\
&= \frac{20}{1} \times \frac{4}{3} \\
&= \frac{80}{3} \text{ or } 26\frac{2}{3} \quad \blacksquare
\end{aligned}$$

EXAMPLE 2 What number is $\frac{2}{3}$ of 600?

SOLUTION

What number is $\frac{2}{3}$ of 600?
$$\Box = \frac{2}{3} \times 600$$

The answer to the multiplication statement is missing, so multiply $\frac{2}{3}$ by 600.

$$\begin{aligned}
\frac{2}{3} \times 600 &= \frac{2}{3} \times \frac{600}{1} \\
&= \frac{2}{\cancel{3}_1} \times \frac{\cancel{3} \times 200}{1} \\
&= 400 \quad \blacksquare
\end{aligned}$$

EXAMPLE 3 What fraction of 48 is 36?

SOLUTION

What fraction of 48 is 36?
$$\Box \times 48 = 36$$

A multiplier is missing, so divide the answer 36 by the multiplier 48 to find the missing multiplier.

$$\begin{aligned}
\text{Missing multiplier} &= \text{answer to the multiplication problem} \div \text{known multiplier} \\
&= 36 \div 48 \\
&= \frac{36}{48} \\
&= \frac{3 \times \cancel{12}}{4 \times \cancel{12}} \\
&= \frac{3}{4} \quad \blacksquare
\end{aligned}$$

Try These Problems

Solve.

7. $12\frac{1}{2}$ is what fraction of 100?
8. $5\frac{2}{5}$ times what number yields 135?
9. $\frac{5}{8}$ equals $\frac{2}{400}$ of what number?
10. $\frac{1}{8}$ of $30\frac{3}{4}$ is what number?
11. The product of $\frac{2}{10}$ and $\frac{15}{17}$ is what number?
12. 900 represents the product of $\frac{7}{4}$ and what number?
13. What fraction of $\frac{1}{4}$ is $\frac{3}{8}$?
14. Find $2\frac{1}{2}$ of 425.

EXAMPLE 4 $16\frac{2}{3}$ represents $\frac{4}{5}$ of what number?

SOLUTION

$16\frac{2}{3}$ represents $\frac{4}{5}$ of what number?

$16\frac{2}{3} = \frac{4}{5} \times \square$

A multiplier is missing, so divide the answer $16\frac{2}{3}$ by the multiplier $\frac{4}{5}$ to find the missing multiplier.

$$\text{Missing multiplier} = \text{answer to the multiplication problem} \div \text{known multiplier}$$

$$= 16\frac{2}{3} \div \frac{4}{5}$$

$$= \frac{50}{3} \times \frac{5}{4}$$

$$= \frac{\cancel{2} \times 25}{3} \times \frac{5}{\cancel{2} \times 2}$$

$$= \frac{125}{6} \text{ or } 20\frac{5}{6} \blacksquare$$

Try Problems 7 through 14.

Special language is used when the number 2 or the number $\frac{1}{2}$ is a multiplier.

Math Symbols	English
2×7	*Twice* 7.
	Double 7.
	Two times seven.
$\frac{1}{2} \times 7$	*Half of* 7.
	Half 7.
	One-half times seven.
	Divide 7 in half.

Now we look at more examples of solving problems by translating.

EXAMPLE 5 Find twice $\frac{3}{5}$.

SOLUTION

$\underline{\text{twice}} \; \frac{3}{5}$

$= 2 \times \frac{3}{5}$ *Twice* means to multiply by 2.

$= \frac{2}{1} \times \frac{3}{5}$

$= \frac{6}{5}$ or $1\frac{1}{5}$ $\blacksquare$

Try These Problems

Solve.
15. Find twice $\frac{6}{7}$.
16. Double $19\frac{1}{4}$.
17. Take half of $\frac{9}{10}$.
18. Divide $6\frac{1}{3}$ in half.
19. Twice what number yields $8\frac{1}{4}$?
20. Half what number yields $\frac{1}{40}$?

EXAMPLE 6   What number is half of $7\frac{1}{3}$?

SOLUTION

$$\text{What number} \quad \underbrace{\text{is}}_{\downarrow} \quad \underbrace{\text{half} \quad \text{of}}_{} \quad 7\frac{1}{3}$$

$$\square = \frac{1}{2} \times 7\frac{1}{3} \qquad \textit{Half of } \text{means to multiply by } \frac{1}{2}.$$

$$= \frac{1}{\cancel{2}} \times \frac{\cancel{22}^{11}}{3}$$

$$= \frac{11}{3} \text{ or } 3\frac{2}{3} \quad \blacksquare$$

EXAMPLE 7   Doubling what number yields $3\frac{1}{4}$?

SOLUTION

$$\underbrace{\text{Doubling}}_{} \text{ what number yields } 3\frac{1}{4}$$

$$2 \times \square = 3\frac{1}{4}$$

A multiplier is missing, so we divide $3\frac{1}{4}$ by 2 to find the missing multiplier.

$$\begin{array}{c}\text{Missing}\\\text{multiplier}\end{array} = \begin{array}{c}\text{answer to the}\\\text{multiplication}\\\text{problem}\end{array} \div \begin{array}{c}\text{known}\\\text{multiplier}\end{array}$$

$$= 3\frac{1}{4} \div 2$$

$$= \frac{13}{4} \times \frac{1}{2}$$

$$= \frac{13}{8} \text{ or } 1\frac{5}{8} \quad \blacksquare$$

 Try Problems 15 through 20.

3   Language That Translates to Division

When reading the division statement $8 \div \frac{2}{3} = 12$, take care to say 8 and $\frac{2}{3}$ in the correct order because $8 \div \frac{2}{3}$ does not equal $\frac{2}{3} \div 8$.

$$8 \div \frac{2}{3} = \frac{\cancel{8}^4}{1} \times \frac{3}{\cancel{2}_1} = 12$$

$$\frac{2}{3} \div 8 = \frac{2}{3} \div \frac{8}{1} = \frac{\cancel{2}^1}{3} \times \frac{1}{\cancel{8}_4}$$

$$= \frac{1}{12}$$

Try These Problems

Solve.
21. Divide $8\frac{1}{2}$ by 3.
22. Divide $8\frac{1}{2}$ into 3.
23. Divide $\frac{1}{3}$ by $20\frac{2}{3}$.
24. Divide $\frac{1}{3}$ into $20\frac{2}{3}$.
25. What number equals $9\frac{3}{4}$ divided into 52?

The number 12 does not equal the number $\frac{1}{12}$. The number 12 is more than 1, but the number $\frac{1}{12}$ is less than 1. The following chart gives correct ways to read the division statement $8 \div \frac{2}{3} = 12$.

Math Symbols	English
$8 \div \frac{2}{3} = 12$	8 *divided by* $\frac{2}{3}$ equals 12.
	$\frac{2}{3}$ *divided into* 8 equals 12.

Note that the word *by* is used when the statement is read from left to right as we read words in a book; however, the word *into* is used if we read the divisor first. The divisor is the number that comes after the division symbol.

Recall that the fraction bar is also used to indicate division. For example,

$$\frac{8}{\frac{2}{3}} = 8 \div \frac{2}{3}$$

The following chart gives correct ways to read the division statement $\frac{8}{\frac{2}{3}} = 12$.

Math Symbols	English
$\frac{8}{\frac{2}{3}} = 12$	8 *divided by* $\frac{2}{3}$ is 12.
	$\frac{2}{3}$ *divided into* 8 is 12.

Note that the word *by* is used when we read the numerator first, but the word *into* is used when we read the denominator first.

 Try Problems 21 through 25.

Suppose a basket contains 5 apples and 8 oranges. Recall from Section 3.6 that we can compare the number of apples to the number of oranges by division. We say, "the **ratio** of apples to oranges is 5 to 8." We write,

$$\text{Ratio of apples to oranges} = \frac{\text{number of apples}}{\text{number of oranges}} = \frac{5}{8}$$

We can also compare the quantities in the reverse order.

$$\text{Ratio of oranges to apples} = \frac{\text{number of oranges}}{\text{number of apples}} = \frac{8}{5}$$

We can also compare either one of the quantities to the total amount of fruit in the basket.

$$\text{Ratio of oranges to the total} = \frac{\text{number of oranges}}{\text{total}}$$
$$= \frac{8}{13}$$

Try These Problems

Solve.
26. Find the ratio of $2\frac{1}{2}$ to 30.
27. Find the ratio of 45 to $3\frac{1}{3}$.

To recognize a comparison by division pay close attention to the language. The word *ratio* indicates to form a fraction, and the quantity that follows the word *to* is placed in the denominator.

EXAMPLE 8 Find the ratio of $9\frac{1}{3}$ to 14. Simplify.

SOLUTION

$$\text{Ratio of } 9\frac{1}{3} \text{ to } 14 = \frac{9\frac{1}{3}}{14}$$ The quantity following the word *to* is placed in the denominator.

$$= 9\frac{1}{3} \div 14$$ Perform the indicated division and simplify.

$$= \frac{28}{3} \times \frac{1}{14}$$

$$= \frac{2 \times 2 \times \not{7}}{3} \times \frac{1}{\underset{1}{\not{2}} \times \not{7}}$$

$$= \frac{2}{3} \blacksquare$$

Try Problems 26 and 27.

Answers to Try These Problems

1. 15   2. 45   3. $\frac{8}{5}$ or $1\frac{3}{5}$ 4. $\frac{1}{30}$ 5. 72   6. 135   7. $\frac{1}{8}$
8. 25   9. 125   10. $\frac{41}{8}$ or $5\frac{1}{8}$ 11. $\frac{3}{17}$ 12. $\frac{3600}{7}$ or $514\frac{2}{7}$
13. $\frac{3}{2}$ or $1\frac{1}{2}$ 14. $\frac{2125}{2}$ or $1062\frac{1}{2}$ 15. $\frac{12}{7}$ or $1\frac{5}{7}$ 16. $\frac{77}{2}$ or $38\frac{1}{2}$
17. $\frac{9}{20}$ 18. $\frac{19}{6}$ or $3\frac{1}{6}$ 19. $\frac{33}{8}$ or $4\frac{1}{8}$ 20. $\frac{1}{20}$ 21. $\frac{17}{6}$ or $2\frac{5}{6}$
22. $\frac{6}{17}$ 23. $\frac{1}{62}$ 24. 62   25. $\frac{16}{3}$ or $5\frac{1}{3}$ 26. $\frac{1}{12}$
27. $\frac{27}{2}$ or $13\frac{1}{2}$

EXERCISES 4.4

Translate each English statement to a multiplication or division statement using math symbols.

1. $\frac{2}{5}$ of 25 is 10.
2. 75 is $\frac{3}{4}$ of 100.
3. The product of $\frac{3}{5}$ and $\frac{5}{6}$ equals $\frac{1}{2}$.
4. $13\frac{1}{2}$ is the product of $4\frac{1}{2}$ and 3.
5. The ratio of 3 to $\frac{1}{4}$ is 12.
6. The ratio of $\frac{1}{4}$ to 3 is $\frac{1}{12}$.
7. Forty-hundredths times 200 gives 80.
8. Fourteen is the result of multiplying $\frac{2}{7}$ by 49.
9. $2\frac{2}{3}$ divided into 4 equals $1\frac{1}{2}$.
10. $2\frac{2}{3}$ divided by 4 equals $\frac{2}{3}$.
11. $4\frac{1}{2}$ is half of 9.
12. Half $\frac{3}{4}$ is $\frac{3}{8}$.
13. $\frac{7}{8}$ is twice as large as $\frac{7}{16}$.
14. 25 is the result of doubling $12\frac{1}{2}$.

Solve.

15. $\frac{2}{5}$ of 900.
16. $\frac{3}{8}$ of $8\frac{4}{5}$.
17. $1\frac{1}{2}$ of 70.
18. 5 of $3\frac{3}{4}$.
19. Find $\frac{8}{9}$ of 81.
20. Three-fifths of 30,100 is what number?
21. Find the product of $5\frac{1}{3}$ and $\frac{3}{4}$.
22. Find the ratio of $5\frac{1}{3}$ to $\frac{3}{4}$.

23. Divide $18\frac{1}{2}$ by 5.
24. Divide 5 by $18\frac{1}{2}$.
25. Divide $\frac{2}{5}$ into 60.
26. Divide 60 into $\frac{2}{5}$.
27. Double $4\frac{1}{4}$.
28. What number is twice $\frac{10}{23}$?
29. Find half of $50\frac{3}{4}$.
30. Half of what number yields $\frac{3}{4}$?
31. $13\frac{1}{2}$ is what number times $\frac{3}{20}$?
32. $\frac{80}{3}$ represents twice what number?
33. One-third of what number is represented by 80?
34. $52\frac{1}{3}$ equals $\frac{5}{7}$ of what number?
35. What number is the ratio of $6\frac{1}{4}$ to 200?
36. What number is the ratio of 200 to $6\frac{1}{4}$?
37. Fifty represents $\frac{7}{20}$ of what number?
38. $\frac{9}{10}$ equals $4\frac{1}{4}$ of what number?
39. 6 is what fraction of 200?
40. $18\frac{2}{3}$ is what fraction of 7?
41. $\frac{3}{5}$ of 9000 is what number?
42. What number times $16\frac{1}{4}$ equals 5?
43. $\frac{9}{40}$ of what number is 180?
44. $\frac{20}{7}$ of $16\frac{1}{3}$ is what number?
45. What fraction of 80 is 200?
46. $\frac{7}{200}$ of what number is 4?

4.5 APPLICATIONS

OBJECTIVES
- Solve an application involving equal parts in a total quantity.
- Solve an application involving equivalent rates.
- Convert a rate to its equivalent rate.
- Solve an application by interpreting a rate as a ratio.
- Solve an application by using a rate as a multiplier.
- Solve an application involving the area of a rectangle.
- Solve an application by using translations.

1 Equal Parts in a Total Quantity

Suppose we have a wire that is 36 inches long. If the wire is cut into 4 equal pieces, then each piece is 9 inches long. Here we picture the situation.

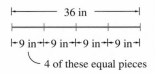

Observe that the numbers are related as follows:

$$\underset{\text{equal parts}}{\text{Number of}} \times \underset{\substack{\text{each part} \\ \text{(in)}}}{\text{Size of}} = \underset{\substack{\text{quantity} \\ \text{(in)}}}{\text{total}}$$

$$4 \times 9 = 36$$

In general, if a situation involves a total quantity that is being split up into a number of equal parts, the following always applies.

$$\begin{array}{c}\text{Number of}\\ \text{equal parts}\end{array} \times \begin{array}{c}\text{Size of}\\ \text{each part}\end{array} = \begin{array}{c}\text{total}\\ \text{quantity}\end{array}$$

A problem can be presented by giving any two of the three quantities and asking for the third quantity. The following examples illustrate this.

EXAMPLE 1 A rope that is 4 feet long is cut into 12 equal pieces. How long is each piece?

SOLUTION We use the formula,

$$\begin{array}{c}\text{Number of}\\ \text{equal parts}\end{array} \times \begin{array}{c}\text{Size of}\\ \text{each part}\\ \text{(ft)}\end{array} = \begin{array}{c}\text{total}\\ \text{quantity}\\ \text{(ft)}\end{array}$$

$$12 \quad \times \quad \boxed{} \quad = \quad 4$$

A multiplier is missing, so we divide 4 by 12 to find the missing multiplier.

$$\begin{array}{c}\text{Size of}\\ \text{each part}\end{array} = \frac{\text{total quantity}}{\text{number of equal parts}}$$

$$= \frac{4}{12}$$

$$= \frac{\cancel{4}^{1}}{3 \times \cancel{4}}$$

$$= \frac{1}{3} \text{ ft}$$

Each piece is $\frac{1}{3}$ foot long. ∎

In Example 1 it is important to note that the total quantity is not always the largest of the three numbers.

EXAMPLE 2 A 14-foot log is cut into pieces that are each $1\frac{3}{4}$ feet long. How many of the smaller pieces can be cut?

SOLUTION We want to know how many of the $1\frac{3}{4}$-foot pieces make up the total 14 feet.

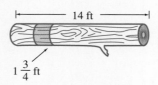

We use the formula,

$$\begin{array}{c}\text{Number of}\\ \text{equal parts}\end{array} \times \begin{array}{c}\text{Size of}\\ \text{each part}\\ \text{(ft)}\end{array} = \begin{array}{c}\text{total}\\ \text{quantity}\\ \text{(ft)}\end{array}$$

$$\boxed{} \quad \times \quad 1\frac{3}{4} \quad = \quad 14$$

A multiplier is missing, so we divide 14 by $1\frac{3}{4}$ to find the missing multiplier.

$$\begin{aligned}\text{Number of} \atop \text{equal parts} &= {\text{total} \atop \text{quantity}} \div {\text{size of} \atop \text{each part}}\\ &= 14 \div 1\tfrac{3}{4}\\ &= \tfrac{14}{1} \div \tfrac{7}{4}\\ &= \tfrac{\overset{2}{\cancel{14}}}{1} \times \tfrac{4}{\underset{1}{\cancel{7}}}\\ &= 8\end{aligned}$$

8 pieces can be cut. ∎

EXAMPLE 3 Ten children share $1\tfrac{1}{2}$ gallons of milk equally. How much milk will each child get?

SOLUTION The $1\tfrac{1}{2}$ gallons of milk is the quantity that is being separated into 10 equal parts. We use the formula,

$$\begin{array}{c}\text{Number of} \\ \text{equal parts}\end{array} \times \begin{array}{c}\text{Size of} \\ \text{each part} \\ \text{(gal)}\end{array} = \begin{array}{c}\text{total} \\ \text{quantity} \\ \text{(gal)}\end{array}$$

$$10 \times \boxed{} = 1\tfrac{1}{2}$$

The total quantity is the quantity that is being split apart, not necessarily the larger number.

A multiplier is missing, so we divide $1\tfrac{1}{2}$ by 10 to find the missing multiplier.

$$\begin{aligned}{\text{Size of} \atop \text{each part}} &= {\text{total} \atop \text{quantity}} \div {\text{number of} \atop \text{equal parts}}\\ &= 1\tfrac{1}{2} \div 10\\ &= \tfrac{3}{2} \times \tfrac{1}{10}\\ &= \tfrac{3}{20} \text{ gal}\end{aligned}$$

Each child receives $\tfrac{3}{20}$ gallon of milk. ∎

EXAMPLE 4 A nurse needs to give out 32 doses of medicine. Each dose contains $1\tfrac{1}{8}$ milliliters. How much medicine does she need?

SOLUTION The 32 doses is the number of equal parts and $1\tfrac{1}{8}$ milliliters is the size of each part. We are missing the total quantity.

Try These Problems

1. Three people share equally 2 pizzas. How much pizza does each person receive?

2. Two people share equally 3 pizzas. How much pizza does each person receive?

3. Thirty milliliters of medicine is to be given out in doses of $1\frac{1}{5}$ milliliters each. How many doses can be given?

4. A board that is $6\frac{3}{10}$ meters long is cut into 6 equal pieces. How long is each piece?

5. Fifteen persons share equally $4\frac{4}{5}$ pounds of cheese. How much cheese does each person receive?

6. A chef needs 50 servings of soup that each contain $14\frac{1}{2}$ ounces. How much soup does the chef need?

$$\begin{aligned}\text{Total quantity (m}\ell\text{)} &= \text{number of equal parts} \times \text{size of each part (m}\ell\text{)} \\ &= 32 \times 1\frac{1}{8} \\ &= \frac{32}{1} \times \frac{9}{8} \\ &= \frac{\cancel{8} \times 4}{1} \times \frac{9}{\cancel{8}} \\ &= 36 \text{ m}\ell\end{aligned}$$

The nurse needs 36 milliliters of medicine.

Try Problems 1 through 6.

2 Equivalent Rates

If you earn 8 dollars per hour, then the phrase *8 dollars per hour,* which means $8 for each 1 hour, indicates the **rate** of pay. A rate is a comparison of two quantities. Here is a list of more phrases that indicate rates.

> *Phrases That Indicate Rates*
> You walked 4 miles per hour.
> Ten nails weigh $\frac{1}{4}$ pound.
> On a map, $\frac{1}{2}$ inch represents 100 miles.
> Socks cost $2 a pair.
> A utility stock costs $28\frac{3}{8}$ dollars per share.

If you earn $8 in 1 hour, then you earn $32 in 4 hours.

Multiply by 4 ⟶ 8 dollars in 1 hour
32 dollars in 4 hours ⟵ Multiply by 4

Observe that multiplying both quantities in a rate by the same nonzero number produces an equivalent rate. This can be used to solve problems as shown in the following examples.

EXAMPLE 5 Stephanie can jog 1 mile in 9 minutes. How long will it take her to jog $4\frac{1}{2}$ miles?

SOLUTION

Multiply by $4\frac{1}{2}$ ⟶ 1 mile in 9 minutes
$4\frac{1}{2}$ miles in ? minutes ⟵ Multiply by $4\frac{1}{2}$

If we multiply each quantity in the given rate by $4\frac{1}{2}$, we obtain how many minutes it takes her to jog $4\frac{1}{2}$ miles.

$$9 \times 4\frac{1}{2} = \frac{9}{1} \times \frac{9}{2}$$
$$= \frac{81}{2} \text{ or } 40\frac{1}{2} \text{ min}$$

It takes Stephanie $40\frac{1}{2}$ minutes to jog $4\frac{1}{2}$ miles.

EXAMPLE 6 A transportation stock sells for $20\frac{1}{4}$ dollars per share. How much do 100 shares of this stock cost?

SOLUTION The phrase *$20\frac{1}{4}$ dollars per share* means $20\frac{1}{4}$ dollars for each 1 share.

Multiply by 100 ⟨ $20\frac{1}{4}$ dollars for 1 share / ? dollars for 100 shares ⟩ Multiply by 100

If we multiply both quantities in the given rate by 100, we obtain the cost of 100 shares of the stock.

$$20\frac{1}{4} \times 100 = \frac{81}{\underset{1}{\cancel{4}}} \times \frac{\overset{25}{\cancel{100}}}{1}$$

$$= 2025$$

It costs $2025 for 100 shares of this stock. ∎

If you walk 28 miles in 7 hours, then you walk 4 miles in 1 hour.

Divide by 7 ⟨ 28 miles in 7 hours / 4 miles in 1 hour ⟩ Divide by 7

Observe that dividing both quantities in a rate by the same nonzero number yields an equivalent rate. This can be used to solve problems as shown in the following examples.

EXAMPLE 7 Fifty nails weigh $2\frac{1}{2}$ pounds. How much does one nail weigh?

SOLUTION

Divide by 50 ⟨ 50 nails weigh $2\frac{1}{2}$ pounds / 1 nail weighs ? pounds ⟩ Divide by 50

Dividing each quantity in the given rate by 50 gives the weight of 1 nail.

$$2\frac{1}{2} \div 50 = \frac{5}{2} \div \frac{50}{1}$$

$$= \frac{\overset{1}{\cancel{5}}}{2} \times \frac{1}{\underset{10}{\cancel{50}}}$$

$$= \frac{1}{20}$$

Each nail weighs $\frac{1}{20}$ pound. ∎

EXAMPLE 8 Betsy can jog $1\frac{1}{4}$ miles in $12\frac{1}{2}$ minutes. How far can she go in 1 minute?

SOLUTION

Divide by $12\frac{1}{2}$ ⟨ $1\frac{1}{4}$ miles in $12\frac{1}{2}$ minutes / ? miles in 1 minute ⟩ Divide by $12\frac{1}{2}$

Try These Problems

7. One pound of bananas costs 56 cents. How much do $1\frac{3}{8}$ pounds cost?

8. A utility stock costs $50\frac{3}{4}$ dollars per share. What is the total cost of 12 shares?

9. Ed can run 4 miles in 27 minutes. How long will it take him to run 1 mile?

10. Dick bought Stephanie some fabric while he was in Paris. He paid $27 for $4\frac{1}{2}$ yards. How much did the fabric cost per yard?

11. Eighty nails weigh $1\frac{3}{4}$ pounds. What does 1 nail weigh?

12. A faucet leaks $2\frac{2}{3}$ ounces of water per hour. How much water has leaked after $5\frac{1}{2}$ hours?

Dividing both quantities in the given rate by $12\frac{1}{2}$ yields how far she can go in 1 minute.

$$1\frac{1}{4} \div 12\frac{1}{2} = \frac{5}{4} \div \frac{25}{2}$$

$$= \frac{\cancel{5}^1}{\cancel{4}_2} \times \frac{\cancel{2}^1}{\cancel{25}_5}$$

$$= \frac{1}{10}$$

Betsy can jog $\frac{1}{10}$ mile in 1 minute. ∎

 Try Problems 7 through 12.

 A Rate as a Ratio

We have learned to recognize phrases that indicate rates, and we know that a rate is a comparison of two quantities. At this time, it will be helpful to recognize that a **rate** is a comparison of two quantities *by division;* thus a rate can be written as a *ratio* of the two quantities. Here is a list of some phrases, their corresponding ratios, and the resulting rates.

English Phrase	*Ratio*	*Rate*
Ann drove 60 miles per hour.	$\dfrac{60 \text{ miles}}{1 \text{ hour}}$	60 miles per hour
Ken paid $100 for 5 shirts.	$\dfrac{100 \text{ dollars}}{5 \text{ shirts}}$	20 dollars per shirt
Water enters the tank at 2 gallons every 3 minutes.	$\dfrac{2 \text{ gallons}}{3 \text{ minutes}}$	$\dfrac{2}{3}$ gallon per minute

First, notice how the ratio units translate to the rate units. The unit in the denominator of the ratio follows the word *per* in the rate unit. That is, the word *per* is used for the fraction bar.

$$\frac{\text{miles}}{\text{hour}} \rightarrow \text{miles per hour}$$

$$\frac{\text{dollars}}{\text{shirts}} \rightarrow \text{dollars per shirt}$$

$$\frac{\text{gallons}}{\text{minutes}} \rightarrow \text{gallons per minute}$$

Second, because the fraction bar indicates to divide the numerator by the denominator, perform this division to obtain the rate quantity.

$$\frac{60 \text{ miles}}{1 \text{ hour}} = 60 \text{ miles per hour} \qquad \text{Divide 60 by 1 to obtain 60.}$$

Chapter 4 Fractions: Multiplication and Division

 Try These Problems

Convert each ratio to a rate. Specify both the quantity and the units.

13. $\dfrac{4 \text{ miles}}{1 \text{ hour}}$

14. $\dfrac{150 \text{ feet}}{6 \text{ seconds}}$

15. $\dfrac{2 \text{ pounds}}{100 \text{ nails}}$

16. $\dfrac{8 \text{ gallons}}{2\frac{1}{2} \text{ minutes}}$

Solve.

17. Jamie drove 90 miles in $1\frac{3}{4}$ hours. How many miles per hour did she drive?

18. It takes Edna $2\frac{1}{2}$ minutes to type 165 words. How many words per minute does she type?

19. A $12\frac{1}{2}$-foot rope weighs $8\frac{1}{3}$ pounds. What does 1 foot of this rope weigh?

20. A chef uses $7\frac{1}{2}$ pounds of butter to make 10 cakes. How much butter is needed for 1 cake?

$\dfrac{100 \text{ dollars}}{5 \text{ shirts}} = 20$ dollars per shirt   Divide 100 by 5 to obtain 20.

$\dfrac{2 \text{ gallons}}{3 \text{ minutes}} = \dfrac{2}{3}$ gallons per minute   Divide 2 by 3 to obtain $\frac{2}{3}$.

 Try Problems 13 through 16.

EXAMPLE 9 A rope weighs $2\frac{1}{2}$ pounds for every 5 feet. How much does the rope weigh per foot?

SOLUTION We want to know the weight per foot, so we need the ratio of pounds to feet.

$$2\frac{1}{2} \text{ pounds for every 5 feet}$$

$$= \dfrac{2\frac{1}{2} \text{ pounds}}{5 \text{ feet}}$$

$$= 2\frac{1}{2} \div 5$$

$$= \dfrac{\overset{1}{\cancel{5}}}{2} \times \dfrac{1}{\underset{1}{\cancel{5}}}$$

$$= \dfrac{1}{2} \text{ pound per foot}$$

We put *pounds* in the numerator and *feet* in the denominator to obtain *pounds per foot*.

The rope weighs $\frac{1}{2}$ pound per foot. ∎

EXAMPLE 10 A $14\frac{1}{2}$-ounce can of beans costs 87 cents. What is the cost of 1 ounce of these beans?

SOLUTION We want to know the cost of 1 ounce; that is, we want the cost per ounce. We need the ratio of the cost to the ounces to obtain the cost per ounce.

$$14\frac{1}{2} \text{ ounces cost 87 cents}$$

$$= \dfrac{87¢}{14\frac{1}{2} \text{ ounces}}$$

$$= 87 \div 14\frac{1}{2}$$

$$= \dfrac{87}{1} \div \dfrac{29}{2}$$

$$= \dfrac{\overset{3}{\cancel{87}}}{1} \times \dfrac{2}{\underset{1}{\cancel{29}}}$$

$$= 6¢ \text{ per ounce}$$

We put *cents* in the numerator and *ounces* in the denominator to obtain *cents per ounce*.

One ounce of these beans costs 6¢. ∎

 Try Problems 17 through 20.

4 More about Rates

If you walk 4 miles per hour for 3 hours, then you walk a total distance of 12 miles. Note these quantities are related in the following way:

$$\frac{\text{miles}}{\text{per hour}} \times \frac{\text{number}}{\text{of hours}} = \frac{\text{total number}}{\text{of miles}}$$
$$4 \times 3 = 12$$

Observe that if you write the rate, 4 miles per hour, as the ratio, $\frac{4 \text{ miles}}{1 \text{ hour}}$, and multiply by 3 hours, the hour units in the denominator cancel with the hour units in the numerator and we are left with the answer in miles.

$$\frac{4 \text{ miles}}{1 \text{ hour}} \times \frac{3 \text{ hours}}{1} = 12 \text{ miles}$$

Here are some more examples of how units cancel when multiplying by a rate.

Situation	Multiplication Statement
Lillian buys 5 baskets at $12 each. The total cost is $60.	$\frac{\$12}{1 \text{ basket}} \times \frac{5 \text{ baskets}}{1} = \$60$
The faucet leaks $2\frac{1}{2}$ ounces of water every minute. After 4 minutes, 10 ounces of water have leaked.	$\frac{2\frac{1}{2} \text{ ounces}}{1 \text{ minute}} \times \frac{4 \text{ minutes}}{1} = 2\frac{1}{2} \times 4$ $= \frac{5}{2} \times \frac{4}{1}$ $= 10 \text{ ounces}$

Paying close attention to the units can help you to set up problems correctly. Here are some examples.

EXAMPLE 11 A car averages 48 miles per hour. How long will it take it to travel 204 miles?

SOLUTION The quantity 48 miles per hour is a rate, thus can be written as a ratio.

$$48 \text{ miles per hour} = \frac{48 \text{ miles}}{1 \text{ hour}}$$

Set up a multiplication statement so that the units in the denominator of the rate cancel.

$$\frac{\text{miles}}{\text{hour}} \times \text{hours} = \text{miles}$$
$$48 \times \boxed{} = 204$$

Because a multiplier is missing, divide 204 by 48 to find the missing multiplier.

$$204 \div 48 = \frac{204}{48} = \frac{51 \times 4}{12 \times 4} = \frac{3 \times 17}{3 \times 4} = \frac{17}{4} \text{ or } 4\frac{1}{4}$$

It will take the car $4\frac{1}{4}$ hours to travel 204 miles. ■

EXAMPLE 12 Coffee costs $6 per pound. How much coffee can you buy for $4?

SOLUTION

METHOD 1 The quantity $6 per pound is a rate, thus can be written as a ratio.

$$\$6 \text{ per pound} = \frac{\$6}{1 \text{ pound}}$$

Set up a multiplication statement so that the units in the denominator of the rate cancel.

$$\frac{\$}{\text{pound}} \times \text{pounds} = \$$$
$$6 \times \boxed{} = 4$$

A multiplier is missing, so divide 4 by 6 to obtain the missing multiplier.

$$4 \div 6 = \frac{4}{6} = \frac{2 \times \cancel{2}}{3 \times \cancel{2}} = \frac{2}{3}$$

You can buy $\frac{2}{3}$ pound coffee for $4.

METHOD 2 The phrase *$6 per pound* is a rate. Before finding the number of pounds that can be bought with $4, we can find the number of pounds that can be bought with $1.

Divide by 6 ⟶ $6 per 1 pound ⟵ Divide by 6
              $1 per ? pound

$$1 \div 6 = \frac{1}{6}$$

Therefore, $1 buys $\frac{1}{6}$ pound of coffee. Now use the rate *$1 per $\frac{1}{6}$ pound* to find how many pounds can be bought with $4.

Multiply by 4 ⟶ $1 per $\frac{1}{6}$ pound ⟵ Multiply by 4
                $4 per ? pound

$$\frac{1}{6} \times 4 = \frac{1}{6} \times \frac{4}{1} = \frac{4}{6} = \frac{\cancel{2} \times 2}{\cancel{2} \times 3} = \frac{2}{3}$$

You can buy $\frac{2}{3}$ pound of coffee for $4. ∎

EXAMPLE 13 Gertrude plans to make a wool jacket. One yard of the wool fabric costs $36. What is the cost of $3\frac{1}{4}$ yards of this fabric?

SOLUTION

METHOD 1 The phrase *one yard costs $36* is indicating a rate, so it can be written as a ratio.

$$\text{one yard costs } \$36 = \frac{\$36}{1 \text{ yard}}$$

Set up a multiplication statement so that the units in the denominator of the rate cancel.

$$\frac{\$}{\text{yard}} \times \text{yards} = \$$$
$$36 \times 3\frac{1}{4} = \boxed{}$$

The answer to the multiplication statement is missing, so multiply.

$$36 \times 3\tfrac{1}{4} = \frac{36}{1} \times \frac{13}{4}$$

$$= \frac{9 \times \cancel{4}}{1} \times \frac{13}{\underset{1}{\cancel{4}}}$$

$$= 117$$

The cost of $3\tfrac{1}{4}$ yards is $117.

METHOD 2 The phrase *one yard costs $36* is a rate.

Multiply by $3\tfrac{1}{4}$ ⟶ 1 yard costs $36
  $3\tfrac{1}{4}$ yards cost $? ⟵ Multiply by $3\tfrac{1}{4}$

Multiply both quantities in the given rate by $3\tfrac{1}{4}$ to obtain the cost of $3\tfrac{1}{4}$ yards.

$$36 \times 3\tfrac{1}{4} = \frac{36}{1} \times \frac{13}{4}$$

$$= \frac{9 \times \cancel{4}}{1} \times \frac{13}{\underset{1}{\cancel{4}}}$$

$$= 117$$

The cost of $3\tfrac{1}{4}$ yards is $117. ■

EXAMPLE 14 Sixty nails weigh $2\tfrac{1}{4}$ pounds. What does one nail weigh?

SOLUTION

METHOD 1 Because we want the weight of 1 nail, we want to know how many pounds per nail. Write the given rate, *60 nails weigh $2\tfrac{1}{4}$ pounds,* as a ratio and perform the indicated division to find how many pounds per nail.

$$\frac{2\tfrac{1}{4} \text{ pounds}}{60 \text{ nails}} = 2\tfrac{1}{4} \div 60$$

$$= \frac{9}{4} \times \frac{1}{60}$$

$$= \frac{\cancel{3} \times 3}{4} \times \frac{1}{\cancel{3} \times 20}$$

$$= \frac{3}{80} \text{ pound per nail}$$

Place *pounds* in the numerator and *nails* in the denominator to get *pounds per nail.*

One nail weighs $\tfrac{3}{80}$ pound.

CHECK We can check this by seeing if the numbers satisfy the following multiplication statement.

$$\frac{\text{pound}}{\text{nail}} \times \text{ nails } = \text{pounds}$$

$$\frac{3}{80} \times 60 \stackrel{?}{=} 2\tfrac{1}{4}$$

$$\frac{3}{4 \times \cancel{20}} \times \frac{3 \times \cancel{20}}{1} \stackrel{?}{=} 2\tfrac{1}{4}$$

$$\frac{9}{4} = 2\tfrac{1}{4} \quad \text{True}$$

Try These Problems

21. A watch gains $2\frac{1}{2}$ minutes every day. How many days will it take the watch to gain 30 minutes?

22. Steve can swim 1 lap in $1\frac{1}{2}$ minutes. How many laps can he swim in 10 minutes?

23. An entertainment stock is selling for $\$20\frac{5}{8}$ per share. What is the cost of 160 shares of this stock?

24. Arthur walks $5\frac{1}{3}$ miles in 60 minutes. How far does he walk in 1 minute?

25. Each nail weighs $\frac{1}{20}$ pound. How much do 120 nails weigh?

26. A fabric sells for $48 per yard. How much fabric can you buy for $516?

27. Martha bought $5\frac{1}{4}$ pounds of coffee for $42. What is the cost per pound?

28. Water enters a tank at the rate of $6\frac{2}{3}$ gallons per minute. How long will it take for 360 gallons of water to enter the tank?

METHOD 2 The phrase *60 nails weigh $2\frac{1}{4}$ pounds* is a rate.

Divide by 60 ⤵ 60 nails weigh $2\frac{1}{4}$ pounds Divide by 60
⤷ 1 nail weighs ? pounds ⤴

Divide both quantities in the given rate by 60 to obtain the weight of 1 nail.

$$2\frac{1}{4} \div 60 = \frac{9}{4} \times \frac{1}{60}$$

$$= \frac{3}{80} \text{ pound}$$

One nail weighs $\frac{3}{80}$ pound. ■

 Try Problems 21 through 28.

5 Area of a Rectangle

Area is a measure of the extent of a region. A square that measures 1 unit on each side is said to have an area of 1 square unit.

1 unit ▢ Area = 1 square unit
1 unit

A rectangle that is 3 units wide and 4 units long contains 12 of the 1-square-unit squares, therefore the area is 12 square units.

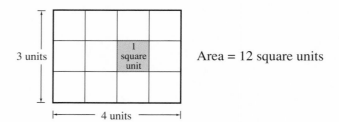

Observe that the area of this rectangle can be obtained by multiplying the length by the width.

$$\text{Area} = \text{length} \times \text{width}$$
$$= 4 \times 3$$
$$= 12 \text{ square units}$$

In general, the area of any rectangle can be computed by multiplying the length by the width.

Area of Rectangle
The area of a rectangle is the length times the width.
Area = length × width

EXAMPLE 15 A rectangular flower garden is $16\frac{1}{2}$ feet long and $4\frac{1}{4}$ feet wide. What is the area of the garden?

SOLUTION

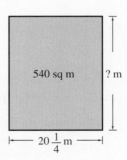

$$\begin{array}{rcl} \text{Area} & = & \text{length} \times \text{width} \\ \text{(sq ft)} & & \text{(ft)} \quad\quad \text{(ft)} \end{array}$$

$$\square = 16\frac{1}{2} \times 4\frac{1}{4}$$

The answer to the multiplication problem is missing, so multiply.

$$16\frac{1}{2} \times 4\frac{1}{4} = \frac{33}{2} \times \frac{17}{4}$$
$$= \frac{561}{8} \quad \text{or} \quad 70\frac{1}{8}$$

The area of the garden is $70\frac{1}{8}$ square feet. ∎

EXAMPLE 16 The area of a vacant lot is 540 square meters. The width of the lot is $20\frac{1}{4}$ meters. How long is the lot?

SOLUTION

$$\begin{array}{rcl} \text{Area} & = & \text{length} \times \text{width} \\ \text{(sq m)} & & \text{(m)} \quad\quad \text{(m)} \end{array}$$

$$540 = \square \times 20\frac{1}{4}$$

A multiplier is missing, so we divide 540 by $20\frac{1}{4}$.

Try These Problems

29. Andrea wants to carpet her bedroom. The floor of the room is $5\frac{2}{3}$ yards wide and $6\frac{1}{3}$ yards long. How many square yards of carpeting will she need to buy?

30. A tablecloth is 27 square feet in area. It is $4\frac{1}{2}$ feet wide. What is the length of the tablecloth?

31. A rectangular garden is $20\frac{1}{4}$ feet long. How wide is the garden if it contains $337\frac{1}{2}$ square feet?

$$540 \div 20\frac{1}{4} = \frac{540}{1} \div \frac{81}{4}$$
$$= \frac{540}{1} \times \frac{4}{81}$$
$$= \frac{\cancel{9} \times 60}{1} \times \frac{4}{\cancel{9} \times 9}$$
$$= \frac{\cancel{3} \times 20}{1} \times \frac{4}{\cancel{3} \times 3}$$
$$= \frac{80}{3} \text{ or } 26\frac{2}{3}$$

The length of the lot is $26\frac{2}{3}$ meters. ■

Try Problems 29 through 31.

6 Problems Involving Translations

In Section 4.4 we saw how translating English phrases to math symbols can be used to solve problems. Now we look at examples involving real-life situations.

EXAMPLE 17 Warren pays $\frac{3}{10}$ of his annual salary in taxes. His annual salary is $70,000. How much does he pay in taxes?

SOLUTION

KEY PHRASE → $\frac{3}{10}$ of his annual salary in taxes
$\frac{3}{10}$ × 70,000 = ☐

The answer to the multiplication statement is missing, so multiply $\frac{3}{10}$ by 70,000.

$$\frac{3}{10} \times 70{,}000 = \frac{3}{10} \times \frac{70000}{1}$$
$$= \frac{3}{\cancel{10}_1} \times \frac{\cancel{10} \times 7000}{1}$$
$$= 21{,}000$$

Warren pays $21,000 in taxes. ■

EXAMPLE 18 A certain college conducted a survey that indicates that $\frac{2}{9}$ of the students are smokers. There are 4000 smokers. What is the total number of students?

SOLUTION

KEY PHRASE → $\frac{2}{9}$ of the students are smokers.
$\frac{2}{9}$ × ☐ = 4000

Try These Problems

32. Mr. Hummer read $\frac{1}{3}$ of a novel last night. If he read 72 pages, how many pages does the novel have?

33. Darlene spends $\frac{3}{20}$ of her income on food. Her monthly income is $2100. How much does she spend on food in one month?

34. Two-thirds of a bottle of medicine weighs $5\frac{1}{2}$ grams. What does a full bottle weigh?

35. The population of a small town in Texas decreased by 350 persons in 5 years. The original population was 1500. The decrease is what fraction of the original population?

A multiplier is missing, so divide 4000 by $\frac{2}{9}$.

$$4000 \div \frac{2}{9} = \frac{4000}{1} \times \frac{9}{2}$$
$$= \frac{\cancel{2} \times 2000}{1} \times \frac{9}{\cancel{2}}$$
$$= 18,000$$

The total number of students is 18,000. ∎

EXAMPLE 19 During last year's football season, Jerry made 81 field goals. The total number attempted was 108. What fractional part of the field goals did he make?

SOLUTION
KEY PHRASE → What fractional part **of** the field goals **did** he make?

☐ × 108 = 81

A multiplier is missing, so divide 81 by 108.

$$81 \div 108 = \frac{81}{108}$$
$$= \frac{\cancel{9} \times 9}{\cancel{9} \times 12}$$
$$= \frac{\cancel{3} \times 3}{\cancel{3} \times 4}$$
$$= \frac{3}{4}$$

Jerry made $\frac{3}{4}$ of his field goals. ∎

Try Problems 32 through 35.

EXAMPLE 20 Mercy spent $3\frac{3}{4}$ hours reading her history lesson. It took Steven $1\frac{1}{5}$ times as long to read the same lesson. How long did it take Steven to read the lesson?

SOLUTION
KEY PHRASE → It took Steven $1\frac{1}{5}$ times as long to read the same lesson.

REPHRASE → Steven's time was $1\frac{1}{5}$ times as long as Marcy's time.

☐ = $1\frac{1}{5}$ × $3\frac{3}{4}$

The answer to the multiplication statement is missing, so multiply $1\frac{1}{5}$ by $3\frac{3}{4}$.

$$1\frac{1}{5} \times 3\frac{3}{4} = \frac{6}{5} \times \frac{15}{4}$$

$$= \frac{\cancel{2} \times 3}{\underset{1}{\cancel{5}}} \times \frac{3 \times \cancel{5}}{\cancel{2} \times 2}$$

$$= \frac{9}{2} \quad \text{or} \quad 4\frac{1}{2}$$

It took Steven $4\frac{1}{2}$ hours to read the lesson. ■

EXAMPLE 21 An auto mechanic earns $\frac{3}{8}$ as much as a computer programmer. If the auto mechanic earns $21,000 annually, what is the annual salary of the computer programmer?

SOLUTION
KEY PHRASE → An auto mechanic earns $\frac{3}{8}$ as much as a computer programmer.

REPHRASE → Auto mechanic's earnings are $\frac{3}{8}$ as much as computer programmer's earnings

$$21{,}000 = \frac{3}{8} \times \square$$

A multiplier is missing, so divide 21,000 by $\frac{3}{8}$.

$$21{,}000 \div \frac{3}{8} = \frac{21000}{1} \times \frac{8}{3}$$

$$= \frac{7000 \times \cancel{3}}{1} \times \frac{8}{\underset{1}{\cancel{3}}}$$

$$= 56{,}000$$

The computer programmer earns $56,000 annually. ■

EXAMPLE 22 The width of a rectangle is $10\frac{3}{4}$ inches. Its length is twice its width. Find the area of the rectangle.

SOLUTION First we find the length of the rectangle.
KEY PHRASE → Length is twice its width.

$$\square = 2 \times 10\frac{3}{4}$$

To find the length, multiply $10\frac{3}{4}$ by 2.

$$2 \times 10\frac{3}{4} = \frac{\overset{1}{\cancel{2}}}{1} \times \frac{43}{\underset{2}{\cancel{4}}}$$

$$= \frac{43}{2} \quad \text{or} \quad 21\frac{1}{2}$$

The length is $\frac{43}{2}$ inches. Now we find the area.

Try These Problems

36. A piano is now worth $2\frac{1}{2}$ times what it was worth 10 years ago. If it is worth $12,000 now, what was it worth 10 years ago?

37. Carl earns $42,000 per year as a computer programmer. José is in business for himself as a tax consultant. José earns $1\frac{1}{4}$ times as much as Carl does. How much does José earn per year?

38. Susan is $2\frac{2}{3}$ feet tall. Her older brother Gary is twice her height. How tall is Gary?

39. The width of a rectangle is $\frac{2}{3}$ as long as the length. If the width is 30 feet, find the area of the rectangle.

$$\begin{aligned}
\text{Area} &= \text{length} \times \text{width} \\
\text{(sq in)} &\quad \text{(in)} \quad \text{(in)} \\
&= \frac{43}{2} \times 10\frac{3}{4} \\
&= \frac{43}{2} \times \frac{43}{4} \\
&= \frac{1849}{8} \text{ or } 231\frac{1}{8}
\end{aligned}$$

The area of rectangle is $231\frac{1}{8}$ square inches. ■

 Try Problems 36 through 39.

EXAMPLE 23 A shipment of fruit contains 500 apples and 850 oranges. What is the ratio of apples to oranges?

SOLUTION

$$\begin{aligned}
\text{Ratio of apples to oranges} &= \frac{\text{number of apples}}{\text{number of oranges}} \\
&= \frac{500}{850} \\
&= \frac{5 \times 10 \times \cancel{10}}{85 \times \cancel{10}} \\
&= \frac{\cancel{5} \times 10}{\cancel{5} \times 17} \\
&= \frac{10}{17}
\end{aligned}$$

The ratio of apples to oranges is $\frac{10}{17}$. ■

EXAMPLE 24 There are 210 women present at a convention of 350 persons. What is the ratio of women to men?

SOLUTION First, determine how many men are at the convention.

$$\begin{aligned}
\text{Number of men} &= \text{Total Number} - \text{Number of women} \\
&= 350 - 210 \\
&= 140
\end{aligned}$$

There are 140 men at the convention.

$$\begin{aligned}
\text{Ratio of women to men} &= \frac{\text{number of women}}{\text{number of men}} \\
&= \frac{210}{140} \\
&= \frac{3 \times \cancel{7} \times \cancel{10}}{2 \times \cancel{7} \times \cancel{10}} \\
&= \frac{3}{2}
\end{aligned}$$

The ratio of women to men is $\frac{3}{2}$. ■

Try These Problems

40. A vending machine contains 30 nickels, 150 dimes, and 75 quarters. What is the ratio of the number of nickels to the total number of coins?

41. Susan is $5\frac{1}{4}$ feet tall and her daughter is $2\frac{1}{2}$ feet tall. What is the ratio of the daughter's height to Susan's height?

42. The sides of a triangle measure $4\frac{5}{8}$ inches, $5\frac{1}{4}$ inches, and $2\frac{1}{8}$ inches. Find the ratio of the longest side to the shortest side.

EXAMPLE 25 A piece of string is cut into two pieces. One piece measures $2\frac{1}{4}$ yards and the other piece measures $4\frac{4}{5}$ yards. What is the ratio of the longer piece to the shorter piece?

SOLUTION

$$\begin{aligned}\text{Ratio of longer piece} \atop \text{to shorter piece} &= \frac{\text{length of longer piece}}{\text{length of shorter piece}} \\ &= \frac{4\frac{4}{5} \text{ yards}}{2\frac{1}{4} \text{ yards}} \quad \text{The yard units cancel and a number without units remains.} \\ &= 4\frac{4}{5} \div 2\frac{1}{4} \\ &= \frac{24}{5} \div \frac{9}{4} \\ &= \frac{24}{5} \times \frac{4}{9} \\ &= \frac{\cancel{3} \times 8}{5} \times \frac{4}{\cancel{3} \times 3} \\ &= \frac{32}{15}\end{aligned}$$

The ratio of the longer piece to the shorter piece is $\frac{32}{15}$. ∎

 Try Problems 40 through 42.

Now we summarize the material in this section by giving guidelines that will help you to decide whether to multiply or divide.

Situations Requiring Multiplication

× 1. You are looking for a total quantity.

$$\text{Total quantity} = \text{size of each part} \times \text{number of equal parts}$$

2. You are looking for a total, as in the following examples.

$$\text{Total miles} = \text{miles per hour} \times \text{number of hours}$$

$$\text{Total cost} = \text{cost per item} \times \text{number of items}$$

3. You are looking for a number that is a certain amount times as large as another number.
4. You are looking for the area of a rectangle.

$$\text{Area} = \text{length} \times \text{width}$$

5. You are looking for a fraction of a number.

4.5 Applications

> **Situations Requiring Division**
>
> ÷ 1. A total quantity is being separated into a number of equal parts.
>
> $$\text{Size of each part} = \text{total quantity} \div \text{number of equal parts}$$
>
> $$= \frac{\text{total quantity}}{\text{number of equal parts}}$$
>
> $$\text{Number of equal parts} = \text{total quantity} \div \text{size of each part}$$
>
> $$= \frac{\text{total quantity}}{\text{size of each part}}$$
>
> 2. You are looking for a missing multiplier in a multiplication statement.
>
> $$\text{Missing multiplier} = \text{answer to the multiplication problem} \div \text{given multiplier}$$
>
> 3. You are looking for an average.
>
> $$\text{Average cost per item} = \text{total cost} \div \text{number of items}$$
>
> $$\text{Average miles per hour} = \text{total miles} \div \text{number of hours}$$
>
> 4. You are looking for what ratio one number is to another number.

▲ Answers to Try These Problems

1. $\frac{2}{3}$ pizza   2. $\frac{3}{2}$ or $1\frac{1}{2}$ pizzas   3. 25 doses   4. $\frac{21}{20}$ or $1\frac{1}{20}$ m
5. $\frac{8}{25}$ lb   6. 725 oz   7. 77¢   8. $609   9. $\frac{27}{4}$ or $6\frac{3}{4}$ min
10. $6   11. $\frac{7}{320}$ lb   12. $\frac{44}{3}$ or $14\frac{2}{3}$ oz   13. 4 mi per hr
14. 25 ft per sec   15. $\frac{1}{50}$ lb per nail   16. $\frac{16}{5}$ or $3\frac{1}{5}$ gal per min
17. $\frac{360}{7}$ or $51\frac{3}{7}$ mi per hr   18. 66 words per min   19. $\frac{2}{3}$ lb
20. $\frac{3}{4}$ lb   21. 12 da   22. $\frac{20}{3}$ or $6\frac{2}{3}$ laps   23. $3300   24. $\frac{4}{45}$ mi
25. 6 lb   26. $\frac{43}{4}$ or $10\frac{3}{4}$ yd   27. $8 per lb   28. 54 min
29. $\frac{323}{9}$ or $35\frac{8}{9}$ sq yd   30. 6 ft   31. $\frac{50}{3}$ or $16\frac{2}{3}$ ft   32. 216 pages
33. $315   34. $\frac{33}{4}$ or $8\frac{1}{4}$ g   35. $\frac{7}{30}$ 36. $4800   37. $52,500
38. $5\frac{1}{3}$ ft   39. 1350 sq ft   40. $\frac{2}{17}$ 41. $\frac{10}{21}$ 42. $\frac{42}{17}$

EXERCISES 4.5

Solve.

1. Eight persons share equally 12 gallons of water. How much water does each person receive?
2. Twelve persons share equally 8 gallons of water. How much water does each person receive?
3. A loaf of bread 35 centimeters long is to be cut into slices that are $1\frac{1}{4}$ centimeters wide. How many slices can be cut?

4. A truck holds $\frac{7}{8}$ ton of sand. The sand is to be unloaded into barrels that hold $\frac{1}{120}$ ton each. How many barrels can be filled?
5. Three-fourths pound of butter is to be divided into 6 equal parts. How much butter is in each part?
6. Ms. Gamez wants to cut a piece of ribbon into 25 equal parts. If the ribbon is $33\frac{1}{3}$ inches long, how long will each part be?
7. For her party, Ms. Crawford wants 150 servings of punch that each contain $6\frac{1}{2}$ ounces. How much punch does she need?
8. Tony is making drapes for his new house. Each pleat uses $2\frac{3}{4}$ inches of material. How wide must the material be to make 16 pleats?
9. One ounce of gold is worth $648. What is $2\frac{1}{3}$ ounces of gold worth?
10. One full bottle contains $6\frac{3}{4}$ liters. You empty out $\frac{2}{3}$ of the bottle. How many liters did you pour out?
11. Asher can jog $3\frac{1}{2}$ miles in 25 minutes. How long will it take him to jog 1 mile?
12. Fifty-two stones weigh $228\frac{4}{5}$ pounds. How much does each stone weigh?
13. Four-fifths of a can holds $5\frac{3}{8}$ gallon. How much does a full can hold?
14. Ninety thumbtacks weigh 1 pound. How many thumbtacks weigh $\frac{2}{3}$ pound?

Convert each ratio to a rate. Specify both the quantity and the units.

15. $\dfrac{6 \text{ feet}}{1 \text{ second}}$
16. $\dfrac{318 \text{ dollars}}{6 \text{ dresses}}$
17. $\dfrac{12 \text{ gallons}}{50 \text{ persons}}$
18. $\dfrac{6\frac{3}{4} \text{ yards}}{15 \text{ minutes}}$

Solve.

19. In a rural county of Montana, 2000 persons live in a 150-square mile region. How many persons per square mile is this?
20. A faucet leaks $1\frac{1}{2}$ ounces every 5 minutes. How many ounces does the faucet leak per minute?
21. It takes Virna $3\frac{1}{2}$ minutes to type 245 words using her computer. How many words per minute does she type?
22. A chef makes 15 dozen doughnuts using $12\frac{1}{2}$ cups of flour. How many cups of flour does he need to make 1 dozen doughnuts?
23. A watch loses $3\frac{1}{2}$ minutes every day. How many minutes has the watch lost in 12 days?
24. A utility stock is selling for $12\frac{5}{8}$ per share. What is the cost of 800 shares of this stock?
25. Onions cost 24¢ per pound. How many pounds of onions can you buy for 78¢?
26. Joan walks $4\frac{1}{2}$ miles per hour. How long will it take her to walk $1\frac{1}{2}$ miles?
27. Three hundred ninety nails weigh 13 pounds. What is the weight of 1 nail?
28. It takes $8\frac{1}{3}$ minutes for 70 gallons of water to enter a reservoir. How many gallons per minute is this?
29. A desktop measures 4 feet by $3\frac{1}{3}$ feet. What is the area of the desktop?
30. The area of a rectangular window is 38 square feet. The width is 6 feet. How long is the window?
31. A rectangular garden contains 275 square feet. Find the width of the garden if the length is $18\frac{1}{3}$ feet.

32. A rectangular floor is $5\frac{1}{4}$ yards wide and $5\frac{1}{3}$ yards long. How many square yards of carpeting are needed to cover this floor?
33. The St. Louis Cardinals won $\frac{3}{4}$ of their baseball games last season. They played a total of 108 games. How many games did they win?
34. Due to a snowstorm, only $\frac{2}{5}$ of the eligible voters in Chicago voted in the election for mayor. A total of 360,000 people voted. How many eligible voters are there?
35. Howard saves $150 out of his monthly income of $500. What fractional part of his income does he save?
36. An oil tank off the coast of Texas contained 450 tons of oil. During a recent spill, $\frac{1}{30}$ of the oil was lost. How much oil spilled?
37. Asher earned $2400 interest on an investment of $12,000 last year. The interest earned is what fractional part of the investment?
38. A solution that contains acid and water has a total volume of $50\frac{1}{4}$ liters. Three-twentieths of the solution is acid. What is the volume of acid?
39. Brenda works on an assembly line at an auto plant in Detroit. She makes $12 per hour. When she works overtime, they pay her time-and-a-half. This means she is paid $1\frac{1}{2}$ times as much as her regular salary. What does Brenda make each hour when working overtime?
40. A recipe calls for $\frac{1}{8}$ teaspoon of pepper. You wish to double the recipe. How much pepper should you use?
41. Due to inflation, the price of gasoline is $2\frac{1}{2}$ times what it was three years ago. The price three years ago was 16 cents per liter. What is the present price of gasoline?
42. The width of a rectangle is $\frac{2}{3}$ as long as the length. If the length is $37\frac{1}{2}$ feet, what is the perimeter of the rectangle?
43. A shipment of fruit contains 750 oranges and twice as many nectarines. What is the ratio of oranges to the total shipment?
44. A rectangle is $4\frac{1}{2}$ meters wide and $5\frac{1}{4}$ meters long. Find the ratio of the length to the width.

A college in California has 36,000 students. The circle graph indicates the breakdown of the students by ethnic background. Use the graph to answer Exercises 45 and 46.

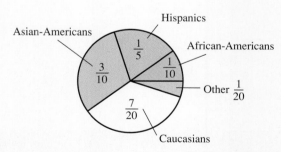

Breakdown of Students by Ethnic Background

45. How many Asian-Americans attend this college?
46. How many Hispanics attend this college?

CHAPTER 4 SUMMARY

KEY WORDS AND PHRASES

reciprocal [4.2]
factor [4.3]
multiplier [4.3]
product [4.3]
twice [4.4]
half [4.4]
double [4.4]
divided by [4.4]
divided into [4.4]
ratio [4.4]
rate [4.5]
area [4.5]

IMPORTANT RULES

How to Multiply Two Fractions [4.1]
- Convert the mixed numerals to improper fractions and convert the whole numerals to fractions.
- Cancel as much as possible. Each time you cancel a factor from a numerator, you must cancel the same factor from a denominator. (Canceling common factors from the numerator and denominator is equivalent to dividing the numerator and denominator by the same nonzero number.)
- Multiply the numerators to obtain the numerator of the answer. Multiply the denominators to obtain the denominator of the answer.
- Check to make sure the answer is simplified. If the answer is an improper fraction, you may leave it as an improper fraction or you may convert it to a mixed numeral.

How to Divide Fractions [4.2]
- Convert the mixed numerals to improper fractions and convert the whole numerals to fractions.
- Change division ($\div$) to multiplication ($\times$) and invert the divisor. The divisor is the second number.
- Cancel as much as possible, then multiply the fractions.
- Check to make sure the answer is simplified. If the answer is an improper fraction, you may leave it as an improper fraction or you may convert it to a mixed numeral.

How to Find the Missing Number in a Multiplication Statement [4.3]
- If you are missing the answer to the multiplication problem (the product), then multiply.
- If you are missing one of the multipliers, then divide the answer (or product) by the known multiplier (or factor) to find the missing multiplier.

How to Find a Fraction of a Number [4.4, 4.5]
To find a fraction of a number, multiply the fraction by the number; that is, the word *of* used in this way translates to multiplication.

Situations that Require Multiplication [4.5]
A summary of these situations appears on page 166.

Situations that Require Division [4.5]
A summary of these situations appears on page 167.

CHAPTER 4 REVIEW EXERCISES

Multiply and simplify.

1. $\frac{1}{4} \times 72$
2. $\frac{1}{3} \times 61$
3. $\frac{2}{5} \times 45$
4. $\frac{3}{8} \times 20$
5. $\frac{1}{6} \times \frac{4}{5}$
6. $\frac{3}{14} \times \frac{5}{3}$
7. $\frac{2}{7} \times 3\frac{1}{4}$
8. $5\frac{2}{3} \times \frac{9}{17}$
9. $2\frac{4}{5} \times 1\frac{7}{8}$
10. $3\frac{3}{5} \times 2\frac{7}{9}$
11. $\frac{36}{20} \times \frac{15}{10}$
12. $\frac{12}{14} \times \frac{7}{21}$

13. $\frac{45}{200} \times 80$ **14.** $12\frac{1}{2} \times \frac{3}{325}$ **15.** $66\frac{2}{3} \times 3120$ **16.** $8\frac{7}{8} \times 4\frac{2}{3}$

17. $\frac{3}{4} \times \frac{11}{12} \times 8$ **18.** $6\frac{1}{2} \times \frac{4}{5} \times \frac{3}{26} \times 15$

19. Don bought $5\frac{3}{4}$ yards of fabric that sells for $12 a yard. How much did he pay for the fabric?

20. Lisa jogged around the lake 5 times. Once around the lake is $\frac{5}{8}$ mile. How far did Lisa jog?

Divide and simplify.

21. $5 \div 9$ **22.** $9 \div 5$ **23.** $\frac{4}{5} \div 3$ **24.** $8\frac{2}{5} \div 5$

25. $13\frac{1}{3} \div 60$ **26.** $\frac{3}{10} \div \frac{3}{4}$ **27.** $\frac{11}{12} \div \frac{1}{8}$ **28.** $\frac{35}{24} \div \frac{42}{30}$

29. $28 \div \frac{4}{7}$ **30.** $3600 \div \frac{90}{12}$ **31.** $15\frac{3}{4} \div \frac{7}{16}$ **32.** $6\frac{1}{4} \div \frac{7}{20}$

33. $7\frac{1}{2} \div 3\frac{1}{3}$ **34.** $\dfrac{\frac{9}{20}}{8}$ **35.** $\dfrac{250}{\frac{4}{5}}$ **36.** $\dfrac{33\frac{1}{3}}{1\frac{1}{9}}$

37. A piece of wire 50 feet long is cut into 75 equal pieces. How long is each piece?

38. A total of $20\frac{1}{4}$ milliliters of medicine is given out in doses of $2\frac{1}{4}$ milliliters each. How many doses can be given?

Find the missing number.

39. $\frac{2}{3} \times 615 = \boxed{}$ **40.** $\frac{1}{4} \times \boxed{} = 8$

41. $\boxed{} \times \frac{4}{5} = \frac{3}{10}$ **42.** $3\frac{1}{2} = \boxed{} \times \frac{3}{4}$

43. $\boxed{} = \frac{3}{7} \times 5\frac{1}{4}$ **44.** $660 = \frac{3}{2} \times \boxed{}$

Solve. Reduce answers to lowest terms.

45. Find $\frac{6}{10}$ of 250.

46. $\frac{9}{4}$ of what number equals 720?

47. What fraction of 25 is $6\frac{1}{4}$?

48. Find $3\frac{1}{2}$ divided by 14.

49. Find half of $\frac{7}{8}$.

50. Find a number that is $4\frac{2}{3}$ times as large as $2\frac{1}{7}$.

51. Find $\frac{4}{5}$ divided into 50.

52. Find the ratio of $\frac{5}{6}$ to $12\frac{1}{2}$.

53. A company conducted a survey indicating that $\frac{3}{20}$ of the workers have young children to care for. There are 240 workers. How many have young children?

54. Barbara paid $\frac{2}{15}$ of her income in federal taxes. She paid $4000 in taxes. What was her income?

55. Cynthia bought a steak for herself that weighed $\frac{7}{8}$ pound. The steak that she bought for her friend John weighed twice as much. How much did John's steak weigh?

56. Debbie bought $2\frac{1}{2}$ pounds of nails for 60 cents. What is the cost of 1 pound of nails?

57. A carpet is $6\frac{3}{4}$ yards long and 5 yards wide. What is the area of the carpet?

58. Five persons share equally $7\frac{1}{2}$ gallons of water. How much water does each person receive?

59. The owner of a small bakery used a total of $72\frac{1}{2}$ pounds of sugar in 7 days. On the average, how many pounds were used each day?

60. The height of a rectangular window is half its width. If the width of the window is $6\frac{1}{2}$ feet, find the area of the window.

61. A car averages 50 miles per hour. How long will it take it to go 175 miles?
62. One yard of fabric costs $28. What is the cost of $6\frac{3}{4}$ yards of this fabric?
63. A tabletop contains $16\frac{5}{8}$ square feet. If the width is $3\frac{1}{2}$ feet, find the length.
64. A piece of string $26\frac{1}{4}$ inches long is cut into two pieces. One piece measures 21 inches and the other piece measures $5\frac{1}{4}$ inches. What is the ratio of the shorter piece to the longer piece?

In Your Own Words

Write complete sentences to discuss each of the following. Support your comments with examples or pictures, if appropriate.

65. Without actually solving the problem, discuss the procedure you would use to find the missing number in the statement $\frac{4}{5} \times 7\frac{1}{2} = \boxed{}$. Assume that you are working without a calculator.
66. Without actually solving the problem, discuss the procedure you would use to find the missing number in the statement $5\frac{1}{4} = \boxed{} \times 14$. Assume that you are working without a calculator.
67. Give at least three examples where the answer to a multiplication problem is more than either of the multipliers (or factors), then give at least three examples where the answer to a multiplication problem is less than one of the multipliers (or factors). Discuss what you observe.
68. Discuss a real-life situation where a total quantity is being split into a number of equal parts. Choose a situation where the number of equal parts is more than the total quantity.
69. Discuss a real-life situation that would involve finding a fraction of a number.

CHAPTER 4 PRACTICE TEST

Multiply and simplify.

1. $\frac{2}{7} \times \frac{3}{5}$
2. $\frac{12}{10} \times \frac{15}{9}$
3. $5\frac{2}{3} \times 900$
4. $12\frac{1}{2} \times 4\frac{3}{5}$
5. $\frac{7}{10} \times \frac{15}{12} \times \frac{2}{21}$

Divide and simplify.

6. $8 \div 20$
7. $\frac{5}{9} \div 60$
8. $\frac{16}{30} \div \frac{6}{5}$
9. $18 \div 8\frac{4}{5}$
10. $\dfrac{7\frac{6}{7}}{3\frac{2}{3}}$

Find the missing number. Simplify.

11. $2\frac{1}{3} \times \boxed{} = 21$
12. $\boxed{} = \frac{3}{20} \times 61{,}500$

Solve.

13. Find 4 divided by $3\frac{1}{2}$.
14. Find a number that is $4\frac{1}{2}$ times as large as 6.
15. Four-fifths of what number is 28?
16. What number is twice $\frac{7}{12}$?
17. Bogard is a chef in a small restaurant. He prepares rice using a recipe that yields 25 cups of cooked rice. How many $\frac{2}{3}$-cup servings will he have?
18. One yard of fabric costs $24. What is the cost of $3\frac{1}{3}$ yards?
19. Earl can run 3 miles in 22 minutes. How long will it take him to run 1 mile?

20. Bette spends $\frac{7}{100}$ of her monthly income on food. Her monthly income is $1200. How much does she spend on food?
21. A wire is $23\frac{1}{3}$ feet long. Bill wants to cut the wire into 35 equal pieces. How long will each piece be?
22. The area of a rectangle is 93 square yards. If the length is $20\frac{2}{3}$ yards, find the width.
23. A solution contains $3\frac{1}{3}$ liters of water and $2\frac{2}{3}$ liters of acid. What is the ratio of acid to water?
24. A faucet leaks $2\frac{1}{3}$ ounces per hour. How long will it take the faucet to leak $8\frac{2}{5}$ ounces?

CHAPTER

Fractions: Addition and Subtraction

5.1 ADDING WITH LIKE DENOMINATORS

OBJECTIVES
- Add fractions with like denominators.
- Add mixed numerals with like denominators.
- Solve an application using addition of fractions.

1 Adding Fractions

If you jog $\frac{4}{8}$ mile and then walk $\frac{3}{8}$ mile, how far have you gone? To solve this problem you need to add $\frac{4}{8}$ and $\frac{3}{8}$ because to add means to total or combine.

$$\frac{4}{8} + \frac{3}{8} = \frac{7}{8}$$

— Add the numerators to obtain the numerator of the answer. $4 + 3 = 7$
— Keep the same denominator.

You have gone a total distance of $\frac{7}{8}$ mile.

Observe that to add fractions with like denominators, add the numerators and keep the same denominator. Here we view a picture of the addition problem $\frac{4}{8} + \frac{3}{8}$; we can see that the result $\frac{7}{8}$ is correct. Assume each separate figure represents 1 whole.

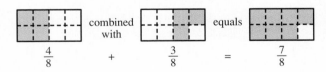

Here are more examples of adding fractions with like denominators.

5.1 Adding with Like Denominators

 Try These Problems

Add and simplify.
1. $\frac{3}{10} + \frac{4}{10}$
2. $\frac{2}{3} + \frac{2}{3}$
3. $\frac{1}{8} + \frac{5}{8}$
4. $\frac{43}{20} + \frac{29}{20}$
5. $\frac{7}{12} + \frac{3}{12} + \frac{2}{12}$
6. $\frac{76}{90} + \frac{29}{90} + \frac{87}{90}$

EXAMPLE 1 Add: $\frac{2}{6} + \frac{5}{6}$

SOLUTION

$\frac{2}{6} + \frac{5}{6} = \frac{7}{6}$ —— Add the numerators. $2 + 5 = 7$.
—— Use the *like* denominator.

or $1\frac{1}{6}$ An answer larger than one whole may be left as an improper fraction or a mixed numeral. ■

EXAMPLE 2 Add: $\frac{7}{12} + \frac{1}{12}$

SOLUTION

$$\frac{7}{12} + \frac{1}{12} = \frac{8}{12}$$

$$= \frac{\cancel{4} \times 2}{\cancel{4} \times 3} \quad \text{Always simplify the answer.}$$

$$= \frac{2}{3} \quad ■$$

EXAMPLE 3 Add: $\frac{25}{27} + \frac{5}{27} + \frac{33}{27}$

SOLUTION

$$\frac{25}{27} + \frac{5}{27} + \frac{33}{27} = \frac{63}{27}$$

$$= \frac{\cancel{9} \times 7}{\cancel{9} \times 3} \quad \text{Simplify the answer.}$$

$$= \frac{7}{3} \quad \text{or} \quad 2\frac{1}{3} \quad \text{The answer may be left as an improper fraction or as a mixed numeral.} \ ■$$

 Try Problems 1 through 6.

2 Adding Mixed Numerals

Now we look at examples of adding mixed numerals.

EXAMPLE 4 Add: $1\frac{3}{5} + 2\frac{1}{5}$

SOLUTION
METHOD 1

$$\begin{aligned} & 1\frac{3}{5} \\ + & 2\frac{1}{5} \\ \hline & 3\frac{4}{5} \end{aligned}$$

It can be convenient to align the numbers vertically.

Add the fractions. $\frac{3}{5} + \frac{1}{5} = \frac{4}{5}$

Add the wholes. $1 + 2 = 3$

Try These Problems

Add and simplify.

7. $5\frac{2}{8} + 2\frac{3}{8}$
8. $17\frac{1}{4} + \frac{3}{4}$
9. $9\frac{7}{12} + 2\frac{3}{12}$
10. $28\frac{25}{36} + 6\frac{20}{36}$
11. $5\frac{13}{40} + 2\frac{28}{40} + 17\frac{39}{40}$
12. $13\frac{22}{24} + 9\frac{41}{24} + 49\frac{25}{24}$

METHOD 2

$$1\frac{3}{5} + 2\frac{1}{5}$$
$$= \frac{8}{5} + \frac{11}{5}$$ Convert each mixed numeral to an improper fraction, then add.
$$= \frac{19}{5} \text{ or } 3\frac{4}{5} \blacksquare$$

From the above example, observe that it is more efficient to add mixed numerals *without* changing them to improper fractions. Method 1 shows the more efficient method. This method is especially easier with problems like $232\frac{2}{8} + 95\frac{1}{8}$, where there are larger numbers.

EXAMPLE 5 Add: $8\frac{10}{11} + \frac{5}{11}$

SOLUTION

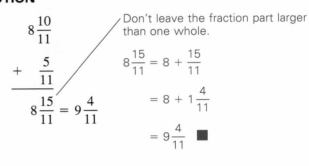

Don't leave the fraction part larger than one whole.

$8\frac{15}{11} = 8 + \frac{15}{11}$

$= 8 + 1\frac{4}{11}$

$= 9\frac{4}{11}$ $\blacksquare$

EXAMPLE 6 Add: $4\frac{16}{21} + 9\frac{12}{21}$

SOLUTION

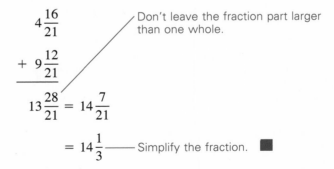

Don't leave the fraction part larger than one whole.

$= 14\frac{1}{3}$ ——— Simplify the fraction.

▲ **Try Problems 7 through 12.**

Here we give a rule for adding fractions with like denominators.

Adding Fractions with Like Denominators

1. Add the numerators and put the answer over the like denominator. (Always reduce the answer to lowest terms by dividing numerator and denominator by the same nonzero number if possible.)
2. To add mixed numbers, add the whole numbers separately from the fractions. (Leave your answer so that the fractional portion is smaller than one and the fraction is reduced to lowest terms.)

Answers to Try These Problems

1. $\frac{7}{10}$ 2. $\frac{4}{3}$ or $1\frac{1}{3}$ 3. $\frac{3}{4}$ 4. $\frac{18}{5}$ or $3\frac{3}{5}$ 5. 1
6. $\frac{32}{15}$ or $2\frac{2}{15}$ 7. $7\frac{5}{8}$ 8. 18 9. $11\frac{5}{6}$ 10. $35\frac{1}{4}$
11. 26 12. $74\frac{2}{3}$

EXERCISES 5.1

Add and simplify.

1. $\frac{2}{8} + \frac{3}{8}$
2. $\frac{4}{15} + \frac{4}{15}$
3. $\frac{4}{9} + \frac{2}{9}$
4. $\frac{12}{25} + \frac{3}{25}$
5. $\frac{9}{20} + \frac{11}{20}$
6. $\frac{8}{7} + \frac{6}{7}$
7. $\frac{11}{18} + \frac{12}{18} + \frac{2}{18}$
8. $\frac{13}{16} + \frac{14}{16} + \frac{10}{16}$
9. $\frac{2}{3} + 3 + \frac{5}{3}$
10. $5 + \frac{1}{6} + \frac{3}{6}$
11. $4\frac{3}{4} + 5$
12. $6 + 14\frac{2}{5}$
13. $\frac{5}{8} + 2\frac{3}{8}$
14. $\frac{7}{10} + 3\frac{1}{10}$
15. $5\frac{5}{24} + 3\frac{15}{24}$
16. $37\frac{3}{7} + 3\frac{4}{7}$
17. $28\frac{17}{32} + 5\frac{20}{32}$
18. $2\frac{7}{9} + 5\frac{5}{9}$
19. $15\frac{27}{40} + 6\frac{29}{40}$
20. $4\frac{19}{21} + 78\frac{16}{21}$
21. $6\frac{3}{7} + 7\frac{6}{7} + 19\frac{5}{7}$
22. $8\frac{11}{12} + 4\frac{8}{12} + 5\frac{9}{12}$
23. $76\frac{2}{39} + 81\frac{24}{39}$
24. $135\frac{27}{42} + 256\frac{22}{42}$

Solve.

25. You plan to go on an overnight hiking trip and want to keep your backpack as light as possible. You have packed items weighing $6\frac{3}{4}$ pounds and $2\frac{1}{4}$ pounds. What is the total weight so far?

26. Don had jogged $2\frac{3}{4}$ miles when he ran into Kimberly, a good friend of his. They talked for a while then jogged together for $4\frac{3}{4}$ miles. How far did Don jog?

27. A carpenter cut a piece of wood $27\frac{3}{8}$ inches long. He realized the piece was too short and he needed it to be $\frac{7}{8}$ inch longer. How long should he cut the piece of wood?

28. Tom is reading a book. He read $\frac{1}{8}$ of it on Monday and $\frac{5}{8}$ of it on Tuesday. What fraction of the book did he read in the two days?

5.2 FINDING THE LEAST COMMON MULTIPLE BY PRIME FACTORING

OBJECTIVES
- Determine whether a number is prime or not prime.
- Write a number as the product of primes.
- Find the least common multiple of two or more numbers by prime factoring.

1 Meaning of the Least Common Multiple

You exercised $\frac{1}{4}$ hour on Wednesday and $\frac{5}{6}$ hour on Thursday. How long did you exercise in the two-day period? To solve this problem you need to add $\frac{1}{4}$ and $\frac{5}{6}$.

$$\frac{1}{4} + \frac{5}{6} = ?$$

In Section 5.1 we learned that it is easy to add fractions with *like* denominators. We need to rewrite these fractions so that they have the same denominator, but we do not want to change the value of the fractions. The number we use for a common denominator must be a multiple of both 4 and 6. The multiples of 4 are

4; 8; **12**; 16; 20; **24**; 28; ...

The multiples of 6 are

6; **12**; 18; **24**; 30; 36; 42; ...

The common multiples of both 4 and 6 are

12; 24; 36; 48; ...

Finally, the **least common multiple (LCM)** of 4 and 6 is 12, and this is the denominator that can be used to add $\frac{1}{4}$ and $\frac{5}{6}$. We also call 12 the **least common denominator (LCD)**.

$$\frac{1}{4} + \frac{5}{6}$$

$$= \frac{1 \times 3}{4 \times 3} + \frac{5 \times 2}{6 \times 2}$$ — What number multiplied by 4 is 12? 3
Multiply the numerator and denominator of the fraction $\frac{1}{4}$ by 3 to obtain $\frac{3}{12}$, a fraction equal to $\frac{1}{4}$.
— What number multiplied by 6 is 12? 2
Multiply the numerator and denominator of the fraction $\frac{5}{6}$ by 2 to obtain $\frac{10}{12}$, a fraction equal to $\frac{5}{6}$.

$$= \frac{3}{12} + \frac{10}{12}$$ —— Now the fractions have a common denominator, so we can see how to add them.

$$= \frac{13}{12} \quad \text{or} \quad 1\frac{1}{12}$$ Add the numerators, $3 + 10 = 13$, to obtain the numerator of the answer. Keep the same denominator, 12.

Therefore, if you exercise $\frac{1}{4}$ hour and $\frac{5}{6}$ hour, you have exercised a total of $\frac{13}{12}$ or $1\frac{1}{12}$ hours.

Try These Problems

Write each as the product of prime numbers.

1. 50
2. 54
3. 49
4. 210
5. 195
6. 297

In general, to add fractions with *unlike* denominators, you need to begin by finding the least common multiple of the denominators. In the previous discussion we found the least common multiple of 4 and 6 by listing multiples of each and observing that 12 is the smallest of the common multiples. There are many techniques for finding the least common multiple of two or more numbers. In this section we discuss finding the least common multiple by prime factoring.

2 Writing Numbers as the Product of Primes

Recall from Section 2.3 that a **prime number** is a whole number larger than 1 that is divisible only by itself and 1. The first ten primes are listed here.

First Ten Primes									
2	3	5	7	11	13	17	19	23	29

Also, recall from Section 2.3 that whole numbers larger than 1 that are *not* prime can be written as the product of primes. For example,

$$36 = 6 \times 6 \qquad 63 = 9 \times 7 \qquad 275 = 5 \times 55$$
$$= 2 \times 3 \times 2 \times 3 \qquad = 3 \times 3 \times 7 \qquad = 5 \times 5 \times 11$$

The process of writing numbers as the product of numbers is called **factoring.** If all of the multipliers (or factors) are prime numbers, then the process is called **prime factoring.** If you need more help with writing numbers as the product of primes, refer to Section 2.3.

Try Problems 1 through 6.

3 Finding the Least Common Multiple of Two or More Numbers

Now we illustrate how to use prime factoring to find the least common multiple (LCM) of two or more numbers.

EXAMPLE 1 Find the least common multiple of 18, 12, and 40.

SOLUTION Write each number as the product of primes.

$$18 = 6 \times 3 \qquad 12 = 4 \times 3 \qquad 40 = 4 \times 10$$
$$= 2 \times 3 \times 3 \qquad = 2 \times 2 \times 3 \qquad = 2 \times 2 \times 2 \times 5$$

Note that the only prime factors appearing in these numbers are 2, 3, and 5.

180 Chapter 5 Fractions: Addition and Subtraction

Here we explain how to decide how many 2s, 3s, and 5s to multiply to obtain the LCM.

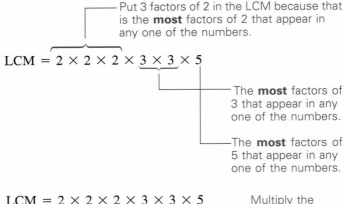

$$\text{LCM} = 2 \times 2 \times 2 \times 3 \times 3 \times 5$$

$$\text{LCM} = 2 \times 2 \times 2 \times 3 \times 3 \times 5$$
$$= 4 \times 6 \times 15$$
$$= 4 \times 90$$
$$= 360$$

Multiply the factors to obtain the LCM. The multiplication can be done in any order. ∎

In the previous example we found 360 to be the least common multiple (LCM) of 18, 12, and 40. This means that 360 is the smallest number that is divisible by 18, 12, and 40. Here we perform each of the divisions in two ways. Observe that there is an alternative to using long division; the division can be done by taking advantage of the prime factoring that has already been done.

Dividing 360 by 18.

$$18\overline{)360} \rightarrow \frac{360}{18} = \frac{2 \times 2 \times \cancel{2} \times \cancel{3} \times \cancel{3} \times 5}{\cancel{2} \times \cancel{3} \times \cancel{3}}$$
$$= 2 \times 2 \times 5$$
$$= 20$$

Dividing 360 by 18 is equivalent to canceling the factors of 18 from 360.

Dividing 360 by 12.

$$12\overline{)360} \rightarrow \frac{360}{12} = \frac{\cancel{2} \times \cancel{2} \times 2 \times 3 \times \cancel{3} \times 5}{\cancel{2} \times \cancel{2} \times \cancel{3}}$$
$$= 2 \times 3 \times 5$$
$$= 30$$

Dividing 360 by 12 is equivalent to canceling the factors of 12 from 360.

Dividing 360 by 40.

$$40 \overline{)360} \rightarrow \frac{360}{40} = \frac{\cancel{2} \times \cancel{2} \times \cancel{2} \times 3 \times 3 \times \cancel{5}}{\cancel{2} \times \cancel{2} \times \cancel{2} \times \cancel{5}}$$

$$= 3 \times 3$$
$$= 9$$

Dividing 360 by 40 is equivalent to canceling the factors of 40 from 360.

Yes, 360 is divisible by the numbers 18, 12, and 40.

Here are more examples of finding the least common multiple by prime factoring.

EXAMPLE 2 Find the least common multiple of 14 and 49.

SOLUTION Write each number as the product of primes.

$$14 = 2 \times 7 \qquad 49 = 7 \times 7$$

Note that the only prime factors that are needed for the LCM are 2s and 7s.

$$\text{LCM} = 2 \times 7 \times 7$$

— Put only one factor of 2 because that is the most factors of 2 that appear in 14 or 49.

— Put 2 factors of 7 because there are 2 factors of 7 in 49 and that is the most factors of 7 that appear in 14 or 49.

$$\text{LCM} = 2 \times 7 \times 7$$
$$= 2 \times 49$$
$$= 98 \blacksquare$$

Multiply the factors in any order to obtain the LCM.

EXAMPLE 3 Find the least common multiple of 55 and 33.

SOLUTION Write each number as the product of primes.

$$55 = 5 \times 11 \qquad 33 = 3 \times 11$$

Note that the only prime factors needed in the LCM are 3s, 5s, and 11s.

$$\text{LCM} = 5 \times 3 \times 11$$
$$= 15 \times 11$$
$$= 165$$

Put only 1 factor of each because none of the factors appear more than once in 55 or 33. ∎

EXAMPLE 4 Find the least common multiple of 7, 72, and 90.

SOLUTION Write each number as the product of primes.

7 is prime

$$72 = 8 \times 9$$
$$= 2 \times 2 \times 2 \times 3 \times 3$$

$$90 = 9 \times 10$$
$$= 3 \times 3 \times 2 \times 5$$

Try These Problems

Find the least common multiple of each group of numbers.
7. 30; 20
8. 28; 49
9. 10; 9
10. 26; 13
11. 8; 12; 9
12. 20; 9; 7
13. 24; 12; 40
14. 81; 18; 8

Note that the only prime factors needed in the LCM are 2s, 3s, 5s, and 7s.

$$LCM = 2 \times 2 \times 2 \times 3 \times 3 \times 5 \times 7$$

- The most factors of 2 appear in 72. There are 3 of them.
- The most factors of 3 appear in 72 and 90. There are 2 of them.
- The only factor of 5 appears in 90.
- The only factor of 7 appears in 7.

$$\begin{aligned} LCM &= 2 \times 2 \times 2 \times 3 \times 3 \times 5 \times 7 \\ &= 8 \times 9 \times 35 \\ &= 72 \times 35 \\ &= 2520 \end{aligned}$$

Now we state a rule for finding the least common multiple of two or more numbers.

> *Finding the Least Common Multiple of Two or More Numbers.*
> 1. Write each number as the product of primes.
> 2. Each kind of prime factor appearing in step 1 must be a factor in the least common multiple. How many of each prime factor? The greatest number of times the factor appears in any one of the original numbers.

 Try Problems 7 through 14.

 Answers to Try These Problems

1. $2 \times 5 \times 5$ 2. $2 \times 3 \times 3 \times 3$ 3. 7×7
4. $2 \times 3 \times 5 \times 7$ 5. $3 \times 5 \times 13$ 6. $3 \times 3 \times 3 \times 11$
7. 60 8. 196 9. 90 10. 26 11. 72 12. 1260
13. 120 14. 648

EXERCISES 5.2

Choose the prime numbers in each group.

1. 2; 6; 7; 9   2. 3; 9; 11; 15   3. 45; 31; 13; 17   4. 27; 43; 51; 19

Write each number as the product of primes. If the number is prime, say prime.

5. 9   6. 13   7. 12   8. 27   9. 11
10. 24   11. 48   12. 19   13. 64   14. 490
15. 125   16. 34   17. 97   18. 84   19. 91
20. 147   21. 245   22. 223

Use prime factoring to find the least common multiple of each group of numbers.

23. 6; 9   **24.** 5; 15   **25.** 30; 24   **26.** 21; 4
27. 28; 12   **28.** 40; 25   **29.** 77; 14   **30.** 39; 52
31. 4; 6; 9   **32.** 10; 12; 15   **33.** 2; 44; 11   **34.** 28; 10; 14
35. 75; 9; 15   **36.** 10; 25; 125   **37.** 65; 25   **38.** 51; 34
39. 52; 68   **40.** 69; 460

5.3 ADDING WITH UNLIKE DENOMINATORS

OBJECTIVES
- Add fractions with unlike denominators.
- Add mixed numerals with unlike denominators.
- Solve an application that involves adding fractions.

1 Working with Easy Denominators

If you complete $\frac{1}{4}$ of a job and your co-worker completes $\frac{3}{10}$ of the job, what fraction of the job have the two of you completed? To solve this problem, we must add $\frac{1}{4}$ and $\frac{3}{10}$.

$$\frac{1}{4} + \frac{3}{10} = ?$$

We must rewrite the fractions so that they have the same denominator. This common denominator is the least common multiple of 4 and 10, which is 20.

$$\frac{1}{4} + \frac{3}{10}$$

$$= \frac{1 \times 5}{4 \times 5} + \frac{3 \times 2}{10 \times 2}$$

Multiply the numerator and denominator of the fraction $\frac{1}{4}$ by 5 to obtain $\frac{5}{20}$, a fraction equal to $\frac{1}{4}$.

Multiply the numerator and denominator of the fraction $\frac{3}{10}$ by 2 to obtain $\frac{6}{20}$, a fraction equal to $\frac{3}{10}$.

$$= \frac{5}{20} + \frac{6}{20}$$

$$= \frac{11}{20}$$

After the fractions have a common denominator, add the numerators and place this over the common denominator.

The two of you completed $\frac{11}{20}$ of the job.

▼ Try These Problems

Add and simplify.

1. $\frac{5}{6} + \frac{3}{8}$
2. $\frac{7}{8} + \frac{3}{40}$
3. $\frac{5}{9} + \frac{11}{6} + \frac{2}{3}$
4. $\frac{1}{2} + \frac{2}{15} + \frac{7}{10}$

Here we look at more examples of adding fractions with unlike denominators.

EXAMPLE 1 Add: $\frac{1}{5} + \frac{7}{15}$

SOLUTION The least common multiple of 5 and 15 is 15, so raise the fraction $\frac{1}{5}$ to higher terms so that it has a denominator of 15.

$$\frac{1}{5} + \frac{7}{15}$$

The fraction $\frac{7}{15}$ already has the denominator 15 so leave it as it is.

Multiply the numerator and denominator of $\frac{1}{5}$ by 3 to obtain $\frac{3}{15}$, a fraction equal to $\frac{1}{5}$.

$$= \frac{1 \times 3}{5 \times 3} + \frac{7}{15}$$

$$= \frac{3}{15} + \frac{7}{15}$$

$$= \frac{10}{15}$$

$$= \frac{\cancel{5} \times 2}{\cancel{5} \times 3}$$

Reduce the answer to lowest terms.

$$= \frac{2}{3} \ \blacksquare$$

EXAMPLE 2 Add: $\frac{5}{6} + \frac{1}{2} + \frac{3}{4}$

SOLUTION Find the least common multiple of 6, 2, and 4.

$$\left.\begin{array}{l} 6 = 2 \times 3 \\ 2 \text{ is prime} \\ 4 = 2 \times 2 \end{array}\right\} \quad \begin{array}{l} \text{LCM} = 2 \times 2 \times 3 \\ = 12 \end{array}$$

Rewrite each fraction so they all have the denominator 12, but do not change the values of the fractions.

$$\frac{5}{6} + \frac{1}{2} + \frac{3}{4}$$

$$= \frac{5 \times 2}{6 \times 2} + \frac{1 \times 6}{2 \times 6} + \frac{3 \times 3}{4 \times 3}$$

Within each individual fraction, multiply the numerator and denominator by the *same* number so that the value of the fraction does not change. Choose the number to multiply by based on the fact that you want each denominator to be 12.

$$= \frac{10}{12} + \frac{6}{12} + \frac{9}{12}$$

$$= \frac{25}{12} \text{ or } 2\frac{1}{12}$$

You may leave the answer as an improper fraction or as a mixed numeral. ∎

▼ Try Problems 1 through 4.

5.3 Adding with Unlike Denominators

EXAMPLE 3   Add: $3\frac{7}{8} + 4\frac{5}{7}$

SOLUTION   Find the least common multiple of 8 and 7.

$$\left.\begin{array}{l} 8 = 4 \times 2 \\ = 2 \times 2 \times 2 \\ \\ 7 \text{ is prime} \end{array}\right\} \begin{array}{l} \text{LCM} = 2 \times 2 \times 2 \times 7 \\ \phantom{\text{LCM}} = 8 \times 7 \\ \phantom{\text{LCM}} = 56 \end{array}$$

The numbers can be left in mixed numeral form but the fractional parts must be written with denominator 56.

$$3\frac{7}{8} = 3\frac{7 \times 7}{8 \times 7} = 3\frac{49}{56}$$

$$+\ 4\frac{5}{7} = 4\frac{5 \times 8}{7 \times 8} = 4\frac{40}{56}$$

$$7\frac{89}{56} = 8\frac{33}{56}$$

— Do not leave the fractional part more than 1.

EXAMPLE 4   Add: $48\frac{2}{9} + 37\frac{10}{21} + 29\frac{6}{7}$

SOLUTION   Find the least common multiple of 9, 21, and 7.

$$\left.\begin{array}{l} 9 = 3 \times 3 \\ 21 = 3 \times 7 \\ 7 \text{ is prime} \end{array}\right\} \text{LCM} = 3 \times 3 \times 7 = 63$$

Rewrite the mixed numerals so that the fractional parts each have the denominator 63.

$$48\frac{2}{9} = 48\frac{2 \times 7}{9 \times 7} = 48\frac{14}{63}$$

Because $63 = \boxed{3 \times 3} \times 7 = 9 \times 7$, multiply numerator and denominator of $\frac{2}{9}$ by 7.

$$37\frac{10}{21} = 37\frac{10 \times 3}{21 \times 3} = 37\frac{30}{63}$$

Because $63 = 3 \times \boxed{3 \times 7} = 3 \times 21$, multiply numerator and denominator of $\frac{10}{21}$ by 3.

$$+\ 29\frac{6}{7} = 29\frac{6 \times 9}{7 \times 9} = 29\frac{54}{63}$$

Because $63 = 3 \times 3 \times \boxed{7} = 9 \times 7$, multiply numerator and denominator of $\frac{6}{7}$ by 9.

$$114\frac{98}{63}$$

— Do not leave the fractional part more than 1.

$$= 115\frac{35}{63}$$

$$= 115\frac{5 \times \cancel{7}}{\cancel{7} \times 9}$$

Reduce the fractional part to lowest terms.

$$= 115\frac{5}{9}$$

Try These Problems

Add and simplify.

5. $8\frac{6}{7} + 13\frac{5}{6}$
6. $3\frac{12}{35} + 9\frac{5}{14}$
7. $25\frac{11}{14} + 38\frac{8}{21} + 52\frac{1}{6}$
8. $5\frac{9}{28} + 3\frac{5}{12} + \frac{1}{21}$

Try Problems 5 through 8.

2 Working with More Difficult Denominators

The next few examples involve more difficult denominators.

EXAMPLE 5   Add: $\frac{7}{25} + \frac{11}{30}$

SOLUTION   Find the LCM of 25 and 30.

$$\left.\begin{array}{l} 25 = 5 \times 5 \\ \\ 30 = 3 \times 10 \\ = 3 \times 2 \times 5 \end{array}\right\} \begin{array}{l} \text{LCM} = 2 \times 3 \times 5 \times 5 \\ \phantom{\text{LCM}} = 6 \times 25 \\ \phantom{\text{LCM}} = 150 \end{array}$$

Rewrite each fraction so that they both have a denominator of 150, but do not change the values of the fractions.

$$\frac{7}{25} + \frac{11}{30}$$

$$= \frac{7 \times 6}{25 \times 6} + \frac{11 \times 5}{30 \times 5}$$

Because $150 = 2 \times 3 \times \boxed{5 \times 5}$ $= 6 \times 25$, multiply numerator and denominator of $\frac{7}{25}$ by 6.

Because $150 = \boxed{2 \times 3 \times 5} \times 5$ $= 30 \times 5$, multiply numerator and denominator of $\frac{11}{30}$ by 5.

$$= \frac{42}{150} + \frac{55}{150}$$

$$= \frac{97}{150}$$

To make sure the fraction is reduced to lowest terms, it is enough to check only the *prime* divisors. We know 2, 3, and 5 are the prime divisors of 150. None of these divide evenly into 97. So the fraction is reduced. ∎

EXAMPLE 6   Add: $\frac{5}{21} + \frac{11}{18} + \frac{2}{63}$

SOLUTION   Find the least common multiple of 21, 18, and 63.

$$\left.\begin{array}{l} 21 = 3 \times 7 \\ 18 = 3 \times 6 \\ = 3 \times 2 \times 3 \\ 63 = 9 \times 7 \\ = 3 \times 3 \times 7 \end{array}\right\} \begin{array}{l} \text{LCM} = 2 \times 3 \times 3 \times 7 \\ \phantom{\text{LCM}} = 6 \times 21 \\ \phantom{\text{LCM}} = 126 \end{array}$$

Convert each fraction to an equal fraction with denominator 126.

5.3 Adding with Unlike Denominators

 **Try These Problems**

Add and simplify.

9. $\frac{11}{32} + \frac{9}{40}$
10. $25\frac{7}{54} + 18\frac{13}{36}$
11. $3\frac{29}{98} + 9\frac{13}{28} + 17\frac{3}{49}$
12. $\frac{23}{50} + \frac{8}{35} + \frac{5}{14}$

$$\frac{5}{21} + \frac{11}{18} + \frac{2}{63}$$

$$= \frac{5 \times 6}{21 \times 6} + \frac{11 \times 7}{18 \times 7} + \frac{2 \times 2}{63 \times 2}$$

Because $126 = 2 \times 3 \times \boxed{3 \times 7} = 6 \times 21$, multiply numerator and denominator by 6.

Because $126 = \boxed{2 \times 3 \times 3} \times 7 = 18 \times 7$, multiply numerator and denominator by 7.

Because $126 = 2 \times \boxed{3 \times 3 \times 7} = 2 \times 63$, multiply numerator and denominator by 2.

$$= \frac{30}{126} + \frac{77}{126} + \frac{4}{126}$$

$$= \frac{111}{126}$$

We already know that the prime divisors of 126 are 2, 3, and 7. Do any of these primes divide evenly into 111? Yes, 3 does because the sum of the digits of 111 is 3. We divide 111 by 3 to find the other factor.

$$= \frac{\cancel{3} \times 37}{\cancel{3} \times 42}$$

$$\begin{array}{r} 37 \\ 3\overline{)111} \\ 9 \\ \overline{21} \\ 21 \end{array} \to 111 = 3 \times 37$$

$$= \frac{37}{42}$$

The fraction is reduced because $42 = 2 \times 3 \times 7$ and none of these prime factors divide evenly into 37. ∎

 Try Problems 9 through 12.

3 Working with Very Difficult Denominators

Now we look at examples of adding fractions where the denominators contain prime factors that are larger than 7.

EXAMPLE 7   Add: $6\frac{16}{39} + 54\frac{5}{9}$

SOLUTION   Find the least common multiple of 39 and 9.

$$\left.\begin{array}{r} 39 = 3 \times 13 \\ 9 = 3 \times 3 \end{array}\right\} \begin{array}{l} \text{LCM} = 3 \times 3 \times 13 \\ \phantom{\text{LCM}} = 9 \times 13 \\ \phantom{\text{LCM}} = 117 \end{array}$$

Try These Problems

Add and simplify.

13. $\frac{17}{22} + \frac{14}{33}$
14. $\frac{5}{68} + \frac{3}{17}$
15. $\frac{7}{26} + \frac{3}{10} + \frac{3}{65}$
16. $2\frac{5}{46} + 5\frac{8}{69} + 8\frac{1}{6}$

Write the fractional parts so that each has a denominator of 117.

$$6\frac{16}{39} = 6\frac{16 \times 3}{39 \times 3} = 6\frac{48}{117}$$
$$+\ 54\frac{5}{9} = 54\frac{5 \times 13}{9 \times 13} = 54\frac{65}{117}$$
$$\phantom{+\ 54\frac{5}{9} = 54\frac{5 \times 13}{9 \times 13} =\ } 60\frac{113}{117}$$

The fraction is reduced to lowest terms because $117 = 3 \times 3 \times 13$ and neither 3 nor 13 divides evenly into 113. ■

EXAMPLE 8   Add: $\frac{13}{55} + \frac{15}{22} + \frac{9}{10}$

SOLUTION   Find the least common multiple of 55, 22, and 10.

$$\left. \begin{array}{l} 55 = 5 \times 11 \\ 22 = 2 \times 11 \\ 10 = 2 \times 5 \end{array} \right\} \begin{array}{l} \text{LCM} = 2 \times 5 \times 11 \\ = 10 \times 11 \\ = 110 \end{array}$$

Convert each fraction to an equal fraction with denominator 110.

$$\frac{13}{55} + \frac{15}{22} + \frac{9}{10}$$

$$= \frac{13 \times 2}{55 \times 2} + \frac{15 \times 5}{22 \times 5} + \frac{9 \times 11}{10 \times 11}$$

Because $110 = 2 \times \boxed{5 \times 11} = 2 \times 55$, multiply numerator and denominator by 2.

Because $110 = \boxed{2} \times 5 \times \boxed{11} = 5 \times 22$, multiply numerator and denominator by 5.

Because $110 = \boxed{2 \times 5} \times 11 = 10 \times 11$, multiply numerator and denominator by 11.

$$= \frac{26}{110} + \frac{75}{110} + \frac{99}{110}$$
$$= \frac{200}{110}$$
$$= \frac{20 \times \cancel{10}}{11 \times \cancel{10}}$$

Cancel the common factor 10 from the numerator and denominator.

$$= \frac{20}{11} \ \text{or} \ 1\frac{9}{11} \ ■$$

 Try Problems 13 through 16.

Now we summarize by stating a rule for adding fractions with unlike denominators.

Adding Fractions with Unlike Denominators

1. Find the least common multiple of the denominators. The number is called the least common denominator.
2. Convert each fraction to an equivalent fraction that has the common denominator.
3. Follow the rules for adding fractions with like denominators.

Answers to Try These Problems

1. $\frac{29}{24}$ or $1\frac{5}{24}$ 2. $\frac{19}{20}$ 3. $\frac{55}{18}$ or $3\frac{1}{18}$ 4. $\frac{4}{3}$ or $1\frac{1}{3}$ 5. $22\frac{29}{42}$
6. $12\frac{7}{10}$ 7. $116\frac{1}{3}$ 8. $8\frac{11}{14}$ 9. $\frac{91}{160}$ 10. $43\frac{53}{108}$ 11. $29\frac{23}{28}$
12. $\frac{183}{175}$ or $1\frac{8}{175}$ 13. $\frac{79}{66}$ or $1\frac{13}{66}$ 14. $\frac{1}{4}$ 15. $\frac{8}{13}$ 16. $15\frac{9}{23}$

EXERCISES 5.3

Add and simplify.

1. $\frac{5}{9} + \frac{7}{6}$
2. $\frac{7}{10} + \frac{11}{15}$
3. $\frac{6}{7} + \frac{29}{56}$
4. $\frac{3}{14} + \frac{9}{10}$
5. $\frac{3}{20} + \frac{1}{15} + \frac{7}{12}$
6. $\frac{3}{8} + \frac{13}{10} + \frac{1}{5}$
7. $\frac{17}{33} + \frac{19}{6}$
8. $\frac{6}{65} + \frac{4}{13}$
9. $\frac{7}{34} + \frac{3}{4} + \frac{5}{17}$
10. $\frac{7}{10} + \frac{4}{55} + \frac{13}{22}$
11. $\frac{6}{35} + \frac{19}{42}$
12. $\frac{17}{45} + \frac{35}{54}$
13. $27\frac{4}{15} + 8\frac{4}{35}$
14. $3\frac{5}{22} + 8\frac{8}{9}$
15. $76\frac{9}{64} + 28\frac{15}{24}$
16. $9\frac{5}{28} + 53\frac{7}{20}$
17. $28\frac{4}{51} + 36\frac{7}{34}$
18. $15\frac{5}{7} + 43\frac{39}{91}$

19. $\frac{5}{6}$
    $\frac{3}{10}$
    $+ \frac{4}{15}$

20. $\frac{4}{7}$
    $\frac{11}{28}$
    $+ \frac{1}{2}$

21. $\frac{3}{7}$
    $\frac{5}{6}$
    $+ \frac{4}{5}$

22. $\frac{1}{6}$
    $\frac{7}{8}$
    $+ \frac{11}{16}$

23. $3\frac{8}{11}$
    $4\frac{5}{33}$
    $+ 6\frac{2}{9}$

24. $5\frac{3}{49}$
    $26\frac{5}{14}$
    $+ \frac{1}{7}$

25. $9\frac{1}{6}$
    $38\frac{3}{7}$
    $+ 66\frac{3}{10}$

26. $10\frac{5}{9}$
    $8\frac{11}{15}$
    $+ 6\frac{4}{7}$

27. $3\frac{17}{24}$
    $7\frac{8}{45}$
    $+ 9\frac{5}{36}$

28. $4\frac{8}{35}$
    $12\frac{40}{63}$
    $+ 7\frac{9}{15}$

29. $13\frac{7}{20}$
    $+ 19\frac{8}{65}$

30. $45\frac{7}{8}$
    $+ 63\frac{5}{34}$

31. A bricklayer worked $4\frac{2}{5}$ hours on Tuesday, $5\frac{1}{4}$ hours on Wednesday, and $6\frac{1}{12}$ hours on Thursday. What was the total time worked for the three-day period?

32. Find the total length of this shaft.

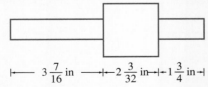

33. A rectangular picture window is $6\frac{3}{8}$ feet wide and $4\frac{2}{3}$ feet long. Find the distance around the window.

34. A business is owned by three women. Joan owns $\frac{7}{12}$ of the business, Roberta owns $\frac{1}{4}$ of the business, and Virginia owns the rest. What fraction of the business is owned by Joan and Roberta?

5.4 COMPARING FRACTIONS

OBJECTIVES
- Compare fractions with like denominators.
- Use the comparing symbols to compare fractions.
- Compare fractions with unlike denominators.
- Solve an application that involves comparing fractions.

1 Comparing Fractions with Like Denominators

In the diagram below there are two figures of the same size. The second figure has a larger portion that is shaded.

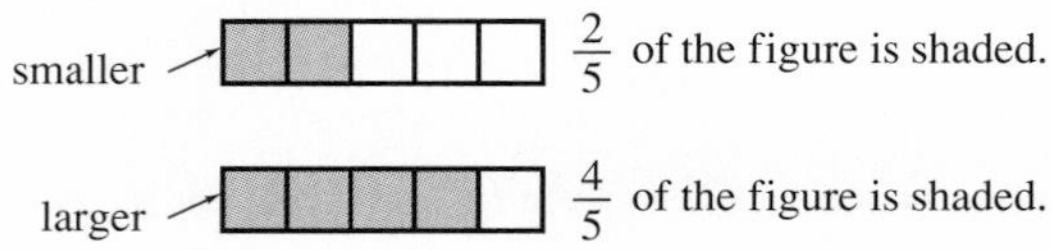

smaller — $\frac{2}{5}$ of the figure is shaded.

larger — $\frac{4}{5}$ of the figure is shaded.

This illustrates that

$\frac{4}{5}$ is larger than $\frac{2}{5}$ or $\frac{2}{5}$ is smaller than $\frac{4}{5}$.

Observe that when two fractions have the same denominator, the larger fraction has the larger numerator.

> *Comparing Fractions with Like Denominators*
> If two fractions have like denominators, then the larger fraction has the larger numerator.

EXAMPLE 1   Name the larger fraction.

a. $\frac{7}{10}, \frac{3}{10}$ **b.** $\frac{3}{3}, \frac{5}{3}$ **c.** $\frac{6}{9}, \frac{7}{9}$

SOLUTION

a. $\frac{7}{10}$ is larger
b. $\frac{5}{3}$ is larger
c. $\frac{7}{9}$ is larger ∎

EXAMPLE 2   Name the smallest fraction.

a. $\frac{3}{4}, \frac{1}{4}, \frac{7}{4}$ **b.** $\frac{5}{7}, \frac{3}{7}, \frac{2}{7}$

SOLUTION

a. $\frac{1}{4}$ is the smallest
b. $\frac{2}{7}$ is the smallest ∎

EXAMPLE 3   List these fractions from smallest to largest.
$\frac{7}{5}, \frac{3}{5}, \frac{2}{5}, \frac{19}{5}$

SOLUTION   Listing from smallest to largest, $\frac{2}{5}, \frac{3}{5}, \frac{7}{5}, \frac{19}{5}$ ∎

5.4 Comparing Fractions   191

 Try These Problems

Name the smaller fraction.
1. $\frac{3}{4}, \frac{1}{4}$
2. $\frac{2}{5}, \frac{18}{5}$
3. $\frac{8}{9}, \frac{7}{9}$

Name the largest fraction.
4. $\frac{1}{5}, \frac{4}{5}, \frac{2}{5}$
5. $\frac{2}{18}, \frac{3}{18}, \frac{11}{18}$

List from smallest to largest.
6. $\frac{6}{8}, \frac{3}{8}, \frac{7}{8}, \frac{2}{8}$
7. $\frac{8}{10}, \frac{3}{10}, \frac{12}{10}, \frac{5}{10}$

Solve.
8. Nanda bought a bag of oranges weighing $6\frac{3}{4}$ pounds. Frank bought a bag weighing $6\frac{1}{4}$ pounds. Who bought the heavier bag of oranges?
9. A recipe calls for $\frac{3}{2}$ cups of milk. Ted has $\frac{1}{2}$ cup of milk. Does Ted have enough milk to prepare this recipe?

Use the symbol $<$, $=$, or $>$ to compare each pair of numbers.
10. $\frac{3}{5} \underset{(<,=,>)}{?} \frac{4}{5}$
11. $\frac{7}{12} \underset{(<,=,>)}{?} \frac{3}{12}$
12. $\frac{11}{5} \underset{(<,=,>)}{?} 2\frac{1}{5}$
13. $4\frac{5}{10} \underset{(<,=,>)}{?} 4\frac{3}{10}$

EXAMPLE 4   Allen ran a race in $5\frac{3}{10}$ minutes and Gregory ran the same race in $5\frac{7}{10}$ minutes. Who won the race?

SOLUTION   Compare $5\frac{3}{10}$ with $5\frac{7}{10}$. The number $5\frac{3}{10}$ is less, so Allen won the race. ■

 Try Problems 1 through 9.

2   The Comparing Symbols

The mathematical symbols $>$ and $<$ are used to compare numbers. The following chart gives examples of how to use these symbols correctly.

English	Math Symbols
2 is smaller than 4	$2 < 4$
4 is larger than 2	$4 > 2$
$\frac{2}{5}$ is smaller than $\frac{4}{5}$	$\frac{2}{5} < \frac{4}{5}$
$\frac{4}{5}$ is larger than $\frac{2}{5}$	$\frac{4}{5} > \frac{2}{5}$

You should make these observations.

1. $\prec$ $\succ$ The pointed side always faces the smaller number.
2. $\prec$ $\succ$ The open side always faces the larger number.

EXAMPLE 5   Use the symbol $<$, $=$, or $>$ to compare each pair of numbers.

a. $\frac{3}{8} \underset{(<,=,>)}{?} \frac{7}{8}$ b. $\frac{4}{4} \underset{(<,=,>)}{?} \frac{2}{4}$ c. $1\frac{1}{16} \underset{(<,=,>)}{?} \frac{17}{16}$

SOLUTION

a. $\frac{3}{8} < \frac{7}{8}$ b. $\frac{4}{4} > \frac{2}{4}$ c. $1\frac{1}{16} = \frac{17}{16}$ ■

Try Problems 10 through 13.

3   Comparing Fractions with Unlike Denominators

In the diagram below there are two figures of the same size. The second figure has a smaller portion that is shaded.

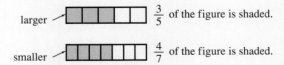

larger — $\frac{3}{5}$ of the figure is shaded.

smaller — $\frac{4}{7}$ of the figure is shaded.

This illustrates that

$$\frac{3}{5} \text{ is larger than } \frac{4}{7} \quad \text{or} \quad \frac{4}{7} \text{ is smaller than } \frac{3}{5}.$$

If we convert each fraction to an equivalent fraction with denominator 35, we can see that $\frac{3}{5}$ is larger than $\frac{4}{7}$ without viewing a picture.

$$\frac{3}{5} = \frac{3 \times 7}{5 \times 7} = \frac{21}{35} \text{ —— larger}$$

$$\frac{4}{7} = \frac{4 \times 5}{7 \times 5} = \frac{20}{35} \text{ —— smaller}$$

Because $\frac{21}{35}$ is more than $\frac{20}{35}$, $\frac{3}{5}$ is more than $\frac{4}{7}$.

> *Comparing Fractions with Unlike Denominators*
> 1. Write each fraction with the same denominator.
> 2. If two fractions have the same denominator, the larger fraction has the larger numerator.

EXAMPLE 6 List from the smallest to largest: $\frac{1}{2}, \frac{4}{9}, \frac{5}{12}, \frac{17}{36}$

SOLUTION Find the least common multiple of 2, 9, 12, and 36.

$$\begin{array}{l} 2 \text{ is prime} \\ 9 = 3 \times 3 \\ 12 = 3 \times 4 \\ = 3 \times 2 \times 2 \\ 36 = 6 \times 6 \\ = 2 \times 3 \times 2 \times 3 \end{array} \Biggr\} \begin{array}{l} \text{LCM} = 2 \times 2 \times 3 \times 3 \\ \phantom{\text{LCM}} = 4 \times 9 \\ \phantom{\text{LCM}} = 36 \end{array}$$

Convert each fraction to an equivalent fraction with denominator 36.

$$\frac{1}{2} = \frac{1 \times 18}{2 \times 18} = \frac{18}{36} \text{ —— largest}$$

$$\frac{4}{9} = \frac{4 \times 4}{9 \times 4} = \frac{16}{36}$$

$$\frac{5}{12} = \frac{5 \times 3}{12 \times 3} = \frac{15}{36} \text{ —— smallest}$$

$$\frac{17}{36}$$

Listing from smallest to largest, $\frac{5}{12}, \frac{4}{9}, \frac{17}{36}, \frac{1}{2}$ ∎

EXAMPLE 7 Beatrice, a mechanic, accidentally picked up a $\frac{7}{16}$-inch wrench to loosen a $\frac{5}{8}$-inch bolt. Was the wrench too large or too small?

Try These Problems

14. Which is larger, $\frac{3}{5}$ or $\frac{7}{12}$?
15. Use the symbol <, =, or > to compare $\frac{1}{5}$ with $\frac{8}{35}$.
16. List from smallest to largest: $\frac{4}{27}, \frac{7}{54}, \frac{1}{6}, \frac{1}{9}$.
17. In a tennis match, Allanah got $\frac{7}{20}$ of her first serves in and Tina got $\frac{2}{5}$. Which of the women got the larger fractional portion of first serves in?
18. Tim is buying a refrigerator for his kitchen. The space in the kitchen is $2\frac{5}{12}$ feet wide. The refrigerator he likes is $2\frac{3}{4}$ feet wide. Does the refrigerator fit in the space?

SOLUTION Which is larger, $\frac{7}{16}$ or $\frac{5}{8}$?

WRENCH $\quad \frac{7}{16}$

BOLT $\quad \frac{5}{8} = \frac{5 \times 2}{8 \times 2} = \frac{10}{16}$ ——— larger

The bolt is larger than the wrench, so the wrench is too small. ■

Try Problems 14 through 18.

Answers to Try These Problems

1. $\frac{1}{4}$ 2. $\frac{2}{5}$ 3. $\frac{7}{9}$ 4. $\frac{4}{5}$ 5. $\frac{11}{18}$ 6. $\frac{2}{8}, \frac{3}{8}, \frac{6}{8}, \frac{7}{8}$ 7. $\frac{3}{10}, \frac{5}{10}, \frac{8}{10}, \frac{12}{10}$
8. Nanda 9. no 10. < 11. > 12. = 13. > 14. $\frac{3}{5}$
15. $\frac{1}{5} < \frac{8}{35}$ 16. $\frac{1}{9}, \frac{7}{54}, \frac{4}{27}, \frac{1}{6}$ 17. Tina 18. no

EXERCISES 5.4

Name the larger fraction.

1. $\frac{2}{3}, \frac{1}{3}$
2. $\frac{7}{12}, \frac{9}{12}$
3. $\frac{2}{3}, \frac{7}{9}$
4. $\frac{3}{10}, \frac{5}{6}$

Name the smaller fraction.

5. $\frac{2}{5}, \frac{4}{5}$
6. $\frac{11}{6}, \frac{7}{6}$
7. $\frac{7}{20}, \frac{2}{5}$
8. $\frac{7}{10}, \frac{7}{9}$

Name the largest fraction in each group.

9. $\frac{2}{12}, \frac{8}{12}, \frac{10}{12}$
10. $\frac{17}{36}, \frac{20}{36}, \frac{15}{36}$
11. $\frac{1}{4}, \frac{1}{5}, \frac{7}{30}$
12. $\frac{5}{6}, \frac{2}{3}, \frac{5}{7}$

Name the smallest fraction in each group.

13. $\frac{7}{6}, \frac{6}{6}, \frac{11}{6}$
14. $\frac{15}{24}, \frac{14}{24}, \frac{20}{24}$
15. $\frac{9}{20}, \frac{1}{2}, \frac{2}{5}$
16. $\frac{5}{8}, \frac{9}{16}, \frac{3}{4}$

List from smallest to largest.

17. $\frac{3}{4}, \frac{1}{4}, \frac{5}{4}, \frac{11}{4}$
18. $\frac{9}{5}, \frac{12}{5}, \frac{4}{5}, \frac{10}{5}$
19. $\frac{3}{5}, \frac{13}{20}, \frac{1}{2}$
20. $\frac{3}{4}, \frac{7}{12}, \frac{2}{3}$
21. $\frac{24}{11}, 3, \frac{11}{24}$
22. $5, \frac{17}{3}, \frac{3}{17}$

Use the symbol >, =, or < to compare each pair of numbers.

23. $2, 7$
24. $11, 9$
25. $\frac{14}{15}, \frac{8}{15}$
26. $\frac{7}{11}, \frac{9}{11}$
27. $3\frac{3}{4}, 3\frac{6}{8}$,
28. $4\frac{5}{9}, 4\frac{1}{2}$
29. $\frac{13}{25}, \frac{3}{5}$
30. $\frac{7}{9}, \frac{3}{4}$
31. $\frac{11}{11}, \frac{12}{13}$
32. $\frac{8}{2}, 4$

Solve.

33. One piece of pipe measures $\frac{5}{8}$ inches in diameter and another piece of pipe measures $\frac{7}{8}$ inches in diameter. Which pipe has the larger diameter?
34. Patricia lives $\frac{8}{10}$ of a mile from school and Agatha lives $\frac{6}{10}$ of a mile from school. Who lives closer to school?
35. Jacqueline receives a weekly salary of $150 and has $10 deducted for health insurance. Paul receives a weekly salary of $100 and has $8 deducted for health insurance. Who pays the larger fractional portion of salary for health insurance?
36. Arthur was comparing his recipe for spaghetti with David's recipe. Arthur's recipe called for $\frac{1}{2}$ cup parsley, whereas David's recipe called for $\frac{1}{4}$ cup parsley. Whose spaghetti recipe contained more parsley?

DEVELOPING NUMBER SENSE #4

COMPARING A FRACTION TO $\frac{1}{2}$

In this feature we look at how to decide if a fraction is less than, equal to, or more than $\frac{1}{2}$. First we look at some fractions that are equal to $\frac{1}{2}$.

$$\frac{1}{2} = \frac{2}{4} = \frac{3}{6} = \frac{4}{8} = \frac{5}{10} = \frac{6}{12} = \frac{7}{14}$$

In each case, observe that the denominator is exactly twice the numerator, or another way to look at it, the numerator is the result of dividing the denominator by 2. For example, in the fraction $\frac{7}{14}$, $14 = 2 \times 7$ or $14 \div 2 = 7$. Using this observation, can you choose the fractions in this list that are exactly equal to $\frac{1}{2}$?

$$\frac{11}{22} \quad \frac{6}{14} \quad \frac{7}{10} \quad \frac{14}{28} \quad \frac{150}{300} \quad \frac{15}{28} \quad \frac{25}{50}$$

The fractions that are equal to $\frac{1}{2}$ are $\frac{11}{22}$, $\frac{14}{28}$, $\frac{150}{300}$, and $\frac{25}{50}$.

Consider the fraction $\frac{6}{14}$. Is it less than or more than $\frac{1}{2}$? Dividing 14 by 2, we see that $\frac{7}{14}$ is equal to $\frac{1}{2}$. Because $\frac{6}{14}$ is less than $\frac{7}{14}$, $\frac{6}{14}$ is less than $\frac{1}{2}$. We write $\frac{6}{14} < \frac{1}{2}$.

Consider the fraction $\frac{14}{25}$. Is it less than or more than $\frac{1}{2}$? Dividing 25 by 2, we see that $\frac{12\frac{1}{2}}{25}$ is exactly equal to $\frac{1}{2}$. Because $\frac{14}{25}$ is more than $\frac{12\frac{1}{2}}{25}$, $\frac{14}{25}$ is more than $\frac{1}{2}$. We write $\frac{14}{25} > \frac{1}{2}$.

Being able to quickly compare a fraction to $\frac{1}{2}$ will help you later when you are rounding off fractions to the nearest whole number.

Number Sense Problems

1. Choose the fractions that are equal to $\frac{1}{2}$.
   - a. $\frac{4}{7}$
   - b. $\frac{10}{20}$
   - c. $\frac{24}{48}$
   - d. $\frac{13}{37}$
   - e. $\frac{114}{360}$
   - f. $\frac{47}{91}$
   - g. $\frac{43}{86}$

2. Choose the fractions that are less than $\frac{1}{2}$.
   - a. $\frac{8}{17}$
   - b. $\frac{24}{47}$
   - c. $\frac{45}{90}$
   - d. $\frac{99}{202}$
   - e. $\frac{87}{180}$
   - f. $\frac{35}{69}$
   - g. $\frac{305}{610}$

3. Choose the fractions that are more than $\frac{1}{2}$.
   - a. $\frac{9}{17}$
   - b. $\frac{17}{30}$
   - c. $\frac{20}{48}$
   - d. $\frac{41}{84}$
   - e. $\frac{46}{92}$
   - f. $\frac{76}{150}$
   - g. $\frac{53}{111}$

Use the symbol $<$, $=$, or $>$ to accurately compare each given fraction with $\frac{1}{2}$.

4. $\frac{1}{4}$ 5. $\frac{5}{8}$ 6. $\frac{10}{18}$ 7. $\frac{18}{36}$ 8. $\frac{34}{70}$ 9. $\frac{42}{87}$

5.5 SUBTRACTING

OBJECTIVES
- Subtract fractions.
- Perform subtraction of mixed numerals that involves borrowing.
- Find the missing number in an addition statement.

Try These Problems

Subtract and simplify.
1. $\frac{11}{8} - \frac{3}{8}$
2. $\frac{1}{2} - \frac{1}{3}$
3. $\frac{7}{12} - \frac{2}{9}$
4. $\frac{9}{5} - \frac{7}{15}$

Solve.
5. A truck was loaded with $\frac{4}{5}$ ton of sand. If $\frac{3}{10}$ ton was unloaded, what fraction of a ton was left?

1 Subtracting Fractions

Suppose that on Monday you walked $\frac{7}{10}$ mile, and on Tuesday you walked only $\frac{3}{10}$ mile. How much farther did you walk on Monday? To solve this problem we need to subtract $\frac{3}{10}$ from $\frac{7}{10}$. Subtraction is the reverse of addition. To subtract means to take away.

$$\frac{7}{10} - \frac{3}{10} = \frac{4}{10}$$ —— Subtract the numerators. $7 - 3 = 4$
   —— Use the common denominator.

$$= \frac{2 \times \cancel{2}}{5 \times \cancel{2}} \quad \text{Simplify the answer.}$$

$$= \frac{2}{5}$$

You walked $\frac{2}{5}$ mile farther on Monday.

Observe that to subtract fractions with like denominators we must subtract the numerators and keep the same denominator. Here we view a picture of the subtraction problem $\frac{7}{10} - \frac{3}{10}$ so that we can see that the result $\frac{2}{5}$ is correct. Assume each separate figure represents 1 whole.

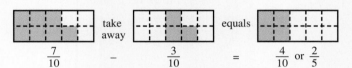

$\frac{7}{10}$ take away $\frac{3}{10}$ equals $\frac{4}{10}$ or $\frac{2}{5}$

Here are more examples of subtracting fractions.

EXAMPLE 1 Subtract: $\frac{3}{4} - \frac{1}{8}$

SOLUTION The least common multiple of 4 and 8 is 8. Convert $\frac{3}{4}$ to an equivalent fraction with denominator 8 before subtracting.

$$\frac{3}{4} - \frac{1}{8}$$

$$= \frac{3 \times 2}{4 \times 2} - \frac{1}{8} \quad \text{Multiply the numerator and denominator of } \frac{3}{4} \text{ by 2 so that it has the common denominator 8.}$$

$$= \frac{6}{8} - \frac{1}{8}$$

$$= \frac{5}{8} \blacksquare$$

 Try Problems 1 through 5.

EXAMPLE 2 Subtract: $9\frac{5}{12} - 2\frac{3}{40}$

SOLUTION Find the least common multiple of 12 and 40.

$$\begin{array}{l} 12 = 3 \times 4 \\ = 3 \times 2 \times 2 \\ 40 = 4 \times 10 \\ = 2 \times 2 \times 2 \times 5 \end{array} \Bigg\} \begin{array}{l} \text{LCM} = 3 \times 2 \times 2 \times 2 \times 5 \\ \phantom{\text{LCM}} = 6 \times 4 \times 5 \\ \phantom{\text{LCM}} = 6 \times 20 \\ \phantom{\text{LCM}} = 120 \end{array}$$

Try These Problems

Subtract and simplify.
6. $4\frac{5}{6} - 1\frac{7}{18}$
7. $42\frac{7}{11} - 17\frac{1}{3}$
8. $15\frac{13}{36} - 9\frac{3}{28}$

Convert each fraction to an equivalent fraction with denominator 120, then subtract.

$$9\frac{5}{12} = 9\frac{5 \times 10}{12 \times 10} = 9\frac{50}{120}$$
$$-2\frac{3}{40} = 2\frac{3 \times 3}{40 \times 3} = 2\frac{9}{120}$$
$$\phantom{-2\frac{3}{40} = 2\frac{3 \times 3}{40 \times 3} =\ } 7\frac{41}{120}$$

Subtract the whole numbers. $9 - 2 = 7$
Subtract the fractions.
$\frac{50}{120} - \frac{9}{120} = \frac{41}{120}$

The answer $7\frac{41}{120}$ is simplified because 2, 3, and 5 are the only prime divisors of 120 and none of these divide evenly into 41. ∎

 Try Problems 6 through 8.

2 Subtraction Involving Borrowing

Art has a tree that is $3\frac{2}{3}$ feet tall. He wants the tree to be 6 feet tall. How much taller must it grow? To solve this problem you need to subtract $3\frac{2}{3}$ from 6.

$$6 - 3\frac{2}{3}$$

Because there is no fraction next to 6 to subtract $\frac{2}{3}$ from, we rewrite 6 as $5\frac{3}{3}$.

$$= 5\frac{3}{3} - 3\frac{2}{3}$$

$$= 2\frac{1}{3}$$

Subtract the whole numbers. $5 - 3 = 2$
Subtract the fractions. $\frac{3}{3} - \frac{2}{3} = \frac{1}{3}$

The tree needs to grow $2\frac{1}{3}$ feet.

Now we view a picture of the subtraction problem $6 - 3\frac{2}{3}$ so that we can see that the result $2\frac{1}{3}$ is correct. Assume each separate figure represents 1 whole.

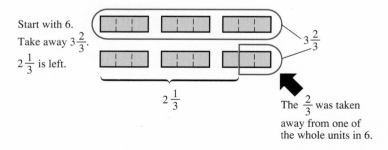

Start with 6.
Take away $3\frac{2}{3}$.
$2\frac{1}{3}$ is left.

$3\frac{2}{3}$

The $\frac{2}{3}$ was taken away from one of the whole units in 6.

Try These Problems

Subtract and simplify.

9. $8 - 2\frac{3}{4}$
10. $19 - 7\frac{9}{20}$
11. $1 - \frac{5}{8}$
12. $1 - \frac{11}{15}$

Solve.

13. A business is owned by two men. Juan owns $\frac{3}{5}$ of the business and Warren owns the rest. What fraction of the business does Warren own?

14. Betsy read $\frac{1}{4}$ of her book last night. What fraction of the book is left?

15. Karl wants to run 10 miles. So far, he has run $5\frac{3}{8}$ miles. How many miles does he have left to go?

16. Sylvia wants to limit her backpack to 8 pounds. She has already packed items weighing $2\frac{1}{2}$ pounds and $4\frac{3}{4}$ pounds. How many more pounds can she pack without going over her limit?

When we rewrite 6 as $5\frac{3}{3}$, we say that we **borrow** 1 from 6 to make 5 and write the 1 as $\frac{3}{3}$. Here we show the problem $6 - 3\frac{2}{3}$ solved in vertical format.

$$\begin{array}{rl} 6 &= 5\frac{3}{3} \\ -3\frac{2}{3} &= 3\frac{2}{3} \\ \hline & 2\frac{1}{3} \end{array}$$

Borrow 1 from 6 to make 5 and write 1 as $\frac{3}{3}$.

EXAMPLE 3 Subtract: $9 - 3\frac{7}{12}$

SOLUTION

$$9 - 3\frac{7}{12}$$
$$= 8\frac{12}{12} - 3\frac{7}{12}$$ Because there is no fraction next to 9 to subtract $\frac{7}{12}$ from, rewrite 9 as $8\frac{12}{12}$.
$$= 5\frac{5}{12}$$ Subtract the whole numbers. $8 - 3 = 5$
Subtract the fractions. $\frac{12}{12} - \frac{7}{12} = \frac{5}{12}$ ■

EXAMPLE 4 Subtract: $1 - \frac{13}{25}$

SOLUTION

$$1 - \frac{13}{25}$$
$$= \frac{25}{25} - \frac{13}{25}$$ Rewrite 1 as $\frac{25}{25}$ so both numbers are in fractional form with denominator 25.
$$= \frac{12}{25}$$ ■

EXAMPLE 5 Dave ate $\frac{1}{3}$ of a pizza. What fraction of the pizza remains?

SOLUTION The whole pizza is represented by the number 1. Subtract $\frac{1}{3}$ from 1.

$$\begin{array}{rl} 1 &= \frac{3}{3} \\ -\frac{1}{3} &= \frac{1}{3} \\ \hline & \frac{2}{3} \end{array}$$

— The whole pizza.
— Fraction of the pizza he ate.
— Fraction of the pizza that remains.

$\frac{2}{3}$ of the pizza remains.

 Try Problems 9 through 16.

Now consider the following subtraction problem.

$$3\frac{1}{6}$$
$$-1\frac{5}{6}$$

Notice that the fraction $\frac{5}{6}$ is larger than the fraction $\frac{1}{6}$ so we cannot subtract as the problem is written. Before working the problem, view the following picture so that we can see what the answer should be.

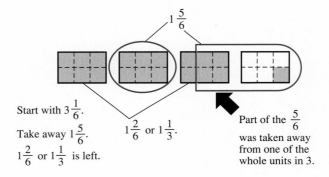

Start with $3\frac{1}{6}$.
Take away $1\frac{5}{6}$.
$1\frac{2}{6}$ or $1\frac{1}{3}$ is left.

$1\frac{2}{6}$ or $1\frac{1}{3}$

Part of the $\frac{5}{6}$ was taken away from one of the whole units in 3.

Now here is a procedure that gives the correct result $1\frac{1}{3}$.

$$3\frac{1}{6} = 2\frac{7}{6}$$
$$-1\frac{5}{6} = 1\frac{5}{6}$$
$$1\frac{2}{6} = 1\frac{1}{3}$$

Observe that $\frac{5}{6}$ is larger than $\frac{1}{6}$, so we cannot subtract the fractions. Borrow 1 from 3 to make 2, and combine the 1 you borrowed with $\frac{1}{6}$.
$1 + \frac{1}{6} = \frac{6}{6} + \frac{1}{6} = \frac{7}{6}$
We are rewriting $3\frac{1}{6}$ as $2\frac{7}{6}$ before subtracting.

EXAMPLE 6 Subtract: $43\frac{2}{9} - 12\frac{7}{9}$

SOLUTION

$$43\frac{2}{9} = 42\frac{11}{9}$$
$$-12\frac{7}{9} = 12\frac{7}{9}$$
$$30\frac{4}{9}$$

Note that $\frac{7}{9}$ is larger than $\frac{2}{9}$ so we cannot subtract the fractions. Borrow 1 from 43 to make 42. Combine the 1 you borrowed with $\frac{2}{9}$.
$1 + \frac{2}{9} = \frac{9}{9} + \frac{2}{9} = \frac{11}{9}$
or $1\frac{2}{9} = \frac{11}{9}$ ∎

EXAMPLE 7 Subtract: $25\frac{5}{21} - 7\frac{8}{9}$

SOLUTION Find a common denominator.

$$\left.\begin{array}{r}21 = 3 \times 7 \\ 9 = 3 \times 3\end{array}\right\} \begin{array}{l}\text{LCM} = 3 \times 3 \times 7 \\ = 63\end{array}$$

Write each fraction with denominator 63 and subtract.

Try These Problems

Subtract and simplify.
17. $5\frac{2}{7} - 3\frac{5}{7}$
18. $8\frac{3}{12} - 2\frac{7}{12}$
19. $6\frac{1}{4} - 3\frac{7}{8}$
20. $23\frac{5}{24} - 7\frac{15}{16}$

Find the missing number.
21. $\boxed{} + \frac{5}{8} = \frac{9}{8}$
22. $\frac{3}{11} + \boxed{} = \frac{9}{11}$
23. $\frac{2}{3} + \frac{3}{4} = \boxed{}$
24. $\boxed{} = \frac{14}{25} + \frac{3}{10}$
25. $4 = 2\frac{1}{2} + \boxed{}$
26. $\frac{2}{3} = \frac{5}{9} + \boxed{}$

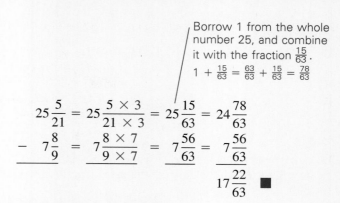

Borrow 1 from the whole number 25, and combine it with the fraction $\frac{15}{63}$.
$1 + \frac{15}{63} = \frac{63}{63} + \frac{15}{63} = \frac{78}{63}$

$$25\frac{5}{21} = 25\frac{5 \times 3}{21 \times 3} = 25\frac{15}{63} = 24\frac{78}{63}$$
$$-\ 7\frac{8}{9} = 7\frac{8 \times 7}{9 \times 7} = 7\frac{56}{63} = 7\frac{56}{63}$$
$$\hspace{6cm} 17\frac{22}{63} \ \blacksquare$$

 Try Problems 17 through 20.

3 The Missing Number in an Addition Statement

The statement $\frac{3}{7} + \frac{2}{7} = \frac{5}{7}$ is an addition statement. If any one of the three numbers is omitted, we can find the missing number. Study the three examples that follow to see how to do this.

1. $\frac{3}{7} + \frac{2}{7} = \boxed{}$ Here the answer to the addition statement is missing, so we *add* to find the missing number.

   The missing number is $\frac{5}{7}$.

2. $\boxed{} + \frac{2}{7} = \frac{5}{7}$ Here one of the terms is missing, so we *subtract* to find the missing number.

   The missing number is $\frac{5}{7} - \frac{2}{7} = \frac{3}{7}$.

3. $\frac{3}{7} + \boxed{} = \frac{5}{7}$ Here one of the terms is missing, so we *subtract* to find the missing number.

   The missing number is $\frac{5}{7} - \frac{3}{7} = \frac{2}{7}$.

 Try Problems 21 through 26.

 Answers to Try These Problems

1. 1   2. $\frac{1}{6}$ 3. $\frac{13}{36}$ 4. $\frac{4}{3}$ or $1\frac{1}{3}$ 5. $\frac{1}{2}$ 6. $3\frac{4}{9}$ 7. $25\frac{10}{33}$
8. $6\frac{16}{63}$ 9. $5\frac{1}{4}$ 10. $11\frac{11}{20}$ 11. $\frac{3}{8}$ 12. $\frac{4}{15}$ 13. $\frac{2}{5}$ of the business
14. $\frac{3}{4}$ of her book   15. $4\frac{5}{8}$ mi   16. $\frac{3}{4}$ lb   17. $1\frac{4}{7}$ 18. $5\frac{2}{3}$
19. $2\frac{3}{8}$ 20. $15\frac{13}{48}$ 21. $\frac{1}{2}$ 22. $\frac{6}{11}$ 23. $\frac{17}{12}$ or $1\frac{5}{12}$ 24. $\frac{43}{50}$
25. $\frac{3}{2}$ or $1\frac{1}{2}$ 26. $\frac{1}{9}$

EXERCISES 5.5

Subtract and simplify:

1. $\frac{5}{6} - \frac{1}{6}$
2. $\frac{13}{9} - \frac{4}{9}$
3. $\frac{7}{8} - \frac{1}{4}$
4. $\frac{10}{21} - \frac{2}{7}$
5. $\frac{7}{9} - \frac{5}{12}$
6. $\frac{21}{10} - \frac{3}{8}$
7. $15\frac{7}{8} - 4\frac{3}{8}$
8. $9\frac{7}{11} - 5\frac{2}{11}$
9. $9\frac{3}{8} - 4$
10. $8\frac{5}{16} - 5$
11. $5\frac{7}{8} - 2\frac{1}{4}$
12. $10\frac{1}{2} - 7\frac{1}{6}$
13. $35\frac{3}{4} - 12\frac{5}{18}$
14. $56\frac{7}{13} - 8\frac{8}{39}$
15. $8 - 3\frac{4}{5}$
16. $72 - 8\frac{1}{3}$

17. $10 - \frac{7}{12}$ **18.** $23 - \frac{8}{15}$ **19.** $1 - \frac{11}{18}$ **20.** $1 - \frac{7}{9}$

21. $7\frac{1}{3}$ **22.** $11\frac{3}{8}$ **23.** $10\frac{5}{14}$ **24.** $15\frac{1}{3}$ **25.** $6\frac{3}{25}$
$-2\frac{2}{3}$ $-6\frac{7}{8}$ $-4\frac{6}{7}$ $-7\frac{2}{5}$ $-2\frac{11}{15}$

26. $72\frac{4}{45}$ **27.** $50\frac{4}{39}$ **28.** $61\frac{4}{33}$ **29.** $12\frac{5}{8}$ **30.** $24\frac{5}{12}$
$-12\frac{25}{27}$ $-\frac{2}{13}$ $-\frac{13}{22}$ $-7\frac{3}{10}$ $-17\frac{8}{21}$

Solve.

31. A carpenter cuts a piece of wood $7\frac{5}{8}$ feet long. She discovers that it is too long by $\frac{3}{8}$ foot. After she cuts off the extra length, how long will the resulting piece be?

32. The value of a certain stock was $22 at the start of the day. During the day it decreased $\$\frac{7}{8}$. What was the value at the end of the day?

33. The string area of the Prince Pro tennis racket is $10\frac{3}{16}$ inches wide; the string area of the Head Pro racket is only $8\frac{5}{16}$ inches wide. What is the difference in the two widths?

34. Mr. Smiegiel purchased a $3\frac{1}{4}$-pound steak at the meat market. After the butcher trimmed the steak of fat, the steak weighed $2\frac{3}{8}$ pounds. How much fat was trimmed from the steak?

35. Cindy is painting the interior of her apartment. She finished $\frac{1}{6}$ of the job yesterday. What fraction of the job remains?

36. A painter finished $\frac{1}{5}$ of the job the first day. What fraction of the job remains?

37. Sara bought a pizza for herself and several friends. Joe ate $\frac{1}{3}$ of the pizza and Carolyn ate $\frac{1}{8}$ of the pizza. What fraction of the pizza remains?

38. Mr. Winetrub started with a full tank of gas. He used $\frac{1}{4}$ of the tank on Friday and $\frac{2}{3}$ of the tank on Saturday. What fraction of the tank is left?

Find the missing number. Simplify the answer.

39. $\frac{5}{9} + \boxed{} = \frac{8}{9}$ **40.** $\boxed{} + \frac{7}{12} = \frac{11}{12}$ **41.** $\boxed{} = \frac{3}{5} + \frac{7}{15}$

42. $\frac{19}{20} = \boxed{} + \frac{1}{4}$ **43.** $5\frac{1}{4} = \boxed{} + 3\frac{2}{3}$ **44.** $16\frac{2}{9} = 5\frac{5}{6} + \boxed{}$

45. $\frac{10}{13} + \boxed{} = 1$ **46.** $4\frac{3}{5} + \boxed{} = 5$

DEVELOPING NUMBER SENSE #5

APPROXIMATING BY ROUNDING

In this feature we look at approximations that involve computations with fractions. For example, suppose you wanted to approximate this addition problem.

$$2\frac{1}{3} + 5\frac{4}{5} + 9\frac{5}{6}$$

One way to obtain an approximation is to round off each number to the nearest whole number, then add.

$2\frac{1}{3} \approx 2$ $2\frac{1}{3}$ is closer to 2 than 3 because $\frac{1}{3}$ is less than $\frac{1}{2}$.

$5\frac{4}{5} \approx 6$ $5\frac{4}{5}$ is closer to 6 than 5 because $\frac{4}{5}$ is more than $\frac{1}{2}$.

$9\frac{5}{6} \approx 10$ $9\frac{5}{6}$ is closer to 10 than 9 because $\frac{5}{6}$ is more than $\frac{1}{2}$.

Adding the numbers 2, 6, and 10, we obtain 18 for the approximation. The actual sum is $17\frac{29}{30}$.

When approximating with fractions, you do not always have to round off to the nearest whole number. If the numbers are larger, you may want to round off to the the nearest ten, hundred, or thousand as you did when working with whole numbers. For example, suppose you want to approximate this multiplication problem.

$$12\frac{7}{8} \times 36\frac{1}{3}$$

One way to obtain an approximation is to round off each factor to the nearest ten, then multiply.

$$12\frac{7}{8} \approx 10$$ Each number is rounded off to the nearest ten.

$$36\frac{1}{3} \approx 40$$

Multiplying 10 by 40, we obtain 400 for the approximation. The actual product is $467\frac{19}{24}$. Of course, if we had rounded off to the nearest whole number the approximation would have been closer to the actual product, $13 \times 36 = 468$.

Number Sense Problems

Round off each of the following numbers to the nearest whole number.

1. $5\frac{1}{3}$ **2.** $7\frac{3}{4}$ **3.** $12\frac{2}{5}$ **4.** $45\frac{2}{3}$ **5.** $\frac{1}{6}$ **6.** $\frac{11}{13}$

Round off each of the following numbers to the nearest ten.

7. 67 **8.** 153 **9.** $18\frac{1}{4}$
10. $14\frac{7}{8}$ **11.** $53\frac{1}{2}$ **12.** $178\frac{12}{25}$

Approximate by first rounding off each number to the nearest whole number.

13. $4\frac{3}{4} + 2\frac{7}{8} + 7\frac{5}{12}$ **14.** $16\frac{2}{7} - 9\frac{8}{9}$ **15.** $16\frac{7}{8} \times 3\frac{2}{9}$
16. $64\frac{2}{7} \div 3\frac{5}{6}$ **17.** $71\frac{2}{3} \times 12\frac{3}{8}$ **18.** $25\frac{7}{20} + 160\frac{5}{12} + 94\frac{5}{8}$

Approximate by first rounding off each number to the nearest ten.

19. $78\frac{1}{2} + 93$ **20.** $174 - 29\frac{1}{4}$ **21.** $13\frac{6}{7} \times 657$ **22.** $692\frac{5}{6} \div 27\frac{4}{5}$

5.6 LANGUAGE

OBJECTIVES
- Translate an English statement involving addition or subtraction to math symbols.
- Solve problems by using translations.

1 Translating English Statements to Math Symbols

The math symbols that we use often have many translations to the English language. Knowing these translations can help in solving application problems. The following chart gives the many English translations for the symbol $=$.

Math Symbol	English
=	equals
	is equal to
	is the same as
	is the result of
	is
	was
	represents
	gives
	makes
	yields
	will be
	were
	are

The addition statement $\frac{3}{4} + \frac{2}{4} = \frac{5}{4}$ is written using math symbols. Some of the ways to read this in English are given in the following chart.

Math Symbols	English
$\frac{3}{4} + \frac{2}{4} = \frac{5}{4}$	Three-fourths plus two-fourths equals five-fourths.
	Three-fourths added to two-fourths gives five-fourths.
	The *sum* of $\frac{3}{4}$ and $\frac{2}{4}$ is $\frac{5}{4}$.
	$\frac{5}{4}$ represents the total when $\frac{3}{4}$ is added to $\frac{2}{4}$.
	$\frac{5}{4}$ is the result of increasing $\frac{3}{4}$ by $\frac{2}{4}$.
	$\frac{3}{4}$ *increased by* $\frac{2}{4}$ makes $\frac{5}{4}$.
	$\frac{2}{4}$ *more than* $\frac{3}{4}$ is $\frac{5}{4}$.

Because $\frac{3}{4} + \frac{2}{4}$ equals $\frac{2}{4} + \frac{3}{4}$, the order in which you read the terms $\frac{3}{4}$ and $\frac{2}{4}$ does not matter; however, when reading the subtraction statement $\frac{5}{4} - \frac{2}{4}$, take care to read $\frac{5}{4}$ and $\frac{2}{4}$ in the correct order, because $\frac{5}{4} - \frac{2}{4}$ does not equal $\frac{2}{4} - \frac{5}{4}$.

$$\frac{5}{4} - \frac{2}{4} = \frac{3}{4} \quad \text{but} \quad \frac{2}{4} - \frac{5}{4} = \frac{-3}{4}$$

The numbers $\frac{3}{4}$ and $\frac{-3}{4}$ are not equal. The number $\frac{-3}{4}$ is a negative number. This book covers negative numbers in Chapter 12. The following chart gives several correct ways to read $\frac{5}{4} - \frac{2}{4} = \frac{3}{4}$.

Try These Problems

Translate each English statement to an addition or subtraction statement using math symbols.

1. $\frac{3}{7}$ subtracted from $\frac{5}{7}$ is $\frac{2}{7}$.
2. 8 increased by $7\frac{1}{2}$ is $15\frac{1}{2}$.
3. $1\frac{1}{4}$ decreased by $\frac{3}{4}$ is $\frac{1}{2}$.
4. The sum of $\frac{1}{3}$ and $\frac{1}{2}$ is $\frac{5}{6}$.

Math Symbols	English
$\frac{5}{4} - \frac{2}{4} = \frac{3}{4}$	Five-fourths minus two-fourths equals three-fourths.
	$\frac{5}{4}$ take away $\frac{2}{4}$ is $\frac{3}{4}$.
	$\frac{5}{4}$ subtract $\frac{2}{4}$ is $\frac{3}{4}$.
	$\frac{2}{4}$ subtracted from $\frac{5}{4}$ is $\frac{3}{4}$.
	$\frac{2}{4}$ less than $\frac{5}{4}$ is $\frac{3}{4}$.
	The *difference* between $\frac{5}{4}$ and $\frac{2}{4}$ is $\frac{3}{4}$.
	$\frac{5}{4}$ *decreased by* $\frac{2}{4}$ gives $\frac{3}{4}$.
	The result of $\frac{2}{4}$ subtracted from $\frac{5}{4}$ is $\frac{3}{4}$.

Note that when reading the symbol — by using the phrases *subtracted from* or *less than*, the numbers $\frac{5}{4}$ and $\frac{2}{4}$ are read in the reverse order than they are written. That is,

$$\frac{2}{4} \text{ subtracted from } \frac{5}{4} \quad \text{means} \quad \frac{5}{4} - \frac{2}{4}$$

and

$$\frac{2}{4} \text{ less than } \frac{5}{4} \quad \text{means} \quad \frac{5}{4} - \frac{2}{4}$$

 Try Problems 1 through 4.

2 Using Translating to Solve Problems

The following examples illustrate how we can use translating to solve problems.

EXAMPLE 1 Sixty represents $27\frac{1}{3}$ plus what number?

SOLUTION Translate the question into math symbols.

204 Chapter 5 Fractions: Addition and Subtraction

Try These Problems

Solve.

5. Find the sum of $2\frac{1}{8}$ and $\frac{7}{16}$.
6. Find the difference between $8\frac{1}{3}$ and $4\frac{5}{9}$.
7. Subtract $\frac{9}{20}$ from $\frac{1}{2}$.
8. Find $\frac{8}{15}$ plus $\frac{7}{9}$.
9. $2\frac{3}{4}$ increased by what number equals 7?
10. The sum of $\frac{7}{22}$ and $\frac{8}{33}$ is what number?
11. What number represents $5\frac{1}{2}$ less than $7\frac{1}{3}$?
12. Seven-eighths subtracted from nine-tenths yields what number?

The missing number is one of the terms in an addition statement, so we subtract the given term $27\frac{1}{3}$ from the total 60 to find the missing term.

$$\begin{array}{rl} 60 & = 59\frac{3}{3} \\ -27\frac{1}{3} & = 27\frac{1}{3} \\ \hline & 32\frac{2}{3} \end{array}$$

The answer is $32\frac{2}{3}$. ∎

EXAMPLE 2 $\frac{8}{25}$ subtracted from $\frac{1}{3}$ yields what number?

SOLUTION   The question translates to math symbols. Observe that since the phrase *subtracted from* is used, we write $\frac{1}{3}$ first in the symbolic statement.

The missing number is the answer to the subtraction problem, $\frac{1}{3} - \frac{8}{25}$.

$$\frac{1}{3} - \frac{8}{25} = \frac{1 \times 25}{3 \times 25} - \frac{8 \times 3}{25 \times 3}$$
$$= \frac{25}{75} - \frac{24}{75}$$
$$= \frac{1}{75}$$

The answer is $\frac{1}{75}$. ∎

Try Problems 5 through 12.

Answers to Try These Problems

1. $\frac{5}{7} - \frac{3}{7} = \frac{2}{7}$ 2. $8 + 7\frac{1}{2} = 15\frac{1}{2}$ 3. $1\frac{1}{4} - \frac{3}{4} = \frac{1}{2}$ 4. $\frac{1}{3} + \frac{1}{2} = \frac{5}{6}$
5. $2\frac{9}{16}$ 6. $3\frac{7}{9}$ 7. $\frac{1}{20}$ 8. $\frac{59}{45}$ or $1\frac{14}{45}$ 9. $4\frac{1}{4}$ or $\frac{17}{4}$ 10. $\frac{37}{66}$
11. $1\frac{5}{6}$ 12. $\frac{1}{40}$

EXERCISES 5.6

Translate each English statement to an addition or subtraction statement using math symbols.

1. The sum of $2\frac{1}{3}$ and $\frac{2}{3}$ equals 3.
2. $\frac{2}{5}$ increased by $\frac{7}{10}$ is $1\frac{1}{10}$.
3. The difference between 7 and $6\frac{1}{4}$ is $\frac{3}{4}$.
4. $\frac{7}{12}$ subtracted from $\frac{2}{3}$ is equal to $\frac{1}{12}$.
5. $\frac{7}{8}$ plus $\frac{1}{8}$ is 1.
6. The result of $5\frac{1}{2}$ decreased by $\frac{3}{4}$ is $4\frac{3}{4}$.

7. One-half plus three-halves equals two.
8. Thirty-two minus one-fourth is thirty-one and three-fourths.
9. $\frac{3}{10}$ is the result of $\frac{1}{2}$ reduced by $\frac{1}{5}$.
10. $5\frac{6}{7}$ less than 7 is $1\frac{1}{7}$.

Solve.

11. Find the sum of 3 and $2\frac{1}{5}$.
12. Find the difference between 3 and $2\frac{1}{5}$.
13. Subtract $\frac{11}{27}$ from $\frac{4}{9}$.
14. Find the result when $\frac{3}{4}$ is decreased by $\frac{1}{4}$.
15. Find the result when $\frac{3}{4}$ is increased by $\frac{1}{4}$.
16. Find $4\frac{4}{5}$ less than $6\frac{1}{4}$.
17. What number plus $\frac{9}{10}$ yields $3\frac{3}{20}$?
18. $6\frac{5}{8}$ represents $4\frac{2}{3}$ increased by what number?
19. What number is $\frac{17}{20}$ less than $\frac{7}{8}$?
20. $\frac{3}{4}$ subtracted from $\frac{5}{6}$ gives what number?
21. The sum of two-thirds and four-fifths equals what number?
22. What number is the result of subtracting five-twelfths from one?

5.7 APPLICATIONS

OBJECTIVES
- Solve an application that involves a total as the sum of parts.
- Find the perimeter of a figure.
- Solve an application that involves increases or decreases.
- Solve an application that involves comparisons.
- Find the average of two or more numbers.
- Solve an application that involves more than one step.

There are many situations in your daily life in which addition and subtraction of fractions is needed. In this section we look at some of these situations and give guidelines on how to decide whether to add or subtract. At the end of the section we also look at applications that require more than one step. These problems may require multiplication and division, as well as addition and subtraction.

1 Problems Involving a Total as the Sum of Parts

We know that to add means to combine or total. In the addition statement

$$\frac{1}{8} + \frac{6}{8} = \frac{7}{8}$$

the sum $\frac{7}{8}$ is the total and the terms $\frac{1}{8}$ and $\frac{6}{8}$ are parts of the total. Recall from Section 5.5 that if any one of the three numbers is missing, we can find it by using the two given numbers. Here we review the procedure.

> *Finding the Missing Number in a Basic Addition Statement*
> 1. If the sum (or total) is missing, add the terms (or parts) to find the missing number.
> 2. If one of the terms (or parts) is missing, subtract the given term from the total to find the missing term.

The following examples illustrate how to use this concept to solve some application problems.

EXAMPLE 1 Ms. Chung purchased a $2\frac{1}{8}$-pound steak at the supermarket. After the butcher trimmed the steak of fat, the steak weighed $1\frac{3}{4}$ pounds. How much fat was trimmed from the original steak?

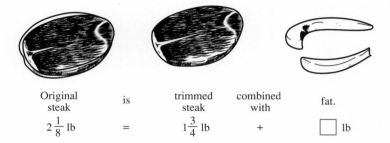

Original steak	is	trimmed steak	combined with	fat.
$2\frac{1}{8}$ lb	=	$1\frac{3}{4}$ lb	+	☐ lb

SOLUTION We are missing part of the total, so we subtract the given term $1\frac{3}{4}$ from the total $2\frac{1}{8}$.

$$2\frac{1}{8} = 2\frac{1}{8} = 1\frac{9}{8}$$
$$-1\frac{3}{4} = 1\frac{6}{8} = 1\frac{6}{8}$$
$$\overline{\frac{3}{8}}$$

$\frac{3}{8}$ pound of fat was trimmed from the steak. ∎

EXAMPLE 2 Find the perimeter of this triangle.

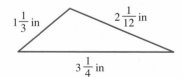

SOLUTION **Perimeter** means the total distance around, so we have the following.

Perimeter = sum of the lengths of the 3 sides

$$\Box = 1\frac{1}{3} + 2\frac{1}{12} + 3\frac{1}{4}$$

We are missing the total, so we add.

5.7 Applications

 Try These Problems

Solve.

1. Ms. Taylor purchased $1\frac{1}{2}$ pounds of Mocha Java coffee beans and $2\frac{1}{2}$ pounds of French Roast coffee beans. What is the total weight of the coffee beans?

2. Sara is hemming drapes for the windows in her apartment. The drapes are now $100\frac{3}{4}$ inches long. They need to be 96 inches long. How wide should she make the hem?

3. Find the distance from B to C.

   A   B   C

   |← $3\frac{1}{2}$ mi →|

   |← $7\frac{3}{5}$ mi →|

4. A desk top is $18\frac{1}{2}$ centimeters wide and $120\frac{1}{4}$ centimeters long. What is the perimeter of the desk top?

5. An oil tanker spills $\frac{3}{8}$ of a load. What fraction of the load remains?

6. Elvira owed money on her credit card. She paid $\frac{3}{10}$ of her debt last month. What fraction of her debt remains?

$$1\frac{1}{3} = 1\frac{1 \times 4}{3 \times 4} = 1\frac{4}{12}$$
$$2\frac{1}{12} = 2\frac{1}{12} = 2\frac{1}{12}$$
$$+\; 3\frac{1}{4} = 3\frac{1 \times 3}{4 \times 3} = 3\frac{3}{12}$$
$$6\frac{8}{12} = 6\frac{\cancel{4} \times 2}{\cancel{4} \times 3} = 6\frac{2}{3}$$

The perimeter of the triangle is $6\frac{2}{3}$ inches. ∎

 Try Problems 1 through 4.

EXAMPLE 3   Peggy spends $\frac{1}{5}$ of her take-home pay for rent. What fraction of her take-home pay is left after she pays the rent?

SOLUTION   All of her take-home pay is represented by the number 1.

$$\text{Fraction spent for rent} + \text{Fraction that remains} = 1$$

$$\frac{1}{5} + \boxed{} = 1$$

We are missing a term in an addition statement, so subtract the given term $\frac{1}{5}$ from the total 1.

$$1 - \frac{1}{5} = \frac{5}{5} - \frac{1}{5}$$
$$= \frac{4}{5}$$

After she pays the rent, $\frac{4}{5}$ of her take-home pay is left. ∎

 Try Problems 5 and 6.

2   Problems Involving Increases and Decreases

The price of a transportation stock decreased from $\$36\frac{1}{2}$ to $\$34$. The amount that the price went down is called the **decrease**. To find the decrease we subtract.

$$\text{Decrease} = \text{higher price} - \text{lower price}$$
$$= \$36\frac{1}{2} - \$34$$
$$= \$2\frac{1}{2}$$

The decrease in price is $\$2\frac{1}{2}$.

Other language can be used to mean decrease. Here we list some situations where a value has gone down and give the corresponding language.

Try These Problems

7. An entertainment stock sold for 28\frac{1}{8}$ per share last week and 24\frac{1}{2}$ per share this week. Find the decrease in price.

8. A biologist was observing closely a rat in an experiment. The rat's weight dropped from $3\frac{2}{3}$ pounds to $2\frac{1}{2}$ pounds in one month. What was the weight loss?

9. The price of a utility stock went from 19\frac{7}{8}$ to 25\frac{1}{2}$. What was the gain in price?

10. A chef increased the amount of chili powder in a chili recipe from $\frac{3}{4}$ teaspoon to 2 teaspoons. What was the increase in the amount of chili powder?

Situation	Words That Indicate That a Value Went Down
Last week Cindy's bird weighed $2\frac{3}{4}$ pounds and this week it weighs only $2\frac{1}{4}$ pounds.	The *loss* in weight is $\frac{1}{2}$ pound. The *reduction* in weight is $\frac{1}{2}$ pound. The *decrease* in weight is $\frac{1}{2}$ pound.
A chef reduced the amount of milk used in a recipe from $4\frac{2}{3}$ cups to 3 cups.	The *reduction* is $1\frac{2}{3}$ cups. The *decrease* is $1\frac{2}{3}$ cups.

▲ **Try Problems 7 and 8.**

A chef increased the amount of paprika used in a recipe from $3\frac{3}{4}$ cups to 4 cups. The amount that the volume of paprika went up is called the **increase** (or **gain**). To find the increase we subtract.

$$\text{Increase} = \frac{\text{larger}}{\text{amount}} - \frac{\text{smaller}}{\text{amount}}$$

$$= 4 - 3\frac{3}{4}$$

$$= 3\frac{4}{4} - 3\frac{3}{4}$$

$$= \frac{1}{4}$$

The increase in the paprika is $\frac{1}{4}$ cup.

▲ **Try Problems 9 and 10.**

3 Problems Involving Comparison

Two unequal numbers, such as 2 and $1\frac{3}{4}$, can be compared using addition and subtraction. Here is a list of several addition and subtraction statements involving 2 and $1\frac{3}{4}$, and the corresponding comparison statements in English.

Comparing Two Numbers	
Math Symbols	**English**
$1\frac{3}{4} + \frac{1}{4} = 2$	$\frac{1}{4}$ more than $1\frac{3}{4}$ is 2.
$2 - \frac{1}{4} = 1\frac{3}{4}$	$\frac{1}{4}$ less than 2 is $1\frac{3}{4}$.
$2 - 1\frac{3}{4} = \frac{1}{4}$	2 and $1\frac{3}{4}$ differ by $\frac{1}{4}$.
	The *difference* between 2 and $1\frac{3}{4}$ is $\frac{1}{4}$.
	$1\frac{3}{4}$ is less than 2 by $\frac{1}{4}$.

▲ Try These Problems

11. Maria uses $2\frac{1}{4}$ cups of Swiss cheese to make a cheese pie that serves 8 persons. José uses half a cup less Swiss cheese when he makes cheese pie for 8. How much cheese does José use in his pie?

12. Manuel studied $3\frac{1}{2}$ hours yesterday. Jamie studied 5 hours. How much longer did Jamie study than Manuel?

13. Stephanie is $1\frac{2}{3}$ feet taller than her younger sister Caitlin. If Stephanie is $5\frac{1}{3}$ feet tall, how tall is Caitlin?

14. Dick has been working at the Stock Investment Company for $4\frac{1}{2}$ years. Rob has worked there $1\frac{1}{2}$ fewer years than Dick. How long has Rob worked at the Stock Investment Company?

Now we solve some problems involving comparisons.

EXAMPLE 4 The length of a rectangle is $3\frac{7}{8}$ feet longer than the width. If the length is 8 feet, find the width.

SOLUTION The first sentence of the problem is comparing the length and width. Translate this sentence to math symbols.

$$\text{The length} \text{ is } 3\frac{7}{8} \text{ ft longer than } \text{the width}.$$

$$8 = 3\frac{7}{8} + \square$$

We are missing one of the terms, so we subtract the given term from the sum to find the missing term.

$$\begin{array}{r} 8 = 7\frac{8}{8} \\ -3\frac{7}{8} = 3\frac{7}{8} \\ \hline 4\frac{1}{8} \end{array}$$

The width of the rectangle is $4\frac{1}{8}$ feet. ■

EXAMPLE 5 Mary jogged $2\frac{1}{2}$ miles yesterday. Alex jogged only $1\frac{3}{4}$ miles. How much farther did Mary jog than Alex?

SOLUTION First we show a picture of the situation.

We want to know the *difference* between $2\frac{1}{2}$ and $1\frac{3}{4}$, so we subtract.

$$\begin{array}{r} 2\frac{1}{2} = 2\frac{2}{4} = 1\frac{6}{4} \\ -1\frac{3}{4} = 1\frac{3}{4} = 1\frac{3}{4} \\ \hline \frac{3}{4} \end{array}$$

Mary jogged $\frac{3}{4}$ mile farther than Alex. ■

▲ **Try Problems 11 through 14.**

4 Averaging

Kathy bought 4 steaks weighing $1\frac{1}{4}$ pounds, $1\frac{3}{4}$ pounds, $\frac{3}{4}$ pound, and $2\frac{1}{4}$ pounds. The total weight of the steaks is 6 pounds.

$$1\frac{1}{4} + 1\frac{3}{4} + \frac{3}{4} + 2\frac{1}{4} = 4\frac{8}{4} = 6$$

What same weight could each of the steaks have and still have a total weight of 6 pounds?

$$6 \div 4 = \frac{6}{4} = \frac{3}{2} \text{ or } 1\frac{1}{2}$$

$$\begin{array}{r} 1\frac{1}{2} \\ 1\frac{1}{2} \\ 1\frac{1}{2} \\ + 1\frac{1}{2} \\ \hline 4\frac{4}{2} = 6 \end{array}\Bigg\}$$ Four weights of $1\frac{1}{2}$ pounds total 6 pounds.

The weight $1\frac{1}{2}$ pounds is called the **average** (or **mean**) of the weights $1\frac{1}{4}$, $1\frac{3}{4}$, $\frac{3}{4}$, and $2\frac{1}{4}$. The average of a set of data is a measure of the middle or center of the data. The average of a collection of numbers can be found by adding the numbers, then dividing by how many numbers there are.

Finding the Average of a Collection of Numbers
1. Add all the numbers in the collection.
2. Divide the sum by how many numbers are in the collection.

EXAMPLE 6 Find the average of $20\frac{1}{2}$, $31\frac{3}{4}$, and $18\frac{1}{4}$.

SOLUTION Add the 3 numbers.

$$\begin{array}{r} 20\frac{1}{2} = 20\frac{2}{4} \\ 31\frac{3}{4} = 31\frac{3}{4} \\ + 18\frac{1}{4} = 18\frac{1}{4} \\ \hline 69\frac{6}{4} = 70\frac{2}{4} = 70\frac{1}{2} \end{array}$$

Divide the sum $70\frac{1}{2}$ by 3 because there are 3 numbers.

$$70\frac{1}{2} \div 3 = \frac{141}{2} \div \frac{3}{1}$$

$$= \frac{\overset{47}{\cancel{141}}}{2} \times \frac{1}{\underset{1}{\cancel{3}}}$$

$$= \frac{47}{2} \text{ or } 23\frac{1}{2}$$

The average is $\frac{47}{2}$ or $23\frac{1}{2}$. ■

Try These Problems

15. Find the average of $\frac{1}{3}, \frac{1}{4}, \frac{3}{4}, \frac{2}{3}$, and $\frac{5}{6}$.

16. Ms. Allen is responsible for making the coffee in the faculty lounge this semester. She recorded the following usage of coffee over a 3-week period.
1st Week $2\frac{1}{2}$ pounds
2nd Week $3\frac{3}{4}$ pounds
3rd Week   3 pounds
What was the average amount of coffee used each week?

EXAMPLE 7   During the last month a small neighborhood bakery used the following amounts of sugar.

Week 1	$10\frac{1}{2}$ pounds
Week 2	$12\frac{3}{4}$ pounds
Week 3	8 pounds
Week 4	$11\frac{1}{4}$ pounds

On the average, how many pounds of sugar were used per week?

SOLUTION   Find the total weight.

$$\begin{aligned} 10\frac{1}{2} &= 10\frac{2}{4} \\ 12\frac{3}{4} &= 12\frac{3}{4} \\ 8 &= 8 \\ +11\frac{1}{4} &= 11\frac{1}{4} \\ \hline &41\frac{6}{4} = 42\frac{2}{4} = 42\frac{1}{2} \end{aligned}$$

The total weight is $42\frac{1}{2}$ pounds. Divide $42\frac{1}{2}$ pounds by 4 because there are 4 weeks.

$$\begin{aligned} 42\frac{1}{2} \div 4 &= \frac{85}{2} \div \frac{4}{1} \\ &= \frac{85}{2} \times \frac{1}{4} \\ &= \frac{85}{8} \text{ or } 10\frac{5}{8} \end{aligned}$$

The bakery used $10\frac{5}{8}$ pounds of sugar per week. ∎

 Try Problems 15 and 16.

5 Problems Involving More Than One Step

Now we put together some of the concepts, language, and procedures we have previously learned to solve problems that involve more than one step. These problems may require multiplication and division as well as addition and subtraction. When solving more complex problems, it is important to label your work clearly with words so that you understand what it is you have obtained at each stage of the problem. Also, if applicable, drawing a picture can be very helpful. Here are some examples.

Try These Problems

17. In the triangle shown here, the longest side is $\frac{7}{8}$ inch longer than the shortest side. The shortest side is $1\frac{3}{8}$ inches long. Find the perimeter of the triangle.

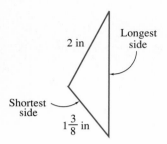

EXAMPLE 8 Shirley is making a rectangular flower bed. The width is $5\frac{1}{3}$ feet. The length is $2\frac{1}{4}$ feet more than the width. Find the perimeter.

SOLUTION Here we view a picture of the situation.

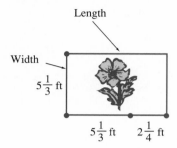

Before finding the perimeter, we must find the length. The third sentence in the problem compares the length and width. Translate this sentence to math symbols.

$$\text{The length} \text{ is } 2\frac{1}{4} \text{ ft more than } \text{the width}.$$

$$\square = 2\frac{1}{4} + 5\frac{1}{3}$$

The length is the sum of $2\frac{1}{4}$ and $5\frac{1}{3}$.

$$\begin{aligned} 5\frac{1}{3} &= 5\frac{4}{12} \\ +\ 2\frac{1}{4} &= 2\frac{3}{12} \\ \hline &\ 7\frac{7}{12} \end{aligned}$$

The length of the rectangle is $7\frac{7}{12}$ feet. The perimeter means the total distance around the rectangle, so we add two lengths and two widths.

$$\begin{aligned} 5\frac{1}{3} &= 5\frac{4}{12} \\ 5\frac{1}{3} &= 5\frac{4}{12} \\ 7\frac{7}{12} &= 7\frac{7}{12} \\ +\ 7\frac{7}{12} &= 7\frac{7}{12} \\ \hline 24\frac{22}{12} &= 25\frac{10}{12} = 25\frac{5}{6} \end{aligned}$$

The perimeter is $25\frac{5}{6}$ feet. ■

 Try Problem 17.

EXAMPLE 9 Steven read $\frac{1}{6}$ of his assignment yesterday and $\frac{2}{5}$ of it today. What fraction of his assignment does he have left to read?

Try These Problems

18. Three persons share the ownership of a building. Don's share is $\frac{3}{10}$, Barbara's share is $\frac{1}{4}$, and David owns the rest. What fraction of the building does David own?

SOLUTION First we find the fraction of the assignment he has read in the 2-day period.

$$\begin{aligned}\frac{1}{6} &= \frac{5}{30}\\ +\frac{2}{5} &= \frac{12}{30}\\ \hline &\frac{17}{30}\end{aligned}$$ —— Fraction he has read

Steven read $\frac{17}{30}$ of his assignment in the two days. The whole assignment is represented by the number 1.

$$\begin{aligned}1 &= \frac{30}{30}\\ -\frac{17}{30} &= \frac{17}{30}\\ \hline &\frac{13}{30}\end{aligned}$$ —— The whole assignment
—— Fraction he has read
—— Fraction left to read

Steven has $\frac{13}{30}$ of his assignment left to read. ■

 Try Problem 18.

EXAMPLE 10 What is the difference in the areas of these two rectangles?

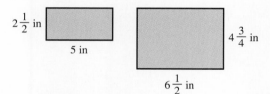

SOLUTION First we find the areas of each of the rectangles.

$$\begin{aligned}\text{Area of the smaller rectangle} &= \text{length} \times \text{width}\\ &= 5 \times 2\frac{1}{2}\\ &= \frac{5}{1} \times \frac{5}{2}\\ &= \frac{25}{2} \text{ or } 12\frac{1}{2} \text{ sq in}\end{aligned}$$

$$\begin{aligned}\text{Area of the larger rectangle} &= \text{length} \times \text{width}\\ &= 6\frac{1}{2} \times 4\frac{3}{4}\\ &= \frac{13}{2} \times \frac{19}{4}\\ &= \frac{247}{8} \text{ or } 30\frac{7}{8} \text{ sq in}\end{aligned}$$

Try These Problems

19. Find the total area of this floor.

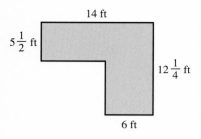

Now to find the *difference* in the areas, we subtract.

$$\begin{array}{c}\text{Difference} \\ \text{in areas} \\ \text{(sq in)}\end{array} = \begin{array}{c}\text{larger} \\ \text{area} \\ \text{(sq in)}\end{array} - \begin{array}{c}\text{smaller} \\ \text{area} \\ \text{(sq in)}\end{array}$$

$$= 30\frac{7}{8} - 12\frac{1}{2}$$
$$= 30\frac{7}{8} - 12\frac{4}{8}$$
$$= 18\frac{3}{8}$$

The difference in the areas is $18\frac{3}{8}$ square inches. ■

 Try Problem 19.

Now we summarize the material presented in this section by giving guidelines that will help you decide whether to add or subtract.

Situations Requiring Addition or Subtraction	
Operation	**Situations**
+	1. You are looking for the total or whole.
	2. You are looking for the result when a quantity has been increased.
−	1. You are looking for one of the parts in a total or whole.
	2. You are looking for the result when a quantity has been decreased.
	3. You are looking for how much larger one quantity is than another.
	4. You are looking for how much smaller one quantity is than another.

 Answers to Try These Problems

1. 4 lb **2.** $4\frac{3}{4}$ in **3.** $4\frac{1}{10}$ mi **4.** $277\frac{1}{2}$ cm **5.** $\frac{5}{8}$
6. $\frac{7}{10}$ **7.** $\$3\frac{5}{8}$ **8.** $1\frac{1}{6}$ lb **9.** $\$5\frac{5}{8}$ **10.** $1\frac{1}{4}$ t **11.** $\frac{7}{4}$ or $1\frac{3}{4}$ cups
12. $1\frac{1}{2}$ hr **13.** $3\frac{2}{3}$ ft **14.** 3 yr **15.** $\frac{17}{30}$ **16.** $\frac{37}{12}$ or $3\frac{1}{12}$ lb
17. $5\frac{5}{8}$ in **18.** $\frac{9}{20}$ **19.** $117\frac{1}{2}$ sq ft

EXERCISES 5.7

Solve.

1. Carlos paid $\frac{1}{3}$ of a debt in June, $\frac{1}{10}$ in July, and $\frac{2}{5}$ in August. What fraction of the debt was paid in these three months?
2. Find the perimeter of a rectangle that is $5\frac{7}{12}$ feet wide and $8\frac{3}{4}$ feet long.

3. A recipe calls for $1\frac{1}{3}$ cups of milk. Bette has only $\frac{3}{4}$ cup of milk. How much more milk does Bette need?

4. Mr. Nguyen wants to drive 380 miles today. He has already driven $192\frac{3}{4}$ miles. How much farther does he have to drive?

5. Lisa spends $\frac{1}{4}$ of her income on rent. What fraction of her income is left?

6. Dave owns $\frac{7}{10}$ of a building and Ann owns the rest. What fraction of the building does Ann own?

7. You set out to walk a distance of $2\frac{1}{2}$ miles. So far, you have gone $\frac{3}{4}$ mile. How much farther do you have to go?

8. Two years ago, Tish planted a tree that she would like to be 6 feet tall. It is now $3\frac{2}{3}$ feet tall. How much more must the tree grow to reach her goal?

9. A spaghetti recipe for 12 persons calls for $2\frac{1}{2}$ cups of parsley. Bruce is especially fond of parsley, so he decided to increase the amount of parsley by $\frac{1}{2}$ cup. What is the total amount of parsley Bruce puts in the spaghetti?

10. An oil stock was selling for $\$100\frac{3}{4}$ per share, then the price decreased by $\$5\frac{1}{2}$. What is the price after the decrease?

11. A puppy's weight dropped from $8\frac{3}{4}$ pounds to $5\frac{7}{16}$ pounds. Find the weight loss.

12. A chef increased the amount of cheese in a recipe from $5\frac{1}{4}$ cups to 8 cups. What is the increase?

13. Today Mildred worked $3\frac{1}{2}$ hours on her project, which is an increase of $1\frac{3}{4}$ hours over yesterday. How long did she work yesterday?

14. Today the price of a utility stock increased by $\$5\frac{7}{16}$. If the price after the increase is $38, what was the price before the increase?

15. A mechanic cut a piece of wire $5\frac{3}{4}$ inches long, then realized the piece was too short. He needs it to be $\frac{7}{16}$ inch longer. How long should he have cut the piece of wire?

16. A carpenter cut a piece of lumber $6\frac{1}{2}$ feet long. She realized the piece was too long by $\frac{3}{4}$ foot. How long should the piece of lumber be?

17. Harvey accidentally picked up a $\frac{1}{2}$-inch wrench to tighten a $\frac{7}{16}$-inch bolt. Was the wrench too large or too small, and by how much?

18. A microwave oven is $20\frac{1}{4}$ inches wide, and the space in Sam's kitchen for the oven is $20\frac{3}{8}$ inches wide. Will the oven fit in the space; and if so, how much space remains?

19. John worked $5\frac{3}{4}$ hours longer than Steve. If John worked $11\frac{1}{3}$ hours, how long did Steve work?

20. The length of a rectangle is $15\frac{2}{3}$ feet longer than the width. If the length is 38 feet, find the width.

21. Find the average of $\frac{5}{8}$ and $\frac{7}{12}$.

22. Find the average of $\frac{1}{2}$, $\frac{3}{8}$, $\frac{2}{3}$, and $\frac{1}{4}$.

23. Sam hiked $2\frac{1}{2}$ miles the first hour, 4 miles the second hour, and $3\frac{3}{4}$ miles the third hour. On the average, how far did he hike each hour?

24. Five boxes weigh $1\frac{1}{4}$ pounds, $2\frac{1}{10}$ pounds, $1\frac{3}{4}$ pounds, 2 pounds and $2\frac{3}{5}$ pounds. What is the averge weight of the 5 boxes?

25. Three rats weigh $2\frac{1}{3}$ pounds each and another rat weighs $3\frac{1}{4}$ pounds. What is the average weight of the 4 rats?

26. Five boxes weigh $7\frac{1}{2}$ pounds each and three boxes weigh $5\frac{2}{3}$ pounds each. What is the average weight of the 8 boxes?
27. Alice has $500. She spends $\frac{2}{5}$ of it for clothes. How much money is left?
28. A math class has 56 students. Five-eighths of the class are men. How many women are in the class?
29. Sally saved $150 out of a monthly income of $2400. Nancy saved $125 out of a monthly income of $1500. Which person saved the larger fractional portion of her income, and what is the difference in these fractions?
30. A stock decreased in price from $30 to 28\frac{3}{4}$. The decrease in price is what fraction of the original price?
31. Find the perimeter around this floor space.

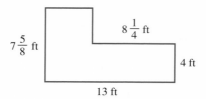

32. Find the total area of this region.

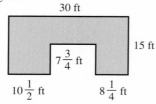

Hannah, Bonnie, Gretchen, and Anna bought some land outside of Washington, D.C. for $72,000. The fractional portion of each person's share is given by the circle graph. Use this information to answer Exercises 33 through 36.

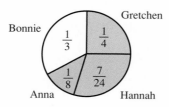

Fractional Parts of Land Valued at $72,000

33. Show that the four fractional parts add up to 1 whole.
34. What fractional part of the land do Gretchen and Hannah together own?
35. How much did Anna pay for her share?
36. How much did Hannah pay for her share?

The bar graph shows the rainfall for a 5-day period in Chicago, Illinois. Use the graph to answer Exercises 37 through 40.

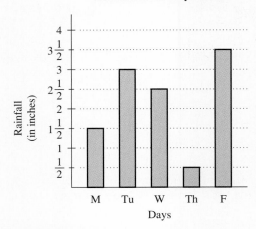

Rainfall for a 5-Day Period

37. Find the total rainfall for the 5-day period.
38. How much more rain fell on Tuesday than on Thursday?
39. On the average, how many inches did it rain each day?
40. Find the ratio of the amount of rain on Friday to the total amount for the 5-day period.

USING THE CALCULATOR # 7

MORE ON THE [a b/c] KEY

A calculator with an [a b/c] key is capable of adding, subtracting, multiplying, and dividing fractions so that the result is a fraction. Here we look at a division problem.

To Compute $6\frac{2}{3} \div \frac{4}{5}$

Enter   6 [a b/c] 2 [a b/c] 3 [÷] 4 [a b/c] 5 [=]

Result   8 ⌐ 1 ⌐ 3.

Therefore, the answer is $8\frac{1}{3}$. Recall that you can obtain the improper fraction form by entering [INV] and [a b/c] when the mixed numeral form is on the display screen.

Next we look at computing an addition of fractions problem by using the [a b/c] key.

To Compute $\frac{5}{27} + \frac{7}{36}$

Enter   5 [a b/c] 27 [+] 7 [a b/c] 36 [=]

Result   41 ⌐ 108.

Therefore, the answer is $\frac{41}{108}$. The result is given in reduced form.

Calculator Problems

Use the [a b/c] key to compute each of the following.

1. $\frac{2}{5} \times \frac{7}{4}$
2. $\frac{12}{25} \div 4\frac{3}{50}$
3. $6\frac{1}{3} - 2\frac{7}{12}$
4. $\frac{1}{20} + \frac{7}{36}$
5. $25\frac{8}{35} + 46\frac{1}{7}$
6. $\frac{4}{51} \times 5\frac{7}{19}$

CHAPTER 5 SUMMARY

KEY WORDS AND PHRASES
least common multiple (LCM) [5.2]
least common denominator (LCD) [5.2]
prime number [5.2]
prime factoring [5.2]
borrowing [5.3]
sum [5.6]
difference [5.6, 5.7]
perimeter [5.7]
decrease [5.7]
increase (gain) [5.7]
average (mean) [5.7]

SYMBOLS
$2 < 4$ means   2 is less than 4. [5.4]
$4 > 2$ means   4 is more than 2. [5.4]

IMPORTANT RULES

How to Add Fractions [5.1, 5.2, 5.3]
- Find the least common multiple (LCM) of the denominators.
- Convert each fraction to an equivalent fraction that has the common denominator.
- Add the numerators to obtain the numerator of the result. The common denominator is the denominator of the result.
- Simplify the answer.

How to Compare Fractions [5.4]
- Find the least common multiple (LCM) of the denominators.
- Convert each fraction to an equivalent fraction that has the common denominator.
- When two fractions have the same denominator, the larger fraction has the larger numerator.

The Perimeter of a Figure [5.7]
The perimeter of a figure is the distance all the way around the figure.

The Average (or Mean) of a Collection of Numbers [5.7]
- Add all of the numbers in the collection.
- Divide the sum by how many numbers are in the collection.

Situations that Require Addition and Subtraction [5.7]
A summary of the situations that require addition and subtraction is on page 214.

CHAPTER 5 REVIEW EXERCISES

Choose the prime numbers in each group.

1. 9, 13, 23, 51, 38   **2.** 2, 7, 17, 37, 57

Write each number as the product of primes.

3. 35   **4.** 72   **5.** 78   **6.** 171

Find the least common multiple of each group of numbers.

7. 6, 10   **8.** 14, 21   **9.** 5, 15, 9
10. 22, 33, 4   **11.** 8, 17   **12.** 9, 4, 5

Add. Simplify if possible.

13. $\frac{3}{4} + \frac{2}{4}$ **14.** $5\frac{1}{6} + 6\frac{5}{6}$ **15.** $\frac{29}{60} + \frac{7}{60}$ **16.** $\frac{2}{3} + \frac{1}{9}$
17. $8\frac{7}{10} + 7\frac{7}{15}$ **18.** $\frac{2}{3} + \frac{1}{7}$ **19.** $13\frac{11}{12} + \frac{4}{9}$ **20.** $\frac{5}{11} + \frac{1}{3}$

21. $6\frac{3}{14} + 29\frac{5}{21}$

22. $\frac{5}{16}$
$\frac{17}{40}$
$+ \frac{1}{10}$

23. $\frac{1}{2}$
$\frac{3}{13}$
$+ \frac{3}{4}$

24. $34\frac{4}{75}$
$18\frac{1}{10}$
$+ 5\frac{1}{6}$

Name the larger fraction.

25. $\frac{3}{9}, \frac{10}{9}$

26. $\frac{3}{11}, \frac{1}{5}$

27. $\frac{3}{4}, \frac{17}{24}$

List from smallest to largest.

28. $\frac{4}{9}, \frac{1}{3}, \frac{8}{27}$

29. $\frac{2}{7}, \frac{2}{3}, \frac{2}{5}$

30. $\frac{19}{20}, 1, \frac{47}{50}$

Use the symbol <, =, or > to accurately compare each pair of fractions.

31. $\frac{5}{7} \underset{(<, =, >)}{?} \frac{13}{21}$

32. $\frac{7}{8} \underset{(<, =, >)}{?} \frac{11}{12}$

33. $6\frac{3}{10} \underset{(<, =, >)}{?} 6\frac{3}{25}$

34. $3\frac{3}{4} \underset{(<, =, >)}{?} 3\frac{2}{3}$

35. $1 \underset{(<, =, >)}{?} \frac{99}{100}$

Subtract. Simplify if possible.

36. $\frac{7}{12} - \frac{3}{12}$

37. $5\frac{7}{8} - 3\frac{1}{8}$

38. $8 - \frac{3}{5}$

39. $17\frac{2}{7} - 5\frac{5}{7}$

40. $\frac{7}{8} - \frac{5}{12}$

41. $12\frac{5}{9} - 3$

42. $6\frac{3}{10}$
$- 2\frac{9}{20}$

43. $10\frac{7}{15}$
$- 3\frac{5}{12}$

44. $28\frac{5}{49}$
$- 8\frac{3}{14}$

Translate each of the following to an addition or subtraction statement using math symbols.

45. The difference between $2\frac{1}{3}$ and 2 is $\frac{1}{3}$.

46. When $\frac{3}{5}$ is increased by $\frac{4}{5}$, the result is $\frac{7}{5}$.

47. The sum of $\frac{2}{3}$ and $\frac{5}{3}$ is $\frac{7}{3}$.

48. 5 diminished by $2\frac{1}{3}$ leaves $2\frac{2}{3}$.

49. $\frac{2}{10}$ more than $\frac{3}{10}$ is $\frac{1}{2}$.

Solve.

50. Find the total when $2\frac{1}{7}$ is added to $\frac{3}{7}$.

51. Find the result of $8\frac{1}{5}$ take away 3.

52. What is $\frac{3}{8}$ subtracted from $\frac{3}{4}$?

53. Find the average of $6\frac{5}{6}$, $3\frac{2}{3}$, and $7\frac{1}{2}$.

54. Steven lives $3\frac{7}{10}$ miles from work and Shirley lives $3\frac{4}{5}$ miles from work. Which one lives farther from work?

55. What is the distance from P to Q?

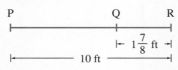

56. Find the perimeter of a rectangle that has a width of $12\frac{1}{3}$ inches and a length of $20\frac{3}{4}$ inches.

57. Marge is half of a foot shorter than Bob. If Bob is 6 feet tall, how tall is Marge?

58. Gloria typed $4\frac{1}{2}$ hours on Wednesday, $6\frac{1}{4}$ hours on Tuesday, and $1\frac{1}{2}$ hours on Monday. What is the total number of hours Gloria typed in the three days?

59. Jean-Louis has a ficus tree in his apartment that is $5\frac{3}{4}$ feet high. He wants the tree to grow to the ceiling, which is 9 feet high. How much higher does the tree need to grow before it reaches the ceiling?

60. Kurt purchased $2\frac{1}{2}$ pounds of ground beef at the meat market. When he got home, he realized he would need 4 pounds to make hamburgers for the barbeque party that evening. How much more meat does he need to buy?

61. Mike was packing his backpack for a hiking trip. He has put in items weighing $5\frac{1}{4}$ pounds, $2\frac{1}{2}$ pounds, and $1\frac{3}{4}$ pounds. What is the total weight so far?

62. An author receives $3000 more of her royalty advance when she has completed $\frac{2}{3}$ of the manuscript. She has completed $\frac{250}{450}$ of the manuscript. Has she completed enough to receive the royalty advance?

63. Ms. Toscano jogged $2\frac{1}{2}$ miles for each of 4 consecutive days, and $4\frac{1}{3}$ miles for each of 3 consecutive days. What was her average daily distance for the 7-day period?

64. In a certain city, $\frac{13}{20}$ of the registered voters are female. What fraction of the voters are male?

65. Doug and Elaine are writing a book. Doug finished $\frac{1}{4}$ of the job and Elaine finished $\frac{2}{3}$ of the job. What fraction of the job is left to do?

66. In a sample of 2700 college students, it was found that $\frac{4}{15}$ of them are smokers. How many students in the sample are nonsmokers?

67. A boat can cruise in still water at the rate of $15\frac{1}{2}$ miles per hour. How fast does the boat go downstream with a current of $2\frac{2}{3}$ miles per hour?

68. The length of a rectangle is $5\frac{3}{8}$ feet more than the width. If the width is $7\frac{1}{2}$ feet, find the area of the rectangle.

69. In a sample of 1500 college students, it was found that $\frac{2}{5}$ of them take public transportation to school. How many students in the sample take public transportation to school?

70. A stock decreased in price from $30 to $$28\frac{3}{4}$. The decrease in price is what fraction of the original price?

✍ In Your Own Words

Write complete sentences to discuss each of the following. Support your comments with examples or pictures, if appropriate.

71. Suppose you have a child in school that is having trouble with fractions. The child incorrectly writes $\frac{9}{10}$ as the answer to the problem $\frac{2}{5} + \frac{7}{5}$. Discuss how you would help the child understand the error made.

72. You have a friend who solved a subtraction problem incorrectly as follows.

$$\begin{aligned} 8\tfrac{1}{3} &= 7\tfrac{11}{3} \\ -\,2\tfrac{2}{3} &= 2\tfrac{2}{3} \\ \hline 5\tfrac{9}{3} &= 8 \end{aligned}$$

Discuss how you would help your friend understand the error made.

73. Discuss the procedure you would use to find out which of the fractions, $\frac{13}{24}$ or $\frac{9}{16}$, is smaller?

74. Discuss several ways to read the subtraction statement $5 - 2\frac{3}{4} = 2\frac{1}{4}$.

75. Discuss a real-life situation that would involve finding the average of several numbers where at least one of the numbers is a fraction or mixed numeral.

CHAPTER 5 PRACTICE TEST

1. Choose the prime numbers in this group. 17, 21, 23, 33, 39

Write each number as the product of primes.

2. 42   **3.** 110

Find the least common multiple of each group of numbers.

4. 15, 20   **5.** 6, 27, 8

Add and simplify.

6. $\frac{5}{6} + \frac{3}{6}$ **7.** $\frac{4}{9} + \frac{5}{18}$ **8.** $6\frac{14}{15} + 8\frac{3}{25}$ **9.** $\frac{3}{4} + \frac{5}{6} + \frac{5}{9}$ **10.** $3\frac{4}{21} + 8\frac{2}{3} + 7\frac{1}{14}$

11. Which is larger, $\frac{7}{18}$ or $\frac{1}{2}$?

12. List from smallest to largest: $\frac{2}{3}, \frac{3}{4}, \frac{7}{12}$

13. Use the symbol $<$, $=$, or $>$ to compare $\frac{17}{24}$ and $\frac{3}{4}$.

Subtract. Simplify if possible.

14. $\frac{3}{4} - \frac{5}{8}$ **15.** $6\frac{4}{15} - \frac{4}{9}$ **16.** $5\frac{7}{11} - 1\frac{3}{22}$

17. $15\frac{5}{12} - 9\frac{1}{8}$ **18.** $10 - 4\frac{2}{3}$

Translate each of the following to an addition or subtraction statement using math symbols.

19. The sum of $\frac{1}{3}$ and $\frac{2}{3}$ is 1.

20. 9 reduced by $1\frac{3}{4}$ is $7\frac{1}{4}$.

Solve.

21. Find the difference between $\frac{2}{3}$ and $\frac{1}{6}$.

22. What is the result when $\frac{4}{5}$ is increased by $\frac{3}{5}$?

23. Find the average of 20, $16\frac{1}{2}$, 12, and $23\frac{3}{4}$.

24. How far is it from B to C?

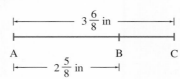

25. Find the perimeter of this rectangle.

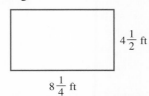

26. Steven is $\frac{3}{4}$ foot taller than Barbara. Barbara is $5\frac{1}{2}$ feet tall. How tall is Steven?
27. Thuy has already put in $2\frac{1}{4}$ cups of milk in the cream sauce she is making. She realizes that she really wants a total of 3 cups of milk in the cream sauce. How much more milk should she put in?
28. A small hose fills a swimming pool at the rate of $1\frac{1}{2}$ gallons per minute. A large hose fills the pool at $5\frac{3}{4}$ gallons per minute. If both hoses are turned on, how fast is the pool being filled?
29. A manufacturer puts $\frac{2}{5}$ of its soft drinks in bottles and the rest in cans. What fraction of the soft drinks are put in cans?
30. The width of a rectangle is $2\frac{7}{8}$ inches less than its length. If the length is $10\frac{1}{8}$ inches, find the area of the rectangle.

CUMULATIVE REVIEW EXERCISES: CHAPTERS 1–5

1. Write the English name for $\frac{13}{100}$.
2. Convert $5\frac{7}{12}$ to an improper fraction.
3. Convert $\frac{160}{15}$ to a mixed numeral.
4. Reduce to lowest terms: $\frac{1800}{210}$
5. List from smallest to largest: $\frac{3}{5}, \frac{2}{3}, \frac{17}{30}$
6. Find an improper fraction with denominator 9 that has the same value as the whole numeral 7.

Write each number as the product of primes.

7. 690
8. 231
9. 1200
10. 3400

Find the least common multiple of each group of numbers.

11. 45; 63
12. 14; 16; 20
13. 80; 32; 48

Perform the indicated operation.

14. $\frac{2}{5} + 7\frac{2}{3}$
15. $\frac{8}{15} \times \frac{45}{60}$
16. $8\frac{1}{3} \div 3\frac{1}{3}$
17. $19 - 6\frac{7}{8}$
18. $65\frac{1}{4} - 48\frac{5}{6}$
19. $\frac{40}{100} \times \frac{25}{36} \times \frac{24}{18}$

Find the missing number.

20. $300 \times \boxed{} = 50$
21. $12\frac{1}{8} + \boxed{} = 25$
22. $6\frac{3}{4} = \boxed{} \times 9\frac{1}{3}$
23. $\boxed{} = \frac{5}{9} + \frac{7}{12}$

Solve. If the result is not a whole numeral, answer with a fraction that is reduced to lowest terms.

24. Find the quotient and remainder when 7895 is divided by 63.
25. Multiply 680 by 96, then add 758 to the product.
26. Subtract 458 from 4020, then multiply the result by 87.
27. Denise has a piece of ribbon that is 84 inches long.
    a. If she cuts the ribbon into 21 equal pieces, how long will each piece be?
    b. If she cuts the ribbon into 20 equal pieces, how long will each piece be?
    c. If she cuts the ribbon into 90 equal pieces, how long will each piece be?
28. A truck contains 24,000 pounds of vegetables, consisting of tomatoes, broccoli, and squash. There are 10,000 pounds of tomatoes and 7200 pounds of broccoli.
    a. How many pounds of squash are on the truck?
    b. The weight of the broccoli is what fraction of the total load?
    c. Find the ratio of the weight of the tomatoes to the weight of the squash.

29. Scott walks $3\frac{3}{4}$ miles per hour.
    a. How far can he walk in 5 hours?
    b. How far can he walk in $6\frac{1}{2}$ hours?
    c. How long will it take him to walk 1 mile?
    d. How long will it take him to walk 4 miles?

30. An oil tank off the coast of Alaska contained 500 tons of oil. During a recent spill, $\frac{2}{25}$ of the oil was lost.
    a. How much oil spilled?
    b. How much oil remained in the tank?
    c. What fraction of the oil remained in the tank?

31. The width of a rectangular floor is $42\frac{1}{2}$ feet. The length is 3 times the width.
    a. Find the length of the floor.
    b. Find the distance all the way around the floor.
    c. Find the area of the floor.
    d. If tile costs $4 per square foot, how much will it cost to tile this floor?

CHAPTER 6

Decimals: Place Value, Addition, and Subtraction

6.1 AN INTRODUCTION TO DECIMALS

OBJECTIVES
- Specify the digit in a decimal numeral that has a certain place value.
- Specify the place value of a certain digit in a decimal numeral.
- Write the expanded form for a decimal numeral.
- Write a decimal numeral, given its expanded form.
- Convert a decimal fraction from decimal form to fractional form and vice versa.
- Write a decimal numeral when given the English name of the numeral.
- Write the English name of a decimal numeral.

1 Decimal Notation and Place Value

Would you rather have $23 or $8.97? Of course you would rather have the $23 because that is more money. Expressing money values is one of the many uses of decimals.

A **decimal** is simply another way of expressing the numbers we have already studied. For example,

$$0.2 = \frac{2}{10} \quad \text{and} \quad 4.25 = 4\frac{25}{100}$$

The decimal system is an extension of the system used to express whole numbers. Here we show a decimal and its **expanded form.**

DECIMAL   EXPANDED FORM

52.469 means $50 + 2 + \frac{4}{10} + \frac{6}{100} + \frac{9}{1000}$

The numbers 5, 2, 4, 6, and 9 are called **digits.** Changing the position of a digit changes the value of the decimal. Each position on either side of the decimal point has a **place value.** The place values for the decimal 52.469 are given in the following diagram.

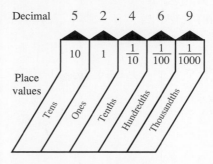

The decimal point is used to separate the whole number portion from the portion that is part of a whole.

$$\underbrace{5\,2}_{\text{whole number}} . \underbrace{4\,6\,9}_{\text{part of a whole}} \leftarrow \text{decimal point}$$

When the decimal has no whole part, we often place a zero in the ones place to bring attention to the decimal point. Here is an example:

$$.18 = 0.18$$

When there is no fractional part to the decimal, we can place one or more zeros past the decimal point to emphasize that there is no fractional part, or we can omit the decimal point and write the number as a whole numeral. Here are some examples:

$$75.00 = 75.0 = 75. = 75$$
$$\$150.00 = \$150. = \$150$$

Here are some more examples of attaching extra zeros to a decimal without changing the value of the decimal.

$$2.6 = 02.6 = 02.600 = 002.6000$$
$$14.709 = 14.70900 = 0014.709$$

Now we look at a diagram that gives more place values. Study this diagram carefully enough so that you make the following observations.

1. As you move from right to left (←), each place value is 10 times the one before it.
2. As you move from left to right (→), each place value is $\frac{1}{10}$ of the one before it.
3. The place value nearest the decimal point on the left is *ones*.
4. The place value nearest the decimal point on the right is *tenths*.
5. The English names for the fractional place values $\frac{1}{10}$, $\frac{1}{100}$, $\frac{1}{1000}$, and so on, all end in *ths*. For example, $\frac{1}{1000}$ is read "one thousand*ths.*"

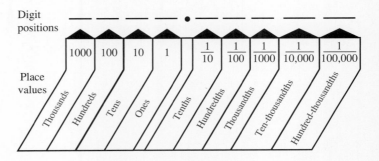

Try These Problems

1. In the decimal 13.82, give the digit in the tenths place.
2. In the decimal 832.754, give the digit in the tens place.
3. Write the decimal for $5000 + 80 + \frac{4}{100} + \frac{7}{1000}$.
4. Write the decimal for $\frac{7}{10} + \frac{2}{1000} + \frac{1}{10000}$.
5. Write 730.04 in expanded form.
6. Write 6.0012 in expanded form.
7. Which of these decimals are equal to 0.07?
   a. 0.007 b. .07
   c. .070   d. 0.0700
8. Which of these money values is the same as $850?
   a. $.850   b. $85.00
   c. $850.00 d. $850,000
9. Which of these decimals are equal to 3.0065?
   a. 03.0065 b. 3.00065
   c. 3.006500 d. 3.6500

Try Problems 1 through 9.

2 Converting Decimal Fractions from Decimal Form to Fractional Form and Vice Versa

Recall that the expanded form for the decimal 52.469 is as follows.

$$52.469 = 50 + 2 + \frac{4}{10} + \frac{6}{100} + \frac{9}{1000}$$

Observe what we get when we add up the numbers in the expanded form.

$$52.469 = 50 + 2 + \frac{4}{10} + \frac{6}{100} + \frac{9}{1000}$$

Each fraction can be written with common denominator 1000.

$$= 50 + 2 + \frac{400}{1000} + \frac{60}{1000} + \frac{9}{1000}$$

$$= 52 + \frac{469}{1000}$$

$$= 52\frac{469}{1000} \quad \text{Mixed numeral}$$

$$= \frac{52,000}{1000} + \frac{469}{1000}$$

$$= \frac{52,469}{1000} \quad \text{Improper fraction}$$

We say "fifty-two and four hundred sixty-nine thousandths" for 52.469 and $52\frac{469}{1000}$. We say "fifty-two thousand, four hundred sixty-nine thousandths" for $\frac{52,469}{1000}$. We call 52.469, $52\frac{469}{1000}$, and $\frac{52,469}{1000}$ **decimal fractions**.

$$52.469 = 52\frac{469}{1000} = \frac{52,469}{1000}$$

decimal form | fractional forms

Observe how the last place value of the decimal form corresponds to the denominator of the fractional form.

$$52.469 = 52\frac{469}{1000} = \frac{52,469}{1000}$$

thousandths

6.1 An Introduction to Decimals

 Try These Problems

Write each in fractional form.
10. 0.003
11. 20.17
12. 0.108
13. 8.0731

Write each in decimal form.
14. $\frac{8}{10}$
15. $\frac{28}{1000}$
16. $\frac{704}{100}$
17. $250\frac{17}{10,000}$

EXAMPLE 1 Write each in fractional form.

a. 0.03   b. 150.0023

SOLUTION

a. $0.03 = \frac{3}{100}$ ← hundredths

b. $150.0023 = 150\frac{23}{10,000}$ or $\frac{1,500,023}{10,000}$ ← ten-thousandths

EXAMPLE 2 Write each in decimal form.

a. $\frac{58}{1000}$ b. $\frac{375}{100}$ c. $26\frac{19}{10,000}$

SOLUTION

a. $\frac{58}{1000} = 0.058$ — The last digit 8 in the numerator must go in the thousandths place.

b. $\frac{375}{100} = 3\frac{75}{100} = 3.75$ — The last digit 5 in the numerator must go in the hundredths place.

c. $26\frac{19}{10,000} = 26.0019$ — The last digit 9 in the numerator must go in the ten-thousandths place.

If you have trouble remembering the place value of a digit by its location in the decimal, here is an observation that can help you to convert decimals to fractions and vice versa.

$0.1 = \frac{1}{10}$ One decimal place in the decimal 0.1. One zero in the denominator 10 of the fraction $\frac{1}{10}$.

$0.08 = \frac{8}{100}$ Two decimal places in the decimal 0.08. Two zeros in the denominator 100 of the fraction $\frac{8}{100}$.

$0.015 = \frac{15}{1000}$ Three decimal places in the decimal 0.015. Three zeros in the denominator of the fraction $\frac{15}{1000}$.

The pattern that you observe here continues. That is, 0.00000567 has 8 decimal places, so the fractional form is $\frac{567}{100,000,000}$, which has 8 zeros in the denominator.

 Try Problems 10 through 17.

3 Reading and Writing Decimals

The following chart gives the English names for several decimal fractions. Study the chart carefully enough so that you understand how to read decimal numerals using the English language.

Try These Problems

Write the decimal numeral for each English name.
18. Eight-tenths
19. Eighteen thousandths
20. Eighty
21. Forty and twelve hundredths
22. Nine hundred four tenths

Write the English name for each.
23. 0.027
24. 5.11
25. 120.0083

Decimal Form	Fraction Form	English Name
0.0002	$\frac{2}{10,000}$	Two ten-thousandths
0.016	$\frac{16}{1000}$	Sixteen thousandths
0.50	$\frac{50}{100}$	Fifty hundredths
7.2	$7\frac{2}{10}$ or $\frac{72}{10}$	Seven and two-tenths or Seventy-two tenths
13.08	$13\frac{8}{100}$ or $\frac{1308}{100}$	Thirteen and eight hundredths or One thousand three hundred eight hundredths
920. or 920	$\frac{920}{1}$ or 920	Nine hundred twenty

Try Problems 18 through 25.

Answers to Try These Problems
1. 8   2. 3   3. 5080.047   4. 0.7021
5. $700 + 30 + \frac{4}{100}$ 6. $6 + \frac{1}{1000} + \frac{2}{10,000}$
7. b. .07   c. .070   d. 0.0700   8. c. $850.00
9. a. 03.0065   c. 3.006500   10. $\frac{3}{1000}$ 11. $20\frac{17}{100}$ or $\frac{2017}{100}$
12. $\frac{108}{1000}$ 13. $8\frac{731}{10,000}$ or $\frac{8731}{10,000}$ 14. 0.8   15. 0.028   16. 7.04
17. 250.0017   18. 0.8   19. 0.018   20. 80   21. 40.12
22. 90.4   23. twenty-seven thousandths
24. five and eleven hundredths
25. one hundred twenty and eighty-three ten-thousandths

EXERCISES 6.1

For the decimal 8073.2459, give the digit with the indicated place value.
1. $\frac{1}{10}$ 2. 1   3. 10   4. $\frac{1}{100}$

For the decimal 9237.0481, give the digit with the indicated place value.
5. thousandths   6. tens   7. ten-thousandths   8. tenths

Write each decimal in expanded form.
9. 0.238   10. 0.035   11. 76.008   12. 3080.704

Write the decimal whose expanded form is given.
13. $\frac{5}{10} + \frac{7}{100}$ 14. $\frac{9}{100} + \frac{1}{1000}$
15. $600 + 30 + \frac{8}{10} + \frac{3}{1000}$ 16. $500 + 7 + \frac{4}{100} + \frac{6}{10,000}$

Write each in fractional form or as a whole numeral. Do not reduce the fractions to lowest terms.
17. 0.3   18. 7.8   19. 0.58
20. 275.   21. 0.006   22. 11.0765

Write a decimal numeral for each.

23. $\frac{4}{100}$
24. $3\frac{5}{10}$
25. $14\frac{32}{1000}$
26. $\frac{45}{10,000}$
27. $\frac{2368}{1000}$
28. $\frac{351}{10}$

Write a decimal numeral for each.

29. Nine-tenths
30. Nineteen hundredths
31. Four hundredths
32. Fourteen ten-thousandths
33. Twelve and three-tenths
34. Forty and thirteen thousandths
35. Fifty and one-hundred-four thousandths
36. One thousand sixty and thirty-six hundredths
37. Five-hundred-three tenths
38. Two-thousand-eight thousandths

Write the English name for each.

39. 0.34
40. 0.237
41. 0.008
42. 0.0002
43. 24.02
44. 405.016
45. 9.0105
46. 7800.0076
47. 9000.
48. 50,630.

6.2 MONEY AND CHECK WRITING

OBJECTIVES
- Write the value of a given piece of currency in both cents and dollars.
- Write the name of a single piece of currency with the indicated value.
- Translate a value expressed using English to math symbols.
- Translate a value in symbols to English.
- Complete the writing of a check by giving a value in the appropriate form.

1 The Pieces of United States Currency and Their Values

A common use of decimals is in expressing the value of money. The currency in circulation in the United States consists of paper bills and metal coins. The value of each piece of currency is based on two different units of measurement, dollars and cents.

The currency worth 1 dollar, also written $1 or $1.00, is called a dollar bill. We show a picture of both sides of a dollar bill.

The currency worth 1 cent, also written 1¢ or $.01, is called a penny. We show a picture of both sides of a penny.

The value of the dollar bill is the same as the value of 100 pennies.

1 dollar = 100 cents

$1 = 100¢

1 cent = 0.01 dollar

1¢ = $0.01

Here is a list of the most common paper bill currencies in circulation. The value of each is given in both dollars and cents.

Paper Bill Currency

Name	Picture	Value in Dollars	Value in Cents
One-dollar bill or dollar bill		$1 or $1.00	100¢
Five-dollar bill		$5 or $5.00	500¢
Ten-dollar bill		$10 or $10.00	1000¢
Twenty-dollar bill		$20 or $20.00	2000¢

Here is a list of the most common coins in circulation. The value of each is given in both cents and dollars.

Try These Problems

*Write the value of the given piece of currency in **a.** cents and **b.** dollars.*
1. ten-dollar bill
2. five-dollar bill
3. quarter
4. nickel

Write the English name of a single piece of currency with the indicated value.
5. $0.10
6. 100¢
7. $20.00
8. 1¢

6.2 Money and Check Writing

Coin Currency

Name	Picture	Value in Cents	Value in Dollars
Penny		1¢	$0.01
Nickel		5¢	$0.05
Dime		10¢	$0.10
Quarter		25¢	$0.25
Half-dollar		50¢	$0.50

A common error is to represent 25 cents by using the symbol .25¢. The value .25¢ means $\frac{25}{100}$ of 1 cent which is $\frac{1}{4}$¢ and is, therefore, less than 1¢. As you see in the chart above, 25 cents is written 25¢ or $.25.

 Try Problems 1 through 8.

 2 Translating Money Values from English to Symbols and Vice Versa

Combining the currencies in various ways creates different money values. You should become familiar with the English language that is used in reading and writing money values. Here we show two ways to read or write $6.25 using English.

METHOD 1
"Six dollars and twenty-five cents." In this method we read the fractional part of the dollar in cents. This method is used most frequently when speaking.

METHOD 2
"Six and twenty-five hundredths dollars." Here the entire value is read in dollars. This method or a slight variation of it is used in check writing.

Try These Problems

Write the symbol for each in dollars.

9. Six cents
10. Fifteen dollars and seventy-five cents
11. Ten and three hundredths dollars.
12. Nine hundred twelve and twenty-three hundredths dollars

Write each value using English in two ways: **a.** *the fractional part of the dollar in cents and* **b.** *the entire value in dollars.*

13. $0.19
14. $7.30
15. $1500.08
16. $97.42

The following chart gives more examples.

Symbolic	English (Fractional part of the dollar in cents)	English (Entire value in dollars)
$0.03	three cents	three hundredths dollars
$0.92	ninety-two cents	ninety-two hundredths dollars
$4.15	four dollars and fifteen cents	four and fifteen hundredths dollars
$28.00	twenty-eight dollars	twenty-eight dollars
$150.07	one hundred fifty dollars and seven cents	one hundred fifty and seven hundredths dollars

 Try Problems 9 through 16.

3 Writing Checks

Many people deposit their money in a checking account at a bank. When they want to make a purchase or pay a bill they write a check for the amount rather than using cash currency.

Jeff Miller has a checking account with National Bank in Palo Alto, California. He wrote a check to pay his gas and electric bill for the month of May. Here is a copy of the check he wrote.

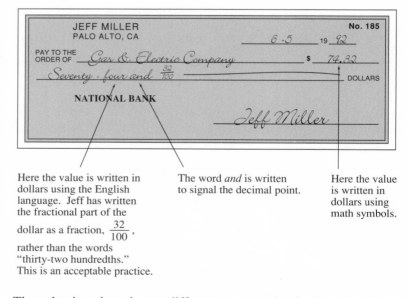

Here the value is written in dollars using the English language. Jeff has written the fractional part of the dollar as a fraction, $\frac{32}{100}$, rather than the words "thirty-two hundredths." This is an acceptable practice.

The word *and* is written to signal the decimal point.

Here the value is written in dollars using math symbols.

The value is written in two different ways on the check so that there is no doubt about the intended value.

Here is another example.

Try These Problems

Give the missing value for each check in the appropriate form.

17. Ms. De Mol wrote a check to pay her water bill. On the check she must express the value in two ways. If she wrote "Thirteen and $\frac{4}{100}$" in the blank that is followed by the word DOLLARS, how must she fill in the blank that follows the dollar symbol $?
$ _____?_____

18. Mr. Moura wrote a check to a financial institution. On the check he must express the value in two ways. If he wrote "10,000.00" in the blank that follows the dollar symbol $, how would he complete the blank that is followed by the word DOLLARS?
_____?_____ DOLLARS

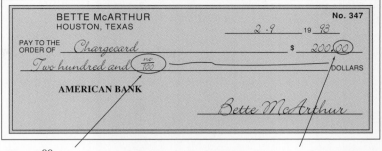

$\frac{no}{100}$ or $\frac{00}{100}$ is written to emphasize there is no fractional part of the dollar.

Two zeros are placed past the decimal point to emphasize there is no fractional part of the dollar.

 Try Problems 17 and 18.

Answers to Try These Problems

1. a. 1000¢ b. $10 or $10.00   2. a. 500¢ b. $5 or $5.00
3. a. 25¢ b. $0.25   4. a. 5¢ b. $0.05   5. dime
6. dollar bill   7. twenty-dollar bill   8. penny   9. $0.06
10. $15.75   11. $10.03   12. $912.23
13. a. nineteen cents
    b. nineteen hundredths dollars
14. a. seven dollars and thirty cents
    b. seven and thirty hundredths dollars
15. a. fifteen hundred dollars and eight cents
    b. fifteen hundred and eight hundredths dollars
16. a. ninety-seven dollars and forty-two cents
    b. ninety-seven and forty-two hundredths dollars
17. 13.04   18. Ten thousand and $\frac{no}{100}$

EXERCISES 6.2

*Write the value of the given piece of currency in **a.** cents and **b.** dollars.*

1. Dollar bill   2. Penny   3. Twenty-dollar bill
4. Nickel   5. Quarter   6. Five-dollar bill

Write the English name of a single piece of currency with the indicated value.

7. 50¢   8. 500¢   9. $0.10   10. $1.00

Write the symbol for each in dollars.

11. Seventeen cents   12. Forty-three cents
13. Forty-six hundredths dollars   14. Sixty-two hundredths dollars
15. Twenty-five dollars and eighty-three cents
16. Two hundred five dollars and seventy-one cents
17. Eight hundred and nine hundredths dollars
18. Ninety-eight and fifty-four hundredths dollars

Write each value using English in two ways: ***a.*** *the fractional part of the dollar in cents and* ***b.*** *the entire value in dollars.*

19. $0.55   **20.** $0.74   **21.** $8.03

22. $728.14   **23.** $67.00   **24.** $5000.00

Complete each check by giving the missing value in the appropriate form.

25.

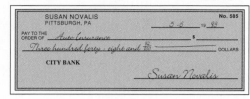

26.

27.

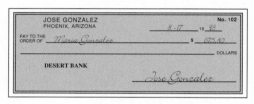

28.

6.3 COMPARING DECIMALS

OBJECTIVES
- Compare decimals.
- Compare decimals using the comparing symbols <, =, or >.

▲ Try These Problems

1. Which of these decimals equal 0.6?
  a. 6.   **b.** .6
  c. .60   **d.** .06
  e. 0.60

2. Which of these decimals equal 82?
  a. .82   **b.** 82.
  c. 820   **d.** 82.0
  e. .082

3. Which of these money values are the same as $200.00?
  a. $200   **b.** $2
  c. $200.   **d.** $20,000

1 Attaching Extra Zeros

Sometimes extra zeros are attached to decimal numerals without changing the value of the decimal. Here are some examples.

.53 = 0.53	fifty-three hundredths
7 or 7. = 7.00	seven
2.6 = 2.6000	two and six-tenths
$40 = $40.00	forty dollars

▲ **Try Problems 1 through 3.**

Try These Problems

4. Which is larger 0.02 or 0.0154?
5. Which is smaller 0.5 or 0.47?
6. List from smallest to largest.
   81.224; 81.2224; 81.2236

2 Comparing Decimals

Now we look at some examples of comparing decimals.

EXAMPLE 1 Which is larger, 0.38 or 0.49?

SOLUTION

$$0.38 = \frac{38}{100}$$

$$0.49 = \frac{49}{100} \text{ —— larger}$$

The decimal 0.49 is larger. ∎

EXAMPLE 2 Which is smaller, 0.004 or 0.0038?

SOLUTION

$$0.004 = \frac{4}{1000}$$ Before comparing, we need to view them with a common denominator.

$$0.0038 = \frac{38}{10000}$$

$$0.004 = 0.0040 = \frac{40}{10,000}$$ Attach a zero here so that each decimal has four digits past the decimal point.

$$0.0038 = 0.0038 = \frac{38}{10,000}$$

$\frac{38}{10,000}$ is smaller than $\frac{40}{10,000}$

The decimal 0.0038 is smaller. ∎

EXAMPLE 3 List from smallest to largest: 30.05; 30.0467; 30.048

SOLUTION It can be helpful to arrange the numbers so that decimal points align vertically.

$$30.05 = 30.0500$$
$$30.0467$$
$$30.048 = 30.0480$$

Attach extra zeros so that each decimal has the same number of decimal places. The numbers are equal out to the tenths place.

Compare 500, 467, and 480, the last three digits of the numbers above. From smallest to largest we have, 467, 480, and 500. Therefore, the decimals listed from smallest to largest are as follows.

$$30.0467;\ 30.0480;\ 30.0500$$

or

$$30.0467;\ 30.048;\ 30.05\ \blacksquare$$

 Try Problems 4 through 6.

Try These Problems

7. Which is larger, 93.0 or 9.30?
8. Which is smaller, 117.2 or 11.72?

Rewrite each statement using the symbol <, =, or >.
9. 15.2 is larger than 15.02
10. 170 is equal to 170.00
11. 0.0342 is smaller than 0.0351

Use the symbol <, =, or > to compare each pair of numbers.
12. 0.05; 0.50
13. 0.4827; 0.6
14. .42; 0.420
15. 17.81; 17.8099

EXAMPLE 4 Which is larger, 8.23 or 82.3?
SOLUTION

If the whole-number portions of the decimals differ, you need only to compare the wholes to determine which decimal is larger.

Since 82 is larger than 8, we conclude that 82.3 is larger. ■

Try Problems 7 and 8.

3 The Comparing Symbols <, =, and >

Special symbols are often used when comparing decimals. Examples of how these symbols are used are given in the following chart.

Math Symbols	English
8 < 20	8 is smaller than 20
20 > 8	20 is larger than 8
20 = 20	20 is equal to 20
0.43 < 0.55	0.43 is smaller than 0.55
0.15 > 0.136	0.15 is larger than 0.136
7.8 = 7.800	7.8 is equal to 7.800

After studying the previous chart, you should make these observations.

1.
   The pointed side always faces the smaller number.
2.
   The open side always faces the larger number.

Try Problems 9 through 15.

Answers to Try These Problems

1. b. .6 c. .60 e. 0.60   2. b. 82. d. 82.0
3. a. $200 c. $200.   4. 0.02   5. 0.47
6. 81.2224; 81.2236; 81.224   7. 93.0   8. 11.72
9. 15.2 > 15.02   10. 170 = 170.00   11. 0.0342 < 0.0351
12. 0.05 < 0.50   13. 0.4827 < 0.6   14. .42 = 0.420
15. 17.81 > 17.8099

EXERCISES 6.3

1. Which of these decimals are equal to 0.43?
   a. .43   b. .043   c. .430   d. 0.430   e. 4.300
2. Which of these decimals are equal to 7.5?
   a. 75.   b. 7.500   c. 7.05   d. 07.50   e. .75

3. Which of these have the same money value as $30?
   a. $3.00   b. $300   c. $30.00   d. $.30   e. $30.0
4. Which of these have the same money value as $9.05.
   a. $9.50   b. $0.95   c. $09.05   d. $9.050   e. $9.005

Which is the smaller decimal?

5. 2.8; 2.9
6. 3.600; 3.489
7. 0.3; 0.03
8. 1.399; 2.04
9. 132.113; 132.1146
10. 5.8607; 58.607

Which is the larger decimal?

11. 8.; .8
12. 0.04; 0.40
13. .132; .138
14. 83.1784; 83.17828

List from smallest to largest.

15. 5; .5; .55
16. .56; 1; 2.3
17. .327; .33; .32
18. 53.112244; 53.114422; 53.113

Use the symbol <, =, or > to compare each pair of numbers.

19. 9.00; 9
20. 8.8; .88
21. 0.8123; 0.814
22. 0.007; 0.000789

6.4 ROUNDING OFF, ROUNDING UP, AND TRUNCATING

OBJECTIVES
- Round off a decimal numeral to a given place value or a given number of decimal places.
- Round up a decimal numeral to a given place value or a given number of decimal places.
- Truncate a decimal numeral to a given place value or a given number of decimal places.

Sometimes the approximate value of a decimal is needed. There are many methods for approximating decimals. We will study three of these methods. The symbol ≈ or ≐ is used to mean *approximately equal to*.

1 Rounding Off

Here are some examples that illustrate what it means to **round off** a decimal to a certain place value or to a given number of decimal places.

EXAMPLE 1 Round off 7.1349 to the nearest hundredths place.

SOLUTION

7.1349 — hundredths place
≈ 7.13   Because the digit 4 is smaller than 5, we drop the 49 and write 7.13 for the approximation.

We are saying that 7.1349 is closer to 7.13 than it is to 7.14. The answer is 7.13. ∎

Try These Problems

1. Round off 0.2317 at three decimal places.
2. Round off 8.045 at two decimal places.
3. Round off 23.3146 at one decimal place.

Round off each to the nearest cent.
4. $0.057
5. $18.912
6. $285.009

Round off each to the nearest dollar.
7. $12.094
8. $180.75
9. $9.82

EXAMPLE 2 Round off 83.68 to the nearest tenths place.
SOLUTION

```
            tenths place
   83.6|8 ── Because the digit 8 is larger than 5,
   83.7       we drop the 8 and write 83.7 for
              the approximation.
```

We are saying that 83.68 is closer to 83.7 than it is to 83.6. The answer is 83.7. ∎

EXAMPLE 3 Round off 132.6935 at three decimal places.
SOLUTION

```
            three decimal places
   132.693|5 ── Because the digit after 3 is 5, we write
       ↓        132.694 as the approximation.
   132.694
```

132.6935 is exactly halfway between 132.693 and 132.694, so we will agree to choose the higher number for the approximation. The answer is 132.694. ∎

 Try Problems 1 through 3.

When approximating money value, special language is sometimes used. The following examples illustrate this.

EXAMPLE 4 Round off $23.168 to the nearest cent.
SOLUTION To the nearest cent means to the nearest hundredth of a dollar. We round off at two decimal places.

```
             cents place
   $23.16|8 ── Because 8 is larger than 5, we increase 6
        ↓      to 7.
   $23.17
```

The answer is $23.17. ∎

EXAMPLE 5 Round off $273.42 to the nearest dollar.
SOLUTION To the nearest dollar means to the nearest one dollar or to the nearest whole dollar. We round off at the ones place.

```
             dollars place
   $ 273|.42
        ↓
   $ 273 ── Because 4 is smaller than 5, we leave 3
            as 3 and drop the 42 cents..
```

The answer is $273. ∎

 Try Problems 4 through 9.

6.4 Rounding Off, Rounding Up, and Truncating

 Try These Problems

10. Round up $132.081 to the nearest cent.
11. Round up $74.60 to the nearest dollar.
12. Round up 0.0342 at three decimal places.
13. Truncate 17.13891 at three decimal places.
14. Truncate 8.0098 at the hundredths place.
15. Truncate 7.3333 at one decimal place.

2 Rounding Up

Suppose the grocery store is selling bananas at 3 pounds for $1. This means that each pound is selling for approximately $0.333. How much will you be charged for 1 pound of these bananas? Most likely, you will be charged $0.34 or 34¢. The store approximates $0.333 as $0.34. Usually, the business world does not follow the rules for rounding off. They approximate in a way that gives them slightly more money. This type of approximating is called **rounding up.** Here are some examples of approximating decimals by rounding up.

EXAMPLE 6 Round up $24.124 to the nearest cent.
SOLUTION

The answer is $24.13. ■

EXAMPLE 7 Round up 18.32 at one decimal place.
SOLUTION

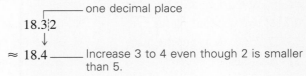

The answer is 18.4. ■

 Try Problems 10 through 12.

3 Truncating

The third method for approximating decimals is called **truncating** (or **rounding down**). To truncate means *to chop off* or *to drop off*. This method of approximating is often used by calculators. For example, if a calculator has a display limited to 8 digits, it might display 0.6666666 for the approximation of the decimal 0.6666666666, instead of displaying 0.6666667. The calculator simply chops off the excess digits. The following example illustrates how to approximate decimals by truncating.

EXAMPLE 8 Truncate 87.367 at two decimal places.
SOLUTION

$$87.36\,|\,7$$
$$\approx 87.36$$

— two decimal places
— Drop this digit and make no other changes no matter how large this digit is.

The answer is 87.36. ■

 Try Problems 13 through 15.

Now we summarize the three types of approximating by making a chart that illustrates all three types. Study the following chart carefully enough so that you know the difference between rounding off, rounding up, and truncating.

Three Types of Rounding

Decimal	Round off At Two Decimal Places	Round up At Two Decimal Places	Truncate At Two Decimal Places
0.362	0.36	0.37	0.36
0.365	0.37	0.37	0.36
0.368	0.37	0.37	0.36

▲ **Answers to Try These Problems**

1. 0.232 2. 8.05 3. 23.3 4. $0.06 5. $18.91 6. $285.01
7. $12 8. $181 9. $10 10. $132.09 11. $75 12. 0.035
13. 17.138 14. 8.00 15. 7.3

EXERCISES 6.4

Round off each decimal at the place indicated.
1. 0.316; two decimal places
2. 7.823; one decimal place
3. 732.013569; thousandths place
4. 832.346; tenths place
5. $17.503; nearest dollar
6. $0.1248; nearest cent

Round up each decimal at the place indicated.
7. $0.431; nearest cent
8. $34.00612; nearest cent
9. 9.01392; four decimal places
10. 8.347; one decimal place
11. 2304.191; hundredths place
12. 0.00828; ten-thousandths place
13. 22.079; nearest whole number
14. $186.63; nearest dollar

Truncate each decimal at the place indicated.
15. 0.23691; three decimal places
16. 1.36898; four decimal places
17. 790.364; tenths place
18. 15.0308; hundredths place
19. 75.59; ones place
20. 100.01; nearest whole number

6.5 ADDITION AND SUBTRACTION

OBJECTIVES
- Add two or more decimals.
- Subtract decimals.
- Find the missing number in an addition statement.

1 Adding

Cynthia needs to pay the following bills:

6.5 Addition and Subtraction

Rent	$300
Gas and electric	$ 31.21
Garbage	$ 15.25
Cable TV	$ 21.50

What is the total amount of money she needs? To solve this problem we need to add the decimals. We add them similarly to the way we add whole numbers; that is, we add like place values.

```
 $300.00
 $ 31.21
 $ 15.25
+$ 21.50
 $367.96
```

Arrange the decimal points in a vertical line so that digits with like place value will be in the same column. Observe that $300 = $300.00

Cynthia will need $367.96 to pay these bills.

Observe that to add decimals, we arrange the numbers vertically so that the decimal points line up in a vertical line. This forces digits with like place value to form a column. Finally, add digits with like place value and place the decimal point in the answer in line with the other decimal points. Here are more examples of adding decimals.

EXAMPLE 1   Add: 27.5 + 2.346 + 0.0018

SOLUTION

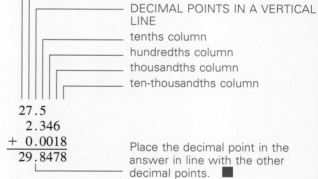

```
  27.5
   2.346
+  0.0018
  29.8478
```

Place the decimal point in the answer in line with the other decimal points. ■

When a column of digits adds up to more than 9, we use carrying as we do when adding whole numbers. Here are some examples:

EXAMPLE 2   Add: 6.5 + 0.846

SOLUTION

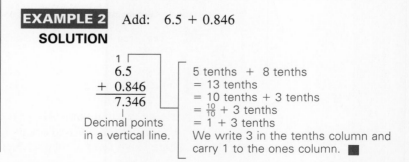

```
   1
   6.5
+  0.846
   7.346
```

Decimal points in a vertical line.

5 tenths + 8 tenths
= 13 tenths
= 10 tenths + 3 tenths
= $\frac{10}{10}$ + 3 tenths
= 1 + 3 tenths

We write 3 in the tenths column and carry 1 to the ones column. ■

Chapter 6 Decimals: Place Value, Addition, and Subtraction

 Try These Problems

Add.
1. $15 + $4.75 + $0.13
2. 703.1 + 0.92 + 2.0008
3. 932.15 + 80.463
4. 56.914 + 870 + 4.061
5. $567.42 + $7.45
6. 0.478 + 2.3 + 7924

EXAMPLE 3 Add: $1.08 + $52.36 + $0.27

SOLUTION

```
    2
  $ 1.08
  $52.36
+ $ 0.27
  $53.71
```
Decimal points in a vertical line.

8 cents + 6 cents + 7 cents
= 21 cents
= 2 dimes + 1 cent
= 2 tenths dollar + 1 hundredth dollar
We write 1 in the hundredths column and carry 2 to the tenths column. ■

EXAMPLE 4 Add: 172.3 + 79.784 + 1392.007

SOLUTION

```
   211  1
    172.3
     79.784
+  1392.007
   1644.091
```
Decimal points in a vertical line ■

EXAMPLE 5 Add: 4.13 + 23 + 63.009

SOLUTION Be careful. The number 23 is a whole numeral. The decimal is placed at the right end. That is, 23 = 23. = 23.000.

```
    4.13
   23.000
+  63.009
   90.139
```
The whole number 23 equals 23. or 23.000.

Now we summarize by writing a procedure for adding decimals.

Adding Decimals
1. If any whole numerals are involved, place the decimal point at the right end.
2. Arrange the numerals vertically for convenience. Line up the decimal points in a vertical line so that digits with like place value form a column.
3. Add each column separately, beginning with the column on the right.
4. When a column adds up to more than 9, use carrying.
5. Place the decimal point in the answer in line with the other decimal points.

 Try Problems 1 through 6.

2 Subtracting

Subtracting decimal numerals is similar to adding decimal numerals in that we line up the decimal points in a vertical line so that digits with like place value form a column. Here are some examples.

6.5 Addition and Subtraction

EXAMPLE 6 Subtract: $7.3 - 0.9$

SOLUTION

$$\begin{array}{r} 6 \\ 7\!\!\!/\,^1 3 \\ -\ 0.9 \\ \hline 6.4 \end{array}$$

We do not subtract 9 from 3 because 9 is larger than 3. We borrow 1 from 7, leaving 6. Write 1 in front of 3 to make 13. The 1 we borrow is really $\frac{10}{10}$ or 10 tenths. Then,
10 tenths + 3 tenths = 13 tenths.

CHECK

$$\begin{array}{r} 1 \\ 0.9 \\ +\ 6.4 \\ \hline 7.3 \end{array}$$

The answer is 6.4. ∎

EXAMPLE 7 Subtract: $18.9 - 0.72$

SOLUTION

$$\begin{array}{r} 8 \\ 1\ 8.\!\!\!/9\,^1 0 \\ -\ \ \ 0.7\ 2 \\ \hline 1\ 8.1\ 8 \end{array}$$

Write in a zero place holder. Borrow 1 from 9, leaving 8. Write 1 in front of 0 to make 10. The 1 we borrow is really
1 tenth = $\frac{1}{10} = \frac{10}{100}$
= 10 hundredths.

CHECK

$$\begin{array}{r} 1 \\ 0.72 \\ +\ 18.18 \\ \hline 18.90 \end{array} = 18.9$$

The answer is 18.18. ∎

EXAMPLE 8 Subtract: $17.6 - 9.358$

SOLUTION

$$\begin{array}{r} 5\ 9 \\ 1\ 7.\!\!\!/6\,\!\!\!/0\,^1 0 \\ -\ \ \ 9.3\ 5\ 8 \\ \hline 8.2\ 4\ 2 \end{array}$$

Write in two zero place holders. Borrow 1 from 60, leaving 59. Write the 1 in front of the zero to make 10. The 1 we borrow is really 1 hundredth = $\frac{1}{100} = \frac{10}{1000}$
= 10 thousandths.

CHECK

$$\begin{array}{r} 11 \\ 9.358 \\ +\ \ 8.242 \\ \hline 17.600 \end{array} = 17.6$$

The answer is 8.242. ∎

Try These Problems

Subtract.
7. 600.24 − 8.013
8. $1436.89 − $78
9. 7.13 − 0.8482
10. 172 − 9.07
11. 1020 − 96.41
12. 5000 − 7.563

EXAMPLE 9 Subtract: $36 − $4.30

SOLUTION Observe that $36 is 36 whole dollars. The decimal point is placed at the right end. That is, $36 = $36. or $36.00.

$$\begin{array}{r} \$\ 3\ \overset{5}{\cancel{6}}\overset{1}{.}0\ 0 \\ -\ \$\quad\ 4\ .\ 3\ 0 \\ \hline \$\ 3\ 1\ .\ 7\ 0 \end{array}$$ ——— $36 = $36.00.

CHECK

$$\begin{array}{r} \overset{1}{}4.30 \\ +\ 31.70 \\ \hline 36.00 \end{array}$$

The answer is $31.70. ■

EXAMPLE 10 Subtract: 9000 − 0.064.

SOLUTION Observe that 9000 is a whole numeral. The decimal point is placed at the right end. That is, 9000 = 9000. = 9000.000.

$$\begin{array}{r} 8\ 9\ 9\ 9\ 9\ 9 \\ \cancel{9}\ \cancel{0}\ \cancel{0}\ \cancel{0}.\cancel{0}\ \cancel{0}{}^{1}0 \\ -\qquad\quad 0\ .0\ 6\ 4 \\ \hline 8\ 9\ 9\ 9\ .9\ 3\ 6 \end{array}$$ ——— 9000 = 9000.000

CHECK

$$\begin{array}{r} 1\ \ 1\ \ 1\ \ 1\ \ 1\ \ 1 \\ 0.0\ 6\ 4 \\ +\ 8\ 9\ 9\ 9\ .9\ 3\ 6 \\ \hline 9\ 0\ 0\ 0\ .0\ 0\ 0 \end{array} = 9000$$

The answer is 8999.936. ■

Now we summarize by writing a procedure for subtracting decimals.

Subtracting Decimals
1. If a whole numeral is involved, place the decimal point at the right end.
2. Arrange the numerals vertically for convenience. Line up the decimal points in a vertical line so that digits with like place value form a column.
3. Subtract each column separately, beginning with the column on the right.
4. When a digit on the bottom is larger than a digit on the top, use borrowing.
5. Place the decimal point in the answer in line with the other decimal points.

 Try Problems 7 through 12.

Try These Problems

Find the missing number.

13. 23.86 + ☐ = 45.3
14. ☐ = 245 + 65.09
15. 304 = 76.34 + ☐
16. ☐ + 0.537 = 50.14

3 The Missing Number in an Addition Statement

The statement 5.8 + 2.1 = 7.9 is an addition statement. If any one of the three numbers is omitted, we can find the missing number. Study the three examples that follow to see how to do this.

1. 5.8 + 2.1 = ☐   Here the answer to the addition statement is missing, so we *add* to find the missing number.

   The missing number is 7.9.

2. ☐ + 2.1 = 7.9   Here one of the terms is missing, so we *subtract* to find the missing number.

   The missing number is 7.9 − 2.1 = 5.8.

3. 5.8 + ☐ = 7.9   Here one of the terms is missing, so we *subtract* to find the missing number.

   The missing number is 7.9 − 5.8 = 2.1.

Try Problems 13 through 16.

Answers to Try These Problems

1. $19.88   2. 706.0208   3. 1012.613   4. 930.975
5. $574.87   6. 7926.778   7. 592.227   8. $1358.89
9. 6.2818   10. 162.93   11. 923.59   12. 4992.437
13. 21.44   14. 310.09   15. 227.66   16. 49.603

EXERCISES 6.5

Add.

1. 43.1   2. 73.4   3. 25.345   4. 456
   + 6.23   + 8.309   56   8.43
                                  + 2.57   + 13.952

5. 0.23 + 9.7 + 728.92   6. 0.8 + 9.07 + 1908.62
7. $346.17 + $250 + $3.74   8. $33.90 + $459.09 + $63
9. 2.8 + 73 + 312.643   10. 135 + 15.007 + 234.72

Solve.

11. Ed ran the first mile in 7.83 minutes and the second mile in 6.7 minutes. What is his total time for the 2-mile run?
12. Tina wrote checks for $23, $17.23, and $426.08. What is the total value of the three checks?

Subtract.

13. 8.749   14. 17.636   15. 154   16. 236
   − 0.68   − 5.9   − 8.32   − 37.065

17. 302.9 − 1.353   18. 85.4 − 3.824
19. $400 − $306.79   20. $670 − $19.46
21. 2004.8 − 32.705   22. 3010 − 259.342
23. 5000 − 1230.67   24. 70,000 − 32.046

Solve.

25. How far is it from C to Y?

```
X          C        Y
•——————————•————————•
|←— 7.3 m —→|
|←———— 12 m ————————→|
```

26. A steak, including the fat, weighs 3.2 pounds. The butcher trims 0.75 pound of fat from the steak. What is the weight of the resulting steak?

Find the missing number.

27. $300 + \boxed{} = 409.67$

28. $\boxed{} + 3.8 = 16.02$

29. $\boxed{} = 35.78 + 9.054$

30. $\boxed{} = 348.7 + 56.238$

31. $2004 = \boxed{} + 805.82$

32. $96.3 + \boxed{} = 101$

6.6 LANGUAGE

OBJECTIVES
- Translate an English phrase to math symbols involving addition and subtraction.
- Solve problems by using translations.

1 Translating

The math symbols that we use often have many translations to the English language. Knowing these translations can help in solving application problems. The following chart gives the many English translations for the symbol $=$.

Math Symbol	English
$=$	equals
	is equal to
	is the same as
	is
	was
	were
	are
	will be
	represents
	gives
	makes
	yields
	is the result of

The addition statement $5.6 + 0.7 = 6.3$ is written using math symbols. Some of the ways to read this in English are given in the following chart.

Try These Problems

Translate each English statement to an addition or subtraction statement using math symbols.

1. The sum of 45.5 and 6.4 is 51.9.
2. The difference between 6 and 1.2 equals 4.8.
3. 67.8 subtracted from 100 yields 32.2.
4. 74.14 minus 56.089 is 18.051.

Solve.

5. Find 4.5 less than 7.2.
6. Find 34.9 increased by 5.02
7. Find 87.6 more than 3.7.
8. Subtract 455.8 from 500.

Math Symbols	English
$5.6 + 0.7 = 6.3$	5.6 plus 0.7 equals 6.3
	5.6 added to 0.7 yields 6.3
	5.6 *increased by* 0.7 is 6.3
	0.7 more than 5.6 is equal to 6.3
	The *sum* of 5.6 and 0.7 equals 6.3
	6.3 is the result of adding 5.6 to 0.7

Because $5.6 + 0.7$ equals $0.7 + 5.6$, the order that you read the terms 5.6 and 0.7 makes no difference; however, when reading the subtraction statement $6.3 - 0.7$, take care to read 6.3 and 0.7 in the correct order, because $6.3 - 0.7$ does not equal $0.7 - 6.3$.

$$6.3 - 0.7 = 5.6 \quad \text{but} \quad 0.7 - 6.3 = -5.6$$

The numbers 5.6 and -5.6 are not equal. The number -5.6 is a negative number. This book covers negative numbers in Chapter 12. The following chart gives several correct ways to read $6.3 - 0.7 = 5.6$.

Math Symbols	English
$6.3 - 0.7 = 5.6$	6.3 minus 0.7 equals 5.6
	6.3 take away 0.7 yields 5.6
	6.3 *decreased by* 0.7 gives 5.6
	6.3 reduced by 0.7 equals 5.6
	6.3 subtract 0.7 is 5.6
	0.7 subtracted from 6.3 is equal to 5.6
	0.7 less than 6.3 gives 5.6
	The *difference* between 6.3 and 0.7 is 5.6

Observe that when reading the symbol $-$ by using the phrases *subtracted from* or *less than*, the numbers 6.3 and 0.7 are read in the reverse order than they are written. That is,

$$0.7 \text{ subtracted from } 6.3 \quad \text{means} \quad 6.3 - 0.7$$

and

$$0.7 \text{ less than } 6.3 \quad \text{means} \quad 6.3 - 0.7$$

 Try Problems 1 through 8.

2 Solving Problems by Using Translations

The following examples illustrate how we can use translating to solve problems.

EXAMPLE 1 The sum of 36.17 and what number is 50.2?

SOLUTION Translate the question into math symbols.

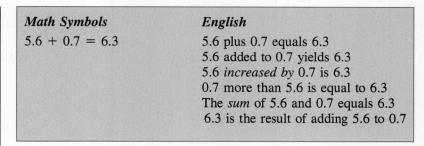

Try These Problems

Solve.

9. 460 represents 348.2 plus what number?
10. What number added to 60.65 yields 120?
11. What number is 75 less than 235.75?
12. What is the difference between 15.1 and 0.35?

The missing number is one of the terms in an addition statement, so we subtract the given term 36.17 from the total 50.2 to find the missing term.

$$\begin{array}{r} \overset{4}{\cancel{5}}0.\overset{1}{2}0 \\ -\ 3\ 6.1\ 7 \\ \hline 1\ 4.0\ 3 \end{array}$$

The answer is 14.03. ■

EXAMPLE 2 What number is 78.4 less than 127.49?

SOLUTION Translate the question into math symbols.

What number is 78.4 less than 127.49

☐ = 127.49 − 78.4

The missing number is the answer to the subtraction problem, 127.49 − 78.4.

$$\begin{array}{r} 1\ \overset{1}{\cancel{2}}7.4\ 9 \\ -\ \ \ 7\ 8.4\ 0 \\ \hline 4\ 9.0\ 9 \end{array}$$

The answer is 49.09. ■

Try Problems 9 through 12.

Answers to Try These Problems

1. 45.5 + 6.4 = 51.9 2. 6 − 1.2 = 4.8
3. 100 − 67.8 = 32.2 4. 74.14 − 56.089 = 18.051
5. 2.7 6. 39.92 7. 91.3 8. 44.2 9. 111.8 10. 59.35
11. 160.75 12. 14.75

EXERCISES 6.6

Translate each statement to an addition or subtraction statement using math symbols.

1. The sum of 0.8 and 0.4 is 1.2.
2. The difference between 41 and 8.3 is 32.7.
3. 18.7 decreased by 7.3 gives 11.4.
4. 7.75 increased by 0.25 yields 8.
5. The total when 72.8 is added to 0.6 is 73.4.
6. 3 is 0.5 more than 2.5.
7. 8.2 is 0.4 less than 8.6.
8. 99.2 is 100.8 subtracted from 200.

Solve.

9. Find the difference between 0.2 and 7.83.
10. Subtract 0.89 from 2.

11. What number is the sum of 4 and 6.82?
12. 12.82 more than 1.6 is what number?
13. Find 100.1 reduced by 79.93.
14. What number increased by 75.3 yields 98?
15. 7.234 plus what number gives 23.65?
16. 18.04 take away 8.7 gives what number?
17. Find the result when 10 is decreased by 3.4.
18. Find the result when 86.24 is increased by 7.

6.7 APPLICATIONS

OBJECTIVES
- Solve an application that involves a total as the sum of parts
- Solve an application that involves an increase or decrease.
- Solve an application that involves comparisons by addition or subtraction.
- Find the perimeter of a figure.
- Solve an application that involves more than one step.

There are many situations in your daily life in which addition and subtraction of decimals is needed. In this section we look at some of these situations and give guidelines on how to decide whether to add or subtract.

1 Problems Involving a Total as the Sum of Parts

We know that to add means to combine or total. In the addition statement

$$15.6 + 9.4 = 25$$

the sum 25 is the total and the terms 15.6 and 9.4 are parts of the total. Recall from Section 6.5 that, if any one of the three numbers is missing, we can find it using the two given numbers. Here we review the procedure.

> *Finding the Missing Number in a Basic Addition Statement*
> 1. If the sum (or total) is missing, add the terms (or parts) to find the missing number.
> 2. If one of the terms (or parts) is missing, subtract the given term from the total to find the missing term.

The following examples illustrate how to use this concept to solve some application problems.

Try These Problems

1. Tom ran a total of 0.3 mile in 2 minutes. The first minute he ran 0.125 mile. How far did he run the second minute?

2. Roberto combined 25.8 gallons of gas with 8.25 gallons of gas. What is the total amount of gas?

3. The total distance from P to N is 7 meters. The distance from P to Q is 3.6 meters. How far is it from Q to N?

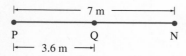

4. Find the perimeter of this triangle.

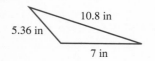

EXAMPLE 1 Arahwana worked on her report 4.5 hours on Monday, 3.75 hours on Tuesday, and 2.25 hours on Wednesday. What is the total time for the three days?

SOLUTION We are looking for the *total* time, so we add.

$$\begin{array}{r} 1\,1\\ 4.5\\ 3.75\\ +\ 2.25\\ \hline 10.50 \end{array} = 10.5$$

She worked a total of 10.5 hours. ■

EXAMPLE 2 The total distance from the bottom of the vase to the top of the flowers is 38 centimeters. The flowers rise 13.4 centimeters above the vase. How tall is the vase?

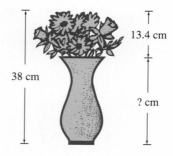

SOLUTION

Total height = height of the vase + height of the flowers above the vase

38 = ☐ + 13.4

We are missing one of the terms in an addition statement, so we subtract 13.4 from 38 to find the missing term.

$$\begin{array}{r} 3\,8.0 \\ -\ 1\,3.4 \\ \hline 2\,4.6 \end{array}$$ —Total height
—Part of the total

The vase is 24.6 centimeters tall. ■

 Try Problems 1 through 4.

2 Problems Involving Increases and Decreases

The price of gasoline increased from $1.20 a gallon to $1.37 during an oil crisis. The amount that the price went up is called the **increase** (or **gain**). What is the increase in price?

Lower price + increase = higher price

$1.20 + ☐ = $1.37

To find the increase we subtract.

Try These Problems

5. On Monday Alan jogged 5.4 miles. On Tuesday he increased his jogging distance by 2.3 miles over Monday's distance. How far did he jog on Tuesday?

6. The monthly service charge for Roger's checking account increased from $13.75 t0 $15. What was the increase?

7. Mr. Wong purchased a car marked $10,840. The state tax increased the price to $11,761.40. What was the amount of the state tax?

8. Last month gas was selling for $1.85 a gallon and this month the price is $1.58 a gallon. What is the reduction in price?

9. An office supply store is having a sale. A felt tip pen that originally sold for $1.25 is now selling for $0.89. What is the savings on one pen?

$$\text{Increase} = \substack{\text{higher} \\ \text{price}} - \substack{\text{lower} \\ \text{price}}$$
$$= \$1.37 - \$1.20$$
$$= \$0.17$$

The increase in price is $0.17.

 Try Problems 5 through 7.

Yesterday lettuce was selling for $1.45 a head. Today a head costs $0.85. The amount that the price went down is called the **decrease.** What is the decrease in price?

$$\text{Decrease} = \substack{\text{higher} \\ \text{price}} - \substack{\text{lower} \\ \text{price}}$$
$$= \$1.45 - \$0.85$$
$$= \$0.60$$

The decrease in price is $0.60.

Other words can be used to mean decrease. Here we list some situations where a value went down and give the corresponding language.

Situation	Words Indicating the Amount the Value Went Down
There is a sale on coats. A $100 value is now selling for $75.50.	The *discount* is $24.50. The *savings* is $24.50 The *reduction* in price is $24.50 The *decrease* in price is $24.50
A dog lost weight. The weight dropped from 10.5 pounds to 8.2 pounds.	The *loss* in weight is 2.3 pounds. The *reduction* in weight is 2.3 pounds. The *decrease* is 2.3 pounds.

 Try Problems 8 and 9.

 3 Problems Involving Comparisons

Two unequal numbers, such as 30 and 18.6, can be compared using addition and subtraction. Here is a list of several addition and subtraction statements involving 30 and 18.6 and the corresponding comparison statements in English.

Comparing Two Numbers

Math Symbols	English
18.6 + 11.4 = 30	11.4 more than 18.6 is 30.
30 − 11.4 = 18.6	11.4 less than 30 is 18.6.
30 − 18.6 = 11.4	30 and 18.6 differ by 11.4.
	The *difference* between 30 and 18.6 is 11.4.
	30 is more than 18.6 by 11.4.
	18.6 is less than 30 by 11.4.

Try These Problems

10. Stephanie has a skirt that is 45.2 centimeters long. Her mother, Tish, wants to lengthen the skirt by 3.5 centimeters. How long will the skirt be after Tish lets the hem out?

11. Betsy mixed a salt solution weighing 50.7 grams. Arthur mixed a salt solution weighing 52 grams. How much lighter is Betsy's solution?

12. Sandy and Leah were in a skiing race. Sandy skied the race in 16.25 seconds. Leah's time was 1.08 seconds less than Sandy's time. How long did it take Leah to ski the race?

Now we look at examples of problems involving comparisons.

EXAMPLE 3 Charlie tried to install a refrigerator that is 62 centimeters wide in a space that is only 61.4 centimeters wide. How much wider is the refrigerator than the space?

SOLUTION We want to know how much larger 62 is than 61.4. We subtract 61.4 from 62.

```
    1
  6 2.0      width of the refrigerator
- 6 1.4      width of the space
  ─────
    0.6      difference in the two widths
```

The refrigerator is 0.6 centimeter wider than the space. ■

EXAMPLE 4 Lou earns $15.50 per hour. Valerie earns $3.75 less than Lou per hour. How much does Valerie earn each hour?

SOLUTION Valerie's hourly wage is $3.75 less than Lou's hourly wage. This statement translates to a subtraction statement.

$$\text{Valerie's hourly wage} = \text{Lou's hourly wage} - \$3.75$$

$$\boxed{} = \$15.50 - \$3.75$$

To find Valerie's hourly wage we subtract $3.75 from $15.50.

```
       4 4
  $ 1 5.5 0
- $   3.7 5
  ─────────
  $ 1 1.7 5
```

Valerie earns $11.75 per hour. ■

 Try Problems 10 through 12.

4 Applications Requiring More than One Operation

Now we look at an application that requires more than one step to solve.

EXAMPLE 5 The length of a rectangle is 15.6 inches more than its width. The width is 45 inches. Find the perimeter of the rectangle.

SOLUTION Here we picture the situation.

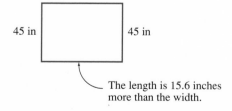

45 in   45 in

The length is 15.6 inches more than the width.

Try These Problems

13. The width of a rectangle is 14.2 centimeters less than the length. If the length is 80 centimeters, find the perimeter of the rectangle.

14. Irene had $412.89 in her checking account. She deposited a check for $356.32 and withdrew $45.79, $124.85, and $68. What is her balance after these transactions?

First, we find the length of the rectangle.

Length is 15.6 inches more than the width
⬚ = 15.6 + 45

To find the length, we must add 15.6 and 45.

```
   15.6
 + 45.0
 ------
   60.6
```

The length of the rectangle is 60.6 inches.

The **perimeter** of the rectangle is the distance all the way around, so we must add two widths and two lengths together to obtain the perimeter.

```
   60.6
   60.6
   45.0
 + 45.0
 ------
  211.2
```

The perimeter is 211.2 inches. ■

 Try Problems 13 and 14.

Now we summarize the material presented in this section by giving guidelines that will help you decide whether to add or subtract.

Situations Requiring Addition and Subtraction

Operation	Situation
+	1. You are looking for the total or whole.
	2. You are looking for the result when a quantity has been increased.
	3. You are looking for the perimeter of a figure.
−	1. You are looking for one of the parts in a total or whole.
	2. You are looking for the result when a quantity has been decreased.
	3. You are looking for how much larger one quantity is than another.
	4. You are looking for how much smaller one quantity is than another.

 Answers to Try These Problems

1. 0.175 mi   **2.** 34.05 gal   **3.** 3.4 m   **4.** 23.16 in
5. 7.7 mi   **6.** $1.25   **7.** $921.40   **8.** $0.27   **9.** $0.36
10. 48.7 cm   **11.** 1.3 g   **12.** 15.17 sec   **13.** 291.6 cm
14. $530.57

EXERCISES 6.7

1. Nell bought a dress for her granddaughter. The dress was marked $15.95. A tax increased the price by $0.80. What was the total price?
2. Cynthia practiced her piano 2.25 hours on Thursday and 1.75 hours on Friday. How long did she practice during the two days?
3. A lamp has a total height of 82.2 centimeters. From the bottom of the base to the bottom of the shade is 42.4 centimeters. Find the height of the shade.

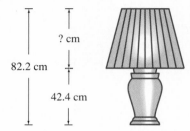

4. Carlos used a 1.2-centimeter screwdriver to loosen a 0.75-centimeter screw. The screwdriver is how much wider than the screw?
5. Find the perimeter of this rectangle.

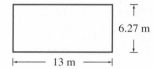

6. Mary bought 10 yards of fabric to make some bicycling clothes for herself and her boyfriend. She used 3.3 yards for herself. How much is left?
7. On a four-day business trip, Ms. Nelle filled her gas tank three times. The amounts were 15.2 gallons, 16 gallons, and 12.9 gallons. How much gasoline did she buy during the trip?
8. Joe bought a portable radio that was marked $85.89. After the tax was added on, it cost $91.37. How much did he pay in tax?
9. Greg jogged a total of 5.4 miles. He jogged 3.6 miles along the highway, then the rest along the river bank. How far did he jog along the river bank?
10. On a two-week trip to Tahiti, Jody encountered the following expenses: $685.75 for air fare, $935.85 for room and board, and $121.60 for miscellaneous other expenses. What was the total cost of the trip?
11. Ms. Taylor bought a microwave oven that was 24.9 inches wide. She put it in a space that was 27.8 inches wide. How much wider was the space than the microwave oven?
12. A cat weighed 5.2 pounds, then lost 1.5 pounds. What was the cat's weight after the weight loss?
13. The price of beans was $1.15 per pound and now is $1.74 per pound. What is the increase in price?
14. Jimmy weighed 8.2 pounds when he was born. Chuck weighed 7.5 pounds at birth. What is the difference in their birth weights?
15. Don is 1.4 feet taller than Richard. If Don is 6.2 feet tall, how tall is Richard?

16. Yesterday the price of oranges was $0.16 per pound less than today. If today's price is $1.25 per pound, what was yesterday's price?
17. The width of a rectangle is 3.5 inches less than the length. If the length is 19.25 inches, find the perimeter of the rectangle.
18. Mr. Stuart has a balance of $500.19 in his checking account. He writes checks for $54.62, $8.74, and $23.91. What is his balance now?
19. Bernice had 450 calories for breakfast, 432 calories for lunch, and 515 calories for supper. That same day, her husband Tom had 520 calories for breakfast, 612 calories for lunch, and 750 calories for supper. How many more calories did Tom have than Bernice for that day?
20. Carlos is taking a 100-mile hike. He hikes 28.5 miles on Monday, 19.8 miles on Tuesday, and 35.2 miles on Wednesday. How much farther does he have to go?
21. What is the difference in the perimeters of these two rectangles?

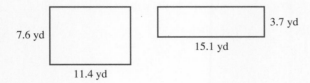

22. A race-car driver wanted to limit total pit-stop time to 50 seconds. He made three pit stops in 16.2 seconds, 15.8 seconds, and 15.6 seconds. How much over or under his desired limit is he?

During an oil crisis the average price per gallon of gasoline fluctuated over a 6-month period as shown in the bar graph. Use the graph to answer Exercises 23 through 26.

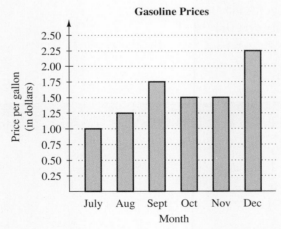

23. What was the average price per gallon for the month of August?
24. For what months was the average price per gallon $1.50?
25. From November to December did the price per gallon increase or decrease, and by how much?
26. From September to October did the price per gallon increase or decrease, and by how much?

CHAPTER 6 SUMMARY

KEY WORDS AND PHRASES
decimal [6.1]
digit [6.1]
expanded form [6.1]
place value [6.1]
decimal fractions [6.1]
round off [6.4]
round up [6.4]
truncate (round down) [6.4]
addition [6.5]
subtraction [6.5]
sum [6.6]
difference [6.6]
increase (gain) [6.7]
decrease [6.7]
perimeter [6.7]

SYMBOLS
3.7 means $3\frac{7}{10}$ [6.1]
$4.12 < 4.3$ means 4.12 is less than 4.3. [6.3]
$4.3 > 4.12$ means 4.3 is more than 4.12. [6.3]
$\approx$ or $\doteq$ means *is approximately equal to*. [6.4]

IMPORTANT RULES
How to Round Off, Round Up, and Truncate Decimals [6.4]
The chart on page 240 summarizes how to approximate decimals using each of these three methods.

How to Add Decimals [6.5]
The rule for adding decimals is on page 242.

How to Subtract Decimals [6.5]
The rule for subtracting decimals is on page 244.

How to Find the Missing Number in an Addition Statement [6.5]
- If the sum (or total) is missing, then add the terms (or parts) to find the sum.
- If one of the terms (or parts) is missing, then subtract the known term from the sum (or total) to obtain the missing term.

The Perimeter of a Figure [6.7]
The perimeter of a figure means the total distance around the figure.

Situations Requiring Addition and Subtraction [6.7]
Guidelines for when to add and when to subtract are given in the chart on page 253.

CHAPTER 6 REVIEW EXERCISES

For the decimal 728.10365, give the digit with the indicated place value.

1. hundredths   **2.** hundreds   **3.** $\frac{1}{10,000}$ **4.** $\frac{1}{10}$

Write the decimal whose expanded form is given.

5. $\frac{6}{10} + \frac{9}{1000}$ **6.** $800 + 4 + \frac{3}{100} + \frac{7}{1000}$

Write each decimal as a fraction, mixed numeral, or whole numeral. Do not reduce the fractions to lowest terms.

7. 0.003   **8.** 7.27

Write a decimal numeral for each of the following.

9. $\frac{19}{10,000}$ **10.** $14\frac{15}{100}$

Write the decimal whose English name is given.
11. Sixteen thousand
12. Sixteen thousandths
13. Ninety and two hundred three ten-thousandths
14. Seven hundred fourteen hundredths

Write the English name for each decimal.
15. 0.05   16. 50.0002   17. 0.304   18. 304,000

*Write the value of the given piece of currency in **a.** cents and **b.** dollars.*
19. nickel   20. quarter   21. dollar bill   22. five-dollar bill

Write the English name of a single piece of currency with the indicated value.
23. $0.10   24. $0.01   25. $10.00   26. 50¢

Write the symbol for each in dollars.
27. Forty-seven cents
28. Sixty-two and twelve hundredths dollars

*Write each value using English in two ways: **a.** with the fractional part of the dollar in cents and **b.** with the entire value in dollars.*
29. $0.61   30. $200.07

31. Which of these decimals are equal to .580?
    a. .58   b. 58.   c. 58.0   d. 0.58   e. 580
32. Which of these have the same money value as $102.00?
    a. $12   b. $102   c. $102.   d. $10.20   e. $10,200

List from smallest to largest.
33. 1; 3.9; 0.39
34. 73.2323; .732323; 73.23222

Choose the correct symbol <, =, or > to compare each pair of numbers.
35. 6.03; 6.003
36. 7.2; 7.20
37. 19.7; 1.97
38. 83.1516; 83.15158

Round off each decimal at the place indicated.
39. 138.137; two decimal places
40. $43.49; nearest dollar

Round up each decimal at the place indicated.
41. 0.23117; thousandths place
42. $204.833; nearest cent

Truncate each decimal at the place indicated.
43. 9.81768; four decimal places
44. 93.1553; hundredths place

Add.
45. 76.7 + 9.682
46. 0.8907 + 28.45
47. 26.03 + 66
48. 968 + 75.45
49. $13.75 + $704 + $8.36
50. $93 + $85.32 + $186.09

Subtract.
51. 76.52 − 5.1
52. $70.13 − $9
53. 511.2 − 83.74
54. 85.7 − 4.308
55. 9 − 3.6
56. 73 − 18.26
57. 3000 − 4.52
58. 80,500 − 751.4

Find the missing number.

59. 34.18 + 7.9 = ☐

60. ☐ + 30.6 = 143

Translate each English statement into an addition or subtraction statement using math symbols.

61. 107 subtracted from 210.8 is 103.8.

62. The sum of 0.7 and 0.6 is 1.3.

63. 4.2 more than 15.8 is 20.

64. 19.7 decreased by 2 gives 17.7.

Solve.

65. Find the difference between 2.1 and 0.75.

66. What number increased by 7.2 yields 10.19?

67. Subtract 19.03 from 168.

68. 15 is more than 3.98 by how much?

69. Sam jogged 5.8 miles today and 3.25 miles yesterday. How much farther did he jog today than yesterday?

70. What is the total distance from A to C?

71. A department store is having a sale on television sets. The price of a TV set was marked down from $500 to $465.85. What was the savings?

72. Jan bought a car marked $20,600. A state tax increased the price to $22,505.80. How much was the state tax?

73. Find the perimeter of this rectangle.

[rectangle: 34 in by 6.75 in]

74. Peter has a rope that is 23.6 meters long. He cuts off a piece that is 4.8 meters long. How long is the remaining piece?

In Your Own Words

Write complete sentences to discuss each of the following. Support your comments with examples or pictures, if appropriate.

75. A student incorrectly writes the numeral 35,000 for "thirty-five thousandths." Discuss how you would help this student understand the error made.

76. Discuss how addition of whole numerals is similar to addition of decimals.

77. Discuss the differences in rounding off, rounding up, and truncating.

78. Discuss a real-life situation where rounding up is used.

79. A sign in the post office reads, "Envelopes for 0.69¢ each." You know that they mean to sell these envelopes for 69 cents each. Discuss what is wrong with the sign. How can you help them to understand that 0.69¢ does not represent 69 cents?

CHAPTER 6 PRACTICE TEST

Write a decimal numeral for each.

1. $\frac{934}{10}$
2. $6\frac{19}{10000}$

Write the decimal whose English name is given.

3. Sixty-five hundredths
4. Forty-eight and seven thousandths

Write the English name for each.

5. 0.0013
6. 7.2
7. Write the value of a quarter in dollars.
8. Write the value of a five-dollar bill in cents.
9. Which of these decimals are equal to 0.90?
   a. 90   b. .900   c. 0.9   d. .9   e. 9.00
10. List from smallest to largest: 1.07; 0.701; 0.079
11. Use the correct symbol <, =, or > to compare 0.007 and 0.00089.
12. Round off 26.0843 to three decimal places.
13. Round up $78.34 to the nearest dollar.
14. Truncate 0.013456 at the ten-thousandths place.

Add.

15. 6.31 + 129.8
16. 864 + 2.7 + 9 + 0.856

Subtract.

17. 13.8 − 2.403
18. $856.34 − $82
19. 232 − 7.44
20. 5000 − 78.6

Solve.

21. Find the sum of 18.6 and 164.82.
22. Find the number that is 2.3 less than 10.
23. Find the perimeter of a rectangle that has width 78.32 feet and length 36.8 feet.
24. A dog weighed 12.4 pounds, then lost 2.8 pounds. What was the dog's weight after the weight loss?
25. Rita paid $7.85 more for a blouse than Britt-Marie. If Rita paid $60.35 for the blouse, what did Britt-Marie pay?
26. Mr. McWilliams was planning to jog a distance of 12 miles. His knee began to hurt after he had gone 8.5 miles, so he walked the rest of the 12 miles. How far did he walk?

CHAPTER 7

Decimals: Multiplication and Division

7.1 MULTIPLICATION

OBJECTIVES
- Multiply two or more decimals.
- Use a shortcut to multiply decimals involving zeros.
- Solve an application that involves multiplication of decimals.

1 Multiplying Decimals

At the supermarket, ground beef sells for $4.20 per pound. How much would you pay for 3.2 pounds of ground beef? We have the following.

$$\text{Cost per pound} \times \text{how many pounds} = \text{total cost}$$

$$\$4.20 \times 3.2 = \boxed{}$$

To find the total cost, we need to multiply $4.20 by 3.2.

```
     4.2 0      2 decimal places
  ×   3.2       1 decimal place
  ─────────
     8 4 0
   1 2 6 0
  ─────────
   1 3.4 4 0    2 + 1 = 3 decimal places
```

You would pay $13.44 for 3.2 pounds of ground beef.

Observe that to multiply the two decimals, begin by multiplying 420 by 32 as if they were whole numbers. This gives 13,440. To place the decimal in the answer, count the number of digits in the multipliers that are to the right of the decimal point. In this case, there are a total of 3 decimal places because there are 2 digits to the right of the decimal point in 4.20 and 1 digit to the right of the decimal point in 3.2. Finally, place the decimal in the answer so that there are 3 digits to the right of the decimal point.

Now we will look carefully at the multiplication problem

$$0.02 \times 1.4 = ?$$

Before doing the problem using the procedure discussed in the previous paragraph, we will do the problem using fractions so that you can see why the previously discussed procedure works.

$$0.02 \times 1.4$$
$$= \frac{2}{100} \times 1\frac{4}{10} \qquad \text{Convert each decimal to a fraction.}$$
$$= \frac{2}{100} \times \frac{14}{10}$$
$$= \frac{28}{1000} \qquad \text{Multiply without canceling so that it will be easy to convert back to a decimal.}$$
$$= 0.028 \qquad \text{Convert back to a decimal.}$$

By solving this problem using fractions, we see that the correct answer should be 0.028. Observe how we can obtain the same answer without using fractions.

```
              Begin by multiplying 2 by 14 to
              obtain 28.
      1 . 4      1 decimal place
    × 0 . 0 2    2 decimal places
      . 0 2 8    1 + 2 = 3 decimal places.
                 Insert a 0 digit in front of the 2 to
                 make 3 decimal places.
```

Here are more examples of multiplying decimals.

EXAMPLE 1 Multiply: .069 × .12

SOLUTION Begin by multiplying 69 by 12 as if the numbers were whole numbers. Place the decimal point as indicated here.

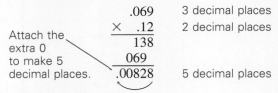

The answer is 0.00828. ∎

Chapter 7 Decimals: Multiplication and Division

 Try These Problems

Multiply.
1. .61 × .3
2. 2.5 × 7.8
3. 45 × 8.16
4. .042 × .9
5. .00082 × 3.4
6. .076 × .45

EXAMPLE 2 Multiply: 7.45 × 356.8

SOLUTION

```
         3 5 6.8      1 decimal place
    ×      7.4 5      2 decimal places
       1 7 8 4 0
       1 4 2 7 2
       2 4 9 7 6
     2 6 5 8.1 6 0    1 + 2 = 3 decimal places
```

The answer is 2658.16. ■

EXAMPLE 3 Multiply: 32 × 20.5

SOLUTION

```
         2 0.5        1 decimal place
    ×      3 2        0 decimal places
           4 1 0
         6 1 5
         6 5 6.0      1 decimal place
```

The answer is 656. ■

EXAMPLE 4 Multiply: .642 × .008

SOLUTION

```
         .6 4 2       3 decimal places
    ×    .0 0 8       3 decimal places
     .0 0 5 1 3 6     3 + 3 = 6 decimal places
                      Attach 2 extra 0s to make 6 decimal
                      places.
```

The answer is 0.005136. ■

Now we summarize by writing a rule for multiplying decimals.

> *Multiplying Decimals*
> 1. Ignore the decimal points for a moment. Multiply the numbers as if they were whole numbers.
> 2. Place the decimal point in the product by adding the number of decimal places in the multipliers to get the number of decimal places in the product.
> 3. You may have to attach extra zeros on the left end of the product to make enough decimal places.

 Try Problems 1 through 6.

2 Recognizing Unnecessary Zero Digits

The following chart illustrates when a zero digit is necessary and when a zero digit is unnecessary. When we say that a zero digit is unnecessary we mean that omitting it does not change the value of the number, where we are discussing values as exact and not approximations.

Try These Problems

Multiply.
7. 0.45 × 0.37
8. .5400 × 0.82
9. $132.00 × 0.90

	Decimal	*Location of the Zero Digit*
Necessary Zero Digits	150. = 150 200. = 200 34,000. = 34,000	Zero digits on the right end but before the decimal point.
	.107 208.7 1006.103	Zero digits between nonzero digits.
	.03 = 0.03 .005 = 0.005 .0017 = 0.0017	Zero digits on the left end but after the decimal point.
Unnecessary Zero Digits	0.5 = .5 00.33 = .33 008.134 = 8.134	Zero digits on the left end but before the decimal point.
	22.00 = 22 .50 = .5 .3600 = .36	Zero digits on the right end but after the decimal point.

When a multiplication problem is presented to you with unnecessary zeros, it is more convenient to drop the extra zeros before multiplying. The following examples illustrate this.

EXAMPLE 5 Multiply: 0.75 × 4.600

SOLUTION Because 0.75 = .75 and 4.600 = 4.6, the problem is .75 × 4.6.

```
      4.6      1 decimal place
    × .7 5     2 decimal places
    ─────
    2 3 0
    3 2 2
    ─────
    3.4 5 0    3 decimal places
```

The answer is 3.45. ∎

EXAMPLE 6 Multiply: $26.00 × 8.40

SOLUTION Because $26.00 = $26 and 8.40 = 8.4, the problem is 26 × 8.4.

```
      2 6      0 decimal places
    × 8.4      1 decimal place
    ─────
    1 0 4
    2 0 8
    ─────
    2 1 8.4    1 decimal place
```

The answer is $218.40. ∎

 Try Problems 7 through 9.

Try These Problems

Multiply.
10. .007 × 81.04
11. 54.006 × 70004
12. 0.807 × 2.0050

3 Using a Shortcut to Multiply Decimals with Necessary Zero Digits

When a zero digit is between two nonzero digits or is between the decimal point and a nonzero digit, then it is a necessary zero digit and cannot be omitted before multiplying. However, there are some steps in the multiplication process that can be omitted. In the following examples, we solve the problem in two ways: first by the long method and then by the shortcut method.

EXAMPLE 7   Multiply: 305 × 32.626

SOLUTION
LONG METHOD

```
      32.626
  ×      305
     163 130
     000 00   —— This row of zeros does not add anything to
     9787 8       the product.
    9950.930
```

SHORTCUT

```
      32.626
  ×      305
     163 130      Omit the row of zeros.
     9787 8   —— Be sure to begin this row directly under
    9950.930      the digit 3 in 305.
```

The answer is 9950.93. ■

EXAMPLE 8   Multiply: 0.0038 × 7.064

SOLUTION   Because 0.0038 = .0038, the problem is .0038 × 7.064.

LONG METHOD

```
       7.064
  ×    .0038
       56 512
      211 92
      000 0    —— These two rows of zeros do not
     0000          add anything to the product.
     .0268 432
```

SHORTCUT

```
       7.064
  ×    .0038
       56512
      21192      Omit the rows of zeros.
     .0268432
```

The answer is 0.268432. ■

 Try Problems 10 through 12.

Try These Problems

Multiply.
13. 10 × 18.07
14. 0.041 × 100
15. 1000 × 6.7
16. 23.8 × 10,000

4 Using a Shortcut to Multiply by a Whole Number Ending in Zero Digits

First we investigate what happens when a decimal is multiplied by 10, 100, 1000, and so on. Here we look carefully at three multiplication problems. First we multiply 3.45 by 10.

$$3.45 \times 10 = 34.5$$

```
    3.45
  ×   10
    0 00
   34 5
   34.50 = 34.5
```

Observe that when 3.45 is multiplied by 10, the decimal point moves 1 place to the right, giving the answer 34.5. Next we multiply 3.45 by 100.

$$3.45 \times 100 = 345. \text{ or } 345$$

```
    3.45
  × 1 00
    0 00
   00 0
  345
  345.00 = 345
```

Observe that when 3.45 is multiplied by 100, the decimal point moves 2 places to the right, giving the answer 345. Finally, we multiply 3.45 by 1000.

$$3.45 \times 1000 = 3450$$

```
     3.45
  × 10 00
     0 00
    00 0
   000
  345
  3450.00  = 3450
```

Observe that when 3.45 is multiplied by 1000, the decimal point moves 3 places to the right, giving the answer 3450. The number of places the decimal moves agrees with the number of zeros in 10, 100, 1000, and so forth. We generalize the previous results by stating a shortcut for multiplying by 10, 100, 1000, and so on.

Multiplying a Decimal by 10, 100, 1000, and So On
Move the decimal point to the right
   One place when multiplying by 10.
   Two places when multiplying by 100.
   Three places when multiplying by 1000.
   Four places when multiplying by 10,000.
The pattern continues. The decimal point moves to the right the same number of places as there are zeros.

 Try Problems 13 through 16.

Next we look at a shortcut that can be used when multiplying a decimal by any whole number ending in zeros. In the examples that

Try These Problems

Multiply.
17. 5.2 × 600
18. 800,000 × .036
19. 4.55 × 28,000

follow, we solve the problem in two ways: first by the long method and then by the shortcut.

EXAMPLE 9   Multiply: 750,000 × .65

SOLUTION
LONG METHOD

```
        7 5 0 0 0 0
      ×         . 6 5
      -------------
        3 7 5 0 0 0 0
      4 5 0 0 0 0
      -------------
      4 8 7 5 0 0 . 0 0
```

The four zero digits at the end of 750,000 caused these four zero digits in the product.

SHORTCUT

```
        7 5 0 0 0 0
      × . 6 5
      ---------
        3 7 5
      4 5 0
      ---------
      4 8 7 5 0 0 . 0 0
```

Bring down the four zeros at the end of 750,000.

The answer is 487,500. ∎

EXAMPLE 10   Multiply: 9000 × 2.607

SOLUTION
LONG METHOD

```
          2 . 6 0 7
        × 9 0 0 0
        -----------
            0 0 0 0
          0 0 0 0
        0 0 0 0
      2 3 4 6 3
      -----------
      2 3 4 6 3 . 0 0 0
```

These three zero digits were caused by the three zero digits at the end of 9000.

SHORTCUT

```
          2 . 6 0 7
        ×   9 0 0 0
        -----------
      2 3 4 6 3 . 0 0 0
```

Bring down the three zeros at the end of 9000.

The answer is 23,463. ∎

▲ **Try Problems 17 through 19.**

5 Multiplying More Than Two Decimals

To multiply more than two numbers, multiply any two of them first, then multiply that result by another one, and continue until all of the numbers have been multiplied. The following examples illustrate the procedure.

Try These Problems

Multiply.
20. 5.09 × 10 × 0.8
21. .008 × 3.6 × 300 × 5

EXAMPLE 11 Multiply: 0.7 × 2.1 × 6

SOLUTION It doesn't matter which two numbers are multiplied first. We begin by multiplying 0.7 by 2.1, then we multiply that result by 6.

```
    2.1        1.47
  ×  .7      ×   6
  1.4 7       8.82
```

The answer is 8.82. ■

EXAMPLE 12 Multiply: 3.75 × 1000 × 4.6 × 3.08

SOLUTION The 4 numbers can be multiplied in any order. We multiply 3.75 by 1000, 4.6 by 3.08, then multiply those two results.

```
3.75 × 1000 = 3750      3.0 8         1 4.1 68
                      ×   4.6       ×    3 750
                        1 8 4 8       7 0 8 400
                        1 2 3 2         9 9 1 7 6
                        1 4.1 6 8     4 2 5 0 4
                                    5 3 1 3 0.000
```

The answer is 53,130. ■

 Try Problems 20 and 21.

 Answers to Try These Problems

1. .183 or 0.183   **2.** 19.50 or 19.5   **3.** 367.20 or 367.2
4. .0378 or 0.0378   **5.** 0.002788   **6.** 0.0342   **7.** 0.1665
8. 0.4428   **9.** $118.80   **10.** 0.56728   **11.** 3,780,636.024
12. 1.618035   **13.** 180.7   **14.** 4.1   **15.** 6700   **16.** 238,000
17. 3120   **18.** 28,800   **19.** 127,400   **20.** 40.72   **21.** 43.2

EXERCISES 7.1

Multiply.

1. .15 × .6   **2.** .07 × 17.1   **3.** 9.1 × 3.6   **4.** 8 × 2.3
5. 8.6 × 409   **6.** 307.9 × 7.1   **7.** 17.378 × 8.79   **8.** .7 × 4034
9. 6943 × .65   **10.** 92.68 × .47

Multiply. Use a shortcut method when appropriate.

11. 0.4 × 7.50   **12.** 23.00 × 0.80   **13.** .1500 × 3.7
14. 07.30 × 52   **15.** 13.008 × .09   **16.** 746 × 500.8
17. 1200 × .045   **18.** 18,000 × 13.2   **19.** 368,000 × .0625
20. 7,030,000 × 4.40   **21.** 100 × 7.813   **22.** .0816 × 100,000
23. 0.360 × 3.007   **24.** 0.0076 × 0.750   **25.** $73.00 × 190,000
26. 50,300 × $80.70   **27.** 0.78 × 5.2 × 3.2   **28.** 4.02 × 0.007 × 40
29. 500 × 45.2 × 20.03 × 0.6   **30.** 0.0052 × 1000 × 6.7 × 0.3

Solve.

31. Chicken breasts are selling for $3.69 per pound. What do 5.5 pounds of these chicken breasts cost? Round up the answer to the nearest cent.

268 Chapter 7 Decimals: Multiplication and Division

32. Pens are selling for $1.39 each. How much would you pay for 400 of these pens?
33. In a certain state, the sales tax is computed by multiplying the decimal 0.065 by the marked price. Find the sales tax for a coat that is marked $145.89. Round up the answer to the nearest cent.
34. Each section of a fence is 2.8 meters long. If 507 of these sections are connected, what is the length of the resulting fence?

7.2 DIVISION

OBJECTIVES
- Divide a decimal or a whole number by a whole number.
- Convert a fraction to decimal form.
- Divide a decimal or whole number by a decimal.
- Round off, round up, or truncate the answer to a division problem to a specified decimal place.
- Divide two places past the decimal point and write the remainder in fractional form.
- Write the answer to a division problem as a repeating decimal when specified.
- Use a shortcut to divide by 10, 100, 1000, and so on.
- Solve an application problem using division of decimals.

A **division** problem can be written in three ways. For example, here are three ways to write *20 divided by 5 is 4,* or *5 divided into 20 is 4.*

$$5\overline{)20}^{\,4} \qquad 20 \div 5 = 4 \qquad \frac{20}{5} = 4 \qquad \text{because} \qquad 5 \times 4 = 20$$

In each of these problems, the number 5 is called the **divisor,** the number 20 is called the **dividend.** In this section, we will learn how to carry out the division when there are decimals involved. We want the procedure to preserve the fact that division is the reverse of multiplication. This means that the answer to the division problem multiplied by the divisor gives the dividend.

1 Dividing Decimals by Whole Numbers

First we look at a couple of easy division problems so that we can see how to place the decimal point when dividing a decimal by a whole number.

$$2\overline{)8.6}^{\,4.3} \qquad \text{because} \qquad 2 \times 4.3 = 8.6$$

$$5\overline{)10.15}^{\,2.03} \qquad \text{because} \qquad 5 \times 2.03 = 10.15$$

Observe that when the divisor is a whole number, the decimal point in the answer is placed directly above the decimal point in the dividend.

Try These Problems

Divide as indicated.
1. $6.5 \div 5$
2. $0.15 \div 6$
3. $\frac{67.54}{22}$
4. $\frac{1584.8}{56}$

Here are more examples to illustrate how to divide a decimal by a whole number.

EXAMPLE 1 Divide: $11.6 \div 4$

SOLUTION $11.6 \div 4$ is the same as $4\overline{)11.6}$
— 4 is the divisor.

$$\begin{array}{r} 2.9 \\ 4\overline{)11.6} \\ \underline{8} \\ 3\,6 \\ \underline{3\,6} \end{array}$$

Place the decimal point in the answer directly above the decimal point in the dividend.

Divide the numbers as whole numbers. Align the digits carefully.

CHECK
$$\begin{array}{r} 2.9 \\ \times\ \ 4 \\ \hline 11.6 \end{array}$$

The answer is 2.9. ■

EXAMPLE 2 Divide: $43.6 \div 8$

SOLUTION $43.6 \div 8$ is the same as $8\overline{)43.6}$
— 8 is the divisor.

$$\begin{array}{r} 5.45 \\ 8\overline{)43.60} \\ \underline{40} \\ 3\,6 \\ \underline{3\,2} \\ 40 \\ \underline{40} \end{array}$$

If you are past the decimal point, extra zeros may be attached to the dividend.
$43.6 = 43.60$

CHECK
$$\begin{array}{r} 5.45 \\ \times\ \ \ 8 \\ \hline 43.60 = 43.6 \end{array}$$

The answer is 5.45. ■

EXAMPLE 3 Divide: $\frac{0.36}{12}$

SOLUTION $\frac{0.36}{12}$ is the same as $12\overline{)0.36}$
— 12 is the divisor.

$$\begin{array}{r} .03 \\ 12\overline{)0.36} \\ \underline{0} \\ 36 \\ \underline{36} \end{array}$$

Zero digits on the left end must be written if you are past the decimal point.

The answer is 0.03. ■

 Try Problems 1 through 4.

In the previous three examples, the division came out evenly. This does not always happen. Here are some examples of various ways to leave the answer when the division does not come out evenly. First we look at approximating the answer at a certain decimal place. You should

Try These Problems

Divide and approximate the result as indicated.

5. 624.3 ÷ 8 (Round off at two decimal places.)
6. 0.4 ÷ 65 (Truncate at 4 decimal places.)
7. $\frac{0.508}{36}$ (Round up at 3 decimal places.)

be familiar with the three types of approximating (**rounding off, rounding up, and truncating**) that were covered in Section 6.4.

EXAMPLE 4 Divide: 12.34 ÷ 7 (Round off at 1 decimal place.)

SOLUTION To round off at 1 decimal place, we will need to see the size of the digit in the 2nd place, so we will divide 2 places past the decimal, then stop and round off as indicated.

$$\begin{array}{r} 1.76 \approx 1.8 \\ 7\overline{)12.34} \\ \underline{7} \\ 5\,3 \\ \underline{4\,9} \\ 44 \\ \underline{42} \\ 2 \end{array}$$

The answer rounded off at 1 decimal place is 1.8. ■

EXAMPLE 5 Divide: $\frac{0.04}{18}$ (Round up at 3 decimal places.)

SOLUTION To round up at 3 decimal places, we will *not* need to look at the size of the digit in the 4th decimal place, so we need only to divide 3 places past the decimal point.

$$\begin{array}{r} .002 \approx 0.003 \\ 18\overline{)0.040} \\ \underline{0} \\ 4 \\ \underline{0} \\ 40 \\ \underline{36} \\ 4 \end{array}$$

The answer rounded up at 3 decimal places is 0.003. ■

 Try Problems 5 through 7.

Another way to leave the answer when the division is not coming out evenly is to stop dividing at some point and write the remainder in a fractional form. The following example illustrates this.

EXAMPLE 6 Divide: 423.1 ÷ 36 (Stop dividing 2 places past the decimal and write the answer in fractional form.)

SOLUTION

$$\begin{array}{r} 1\,1.7\,5\,\tfrac{10}{36} = 11.75\tfrac{5}{18} \\ 36\overline{)4\,2\,3.1\,0} \\ \underline{3\,6} \\ 6\,3 \\ \underline{3\,6} \\ 2\,7\,1 \\ \underline{2\,5\,2} \\ 1\,9\,0 \\ \underline{1\,8\,0} \\ 1\,0 \end{array}$$

Fraction = $\frac{\text{remainder}}{\text{divisor}}$

Try These Problems

Divide. Stop dividing 2 places past the decimal and write the remainder in fractional form.

8. $2.34 \div 28$
9. $19.8 \div 16$

Divide. Write the answer as a repeating decimal.

10. $10.4 \div 9$
11. $0.4 \div 33$

The notation $11.75\frac{5}{18}$ is ambiguous. Although it is often used in business, it is seldom used by mathematicians. Here is what it means.

$$11.75\frac{5}{18} \quad \text{means} \quad 11\frac{75\frac{5}{18}}{100}$$

The fraction $\frac{5}{18}$ is not in the thousandths place, it is associated with the digit 5 that is in the hundredths place.

 Try Problems 8 and 9.

The answer to a division problem can be a repeating decimal. Sometimes the repeating pattern is easy to see after a few division steps. Here is an example:

EXAMPLE 7 Divide: $\frac{5.5}{6}$ (Write the answer as a repeating decimal.)

SOLUTION $\frac{5.5}{6}$ is the same as $6 \overline{)5.5}$

```
        . 9 1 6 6  = .916̄
    6)5.5 0 0 0
       5 4
         1 0
            6
            4 0
            3 6
              4 0
              3 6
                4
```

The digit 6 continues to repeat no matter how far we carry out the division.

A bar written above the 6 indicates this digit repeats.

The answer is $0.91\overline{6}$. ∎

 Try Problems 10 and 11.

2 Dividing a Whole Number by a Whole Number

Recall from Chapter 6 that the decimal point in a whole numeral is placed at the right end. For example,

$$5 = 5. = 5.00 \qquad \text{Five}$$
$$40 = 40. = 40.000 \qquad \text{Forty}$$
$$128 = 128. = 128.0 \qquad \text{One hundred twenty-eight}$$

Any number of zeros can be placed past the decimal point without changing the value of the whole number.

Now we look at examples of dividing a whole number by a whole number, where we write the answer as a decimal.

EXAMPLE 8 Divide: $32 \div 5$

SOLUTION $32 \div 5$ is the same as $5 \overline{)32}$

```
       6.4
   5)32.0
     30
      2 0
      2 0
```

Every whole number can be written as a decimal. The decimal point is placed at the right end.
$32 = 32. = 32.0$

Try These Problems

Divide as indicated.
12. 15 ÷ 6
13. 6 ÷ 15
14. 7 ÷ 125
15. 125 ÷ 7 (Truncate at 2 decimal places.)

CHECK
$$\begin{array}{r} 6.4 \\ \times\ \ 5 \\ \hline 32.0 = 32 \end{array}$$

The answer is 6.4. ■

EXAMPLE 9 Divide: 5 ÷ 32

SOLUTION 5 ÷ 32 is the same as $32\overline{)5}$

In this book, you will be expected to continue dividing until the division comes out evenly unless told otherwise.

```
         .1 5 6 2 5
    32)5.0 0 0 0 0
        3 2
        ───
        1 8 0
        1 6 0
        ─────
          2 0 0
          1 9 2
          ─────
            8 0
            6 4
            ───
            1 6 0
            1 6 0
```

If you are past the decimal point, attach as many zeros as needed to continue the division.
5 = 5. = 5.00000

CHECK
$$\begin{array}{r} .15625 \\ \times\ \ \ \ 32 \\ \hline 31250 \\ 4\ 6875 \\ \hline 5.00000 = 5 \end{array}$$

The answer is 0.15625. ■

Compare Examples 8 and 9. In Example 8, the divisor is 5 and the dividend is 32; however, in Example 9 we have the reverse situation. Observe that when using the division symbol ÷, the second number is always the divisor.

 Try Problems 12 through 15.

3 Converting Fractions to Decimals

Now that we can divide whole numbers by whole numbers, we can convert fractions to decimals, because the fraction bar means division.

EXAMPLE 10 Convert $\frac{5}{8}$ to a decimal.

SOLUTION $\frac{5}{8}$ means $8\overline{)5}$

The denominator is always the divisor.

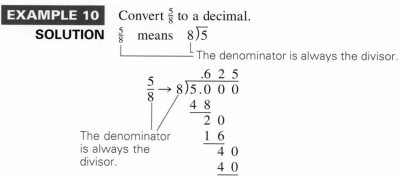

The decimal form is 0.625. ■

Try These Problems

Convert each of these fractions to a decimal.

16. $\frac{4}{5}$
17. $\frac{5}{4}$
18. $\frac{3}{40}$
19. $\frac{40}{3}$ (Write the answer as a repeating decimal.)

EXAMPLE 11 Convert $\frac{70}{33}$ to a decimal.

SOLUTION $\frac{70}{33}$ means $33\overline{)70}$ — The denominator is always the divisor.

$$\frac{70}{33} \rightarrow 33\overline{)70.0000} 2.1212 = 2.\overline{12}$$

- The digit pattern 12 continues to repeat no matter how far we carry out the division.
- A bar written above 12 indicates that the digit pattern 12 repeats.

The decimal form is $2.\overline{12}$. ∎

EXAMPLE 12 Convert $\frac{4}{53}$ to a decimal. Truncate at 4 decimal places.

SOLUTION $\frac{4}{53}$ means $53\overline{)4}$ — The denominator is always the divisor.

To truncate at 4 decimal places, we do *not* need to see the size of the digit in the 5th decimal place, so we divide 4 places past the decimal point.

$$53\overline{)4.0000} = .0754 \approx 0.0754$$

The decimal truncated at 4 decimal places is 0.0754. ∎

 Try Problems 16 through 19.

4 Dividing a Decimal by a Decimal

So far we have divided whole numbers and decimals by *whole numbers*. How do we divide by *decimals*? Here we look carefully at the following problem.

$$0.48 \div 0.2 \quad \text{or} \quad 0.2\overline{)0.48}$$

The divisor 0.2 is not a whole number, but we can convert this division problem to an equivalent division problem where the divisor is a whole number. First we view the division problem as a fraction.

$$0.48 \div 0.2 = \frac{0.48}{0.2}$$

We know that if we multiply the numerator and denominator of a fraction by the same nonzero number, we do not change the value of the fraction. So if we multiply the numerator and denominator by 10, we will have a division problem that has the divisor a whole number.

$$0.48 \div 0.2 = \frac{0.48}{0.2}$$
$$= \frac{0.48 \times 10}{0.2 \times 10}$$
$$= \frac{4.8}{2} \qquad \text{Multiplying by 10 shifts the decimal point 1 place to the right.}$$

Therefore, dividing 0.48 by 0.2 is the same as dividing 4.8 by 2. Our procedure could have been as follows.

$0.2 \overline{)0.4\,8}$

Step 1: Make the divisor a whole number by moving the decimal point 1 place to the right.

Step 2: Move the decimal point in the dividend, 0.48, 1 place to the right also.

$$\begin{array}{r} 2.4 \\ 02.\overline{)04.8} \end{array}$$

Step 3: Place the decimal point in the quotient directly above the new decimal point in the dividend.

Step 4: Divide the numbers as whole numbers. Align the digits carefully.

The result is 2.4.

Now we look at more examples where both the divisor and the dividend contain a decimal point.

EXAMPLE 13 Divide: $50.75 \div 2.5$

SOLUTION The second number 2.5 is the divisor. It is placed outside the division symbol $\overline{)}$.

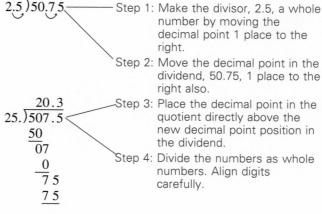

The answer is 20.3.

Try These Problems

Divide as indicated.
20. $0.756 \div .06$
21. $0.171 \div 4.5$
22. $\frac{10.5}{.015}$
23. $\frac{302.8}{.009}$ (Round off at 3 decimal places.)

EXAMPLE 14 Divide: $0.185 \div .37$

SOLUTION The second number .37 is the divisor. It is placed outside the division symbol $\overline{)}$.

```
          . 0 5
    .37 ).0 1 8 5
            0
          1 8 5
          1 8 5
```
Zero digits on the left end must be written if you are past the decimal point.

The answer is 0.05. ∎

EXAMPLE 15 Divide: $\frac{4.2}{2.03}$ (Round off at 2 decimal places.)

SOLUTION The denominator 2.03 is the divisor. It is placed outside the division symbol $\overline{)}$.

```
              2.0 6 8  ≈ 2.0 7
   2.03 )4.2 0 0 0 0
           4 0 6
             1 4 0
                 0
             1 4 0 0
             1 2 1 8
               1 8 2 0
               1 6 2 4
                 1 9 6
```
Extra zeros may be attached to the dividend if you are past the decimal point.

The answer rounded off at 2 decimal places is 2.07. ∎

Try Problems 20 through 23.

5 Dividing a Whole Number by a Decimal

Remember that the decimal point in a whole numeral is placed at the right end. For example,

$$62 = 62. = 62.00$$
$$450 = 450. = 450.000$$
$$3482 = 3482. = 3482.0$$

Any number of zeros can be placed after the decimal point without changing the value of the whole number. Here are some examples of dividing a whole number by a decimal.

EXAMPLE 16 Divide: $23 \div 9.2$

SOLUTION Write the whole numeral 23 as a decimal before beginning the division process. $23 = 23. = 23.00$

```
           2.5
   9.2 )2 3.0 0
         1 8 4
           4 6 0
           4 6 0
```
Write the whole number 23 as a decimal before moving the decimal point.
$23 = 23. = 23.00$

The answer is 2.5. ∎

276 Chapter 7 Decimals: Multiplication and Division

 Try These Problems

Divide as indicated.
24. 27 ÷ .08
25. 153 ÷ 1.7
26. $\frac{8}{.24}$ (Write the answer as a repeating decimal.)
27. $\frac{801}{0.028}$ (Round up at 2 decimal places.)

EXAMPLE 17 Divide: $\frac{7}{.033}$ (Write the answer as a repeating decimal.)

SOLUTION Write the whole numeral 7 as a decimal before beginning the division process. 7 = 7. = 7.000.

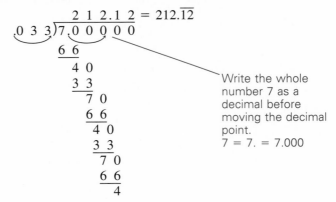

The repeating decimal is 212.$\overline{12}$. ∎

 Try Problems 24 through 27.

6 Paying Special Attention to Zeros in the Quotient

A common error that students make is to omit necessary zero digits in the quotient. Here are some examples to help you with this problem.

EXAMPLE 18 Divide: 195,000 ÷ 2.6

SOLUTION Write the whole numeral 195,000 as a decimal before beginning the division process.
195,000 = 195,000. = 195,000.0

```
            7 5 0 0 0.
      2.6)1 9 5 0 0 0.0
          1 8 2
            1 3 0
            1 3 0
```
Be sure to include these zero digits.

The answer is 75,000. ∎

EXAMPLE 19 Divide: 0.027 ÷ 45

SOLUTION

```
            .0 0 0 6
        4 5).0 2 7 0
              2 7 0
```
Be sure to include these zero digits.

The answer is 0.0006. ∎

Try These Problems

Divide as indicated.
28. 7.2 ÷ .008
29. 11,100 ÷ .74
30. .00225 ÷ 12.5
31. 1002 ÷ 500
32. $\frac{21}{400.2}$ (Truncate at 3 decimal places.)
33. $\frac{53{,}352}{.76}$

EXAMPLE 20 Divide: $\frac{240.123}{.06}$

SOLUTION

```
           4 0 0 2.0 5
    .0 6 ) 2 4 0.1 2 3 0
           2 4
           0 0 1 2
               1 2
               0 3 0
                 3 0
```
Be sure to include these zero digits.

The answer is 4002.05. ∎

 Try Problems 28 through 33.

7 Using a Shortcut to Divide by 10, 100, or 1000

Here we show the result when dividing the numbers 25 and 3.7 by 10.

```
       2.5              .37
  10)25.0          10)3.70
     20               3 0
      5 0              70
      5 0              70
```

Observe that in each case, dividing a number by 10 causes the decimal point to shift 1 place to the left. Here we show the results.

$$25 \div 10 = 2.5 \qquad 3.7 \div 10 = 0.37$$

Similarly, dividing a decimal by 100 causes the decimal point to shift 2 places to the left. Here we show the results of dividing the numbers 25 and 3.7 by 100.

$$25 \div 100 = 0.25 \qquad 3.7 \div 100 = 0.037$$

As you might predict, dividing a decimal by 1000 causes the decimal point to shift 3 places to the left. Here we show the results of dividing the numbers 25 and 3.7 by 1000. This time we use the fraction bar to indicate the division.

$$\frac{25}{1000} = 0.025 \qquad \frac{3.7}{1000} = 0.0037$$

We generalize the previous results by stating a shortcut for dividing by 10, 100, 1000 and so on.

Dividing a Decimal by 10, 100, 1000, and So On

Move the decimal point to the left
  one place when dividing by 10,
  two places when dividing by 100,
  three places when dividing by 1000,
  four places when dividing by 10,000.
The pattern continues. The decimal point moves to the left the same number of places as there are zeros.

Try These Problems

Divide as indicated without using long division. Answer with a decimal.
34. 673.8 ÷ 10
35. 0.9 ÷ 100
36. $\frac{728}{100}$
37. $\frac{9.76}{1000}$

 Try Problems 34 through 37.

Answers To Try These Problems

1. 1.3   2. 0.025   3. 3.07   4. 28.3   5. 78.04
6. 0.0061   7. 0.015   8. $0.08\frac{5}{14}$ 9. $1.23\frac{3}{4}$ 10. $1.1\overline{5}$
11. $0.0\overline{12}$ 12. 2.5   13. 0.4   14. 0.056   15. 17.85
16. 0.8   17. 1.25   18. 0.075   19. $13.\overline{3}$ 20. 12.6
21. 0.038   22. 700   23. 33,644.444   24. 337.5   25. 90
26. $33.\overline{3}$ 27. 28,607.15   28. 900   29. 15,000
30. 0.00018   31. 2.004   32. 0.052   33. 70,200
34. 67.38   35. 0.009   36. 7.28   37. 0.00976

EXERCISES 7.2

Divide. Continue dividing until the remainder is zero.

1. 8.4 ÷ 6
2. 6.7 ÷ 50
3. 8 ÷ 50
4. 3 ÷ 8
5. 0.65 ÷ 0.5
6. 0.11 ÷ 5.5
7. 6.3 ÷ 0.15
8. 5.7 ÷ 0.125
9. 21 ÷ 0.06
10. 3 ÷ 0.12
11. 1244 ÷ .625
12. 555 ÷ .185
13. 6030 ÷ .09
14. 34.8 ÷ .058
15. 107.92 ÷ .0568
16. 16.2024 ÷ .03
17. 45,093 ÷ 15
18. 20.3203 ÷ 4.06
19. .244 ÷ 8
20. 3624 ÷ 300
21. $\frac{30400.2}{50}$
22. $\frac{562.5}{.0625}$
23. $\frac{.02197}{8.45}$
24. $\frac{1}{32}$
25. $\frac{14}{.035}$
26. $\frac{31}{12.4}$

Divide. Write the answer as a repeating decimal.

27. 26.8 ÷ 3
28. 95 ÷ 12
29. 125 ÷ .9
30. $\frac{2}{.003}$
31. $\frac{1}{90}$
32. $\frac{21.3}{.99}$

Convert each of these fractions to decimal form.

33. $\frac{7}{8}$
34. $\frac{3}{16}$
35. $\frac{7}{400}$
36. $\frac{13}{250}$
37. $\frac{50}{48}$
38. $\frac{458}{90}$
39. $\frac{3}{11}$
40. $\frac{5}{27}$

Divide. Approximate the answer as indicated.

41. 82.9 ÷ 6   (Round off at 3 decimal places.)
42. 930 ÷ 64   (Round off at 1 decimal place.)
43. 63.13 ÷ 8.3   (Truncate at 3 decimal places.)
44. 315 ÷ .073   (Truncate at 2 decimal places.)
45. 3 ÷ 360   (Round up at 3 decimal places.)
46. .01 ÷ .62   (Round up at 3 decimal places.)
47. $\frac{.0034}{7.7}$ (Round off at 4 decimal places.)
48. $\frac{2990.069}{4.6}$ (Truncate at 2 decimal places.)
49. $\frac{63}{4.03}$ (Round up at 2 decimal places.)
50. $\frac{40}{13}$ (Truncate at 2 decimal places.)

Divide. Stop dividing 2 places past the decimal point and write the remainder in fractional form.

51. $18.3 \div 7$ **52.** $40 \div 62$ **53.** $2.15 \div .03$
54. $\frac{8}{.28}$ **55.** $\frac{18}{6.3}$ **56.** $\frac{31}{7}$

Use a shortcut to do each indicated division. Answer with a decimal.

57. $4.56 \div 100$ **58.** $0.08 \div 10$ **59.** $8976 \div 1000$
60. $45.3 \div 1000$ **61.** $\frac{0.075}{1000}$ **62.** $\frac{45}{10}$

Solve. Write the answer as a decimal.

63. A string is 30.8 inches long. You cut the string into 8 equal pieces. How long is each piece?

64. In San Francisco, the sales tax is computed by multiplying the decimal 0.085 by the marked price of an item. If the sales tax on an item was $52.70, what was the marked price of this item?

65. The length of a rectangle is 12.8 meters. If the area is 83.2 square meters, how wide is the rectangle?

66. A wire is 3.8 feet long. Sasha wants to cut the wire into 0.04-foot pieces. How many pieces can she cut?

DEVELOPING NUMBER SENSE #6

MULTIPLYING BY 0.1, 0.01, AND 0.001 MENTALLY

Multiplying by the decimal 0.1 is the same as multiplying by the fraction $\frac{1}{10}$, and thus is equivalent to dividing by 10. Therefore, this should cause the decimal point to move 1 place to the left. Here we show some examples:

$$36 \times 0.1 = 3.6$$
$$75.8 \times 0.1 = 7.58$$
$$0.07 \times 0.1 = 0.007$$

Similarly, multiplying by the decimal 0.01 is the same as multiplying by the fraction $\frac{1}{100}$, and thus is equivalent to dividing by 100. Therefore, this should cause the decimal point to move 2 places to the left. Here are some examples:

$$36 \times 0.01 = 0.36$$
$$75.8 \times 0.01 = 0.758$$
$$0.07 \times 0.01 = 0.0007$$

The pattern observed above continues. Multiplying by 0.001 causes the decimal point to move 3 places to the left.

Number Sense Problems

Compute each of the following mentally.

1. 0.1×340 **2.** 0.1×78.4 **3.** 0.1×0.659 **4.** 0.1×9.2
5. 0.01×84 **6.** 0.01×8.75 **7.** 0.01×0.006 **8.** 0.01×12.45
9. 0.001×3 **10.** 0.001×57 **11.** 0.001×45.7 **12.** 0.001×0.5
13. 0.01×3.67 **14.** 0.1×0.786 **15.** 0.01×80 **16.** 0.1×1200

7.3 THE MISSING NUMBER IN A MULTIPLICATION STATEMENT

OBJECTIVE ▌ Find the missing number in a multiplication statement.

In the multiplication statement

$$9 \times 3 = 27 \quad \text{or} \quad 27 = 9 \times 3$$

the numbers 9 and 3 are called **multipliers** or **factors.** The answer to the multiplication problem, 27, is called the **product** of 9 and 3. Note that the statement can be written with the product located after the equality symbol or before the equality symbol.

Three questions can be asked by omitting any one of the three numbers.

1. $9 \times 3 = \boxed{}$
   Missing number is 27.
   We are missing the answer to a multiplication problem, so we *multiply* to find the missing number.

2. $9 \times \boxed{} = 27$
   $9 \overline{)27} 3$
   known multiplier — answer to the multiplication problem
   We are missing one of the multipliers, so we divide to find the missing number.

3. $\boxed{} \times 3 = 27$
   $3 \overline{)27} 9$
   known multiplier — answer to the multiplication problem
   We are missing one of the multipliers, so we *divide* to find the missing number.

In the multiplication statement $9 \times 3 = 27$, the product 27 is larger than either one of the multipliers. Do not expect this to always be true. Here we look at more examples.

$2.5 \times 400 = 1000$ — Here the product 1000 is larger than either multiplier because both multipliers are larger than 1.

$0.5 \times 400 = 200$ — Here the product 200 is less than the multiplier 400 because the other multiplier, 0.5, is less than 1.

EXAMPLE 1 $\quad 0.85 \times 908 = \boxed{}$

SOLUTION We are missing the answer to the multiplication problem, so we multiply to find the missing number.

$$\begin{array}{r} 9\,08 \\ \times\ .85 \\ \hline 45\,40 \\ 726\,4 \\ \hline 771.80 \end{array} = 771.8$$

The missing product is 771.8. ▌

Try These Problems

Find the missing number. Write the answer as a decimal.

1. $70 \times 0.3 = \boxed{}$
2. $\boxed{} = 16 \times 0.83$
3. $0.72 = 18 \times \boxed{}$
4. $\boxed{} \times 700 = 35$
5. $\boxed{} \times 82.5 = 0.66$
6. $120{,}000 = 0.375 \times \boxed{}$

Note in Example 1 that we expected the product to be less than 908 because 0.85 is less than 1.

Try Problems 1 and 2.

EXAMPLE 2 $\quad 12 = 0.25 \times \boxed{}$

SOLUTION One of the multipliers is missing, so we divide to find the missing number.

$$\underset{\text{multiplier}}{\text{Missing}} = \underset{\text{problem}}{\underset{\text{multiplication}}{\text{answer to the}}} \div \underset{\text{multiplier}}{\text{known}}$$

$$\boxed{} = 12 \div 0.25$$

$$0.25 \overline{)12.00} \quad \begin{array}{r} 48. \\ \underline{10\ 0} \\ 2\ 00 \\ \underline{2\ 00} \end{array} = 48 \longleftarrow \text{missing multiplier}$$

known multiplier ⟶ ⟵ answer to the multiplication problem

CHECK
$$\begin{array}{r} 48 \\ \times\ .25 \\ \hline 2\ 40 \\ 9\ 6 \\ \hline 12.00 = 12 \end{array}$$

The missing multiplier is 48. ∎

In Example 2 note that the multiplier we found, 48, is more than the product 12. This happened because the other multiplier, 0.25, is less than 1.

EXAMPLE 3 $\quad 6.8 \times \boxed{} = 20.74$

SOLUTION One of the multipliers is missing, so we divide to find the missing number.

$$\underset{\text{multiplier}}{\text{Missing}} = \underset{\text{problem}}{\underset{\text{multiplication}}{\text{answer to the}}} \div \underset{\text{multiplier}}{\text{known}}$$

$$\boxed{} = 20.74 \div 6.8$$

$$6.8 \overline{)20.740} \quad \begin{array}{r} 3.05 \\ \underline{20\ 4} \\ 34 \\ \underline{0} \\ 3\ 40 \\ \underline{3\ 40} \end{array}$$

The missing multiplier is 3.05. ∎

Try Problems 3 through 6.

Now we summarize the procedure for finding the missing number in a multiplication statement.

> *Finding the Missing Number in a Multiplication Statement*
> 1. If you are missing the answer to the multiplication problem, then **multiply**.
> 2. If you are missing one of the multipliers, then **divide**. Take care to divide in the correct order. Here is how it works.
>
> known multiplier ⟌ missing multiplier / answer to the multiplication problem

 Answers to Try These Problems

1. 21 2. 13.28 3. 0.04 4. 0.05 5. 0.008 6. 320,000

EXERCISES 7.3

Find the missing number. Write the answer as a decimal.

1. 0.065 × ☐ = 1.508
2. 72.45 = 0.35 × ☐
3. 1800 × ☐ = 36.9
4. 48 = 0.6 × ☐
5. 76.2 × 0.05 = ☐
6. ☐ = 2.3 × 0.047
7. 4.75 = ☐ × 380
8. ☐ × 5200 = 16,380
9. 900 = 2.5 × ☐
10. 32.4 × ☐ = 200.88
11. ☐ = 150.20 × 0.065
12. 600 × 8.3 = ☐
13. 91 = 7 × ☐
14. ☐ × 6000 = 15
15. 985.32 = 122.4 × ☐
16. 5000 = 3.125 × ☐
17. 0.95 × ☐ = 0.9785
18. 30.4 × 720 = ☐
19. 65 = ☐ × 0.13
20. 31.59 = 4050 × ☐

DEVELOPING NUMBER SENSE #7

MULTIPLYING AND DIVIDING BY A DECIMAL LESS THAN, EQUAL TO, OR MORE THAN 1

Here we look at the number 480 multiplied by 2.4, 1, and 0.75. Observe how the answers compare with 480.

$$480 \times 2.4 = 1152$$
$$480 \times 1 = 480$$
$$480 \times 0.75 = 360$$

The first result 1152 is more than 480 because the factor 2.4 is more than 1. The second result 480 is equal to 480 because the factor being multiplied by 480 is 1. The third result 360 is less than 480 because the factor 0.75 is less than 1. This property of numbers was also observed in Developing Number Sense #3 with relation to fractions.

What happens with *division* by numbers less than, equal to, or more than 1? Here we divide the number 480 by the numbers 2.4, 1, and 0.75. Observe how the answers compare with 480.

$$480 \div 2.4 = 200$$
$$480 \div 1 = 480$$
$$480 \div 0.75 = 640$$

The first result 200 is less than 480 because the divisor 2.4 is more than 1. The second result is 480 because the divisor is 1. The third result 640 is more than 480 because the divisor 0.75 is less than 1.

Cont. page 283

Number Sense Problems

Without computing, decide whether the result is less than, equal to, or more than 850.

1. 850 × 7.5
2. 850 × 0.98
3. 0.005 × 850
4. 1.86 × 850
5. 850 ÷ 1
6. 850 ÷ 1.7
7. 850 ÷ 0.64
8. 850 ÷ 0.005

Without computing, decide whether the result is less than, equal to, or more than 47.2.

9. 47.2 ÷ 0.16
10. 47.2 × 0.16
11. 47.2 × 56.8
12. 47.2 ÷ 56.8

7.4 LANGUAGE

OBJECTIVES
- Find a decimal of a number.
- Translate an English phrase to math symbols involving multiplication and division.
- Solve problems by using translations.

 Try These Problems

Solve.
1. 0.25 of 68.5
2. 4.5 of 4600
3. 0.076 of 30.6
4. 5.09 of 0.7

1 Finding a Decimal of a Number

Recall from Section 4.4 that to find a fraction of a number we multiply the fraction by the number. For example,

$$\frac{3}{4} \text{ of } 20$$
$$\downarrow$$
$$= \frac{3}{\underset{1}{4}} \times \frac{\overset{5}{20}}{1} \quad \text{The word } of \text{ used in this way translates to multiplication.}$$
$$= 15$$

Also to find a decimal of a number we multiply the decimal by the number. For example,

$$0.75 \text{ of } 20$$
$$\downarrow$$
$$= 0.75 \times 20 \quad \text{The word } of \text{ used in this way translates to multiplication.}$$
$$= 15.00$$
$$= 15$$

> **Finding a Decimal of a Number**
> To find a decimal of a number, multiply the decimal by the number, that is, the word *of* used in this way translates to multiplication.

 Try Problems 1 through 4.

2 Translating English to Multiplication

Here is a review of the many English translations for the symbol =.

Math Symbol	English
=	equals
	is equal to
	is the same as
	is the result of
	is
	are
	was
	were
	will be
	represents
	gives
	makes
	yields

The multiplication statement $1.5 \times 3 = 4.5$ or $4.5 = 1.5 \times 3$ is written with math symbols. Some of the ways to read this statement in English are given in the following chart.

Math Symbols	English
$1.5 \times 3 = 4.5$	1.5 times 3 equals 4.5.
$4.5 = 1.5 \times 3$	1.5 multiplied by 3 is 4.5.
	1.5 of 3 yields 4.5.
	4.5 is 3 times as large as 1.5.
	The *product* of 1.5 and 3 is 4.5.

Because 1.5×3 equals 3×1.5, the order that you say the multipliers does not matter.

The following examples illustrate how to use translating to solve problems.

EXAMPLE 1 What number represents the product of 4.05 and 60?

SOLUTION The question translates to a multiplication statement.

What number | represents | the product of 4.05 and 60 | ?
□ | = | 4.05×60 |

The answer to the multiplication statement is missing, so we multiply to find the missing number.

$$\begin{array}{r} 4.05 \\ \times\ \ 60 \\ \hline 243.00 \end{array} = 243$$

The missing product is 243. ■

Try These Problems

5. What number is 5 times as large as 43.2?
6. 65 represents the product of 12.5 and what number?
7. What decimal of 700.5 is 5.604?
8. 15.7 times 50.9 yields what number?

EXAMPLE 2 0.75 of what number is 300?

SOLUTION The question translates to a multiplication statement.

$$0.75 \text{ of what number is } 300?$$
$$0.75 \times \boxed{} = 300$$

A multiplier is missing, so divide the answer 300 by the given multiplier 0.75 to find the missing multiplier.

$$0.75 \overline{)300}$$ → 4, remainder 300

The missing number is 4. ■

 Try Problems 5 through 8.

Special language is used when the number 2 or the number 0.5 is a multiplier.

Math Symbols	*English*
$2 \times 8.6 = 17.2$	*Twice* 8.6 is 17.2. Doubling 8.6 gives 17.2. 2 times 8.6 equals 17.2.
$0.5 \times 8.6 = 4.3$	*Half of* 8.6 is 4.3. 0.5 times 8.6 gives 4.3.

EXAMPLE 3 What number is half of 0.35?

SOLUTION The question translates to a multiplication statement.

$$\text{What number is half of } 0.35?$$
$$\boxed{} = 0.5 \times 0.35$$

The answer to the multiplication statement is missing, so we multiply to find the missing number.

$$\begin{array}{r} 0.35 \\ \times\ 0.5 \\ \hline .175 \end{array}$$ ■

EXAMPLE 4 1.21 equals twice what number?

SOLUTION The question translates to a multiplication statement.

$$1.21 \text{ equals twice what number?}$$
$$1.21 = 2 \times \boxed{}$$

A multiplier is missing, so divide the answer 1.21 by the given multiplier 2 to find the missing multiplier.

Try These Problems

Solve. Answer with a decimal.

9. 8.15 is half of what number?
10. What is the result when 3.8 is doubled?
11. Find twice 0.056.
12. Find half of 45.9.
13. Divide 500 by 3.2.
14. Divide 500 into 3.2.
15. Divide 1.2 into 0.15.
16. Divide 1.2 by 0.15.
17. Find the ratio of 50 to 0.32.
18. Find the ratio of 0.32 to 50.

$$\begin{array}{r}.605\\2\overline{)1.210}\\\underline{1\,2}\\1\\\underline{0}\\10\\\underline{10}\end{array}$$

 Try Problems 9 through 12.

3 Translating English to Division

When reading the division statement $80 \div 0.4$, take care to say 80 and 0.4 in the correct order because $80 \div 0.4$ does not equal $0.4 \div 80$. Here we illustrate the division done in both ways.

$$80 \div 0.4 \rightarrow 0.4\overline{)80.0} \quad \underset{}{200.} = 200$$

$$0.4 \div 80 \rightarrow 80\overline{)0.400} \quad \underset{}{.005}$$
$$\phantom{0.4 \div 80 \rightarrow 80\overline{)0.}}\underline{400}$$

Observe that the results are different. The following chart gives correct ways to read the division statement $80 \div 0.4 = 200$.

Math Symbols	English
$80 \div 0.4 = 200$	80 *divided by* 0.4 equals 200. 0.4 *divided into* 80 is 200. The *ratio* of 80 to 0.4 is 200.

Note that the word *by* is used when the statement is read from left to right as you read words in a book; however, the word *into* is used if you read the divisor first. The divisor is the number after the division symbol.

Recall that the fraction bar is also used to indicate division. For example,

$$\frac{80}{0.4} = 80 \div 0.4$$

The following chart gives correct ways to read the division statement $\frac{80}{0.4} = 200$.

Math Symbols	English
$\frac{80}{0.4} = 200$	80 *divided by* 0.4 is 200. 0.4 *divided into* 80 equals 200. The *ratio* of 80 to 0.4 is 200.

Note that the word *by* is used when the numerator is read first, but the word *into* is used when the denominator is read first.

 Try Problems 13 through 18.

Answers to Try These Problems

1. 17.125   **2.** 20,700   **3.** 2.3256   **4.** 3.563   **5.** 216
6. 5.2   **7.** 0.008   **8.** 799.13   **9.** 16.3   **10.** 7.6   **11.** 0.112
12. 22.95   **13.** 156.25   **14.** 0.0064   **15.** 0.125   **16.** 8
17. 156.25   **18.** 0.0064

EXERCISES 7.4

Translate each of these statements to a multiplication or division statement using math symbols.

1. The product of 0.3 and 0.2 equals 0.06.
2. 0.5 divided by 0.2 equals 2.5.
3. 45.2 is 5 times as large as 9.04.
4. 19.5 is half of 39.
5. 41.6 is twice 20.8.
6. 450 represents 0.75 of 600.

Solve.

7. Find the product of 4.7 and 200.
8. 0.875 is the product of 2.5 and what number?
9. Divide 0.4 by 16.
10. Divide 0.4 into 16.
11. Find half of 30.5.
12. Half of what number is 67.5?
13. 9.45 represents 0.07 of what number?
14. Find 0.035 of 46,000.
15. What decimal of 60.4 is 392.6?
16. 32,500 equals 1.25 of what number?
17. Divide 0.075 into 37.5.
18. Divide 0.075 by 37.5.
19. What number is twice 18.09?
20. Twice what number is 1.05?
21. Find the ratio of 81 to 4.05.
22. What number is 8.5 times as large as 55?
23. 300 is what decimal of 3750?
24. Find the ratio of 18 to 0.036.

7.5 APPLICATIONS

OBJECTIVES
- Solve an application involving equal parts in a total quantity.
- Solve an application involving equivalent rates.
- Convert a rate to an equivalent rate.
- Solve an application by interpreting a rate as a ratio.
- Solve an application by using a rate as a multiplier.
- Solve an application involving the area of a rectangle.
- Solve an application by using translations.
- Find the average (or mean) of a collection of numbers.

1 Solving Problems Involving Equal Parts in a Total Quantity

Suppose you have a string that is 24 feet long. If you cut the string into 3 equal pieces, then each piece is 8 feet. Here is a picture of the situation.

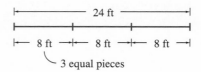

Observe that the numbers are related as follows.

$$\begin{array}{c} \text{Size of} \\ \text{each part} \\ \text{(ft)} \end{array} \times \begin{array}{c} \text{number of} \\ \text{equal parts} \end{array} = \begin{array}{c} \text{total} \\ \text{quantity} \\ \text{(ft)} \end{array}$$

$$8 \quad \times \quad 3 \quad = \quad 24$$

In general, if a situation involves a total quantity being split into a number of equal parts, the following always applies.

$$\begin{array}{c} \textbf{Size of} \\ \textbf{each part} \end{array} \times \begin{array}{c} \textbf{number of} \\ \textbf{equal parts} \end{array} = \begin{array}{c} \textbf{total} \\ \textbf{quantity} \end{array}$$

A problem can be presented by giving any two of the three quantities and asking for the third quantity. The following examples illustrate this.

EXAMPLE 1 A board 6 feet long is cut into 24 equal pieces. How long is each piece?

SOLUTION Here we picture the situation.

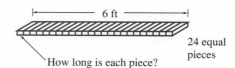

We use the formula

$$\begin{array}{c} \text{Size of} \\ \text{each part} \\ \text{(ft)} \end{array} \times \begin{array}{c} \text{number of} \\ \text{equal parts} \end{array} = \begin{array}{c} \text{total} \\ \text{quantity} \\ \text{(ft)} \end{array}$$

$$\boxed{} \quad \times \quad 24 \quad = \quad 6$$

A multiplier is missing, so we divide 6 by 24 to find the missing multiplier.

```
              .25   ── length of each piece in feet
          24)6.00   ── total length in feet
             48
             ───
             120
             120
```
number of equal pieces

Each piece is 0.25 foot long. ■

Try These Problems

Solve.

1. A rope that is 45 inches long is cut into 36 equal pieces. How long is each piece?

2. A rope that is 36 inches long is cut into 45 equal pieces. How long is each piece?

3. Medicine is given out in doses that each contain 1.8 centiliters. How many doses can be given from a bottle that contains 45 centiliters?

4. Twelve children share equally 2.4 gallons of milk. How much milk does each child receive?

5. Mr. Henry pays off a debt by making 40 equal payments. If the total debt is $4370, what is each payment?

6. Linda is preparing punch for a party. She wants to have 75 servings that each contain 6.5 ounces. How many ounces of punch does she need?

EXAMPLE 2 A slicing machine for a baking company slices a 24-centimeter loaf of bread into slices that are each 1.5 centimeters wide. How many slices can be cut from the loaf of bread?

SOLUTION Here we picture the situation.

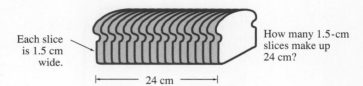

We use the formula

$$\begin{array}{c}\text{Size of}\\\text{each part}\\\text{(cm)}\end{array} \times \begin{array}{c}\text{number of}\\\text{equal parts}\end{array} = \begin{array}{c}\text{total}\\\text{quantity}\\\text{(cm)}\end{array}$$

$$1.5 \quad \times \quad \boxed{} \quad = \quad 24$$

A multiplier is missing, so we divide 24 by 1.5 to find the missing multiplier.

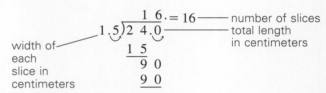

16 slices can be cut from the loaf. ∎

EXAMPLE 3 Theresa pays off a debt by making 60 equal payments of $230.70. What is the total debt?

SOLUTION We use the formula

$$\begin{array}{c}\text{Size of}\\\text{each part}\\\text{(\$)}\end{array} \times \begin{array}{c}\text{number of}\\\text{equal parts}\end{array} = \begin{array}{c}\text{total}\\\text{quantity}\\\text{(\$)}\end{array}$$

$$230.70 \quad \times \quad 60 \quad = \quad \boxed{}$$

We are missing the answer to the multiplication problem, so we multiply to find the total debt.

$$\begin{array}{r}230.7\\ \times60\\ \hline 1384\,2.0 = 13{,}842\end{array}$$

The total debt is $13,842. ∎

 **Try Problems 1 through 6.**

2 Generating Equivalent Rates

If you earn 15 dollars per hour, then the phrase *15 dollars per hour,* (that means $15 for each 1 hour) indicates the **rate** of pay. A rate is a comparison of two quantities. Here is a list of more phrases that indicate rates.

> **Phrases that Indicate Rates**
> The faucet leaks 3 ounces every hour.
> Shirts cost $35.69 each.
> A stock costs $89.70 per share.
> Twenty nails weigh 0.9 pound.
> He bicycles 1.8 miles in 3 minutes.

If you earn $15 in 1 hour, then you earn $45 in 3 hours.

$$\text{Multiply by 3} \left(\begin{array}{l} \text{15 dollars in 1 hour} \\ \text{45 dollars in 3 hours} \end{array} \right) \text{Multiply by 3}$$

Observe that multiplying both quantities in a rate by the same nonzero number produces an equivalent rate. This can be used to solve problems as shown in the following examples.

EXAMPLE 4 Charles earns $212.40 each week. How much money does he earn in 16 weeks?

SOLUTION The phrase *$212.40 each week* means $212.40 for 1 week.

$$\text{Multiply by 16.} \left(\begin{array}{l} \$212.40 \text{ in 1 week} \\ ? \text{ in 16 weeks} \end{array} \right) \text{Multiply by 16.}$$

If we multiply each quantity in the given rate by 16, we obtain the money earned in 16 weeks.

```
     212.40
  ×      16
   1274 40
   2124 0
   3398.40
```

Charles earns $3398.40 in 16 weeks. ■

EXAMPLE 5 Frances can jog 1 mile in 8.5 minutes. How long will it take her to jog 5.2 miles?

SOLUTION The phrase *1 mile in 8.5 minutes* is the given rate. We want an equivalent rate that involves 5.2 miles.

$$\text{Multiply by 5.2.} \left(\begin{array}{l} \text{1 mile in 8.5 minutes} \\ \text{5.2 miles in ? minutes} \end{array} \right) \text{Multiply by 5.2.}$$

If we multiply both quantities in the given rate by 5.2, we obtain the number of minutes it takes Frances to jog 5.2 miles.

$$8.5 \times 5.2 = 44.20 = 44.2$$

It will take Frances 44.2 minutes to jog 5.2 miles. ■

If 3 pounds of coffee cost $18, then 1 pound of coffee costs $6.

$$\text{Divide by 3.} \left(\begin{array}{l} \text{3 pounds cost \$18} \\ \text{1 pound costs \$6} \end{array} \right) \text{Divide by 3.}$$

Observe that dividing both quantities of a rate by the same nonzero number yields an equivalent rate. This can be used to solve problems as shown in the following examples.

EXAMPLE 6 Twenty nails weigh 16 ounces. What is the weight of 1 nail?

SOLUTION The phrase *20 nails weigh 16 ounces* is the given rate. We want an equivalent rate that involves 1 nail.

Divide by 20. ⟶ 20 nails weigh 16 ounces / 1 nail weighs ? ounces ⟵ Divide by 20.

Dividing each quantity in the given rate by 20 gives the weight of 1 nail.

$$20\overline{)16.0}$$
$$\underline{16\ 0}$$
$$.8$$

One nail weighs 0.8 ounces. ■

EXAMPLE 7 An airplane travels 3510 miles in 5.4 hours. How far can the plane go in 1 hour?

SOLUTION The phrase *3510 miles in 5.4 hours* is the given rate. We want an equivalent rate that involves 1 hour.

Divide by 5.4. ⟶ 3510 miles in 5.4 hours / ? miles in 1 hour ⟵ Divide by 5.4.

Dividing both quantities in the given rate by 5.4 gives the number of miles the plane goes in 1 hour.

$$5.4\overline{)3510.0} = 650$$

with quotient 6 5 0., and steps 3 2 4, 2 7 0, 2 7 0, 0, 0.

The plane can go 650 miles in 1 hour. ■

EXAMPLE 8 The cost of 5.5 pounds of bananas is $3.74. What is the cost per pound?

SOLUTION The phrase *5.5 pounds cost $3.74* is the given rate. The phrase *cost per pound* means cost for 1 pound. We want to convert the given rate to an equivalent rate that involves 1 pound.

Divide by 5.5. ⟶ 5.5 pounds cost $3.74 / 1 pound costs ? ⟵ Divide by 5.5.

Dividing both quantities in the given rate by 5.5 gives the cost of 1 pound of bananas.

$$5.5\overline{)3.740}$$
with quotient .6 8, steps 3 3 0, 4 4 0, 4 4 0.

One pound of bananas costs $0.68. ■

Try These Problems

7. One nail weighs 0.03 pound. What do 75 nails weigh?

8. One share of utility stock sells for $35.40. What is the cost of 27 shares?

9. Rick paid $42.60 for 7.5 yards of fabric. What is the cost of 1 yard of fabric?

10. A watch loses 3 minutes every 4.8 hours. How many minutes does the watch lose in 1 hour?

11. A faucet leaks 3.4 ounces per hour. How many ounces does the faucet leak in 8.6 hours?

12. It takes you 35 minutes to drive 21 miles. How far do you drive per minute?

Convert each ratio to a rate. Specify both the quantity and the units.

13. $\dfrac{5 \text{ pounds}}{1 \text{ week}}$

14. $\dfrac{9 \text{ gallons}}{4 \text{ hours}}$

15. $\dfrac{15.6 \text{ ¢}}{2.4 \text{ ounces}}$

16. $\dfrac{4 \text{ dollars}}{3.2 \text{ pounds}}$

We summarize the previous examples by stating a rule about rates.

> *Generating Equivalent Rates*
> 1. Both quantities in a rate can be multiplied by the same nonzero number to produce an equivalent rate.
> 2. Both quantities in a rate can be divided by the same nonzero number to produce an equivalent rate.

 Try Problems 7 through 12.

3 A Rate as a Ratio

Recall from Sections 3.6 and 4.5 that a **ratio** is a fraction or a comparison by division. For example,

$$\text{the ratio of 3 to 5} = \frac{3}{5}$$

Recall also from Section 4.5 that a **rate** is a comparison of two quantities *by division,* thus a rate can be written as a ratio of two quantities. Here are some phrases, their corresponding ratios, and the resulting rates.

English Phrase	Ratio	Rate
Sue earns $25 per hour.	$\dfrac{25 \text{ dollars}}{1 \text{ hour}}$	25 dollars per hour
Dave lost 7 pounds in 2 weeks.	$\dfrac{7 \text{ pounds}}{2 \text{ weeks}}$	$\dfrac{7}{2} = 3.5$ pounds per week
A fish swims 72 feet every 90 seconds.	$\dfrac{72 \text{ feet}}{90 \text{ seconds}}$	$\dfrac{72}{90} = 0.8$ feet per second

First, notice how the ratio units translate to the rate units. The unit in the denominator of the ratio follows the word *per* in the rate unit. That is, the word *per* is used for the fraction bar. Here we look at only the units.

$$\frac{\text{dollars}}{\text{hour}} \longrightarrow \text{dollars per hour}$$

$$\frac{\text{pounds}}{\text{week}} \longrightarrow \text{pounds per week}$$

$$\frac{\text{feet}}{\text{second}} \longrightarrow \text{feet per second}$$

Second, because the fraction bar indicates to divide the numerator by the denominator, perform this division to obtain the rate quantity. For example,

$$\frac{72 \text{ feet}}{90 \text{ seconds}} = 0.8 \text{ feet per second}$$

Divide 72 by 90 to obtain 0.8.

 Try Problems 13 through 16.

7.5 Applications

▲ Try These Problems

Solve. Answer with a decimal.

17. A particle travels 3 meters in 24 seconds. How far does the particle travel in 1 second?

18. Two hundred fifty nails weigh 5 pounds. What is the weight per nail?

19. Ms. Yeh earns $442.50 in 30 hours. How much does she earn per hour?

20. A 15.2-ounce can of corn costs $1.29. What is the cost of 1 ounce? (Round off the answer to the nearest cent.)

Interpreting a rate as a ratio can help in solving problems. Here we show some examples.

EXAMPLE 9 A car can go 85.4 miles on 2.8 gallons of gas. How many miles per gallon is this?

SOLUTION We want to know the number of miles per gallon, so we need the ratio of miles to gallons.

85.4 miles on 2.8 gallons

$$= \frac{85.4 \text{ miles}}{2.8 \text{ gallons}}$$

We put miles in the numerator and gallons in the denominator to obtain miles per gallon.

We need to divide 85.4 by 2.8 to obtain the number of miles per gallon.

$$\begin{array}{r} 30.5 \\ 2.8{\overline{\smash{\big)}\,85.40}} \\ \underline{84} \\ 1\,4 \\ \underline{0} \\ 1\,4\,0 \\ \underline{1\,4\,0} \end{array}$$

The car gets 30.5 miles per gallon.

EXAMPLE 10 In 6.25 minutes a tank leaks 2 liters of water. How much does the tank leak in 1 minute?

SOLUTION We want the number of liters in 1 minute; that is, we want the number of liters per minute. We need the ratio of liters to minutes. In 6.25 minutes the tank leaks 2 liters

$$= \frac{2 \text{ liters}}{6.25 \text{ minutes}}$$

We put liters in the numerator and minutes in the denominator to obtain liters per minute.

We divide 2 by 6.25 to obtain the number of liters per minute.

$$\begin{array}{r} 0.32 \\ 6.25{\overline{\smash{\big)}\,2.0000}} \\ \underline{1\,875} \\ 1\,2\,5\,0 \\ \underline{1\,2\,5\,0} \end{array}$$

The tank leaks 0.32 liter in 1 minute.

 Try Problems 17 through 20.

4 Using a Rate as a Multiplier

If you drive 50 miles per hour for 3 hours, then you travel a total distance of 150 miles. We have the following.

$$\frac{\text{miles}}{\text{per hour}} \times \frac{\text{number}}{\text{of hours}} = \frac{\text{total number}}{\text{of miles}}$$

$$50 \quad \times \quad 3 \quad = \quad 150$$

Observe that if you write the rate, 50 miles per hour, as the ratio, $\frac{50 \text{ miles}}{1 \text{ hour}}$, and multiply by 3 hours, the hour units in the denominator cancel with the hour units in the numerator and we are left with the answer in miles.

$$\frac{50 \text{ miles}}{1 \text{ hour}} \times \frac{3 \text{ hours}}{1} = 150 \text{ miles}$$

Here are some more examples of how units cancel when multiplying by a rate.

$$\frac{\text{feet}}{\text{second}} \times \text{seconds} = \text{feet}$$

$$\frac{\text{cents}}{\text{pound}} \times \text{pounds} = \text{cents}$$

Paying close attention to the units can help you to set up problems correctly. Here are some examples.

EXAMPLE 11 Canned apricots cost $0.13 per ounce. What is the cost of 15 ounces?

SOLUTION The phrase *$0.13 per ounce* indicates a rate, so it can be written as a ratio.

$$\$0.13 \text{ per ounce} = \frac{\$0.13}{1 \text{ ounce}} = \frac{0.13 \text{ dollars}}{1 \text{ ounce}}$$

Set up a multiplication statement using this rate.

$$\frac{\text{dollars}}{\text{ounce}} \times \text{ounces} = \text{dollars}$$

$$0.13 \quad \times \quad 15 \quad = \quad \boxed{}$$

The answer to the multiplication problem is missing, so we multiply to find the missing number.

$$0.13 \times 15 = 1.95$$

The cost of 15 ounces is $1.95. ■

EXAMPLE 12 Emerald, a college student, earns $12.50 per hour. How many hours must she work to earn $600?

SOLUTION
METHOD 1 The phrase *$12.50 per hour* indicates a rate so can be written as a ratio.

$$\$12.50 \text{ per hour} = \frac{\$12.50}{1 \text{ hour}} = \frac{12.5 \text{ dollars}}{1 \text{ hour}}$$

Set up a multiplication statement using this rate.

$$\frac{\text{dollars}}{\text{hour}} \times \text{hours} = \text{dollars}$$

$$12.5 \quad \times \quad \boxed{} \quad = \quad 600$$

A multiplier is missing, so we divide 600 by 12.5 to find the missing multiplier.

$$12.5 \overline{)600.0} = 48 \quad \text{— Number of hours it takes her to earn \$600.}$$

```
         4 8. = 48
 1 2.5 )6 0 0.0
        5 0 0
        1 0 0 0
        1 0 0 0
```

She must work 48 hours to earn $600.

METHOD 2 The given rate *$12.50 per hour* means $12.50 for 1 hour. We want to convert this rate to an equivalent rate involving $600.

$12.50   for   1   hour
$600   for   ?   hours

It is not obvious how to convert the given rate to an equivalent rate, so we find out how long she works to earn $1.

Divide by $12.50 for 1 hour   Divide by
$12.50. $1 for ? hour $12.50.

Dividing both quantities in the given rate by 12.5 produces an equivalent rate involving $1.

```
         . 0 8     — Fractional part of an hour it
 1 2.5 )1.0 0 0       takes her to earn $1.
         1 0 0 0
```

She earns $1 in 0.08 hour. Now we convert this rate to an equivalent rate involving $600.

Multiply $1 in 0.08 hour   Multiply
by 600. $600 in ? hours   by 600.

Multiplying both quantities in the known rate by 600 gives an equivalent rate involving $600.

$$0.08 \times 600 = 48$$

She must work 48 hours to earn $600. ■

EXAMPLE 13 Mr. Wong types 420 words in 10.5 minutes. How long will it take him to type 750 words?

SOLUTION

METHOD 1 The phrase *420 words in 10.5 minutes* indicates the rate Mr. Wong types. We can write this rate as a ratio.

$$420 \text{ words in } 10.5 \text{ minutes} = \frac{420 \text{ words}}{10.5 \text{ minutes}}$$

Dividing 420 by 10.5 gives us how many words Mr. Wong types per minute.

```
          4 0.    — Number of words he
 1 0.5 )4 2 0.0      can type in 1 minute.
        4 2 0
              0
```

Try These Problems

Solve. Answer with a decimal.
21. You drive 0.6 mile each minute. How far can you drive in 85 minutes?
22. Arthur paid $1.25 per pound for oranges. How many pounds can he buy for $10.75?
23. A printer prints pages at the rate of 8 pages per minute. How long will it take the printer to print 1500 pages?
24. Bob stuffs 24 envelopes in 60 seconds. How long does it take him to stuff 50 envelopes?
25. A cable that is 5.2 feet long weighs 26 pounds. What is the weight of 3.9 feet of this cable?

Mr. Wong types 40 words per minute. Finally, we set up a multiplication statement using this rate so that we can find out how long it takes him to type 750 words.

$$\frac{\text{words}}{\text{minute}} \times \text{minutes} = \text{words}$$

$$40 \times \boxed{} = 750$$

A multiplier is missing, so we divide the answer 750 by the given multiplier 40 to find the missing multiplier.

```
          18.75  ——— Number of minutes
    40)750.00        it takes him to type 750 words.
       40
       ---
       350
       320
       ---
        30 0
        28 0
        ----
         2 00
         2 00
```

It takes him 18.75 minutes to type 750 words.

METHOD 2 The phrase *420 words in 10.5 minutes* indicates a rate. We want an equivalent rate that involves 750 words.

420 words in 10.5 minutes

750 words in ? minutes

It is not obvious how to convert the given rate to an equivalent rate involving 750 words. Therefore, before finding how long it takes him to type 750 words, we find how long it takes him to type 1 word.

Divide by 420. ⌒ 420 words in 10.5 minutes ⌐ Divide
                ↘ 1 word in ? minutes   ↙ by 420.

Dividing both quantities in the given rate by 420 gives how many minutes it takes him to type 1 word.

```
            .0 2 5 ——— Fractional part of a minute it
    4 2 0)1 0.5 0 0       takes him to type 1 word.
          8 4 0
          -----
          2 1 0 0
          2 1 0 0
```

It takes him 0.025 minute to type 1 word. Now we convert the rate, 1 word in 0.025 minute to an equivalent rate involving 750 words.

Multiply by 750. ⌒ 1 word in 0.025 minute ⌐ Multiply
                  ↘ 750 words in ? minutes ↙ by 750.

Multiplying both quantities in the known rate by 750 gives an equivalent rate involving 750 words.

$$750 \times 0.025 = 18.75$$

It takes Mr. Wong 18.75 minutes to type 750 words. ■

 Try Problems 21 through 25.

5 Area of a Rectangle

The **area** of a rectangle is a measure of the extent of the region inside the boundary. A square that measures 1 unit on each side is said to have an area of 1 square unit.

1 unit Area = 1 square unit
1 unit

A rectangle that is 3 units wide and 4 units long contains 12 of the 1-square-unit squares, therefore the area is 12 square units.

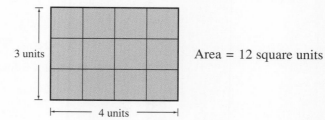

3 units   Area = 12 square units

4 units

Observe that the area of this rectangle can be obtained by multiplying the length by the width.

$$\text{Area} = \text{length} \times \text{width}$$
$$= 4 \times 3$$
$$= 12 \text{ square units}$$

In general, the area of any rectangle can be computed by multiplying the length by the width.

Area of a Rectangle
The area of a rectangle is the length times the width.
Area = length × width

EXAMPLE 14 Find the area of a rectangle that has length 65.4 feet and width 40.5 feet.

SOLUTION We use the following formula.

$$\underset{(\text{sq ft})}{\text{Area}} = \underset{(\text{ft})}{\text{length}} \times \underset{(\text{ft})}{\text{width}}$$
$$\boxed{} = 65.4 \times 40.5$$

We multiply 65.4 by 40.5 to find the area.

```
      6 5.4
   ×  4 0.5
     3 2 7 0
   2 6 1 6
   2 6 4 8.7 0 = 2648.7
```

The area is 2648.7 square feet. ∎

Try These Problems

26. The area of a plot of land is 1.76 square miles. The length is 2.2 miles. How wide is the plot?

27. The rectangular floor of a room measures 7.5 yards by 8.4 yards. How many square yards does this floor contain?

EXAMPLE 15 The area of a rectangle is 54.4 square meters. The width is 6.4 meters. Find the length.

SOLUTION We use the following formula.

$$\begin{array}{c} \text{Area} \\ \text{(sq m)} \end{array} = \begin{array}{c} \text{length} \\ \text{(m)} \end{array} \times \begin{array}{c} \text{width} \\ \text{(m)} \end{array}$$

$$54.4 = \square \times 6.4$$

A multiplier is missing, so we divide the answer 54.4 by the given multiplier 6.4 to find the missing multiplier.

$$\begin{array}{r} 8.5 \\ 6.4\overline{)54.40} \\ \underline{512} \\ 320 \\ \underline{320} \end{array} \text{—length in meters}$$

The length is 8.5 meters. ■

 Try Problems 26 and 27.

6 Solving Problems Involving Translations

In Section 7.4 we saw how translating English phrases to math symbols can be used to solve problems. Now we look at examples involving real-life situations.

EXAMPLE 16 This week apples cost 1.5 times as much as they did last week. This week apples cost $0.89 a pound. What was the cost last week? (Round off to the nearest cent.)

SOLUTION The first sentence translates to a multiplication statement.

$$\begin{array}{ccccc} \text{Cost} & & & & \text{cost} \\ \text{this week} & \text{is} & 1.5 & \text{times} & \text{last week} \\ (\$) & \downarrow & \downarrow & & (\$) \\ 0.89 & = & 1.5 & \times & \square \end{array}$$

A multiplier is missing, so we divide the answer, 0.89, by the given multiplier, 1.5, to find the missing multiplier.

$$\begin{array}{r} .593 \approx 0.59 \\ 1.5\overline{)0.900} \\ \underline{75} \\ 140 \\ \underline{135} \\ 50 \\ \underline{45} \\ 5 \end{array}$$

The apples cost $0.59 a pound last week. ■

7.5 Applications

EXAMPLE 17 Margaret pays 0.35 of her take-home pay for rent. How much does she pay for rent if her take-home pay is $800?

SOLUTION The first sentence translates to a multiplication statement.

0.35	of	her take-home pay ($)	is	rent ($)
↓		↓		
0.35	×	800	=	☐

The answer to the multiplication problem is missing, so we multiply to find the rent.

$$\begin{array}{r} 800 \\ \times\ 0.35 \\ \hline 40\ 00 \\ 240\ 0\ \ \\ \hline 280.00 = 280 \end{array}$$

Margaret pays $280 in rent. ■

EXAMPLE 18 At City College, 0.15 of the students are over 25 years old. If 2250 students are over the age of 25, how many students does the college have?

SOLUTION The first sentence translates to a multiplication statement.

0.15	of	all the students	are	over 25 years old
↓		↓		↓
0.15	×	☐	=	2250

A multiplier is missing, so we divide the answer 2250 by the given multiplier to find the missing multiplier.

$$0.15 \overline{)2\,2\,5\,0.0\,0}\ \ \begin{array}{c} 1\,5\,0\,0.\ = 15{,}000 \\ \underline{1\,5}\ \ \ \ \ \ \ \ \ \ \\ 7\,5\ \ \ \ \ \ \\ \underline{7\,5}\ \ \ \ \ \ \end{array}$$

There are 15,000 students. ■

EXAMPLE 19 In a recent football game, a quarterback completed 20 out of 32 passes. The number of completions is what decimal of the total number of passes?

SOLUTION The last sentence translates to a multiplication statement.

Number of completions	is	what decimal	of	total passes
↓				↓
20	=	☐	×	32

A multiplier is missing, so we divide the answer 20 by the given multiplier 32 to obtain the missing multiplier.

Try These Problems

Solve. Answer with a decimal.

28. Mr. Stribolt cut two pieces of wire. The shorter piece is 0.75 times as long as the longer piece. The shorter piece is 12.3 meters long. What is the length of the longer piece?

29. The price of gasoline is 1.2 times as much as it was last year. Last year the price was $0.54 per liter. What is the price this year? (Truncate to the nearest cent.)

30. A company conducted a survey indicating that 0.4 of the workers have young children at home to care for. There are 1600 workers. How many have young children?

31. Ms. Cruz pays 0.30 of her income in federal taxes. If she pays $7680 in taxes, what is her income?

32. A baseball player was up to bat 150 times. He got 33 hits. His hits are what decimal part of his times at bat? (The answer is called his batting average.)

33. Six-tenths of a bottle of medicine weighs 3.45 grams. How much does the full bottle weigh?

34. Find the average of 6.25; 8.75; 10.5; and 12.95.

35. Mr. Lau earns $45.30 per hour, Ms. Tomm earns $52.65 per hour, and Mr. Pearson earns $80.75 per hour. What is the average of these three rates of pay? (Truncate to the nearest cent.)

```
      0.625
32)20.000
   192
    80
    64
   160
   160
```

He completed 0.625 of his passes. ∎

 Try Problems 28 through 33.

Averaging

Recall from Sections 2.5 and 5.7 that the **average** (or **mean**) of a set of data is a measure of the middle or center of the data. The average of a collection of numbers can be found by adding the numbers, then dividing by how many numbers there are.

> *Finding the Average of a Collection of Numbers*
> 1. Add all the numbers in the collection.
> 2. Divide the sum by how many numbers are in the collection.

EXAMPLE 20 Find the average of 34.2, 65.85, and 56. (Round up at 2 decimal places.)

SOLUTION First add the numbers, then divide the sum by 3.

```
                  52.01 ≈ 52.02
   34.20       3)156.05
   65.85         15
 + 56.00          6
  156.05          6
                  05
                   3
                   2
```

The average is 52.02. ∎

EXAMPLE 21 The rainfall in Atlanta for a 5-day period was as follows.

1.7 in   2.4 in   0.65 in   0.84 in   2.1 in

What was the average rainfall for the 5-day period? (Truncate at 1 decimal place.)

SOLUTION Add the 5 numbers, then divide the total by 5.

```
            1.5
   1.7   5)7.69     Because we want the
   2.4     5        answer truncated at 1
   0.65    2 6      decimal place, we do
   0.84    2 5      not need to divide any
 + 2.1       1      further.
   7.69
```

The average rainfall is 1.5 inches. ∎

 Try Problems 34 and 35.

Now we summarize the material in this section by giving guidelines that will help you to decide whether to multiply or divide.

> **Situations Requiring Multiplication**
>
> ×   1. You are looking for a total quantity.
>
> $$\text{Total quantity} = \text{size of each part} \times \text{number of equal parts}$$
>
> 2. You are looking for a total, as in the following examples.
>
> $$\text{Total miles} = \text{miles per hour} \times \text{number of hours}$$
>
> $$\text{Total cost} = \text{cost per item} \times \text{number of items}$$
>
> 3. You are looking for a number that is a certain amount times as large as another number.
> 4. You are looking for the area of a rectangle.
>
> $$\text{Area} = \text{length} \times \text{width}$$
>
> 5. You are looking for a decimal of a number.

> **Situations Requiring Division**
>
> ÷   1. A total quantity is being separated into a number of equal parts.
>
> $$\text{Size of each part} = \text{total quantity} \div \text{number of equal parts}$$
> $$= \frac{\text{total quantity}}{\text{number of equal parts}}$$
>
> $$\text{Number of equal parts} = \text{total quantity} \div \text{size of each part}$$
> $$= \frac{\text{total quantity}}{\text{size of each part}}$$
>
> 2. You are looking for a missing multiplier in a multiplication statement.
>
> $$\text{Missing multiplier} = \text{answer to the multiplication problem} \div \text{given multiplier}$$
>
> 3. You are looking for an average.
>
> $$\text{Average cost per item} = \text{total cost} \div \text{number of items}$$
>
> $$\text{Average miles per hour} = \text{total miles} \div \text{number of hours}$$
>
> $$\text{Average of a collection of numbers} = \text{sum of the numbers} \div \text{how many numbers}$$
>
> 4. You are looking for what ratio one number is to another number.

Answers to Try These Problems

1. 1.25 in 2. 0.8 in 3. 25 doses 4. 0.2 gal
5. $109.25 6. 487.5 oz 7. 2.25 lb 8. $955.80
9. $5.68 10. 0.625 min 11. 29.24 oz 12. 0.6 mi
13. 5 lb per wk 14. 2.25 gal per hr 15. 6.5¢ per oz
16. $1.25 per lb 17. 0.125 m 18. 0.02 lb 19. $14.75
20. $0.08 21. 51 mi 22. 8.6 lb 23. 187.5. min 24. 125 sec
25. 19.5 lb 26. 0.8 mi 27. 63 sq yd 28. 16.4 m
29. $0.64 per ℓ 30. 640 workers 31. $25,600 32. 0.22 or 0.220
33. 5.75 g 34. 9.6125 35. $59.56 per hr

EXERCISES 7.5

Solve.

1. Cathy earned $3829.50 over a 9-month period. How much did she earn each month?
2. Four persons share an apartment that rents for $323 per month. If they split the rent equally, what does each pay per month?
3. One hundred twenty strips of metal are laid one right after the other. If each strip is 4.05 meters long, what is the resulting total length?
4. A large container of milk is distributed equally among 24 persons. If each person receives 6.5 ounces, how much milk was in the original large container?
5. How many 1.2-foot pieces of wood can be cut from a board that is 21.6 feet long?
6. A chef made 975 ounces of spaghetti sauce. How many 7.5-ounce servings is this?
7. Find the cost of 4 pounds of meat at $4.15 per pound.
8. Lumber costs $0.82 per foot. What is the cost of 36.5 feet of this lumber?
9. Nancy can go 148.8 miles in 3 hours. How far can she go in 1 hour?
10. Jack paid $4.95 for 5 pounds of bananas. How much does 1 pound cost?
11. Rene jogs 3.5 miles in 32.2 minutes. How long does it take her to jog 1 mile?
12. A jar of mushrooms containing 8.6 ounces costs $0.69. What is the cost per ounce? (Truncate to the nearest cent.)
13. If sales tax is 0.055 of the marked price, what is the sales tax on a TV set marked $495? (Round off to the nearest cent.)
14. Bill's take-home pay for one month is $1475. He pays $354 in rent. What decimal part of his take-home pay is his rent?
15. Today radishes cost 0.8 times as much as they did yesterday. Today the cost is $0.85 a bunch. What was the cost yesterday? (Truncate to the nearest cent.)
16. Susan has a charge account at a department store. The store charges a service charge which is 0.015 multiplied by the unpaid balance. What is the service charge if her unpaid balance is $180?
17. A baseball player had 32 hits out of 95 times at bat. What is the ratio of hits to times at bat? (Round off at 3 decimal places.)

18. A solution contains 960 centiliters of water and 4.5 centiliters of acid. What is the ratio of acid to water? (Truncate at 4 decimal places.)
19. The area of a rectangle is 70.68 square feet. If the length is 7.6 feet, find the width.
20. The length of a rectangle is 12.03 centimeters and the width is 4.8 centimeters. Find the area of the rectangle.
21. Gordon bought three suits at the following prices: $475.80, $395.60, and $419.20. What is the average price of the suits?
22. A statistician measured the weight of drained peaches in 5 randomly selected 15-ounce cans packed by the Sunny Fruit Company. The weights (in ounces) were as follows.
    10.9   11.4   11.3   12.4   10.8
    Compute the average of these 5 weights.
23. A bottle of medicine contains 36 centiliters. A nurse needs to give out doses of this medicine, each containing 2.4 centiliters. How many doses can be given from the bottle?
24. A pile of sand weighs 0.5 ton. The sand is to be relocated using barrels that each hold 0.025 ton of sand. How many barrels are needed?
25. Susan's car averages 35.8 miles per gallon. How far can she go on a full tank of 15 gallons?
26. If Joe bicycles 0.4 mile each minute, how far can he bicycle in 12 minutes?
27. Gary wants to drive from Houston to New Orleans, which is a distance of 359 miles. If his car averages 24.6 miles per gallon, how many gallons of gas does he need to make the one-way trip? (Round up at 1 decimal place.)
28. Peter walks at the rate of 5.2 miles per hour. How long will it take him to walk 160 miles? (Truncate at 1 decimal place.)
29. Five pens weigh 32 grams. What is the weight of 19 pens?
30. Fifteen ounces of a product cost $1.20. What is the cost of 24 ounces?
31. In a tennis match, Tracy got 0.65 of her first serves in. If she got 78 first serves in, how many first serves did she attempt?
32. Today the price of oil is 0.75 of what it was last year. If the price today is $36 a barrel, what was the price last year?
33. You buy 1.5 pounds of mushrooms for $3.27. What is the cost per pound?
34. A watch loses 4 seconds every 25 hours. How much time does the watch lose per hour?
35. Betsy cuts a 4.5-meter wire into 30 equal pieces. How long is each piece?
36. Chuck wants to practice his piano for 24 hours during the next 15 days. If he spreads the time equally over the 15 days, how many hours will he practice each day?
37. The width of a rectangle is 45.3 feet. The length is twice the width. Find the area of the rectangle.
38. A rectangular floor contains 400 square feet. If the length is 27.5 feet, find the width. (Round off at 1 decimal place.)
39. Lou bought 3 steaks. The prices were $3.75, $4.20, and $5.15. What was the average price? (Round off to the nearest cent.)
40. The hourly wages of 5 workers were as follows.
    $16.75 $18.45 $20.75 $32.60 $17.50
    What is the average of these wages?

41. A basketball team won 0.75 of the total games played. If they won 96 games, how many games have they played?

42. The price of a computer decreased from $4200 to $3360. The decrease is what decimal of the original price?

The cost per pound for fresh asparagus fluctuated during a 5-month period as shown in the bar graph. Use the graph to answer Exercises 43 through 46.

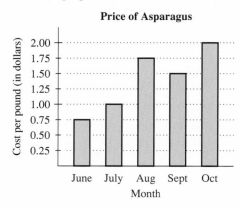

43. What was the cost per pound for asparagus in August?

44. What was the cost of 20 pounds of asparagus in June?

45. Find the average cost per pound for the 5-month period.

46. The decrease in price from August to September is what decimal of the August price? (Round off at 2 decimal places.)

CHAPTER 7 SUMMARY

KEY WORDS AND PHRASES

multiplication [7.1]
division [7.2]
divisor [7.2]
dividend [7.2]
quotient [7.2]
round off [7.2]
round up [7.2]

truncate [7.2]
repeating decimal [7.2]
multiplier [7.3]
factor [7.3]
product [7.3, 7.4]
twice [7.4]
half of [7.4]

divided by [7.4]
divided into [7.4]
ratio [7.4, 7.5]
rate [7.5]
area of a rectangle [7.5]
average (mean) [7.5]

SYMBOLS

$8.4\overline{56}$ means $8.456565656\ldots$

$0.33\tfrac{1}{3}$ means $\dfrac{33\tfrac{1}{3}}{100}$

There are three ways to indicate the division problem 42 divided by 6.

$42 \div 6 = 7$ $6\overline{)42}^{\,7}$ $\dfrac{42}{6} = 7$

IMPORTANT RULES

How to Multiply Two Decimals [7.1]

- Ignore the decimal points and multiply the numbers as if they were whole numbers.
- Place the decimal point in the product by adding the number of decimal places in the multipliers to get the number of decimal places in the product.

How to Multiply a Decimal by 10, 100, 1000, and so on [7.1]
The rule for how to multiply a decimal by 10, 100, 1000, and so on is on page 265.

How to Divide Decimals [7.2]
- If the divisor is a whole number, place the decimal point in the quotient directly above the decimal point in the dividend.
- If the divisor is not a whole number, move the decimal point in the divisor to the right as many places as it takes to make a whole number. Move the decimal in the dividend the same number of places as was moved in the divisor. Place the decimal point in the quotient directly above the new decimal point in the dividend.

How to Divide a Decimal by 10, 100, 1000, and so on [7.2]
The rule for how to divide a decimal by 10, 100, 1000, and so on is on page 277.

How to find the Missing Number in a Multiplication Statement [7.3]
- If you are missing the answer to the multiplication problem (the product), then multiply.
- If you are missing one of the multipliers, then divide the answer (or product) by the known multiplier (or factor) to find the missing multiplier.

How to find a Decimal of a Number [7.4, 7.5]
To find a decimal of a number, multiply the decimal by the number, that is, the word *of* used in this way translates to multiplication.

Situations Requiring Multiplication [7.5]
A summary of these situations appears on page 301.

Situations Requiring Division [7.5]
A summary of these situations appears on page 301.

CHAPTER 7 REVIEW EXERCISES

Multiply.

1. $.75 \times .8$
2. $.12 \times .64$
3. 35×6.3
4. 307×8.4
5. 0.83×54.0
6. 92.40×0.70
7. 593.00×0.065
8. 12×8.20
9. $.600 \times 0.360$
10. $.090 \times 705$
11. 3050×60.5
12. 60.70×0.103
13. $950{,}000 \times 2.50$
14. $3600 \times .045$
15. 9000×11.2
16. $120{,}000 \times .009$
17. $.086 \times 1000$
18. $0.78 \times 100{,}000$
19. $700 \times 4.5 \times 0.5$
20. $506 \times 3.4 \times 0.23 \times 6$

Divide. Continue dividing until the remainder is zero.

21. $48.6 \div 120$
22. $0.185 \div 74$
23. $17 \div 5$
24. $16 \div 64$
25. $2.268 \div 40.5$
26. $0.1456 \div 0.07$
27. $83.12 \div 0.004$
28. $2737.5 \div 0.375$
29. $4440 \div 0.12$
30. $\frac{90}{3.2}$
31. $\frac{2.22}{0.6}$
32. $\frac{3}{400}$

Divide. Express the answer as a repeating decimal.

33. $5 \div 9$
34. $26 \div 3$
35. $3.85 \div 3.3$
36. $43 \div 0.075$

Divide two places past the decimal and write the remainder in fractional form.

37. $468.3 \div 56$
38. $620 \div 6$
39. $\frac{0.4}{6.5}$
40. $\frac{50}{0.48}$

Divide. Approximate the answer as indicated.
41. 490.3 ÷ 16   (Truncate at 3 decimal places.)
42. 4 ÷ 6.9   (Truncate at 2 decimal places.)
43. 7 ÷ 26   (Round off at 4 decimal places.)
44. 5.2 ÷ 20.5   (Round off at 3 decimal places.)
45. $\frac{35}{6}$ (Round up at 3 decimal places.)
46. $\frac{7.08}{0.38}$ (Round up at 2 decimal places.)

Convert each of these fractions to a decimal.
47. $\frac{3}{4}$ 48. $\frac{210}{8}$ 49. $\frac{62}{25}$ 50. $\frac{7}{40}$
51. $\frac{53}{6}$ 52. $\frac{5}{120}$ 53. $\frac{19}{27}$ 54. $\frac{93}{11}$

Find the missing number. Write the answer in decimal form.
55. 8.5 × 9 = ☐
56. 8 × ☐ = 201
57. 33 = 0.48 × ☐
58. 299 = 0.065 × ☐
59. ☐ × 6.4 = 0.08
60. 0.35 × 27,000 = ☐

Solve. Write the answer in decimal form.
61. Find 0.015 of 650.
62. Divide 3.51 by 2.7.
63. Divide 2500 into 93.
64. What decimal of 2400 is 144?
65. Find twice 8.6.
66. What number is half of 17.73?
67. Find the ratio of 4.5 to 12.
68. Find the average of 0.17 and 2.89.
69. Twelve families share equally 8.7 pounds of cheese. How much cheese does each family receive?
70. How many 0.8-foot pieces of wood can be cut from a log that is 10.4 feet long?
71. A program consists of 9 segments that are each 3.75 minutes long. What is the total length of the program?
72. Find the cost of 7.2 pounds of chicken at $3.28 per pound. (Round up to the nearest cent.)
73. Six loads of dirt weigh 0.09 ton. What is the weight of 1 load?
74. Ching earns $24.75 per hour. At this rate, how many hours must he work to earn $1980?
75. Water enters a tank at the rate of 56 gallons every 3.5 minutes. How many gallons per minute is this?
76. If sales tax is 0.085 of the marked price, what is the sales tax on a microwave oven that is marked $380?
77. Ms. Gomez has a charge account at a department store. The store charges a service charge which is 0.015 of the unpaid balance. If the service charge is $6.12, what is the unpaid balance?
78. At an auto plant outside of Detroit, 76 out of 3800 workers called in sick. What decimal of the workers called in sick?
79. The price of sugar has doubled in the past ten years. If the price now is $5.68 per pound, what was the price ten years ago?
80. Stanley, an auto mechanic, earns $42 per hour. His girlfriend Edna is a lawyer making 3.5 times as much as Stanley. What does Edna make per hour?

81. One side of a box is 35.8 centimeters wide and 60 centimeters long. What is the area of this one side?
82. A rectangular wall is 9.5 feet high and contains 200 square feet. How wide is the wall? (Round up at 1 decimal place.)
83. The hourly wages for 4 workers are as follows.
    $13.80 $15 $16.75 $20.50
    Find the average of 4 hourly wages. (Round off to the nearest cent.)
84. The price of gas decreased from $1.89 per gallon to $1.50 per gallon in one week. Find the ratio of the decrease in price to the original price. (Round off at 2 decimal places.)
85. A watch loses 4.8 minutes in 8 hours. How much time does the watch lose in 12 hours?
86. On a map 1.25 inches represents 20 miles. How many miles does 2 inches represent?

In Your Own Words

Write complete sentences to discuss each of the following. Support your comments with examples or pictures, if appropriate.

87. Discuss how the following two problems are similar and how they differ.
    a. $2.5 \times \boxed{} = 0.4$ **b.** $0.4 \times \boxed{} = 2.5$
88. Discuss how you decide whether a decimal is less than or more than the number 1.
89. A rectangle is 8.5 feet long and 4.2 feet wide. Discuss how the procedure for finding the perimeter of this rectangle is different than the procedure for finding the area.
90. Give examples of at least three different rates that you experience in your life. Choose examples that involve decimals.
91. Discuss how you can always find the decimal representation for a fraction.

CHAPTER 7 PRACTICE TEST

Multiply.

1. 7.2×8.6
2. 208.30×0.603
3. $60{,}000 \times 7.5$
4. 0.0608×1000
5. 0.0013×9
6. $0.8 \times 3.4 \times 50$

Divide. Continue dividing until the remainder is zero.

7. $230 \div 80$
8. $2.1654 \div 5.4$
9. $\frac{1404}{0.004}$

Divide. Express the answer as a repeating decimal.

10. $2 \div 3$
11. $21.4 \div 12$
12. $\frac{1300}{2.7}$

Divide two places past the decimal and write the remainder in fractional form.

13. $7.82 \div 0.3$
14. $1.2 \div 64$
15. $\frac{18.9}{0.81}$

Divide. Approximate the answer as indicated.

16. $28.023 \div 0.7$ (Round off at 3 decimal places.)
17. $8 \div 4.25$ (Truncate at 4 decimal places.)
18. $\frac{2.3}{0.72}$ (Round up at 3 decimal places.)

Convert each of these fractions to a decimal.

19. $\frac{5}{8}$
20. $\frac{43}{25}$
21. $\frac{5}{18}$
22. $\frac{208}{33}$

Find the missing number. Write the answer as a decimal.

23. $16 \times \boxed{} = 8$
24. $\boxed{} \times 6.4 = 32$
25. $\boxed{} = 0.07 \times 415$
26. $128 = 0.08 \times \boxed{}$

Solve. Write the answer as a decimal.

27. What is 12 divided by 0.03?
28. Find 6.2 of 500.
29. Divide 16 into 0.05.
30. What decimal of 125 is 8.2?
31. Kathy earns $250.60 each week. How much money will she earn in 12 weeks?
32. Bob pays 0.08 of his take-home pay for food each month. He spends $60 on food. How much is his take-home pay?
33. Lieu bought an unsliced loaf of bread from the neighborhood bakery. It was 21 centimeters long. She wants to cut slices that are 1.4 centimeters wide. How many slices can she cut from the loaf of bread?
34. The length of a rectangle is 80 feet and the width is 17.3 feet. What is the area of the rectangle?
35. Anthony drove a total of 238 miles in 5 hours. How many miles per hour is this?
36. An airplane flies 2478 miles in 3.5 hours. How far does it go in 6.7 hours?
37. The amount of snow that fell in a 5-day period at a ski resort in Utah was as follows.
    6.8 in   10.9 in   24.5 in   36.6 in   48 in
    What was the average amount for the 5-day period? (Round off at 1 decimal place.)

CUMULATIVE REVIEW EXERCISES: CHAPTERS 1–7

Convert each of these decimals to a fraction. Reduce to lowest terms.

1. 3.6
2. 0.85
3. 0.0025
4. 2.125

Convert each of these fractions to a decimal.

5. $\frac{3}{8}$
6. $\frac{11}{200}$
7. $\frac{302}{4}$
8. $\frac{354}{15}$

Find the missing number. Answer with a fraction.

9. $4\frac{2}{3} + \boxed{} = 24$
10. $\boxed{} \times \frac{7}{100} = 140$
11. $\boxed{} = \frac{27}{45} \times \frac{65}{26}$
12. $\frac{3}{14} + \frac{7}{8} + \frac{11}{28} = \boxed{}$

Solve.

13. Find the difference between 20.1 and 5.76.
14. Find the average of 4.3, 12.8, and 6.
15. Twice what number yields $7\frac{1}{2}$?
16. Find the ratio of $12\frac{1}{2}$ to 100.
17. Find the product of 39.78 and 10,000.
18. Divide 400 into 2.5.
19. Divide 400 by 2.5.

20. Aritha purchased items costing $4.58, $12.96, and $23.79. How much change does she receive from a $100 bill?

21. Bernice and Tom planned to work on a project for 20 hours this week. On Monday they worked $5\frac{1}{2}$ hours; on Tuesday, they worked $4\frac{1}{4}$ hours; and on Wednesday, they worked $3\frac{3}{4}$ hours. How many more hours do they need to work to reach their goal?

22. A rectangular piece of land is 40 feet wide and 65 feet long.
    a. What is the area of this plot?
    b. What is the perimeter of this plot?
    c. If 64 fluid ounces of fertilizer are needed to fertilize each 1000 square feet of land, how many fluid ounces of fertilizer are needed to fertilize this plot?
    d. If you build a fence around the plot at $24.75 per foot, what is the total cost of the fence?

23. An elderly couple inherits $60,000. They invest $\frac{4}{5}$ of the money and divide the rest equally among their 8 grandchildren. How much does each grandchild receive?

CHAPTER 8

Problem Solving: Whole Numbers, Fractions, and Decimals

8.1 THE NUMBER LINE

OBJECTIVES

- Name whole numbers that are associated with points on a number line.
- Name fractions that are associated with points on a number line.
- Name decimals that are associated with points on a number line.

1 Whole Numbers on the Number Line

We picture numbers associated with points along a line. Here is a **number line** containing some whole numbers.

The arrow at the right end indicates the direction in which the numbers get larger. On this number line the numbers are increasing by 1 unit as we move to the right so that the length of each of the smaller segments is 1. Observe that the **distance** between any two points can be found by subtracting the numbers associated with these points. Subtract the smaller number from the larger number to obtain the distance between them.

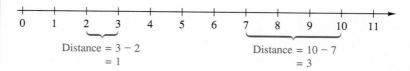

Try These Problems

Give the numbers associated with points A and B.

1.
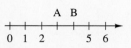

2.
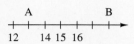

3.
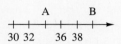

4.
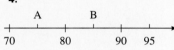

EXAMPLE 1 Give the numbers associated with points A, B, and C.

SOLUTION First we need to find the length of each of the smaller segments. Because the distance between 202 and 203 is 1 unit, each of the smaller segments has length 1 unit.

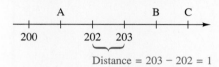

Now we fill in the missing numbers. As we move from left to right, we add 1 unit.

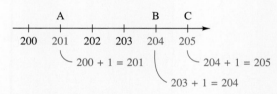

A. 201 **B.** 204 **C.** 205 ■

EXAMPLE 2 Give the numbers associated with points A, B, and C.

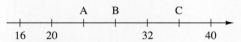

SOLUTION First we find the length of each of the smaller segments. Because the distance between 16 and 20 is 4, each of the smaller segments has a length of 4 units.

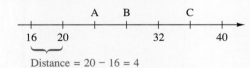

Now we fill in the missing numbers by adding 4 as we move from left to right.

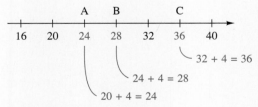

A. 24 **B.** 28 **C.** 36 ■

Try Problems 1 through 4.

EXAMPLE 3 Give the numbers associated with points A and B.

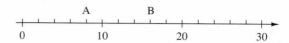

SOLUTION First find the distance between two of the given numbers. For example, the distance between 20 and 30 is 10 units.

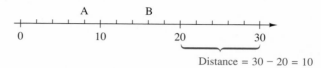

Now count the number of segments between 20 and 30. There are 5 segments between 20 and 30. If a 10-unit length is divided into 5 equal parts then each part has length 2 units. Thus each of the smaller segments has length 2 units.

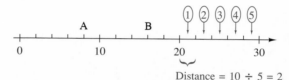

Now fill in the missing numbers by adding 2 as you move from left to right from one tick mark to the next.

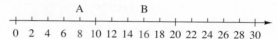

A. 8   **B.** 16   ∎

EXAMPLE 4 Give the numbers associated with points A, B, and C.

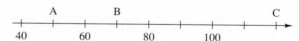

SOLUTION First, find the distance between two of the given numbers. For example, the distance between 80 and 100 is 20.

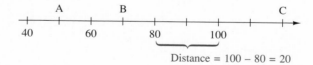

Second, count the number of segments between 80 and 100. There are 2 segments between 80 and 100. If a 20-unit length is divided into 2 equal parts then each part has length 10 units. Thus each of the smaller segments has length 10 units.

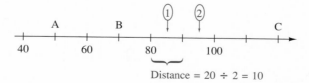

▲ **Try These Problems**

Give the numbers associated with points A and B.

5.
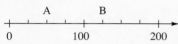

6.
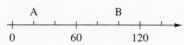

7.
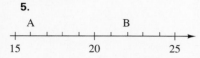

8.
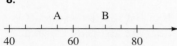

Finally, we fill in the missing numbers by adding 10 as we move left to right from one tick mark to the next.

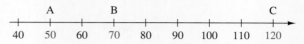

A. 50   **B.** 70   **C.** 120   ■

▲ **Try Problems 5 through 8.**

2   Fractions on the Number Line

We also picture fractions on a number line. Here we show a number line where we look at some fractions with denominator 2.

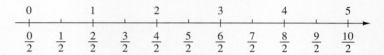

The fractions get larger as we move in the direction of the arrow. On this number line, the numbers are increasing by $\frac{1}{2}$ unit as we move from left to right, so that the length of each of the smaller segments is $\frac{1}{2}$ unit. Observe that the distance between any two points can be found by subtracting the numbers associated with these points. Subtract the smaller number from the larger number to obtain the distance.

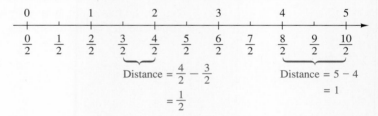

EXAMPLE 5   Give the fractions associated with points A, B, and C.

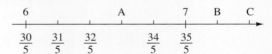

SOLUTION   We can see that each of the smaller segments has length $\frac{1}{5}$ by subtracting any two numbers that are associated with adjacent tick marks. For example, the distance between $\frac{30}{5}$ and $\frac{31}{5}$ is $\frac{1}{5}$.

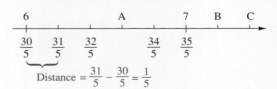

Now fill in the missing numbers by adding $\frac{1}{5}$ as you move from left to right from one tick mark to the next.

Try These Problems

Give the fractions associated with points A and B.

9.
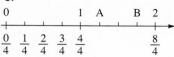

10.
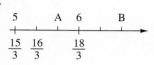

11.
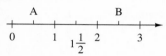

12.
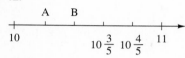

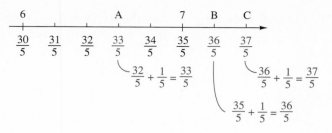

A. $\frac{33}{5}$ or $6\frac{3}{5}$ **B.** $\frac{36}{5}$ or $7\frac{1}{5}$ **C.** $\frac{37}{5}$ or $7\frac{2}{5}$ ■

EXAMPLE 6   Give the fractions associated with points A, B, and C.

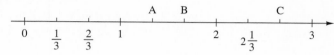

SOLUTION   We can see that each of the smaller segments has length $\frac{1}{3}$ by subtracting any two numbers that are associated with adjacent tick marks. For example, the distance between 2 and $2\frac{1}{3}$ is $\frac{1}{3}$.

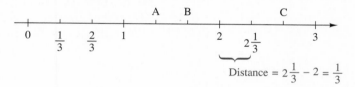

Fill in the missing numbers by adding $\frac{1}{3}$ as you move from left to right from one tick mark to the next.

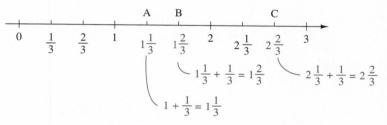

A. $1\frac{1}{3}$ or $\frac{4}{3}$ **B.** $1\frac{2}{3}$ or $\frac{5}{3}$ **C.** $2\frac{2}{3}$ or $\frac{8}{3}$ ■

 Try Problems 9 through 12.

EXAMPLE 7   Give the fractions associated with points A and B.

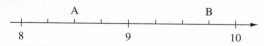

SOLUTION   First, find the distance between two given numbers. For example, the distance between 8 and 9 is 1.

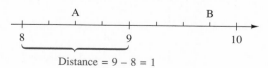

Second, count the number of segments between 8 and 9. There are 4 segments between 8 and 9. If a 1-unit length is divided into 4 equal parts then each part has length $\frac{1}{4}$ unit. Thus each of the smaller segments has length $\frac{1}{4}$ unit.

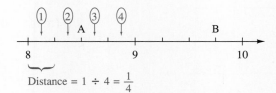

Finally, fill in the missing numbers by adding $\frac{1}{4}$ as you move from left to right from one tick mark to the next.

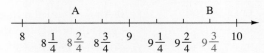

A. $8\frac{1}{2}$ or $\frac{17}{2}$ **B.** $9\frac{3}{4}$ or $\frac{39}{4}$ ■

EXAMPLE 8   Give the fractions associated with points A and B.

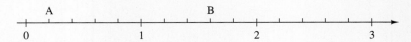

SOLUTION   First, find the distance between two given numbers. For example, the distance between 2 and 3 is 1.

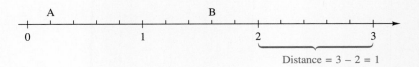

Second, count the number of segments between 2 and 3. There are 5 segments between 2 and 3. If a 1-unit length is divided into 5 equal parts then each part has length $\frac{1}{5}$ unit. Thus each of the smaller segments has length $\frac{1}{5}$ unit.

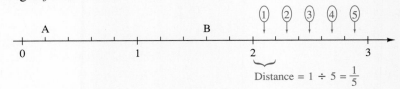

Finally, fill in the missing numbers by adding $\frac{1}{5}$ as you move from left to right from one tick mark to the next.

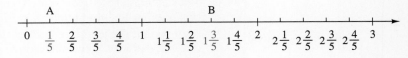

A. $\frac{1}{5}$ **B.** $1\frac{3}{5}$ or $\frac{8}{5}$ ■

316 Chapter 8 Problem Solving: Whole Numbers, Fractions, and Decimals

 Try These Problems

Give the fractions associated with points A and B.

13.
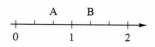

14.
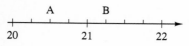

15.
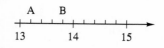

16.
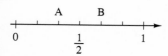

EXAMPLE 9 Give the fractions associated with points B and C.

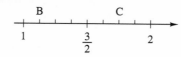

SOLUTION First, find the distance between two given numbers. For example, the distance between 1 and 2 is 1.

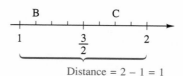

Second, count the number of segments between 1 and 2. There are 8 segments between 1 and 2. If a 1-unit length is divided into 8 equal parts then each part has length $\frac{1}{8}$ unit. Thus each of the smaller segments has length $\frac{1}{8}$ unit.

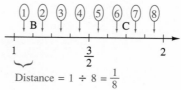

Finally, fill in the missing numbers by adding $\frac{1}{8}$ as you move from left to right from one tick mark to the next.

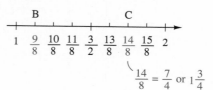

B. $1\frac{1}{8}$ or $\frac{9}{8}$ **C.** $1\frac{3}{4}$ or $\frac{7}{4}$ ■

 Try Problems 13 through 16.

3 Decimals on the Number Line

We also picture decimals on a number line. Here is a number line with some decimals between 0 and 1.

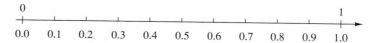

The decimals get larger as we move in the direction of the arrow. On this number line the numbers are increasing by 0.1 as we move to the right so that the length of each of the smaller segments is 0.1. Observe that the distance between any two points can be found by subtracting the numbers associated with these points. Subtract the smaller number from the larger number to obtain the distance.

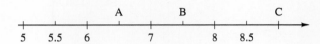

EXAMPLE 10 Give the decimals associated with points A, B, and C.

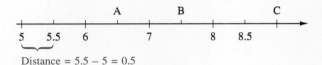

SOLUTION We can see that each of the smaller segments has length 0.5 by subtracting any two numbers that are associated with adjacent tick marks. For example, the distance between 5 and 5.5 is 0.5.

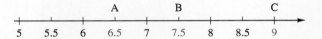

Now fill in the missing numbers by adding 0.5 as you move from left to right from one tick mark to the next.

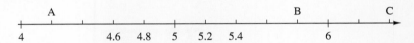

A. 6.5   **B.** 7.5   **C.** 9 ■

EXAMPLE 11 Give the decimals associated with points A, B, and C.

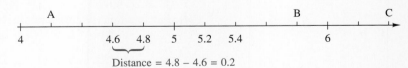

SOLUTION We can see that each of the smaller segments has length 0.2 by subtracting any two numbers that are associated with adjacent tick marks. For example, the distance between 4.6 and 4.8 is 0.2.

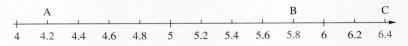

Now fill in the missing numbers by adding 0.2 as you move from left to right from one tick mark to the next.

A. 4.2   **B.** 5.8   **C.** 6.4 ■

Try These Problems

Give the decimals associated with points A and B.

17.
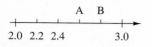

18.
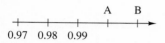

19.
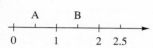

20.
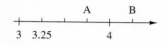

 Try Problems 17 through 20.

EXAMPLE 12 Give the decimals associated with points A, B, and C.

SOLUTION First, find the distance between two given numbers. For example, the distance between 0 and 1 is 1.

Second, count how many segments are between 0 and 1. There are 4 segments between 0 and 1. If a 1-unit segment is divided into 4 equal parts then each part has length $\frac{1}{4}$ or 0.25 unit.

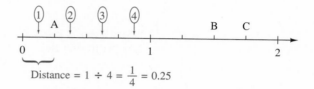

Now fill in the missing numbers by adding 0.25 as you move from left to right from one tick mark to the next.

A. 0.25 **B.** 1.50 or 1.5 **C.** 1.75 ∎

EXAMPLE 13 Give the decimals associated with points A, B, and C.

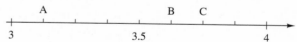

SOLUTION First, find the distance between two given numbers. For example, the distance between 3 and 4 is 1.

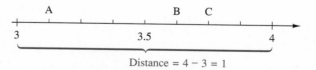

Second, count the number of segments between 3 and 4. There are 8 segments between 3 and 4. If a 1-unit length is divided into 8 equal parts then each part has length $\frac{1}{8}$ or 0.125. Thus each of the smaller segments has length 0.125.

Try These Problems

Give the decimals associated with points A and B.
21.

22.
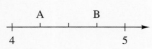

23.
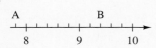

24.
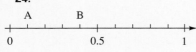

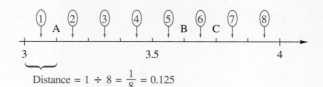

$$\text{Distance} = 1 \div 8 = \frac{1}{8} = 0.125$$

Now fill in the missing numbers by adding 0.125 as you move from left to right from one tick mark to the next.

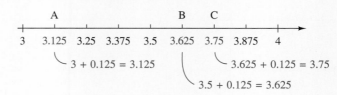

A. 3.125 **B.** 3.625 **C.** 3.75

Try Problems 21 through 24.

Answers to Try These Problems

1. **A.** 3 **B.** 4 2. **A.** 13 **B.** 18
3. **A.** 34 **B.** 40 4. **A.** 75 **B.** 85
5. **A.** 16 **B.** 22 6. **A.** 50 **B.** 125
7. **A.** 20 **B.** 100 8. **A.** 55 **B.** 70
9. **A.** $1\frac{1}{4}$ or $\frac{5}{4}$ **B.** $1\frac{3}{4}$ or $\frac{7}{4}$
10. **A.** $\frac{17}{3}$ or $5\frac{2}{3}$ **B.** $\frac{20}{3}$ or $6\frac{2}{3}$
11. **A.** $\frac{1}{2}$ **B.** $2\frac{1}{2}$ or $\frac{5}{2}$ 12. **A.** $10\frac{1}{5}$ **B.** $10\frac{2}{5}$
13. **A.** $\frac{2}{3}$ **B.** $1\frac{1}{3}$ or $\frac{4}{3}$ 14. **A.** $20\frac{1}{2}$ **B.** $21\frac{1}{4}$
15. **A.** $13\frac{1}{5}$ **B.** $13\frac{4}{5}$ 16. **A.** $\frac{2}{6}$ or $\frac{1}{3}$ **B.** $\frac{4}{6}$ or $\frac{2}{3}$
17. **A.** 2.6 **B.** 2.8 18. **A.** 1.00 or 1 **B.** 1.01
19. **A.** 0.5 **B.** 1.5 20. **A.** 3.75 **B.** 4.25
21. **A.** 1.5 **B.** 3.5 22. **A.** 4.25 **B.** 4.75
23. **A.** 7.8 **B.** 9.4 24. **A.** 0.1 **B.** 0.4

EXERCISES 8.1

Give the numbers associated with points A and B.

1.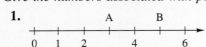

2.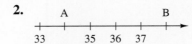

3.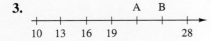

4.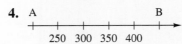

5.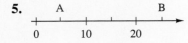

6.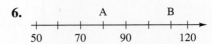

7.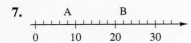

8.

Give the fractions associated with points A and B.

9.
```
    A         B
+---+---+---+---+---+---+---+---+→
0   1/8     3/8 4/8 5/8 6/8     1
```

10.
```
  5   A   6       B       8
+---+---+---+---+---+---+---+---+→
   10/2     12/2 13/2        16/2
```

11.
```
    A       B
+---+---+---+---+---+→
4       5       6
```

12.
```
A       B
+---+---+---+---+→
    16      17
```

13.
```
    A       B
+---+---+---+---+→
0               1
```

14.
```
  A     B
+---+---+---+---+→
0               2
```

15.
```
      A       B
+---+---+---+---+---+→
2   2½     3   3½
```

16.
```
A       B
+---+---+---+---+---+→
0   ½      1   1½
```

Give the decimals associated with points A and B.

17.
```
   A B
+---+---+---+---+---+→
2.8     3.1 3.2 3.3
```

18.
```
           A   B
+---+---+---+---+---+→
0.70 0.75 0.80    0.95
```

19.
```
  A         B
+---+---+---+---+→
8       9       10
```

20.
```
    A   B
+---+---+---+---+→
0   1   2   3
```

21.
```
  A B
+---+---+---+---+→
13      14      15
```

22.
```
A       B
+---+---+---+---+→
0               1
```

23.
```
A       B
+---+---+---+→
4   4.5     5
```

24.
```
A               B
+---+---+---+---+→
50              51
```

The bar graph gives the fraction of administrative positions held by women at a large university for each of four years. Use the graph to answer Exercises 25 through 28.

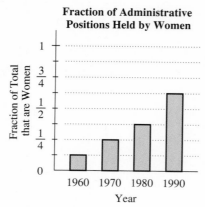

Fraction of Administrative Positions Held by Women

25. In 1960 what fraction of the administrative positions were held by women?
26. In 1990 what fraction of the administrative positions were held by women?
27. In 1970 what fraction of the administrative positions were held by men?
28. In 1980 what fraction of the administrative positions were held by men?

8.2 COMPARING NUMBERS

OBJECTIVES
- Convert fractions, decimals, and whole numbers from one form to another.
- Compare decimals with decimals.
- Compare fractions with fractions.
- Compare fractions with decimals.

1 Equality of Numbers

A number can be written in many forms without changing its value. For example,

$$\frac{1}{2} = \frac{2}{4} = \frac{3}{6} = \frac{4}{8} = \frac{5}{10} = 0.5 = 0.50$$

Here we picture the number $\frac{1}{2}$ in various forms on the number lines that follow.

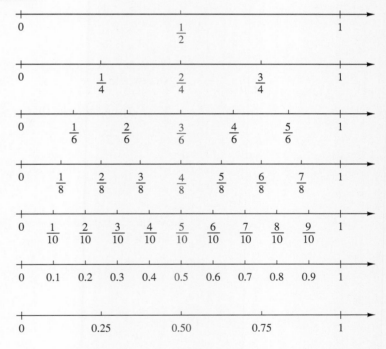

Observe that the position of the number $\frac{1}{2}$ remains halfway between the numbers 0 and 1 regardless of its form. In general, equal numbers occupy the same position on the number line.

When problem solving, it is often necessary to convert numbers from one form to another. Here is a review of the various techniques of **converting** a number; that is, changing its form without changing its value.

CONVERTING NUMBERS FROM ONE FORM TO ANOTHER

1. Reduce a fraction to lower terms by dividing the numerator and denominator by the same nonzero number, or by canceling common factors from the numerator and denominator.

   Example: $\dfrac{21}{28} = \dfrac{3 \times \cancel{7}}{4 \times \cancel{7}} = \dfrac{3}{4}$

2. Raise a fraction to higher terms by multiplying the numerator and denominator by the same nonzero number.

   Example: $\dfrac{2}{3} = \dfrac{2 \times 4}{3 \times 4} = \dfrac{8}{12}$

3. Convert a mixed numeral to an improper fraction.

   Example: $3\dfrac{1}{5} = 3 + \dfrac{1}{5} = \dfrac{15}{5} + \dfrac{1}{5} = \dfrac{16}{5}$

4. Convert an improper fraction to a mixed numeral.

   Example: $\dfrac{18}{7} \rightarrow 7\overline{)18}$ with quotient 2 remainder 4. Therefore, $\dfrac{18}{7} = 2\dfrac{4}{7}$

5. Write a mixed numeral so that the fractional part is smaller than one whole.

   Example: $9\underbrace{\dfrac{8}{5}}_{\text{Fractional part is larger than one whole.}} = 9 + \dfrac{8}{5} = 9 + 1\dfrac{3}{5} = 10\underbrace{\dfrac{3}{5}}_{\text{Fractional part is smaller than one whole.}}$

6. Write a mixed numeral so that the fractional part is larger than one whole.

   Example: $7\underbrace{\dfrac{3}{8}}_{\text{Fractional part is smaller than one whole.}} = 6 + 1\dfrac{3}{8} = 6 + \dfrac{11}{8} = 6\underbrace{\dfrac{11}{8}}_{\text{Fractional part is larger than one whole.}}$

7. Convert a fraction to a decimal.

   Example: $\dfrac{9}{20} \rightarrow 20\overline{)9.00}$ gives $.45$. Therefore, $\dfrac{9}{20} = 0.45$

8. Convert a decimal to a fraction.

$$\textit{Example:} \quad 0.125 = \frac{125}{1000} = \frac{5 \times \cancel{25}}{40 \times \cancel{25}} = \frac{5}{40} = \frac{1 \times \cancel{5}}{8 \times \cancel{5}} = \frac{1}{8}$$

9. Convert a whole numeral to a decimal or vice versa.

Examples: $\quad 3 = 3.$
$\qquad\qquad 182 = 182.$
$\qquad\qquad 75. = 75$

10. Attach extra zeros to a decimal or remove extra zeros.

Examples: $\quad 6.5 = 6.50$
$\qquad\qquad .823 = 0.823000$
$\qquad\qquad 4 = 4.00$
$\qquad\qquad 12.300 = 12.3$

The symbol $=$ is used to indicate that two numbers are equal and the symbol $\neq$ is used to indicate that two numbers are not equal. For example,

$$0.3 = 0.30 \quad \text{0.3 equals 0.30}$$
$$3 \neq 0.3 \quad \text{3 does not equal 0.3}$$

EXAMPLE 1 Which of these numbers are equal to $\frac{3}{20}$?

a. $6\frac{2}{3}$ **b.** 0.203 **c.** $\frac{27}{180}$

SOLUTION

a. Does $6\frac{2}{3} = \frac{3}{20}$? $6\frac{2}{3}$ is larger than 1 and $\frac{3}{20}$ is smaller than 1.

$$6\frac{2}{3} \neq \frac{3}{20}$$

b. Does $0.203 = \frac{3}{20}$? Convert $\frac{3}{20}$ to a decimal.

$$\frac{3}{20} \rightarrow 20\overline{)3.00} \quad \text{Therefore, } \frac{3}{20} = \underset{\text{Smaller than 0.203}}{0.15}$$
$$\phantom{\frac{3}{20} \rightarrow 20)}\underline{2\,0}$$
$$\phantom{\frac{3}{20} \rightarrow 20)}\,1\,00$$
$$\phantom{\frac{3}{20} \rightarrow 20)}\,\underline{1\,00}$$

$$\frac{3}{20} \neq 0.203$$

c. Does $\frac{27}{180} = \frac{3}{20}$? Reduce $\frac{27}{180}$ to lowest terms.

$$\frac{27}{180} = \frac{3 \times \cancel{9}}{20 \times \cancel{9}} = \frac{3}{20}$$

$$\frac{27}{180} = \frac{3}{20}$$

Answer **c.** $\frac{27}{180}$ ∎

Try These Problems

1. Which of these numbers are equal to $\frac{3}{8}$?
   a. $\frac{12}{32}$ b. $\frac{8}{3}$
   c. 0.375 d. 2.66

2. Which of these numbers are equal to $12\frac{2}{5}$?
   a. $\frac{29}{5}$ b. $10\frac{12}{5}$
   c. 12.4 d. $\frac{62}{5}$

3. Which of these numbers are equal to 8.3?
   a. $8\frac{1}{3}$ b. 0.083
   c. $\frac{83}{10}$ d. 8.300

4. Which of these numbers are equal to 0.008?
   a. $\frac{8}{100}$ b. 0.0080
   c. $\frac{8}{1000}$ d. $\frac{4}{500}$

EXAMPLE 2 Which of these numbers are equal to 7.8?

a. $6\frac{18}{10}$ b. $7\frac{4}{5}$ c. $\frac{115}{15}$

SOLUTION

a. Does $6\frac{18}{10} = 7.8$? Write $6\frac{18}{10}$ so that the fraction part is smaller than one whole, then convert to a decimal.

$$6\frac{18}{10} = 6 + \frac{18}{10} = 6 + 1\frac{8}{10} = 7\frac{8}{10} = 7.8$$

$6\frac{18}{10} = 7.8$

b. Does $7\frac{4}{5} = 7.8$? Convert 7.8 to a mixed numeral.

$$7.8 = 7\frac{8}{10} = 7\frac{4 \times 2}{5 \times 2} = 7\frac{4}{5}$$

$7\frac{4}{5} = 7.8$

c. Does $\frac{115}{15} = 7.8$? Convert $\frac{115}{15}$ to a decimal.

$$\begin{array}{r} 7.66 \\ 15{\overline{\smash{\big)}\,115.00}} \\ \underline{105} \\ 10\,0 \\ \underline{9\,0} \\ 1\,00 \\ \underline{90} \\ 10 \end{array} \rightarrow \frac{115}{15} = 7.\overline{6}$$

Smaller than 7.8

$\frac{115}{15} \ne 7.8$

Answer a. $6\frac{18}{10}$ b. $7\frac{4}{5}$ ■

 Try Problems 1 through 4.

2 Comparing Decimals

Decimals on the number line are arranged in order. As you move from left to right the decimals get larger. For example, this picture illustrates that 0.6 is more than 0.58.

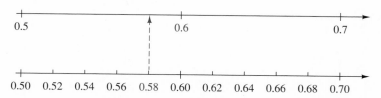

The symbols $<$ and $>$ are used to **compare** numbers that are not equal. For example,

$0.6 > 0.58$ 0.6 is more than 0.58

$0.58 < 0.6$ 0.58 is less than 0.6

Remember that the pointed side of the symbol always points toward the smaller number.

Try These Problems

In Problems 5 and 6, use the symbol < or > to compare each pair of numbers.

5. 0.3, 0.256
6. 5.14, 5.014
7. Which is less, 0.00786 or 0.007849?
8. Which is more, 8.508 or 8.50799?
9. Which is less, 0.3$\overline{52}$ or 0.3$\overline{5}$?
10. Which is more, 13.45 or 13.$\overline{435}$?

One method of comparing decimals is to write each of them with the same number of decimal places, then compare. Here we compare 0.6 with 0.58 using this method.

$$0.6 = 0.60 = \frac{60}{100} \quad \text{60 hundredths}$$

$$0.58 = 0.58 = \frac{58}{100} \quad \text{58 hundredths}$$

Because $\frac{60}{100}$ is more than $\frac{58}{100}$, 0.6 is more than 0.58. This method for comparing decimals is discussed carefully in Section 6.3.

 Try Problems 5 and 6.

It is possible to compare decimals without writing each decimal with the same number of decimal places. Here are examples to illustrate the technique.

EXAMPLE 3 Which is less, 7.132876 or 7.1345?

SOLUTION Arrange the numbers so that decimal points are aligned vertically and digits with the same place value form a column. Compare digits with like place value, starting on the left.

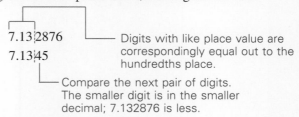

The decimal 7.132876 is less. ■

 Try Problems 7 and 8.

EXAMPLE 4 Which is more, 12.438 or 12.43$\overline{8}$?

SOLUTION Arrange the numbers so that decimal points are aligned vertically, and digits with the same place value form a column. Compare digits with like place value, starting on the left.

12.438   Digits with like place value are
12.43$\overline{8}$ correspondingly equal out to the
         thousandths place.

After the thousandths place the digits will not be equal so write the decimals so that you can view these digits carefully.

$$12.438 = 12.438|0$$
$$12.43\overline{8} = 12.438|888\ldots$$

Compare the next pair of digits. The larger digit is in the larger decimal. 12.43$\overline{8}$ is more.

The decimal 12.43$\overline{8}$ is more. ■

 Try Problems 9 and 10.

Try These Problems

List from smallest to largest.
11. 0.75; 7.52; 7.5; 7.499
12. 22.36; 22.$\overline{3}$; 2.236; 2.$\overline{23}$

Use the symbol < or > to compare each pair of numbers.
13. $\frac{9}{20}$, $\frac{3}{5}$
14. $2\frac{1}{3}$, $2\frac{2}{7}$

EXAMPLE 5 List from smallest to largest: $3.\overline{7}$, 3.76, $3.\overline{75}$

SOLUTION First, recall that $3.\overline{7}$ means $3.7777777\ldots$ and $3.\overline{75}$ means $3.7575757575\ldots$

When there are more than two decimals to compare, it can be helpful to picture the approximate positions of the decimals on the number line.

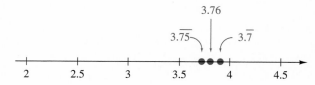

From smallest to largest the numbers are; $3.\overline{75}$, 3.76, and $3.\overline{7}$. ■

 Try Problems 11 and 12.

3 Comparing Fractions

Fractions on the number line are arranged in order. As you move from left to right the fractions get larger. For example, this picture illustrates that $\frac{7}{10}$ is less than $\frac{4}{5}$.

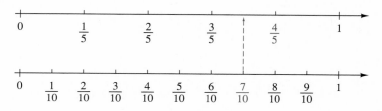

The symbols < and > can be used to compare fractions. For example,

$$\frac{7}{10} < \frac{4}{5} \qquad \text{$\frac{7}{10}$ is less than $\frac{4}{5}$}$$

$$\frac{4}{5} > \frac{7}{10} \qquad \text{$\frac{4}{5}$ is more than $\frac{7}{10}$}$$

One method of comparing fractions is to write them with the same denominator, then compare. Here we compare $\frac{7}{10}$ and $\frac{4}{5}$ using this method. The least common multiple of 5 and 10 is 10, so we convert $\frac{4}{5}$ to a fraction with denominator 10.

$$\frac{4}{5} = \frac{4 \times 2}{5 \times 2} = \frac{8}{10} \qquad \text{8 tenths}$$

$$\frac{7}{10} \phantom{= \frac{4 \times 2}{5 \times 2}} = \frac{7}{10} \qquad \text{7 tenths}$$

Because $\frac{8}{10}$ is more than $\frac{7}{10}$ and $\frac{4}{5} = \frac{8}{10}$, then $\frac{4}{5}$ is more than $\frac{7}{10}$. Using symbols, we write $\frac{4}{5} > \frac{7}{10}$. This method of comparing fractions is discussed thoroughly in Section 5.4.

 Try Problems 13 and 14.

Try These Problems

15. Which is less, $\frac{1}{2}$ or $\frac{2}{5}$?
16. Which is more, $\frac{8}{10}$ or $\frac{31}{39}$?
17. Which is less, $3\frac{7}{11}$ or $3\frac{19}{30}$?
18. Which is more, $5\frac{2}{17}$ or $5\frac{29}{250}$?

Another way to compare fractions is change each to a decimal, then compare. Here are examples to illustrate the technique.

EXAMPLE 6 Which is less, $\frac{9}{20}$ or $\frac{21}{50}$?

SOLUTION Convert each fraction to a decimal.

$$20\overline{)9.00}^{.45} \rightarrow \frac{9}{20} = 0.45$$
$$\underline{8\ 0}$$
$$1\ 00$$
$$\underline{1\ 00}$$

$$50\overline{)21.00}^{.42} \rightarrow \frac{21}{50} = 0.42 \text{ —— This number is less.}$$
$$\underline{20\ 0}$$
$$1\ 00$$
$$\underline{1\ 00}$$

The fraction $\frac{21}{50}$ is less. ■

 Try Problems 15 and 16.

EXAMPLE 7 Which is more, $5\frac{3}{8}$ or $5\frac{7}{18}$?

SOLUTION Convert each mixed numeral to a decimal.

$$8\overline{)3.000}^{.375} \rightarrow 5\frac{3}{8} = 5.375$$
$$\underline{2\ 4}$$
$$60$$
$$\underline{56}$$
$$40$$
$$\underline{40}$$

$$18\overline{)7.000}^{.388} \rightarrow 5\frac{7}{18} = 5.3\overline{8} \text{ —— This number is more.}$$
$$\underline{5\ 4}$$
$$1\ 60$$
$$\underline{1\ 44}$$
$$160$$
$$\underline{144}$$
$$16$$

The number $5\frac{7}{18}$ is more. ■

 Try Problems 17 and 18.

EXAMPLE 8 List from smallest to largest: $\frac{3}{2}, \frac{5}{7}, \frac{3}{4}$

SOLUTION Convert each of the fractions to a decimal, then compare.

$$2\overline{)3.0}^{1.5} \rightarrow \frac{3}{2} = 1.5$$
$$\underline{2}$$
$$1\ 0$$
$$\underline{1\ 0}$$

Try These Problems

In Problems 19 and 20, list from smallest to largest.

19. $\frac{180}{200}, \frac{3}{8}, \frac{7}{5}$

20. $\frac{211}{40}, \frac{79}{13}, \frac{50}{9}, \frac{61}{10}$

21. Which is more, $\frac{8}{13}$ or 0.62?

22. Which is less, 5.8724 or $5\frac{7}{8}$?

$$\begin{array}{r} .71 \\ 7{\overline{\smash{\big)}\,5.00}} \\ \underline{4\ 9} \\ 10 \\ \underline{7} \\ 3 \end{array} \rightarrow \frac{5}{7} \approx 0.71$$

$$\begin{array}{r} .75 \\ 4{\overline{\smash{\big)}\,3.00}} \\ \underline{2\ 8} \\ 20 \\ \underline{20} \end{array} \rightarrow \frac{3}{4} \approx 0.75$$

When comparing more than two numbers, it can be helpful to picture the numbers on the number line.

From smallest to largest the numbers are $\frac{5}{7}, \frac{3}{4}$, and $\frac{3}{2}$. ∎

Try Problems 19 and 20.

4 Comparing Fractions to Decimals

If you need to compare numbers where some of them are written as fractions and others are written as decimals, write them all as decimals then compare. The following examples illustrate this.

EXAMPLE 9 Which is more, $\frac{7}{11}$ or 0.6367?

SOLUTION Convert $\frac{7}{11}$ to a decimal.

$$\begin{array}{r} .6363 \\ 11{\overline{\smash{\big)}\,7.0000}} \\ \underline{6\ 6} \\ 40 \\ \underline{33} \\ 70 \\ \underline{66} \\ 40 \\ \underline{33} \\ 7 \end{array} \rightarrow \frac{7}{11} = 0.\overline{63}$$

Now compare the two decimals carefully. It helps to arrange them so that the decimal points align vertically and digits with like place value form a column.

$$\frac{7}{11} = 0.\overline{63} = 0.636{\vert}363\ldots$$

$$0.6367 = 0.636{\vert}700 \text{—— This number is more.}$$

The number 0.6367 is more. ∎

Try Problems 21 and 22.

▲ **Try These Problems**

23. List from smallest to largest: 1.09, 0.916, $\frac{11}{12}$, $\frac{9}{10}$

EXAMPLE 10 List from smallest to largest: $\frac{2}{9}$, 0.22, 1.2, $\frac{1}{4}$

SOLUTION Convert $\frac{2}{9}$ and $\frac{1}{4}$ to decimals.

$$9\overline{)2.00}^{.22} \rightarrow \frac{2}{9} = 0.\overline{2}$$
$$\underline{1\ 8}$$
$$20$$
$$\underline{18}$$
$$2$$

$$4\overline{)1.00}^{.25} \rightarrow \frac{1}{4} = 0.25$$
$$\underline{8}$$
$$20$$
$$\underline{20}$$

When there are more than two numbers it helps to picture the numbers on the number line.

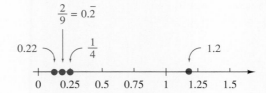

From smallest to largest the numbers are 0.22, $\frac{2}{9}$, $\frac{1}{4}$, and 1.2. ∎

▲ **Try Problem 23.**

▲ **Answers to Try These Problems**

1. a. $\frac{12}{32}$ **c.** 0.375   **2. b.** $10\frac{12}{5}$ **c.** 12.4 **d.** $\frac{62}{5}$ **3. c.** $\frac{83}{10}$ **d.** 8.300
4. b. 0.0080 **c.** $\frac{8}{1000}$ **d.** $\frac{4}{500}$ **5.** >   **6.** >   **7.** 0.007849
8. 8.508   **9.** 0.35$\overline{2}$ **10.** 13.45   **11.** 0.75; 7.499; 7.5; 7.52
12. 2.2$\overline{3}$; 2.236; 22.$\overline{3}$; 22.36   **13.** <   **14.** >   **15.** $\frac{2}{5}$
16. $\frac{8}{10}$ **17.** $3\frac{19}{30}$ **18.** $5\frac{2}{17}$ **19.** $\frac{3}{8}, \frac{180}{200}, \frac{7}{5}$
20. $\frac{211}{40}, \frac{50}{9}, \frac{79}{13}, \frac{61}{10}$ **21.** 0.62   **22.** 5.8724
23. $\frac{9}{10}$, 0.916, $\frac{11}{12}$, 1.09

EXERCISES 8.2

Use the symbol <, =, or > to compare each pair of numbers.

1. 23.54, 23.268
2. 0.0076, 0.012
3. $\frac{5}{12}, \frac{3}{4}$
4. $8\frac{7}{50}, 8\frac{2}{10}$
5. $14\frac{12}{4}$, 17
6. $9\frac{6}{4}, 10\frac{1}{2}$
7. 0.78, $\frac{7}{9}$
8. $\frac{2}{13}$, 0.154
9. 16.25, $16\frac{11}{48}$
10. $\frac{104}{5}$, 20.8

11. Which of these numbers are equal to $\frac{7}{25}$?
   a. $\frac{28}{100}$ **b.** 0.725   **c.** $\frac{35}{125}$ **d.** 0.28

12. Which of these numbers are equal to $9\frac{3}{16}$?
   a. $\frac{147}{16}$ **b.** 9.316   **c.** $8\frac{19}{16}$ **d.** 9.1875

13. Which of these numbers are equal to 0.06?
    a. $\frac{3}{500}$ b. $\frac{3}{50}$ c. 0.0600   d. $\frac{60}{1000}$
14. Which of these numbers are equal to 7.4?
    a. 7.400   b. $\frac{37}{5}$ c. $\frac{74}{100}$ d. $\frac{296}{40}$
15. Which of these numbers are equal to 12?
    a. $\frac{60}{5}$ b. $\frac{2}{24}$ c. 0.12   d. 12.0
16. Which of these numbers are equal to 35?
    a. $34\frac{6}{6}$ b. $\frac{1220}{4}$ c. 35.00   d. $\frac{420}{12}$

Which number is less?
17. 8.16394, 8.1638109
18. $\frac{3}{1000}$, $\frac{1}{33}$
19. 1.75, $1\frac{13}{18}$
20. $\frac{5}{34}$, 0.148

Which number is greater in value?
21. $8\frac{1}{12}$, $8\frac{2}{25}$
22. $25.\overline{3}$, 25.3378
23. $\frac{52}{50}$, 1.3
24. $\frac{5}{17}$, 0.29

List from smallest to largest.
25. 21.5; 2.1; 20.75; 21.6
26. $7.\overline{7}$; 7.7; 0.7; $0.\overline{7}$
27. $1\frac{1}{4}$, $\frac{3}{4}$, $\frac{4}{3}$, 1
28. $\frac{14}{33}$, $\frac{2}{50}$, $\frac{2}{5}$, $\frac{5}{2}$
29. 0.64, $\frac{3}{5}$, $\frac{23}{40}$, $0.63\overline{7}$
30. 3.6, 3.628, $3\frac{2}{3}$, $3\frac{17}{27}$
31. $20\frac{11}{18}$, 20.6, 2.06, $20\frac{2}{3}$
32. 0.114, $\frac{1}{10}$, $\frac{10}{1}$, 9.8

8.3 LANGUAGE

OBJECTIVE ■ Solve problems by using translations.

In this section we review the language that has been introduced throughout this book and look at some examples of solving problems by using translations. The first chart gives many English words or phrases that often translate to the equality symbol =.

Math Symbol	English
=	equals is equal to is the same as is the result of is are was were will be represents gives makes yields

1 Translating English to Addition and Subtraction

The addition statement $3.2 + 0.4 = 3.6$ is written using math symbols. Some of the ways to read this statement using English are given in the chart that follows.

Math Symbols	English
$3.2 + 0.4 = 3.6$	3.2 plus 0.4 equals 3.6.
	3.2 added to 0.4 gives 3.6.
	3.2 *increased by* 0.4 yields 3.6.
	The *sum* of 3.2 and 0.4 is 3.6.
	0.4 more than 3.2 is 3.6.
	3.6 is the result of adding 3.2 and 0.4.

The subtraction statement $3.2 - 0.4 = 2.8$ is written using math symbols. Some of the ways to read this statement using English are given in the following chart.

Math Symbols	English
$3.2 - 0.4 = 2.8$	3.2 minus 0.4 is 2.8
	3.2 take away 0.4 equals 2.8.
	3.2 *decreased by* 0.4 gives 2.8.
	0.4 subtracted from 3.2 is 2.8.
	0.4 less than 3.2 is equal to 2.8.
	The *difference* between 3.2 and 0.4 is 2.8.
	2.8 is the result of subtracting 0.4 from 3.2.

Observe that when reading the subtraction symbol $-$ using the phrases *subtracted from* or *less than*, the numbers are read in the reverse order than they are written. Here is a closer look at translations that involve these phrases.

$$\underset{3.2\ -\ 0.4}{\text{0.4 subtracted from 3.2}} \qquad \underset{3.2\ -\ 0.4}{\text{0.4 less than 3.2}}$$

Here is an example of using translations to help with problem solving.

EXAMPLE 1   What number increased by $\frac{3}{4}$ gives $7\frac{5}{18}$?

SOLUTION   The question translates to an addition statement.

What number $\underbrace{\text{increased by}}$ $\frac{3}{4}$ gives $7\frac{5}{18}$?

$\square \quad + \quad \frac{3}{4} \quad = \quad 7\frac{5}{18}$

Try These Problems

Solve.
1. Find the difference between 230 and 1.98.
2. Find the sum of $3\frac{1}{12}$ and $6\frac{3}{20}$.
3. What number plus 0.86 yields 9.7?
4. 16 represents $12\frac{3}{5}$ more than what number?

One of the parts is missing so we subtract to find the missing number.

$$7\frac{5}{18} - \frac{3}{4}$$
$$= 7\frac{5 \times 2}{18 \times 2} - \frac{3 \times 9}{4 \times 9}$$ Convert each fraction to an equivalent fraction with denominator 36.
$$= 7\frac{10}{36} - \frac{27}{36}$$
$$= 6\frac{46}{36} - \frac{27}{36}$$
$$= 6\frac{19}{36}$$

The missing number is $6\frac{19}{36}$. ■

If you need to see more examples of translation problems involving addition and subtraction, refer to Sections 1.3, 5.6, and 6.6.

 Try Problems 1 through 4.

2 Translations Involving Multiplication and Division

The multiplication statement $\frac{3}{4} \times 200 = 150$ is written using math symbols. The chart that follows illustrates many correct ways to read the statement using English.

Math Symbols	English
$\frac{3}{4} \times 200 = 150$	$\frac{3}{4}$ times 200 is 150.
	$\frac{3}{4}$ of 200 equals 150.
	$\frac{3}{4}$ multiplied by 200 gives 150.
	The *product* of $\frac{3}{4}$ and 200 is 150.
	150 represents $\frac{3}{4}$ of 200.
	150 is the result of multiplying $\frac{3}{4}$ by 200.

EXAMPLE 2 What fraction of 60 is 2.4?

SOLUTION The question translates to a multiplication statement.

What fraction of 60 is 2.4?
↓   ↓
☐ × 60 = 2.4

Try These Problems

Solve.
5. Find 0.15 of 460.
6. Find the product of $\frac{1}{30}$ and $8\frac{1}{3}$.
7. 115 represents 0.005 multiplied by what number?
8. What fraction of $45\frac{1}{2}$ is $9\frac{1}{10}$?
9. Find the ratio of 0.45 to 30.
10. Find the result when $3\frac{3}{4}$ is divided by $1\frac{1}{2}$.
11. Divide 0.125 into 16.
12. Divide $\frac{2}{5}$ into 80.

A multiplier is missing, so we divide 2.4 by 60 to find the missing multiplier.

$$\begin{array}{r} .04 \\ 60{\overline{\smash{\big)}\,2.40}} \\ \underline{0} \\ 2\,40 \\ \underline{2\,40} \end{array}$$

The product 2.4 is written inside the division symbol.

The answer is 0.04 or $\frac{4}{100}$. ■

Now that we have studied both fractions and decimals, it is correct to answer the question, "what fraction of," with either a fraction or decimal.

If you need to see more examples of problems that translate to a multiplication statement, refer to Sections 2.4, 4.4, and 7.4.

 Try Problems 5 through 8.

When reading the division statement $12 \div 3$, take care to read the numbers in the correct order, because $12 \div 3$ does not equal $3 \div 12$.

$$12 \div 3 = \frac{12}{3} = 4 \quad \text{but} \quad 3 \div 12 = \frac{3}{12} = \frac{1}{4} \text{ or } 0.25$$

Larger than 1.   Smaller than 1.

The following chart gives several correct ways to read each of these division statements using English.

Math Symbols	English
$12 \div 3 = 4$ or $\frac{12}{3} = 4$	12 *divided by* 3 is 4. The *ratio* of 12 to 3 is 4. 3 *divided into* 12 equals 4. 4 is the result of dividing 12 by 3.
$3 \div 12 = \frac{1}{4}$ or 0.25 $\frac{3}{12} = \frac{1}{4}$ or 0.25	3 *divided by* 12 is 0.25. The *ratio* of 3 to 12 is $\frac{1}{4}$. 12 *divided into* 3 is $\frac{1}{4}$. 0.25 is the result of dividing 3 by 12.

If you need more help with the language related to division, refer to Sections 2.4, 4.4, and 7.4.

 Try Problems 9 through 12.

 Answers to Try These Problems

1. 228.02   2. $9\frac{7}{30}$ 3. 8.84   4. $3\frac{2}{5}$ 5. 69   6. $\frac{5}{18}$
7. 23,000   8. $\frac{1}{5}$ or 0.2   9. 0.015   10. $\frac{5}{2}$ or $2\frac{1}{2}$ or 2.5
11. 128   12. 200

EXERCISES 8.3

Solve. Express the answer as a fraction or a decimal.

1. Find the product of 4.5 and 300.
2. Find the sum of 23.9 and 360.
3. Find the difference between $2\frac{1}{3}$ and $\frac{3}{4}$.
4. Find the ratio of $3\frac{1}{2}$ to $10\frac{1}{2}$.
5. Divide 40 into 3.4.
6. Subtract 9.45 from 82.
7. What number is $5\frac{5}{8}$ more than $7\frac{3}{4}$?
8. Find $\frac{7}{8}$ of 320.
9. 50 equals 1.6 times what number?
10. Divide 8 by 3.2.
11. What number is $3\frac{8}{9}$ less than $8\frac{13}{20}$?
12. What number plus $\frac{4}{5}$ yields $\frac{13}{7}$?
13. What fraction of 84 is 2.1?
14. 0.02 of what number gives 0.15?
15. 14 is what fraction of 56?
16. Find the ratio of 13 to 104.
17. $12\frac{5}{6}$ decreased by $4\frac{2}{3}$ yields what number?
18. 36 represents $2\frac{1}{2}$ of what number?
19. Divide $3\frac{1}{5}$ by 200.
20. Divide $3\frac{1}{5}$ into 200.

USING THE CALCULATOR # 8

MULTIPLYING AND DIVIDING FRACTIONS AS DECIMALS

When we use a calculator to compute mathematics problems, we are often satisfied with the answer in decimal form even if the original problem was given in fractional form and even if the answer is only an approximation. In this feature, we look at operating with fractions without the use of the [a b/c] key.

Of course, you can convert a fraction to a decimal on the calculator by dividing the numerator by the denominator. For example, here we convert $\frac{3}{4}$ to a decimal.

To Convert to a Decimal	$\frac{3}{4}$
Enter	3 [÷] 4 [=]
Result	0.75

Therefore, the fraction $\frac{3}{4}$ is equal to the decimal 0.75. The procedure for converting a fraction to a decimal is the same for both the scientific and the basic calculator.

Next we look at multiplying two fractions on the calculator. The procedure is the same for both the scientific and the basic calculator. For example, here we compute $\frac{5}{8} \times \frac{2}{5}$.

To Compute	$\frac{5}{8} \times \frac{2}{5}$
Enter	5 [÷] 8 [×] 2 [÷] 5 [=]
Result	0.25

Dividing two fractions on the calculator is not as straightforward as multiplying two fractions. If you enter the [÷] key instead of the [×] key in the previous procedure, you do not get the result of $\frac{5}{8} \div \frac{2}{5}$. The reason has to do with the order in which the calculator is doing the repeated division.

Probably the safest way for you to divide two fractions on the calculator is first to change the problem to multiplication, then use the procedure for multiplication. Here we show how to compute $\frac{5}{8} \div \frac{2}{5}$.

To Compute	$\frac{5}{8} \div \frac{2}{5}$
Change to	$\frac{5}{8} \times \frac{5}{2}$
Enter	5 [÷] 8 [×] 5 [÷] 2 [=]
Result	1.5625

Cont. page 338.

Calculator Problems

Convert each fraction to a decimal. If the decimal representation goes beyond 3 decimal places, round off at 3 decimal places.

1. $\frac{7}{8}$ 2. $\frac{8}{7}$ 3. $\frac{35}{13}$ 4. $\frac{13}{35}$ 5. $\frac{250}{6000}$ 6. $\frac{6000}{250}$

Evaluate each without using the $\boxed{a^b/c}$ *key. If the decimal representation goes beyond 3 decimal places, round off at 3 decimal places.*

7. $\frac{5}{6} \times \frac{3}{4}$ 8. $\frac{15}{38} \times \frac{75}{41}$ 9. $678 \times \frac{97}{186}$ 10. $\frac{85}{300} \times 45$
11. $\frac{5}{6} \div \frac{3}{4}$ 12. $\frac{15}{38} \div \frac{75}{41}$ 13. $678 \div \frac{97}{186}$ 14. $\frac{85}{300} \div 45$

Convert the mixed numerals to improper fractions, then use the procedure discussed in this feature to evaluate. If the decimal representation goes beyond 3 decimal places, round off at 3 decimal places.

15. $8\frac{4}{5} \times 3\frac{6}{7}$ 16. $8\frac{4}{5} \div 3\frac{6}{7}$ 17. $78\frac{17}{456} \times 27\frac{45}{88}$ 18. $78\frac{17}{456} \div 27\frac{45}{88}$

8.4 APPLICATIONS: CHOOSING THE CORRECT OPERATION

OBJECTIVES
- Solve an application that involves a total as the sum of parts.
- Solve an application that involves equal parts in a total quantity.
- Solve an application that involves the perimeter of a figure.
- Solve an application that involves the area of a rectangle.
- Solve an application that involves translating.
- Solve an application that involves rates or ratios.

In this section we look at the various types of applications that have been introduced throughout the text thus far. Also we refer you to the appropriate previous sections where more detailed instruction is given on a particular type of problem.

1 Problems Involving a Total as the Sum of Parts

EXAMPLE 1 Dick wanted to jog a total distance of 8 miles. After jogging 6.4 miles, his foot began to hurt and he walked the rest of the way. How far did he walk?

SOLUTION First picture the situation.

Observe that we are missing one of the parts in a total quantity, so we have the following.

336 Chapter 8 Problem Solving: Whole Numbers, Fractions, and Decimals

▲ Try These Problems

Solve.

1. Ms. Rodriquez, the vice-president of finance for a small company, wrote checks for $456.96, $365.75, and $1087.54. What is the total amount of the checks?

2. Doug has a piece of wire that is 18.2 meters long. He cuts off a piece that is 9.75 meters long. How long is the remaining piece?

3. On a vacation Bette spent $\frac{1}{3}$ of her money on air fare. What fraction of her money was left for other expenses?

4. Jose received a bonus from his employer. He invested $\frac{2}{5}$ of it in a money market fund, $\frac{1}{4}$ of it in municipal bonds, and spent the rest on furniture for his new home. What fraction of his bonus did he invest?

To find the missing part, subtract the given part 6.4 from the total 8.

$$\begin{array}{rl} 8.0 & \text{Total} \\ -6.4 & \text{Jogging part} \\ \hline 1.6 & \text{Walking part} \end{array}$$

Dick walked 1.6 miles. ■

EXAMPLE 2   An engineer completed $\frac{2}{15}$ of a job. What fraction of the job is left?

SOLUTION   The whole job is represented by the number 1, thus we have the following.

$$\begin{array}{c} \text{Fraction} \\ \text{completed} \end{array} + \begin{array}{c} \text{fraction} \\ \text{left} \end{array} = \text{total}$$

$$\frac{2}{15} + \boxed{} = 1$$

To find the missing part, subtract the given part $\frac{2}{15}$ from the total 1.

$$1 - \frac{2}{15}$$
$$= \frac{15}{15} - \frac{2}{15} \qquad \text{Convert 1 to } \tfrac{15}{15}.$$
$$= \frac{13}{15}$$

$\frac{13}{15}$ of the job is left. ■

If you need to see additional examples of problems involving a total as the sum of parts; the list that follows gives the section of the text where the specified examples can be found.

Section	Examples
1.4	1, 2, and 3
5.7	1, 2, and 3
6.7	1 and 2

▲ Try Problems 1 through 4.

2   Problems Involving Equal Parts in a Total Quantity

Suppose you have a rope that is 3 feet in length. If you cut the rope into 12 equal pieces, then the length of each piece is $\frac{1}{4}$ or 0.25 foot. Here is a picture of the situation.

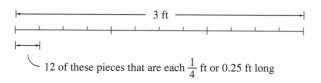

12 of these pieces that are each $\frac{1}{4}$ ft or 0.25 ft long

Observe that the numbers are related as follows.

$$\begin{array}{c} \text{Size of} \\ \text{each part} \\ \text{(ft)} \end{array} \times \begin{array}{c} \text{number of} \\ \text{equal parts} \end{array} = \begin{array}{c} \text{total} \\ \text{quantity} \\ \text{(ft)} \end{array}$$

$$0.25 \quad \times \quad 12 \quad = \quad 3$$

In general, if a situation involves a total quantity being split into a number of equal parts, the following always applies.

$$\begin{array}{c} \textbf{Size of} \\ \textbf{each part} \end{array} \times \begin{array}{c} \textbf{number of} \\ \textbf{equal parts} \end{array} = \begin{array}{c} \textbf{total} \\ \textbf{quantity} \end{array}$$

When using this formula be careful not to assume that the total quantity is always the largest of the three numbers. Notice in the situation above, the total quantity is 3, and the number of equal parts is a larger number, 12. The total quantity is always larger than the size of each part but may or may not be larger than the number of equal parts.

A problem can be presented by giving any two of the three quantities and asking for the third quantity. The following examples illustrate this.

EXAMPLE 3   A nurse needs to give a patient 42 doses of medicine. Each dose contains 2.7 centiliters. How much medicine does the nurse need for this patient?

SOLUTION   This problem involves a total quantity that is made up of a number of equal parts. The following formula applies.

$$\begin{array}{c} \text{Size of} \\ \text{each part} \\ (c\ell) \end{array} \times \begin{array}{c} \text{number of} \\ \text{equal parts} \end{array} = \begin{array}{c} \text{total} \\ \text{quantity} \\ (c\ell) \end{array}$$

$$2.7 \quad \times \quad 42 \quad = \quad \boxed{}$$

To find the total amount of medicine, multiply the number of doses 42 by the size of each dose, 2.7 centiliters.

```
    4 2          Number of doses
  × 2.7          Size of each dose in centiliters
  ─────
   29 4
   84
  ─────
  113.4          Total amount in centiliters.
```

The nurse needs a total of 113.4 centiliters.   ■

EXAMPLE 4   A chef prepared 174 cups of pudding. How many $\frac{2}{3}$-cup portions can she serve?

SOLUTION   We want to know how many $\frac{2}{3}$-cup servings are in the total quantity, 174 cups. The following formula applies.

$$\begin{array}{c} \text{Size of} \\ \text{each part} \\ \text{(cups)} \end{array} \times \begin{array}{c} \text{number of} \\ \text{equal parts} \end{array} = \begin{array}{c} \text{total} \\ \text{quantity} \\ \text{(cups)} \end{array}$$

$$\frac{2}{3} \quad \times \quad \boxed{} \quad = \quad 174$$

Try These Problems

5. A bookshelf is 57 inches wide. Books that are each 0.6 inch thick are to stand on the shelf. How many books fit on the shelf?

6. The cost of putting in a new road through a residential area was $19,800. The 55 homeowners who live in the area shared the cost equally. What was each homeowner's share?

7. Janet is making drapes for a very wide window. If she sews together 12 panels that are each $3\frac{5}{8}$ feet wide, what is the total width of the resulting drapes?

A multiplier is missing, so we divide 174 by $\frac{2}{3}$ to find the missing multiplier.

$$174 \div \frac{2}{3} = \frac{\overset{87}{\cancel{174}}}{1} \times \frac{3}{\underset{1}{\cancel{2}}} = 261$$

Total amount in cups / Size of each serving in cups / Number of servings

The chef can serve 261 of the $\frac{2}{3}$-cup portions. ∎

If you need to see additional examples involving equal parts in a total quantity, the following list gives the section of the text where the specified examples can be found.

Section	Examples
2.5	1, 2, and 3
3.6	1, 2, and 3
4.5	1, 2, 3 and 4
7.5	1, 2, and 3

 Try Problems 5 through 7.

3 Perimeter and Area

The segment shown here has a length of 1 centimeter.

——— Length = 1 centimeter

Here is a square that has each side measuring 1 centimeter.

Area = 1 square centimeter
Perimeter = 4 centimeters

We say that the **area** of this square measures 1 square centimeter. However, the **perimeter** of this square is 4 centimeters. Area is the measure of the extent of a region on a flat surface, while perimeter is the total distance around the boundary of the region. Consider a rectangular region that has a width of 3 centimeters and a length of 4 centimeters.

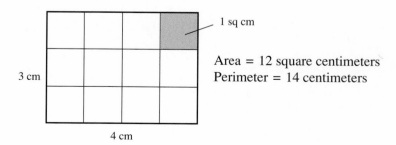

Area = 12 square centimeters
Perimeter = 14 centimeters

The area of each of the smaller squares inside measures 1 square centimeter, and there are 12 of these squares inside. Therefore, the area of the entire rectangle is 12 square centimeters. Because the distance around the rectangle is 14 centimeters, the perimeter of the rectangle is 14 centimeters. Observe that the area of this rectangle can be found by multiplying the width by the length.

$$\text{Area} = 3 \text{ cm} \times 4 \text{ cm} = 12 \text{ sq cm}$$

Also, observe that the perimeter of this rectangle can be found by adding the lengths of all four sides.

$$\text{Perimeter} = 3 \text{ cm} + 3 \text{ cm} + 4 \text{ cm} + 4 \text{ cm}$$
$$= 14 \text{ cm}$$

In general the area of any rectangle can be found by multiplying the width by the length. Before multiplying, the distances must be written using the same unit of measurement. The perimeter of a rectangle can be found by adding the lengths of all four sides.

Area of a Rectangle

The area of a rectangle is the length times the width.
Area = length × width

Perimeter of a Rectangle

Perimeter = length + length + width + width
= twice the length + twice the width

EXAMPLE 5 A rectangular window is $6\frac{1}{2}$ feet wide and $3\frac{3}{4}$ feet high.
a. Find the perimeter of the window.
b. Find the area of the window.

SOLUTION

a. Perimeter (ft) $= 6\frac{1}{2} + 6\frac{1}{2} + 3\frac{3}{4} + 3\frac{3}{4}$

$= 6\frac{2}{4} + 6\frac{2}{4} + 3\frac{3}{4} + 3\frac{3}{4}$

$= 18\frac{10}{4}$

$= 20\frac{2}{4}$

$= 20\frac{1}{2}$ or 20.5

b. Area (sq ft) $= 6\frac{1}{2} \times 3\frac{3}{4}$

$= \frac{13}{2} \times \frac{15}{4}$

$= \frac{195}{8}$ or $24\frac{3}{8}$ or 24.375

The perimeter is $20\frac{1}{2}$ feet and the area is $24\frac{3}{8}$ square feet.

Try These Problems

8. A rectangular piece of fabric is $17\frac{3}{8}$ inches wide and $35\frac{1}{3}$ inches long.
   a. Find the perimeter of the piece.
   b. Find the area of the piece.

9. Find the perimeter of a triangle whose sides have lengths of 28 centimeters, 5.7 centimeters, and 18.73 centimeters.

10. The length of a sheet of metal is 8.6 meters. The sheet consists of 54.61 square meters.
    a. What is the width of the sheet?
    b. What is the perimeter of the sheet?

EXAMPLE 6 A rectangular lot contains 1020 square yards. If the width of the lot is 25 yards, find the length.

SOLUTION The quantity *1020 square yards* is the area of the lot. This should be obvious because of the units *square yards* attached to the 1020. We have the following:

$$\begin{array}{ccc} \text{Area} & = \text{length} \times \text{width} \\ \text{(sq yd)} & \text{(yd)} \quad \text{(yd)} \\ 1020 & = \boxed{} \times 25 \end{array}$$

A multiplier is missing, so we divide 1020 by 25 to find the missing multiplier.

$$\begin{array}{r} 40.8 \\ 25 \overline{\smash{)}1020.0} \\ \underline{100} \\ 20 \\ \underline{0} \\ 20\,0 \\ \underline{20\,0} \end{array}$$

Length in yards
Area in square yards
Width in yards

The length of the lot is 40.8 yards. ∎

If you need to see more examples related to perimeter or area, the following list gives the location of these examples.

Section	Examples
1.4	2
2.5	11, 12, 17, and 18
4.5	15 and 16
5.7	2, 8, and 10
6.7	5
7.5	14 and 15

 Try Problems 8 through 10.

4 Problems Involving Translations

EXAMPLE 7 A certain stock opened for $\$18\frac{3}{8}$ a share. At the end of the day the price was $\$19\frac{5}{8}$. What was the increase in price?

SOLUTION $\$18\frac{3}{8}$ is the original price and $\$19\frac{5}{8}$ is the price after the increase. We have the following:

$$\begin{array}{ccc} \text{Original} & \text{Increase} & \text{final} \\ \text{price} & + \text{in price} & = \text{price} \\ (\$) & (\$) & (\$) \\ 18\frac{3}{8} & + \boxed{} & = 19\frac{5}{8} \end{array}$$

We are missing one of the terms, so we subtract to find the missing term.

$$19\frac{5}{8} - 18\frac{3}{8}$$
$$= 19\frac{5}{8} - 18\frac{3}{8}$$
$$= 1\frac{2}{8}$$
$$= 1\frac{1}{4}$$

The increase in price is $1\frac{1}{4}$ or \$1.25. ∎

EXAMPLE 8 Dorothy worked for 3.5 hours on Tuesday. Steve worked 2.25 times as long as Dorothy. How long did Steve work?

SOLUTION The sentence, "Steve worked 2.25 times as long as Dorothy," really means that Steve's time is 2.25 times as long as Dorothy's time, and can be translated to a multiplication statement.

Steve's time	is	2.25	times as long as	Dorothy's time
↓	↓			
☐	=	2.25	×	3.5

The answer to the multiplication problem is missing, so we multiply to find the answer.

$$\begin{array}{r} 2.25 \\ \times\ \ 3.5 \\ \hline 1125 \\ 675\ \ \\ \hline 7.875 \end{array}$$

Steve worked 7.875 hours. ∎

EXAMPLE 9 Ms. Tang pays 0.4 of her annual gross salary in taxes. She paid \$24,000 in taxes last year. What is her annual gross salary?

SOLUTION The phrase *0.4 of her annual gross salary in taxes* really means that 0.4 of her annual gross salary is the amount paid in taxes, and can be translated to a multiplication statement.

0.4	of	her salary (\$)	is	taxes (\$)
↓		↓		
0.4	×	☐	=	24,000

A multiplier is missing, so we divide to find the missing multiplier.

$$0.4\overline{)24000.0}\ \ \ \text{quotient } 60000.$$

Her salary was \$60,000. ∎

Try These Problems

11. A coat is marked $165. Tax increases the price by $9.90. What is the total price of the coat?

12. Kathy earns 0.04 of her total sales in commissions. She sells $2600 worth of merchandise. How much does she earn in commissions?

13. A carpenter has one board that is $5\frac{3}{8}$ feet long and another board that is $12\frac{2}{3}$ feet long.
    a. What is the difference in the lengths of these two boards?
    b. What is the ratio of the shorter board to the longer board?

14. Cyrus earns 1.8 times as much as Betsy earns. Cyrus earns $900 each month. What is Betsy's monthly salary?

15. Seven-thirtieths of the workers in a factory were out with the flu last week. There were 490 workers out with the flu. How many workers are employed by this factory?

16. Ms. Moody walks $4\frac{1}{2}$ miles every 3 hours. How many miles per hour is this?

17. Eight-tenths ounce of gold is worth $450. How much is the gold worth per ounce?

If you need additional examples related to solving problems using translations, the following list gives the location of these examples.

Section	Examples
1.4	4, 5
2.5	13, 14
3.6	4–8
4.5	17–25
5.7	4, 5
6.7	3, 4, 5
7.5	16–19
8.3	1, 2

 Try Problems 11 through 15.

5 Problems Involving Rates

A **rate** is a comparison by division; a rate can therefore be written as the ratio of two quantities. Here are examples of phrases that indicate rates.

$$70 \text{ words per minute} \rightarrow \frac{70 \text{ words}}{1 \text{ minute}}$$

$$60 \text{ feet in 2 seconds} \rightarrow \frac{60 \text{ feet}}{2 \text{ seconds}} = 30 \text{ feet per second}$$

$$8 \text{ pounds every 10 feet} \rightarrow \frac{8 \text{ pounds}}{10 \text{ feet}} = 0.8 \text{ pound per foot}$$

Observe that words like *per, in,* and *every* that are used to indicate rates translate to the fraction bar or division.

EXAMPLE 10 On a map, $2\frac{1}{2}$ inches represent 400 miles. Express this rate in miles per inch.

SOLUTION Write the rate $2\frac{1}{2}$ *inches represent 400 miles* as a ratio. Put inches in the denominator because we want miles per inch.

$$2\frac{1}{2} \text{ inches represent 400 miles}$$

$$= \frac{400 \text{ miles}}{2\frac{1}{2} \text{ inches}}$$

To obtain *miles per inch* put inches in the denominator.

Divide 400 by $2\frac{1}{2}$ to obtain how many miles per inch.

$$\frac{400}{2\frac{1}{2}} = 400 \div 2\frac{1}{2} = \frac{400}{1} \div \frac{5}{2}$$

$$= \frac{\overset{80}{\cancel{400}}}{1} \times \frac{2}{\underset{1}{\cancel{5}}}$$

$$= 160$$

The rate is 160 miles per inch. ∎

 Try Problems 16 and 17.

8.4 Applications: Choosing the Correct Operation

If you type 70 words per minute, then you can type 350 words in 5 minutes. We have the following:

$$\frac{70 \text{ words}}{1 \text{ minute}} \times \frac{5 \text{ minutes}}{1} = 350 \text{ words}$$

Observe that if you multiply the rate $\frac{70 \text{ words}}{1 \text{ minute}}$ by minutes so that the minute units cancel, the resulting units must be words. Here are more examples of how units cancel when multiplying by a rate.

$$\frac{60 \text{ feet}}{2 \text{ seconds}} \times \frac{9.5 \text{ seconds}}{1} = 30 \times 9.5 \text{ feet}$$
$$= 285 \text{ feet}$$

$$\frac{8 \text{ pounds}}{10 \text{ feet}} \times \frac{5.25 \text{ feet}}{1} = \frac{4 \times 5.25}{5} \text{ pounds}$$
$$= \frac{21}{5} \text{ pounds}$$
$$= 4.2 \text{ pounds}$$

EXAMPLE 11 Mary needs $1\frac{1}{3}$ yards of fabric to make a blouse. The fabric costs $3.99 per yard. What is the total cost of the fabric?

SOLUTION The phrase *$3.99 per yard* is a rate, and can be written as a ratio.

$$\$3.99 \text{ per yard} \rightarrow \frac{3.99 \text{ dollars}}{1 \text{ yard}}$$

Because yards is in the denominator, we can set up a multiplication statement involving this rate so that yards cancel.

$$\frac{\text{dollars}}{\text{yard}} \times \text{yards} = \text{dollars}$$

$$3.99 \times 1\frac{1}{3} = \boxed{}$$

The answer to the multiplication statement is missing, so we multiply to find the missing number.

$$3.99 \times 1\frac{1}{3}$$

$$= \frac{\overset{1.33}{3.99}}{1} \times \frac{4}{\underset{1}{\cancel{3}}} \qquad \text{Divide both 3.99 and 3 by 3.}$$

$$= 5.32$$

The total cost is $5.32. ∎

EXAMPLE 12 A faucet leaks 3.2 ounces of water every 8 minutes. How long does it take the faucet to leak 7 ounces?

Try These Problems

18. Mr. Garcia earns $32.46 per hour. How much does he earn in 6.5 hours?
19. Water enters a tank at the rate of 3.2 gallons per minute. How many gallons have entered the tank after 7.5 minutes?
20. A fish swims 60 feet in 2 seconds. How far can the fish swim in 9 seconds?
21. If Darby jogs $3\frac{1}{3}$ miles in 25 minutes, how long will it take him to go 20 miles?

SOLUTION The phrase *3.2 ounces every 8 minutes* is a rate and can be written as a ratio.

$$3.2 \text{ ounces every 8 minutes}$$
$$= \frac{3.2 \text{ ounces}}{8 \text{ minutes}}$$
$$= \frac{0.4 \text{ ounces}}{1 \text{ minute}} \quad \text{Divide 3.2 by 8 to obtain 0.4.}$$

Therefore, the faucet leaks 0.4 ounces per minute. Since minutes is in the denominator, set up a multiplication statement involving this rate so that minutes cancel.

$$\frac{\text{ounces}}{\cancel{\text{minutes}}} \times \cancel{\text{minutes}} = \text{ounces}$$
$$0.4 \times \boxed{} = 7$$

A multiplier is missing, so we divide 7 by 0.4 to find the missing multiplier.

```
        1 7.5
0.4 ) 7.0 0
        4
        3 0
        2 8
          2 0
          2 0
```

It takes the faucet 17.5 minutes to leak 7 ounces. ∎

If you need to see additional examples involving rates, the following list gives the location of the material. Also, solving equivalent rate problems by using proportions is discussed in Section 8.6.

Section	Example
2.5	5–10
4.5	5–14
7.5	4–13

 Try Problems 18 through 21.

If you still need additional guidelines on choosing the correct operation in an application problem, refer to Table 5 at the end of the book, which reviews the operations that correspond to various situations.

 Answers to Try These Problems

1. $1910.25   2. 8.45 m   3. $\frac{2}{3}$ 4. $\frac{13}{20}$ 5. 95 books
6. $360   7. $\frac{87}{2}$ or $43\frac{1}{2}$ ft
8. a. $105\frac{5}{12}$ in   b. $\frac{7367}{12}$ or $613\frac{11}{12}$ sq in   9. 52.43 cm
10. a. 6.35 cm   b. 29.9 cm   11. $174.90   12. $104
13. a. $7\frac{7}{24}$ ft   b. $\frac{129}{304}$ 14. $500   15. 2100 workers
16. $1\frac{1}{2}$ mi per hr   17. $562.50 per oz   18. $210.99
19. 24 gal   20. 270 ft   21. 150 min

EXERCISES 8.4

Solve.

1. How far is it from A to B?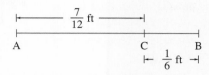

2. How far is it from X to Y?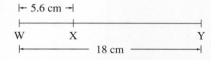

3. A painter used $\frac{3}{4}$ of a bucket of paint. What fraction of the paint remained?

4. Roger, Ted, and Claudia own an apartment building together. Ted owns $\frac{1}{3}$ of the property and Claudia owns $\frac{4}{9}$ of it. What fraction of the property do Ted and Claudia together own?

5. Four hundred forty pounds of candy are packaged in boxes each weighing 0.8 pound. How many boxes of candy can be packaged?

6. A wire that is 3 meters long is cut into 12 equal pieces. How long is each piece?

7. How many 6.5-ounce servings can a chef get from 200 ounces of rice? (Truncate to the nearest whole number.)

8. Mr. Wilson built a fence along one side of his backyard by joining together 15 segments of fencing that were each $7\frac{1}{3}$ feet. What is the total length of this fence?

9. A rectangular piece of paper is $8\frac{1}{2}$ inches wide and 11 inches long.
   a. What is the area of the piece of paper?
   b. What is the perimeter of the piece of paper?

10. The width of a rectangular garden is 12.8 feet. The garden contains 262.4 square feet.
    a. Find the length of the garden.
    b. Find the perimeter of the garden.

11. Find the perimeter of this triangle.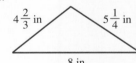

12. The width of a rectangle is 0.6 of the length. The length is 80 meters.
    a. Find the width of the rectangle.
    b. Find the area of the rectangle.

13. Because of inflation, the price of steak has risen from $2.75 per pound to $5.29 per pound. What is the increase in the price per pound?

14. A construction worker earned $17.50 an hour last year. This year she got a raise of $1.65 an hour. What is her hourly wage now?

15. A plane flew to New York and back. Because of weather conditions, the return trip took 1.2 times as long as the trip to New York. The return trip took 6.3 hours. How long was the flight to New York?

16. In a certain state, sales tax is 0.056 of the marked price. The tax paid on a textbook was $1.96. Find the marked price of the textbook.
17. A student got 168 questions correct out of 180.
    a. What fraction of the questions did the student get correct?
    b. How many questions did the student get wrong?
18. Karl traveled a total distance of 5800 miles. Eight-tenths of the distance was by airplane. How far did he travel by plane?
19. A factory employs 3500 workers. There are 500 African-Americans and 800 Hispanics.
    a. What is the ratio of African-Americans to total workers?
    b. What is the ratio of Hispanics to African-Americans?
20. A chemist mixes 2.5 liters of acid with 8 liters of water.
    a. What is the total volume of this solution?
    b. Find the ratio of the acid volume to the total volume. (Round off at 2 decimal places.)
21. Jeff types 208 words in 4 minutes. On average, how many words per minute does he type?
22. Carpeting sells for $23.50 per square yard. What is the cost of $7\frac{1}{4}$ square yards of carpeting? (Round up to the nearest cent.)
23. You buy 8 heads of lettuce that sell for $0.89 per head. What is the total cost of the lettuce?
24. John pays $107.10 for 8.5 yards of fabric. What is the cost per yard?
25. On a map $\frac{3}{4}$ inch represents 25 miles. How many miles do 6 inches represent?
26. A watch loses 2 minutes every 12 hours. At this rate, how much time does the watch lose in 51 hours?

USING THE CALCULATOR #9

ADDING AND SUBTRACTING FRACTIONS AS DECIMALS

Now we look at adding and subtracting fractions on the calculator without using the [a b/c] key. There is a procedure that will work on both scientific and basic calculators. First we use this procedure to compute $\frac{3}{4} + \frac{1}{2}$.

Before beginning, be sure that the memory is clear, that is, that it contains the number 0. If your calculator does not have a clear-the-memory key, you can enter [MR] then enter [M−] to clear the memory. On a scientific calculator, you can clear the memory by entering the number 0, then enter [Min] or [x→M].

To Compute $\frac{3}{4} + \frac{1}{2}$

Enter 3 [÷] 4 [M+] 1 [÷] 2 [M+] [MR]

Result 1.25

Now we do a subtraction problem. We compute $\frac{3}{4} - \frac{1}{2}$.

To Compute $\frac{3}{4} - \frac{1}{2}$

Enter 3 [÷] 4 [M+] 1 [÷] 2 [M−] [MR]

Result 0.25

Observe that you enter the [M−] key after the number you are subtracting. You still enter the [M+] key after the number you are subtracting from.

To compute the above problems on a scientific calculator, you do not need to use the memory. Here we show how to do the addition problem.

Cont. page 347.

ON A SCIENTIFIC CALCULATOR

To Compute $\dfrac{3}{4} + \dfrac{1}{2}$

Enter 3 ÷ 4 + 1 ÷ 2 =

Result 1.25

A subtraction problem would be done similarly by entering the − key instead of the + key.

Calculator Problems

Evaluate each without using the a b/c key. If the decimal representation goes beyond 3 decimal places, round off at 3 decimal places.

1. $\dfrac{2}{5} + \dfrac{3}{8}$
2. $\dfrac{2}{5} - \dfrac{3}{8}$
3. $\dfrac{13}{35} + \dfrac{7}{41}$
4. $\dfrac{13}{35} - \dfrac{7}{41}$
5. $348 + \dfrac{16}{77}$
6. $348 - \dfrac{16}{77}$
7. $\dfrac{881}{23} + 21$
8. $\dfrac{881}{23} - 21$

Convert each mixed numeral to an improper fraction, then use a procedure discussed in this feature to evaluate each. If the decimal representation goes beyond 3 decimal places, round off at 3 decimal places.

9. $4\dfrac{1}{5} + 3\dfrac{2}{3}$
10. $4\dfrac{1}{5} - 3\dfrac{2}{3}$
11. $18\dfrac{5}{13} + 7\dfrac{15}{23}$
12. $18\dfrac{5}{13} - 7\dfrac{15}{23}$

8.5 APPLICATIONS: MORE THAN ONE STEP

OBJECTIVES
- Solve an application requiring more than one step that involves addition and subtraction.
- Solve an application requiring more than one step that involves multiplication and division.
- Solve an application requiring more than one step that involves multiplication and addition.
- Solve an application requiring more than one step that involves a combination of any of the four operations: multiplication, division, addition, or subtraction.

In this section we look at applications that require more than one step to solve. We will present these problems in four different categories as stated in the objectives. When solving more complex problems, it is important to label your work clearly with words so that you understand what it is you have obtained at each stage of the problem. Also, if applicable, drawing a diagram can be very helpful. Here are some examples.

Try These Problems

Solve.

1. Your checking account shows a balance of $538.24. What is your checking account balance after you write checks for $28.30, $95.64, and $260?

2. At the hardware store Henry buys items marked $3.98, $2.76, and $13.89. A tax of $1.34 is added to the bill. Henry gives the clerk $25. How much change does he receive?

3. A triangular plot of land has sides measuring $2\frac{3}{4}$ feet, $4\frac{1}{4}$ feet, and $5\frac{3}{4}$ feet. Mr. Nelson buys 14 feet of fencing to put around the border of the plot. How much fencing is left over?

4. Carmen, Patricia, and Tosca own a small business together. Carmen owns $\frac{1}{4}$ of it, Tosca owns $\frac{2}{5}$ of it, and Patricia owns the rest. What fraction of the business does Patricia own?

1 Applications Involving Addition and Subtraction

EXAMPLE 1 Carol buys the following items at the grocery store:

Milk	$2.39
Potatoes	$4.90
Meat	$9.64

If we assume no tax is added, how much change does she receive from a $20 bill?

SOLUTION First find the total cost of the three items by adding.

```
  2.39
  4.90
+ 9.64
------
 16.93   —— Total cost
            in dollars
```

Carol paid a total of $16.93 for the three items. To find how much change she received from a $20 bill, subtract 16.93 from 20.

```
  20.00
- 16.93
-------
   3.07   —— Change she received
             in dollars
```

Carol received $3.07 in change. ∎

EXAMPLE 2 One rectangle has width $19\frac{1}{3}$ feet and length $20\frac{2}{3}$ feet. Another rectangle has width $16\frac{1}{2}$ feet and length 18 feet. What is the difference in the perimeters of these two rectangles?

SOLUTION First find the perimeter of each of the rectangles.

$$\text{Perimeter of rectangle \#1 (ft)} = 19\frac{1}{3} + 19\frac{1}{3} + 20\frac{2}{3} + 20\frac{2}{3}$$
$$= 78\frac{6}{3}$$
$$= 80$$

$$\text{Perimeter of rectangle \#2 (ft)} = 16\frac{1}{2} + 16\frac{1}{2} + 18 + 18$$
$$= 68\frac{2}{2}$$
$$= 69$$

Now subtract to find the difference in the perimeters.

$$80 - 69 = 11$$

The difference in the two perimeters is 11 feet. ∎

 Try Problems 1 through 4.

2 Applications Involving Multiplication and Division

EXAMPLE 3   Bob can jog 4 miles in 38 minutes. How long does it take him to jog 6 miles?

SOLUTION

METHOD 1   The phrase *4 miles in 38 minutes* is a rate and can be written as a ratio.

$$4 \text{ miles in } 38 \text{ minutes} = \frac{4 \text{ miles}}{38 \text{ minutes}}$$

$$= \frac{2 \text{ miles}}{19 \text{ minutes}}$$

Now set up a multiplication statement using this rate so that units in the denominator cancel.

$$\frac{\text{miles}}{\cancel{\text{minute}}} \times \cancel{\text{minutes}} = \text{miles}$$

$$\frac{2}{19} \times \boxed{} = 6$$

A multiplier is missing, so divide 6 by $\frac{2}{19}$ to find the missing multiplier.

$$6 \div \frac{2}{19} = \frac{\overset{3}{\cancel{6}}}{1} \times \frac{19}{\underset{1}{\cancel{2}}}$$

$$= 57$$

It takes Bob 57 minutes to jog 6 miles.

METHOD 2   The phrase *4 miles in 38 minutes* can be written as a ratio so that miles are in the denominator instead of minutes.

$$4 \text{ miles in } 38 \text{ minutes} = \frac{38 \text{ minutes}}{4 \text{ miles}}$$

$$= \frac{9.5 \text{ minutes}}{1 \text{ mile}}$$

Now set up a multiplication statement using this rate so that units in the denominator cancel.

$$\frac{\text{minutes}}{\cancel{\text{mile}}} \times \cancel{\text{miles}} = \text{minutes}$$

$$9.5 \times 6 = \boxed{}$$

The answer to the multiplication statement is missing so multiply 6 by 9.5 to obtain the answer.

$$9.5 \times 6 = 57.0 = 57$$

It takes Bob 57 minutes to jog 6 miles. ■

Try These Problems

5. Seventy-five nails weigh 15 ounces. What is the weight of 101 nails?

6. The dosage of a certain medicine is $\frac{2}{3}$ ounce for each 100 pounds of body weight. How much medicine is required for a person who weighs 240 pounds?

7. Mohammed wants to cover a rectangular floor with tile that costs $12.70 per square foot. If the floor is 10.2 feet wide and 14.5 feet long, what is the total cost of the tile? Do not include tax.

8. A real estate agent sells a piece of property for $250,000. She receives 0.06 of the selling price. She then must turn over 0.25 of her money to the company she works for. How much money does the agent give to the company?

EXAMPLE 4 The floor of a rectangular room measures 4.75 yards by 6 yards. You want to carpet the floor with carpeting that costs $11.95 per square yard. Without including tax, what is the total cost of carpeting the floor? (Round up the answer to the nearest cent.)

SOLUTION The carpeting sells for $11.95 *per square yard*. We need to find out how many square yards this floor covers; that is, we need to find out the area of the floor.

$$
\begin{aligned}
\text{Area} &= \text{length} \times \text{width} \\
(\text{sq yd}) & \quad (\text{yd}) \quad\quad (\text{yd}) \\
&= 6 \times 4.75 \\
&= 28.5
\end{aligned}
$$

The floor covers an area of 28.5 square yards. Set up a multiplication statement using the rate *$11.95 per square yard*.

$$
\begin{aligned}
\text{total cost (\$)} &= \frac{\text{cost (\$)}}{\text{square yard}} \times \text{square yards} \\
&= 11.95 \times 28.5 \\
&= 340.575 \\
&\approx 340.58
\end{aligned}
$$

The total cost is $340.58. ■

EXAMPLE 5 A machine fills and caps 250 bottles of mineral water each hour. After 24 hours, how many 6-packs of mineral water have been filled and capped?

SOLUTION The phrase *250 bottles every hour* is a rate. Set up a multiplication statement using this rate.

$$
\begin{aligned}
\text{Total number of bottles} &= \frac{\text{bottles}}{\text{hour}} \times \text{number of hours} \\
&= 250 \times 24 \\
&= 6000
\end{aligned}
$$

In 24 hours the machine has filled and capped a total of 6000 bottles. Now we want to know how many 6-packs can be made from the 6000 bottles. That is, we want to separate 6000 into a number of equal parts. The size of each part is 6.

$$
\begin{aligned}
\text{Total quantity} &= \text{size of each part} \times \text{number of equal parts} \\
6000 &= 6 \times \boxed{}
\end{aligned}
$$

A multiplier is missing so we divide 6000 by 6 to find the missing multiplier.

$$6000 \div 6 = 1000$$

In 24 hours 1000 6-packs can be filled and capped. ■

 Try Problems 5 through 8.

Try These Problems

9. Nanda bought 7 yards of fabric selling for $9.69 per yard and $5\frac{1}{2}$ yards of fabric selling for $7.98 per yard. What is the total cost of the fabric? Do not include tax.

10. The marked price of a tennis racket is $89.95. The store adds a sales tax, which is 0.065 of the marked price. What is the total cost of the tennis racket? (Round up the answer to the nearest cent.)

11. Chuck is reading a book that has 860 pages. He read $\frac{1}{4}$ of the book on Saturday and $\frac{2}{5}$ of the book on Sunday. How many pages has he read?

12. A taxi charges $1.20 in addition to $0.85 per mile. What is the cost of a 5-mile taxi ride?

3 Applications Involving Multiplication and Addition

EXAMPLE 6 David bought two shirts at $26.50 each and three ties at $22.85 each. What is the total cost of these items?

SOLUTION First find the total cost of the shirts.

$$\begin{aligned}\text{Total cost of the shirts (\$)} &= \text{cost per shirt (\$)} \times \text{how many shirts} \\ &= 26.50 \times 2 \\ &= 53\end{aligned}$$

Next find the total cost of the ties.

$$\begin{aligned}\text{Total cost of the ties (\$)} &= \text{cost per tie (\$)} \times \text{how many ties} \\ &= 22.85 \times 3 \\ &= 68.55\end{aligned}$$

Last find the total cost of the shirts and the ties.

$$\begin{aligned}\text{Total cost (\$)} &= \text{cost of the shirts (\$)} + \text{cost of the ties (\$)} \\ &= 53 + 68.55 \\ &= 121.55\end{aligned}$$

The total cost of all the items is $121.55. ■

EXAMPLE 7 The marked price of a pair of running shoes is $65.80. The store adds a sales tax, which is 0.065 of the marked price. What is the total cost of the shoes? (Round up the answer to the nearest cent.)

SOLUTION First find the sales tax. The problem states the following.

$$\begin{aligned}\text{Sales tax (\$)} &\text{ is } 0.065 \text{ of the marked price (\$)} \\ &= 0.065 \times 65.80 \\ &= 4.277 \quad \text{Do not round until the end.}\end{aligned}$$

The sales tax is $4.277. The sales tax is added to the marked price to obtain the final selling price.

$$\begin{aligned}\text{Total cost (\$)} &= \text{marked price (\$)} + \text{sales tax (\$)} \\ &= 65.80 + 4.277 \\ &= 70.077 \\ &\approx 70.08\end{aligned}$$

The total cost of the shoes is $70.08. ■

 Try Problems 9 through 12.

EXAMPLE 8 A carpenter wants to put weather stripping around the border of a rectangular window that is 3.4 feet wide and 2.5 feet high. If the weather stripping sells for $2.45 per foot, what is the total cost of the weather stripping? Do not include tax.

SOLUTION The weather stripping sells for $2.45 per foot. First we must find out how many feet are around the border of the window; that is, we want the perimeter of the window.

$$\text{Perimeter (ft)} = 3.4 + 3.4 + 2.5 + 2.5$$
$$= 11.8$$

The total distance around the window is 11.8 feet. Because each foot of weather stripping costs $2.45, we multiply 11.8 by $2.45 to find the total cost.

$$\text{Total cost (\$)} = \text{cost per foot (\$)} \times \text{how many feet}$$
$$= 2.45 \times 11.8$$
$$= 28.91$$

The total cost is $28.91. ∎

EXAMPLE 9 Find the total area of this region.

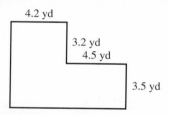

SOLUTION View the region as two rectangular regions.

Observe that the length of rectangle #1 is not given but we can find it by adding 3.2 yards to 3.5 yards. Now find the areas of each of the two rectangular regions.

$$\text{Area of rectangle \#1 (sq yd)} = \text{length (yd)} \times \text{width (yd)}$$
$$= 6.7 \times 4.2$$
$$= 28.14$$

$$\text{Area of rectangle \#2 (sq yd)} = \text{length (yd)} \times \text{width (yd)}$$
$$= 4.5 \times 3.5$$
$$= 15.75$$

Try These Problems

13. Carlos wants to put a fence around a rectangular piece of land that is 8.5 feet wide and 20.8 feet long. The fencing material costs $4.59 per foot. What is the total cost of the fence? (Round up the answer to the nearest cent.)

14. Find **a.** the perimeter and **b.** the area of this region.

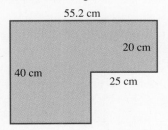

15. Joanne jogged these distances during the last five days:

    Monday   3.5 mi
    Tuesday   4.7 mi
    Wednesday   4.6 mi
    Thursday   5.2 mi
    Friday   6.2 mi

    What is her average daily jogging distance? (Round off at one decimal place.)

To find the total area of the region, add the area of rectangle #1 to the area of rectangle #2.

$$\begin{array}{c}\text{Total area}\\ \text{of the region}\\ \text{(sq yd)}\end{array} = \begin{array}{c}\text{area of}\\ \text{rectangle \#1}\\ \text{(sq yd)}\end{array} + \begin{array}{c}\text{area of}\\ \text{rectangle \#2}\\ \text{(sq yd)}\end{array}$$

$$= 28.14 + 15.75$$
$$= 43.89$$

The total area of the region is 43.89 square yards. ∎

▲ **Try Problems 13 and 14.**

4 Applications Involving Multiplication, Division, Addition, and Subtraction

EXAMPLE 10 Lula bowled four games with the following scores.

$$132 \quad 146 \quad 127 \quad 137$$

What was her average score? (Round off to the nearest whole number.)

SOLUTION Find the sum of the 4 scores.

$$132 + 146 + 127 + 137 = 542$$

Divide the sum 542 by 4 to find the average of the 4 scores.

$$\begin{array}{r} 135.5 \approx 136 \\ 4\overline{)542.0} \\ \underline{4} \\ 14 \\ \underline{12} \\ 22 \\ \underline{20} \\ 2\,0 \\ \underline{2\,0} \end{array}$$

The average is 136. ∎

▲ **Try Problem 15.**

EXAMPLE 11 The Anderson Food Company sells 7.2 ounces of canned apricots for $2.81. The Justrite Food Company sells 6.5 ounces of canned apricots for $2.60. Which company has the better price per ounce? (Round off to the nearest cent.)

SOLUTION Find the price per ounce for each. The lower rate is the better price per ounce.

ANDERSON FOOD COMPANY

7.2 ounces for $2.81

$$= \frac{\$2.81}{7.2 \text{ ounces}}$$

$$\approx \$0.39 \text{ per ounce}$$

Put the price in the numerator and ounces in the denominator to find the price per ounce.

Divide 2.81 by 7.2 to obtain 0.39.

Try These Problems

16. James earned $235 in 20 hours and Rebecca earned $188.10 in 15 hours. Who earns more money per hour?

17. Last week Olga purchased 6 ounces of gold for $2604. This week she purchased 5.5 ounces for $2365. Did the price per ounce increase or decrease during the week and by how much?

18. The price of gasoline increased from $1.20 per gallon to $1.50 per gallon. The increase in price is what fraction of the original price?

19. A utility stock decreased in price from $20 per share to $18\frac{3}{4}$ per share. The decrease in price is what fraction of the original price?

JUSTRITE FOOD COMPANY

$$\frac{6.5 \text{ ounces for } \$2.60}{} = \frac{\$2.60}{6.5 \text{ ounces}}$$

$$= \$0.40 \text{ per ounce}$$

Put the price in the numerator and ounces in the denominator to find the price per ounce. Divide 2.60 by 6.5 to obtain 0.40.

The Anderson Food Company has the better price per ounce at approximately $0.39 per ounce. ∎

 Try Problems 16 and 17.

EXAMPLE 12 During a sale, the price of a pair of shoes is reduced from $85 to $50. The decrease is what fraction of the original price?

SOLUTION First find out how much the price decreased by subtracting $50 from $85.

$$\text{Decrease in price} = \$85 - \$50 = \$35$$

The question is, "The decrease is what fraction of the original price?" This sentence can be translated to a multiplication statement.

The decrease	is	what fraction	of	original price
↓		↓		
35	=	☐	×	85

A multiplier is missing, so divide 35 by 85 to find the missing multiplier.

$$35 \div 85 = \frac{35}{85} = \frac{\cancel{5} \times 7}{\cancel{5} \times 17} = \frac{7}{17}$$

The decrease in price is $\frac{7}{17}$ of the original price. ∎

 Try Problems 18 and 19.

EXAMPLE 13 A large tank is leaking water out of two different holes. The larger hole is leaking water at the rate of 2 gallons each hour. The smaller hole is leaking water at the rate of $\frac{1}{2}$ gallon each hour. After $5\frac{1}{2}$ hours, how many more gallons of water have leaked through the larger hole than through the smaller hole?

SOLUTION First find out how many gallons of water have leaked through each of the holes in $5\frac{1}{2}$ hours.

LARGER HOLE

$$\begin{aligned}
\text{Total gallons} &= \text{gallons per hour} \times \text{number of hours} \\
&= 2 \times 5\frac{1}{2} \\
&= \frac{2}{1} \times \frac{11}{2} \\
&= 11
\end{aligned}$$

▲ Try These Problems

20. Two airplanes started from the same airport and flew in different directions. One plane flew for 3 hours at a rate of 350 miles per hour. The other plane flew for $5\frac{1}{2}$ hours at a rate of 470 miles per hour. How much farther did the second plane fly?

SMALLER HOLE

$$\begin{aligned}\text{Total gallons} &= \text{gallons per hour} \times \text{number of hours} \\ &= \frac{1}{2} \times 5\frac{1}{2} \\ &= \frac{1}{2} \times \frac{11}{2} \\ &= \frac{11}{4} = 2\frac{3}{4}\end{aligned}$$

After $5\frac{1}{2}$ hours, 11 gallons of water have leaked through the larger hole and $2\frac{3}{4}$ gallons have leaked through the smaller hole. Subtract $2\frac{3}{4}$ from 11 to find out how many more gallons have leaked through the larger hole.

$$11 - 2\frac{3}{4} = 10\frac{4}{4} - 2\frac{3}{4} = 8\frac{1}{4}$$

$8\frac{1}{4}$ gallons more have leaked through the larger hole. ■

▲ Try Problem 20.

▲ Answers to Try These Problems

1. $154.30 2. $3.03 3. $1\frac{1}{4}$ ft 4. $\frac{7}{20}$ 5. 20.2 oz
6. $1\frac{3}{5}$ or 1.6 oz 7. $1878.33 8. $3750 9. $111.72
10. $95.80 11. 559 pages 12. $5.45 13. $268.98
14. a. 190.4 cm b. 1708 sq cm 15. 4.8 mi
16. Rebecca at $12.54 per hr 17. decreased by $4 per oz
18. $\frac{1}{4}$ or 0.25 19. $\frac{1}{16}$ or 0.0625 20. 1535 mi

EXERCISES 8.5

Solve.

1. Your checking account shows a balance of $1290.45. After you write checks for $235.15 and $76.98 and you deposit your paycheck of $800.35, what is your balance?

2. How far is it from A to B?

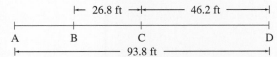

3. Cynthia bought five pens at $0.69 each and two tablets at $1.29 each. She is charged a sales tax that is 0.06 times the marked price. How much change does she receive from a 10-dollar bill? (Round off to the nearest cent.)

4. Lucinda completed $\frac{1}{3}$ of the job yesterday and $\frac{1}{4}$ of the job today. What fraction of the job does she have left to do?

5. A watch loses $3\frac{3}{4}$ minutes every 24 hours. At this rate, how much time does the watch lose in 40 hours?

6. Five pounds of oranges cost $2.45. How much do $3\frac{1}{2}$ pounds cost? (Round up the answer to the nearest cent.)

7. Yesterday Mr. Yee paid $4062.50 for 250 shares of an energy stock. Today he paid $7000 for 400 shares of the same stock. Did the price per share increase or decrease and by how much?

8. Machine A can print 460 pages in 50 minutes and Machine B can print 540 pages in 60 minutes. Which machine is faster; that is, which one prints more pages per minute?

9. The length of a rectangle is 1.8 times as long as the width. The width is 38.7 feet. What is **a.** the perimeter, and **b.** the area of the rectangle?

10. What is the total cost of tiling this floor if tile sells for $3.79 per square foot? (Round up the answer to the nearest cent.)

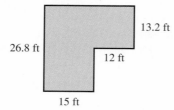

11. A stock increased in value from $35\frac{3}{8}$ dollars to $36\frac{1}{2}$ dollars. The increase in value is what fraction of the original value?

12. The price of a computer decreased from $4000 to $3400. What fraction of the original is the decrease?

13. A Tennessee farmer has a total of 2500 acres to plant some rice, cotton, and corn. He uses $\frac{1}{4}$ of the land for rice, $\frac{3}{5}$ of it for cotton, and the rest for corn. How many acres of corn did he plant?

14. A salesperson earns a commission that is $\frac{6}{100}$ of the first $3000 in sales combined with $\frac{9}{100}$ of the sales over $3000. How much commission does the person receive for sales of $5400?

15. Three suitcases weigh $82\frac{3}{4}$ pounds, $50\frac{1}{2}$ pounds, and $63\frac{1}{4}$ pounds. What is the average weight of the suitcases?

16. Anna has 6 quarters and 12 nickels. How much money does Anna have?

17. Ms. Truong left on a trip at 8 AM in the morning. The odometer on her car read 55140. At 1 PM, when she stopped for lunch, the odometer read 55409. How many miles per hour did she average?

18. Mr. Hsu sells flowers. He put 1545 roses in bunches with 18 roses in each bunch. After making as many bunches as possible, he took home the leftover roses and distributed them equally among his three children. How many roses did each child receive?

19. A parking lot charges $1.50 for the first hour, then $0.65 for each additional half hour. You enter the lot at a 9 AM and leave at 2:30 PM. How much is your parking fee?

20. Inez wants to put a fence around the border of this triangular plot of land. If the fencing material costs $3.85 per foot, find the total cost of the fence.

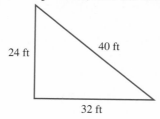

A statistician for a local car dealer made this double-bar graph that indicates the number of new cars sold during each quarter of 1992 and 1993. Use the graph to answer Exercises 21 through 24.

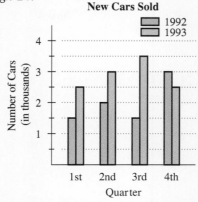

21. How many new cars did the company sell during the 3rd quarter of 1993?
22. How many new cars did the company sell during the 4th quarter of 1992?
23. Find the average number of new cars sold per quarter in 1993.
24. Find the average number of new cars sold per quarter in 1992.

8.6 SOLVING PROPORTIONS (OPTIONAL)

OBJECTIVES
- Test the equality of two fractions, ratios, or rates by using cross products.
- Find the missing number in a proportion.
- Use a proportion to solve an equivalent-rate application problem.

1 Testing the Equality of Two Fractions, Ratios, or Rates

As we discussed in Sections 5.4 and 8.2, one way to compare fractions is to write them with a common denominator and compare the numerators. For example, how do the fractions $\frac{3}{4}$ and $\frac{9}{12}$ compare? The least common denominator is 12, so we write each fraction with denominator 12.

$$\frac{3}{4} = \frac{3 \times 3}{4 \times 3} = \frac{9}{12}$$

$\frac{9}{12}$ is already written with denominator 12.

After writing each with denominator 12, we see they have the same numerator. Therefore, $\frac{3}{4}$ is equal to $\frac{9}{12}$.

There is another way to see that two fractions are equal. Here we look at the same pair of equal fractions.

$$4 \times 9 = 36$$
$$\frac{3}{4} = \frac{9}{12}$$
$$3 \times 12 = 36$$

Try These Problems

Are the given ratios equal? Answer yes or no.

1. $\dfrac{8}{10}, \dfrac{20}{25}$
2. $\dfrac{12}{90}, \dfrac{300}{1500}$
3. $\dfrac{2\frac{1}{2}}{10}, \dfrac{1\frac{1}{2}}{6}$
4. $\dfrac{18.4}{240}, \dfrac{2.3}{30}$

Observe that 3 times 12 is equal to 4 times 9. Both products are 36. These products are often referred to as **cross products.** In general, if two fractions are equal their cross products are equal. Also, conversely, if the cross products of two fractions are equal, the fractions are equal.

Here are more examples of using cross products to test the equality of fractions.

EXAMPLE 1 Are the fractions $\frac{2}{3}$ and $\frac{10}{15}$ equal?

SOLUTION

$$3 \times 10 = 30$$
$$\frac{2}{3} \stackrel{?}{=} \frac{10}{15}$$
$$2 \times 15 = 30$$

The cross products are each equal to 30.
Yes, $\frac{2}{3} = \frac{10}{15}$. ∎

EXAMPLE 2 Are the fractions $\frac{\frac{1}{2}}{35}$ and $\frac{3}{270}$ equal?

SOLUTION

$$35 \times 3 = 105$$
$$\frac{\frac{1}{2}}{35} \stackrel{?}{=} \frac{3}{270}$$
$$\frac{1}{2} \times 270 = 135$$

The cross products are not equal.
No, $\frac{\frac{1}{2}}{35} \neq \frac{3}{270}$. ∎

 Try Problems 1 through 4.

Recall that a ratio or rate is a comparison by division, and thus can be written in a fractional form. For example, the rate 50 miles per hour can be written

$$50 \text{ miles per hour} = \frac{50 \text{ miles}}{1 \text{ hour}}$$

When dealing with ratios or rates there are often units involved, such as miles per hour, persons per square foot, or dollars per year. If you want to test the equality of two ratios or rates, first make sure the units associated with each of the ratios are the same. If the units are the same, then you can use cross products or finding a common denominator to test the equality of the ratios or rates. For example, is the rate of pay, $12 per hour, equal to the rate, $96 every 8 hours?

$$\frac{\$12}{1 \text{ hr}} \stackrel{?}{=} \frac{\$96}{8 \text{ hr}}$$ The rate unit in each case is dollars per hour.

Try These Problems

Are the given rates equal? Answer yes or no.

5. $\dfrac{50 \text{ miles}}{2 \text{ gallons}}$, $\dfrac{75 \text{ miles}}{3 \text{ gallons}}$

6. $\dfrac{680 \text{ trees}}{8 \text{ acres}}$, $\dfrac{510 \text{ trees}}{6 \text{ acres}}$

7. $\dfrac{\$15.60}{24 \text{ ounces}}$, $\dfrac{7.8 \text{ ounces}}{\$12}$

8. $\dfrac{40 \text{ feet}}{2\frac{1}{3} \text{ minutes}}$, $\dfrac{60 \text{ feet}}{3\frac{1}{2} \text{ minutes}}$

The units agree, so we can use cross products to check the equality of these two rates.

$$1 \times 96 = 96$$

$$\dfrac{\text{dollars}}{\text{hour}} \quad \dfrac{12}{1} \stackrel{?}{=} \dfrac{96}{8}$$

$$12 \times 8 = 96$$

Because the units agree and the cross products are equal, the two rates are equal. That is, earning $12 per hour is equivalent to earning $96 every 8 hours.

Here are more examples.

EXAMPLE 3 Are the rates $\dfrac{2.8 \text{ oz}}{5 \text{ servings}}$ and $\dfrac{5.6 \text{ oz}}{10 \text{ servings}}$ equal?

SOLUTION The rate unit in each case is ounces per serving. Therefore, we can use cross products to test the equality of the two rates.

$$5 \times 5.6 = 28$$

$$\dfrac{\text{ounces}}{\text{serving}} \quad \dfrac{2.8}{5} \stackrel{?}{=} \dfrac{5.6}{10}$$

$$2.8 \times 10 = 28$$

The units agree and the cross products are equal.

Yes, $\dfrac{2.8 \text{ oz}}{5 \text{ servings}} = \dfrac{5.6 \text{ oz}}{10 \text{ servings}}$. ■

▲ **Try Problems 5 through 8.**

2 Solving Proportions

A **proportion** is a statement that two fractions, ratios, or rates are equal. For example, the statements

$$\dfrac{1}{2} = \dfrac{50}{100} \quad \text{and} \quad \dfrac{30 \text{ miles}}{1 \text{ gallon}} = \dfrac{450 \text{ miles}}{15 \text{ gallons}}$$

are proportions.

If any one of the four numbers that make up a proportion is missing, we can find that number. Suppose we are missing the number 50 in the first proportion above. If so, we have the following.

$$\dfrac{1}{2} = \dfrac{\Box}{100}$$

To find this missing number we can proceed as follows. Remember we want a number that makes the two fractions equal.

$$\dfrac{1}{2} = \dfrac{\Box}{100}$$

$$1 \times 100 = 2 \times \Box \qquad \text{Set the cross products equal to each other because, if two fractions are equal, their cross products are equal.}$$

Try These Problems

Solve each of the following proportions.

9. $\dfrac{\square}{5} = \dfrac{16}{40}$

10. $\dfrac{14}{49} = \dfrac{\square}{84}$

11. $\dfrac{45}{\square} = \dfrac{60}{1600}$

$100 = 2 \times \square$ — Now this is a multiplication statement with a missing multiplier, so divide 100 by 2 to find the missing number.

$100 \div 2 = \square$

$50 = \square$

The missing number is 50.

The process of finding a missing number in a proportion is called **solving the proportion.** Here are more examples of solving proportions.

EXAMPLE 4 Solve the proportion: $\dfrac{\square}{25} = \dfrac{24}{40}$

SOLUTION

$\dfrac{\square}{25} = \dfrac{24}{40}$

$\square \times 40 = 25 \times 24$ — Set the cross products equal to each other because if two fractions are equal their cross products are equal.

$\square \times 40 = 600$ — This is a multiplication statement with a missing multiplier, so divide 600 by 40 to find the missing number.

$\square = 600 \div 40$

$\square = 15$

The missing number is 15. ■

Try Problems 9 through 11.

EXAMPLE 5 Solve the proportion: $\dfrac{16}{\square} = \dfrac{70}{3.5}$

SOLUTION

$\dfrac{16}{\square} = \dfrac{70}{3.5}$

$16 \times 3.5 = \square \times 70$ — Set the cross products equal to each other.

$56 = \square \times 70$ — A multiplier is missing in this multiplication statement.

$56 \div 70 = \square$ — Divide 56 by 70 to find the missing multiplier.

$$\begin{array}{r} .8 \\ 70\overline{)56.0} \\ \underline{56\ 0} \end{array}$$

The missing number is 0.8. ■

Try These Problems

Solve each of the following proportions.

12. $\dfrac{\boxed{}}{1} = \dfrac{13.5}{9}$

13. $\dfrac{12.18}{\boxed{}} = \dfrac{20.30}{5}$

14. $\dfrac{26}{46.8} = \dfrac{3.9}{\boxed{}}$

15. $\dfrac{2\frac{2}{3}}{90} = \dfrac{\boxed{}}{45}$

EXAMPLE 6 Solve the proportion: $\dfrac{\frac{3}{4}}{200} = \dfrac{12}{\boxed{}}$

SOLUTION

$\dfrac{\frac{3}{4}}{200} = \dfrac{12}{\boxed{}}$

$\dfrac{3}{4} \times \boxed{} = 200 \times 12$ Set the cross products equal to each other.

$\dfrac{3}{4} \times \boxed{} = 2400$ A multiplier is missing in this multiplication statement.

$\boxed{} = 2400 \div \dfrac{3}{4}$ Divide 2400 by $\frac{3}{4}$ to find the missing multiplier.

$\boxed{} = \dfrac{\overset{800}{\cancel{2400}}}{1} \times \dfrac{4}{\underset{1}{\cancel{3}}}$ Dividing by $\frac{3}{4}$ is the same as multiplying by $\frac{4}{3}$.

$\boxed{} = 3200$

The missing number is 3200. ■

▲ Try Problems 12 through 15.

3 Using Proportions to Solve Equivalent-Rate Problems

We have discussed equivalent-rate problems in Sections 2.5, 4.5, 7.5, 8.4, and 8.5. Now we look at solving these types of problems by using proportions. Here are some examples.

EXAMPLE 7 Is 15 pounds of beef for 20 servings equivalent to 9 pounds of beef for 12 servings?

SOLUTION Two rates are given and we want to know if they are equal.

$$\dfrac{15 \text{ lb}}{20 \text{ servings}} \stackrel{?}{=} \dfrac{9 \text{ lb}}{12 \text{ servings}}$$

The rate unit associated with each is pounds per serving. We use cross products to test the equality of the rates.

$\dfrac{\text{pounds}}{\text{serving}}$ $\dfrac{15}{20} \stackrel{?}{=} \dfrac{9}{12}$

$20 \times 9 = 180$

$15 \times 12 = 180$

The units agree and the cross products are equal.

Yes, $\dfrac{15 \text{ lb}}{20 \text{ servings}} = \dfrac{9 \text{ lb}}{12 \text{ servings}}$. ■

EXAMPLE 8 At a college, there were 105 women and 130 men who had health related majors. Are the number of men to the number of women in the ratio 5 to 4?

362 Chapter 8 Problem Solving: Whole Numbers, Fractions, and Decimals

 Try These Problems

Answer each yes or no.

16. A shipment consists of 450 televisions and 600 computers. Is the ratio of the number of televisions to the number of computers 3 to 4?

17. A car is said to get 35 miles per gallon of gas. Is this the same as getting 100 miles on 3 gallons of gas?

18. Jesse earned $132 in 8 hours. Is this rate of pay equivalent to $16.50 per hour?

19. On a map $3\frac{3}{4}$ inches represents 100 miles. At this rate, does $9\frac{3}{8}$ inches represent 250 miles?

Set up a proportion and solve.

20. Penny lost 4 pounds in 6 weeks. At this rate, how many pounds can she lose is 15 weeks?

21. In a certain town, the ratio of the number of African-Americans to the number of Hispanics is 2 to 15. If there are 7800 African-Americans, how many Hispanics are there?

SOLUTION The second sentence translates as follows.

$$\frac{\text{number of men}}{\text{number of women}} \stackrel{?}{=} \frac{5}{4}$$

The number of men must go in the numerator and the number of women in the denominator.

$$\frac{\text{men}}{\text{women}} \quad \frac{130}{105} \stackrel{?}{=} \frac{5}{4}$$

$105 \times 5 = 525$

$130 \times 4 = 520$

After comparing in the correct order, men to women, the cross products are not equal. No, the ratio of men to women is not 5 to 4. ∎

 Try Problems 16 through 19.

EXAMPLE 9 A department store is having a sale on T-shirts. You can buy 2 T-shirts for $15. At this rate, what do 7 T-shirts cost?

SOLUTION Set up a proportion that involves the given rate, *2 T-shirts for $15*. Be sure that units are the same for each rate. We choose to put the number of T-shirts in the numerator and the number of dollars in the denominator.

$$\frac{2 \text{ T-shirts}}{\$15} = \frac{7 \text{ T-shirts}}{\$\boxed{}}$$

The rate unit for each is T-shirts per dollar, so we can write the proportion without the units and solve.

$$\frac{2}{15} = \frac{7}{\boxed{}} \quad \frac{\text{T-shirts}}{\$}$$

$2 \times \boxed{} = 15 \times 7$ Set the cross products equal to each other.

$2 \times \boxed{} = 105$ This is a multiplication statement with a multiplier missing.

$\boxed{} = 105 \div 2$ Divide 105 by 2 to find the missing multiplier.

$\boxed{} = 52.5$

The cost of 7 T-shirts is $52.50. ∎

 Try Problems 20 and 21.

EXAMPLE 10 Kathy bicycles 2.8 miles in 14 minutes. At this rate, how far can she go in 25 minutes?

SOLUTION Set up a proportion that involves the given rate, *2.8 miles in 14 minutes*. Be sure the units are the same for each rate. We choose to put the number of miles in the numerator and the number of minutes in the denominator.

$$\frac{2.8 \text{ miles}}{14 \text{ minutes}} = \frac{\boxed{} \text{ miles}}{25 \text{ minutes}}$$

▲ Try These Problems

Set up a proportion and solve.

22. Fresh mushrooms are selling for $1.35 per 0.5 pound. At this rate, how many pounds of mushrooms can you buy for $21.60?

23. The dosage for a medication is 3.3 centigrams for every 60 pounds of body weight. At this rate, how many centigrams of this medication are required for a child who weighs 55 pounds?

The rate unit for each is miles per minute, so we can write the proportion without the units and solve.

$$\frac{2.8}{14} = \frac{\boxed{}}{25} \quad \frac{\text{miles}}{\text{minutes}}$$

$2.8 \times 25 = 14 \times \boxed{}$ Set the cross products equal to each other.

$70 = 14 \times \boxed{}$ This is a multiplication statement with a multiplier missing.

$70 \div 14 = \boxed{}$ Divide 70 by 14 to find the missing multiplier.

$5 = \boxed{}$

Kathy can bicycle 5 miles in 25 minutes. ■

▲ Try Problems 22 and 23.

EXAMPLE 11 The label on a bottle of rug shampoo says to mix shampoo and water in the ratio 1 to 12. How much water should be mixed with $1\frac{1}{3}$ cups of shampoo?

SOLUTION Set up a proportion that involves the given ratio, $\frac{1}{12}$. The first sentence implies that the numerator, 1, corresponds to the amount of shampoo, and the denominator, 12, corresponds to the amount of water.

$$\frac{\text{cups of shampoo}}{\text{cups of water}} = \frac{1}{12}$$

We are given that there are $1\frac{1}{3}$ cups of shampoo and this quantity goes in the numerator.

$$\frac{1\frac{1}{3} \text{ cups}}{\boxed{} \text{ cups}} = \frac{1}{12} \quad \text{Units cancel}$$

Because the units in the numerator and denominator are the same in the first ratio, they cancel and that ratio is really a fraction with no units, as is the ratio $\frac{1}{12}$.

Now we can write the proportion without any units and solve.

$$\frac{1\frac{1}{3}}{\boxed{}} = \frac{1}{12}$$

$1\frac{1}{3} \times 12 = \boxed{} \times 1$ Set the cross products equal to each other.

$\frac{4}{3} \times \frac{\cancel{12}^{4}}{1} = \boxed{} \times 1$ Convert $1\frac{1}{3}$ to $\frac{4}{3}$ and multiply by 12.

$16 = \boxed{} \times 1$ This is a multiplication statement with a missing multiplier.

$16 \div 1 = \boxed{}$ Divide 16 by 1 to find the missing multiplier.

$16 = \boxed{}$

16 cups of water need to be mixed with the $1\frac{1}{3}$ cups of shampoo. ■

364 Chapter 8 Problem Solving: Whole Numbers, Fractions, and Decimals

▲ Try These Problems
Set up a proportion and solve.

24. A recipe calls for $6\frac{2}{3}$ cups of flour for 20 servings. At this rate, how many cups of flour are needed for 12 servings?

25. A chemist is mixing acid and water in the ratio 3 to 20. How many ounces of water does she need to mix with $8\frac{3}{4}$ ounces of acid?

▲ Try Problems 24 and 25.
▲ Answers to Try These Problems

1. yes 2. no 3. yes 4. yes 5. yes 6. yes 7. no
8. yes 9. 2 10. 24 11. 1200 12. 1.5 13. 3 14. 7.02
15. $1\frac{1}{3}$ 16. yes 17. no 18. yes 19. yes 20. 10 lb
21. 58,500 22. 8 lb 23. 3.025 cg 24. 4 c 25. $58\frac{1}{3}$ oz

EXERCISES 8.6

Are the given ratios or rates equal? Answer yes or no.

1. $\frac{2}{3}, \frac{4}{5}$
2. $\frac{3}{13}, \frac{15}{65}$
3. $\frac{280}{400}, \frac{350}{500}$
4. $\frac{12}{1800}, \frac{10}{755}$
5. $\frac{4}{24\frac{2}{3}}, \frac{6}{37}$
6. $\frac{8\frac{1}{3}}{125}, \frac{3\frac{2}{3}}{55}$
7. $\frac{1.2}{5}, \frac{6}{26}$
8. $\frac{0.08}{300}, \frac{2}{7500}$
9. $\frac{15 \text{ miles}}{4 \text{ hours}}, \frac{60 \text{ miles}}{16 \text{ hours}}$
10. $\frac{200 \text{ bricks}}{4 \text{ feet}}, \frac{1500 \text{ bricks}}{30 \text{ feet}}$
11. $\frac{2 \text{ pounds}}{5 \text{ weeks}}, \frac{4 \text{ weeks}}{10 \text{ pounds}}$
12. $\frac{24 \text{ persons}}{3 \text{ square miles}}, \frac{200 \text{ persons}}{25 \text{ square miles}}$
13. $\frac{\$24}{1 \text{ hour}}, \frac{\$72}{2.5 \text{ hours}}$
14. $\frac{\$1.53}{0.75 \text{ pound}}, \frac{\$4.16}{2 \text{ pounds}}$
15. $\frac{1\frac{1}{4} \text{ inches}}{20 \text{ miles}}, \frac{5 \text{ miles}}{80 \text{ inches}}$
16. $\frac{6\frac{3}{4} \text{ feet}}{15 \text{ seconds}}, \frac{3 \text{ feet}}{6\frac{2}{3} \text{ seconds}}$

Solve each of the following proportions.

17. $\frac{\Box}{24} = \frac{5}{30}$
18. $\frac{16}{56} = \frac{\Box}{21}$
19. $\frac{450}{\Box} = \frac{750}{100}$
20. $\frac{27}{3600} = \frac{18}{\Box}$
21. $\frac{\Box}{36} = \frac{6}{40}$
22. $\frac{25}{\Box} = \frac{300}{33}$
23. $\frac{\Box}{1} = \frac{9}{2.4}$
24. $\frac{0.04}{50} = \frac{\Box}{750}$
25. $\frac{18}{\Box} = \frac{7.5}{3}$
26. $\frac{1000}{0.2} = \frac{6}{\Box}$
27. $\frac{3\frac{1}{3}}{12} = \frac{\Box}{126}$
28. $\frac{9}{\Box} = \frac{\frac{3}{20}}{7}$
29. $\frac{\Box}{100} = \frac{\frac{8}{3}}{4}$
30. $\frac{20}{1\frac{1}{2}} = \frac{70}{\Box}$

Answer each yes or no.

31. The property tax on a house worth $50,000 is $350. Is this the same rate as paying $420 in property tax for a house worth $60,000?

32. A painter mixes 6 gallons of red paint with 10 gallons of blue paint. Is he mixing red and blue paint in the ratio 3 to 5?

33. Five pounds of carrots cost $3.45. Is this rate equivalent to 8 pounds for $4.92?

34. Mr. Clinton drove 225 miles in 4.5 hours. Is this rate equivalent to 50 miles per hour?

35. A recipe that serves 6 persons calls for $\frac{3}{4}$ cup of sugar. Is the ratio of the cups of sugar to the number of servings 8 to 1?

36. Tipper drives 320 miles in $\frac{1}{2}$ day. At this rate, will she drive 3840 miles in 6 days?

Set up a proportion and solve.

37. An inspector finds 3 defective light bulbs in a sample of 400 bulbs. At this rate, how many defective bulbs can be expected in a shipment of 10,000 bulbs?
38. The administration wants to keep the ratio of teachers to students at 1 to 25. How many students can be served with 26 teachers?
39. The ratio of women to men in the fire department is 3 to 40. If there are 18 women in the department, how many men are there?
40. A car went 840 miles on 20 gallons of gas. How many miles per gallon is this?
41. If 5 pounds of potatoes cost $1.25, how many pounds can be bought for $12?
42. A cable that is 2 meters in length weighs 5.4 pounds. At this rate, what is the weight of 3.6 meters of this cable?
43. One dose of a medication is 1.5 ounces for every 40 pounds of body weight. At this rate, how many ounces are required for a dose of this medication for a woman who weighs 136 pounds?
44. A carpenter is paid $28.75 per hour. How many hours does he need to work to earn $4025?
45. Chuck, an architect, made a drawing with a scale of 1 inch representing 8 feet. How wide is a room that measures $5\frac{1}{2}$ inches on the drawing?
46. A bartender mixes champagne and orange juice in the ratio 2 to 3. How much orange juice does she mix with $8\frac{1}{2}$ ounces of champagne?

CHAPTER 8 SUMMARY

KEY WORDS AND PHRASES

number line [8.1]
distance [8.1]
converting numbers [8.2]
comparing numbers [8.2]
increased by [8.3]
sum [8.3]
decreased by [8.3]
difference [8.3]
product [8.3]
ratio [8.3, 8.6]
divided by [8.3]
divided into [8.3]
area [8.4]
perimeter [8.4]
rate [8.4, 8.5, 8.6]
cross products [8.6]
proportion [8.6]
solving a proportion [8.6]

SYMBOLS

$0.9 < 5$ means 0.9 is less than 5
$\frac{3}{4} > \frac{1}{4}$ means $\frac{3}{4}$ is more than $\frac{1}{4}$
$19.\overline{6}$ means $19.66666666\ldots$

IMPORTANT RULES

Converting Numbers From One Form to Another [8.2]

A summary of the different techniques for changing a number from one form to another is on pages 322 and 323.

Area of a Rectangle [8.4]

The area of a rectangle measures the number of square units inside the rectangle. The area can be found by multiplying the length by the width.
Area = length × width

Perimeter of a Rectangle [8.4]
The perimeter of a rectangle is the distance all the way around it. The perimeter can be found by adding the lengths of the four sides.
Perimeter = length + length + width + width

Choosing the Correct Operation [8.4]
Table 5 at the end of the book reviews the operations that correspond to various situations.

CHAPTER 8 REVIEW EXERCISES

Give the number associated with the points A and B.

1.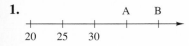
2.
3.
4.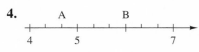
5.
6.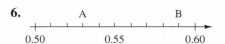

7. Which of these numbers are equal to $\frac{9}{50}$?
   a. $5.\overline{5}$ b. 0.18 c. 0.1800 d. $\frac{27}{150}$
8. Which of these numbers are equal to 7.8?
   a. 7.08 b. $\frac{39}{5}$ c. $7\frac{8}{100}$ d. $7\frac{32}{40}$
9. Which of these numbers are equal to 25?
   a. $\frac{3}{75}$ b. $\frac{75}{3}$ c. 25.00 d. 0.25

Use the symbol $<$, $=$, or $>$ to compare each pair of numbers.

10. $13\frac{4}{9}\underset{(<,\,=,\,>)}{\underline{\quad?\quad}}13.45$
11. $8.2\overline{3}\underset{(<,\,=,\,>)}{\underline{\quad?\quad}}8.\overline{23}$
12. $\frac{5}{24}\underset{(<,\,=,\,>)}{\underline{\quad?\quad}}\frac{1}{4}$
13. $\frac{7}{16}\underset{(<,\,=,\,>)}{\underline{\quad?\quad}}0.4375$

List from smallest to largest.

14. $\frac{7}{8}$, $0.8\overline{7}$, $\frac{31}{36}$
15. $4\frac{2}{5}$, 0.44, 0.044, $\frac{4}{90}$
16. $8\frac{5}{6}$, $8\frac{2}{3}$, $8\frac{3}{4}$
17. 0.58, $\frac{5}{8}$, 0.5625

Solve.

18. Find the difference between 32 and 4.8.
19. Find the ratio of $3\frac{1}{5}$ to 8.
20. Divide $4\frac{1}{2}$ into $\frac{2}{3}$.
21. Find $\frac{5}{6}$ of $7\frac{1}{2}$.
22. 0.075 of what number is 30?
23. What number increased by $22\frac{1}{3}$ yields 50?
24. 12 is what decimal of 750?
25. The profit for a company was $450,000 last year. This year the profit is 0.7 times as large as last year. Find the profit this year.

26. Pedro takes home $1512 each month. He spends $630 on rent. What fraction of his take-home pay is spent on rent? Simplify your answer.
27. A painter has $6\frac{1}{2}$ gallons of paint. He needs an additional $3\frac{3}{4}$ gallons to do his job. How much paint does it take to do the job?
28. A bakery makes a loaf of bread that is 34 centimeters long. How many slices can be cut from one loaf if each slice is 1.7 centimeters wide?
29. Sandra spent $\frac{2}{3}$ of her vacation in Hawaii. She spent the rest in Vancouver. What fraction of her vacation was spent in Vancouver?
30. A picture is 28.2 inches wide and 36.5 inches long.
    a. Find the perimeter around the picture.
    b. Find the area of the picture.
31. Water enters a tank at the rate of 4.5 gallons per hour. How much water has entered the tank after 7.5 hours?
32. A fish swims 210 yards in 50 seconds. How many yards per second is this?
33. A rectangular plot of land is 34.6 feet wide and 60.8 feet long. Jack buys 200 feet of fencing to put around the lot. How much fencing is left over?
34. Janet buys $4\frac{1}{3}$ yards of fabric at $12.60 per yard and 5 yards of fabric at $7.25 per yard. The store adds a sales tax that is 0.058 of the marked price. Find the total cost of the fabric. (Round up at two decimal places.)
35. Due to a sale the price of a television decreased from $600 to $450. The decrease in price is what fraction of the original price?
36. The width of a rectangle is 6.8 feet less than the length. If the length is 12 feet, what is the area of the rectangle?
37. A rectangular mirror is 4.5 feet wide and 3 feet high. You put a frame around the mirror that costs $5.75 per foot. What is the total cost of the frame?
38. Find the total cost of carpeting this floor if the carpeting costs $17.95 per square yard.

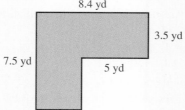

39. Bob saved the following amounts the last four months:
    April $300
    May $260
    June $400
    July $150
    On average, how much did he save each month?
40. A parking lot charges $1.00 for the first half hour, then $0.80 for each additional half hour. You enter the garage at 11 AM and leave at 2 PM. How much is your parking fee?
41. A watch loses 6.5 seconds every 5 hours. At this rate, how much time does the watch lose in 24 hours?
42. Blanca types 276 words in 5 minutes and Natalie types 222 words in 4 minutes. Who types faster and at what rate?

43. If Stephanie can walk 13 miles in 3 hours, how long will it take her to walk 52 miles?
44. The area of a rectangle is 32 square yards and the length is $13\frac{1}{3}$ yards. Find the perimeter of the rectangle.

The rainfall for the first 5 months of a year for a certain city is given in the bar graph. Use the graph to answer Exercises 45 through 50.

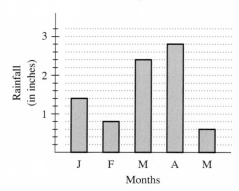

Monthly Rainfall

45. How much rain fell in February?
46. How much rain fell in March?
47. What was the total rainfall for the 5-month period?
48. What was the average rainfall per month for the 5-month period?
49. The amount of rainfall in May is what fraction of the total rainfall for the 5-month period?
50. Which month had the least amount of rain?

In Your Own Words

Write complete sentences to discuss each of the following. Support your comments with examples or pictures, if appropriate.

51. Discuss at least two ways of showing that the decimal 12.4 is equal to the fraction $\frac{620}{50}$.
52. Matthew is building a fence around a rectangular plot of land. Discuss whether he should find the perimeter or the area of this rectangular plot before buying the fencing materials.
53. A committee consists of 4 men and 6 women. Discuss at least two ways to compare the number of women and men by using addition or subtraction. Also, discuss at least two ways to compare the number of men and women by using division or a ratio.
54. Discuss how you can decide, before dividing, how the answers to these two problems should compare with the number 200.
    a. $200 \div 2.5$ b. $200 \div 0.25$
55. Discuss how you would decide which one of the following is the better buy.
    a. 14.5-ounce can of beans for $1.19
    b. 16-ounce can of the same beans for $1.29

CHAPTER 8 PRACTICE TEST

Give the numbers associated with the points A and B.

1.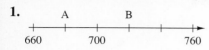
2.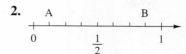
3.
   ```
   +----A--B----+
   9.0  9.4   11.0
   ```

4. Which of these numbers are equal to $9\frac{7}{25}$?
   a. 9.14   b. 9.357   c. $\frac{232}{25}$ d. $8\frac{32}{25}$

Use the symbol <, =, or > to compare each pair of numbers.

5. $12\frac{5}{8}$, 12.625
6. 0.65, $0.\overline{6}$
7. List from smallest to largest: $9\frac{1}{4}$, $\frac{46}{5}$, 9.14
8. List from smallest to largest: $\frac{83}{10}$, $7\frac{4}{3}$, 0.95, 9.2

Solve.

9. Find the sum of $5\frac{1}{6}$ and $2\frac{2}{3}$.
10. Find the ratio of 5.2 to 12.5.
11. Sales tax is 0.058 of the marked price. If the marked price of a stereo is $450, how much is the sales tax?
12. A chef prepared 57 ounces of rice. How many $\frac{3}{4}$-ounce portions can he serve?
13. A rectangular sheet of metal measures 7.9 feet by 18.7 feet. What is the area of the sheet of metal?
14. Jennifer filled $\frac{2}{5}$ of a bucket with water. What fraction of the bucket has no water?
15. Cindy studied for $5\frac{3}{4}$ hours and Roger studied for $7\frac{1}{4}$ hours. How much longer did Roger study than Cindy?
16. Ms. Wilson pays $\frac{1}{3}$ of her take-home pay in mortgage payments each month. If her monthly mortgage payment is $725.50, how much is her take-home pay?
17. Four pounds of coffee cost $21. How much of this coffee can you buy for $45.15?
18. A watch loses 4 seconds every 6 hours. At this rate, how many seconds does the watch lose in 10 hours?
19. The area of a rectangle is 170 square meters. The width is 8.5 meters. Find the perimeter of this rectangle.
20. The price of a certain stock increased from $13\frac{1}{8}$ to $15\frac{3}{4}$. The increase in price is what fraction of the original price?
21. You buy 6 pairs of socks at $5.75 each and 5 scarves at $12.50 each. If $5.82 in tax is added to the cost, how much change do you receive from $120?
22. What is the total cost of tiling this floor space if the tile costs $8.60 per square foot?

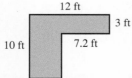

CHAPTER 9

Percent

9.1 CONVERTING PERCENTS TO DECIMALS AND FRACTIONS

OBJECTIVES
- Recognize whether a percent is less than, equal to, or more than 1.
- Convert a percent to a decimal.
- Convert a percent to a fraction.

1 Introduction

Most people encounter **percents** in their everyday life. You probably have seen a sign similar to this one in a department store.

SWEATER SALE
40% OFF

The sign indicates that the store management has reduced the price of some sweaters. Percents are being used to tell the customer how much the price has been reduced. We will take a closer look at this and other applications later in the chapter.

Why do we need percents? Percents were introduced in an attempt to standardize the arithmetic numbers by thinking of all of them with the same denominator so that they are easier to compare. The number 100 was chosen as the common denominator. Why was 100 chosen? Perhaps because it is convenient to work with and fits well with the decimal system. The symbol, %, read "percent," was introduced and is used in the place of the fraction bar and the denominator 100.

$$\frac{50}{100} = 50\%$$

Observe that the symbol % was chosen very cleverly. The slash (/) between the two 0s symbolizes the fraction bar or division. The 0s on each side of the slash symbolize the two 0s in the denominator 100.

The word *percent* comes from the Latin phrase *per centum*.

Per means **For each** or **Divide by**

Centum means **100** or **One hundred**

Therefore, percent means *per 100* or *divide by 100*.

The percent, 50%, is simply another way to write the number $\frac{50}{100}$ or 0.50 or $\frac{1}{2}$. Study the following chart to see how $\frac{1}{2}$ or .50 changes to percent.

Fraction to Percent	Decimal to Percent
$\frac{1}{2}$ one-half	0.5 five-tenths
$= \frac{50}{100}$ fifty hundredths	$= 0.50$ fifty hundredths
$= 50$ per 100	$= 50$ per 100
$= 50$ percent	$= 50$ percent
$= 50\%$	$= 50\%$

All numbers can be written in percent form. We now have three ways to write each number. Each number can be written as a fraction, as a decimal, or as a percent. A few examples are given in the table below.

Fraction	Decimal	Percent
$\frac{1}{100}$	0.01	1%
$\frac{1}{10} = \frac{10}{100}$	$0.1 = 0.10$	10%
$\frac{1}{4} = \frac{25}{100}$	0.25	25%
$\frac{1}{2} = \frac{50}{100}$	$0.5 = 0.50$	50%
$\frac{3}{4} = \frac{75}{100}$	0.75	75%
$1 = \frac{1}{1} = \frac{100}{100}$	$1 = 1.00$	100%
$1\frac{1}{2} = \frac{3}{2} = \frac{150}{100}$	$1.5 = 1.50$	150%
$2 = \frac{2}{1} = \frac{200}{100}$	$2 = 2.00$	200%
$3\frac{1}{4} = \frac{13}{4} = \frac{325}{100}$	3.25	325%

Observe from the previous table that the number 1 is the same as 100%, that a number less than 1 is less than 100%, and a number more than 1 is more than 100%.

Try These Problems

Indicate whether each percent is less than, equal to, or more than the number 1.
1. 35%
2. 180%
3. 6%
4. 100%

Convert each percent to a decimal.
5. 4%
6. 39%
7. 100%
8. 375%

A Useful Observation
1. 100% = 1
2. A percent that is less than 100% represents a number less than 1.
3. A percent that is more than 100% represents a number more than 1.

 Try Problems 1 through 4.

2 Converting Percents to Decimals

The percent symbol, %, means *per 100* or *divide by 100*. Therefore to convert a percent to a decimal, drop the percent symbol and divide by 100. For example,

$$50\% = 50 \div 100 = \frac{50}{100} = 0.50$$

Converting a Percent to a Decimal
To convert a percent to a decimal, drop the percent symbol and divide by 100.

Here are more examples that illustrate how to use this rule to convert percents to decimals.

EXAMPLE 1 Convert 3% to a decimal.
SOLUTION

$3\% $	$= 3 \div 100$	% means divided by 100.
	$= 3. \div 100$	Because 3 is a whole number, the decimal point is placed at the right end. 3. = 3
	$= .03$	Dividing a decimal by 100 moves the decimal point 2 places to the left.
	$= .03$ or 0.03	The digit 0 before the decimal point is not necessary but helps bring attention to the decimal point. ∎

EXAMPLE 2 Convert 425% to a decimal.
SOLUTION

425%	$= 425 \div 100$	% means divided by 100.
	$= 425. \div 100$	
	$= 4.25$	Dividing the decimal by 100 moves the decimal point 2 places to the left. ∎
	$= 4.25$	

 Try Problems 5 through 8.

9.1 Converting Percents to Decimals and Fractions

 Try These Problems

Convert each percent to a decimal.
9. 8.05%
10. 0.25%
11. 62.5%
12. 200.8%
13. $\frac{1}{2}$%
14. $\frac{5}{80}$%

EXAMPLE 3 Convert 5.2% to a decimal.

SOLUTION

$5.2\% = 5.2 \div 100$ % means divided by 100.

$ = .052$ Dividing a decimal by 100 moves the decimal point 2 places to the left.

$ = .052$ or 0.052 The digit 0 before the decimal point is not necessary but helps bring attention to the decimal point. ∎

EXAMPLE 4 Convert 0.06% to a decimal.

SOLUTION

$0.06\% = 0.06 \div 100$ % means divided by 100.

$ = .0006$ Dividing a decimal by 100 moves the decimal point 2 places to the left.

$ = .0006$ or 0.0006 The digit 0 before the decimal point is not necessary but helps bring attention to the decimal point. ∎

 Try Problems 9 through 12.

EXAMPLE 5 Convert $\frac{1}{4}$% to a decimal.

SOLUTION Be careful here. The percent, $\frac{1}{4}$%, is not the same as the number $\frac{1}{4}$.

$\frac{1}{4}\% = 0.25\%$ Convert $\frac{1}{4}$ to a decimal before dropping the percent symbol and dividing by 100.

$$4\overline{)1.00}^{\,.25} \rightarrow \frac{1}{4} = 0.25$$
$$\underline{8}$$
$$20$$
$$\underline{20}$$

$\phantom{\frac{1}{4}\%} = 0.25 \div 100$ Drop the percent symbol and divide by 100.

$\phantom{\frac{1}{4}\%} = .0025$ Dividing a decimal by 100 moves the decimal point 2 places to the left.

$\phantom{\frac{1}{4}\%} = .0025$ or 0.0025 ∎

 Try Problems 13 and 14.

374 Chapter 9 Percent

 Try These Problems

Convert each percent to a decimal.
15. $12\frac{1}{2}\%$
16. $4\frac{7}{8}\%$
17. $66\frac{2}{3}\%$
18. $5\frac{1}{8}\%$
19. $\frac{1}{3}\%$
20. $4\frac{7}{11}\%$

EXAMPLE 6 Convert $6\frac{3}{8}\%$ to a decimal.
SOLUTION

$6\frac{3}{8}\% = 6.375\%$ Convert $6\frac{3}{8}$ to a decimal before dropping the percent symbol and dividing by 100.

$$\begin{array}{r} .375 \\ 8\overline{)3.000} \\ \underline{2\ 4} \\ 60 \\ \underline{56} \\ 40 \\ \underline{40} \end{array} \rightarrow \frac{3}{8} = 0.375$$

$= 6.375 \div 100$ Drop the percent symbol and divide by 100.

$= .06375$ Dividing a decimal by 100 moves the decimal point 2 places to the left.

$= .06375$ or 0.06375

 Try Problems 15 and 16.

EXAMPLE 7 Convert $33\frac{1}{3}\%$ to a decimal.
SOLUTION
METHOD 1

$33\frac{1}{3}\% = 33.\overline{3}\%$ Convert $33\frac{1}{3}$ to a decimal before dropping the percent symbol and dividing by 100.

$$\begin{array}{r} .333 \\ 3\overline{)1.000} \\ \underline{9} \\ 10 \\ \underline{9} \\ 10 \\ \underline{9} \\ 1 \end{array} \rightarrow \begin{array}{l} \frac{1}{3} = 0.33333\ldots \\ \frac{1}{3} = 0.\overline{3} \end{array}$$

$= 33.\overline{3} \div 100$ Drop the percent symbol and divide by 100.

$= .33\overline{3}$ Dividing a decimal by 100 moves the decimal point 2 places to the left.

$= 0.\overline{3}$

METHOD 2 When the percent converts to a repeating decimal, we often write it as a decimal with fraction rather than a repeating decimal.

$$33\frac{1}{3}\% = \frac{33\frac{1}{3}}{100} = 0.33\frac{1}{3}$$

Repeating decimals like $0.\overline{3}$ and decimals with fractions like $0.33\frac{1}{3}$ are not convenient to work with. As you will see later in the chapter, it is better to convert percents like $33\frac{1}{3}\%$ to fractions rather than decimals.

 Try Problems 17 through 20.

Try These Problems

Convert each percent to a fraction.
21. 2%
22. 80%
23. 420%

3 Converting Percents to Fractions

To convert a percent to a decimal, we dropped the percent symbol and divided by 100. This same rule can be used to convert a percent to a fraction. For example,

$$50\% = 50 \div 100 = \frac{50}{100} = \frac{1}{2}$$

> ### Converting a Percent to a Fraction
> To convert a percent to a fraction, drop the percent symbol and divide by 100.

Here are some examples to illustrate how to use this rule.

EXAMPLE 8 Convert 35% to a fraction.
SOLUTION

$$35\% = 35 \div 100 \quad \text{Drop the percent symbol and divide by 100.}$$
$$= \frac{35}{100}$$
$$= \frac{\cancel{5} \times 7}{\cancel{5} \times 20} \quad \text{Reduce the fraction to lowest terms.}$$
$$= \frac{7}{20} \blacksquare$$

EXAMPLE 9 Convert 250% to a fraction.
SOLUTION

$$250\% = 250 \div 100 \quad \text{Drop the percent symbol and divide by 100.}$$
$$= \frac{250}{100}$$
$$= \frac{25 \times \cancel{10}}{10 \times \cancel{10}}$$
$$= \frac{25}{10}$$
$$= \frac{5 \times \cancel{5}}{2 \times \cancel{5}} \quad \text{Reduce the fraction to lowest terms.}$$
$$= \frac{5}{2} \text{ or } 2\frac{1}{2} \quad \text{The answer may be expressed as an improper fraction or as a mixed numeral.}$$

 Try Problems 21 through 23.

Try These Problems

Convert each percent to a fraction.
24. $\frac{1}{4}\%$
25. $\frac{3}{10}\%$
26. $3\frac{1}{5}\%$
27. $6\frac{2}{3}\%$
28. $12\frac{1}{2}\%$
29. $10\frac{2}{3}\%$

EXAMPLE 10 Convert $\frac{3}{4}\%$ to a fraction.

SOLUTION Be careful here. The percent, $\frac{3}{4}\%$, is not the same as the number $\frac{3}{4}$.

$$\frac{3}{4}\% = \frac{3}{4} \div 100 \qquad \text{Drop the percent symbol and divide by 100.}$$

$$= \frac{3}{4} \times \frac{1}{100} \qquad \text{Dividing by 100 is the same as multiplying by } \tfrac{1}{100}.$$

$$= \frac{3}{400} \quad\blacksquare$$

 Try Problems 24 and 25.

EXAMPLE 11 Convert $8\frac{1}{3}\%$ to a fraction.

SOLUTION

$$8\frac{1}{3}\% = 8\frac{1}{3} \div 100 \qquad \text{Drop the percent symbol and divide by 100.}$$

$$= \frac{25}{3} \times \frac{1}{100} \qquad \text{Dividing by 100 is the same as multiplying by } \tfrac{1}{100}.$$

$$= \frac{\cancel{25}^{1}}{3} \times \frac{1}{\cancel{25} \times 4}$$

$$= \frac{1}{12} \quad\blacksquare$$

 Try Problems 26 and 27.

EXAMPLE 12 Convert $66\frac{2}{3}\%$ to a fraction.

SOLUTION

$$66\frac{2}{3}\% = 66\frac{2}{3} \div 100 \qquad \text{Drop the percent symbol and divide by 100.}$$

$$= \frac{\cancel{200}^{2}}{3} \times \frac{1}{\cancel{100}_{1}} \qquad \text{Dividing by 100 is the same as multiplying by } \tfrac{1}{100}.$$

$$= \frac{2}{3} \quad\blacksquare$$

Each of the percents in Examples 11 and 12 convert to a repeating decimal or to a decimal with fraction. That is, $8\frac{1}{3}\% = 0.08\overline{3}$ or $0.08\frac{1}{3}$ and $66\frac{2}{3}\% = 0.\overline{6}$ or $0.66\frac{2}{3}$. These percents can also be approximated as terminating decimals, $8\frac{1}{3}\% \approx 0.083$ and $66\frac{2}{3}\% \approx 0.66$ or 0.67. When you work with percents like $8\frac{1}{3}\%$ and $66\frac{2}{3}\%$ be sure to convert them to fractions if you want an exact answer. Using the decimal approximations will result in only an approximate answer. Also, do not try to calculate with a repeating decimal or a decimal with fraction because they are much harder to work with than fractions.

 Try Problems 28 and 29.

9.1 Converting Percents to Decimals and Fractions   **377**

 Try These Problems

Convert each percent to a fraction
30. 2.5%
31. 5.6%
32. 87.5%
33. 0.6%

EXAMPLE 13   Convert 3.5% to a fraction.
SOLUTION

$$3.5\% = 3.5 \div 100$$ Drop the percent symbol and divide by 100.

$$= .035$$ Dividing a decimal by 100 moves the decimal point 2 places to the left.

$$= 0.035$$ This decimal is 35 thousandths.

$$= \frac{35}{1000}$$ Convert the decimal to a fraction.

$$= \frac{\cancel{5} \times 7}{\cancel{5} \times 200}$$ Reduce the fraction to lowest terms.

$$= \frac{7}{200} \ \blacksquare$$

 Try Problems 30 and 31.

EXAMPLE 14   Convert 37.5% to a fraction.
SOLUTION

$$37.5\% = 37.5 \div 100$$ Drop the percent symbol and divide by 100.

$$= .375$$ Dividing a decimal by 100 moves the decimal point 2 places to the left.

$$= 0.375$$ This decimal is 375 thousandths.

$$= \frac{375}{1000}$$ Convert the decimal to a fraction.

$$= \frac{25 \times 15}{25 \times 40}$$

$$= \frac{15}{40}$$

$$= \frac{3 \times \cancel{5}}{\cancel{5} \times 8}$$ Reduce the fraction to lowest terms.

$$= \frac{3}{8} \ \blacksquare$$

 Try Problems 32 and 33.

EXAMPLE 15   Convert 0.04% to a fraction.
SOLUTION   Be careful here. The percent, 0.04%, is not the same as the number 0.04.

$$0.04\% = 0.04 \div 100$$ Drop the percent symbol and divide by 100.

$$= .0004$$ Dividing a decimal by 100 moves the decimal point 2 places to the left.

$$= 0.0004$$ This decimal is 4 ten-thousandths.

378 Chapter 9 Percent

 Try These Problems

Convert each percent to a fraction.
34. 0.05%
35. 0.12%
36. 3.75%
37. 8.25%

$$= \frac{4}{10,000}$$ Convert the decimal to a fraction.

$$= \frac{\cancel{4}^1}{\cancel{4} \times 2500}$$ Reduce the fraction to lowest terms.

$$= \frac{1}{2500} \blacksquare$$

 Try Problems 34 and 35.

EXAMPLE 16 Convert 6.25% to a fraction.
SOLUTION

$$6.25\% = 6.25 \div 100$$ Drop the percent symbol and divide by 100.

$$= .0625$$ Dividing a decimal by 100 moves the decimal point 2 places to the left.

$$= 0.0625$$ This decimal is 625 ten-thousandths.

$$= \frac{625}{10,000}$$ Convert the decimal to a fraction.

$$= \frac{25 \times \cancel{25}}{\cancel{25} \times 400}$$

$$= \frac{25}{400}$$

$$= \frac{\cancel{25}^1}{16 \times \cancel{25}}$$ Reduce the fraction to lowest term.

$$= \frac{1}{16} \blacksquare$$

Try Problems 36 and 37.

Answers to Try These Problems

1. less than 1 2. more than 1 3. less than 1 4. equal to 1
5. 0.04 6. 0.39 7. 1 8. 3.75 9. 0.0805 10. 0.0025
11. 0.625 12. 2.008 13. 0.005 14. 0.000625 15. 0.125
16. 0.04875 17. $0.\overline{6}$ or $0.66\frac{2}{3}$ 18. $0.051\overline{6}$ or $0.05\frac{1}{6}$
19. $0.00\overline{3}$, $0.003\frac{1}{3}$ 20. $0.04\overline{63}$, $0.04\frac{7}{11}$ 21. $\frac{1}{50}$ 22. $\frac{4}{5}$
23. $\frac{21}{5}$ or $4\frac{1}{5}$ 24. $\frac{1}{400}$ 25. $\frac{3}{1000}$ 26. $\frac{4}{125}$ 27. $\frac{1}{15}$ 28. $\frac{1}{8}$ 29. $\frac{8}{75}$
30. $\frac{1}{40}$ 31. $\frac{7}{125}$ 32. $\frac{7}{8}$ 33. $\frac{3}{500}$ 34. $\frac{1}{2000}$ 35. $\frac{3}{2500}$ 36. $\frac{3}{80}$
37. $\frac{33}{400}$

EXERCISES 9.1

Indicate whether each percent is less than, equal to, or more than the number 1.

1. 50%   **2.** 300%   **3.** 150%   **4.** 8%   **5.** 100%   **6.** 1%

Convert each percent to a decimal. If the decimal is repeating, express the answer as both a repeating decimal and as a decimal with fraction.

7. 6% **8.** 5% **9.** 10% **10.** 25% **11.** 100%
12. 125% **13.** 350% **14.** 275% **15.** 4.2% **16.** 8.7%
17. 15.32% **18.** 9.02% **19.** 0.9% **20.** 0.75% **21.** 0.145%
22. 0.025% **23.** 100.5% **24.** 230.6% **25.** $\frac{1}{4}$% **26.** $\frac{2}{5}$%
27. $2\frac{1}{2}$% **28.** $6\frac{2}{5}$% **29.** $37\frac{3}{4}$% **30.** $50\frac{1}{4}$% **31.** $\frac{1}{40}$%
32. $\frac{3}{100}$% **33.** $\frac{5}{6}$% **34.** $2\frac{2}{3}$% **35.** $33\frac{1}{3}$% **36.** $6\frac{1}{6}$%

Convert each percent to a fraction. Simplify.

37. 1% **38.** 8% **39.** 75% **40.** 12% **41.** 36%
42. 40% **43.** 175% **44.** 200% **45.** 325% **46.** 260%
47. $\frac{2}{5}$% **48.** $\frac{2}{3}$% **49.** $33\frac{1}{3}$% **50.** $62\frac{1}{2}$% **51.** $3\frac{1}{3}$%
52. $7\frac{1}{2}$% **53.** $\frac{1}{2}$% **54.** $\frac{2}{5}$% **55.** $\frac{7}{10}$% **56.** $\frac{1}{30}$%
57. $43\frac{3}{4}$% **58.** $16\frac{2}{5}$% **59.** $9\frac{2}{3}$% **60.** $4\frac{3}{5}$% **61.** $\frac{3}{400}$%
62. $\frac{9}{200}$% **63.** 2.5% **64.** 7.5% **65.** 4.5% **66.** 6.4%
67. 37.5% **68.** 62.5% **69.** 0.8% **70.** 0.4% **71.** 0.09%
72. 0.02% **73.** 0.25% **74.** 0.16% **75.** 4.25% **76.** 8.75%

9.2 FINDING A PERCENT OF A NUMBER

OBJECTIVES
- Find a percent of a number by using decimals.
- Find a percent of a number by using fractions.

1 Finding a Percent of a Number in Two Ways: Using Decimals and Using Fractions

Suppose your take-home pay for one month is $2000 and you pay 25% of the take-home pay in rent. How much do you pay in rent? We have the following.

Amount paid in rent = 25% of take-home pay
 = 25% × $2000

The word *of* used in this way translates to multiplication. To find 25% of $2000 we must multiply 25% by $2000. Before multiplying, convert 25% to a fraction or a decimal. Here we show the calculation done in both ways.

USING DECIMALS

Amount paid in rent ($) = 25% of 2000
= 0.25 × 2000
= 500.00
= 500

USING FRACTIONS

$$\text{Amount paid in rent (\$)} = 25\% \text{ of } 2000$$

$$= \frac{25}{100} \times 2000$$

$$= \frac{25}{\cancel{100}_{1}} \times \frac{\cancel{2000}^{20}}{1}$$

$$= 500$$

Observe that 25% of 2000 is 500. Therefore, the amount paid in rent is $500.

In the above problem, it did not matter whether we chose to do the problem using fractions or decimals. One method was as easy as the other and both methods gave the precise answer. Sometimes, however, choosing decimals is easier, and at other times, choosing fractions is easier. Also, when the percent converts to a repeating decimal, you need to choose fractions in order to obtain an answer that is not an approximation. In this section we learn how to choose the most convenient method or the method that leads to a precise answer. First we look at more examples where one method is as easy as the other and both are precise.

EXAMPLE 1 Find 5% of 60.

SOLUTION
USING DECIMALS

$$5\% \text{ of } 60$$
$$\downarrow$$
$$= 0.05 \times 60$$
$$= 3.00$$
$$= 3$$

Convert 5% to a decimal.
$5\% = 5 \div 100 = 0.05$
Multiply 0.05 by 60.
$$\begin{array}{r} 0.05 \\ \times 60 \\ \hline 3.00 \end{array}$$

USING FRACTIONS

$$5\% \text{ of } 60$$
$$\downarrow$$
$$= \frac{1}{20} \times \frac{60}{1}$$

Convert 5% to a fraction.
$$5\% = 5 \div 100 = \frac{5}{100} = \frac{\cancel{5}^{1}}{\cancel{5} \times 20} = \frac{1}{20}$$

$$= \frac{1}{\cancel{20}_{1}} \times \frac{3 \times \cancel{20}}{1}$$

Cancel common factors from the numerator and denominator before multiplying the fractions.

$$= 3$$

5% of 60 is 3. ∎

9.2 Finding a Percent of a Number 381

 Try These Problems

Solve each problem in two ways, by using decimals and by using fractions.

1. 3% of 400
2. 50% of 36
3. $7\frac{1}{2}$% of 620
4. $37\frac{1}{2}$% of 20

In Example 1 note that the result 3 is less than 60 because 5% is less than 100% or 1.

 Try Problems 1 and 2.

EXAMPLE 2 Find $3\frac{1}{2}$% of 2600.

SOLUTION
USING DECIMALS

$$3\frac{1}{2}\% \text{ of } 2600$$
$$\downarrow$$
$$= 0.035 \times 2600$$

Convert $3\frac{1}{2}$% to a decimal.
$3\frac{1}{2}\% = 3.5\% = 3.5 \div 100$
$\phantom{3\frac{1}{2}\%} = 0.035$

$= 91.000$

Multiply 0.035 by 2600.

$$\begin{array}{r} 0.035 \\ \times 2600 \\ \hline 21\,0 \\ 70 \\ \hline 91.000 \end{array}$$

$= 91$

USING FRACTIONS

$$3\frac{1}{2}\% \text{ of } 2600$$
$$\downarrow$$
$$= \frac{7}{200} \times \frac{2600}{1}$$

Convert $3\frac{1}{2}$% to a fraction.
$3\frac{1}{2}\% = 3\frac{1}{2} \div 100$
$\phantom{3\frac{1}{2}\%} = \frac{7}{2} \times \frac{1}{100} = \frac{7}{200}$

$$= \frac{7}{\cancel{200}} \times \frac{13 \times \cancel{200}}{1}$$

$= 91$

$3\frac{1}{2}$% of 2600 is 91. ∎

 Try Problems 3 and 4.

EXAMPLE 3 Find 125% of 86.

SOLUTION
USING DECIMALS

$$125\% \text{ of } 86$$
$$\downarrow$$
$$= 1.25 \times 86$$

Convert 125% to a decimal.
$125\% = 125 \div 100 = 1.25$

$= 107.50$

Multiply 1.25 by 86.

$= 107.5$

Try These Problems

Solve each problem in two ways, by using decimals and by using fractions.

5. 175% of 1600
6. 250% of 47
7. $\frac{2}{5}$% of 2500
8. 0.03% of 5000

USING FRACTIONS

$$125\% \text{ of } 86$$
$$\downarrow$$
$$= \frac{5}{4} \times \frac{86}{1}$$
$$= \frac{5}{2 \times 2} \times \frac{2 \times 43}{1}$$
$$= \frac{215}{2} \text{ or } 107\frac{1}{2}$$

Convert 125% to a fraction.
$125\% = \frac{125}{100} = \frac{5 \times 25}{4 \times 25} = \frac{5}{4}$

125% of 86 is 107.5 or $107\frac{1}{2}$. ■

In Example 3 note that the result 107.5 is more than 86, because 125% is more than 100% or 1.

 Try Problems 5 and 6.

EXAMPLE 4 Find $\frac{1}{4}$% of 3200.

SOLUTION
USING DECIMALS

$$\frac{1}{4}\% \text{ of } 3200$$
$$\downarrow$$
$$= 0.0025 \times 3200$$

Convert $\frac{1}{4}$% to a decimal.
$\frac{1}{4}\% = 0.25\%$
$\quad\quad = 0.25 \div 100 = 0.0025$

$= 8.0000$
$= 8$

Multiply 0.0025 by 3200.

```
    0.0025
  ×   3200
      50
     75
   8.0000
```

USING FRACTIONS

$$\frac{1}{4}\% \text{ of } 3200$$
$$\downarrow$$
$$= \frac{1}{400} \times \frac{3200}{1}$$
$$= \frac{1}{4 \times 100} \times \frac{32 \times 100}{1}$$
$$= \frac{32}{4}$$
$$= 8$$

Convert $\frac{1}{4}$% to a fraction.
$\frac{1}{4}\% = \frac{1}{4} \div 100$
$\quad\quad = \frac{1}{4} \times \frac{1}{100} = \frac{1}{400}$

$\frac{1}{4}$% of 3200 is 8. ■

In Example 4 we should expect $\frac{1}{4}$% of 3200 to be less than 1% of 3200, because $\frac{1}{4}$% is less than 1%. This means we should expect the result, 8, to be less than 1% of 3200 which is 32.

 **Try Problems 7 and 8.**

Try These Problems

Solve each of these problems by using decimals.

9. 9.3% of 780
10. 275% of 13.6
11. 0.7% of $59.30 (Round off to the nearest cent.)

2 Choosing Decimals over Fractions to Solve Problems

Now that you have seen that we can use decimals or fractions to find a percent of a number, we want to look at some situations where it is better to choose the decimal method.

EXAMPLE 5   Find 8% of 37.9.

SOLUTION   Because the number 37.9 is already written in decimal form, and it is easy to convert 8% to a decimal, we choose to use decimals to solve this problem.

$$8\% \text{ of } 37.9$$
$$\downarrow$$
$$= 0.08 \times 37.9 \qquad \text{Convert 8\% to a decimal.}$$
$$\qquad\qquad\qquad\qquad 8\% = 8 \div 100 = 0.08$$
$$= 3.032 \qquad\qquad \text{Multiply 0.08 by 37.9.}$$

8% of 37.9 is 3.032. ■

EXAMPLE 6   Find 0.75% of $65.25. (Round off the answer to the nearest cent.)

SOLUTION   Because the number 65.25 is already written in decimal form, and the percent 0.75% is easy to convert to a decimal, we choose to use decimals to solve this problem.

$$0.75\% \text{ of } \$65.25$$
$$\downarrow$$
$$= 0.0075 \times \$65.25 \qquad \text{Convert 0.75\% to a decimal.}$$
$$\qquad\qquad\qquad\qquad\quad 0.75\% = 0.75 \div 100 = 0.0075$$
$$= \$0.489375 \qquad\qquad \text{Multiply 0.0075 by 65.25.}$$
$$\approx \$0.49$$

$$\begin{array}{r} 65.25 \\ \times\; 0.0075 \\ \hline 32625 \\ 45675 \\ \hline .489375 \approx 0.49 \end{array}$$

0.75% of $65.25 is approximately $0.49. ■

In Example 6 we expect the result to be less than 1% of $65.25 because 0.75% is less than 1%. This means the result $0.49 should be less than 1% of $65.25 which is approximately $0.65.

 Try Problems 9 through 11.

3 Choosing Fractions over Decimals to Solve Problems

Sometimes fractions convert to repeating decimals. Here are some examples.

$$\frac{1}{3} = 0.333\ldots \qquad \text{Divide 1 by 3.}$$
$$= 0.\overline{3}$$

$$\begin{array}{r} .333 \\ 3{\overline{\smash{\big)}\,1.000}} \\ \underline{9} \\ 10 \\ \underline{9} \\ 10 \\ \underline{9} \\ 1 \end{array}$$

Try These Problems

Solve each of these problems by using fractions.

12. $66\frac{2}{3}\%$ of 900

13. $10\frac{5}{6}\%$ of 75

$$\frac{2}{3} = 0.666\ldots$$
$$= 0.\overline{6}$$

Divide 2 by 3.

```
    .666
3)2.000
  1 8
   20
   18
    20
    18
     2
```

$$\frac{5}{11} = 0.454545\ldots$$
$$= 0.\overline{45}$$

Divide 5 by 11.

```
     .4545
11)5.0000
   4 4
    60
    55
     50
     44
      60
      55
       5
```

If a fraction or percent converts to a repeating decimal, and you need to multiply or divide by this number, you should avoid using decimals. Using a decimal approximation for a repeating decimal will result in an answer that is not precise. *Use fractions when solving problems that involve repeating decimals.* Here are some examples.

EXAMPLE 7   Find $33\frac{1}{3}\%$ of 15.

SOLUTION   Because $33\frac{1}{3}\%$ converts to the repeating decimal $0.\overline{3}$, we must use fractions to solve this problem.

$$33\frac{1}{3}\% \text{ of } 15$$
$$= \frac{1}{3} \times \frac{15}{1}$$
$$= \frac{15}{3} \text{ or } 5$$

Convert $33\frac{1}{3}\%$ to a fraction.
$$33\frac{1}{3}\% = 33\frac{1}{3} \div 100$$
$$= \frac{\overset{1}{\cancel{100}}}{3} \times \frac{1}{\underset{1}{\cancel{100}}} = \frac{1}{3}$$

$33\frac{1}{3}\%$ of 15 is 5.

In the previous example, let's look at what would happen if we used a decimal approximation for $33\frac{1}{3}\%$, instead of the fraction $\frac{1}{3}$, to calculate $33\frac{1}{3}\%$ of 15. Here we show the results when converting $33\frac{1}{3}\%$ to 0.3, 0.33, or 0.333 to compute $33\frac{1}{3}\%$ of 15.

$$0.3 \times 15 = 4.5$$
$$0.33 \times 15 = 4.95$$
$$0.333 \times 15 = 4.995$$

These three results—4.5, 4.95, and 4.995—are only approximations of the precise answer 5. When a percent converts to a repeating decimal, you must convert the percent to a fraction before calculating with it. Using the fraction will produce the precise answer, while using a decimal approximation will not produce the precise answer.

Try Problems 12 and 13.

Try These Problems

Solve each of these problems by using fractions.

14. $12\frac{1}{2}\%$ of $8\frac{1}{3}$
15. 125% of $\frac{5}{12}$
16. $\frac{2}{7}\%$ of 2100
17. $\frac{1}{4}\%$ of $66\frac{2}{3}$

EXAMPLE 8 Find 250% of $10\frac{2}{3}$.

SOLUTION Because $10\frac{2}{3}$ converts to the repeating decimal $10.\overline{6}$, we must use fractions to solve this problem.

$$250\% \text{ of } 10\frac{2}{3}$$
$$= \frac{\overset{1}{\cancel{5}}}{\cancel{2}} \times \frac{\overset{16}{\cancel{32}}}{3}$$

Convert 250% to a fraction.
$250\% = \frac{250}{100} = \frac{5 \times \cancel{50}}{2 \times \cancel{50}} = \frac{5}{2}$

$$= \frac{80}{3} \text{ or } 26\frac{2}{3}$$

250% of $10\frac{2}{3}$ is $26\frac{2}{3}$. ■

▲ Try Problems 14 and 15.

EXAMPLE 9 Find $\frac{3}{5}\%$ of $4\frac{1}{6}$.

SOLUTION Because $4\frac{1}{6}$ converts to the repeating decimal $4.1\overline{6}$, we must use fractions to solve this problem.

$$\frac{3}{5}\% \text{ of } 4\frac{1}{6}$$
$$= \frac{3}{500} \times \frac{25}{6}$$

Convert $\frac{3}{5}\%$ to a fraction.
$\frac{3}{5}\% = \frac{3}{5} \div 100 = \frac{3}{5} \times \frac{1}{100} = \frac{3}{500}$

$$= \frac{\overset{1}{\cancel{3}}}{\cancel{25} \times 20} \times \frac{\overset{1}{\cancel{25}}}{\underset{2}{\cancel{6}}}$$

$$= \frac{1}{40}$$

$\frac{3}{5}\%$ of $4\frac{1}{6}$ is $\frac{1}{40}$. ■

▲ Try Problems 16 and 17.

At this time we summarize the procedure for finding the percent of a number.

Finding a Percent of a Number

1. Convert the percent to a fraction or decimal. Use fractions if repeating decimals are involved. Otherwise, choose whichever seems more convenient.
2. Translate the word *of* to *multiply*.
3. Follow the rules for multiplying decimals or fractions when carrying out the multiplication.

▲ Answers to Try These Problems.

1. 12   2. 18   3. 46.5 or $46\frac{1}{2}$ or $\frac{93}{2}$ 4. 7.5 or $7\frac{1}{2}$ or $\frac{15}{2}$
5. 2800   6. 117.5 or $117\frac{1}{2}$ or $\frac{235}{2}$ 7. 10
8. 1.5 or $1\frac{1}{2}$ or $\frac{3}{2}$ 9. 72.54   10. 37.4   11. $0.42   12. 600
13. $\frac{65}{8}$ or $8\frac{1}{8}$ 14. $\frac{25}{24}$ or $1\frac{1}{24}$ 15. $\frac{25}{48}$ 16. 6   17. $\frac{1}{6}$

EXERCISES 9.2

Solve these problems in two ways, by using decimals and by using fractions.

1. 1% of 3000
2. 10% of 250
3. 25% of 72
4. 50% of 700
5. 350% of 40
6. 275% of 64
7. $6\frac{1}{4}$% of 200
8. $12\frac{1}{2}$% of 680
9. 0.2% of 1500
10. $\frac{3}{4}$% of 2400

Use decimals to solve these problems.

11. 8.7% of 162
12. 36.4% of 95
13. 42% of 17.8
14. 9% of 3.14
15. 100% of 12.3
16. 100% of 7.98
17. 120% of 130.7
18. 250% of 9400
19. 162% of 8
20. 215% of 0.6
21. 0.6% of 42.5
22. 0.45% of 906
23. 0.064% of 5600
24. 0.03% of 8000
25. $\frac{1}{2}$% of 15.60
26. $5\frac{1}{4}$% of 76.8

Use fractions to solve these problems.

27. $33\frac{1}{3}$% of 450
28. $6\frac{2}{3}$% of 1800
29. $7\frac{1}{2}$% of $22\frac{2}{9}$
30. $87\frac{1}{2}$% of $13\frac{1}{3}$
31. 100% of $\frac{7}{12}$
32. 100% of $9\frac{5}{6}$
33. 240% of $14\frac{1}{6}$
34. 350% of $2\frac{2}{7}$
35. $\frac{2}{5}$% of $10\frac{5}{12}$
36. $\frac{1}{6}$% of 540
37. 1.5% of $16\frac{2}{3}$
38. 0.3% of $18\frac{1}{3}$

Solve.

39. Gloria pays 35% of her annual income in taxes. If her annual income is $56,000, how much does she pay in taxes?

40. Sales tax in a certain state is 5.2% of the marked price. How much tax is paid on a stereo that is marked $580?

A used-car dealer has a total of 460 automobiles on his lot to sell. The circle graph gives the percent of the total that are trucks, 4-door sedans, and 2-door sedans, respectively. Use the graph to answer Exercises 41 and 42.

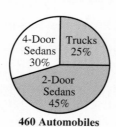

460 Automobiles

41. How many trucks are on the lot?
42. How many 2-door sedans are on the lot?

9.3 CONVERTING DECIMALS AND FRACTIONS TO PERCENT

OBJECTIVES
- Convert a decimal to a percent.
- Convert a fraction to a percent.
- Given a fraction, decimal, or percent, find the other two forms of the number.

1 Converting Decimals to Percents

In Section 9.1 you learned how to convert a percent to a decimal. For example,

Percent	→	Decimal
50%	=	0.50

Drop the percent symbol and divide by 100.

Now we want to reverse this process. We want to be able to convert a decimal to a percent. For example,

Decimal	→	Percent
0.50	=	50%

Multiply by 100 and attach the percent symbol, %.

Observe that if we multiply 0.50 by 100, then attach the percent symbol, we get 50% which we know to be the correct result. Here we show the procedure carefully.

$$0.50 = 0.50 \times 100\%$$ Multiply by 100 and attach the percent symbol, %.
$$= 050.\ \%$$ Multiplying a decimal by 100 moves the decimal point 2 places to the right.
$$= 50\%$$

Here we have really multiplied 0.50 by 100% to obtain 50%. Recall that 100% is equal to 1. Multiplying a number by 1 does not change that number, so multiplying a number by 100% does not alter the value of that number. Therefore, 0.50 is equal to 50%.

The above discussion suggests the following rule.

Converting a Decimal to a Percent
To convert a decimal to a percent, multiply by 100 and attach the percent symbol, %.

Here are some examples that illustrate how to use this rule to convert decimals to percents.

EXAMPLE 1 Convert 0.3 to a percent.

SOLUTION

$$0.3 = 0.3 \times 100\%$$ Multiply by 100 and attach the percent symbol, %.
$$= 030.\ \%$$ Multiplying a decimal by 100 moves the decimal point 2 places to the right.
$$= 30\%$$

EXAMPLE 2 Convert 0.065 to a percent.

SOLUTION

$$0.065 = 0.065 \times 100\%$$ Multiply by 100 and attach the percent symbol, %.
$$= 006.5\ \%$$ Multiplying a decimal by 100 moves the decimal point 2 places to the right.
$$= 6.5\%$$

Try These Problems

Convert each decimal to a percent.
1. 0.68
2. 0.8
3. 0.075
4. 7
5. 1.6
6. 0.0004

EXAMPLE 3 Convert 2 to a percent.
SOLUTION

$2 = 2 \times 100\%$ Multiply by 100 and attach the percent symbol, %.

$= 200\%$ ■

EXAMPLE 4 Convert 0.0054 to a percent.
SOLUTION

$0.0054 = 0.0054 \times 100\ \%$ Multiply by 100 and attach the percent symbol, %.

$= 000.54\ \%$ Multiplying a decimal by 100 moves the decimal point 2 places to the right.
$= 0.54\%$ ■

 Try Problems 1 through 6.

2 Converting Fractions to Percents

You have just learned how to convert a decimal to a percent. For example,

Decimal → Percent
0.04 = 4% Multiply by 100 and attach the percent symbol, %.

The same rule should work for fractions. For example,

Fraction → Percent
$\frac{4}{100}$ = 4% Multiply by 100 and attach the percent symbol, %.

Observe that if we multiply the fraction $\frac{4}{100}$ by 100, then attach the percent symbol, we get 4% which we know to be the correct result. Here we show the procedure.

$\frac{4}{100} = \frac{4}{100} \times 100\ \%$ Multiply by 100 and attach the percent symbol, %.

$= \frac{4}{\underset{1}{\cancel{100}}} \times \frac{\overset{1}{\cancel{100}}}{1}\%$ Cancel the common factor 100.

$= 4\%$

Here we have really multiplied $\frac{4}{100}$ by 100% to obtain 4%. Recall that 100% equals 1. Multiplying a number by 1 does not change the value of the number, so multiplying a number by 100% does not change the value of that number. Therefore, the fraction $\frac{4}{100}$ is equal to 4%.

The above discussion suggests the following rule.

Converting a Fraction to a Percent

To convert a fraction to a percent, multiply by 100 and attach the percent symbol, %.

Try These Problems

Convert each fraction to a percent.
7. $\frac{2}{5}$
8. $\frac{3}{20}$

Here are some examples that illustrate how to use this rule to convert fractions to percents.

EXAMPLE 5 Convert $\frac{3}{4}$ to a percent.

SOLUTION
METHOD 1

$$\frac{3}{4} = \frac{3}{4} \times 100 \,\%$$ Multiply by 100 and attach the percent symbol, %.

$$= \frac{3}{4} \times \frac{100}{1}\,\%$$

$$= \frac{3}{\underset{1}{\cancel{4}}} \times \frac{\cancel{4} \times 25}{1}\,\%$$ Cancel the common factor 4.

$$= 75\%$$

METHOD 2 It is also possible to convert the fraction to a decimal, then multiply by 100 and attach the percent symbol, %.

$$\frac{3}{4} = 0.75$$ Convert $\frac{3}{4}$ to a decimal.

$$\begin{array}{r} .75 \\ 4\overline{)3.00} \\ \underline{2\,8} \\ 20 \\ \underline{20} \end{array}$$

$$= 0.75 \times 100 \,\%$$ Multiply by 100 and attach the percent symbol, %.

$$= 075. \,\%$$ Multiplying a decimal by 100 moves the decimal point 2 places to the right.

$$= 75\%$$

 Try Problems 7 and 8.

EXAMPLE 6 Convert $3\frac{1}{2}$ to a percent.

SOLUTION
METHOD 1

$$3\frac{1}{2} = 3\frac{1}{2} \times 100 \,\%$$ Multiply by 100 and attach the percent symbol, %.

$$= \frac{7}{2} \times \frac{100}{1}\,\%$$ Convert $3\frac{1}{2}$ to an improper fraction. $3\frac{1}{2} = \frac{7}{2}$

$$= \frac{7}{\underset{1}{\cancel{2}}} \times \frac{\cancel{2} \times 50}{1}\,\%$$ Cancel the common factor 2.

$$= 350\%$$

METHOD 2 Here we convert $3\frac{1}{2}$ to a decimal, then multiply by 100 and attach the percent symbol, %.

$$3\frac{1}{2} = 3.5$$ Convert $3\frac{1}{2}$ to a decimal.

$$= 3.5 \times 100 \,\%$$ Multiply by 100 and attach the percent symbol, %.

$$= 350. \,\%$$ Multiplying a decimal by 100 moves the decimal point 2 places to the right.

$$= 350\%$$

Try These Problems

Convert each fraction to a percent.
9. $2\frac{1}{4}$
10. $1\frac{3}{5}$
11. $\frac{7}{8}$
12. $\frac{3}{40}$

 Try Problems 9 and 10.

EXAMPLE 7 Convert $\frac{7}{40}$ to a percent.

SOLUTION
METHOD 1

$$\frac{7}{40} = \frac{7}{40} \times 100 \ \%$$ Multiply by 100 and attach the percent symbol, %.

$$= \frac{7}{40} \times \frac{100}{1} \%$$

$$= \frac{7}{2 \times 20} \times \frac{5 \times 20}{1} \%$$ Cancel the common factor 20.

$$= \frac{35}{2} \%$$ Convert $\frac{35}{2}$ to a mixed numeral or a decimal.

$$= 17\frac{1}{2}\%$$

or 17.5%

METHOD 2 Here we convert $\frac{7}{40}$ to a decimal, then multiply by 100 and attach the percent symbol, %.

$$\frac{7}{40} = 0.175$$ Convert $\frac{7}{40}$ to a decimal.

$$\begin{array}{r} .175 \\ 40\overline{)7.000} \\ \underline{4\ 0} \\ 3\ 00 \\ \underline{2\ 80} \\ 200 \\ \underline{200} \end{array}$$

$$= 0.175 \times 100 \ \%$$ Multiply by 100 and attach the percent symbol, %.

$$= 017.5 \ \%$$ Multiplying a decimal by 100 moves the decimal point 2 places to the right.

$$= 17.5\%$$

 Try Problems 11 and 12.

EXAMPLE 8 Convert $\frac{1}{3}$ to a percent.

SOLUTION
METHOD 1

$$\frac{1}{3} = \frac{1}{3} \times 100 \ \%$$ Multiply by 100 and attach the percent symbol, %.

$$= \frac{1}{3} \times \frac{100}{1} \%$$

$$= \frac{100}{3} \%$$ Convert $\frac{100}{3}$ to a mixed numeral.

$$= 33\frac{1}{3}\%$$ $\frac{100}{3} \rightarrow 3\overline{)100} \rightarrow \frac{100}{3} = 33\frac{1}{3}$
$\phantom{= 33\frac{1}{3}\%\ \ \ \ \frac{100}{3} \rightarrow}\begin{array}{r}33\\\underline{9}\\10\\\underline{9}\\1\end{array}$

METHOD 2 Here we convert $\frac{1}{3}$ to a decimal, then multiply by 100 and attach the percent symbol, %.

$$\frac{1}{3} = 0.33\frac{1}{3}$$

Convert $\frac{1}{3}$ to a decimal. Because the division does not come out evenly, divide two places past the decimal point and write the remainder in fractional form.

$$\frac{1}{3} \rightarrow 3\overline{)1.00}^{\,.33\frac{1}{3}} \rightarrow \frac{1}{3} = 0.33\frac{1}{3}$$
$$\phantom{\frac{1}{3} \rightarrow 3\overline{)1.}}\underline{9}$$
$$\phantom{\frac{1}{3} \rightarrow 3\overline{)1.0}}10$$
$$\phantom{\frac{1}{3} \rightarrow 3\overline{)1.00}}\underline{9}$$
$$\phantom{\frac{1}{3} \rightarrow 3\overline{)1.00}}1$$

$$= 0.33\frac{1}{3} \times 100\ \%$$

Multiply by 100 and attach the percent symbol, %.

$$= 033.\frac{1}{3}\ \%$$

Multiplying a decimal by 100 moves the decimal point 2 places to the right.

$$= 33\frac{1}{3}\ \% \quad \blacksquare$$

EXAMPLE 9 Convert $\frac{5}{7}$ to a percent.

SOLUTION
METHOD 1

$$\frac{5}{7} = \frac{5}{7} \times 100\ \%$$

Multiply by 100 and attach the percent symbol, %.

$$= \frac{5}{7} \times \frac{100}{1}\ \%$$

$$= \frac{500}{7}\ \%$$

Convert $\frac{500}{7}$ to a mixed numeral.

$$= 71\frac{3}{7}\ \%$$

$$\frac{500}{7} \rightarrow 7\overline{)500}^{\,71} \rightarrow \frac{500}{7} = 71\frac{3}{7}$$
$$\phantom{\frac{500}{7} \rightarrow 7)}\underline{49}$$
$$\phantom{\frac{500}{7} \rightarrow 7)5}10$$
$$\phantom{\frac{500}{7} \rightarrow 7)5}\underline{7}$$
$$\phantom{\frac{500}{7} \rightarrow 7)5}3$$

METHOD 2 Here we convert $\frac{5}{7}$ to a decimal, then multiply by 100 and attach the percent symbol, %.

$$\frac{5}{7} = 0.71\frac{3}{7}$$

Convert $\frac{5}{7}$ to a decimal. Because the division does not come out evenly, divide two places past the decimal point and write the remainder in fractional form.

$$\frac{5}{7} \rightarrow 7\overline{)5.00}^{\,.71\frac{3}{7}} \rightarrow \frac{5}{7} = 0.71\frac{3}{7}$$
$$\phantom{\frac{5}{7} \rightarrow 7)}\underline{4\ 9}$$
$$\phantom{\frac{5}{7} \rightarrow 7)5.}10$$
$$\phantom{\frac{5}{7} \rightarrow 7)5.0}\underline{7}$$
$$\phantom{\frac{5}{7} \rightarrow 7)5.0}3$$

▲ **Try These Problems**

Convert each fraction to a percent.
13. $\frac{2}{3}$
14. $\frac{5}{6}$
15. $\frac{5}{18}$
16. $\frac{7}{6}$
17. $\frac{1}{200}$
18. $\frac{7}{4000}$

$$= 0.71\frac{3}{7} \times 100 \; \%$$ Multiply by 100 and attach the percent symbol, %.

$$= 071.\frac{3}{7} \%$$ Multiplying a decimal by 100 moves the decimal point 2 places to the right.

$$= 71\frac{3}{7} \% \quad \blacksquare$$

▲ **Try Problems 13 through 16.**

EXAMPLE 9 Convert $\frac{3}{800}$ to a percent.

SOLUTION
METHOD 1

$$\frac{3}{800} = \frac{3}{800} \times 100 \; \%$$ Multiply by 100 and attach the percent symbol, %.

$$= \frac{3}{800} \times \frac{100}{1} \%$$

$$= \frac{3}{8 \times \cancel{100}} \times \frac{\cancel{100}^1}{1} \%$$ Cancel the common factor 100.

$$= \frac{3}{8} \%$$ The number in front of the percent symbol may be written as a fraction or a decimal.

or 0.375%

$$\frac{3}{8} \rightarrow 8\overline{)3.000}^{.375} \rightarrow \frac{3}{8} = 0.375$$
$$\phantom{\frac{3}{8} \rightarrow 8)}\underline{2\;4}$$
$$\phantom{\frac{3}{8} \rightarrow 8)}\;\;60$$
$$\phantom{\frac{3}{8} \rightarrow 8)}\;\;\underline{56}$$
$$\phantom{\frac{3}{8} \rightarrow 8)}\;\;\;\;40$$
$$\phantom{\frac{3}{8} \rightarrow 8)}\;\;\;\;\underline{40}$$

METHOD 2 Here we convert $\frac{3}{800}$ to a decimal, then multiply by 100 and attach the percent symbol, %.

$$\frac{3}{800} = 0.00375$$ Convert $\frac{3}{800}$ to a decimal.

$$\frac{3}{800} \rightarrow 800\overline{)3.00000}^{.00375} \rightarrow \frac{3}{800} = 0.00375$$
$$\phantom{\frac{3}{800} \rightarrow 800)}\underline{2\;400}$$
$$\phantom{\frac{3}{800} \rightarrow 800)}\;\;6000$$
$$\phantom{\frac{3}{800} \rightarrow 800)}\;\;\underline{5600}$$
$$\phantom{\frac{3}{800} \rightarrow 800)}\;\;\;\;4000$$
$$\phantom{\frac{3}{800} \rightarrow 800)}\;\;\;\;\underline{4000}$$

$$= 0.00375 \times 100 \; \%$$ Multiply by 100 and attach the percent symbol, %.

$$= 000.375 \; \%$$ Multiplying a decimal by 100 moves the decimal point 2 places to the right.

$$= 0.375\% \quad \blacksquare$$

▲ **Try Problems 17 and 18.**

3 Three Ways to Write a Number

In Section 9.1 we converted percents to fractions and decimals, and in this section we converted fractions and decimals to percents. In Section 6.1 we converted decimals to fractions, and in Section 7.2 we converted fractions to decimals. Now it should be clear to you that each number can be written in three ways: fraction, decimal, and percent. You want to be good at converting a number from one form to the other. Here we summarize the procedures.

CONVERTING NUMBERS FROM ONE FORM TO ANOTHER

Percent → Fraction or Decimal
To convert a percent to a fraction or decimal, drop the percent symbol, %, and divide by 100.

$$\textit{Example}: 5\% = 5 \div 100 = 0.05$$

$$5\% = 5 \div 100 = \frac{5}{100} = \frac{\cancel{5}^{1}}{\cancel{5} \times 20} = \frac{1}{20}$$

Fraction or Decimal → Percent
To convert a fraction or decimal to a percent, multiply by 100 and attach the percent symbol, %.

$$\textit{Example}: 0.6 = 0.6 \times 100\% = 60\%$$

$$\frac{3}{5} = \frac{3}{5} \times 100\% = \frac{3}{\underset{1}{\cancel{5}}} \times \frac{\cancel{5} \times 20}{1}\% = 60\%$$

Fraction → Decimal
To convert a fraction to a decimal, divide the numerator by the denominator.

$$\textit{Example}: \frac{3}{8} \rightarrow 8\overline{)3.000} \rightarrow \frac{3}{8} = 0.375$$

(long division showing .375, 24, 60, 56, 40, 40)

Decimal → Fraction
To convert a *terminating* decimal to a fraction, think of the formal English name for the decimal using tenths, hundredths, thousandths, and so on, then write the fraction with the same English name.

$$\textit{Example}: 0.004 = \frac{4}{1000} \quad \text{Four thousandths}$$

$$= \frac{\cancel{4}^{1}}{\cancel{4} \times 250}$$

$$= \frac{1}{250}$$

Converting a *repeating* decimal to a fraction requires algebra, so we do not discuss this procedure in this text.

Try These Problems

19. Convert 35% to **a.** a decimal and **b.** a fraction.
20. Convert 0.025 to **a.** a fraction and **b.** a percent.
21. Convert $\frac{7}{2}$ to **a.** a decimal and **b.** a percent.

Try Problems 19 through 21.

Answers to Try These Problems

1. 68%   2. 80%   3. 7.5%   4. 700%   5. 160%   6. 0.04%
7. 40%   8. 15%   9. 225%   10. 160%   11. $87\frac{1}{2}$% or 87.5%
12. $7\frac{1}{2}$% or 7.5%   13. $66\frac{2}{3}$%   14. $83\frac{1}{3}$%   15. $27\frac{7}{9}$%
16. $116\frac{2}{3}$%   17. $\frac{1}{2}$% or 0.5%   18. $\frac{7}{40}$% or 0.175%
19. **a.** 0.35 **b.** $\frac{7}{20}$ 20. **a.** $\frac{1}{40}$ **b.** 2.5% or $2\frac{1}{2}$%
21. **a.** 3.5 **b.** 350%

EXERCISES 9.3

Convert each decimal to a percent.

1. 0.45   2. 0.62   3. 0.9   4. 0.3   5. 0.053
6. 0.015   7. 1   8. 3   9. 2.4   10. 1.7
11. 1.08   12. 3.24   13. 0.005   14. 0.0072   15. 0.00038
16. 0.0009   17. 0.672   18. 1.456   19. 0.0746   20. 0.0805

Convert each fraction to a percent.

21. $\frac{1}{2}$ 22. $\frac{3}{5}$ 23. $\frac{7}{20}$ 24. $\frac{3}{25}$ 25. $1\frac{4}{5}$
26. $3\frac{1}{4}$ 27. $\frac{13}{10}$ 28. $\frac{69}{50}$ 29. $\frac{3}{8}$ 30. $\frac{11}{80}$
31. $\frac{5}{16}$ 32. $\frac{1}{40}$ 33. $\frac{1}{3}$ 34. $\frac{5}{12}$ 35. $\frac{4}{15}$
36. $\frac{7}{300}$ 37. $\frac{5}{3}$ 38. $\frac{11}{6}$ 39. $3\frac{1}{6}$ 40. $2\frac{2}{3}$
41. $\frac{3}{500}$ 42. $\frac{7}{200}$ 43. $\frac{3}{1600}$ 44. $\frac{1}{800}$

Complete the following chart. Simplify the fractions when possible.

	Fraction	Decimal	Percent
45.	$\frac{3}{100}$		
46.		0.25	
47.			80%
48.	$1\frac{1}{5}$		
49.		2.06	
50.			375%
51.	$\frac{5}{8}$		
52.		0.125	
53.			$37\frac{1}{2}$% or 37.5%
54.	$\frac{1}{200}$		
55.		0.006	
56.			$\frac{3}{4}$% or 0.75%
57.	$\frac{1}{6}$		
58.		$0.83\frac{1}{3}$	
59.			$66\frac{2}{3}$%
60.	$\frac{4}{3}$		

9.4 THE MISSING NUMBER IN A PERCENT STATEMENT

OBJECTIVES
- Find the missing number in a basic percent statement by using decimals.
- Find the missing number in a basic percent statement by using fractions.

1 Using Fractions or Decimals

The statement *60 is 50% of 120* is a percent statement. Observe that the statement translates to a multiplication statement.

$$60 \text{ is } 50\% \text{ of } 120$$
$$60 = 0.5 \times 120$$

Three types of percent problems can be obtained by omitting any one of the three numbers involved in the basic percent statement. For example, from the statement *60 is 50% of 120*, three questions can arise.

1. What number is 50% of 120?

   $\boxed{} = 0.5 \times 120$ Here we are missing the answer to a multiplication problem, so we multiply to find the missing number 60.

2. 60 is 50% of what number?

   $60 = 0.5 \times \boxed{}$ Here we are missing a multiplier, so we divide 60 by 0.5 to find the missing multiplier 120.

3. 60 is what percent of 120?

   $60 = \boxed{}\% \times 120$ Here we are missing a multiplier, so we divide 60 by 120 to find the missing multiplier 0.5. Finally, we convert the decimal 0.5 to the percent 50%.

When solving these problems without a calculator, you will have to decide whether to use fractions or decimals. In some cases the problems can be done either way, with one method as convenient as the other. The first examples shown here are of this type.

EXAMPLE 1   150 is 25% of what number?

SOLUTION   This problem can be solved by using either decimals or fractions. We show the problems done both ways.
USING DECIMALS

$$150 \text{ is } 25\% \text{ of what number?}$$
$$150 = 0.25 \times \boxed{}$$

Translate to a multiplication statement and convert the percent to a decimal.

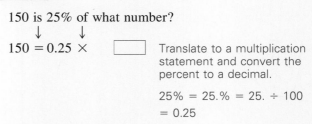

Try These Problems

Solve these problems in two ways
a. *by using* decimals, *and* **b.** *by using* fractions.

1. 720 is 75% of what number?
2. 40% of what number is 80?

A multiplier is missing, so divide 150 by 0.25 to find the missing multiplier.

$$0.25\overline{)150.00} = 600$$
$$150$$

The answer to the multiplication statement goes inside the division symbol.

USING FRACTIONS

150 is 25% of what number?

$$150 = \frac{1}{4} \times \square$$

Translate to a multiplication statement and convert the percent to a fraction.

$$25\% = \frac{25}{100}$$

$$= \frac{\overset{1}{\cancel{25}}}{\cancel{25} \times 4} = \frac{1}{4}$$

A multiplier is missing, so divide 150 by $\frac{1}{4}$ to find the missing multiplier.

$$150 \div \frac{1}{4}$$

The answer to the multiplication statement is written first.

$$= \frac{150}{1} \times \frac{4}{1}$$

Dividing by $\frac{1}{4}$ is the same as multiplying by 4.

$$= 600$$

The answer is 600. ■

Try Problems 1 and 2.

EXAMPLE 2 4% of 200 is what number?

SOLUTION This problem can be solved by using either decimals or fractions. We show the problems done both ways.

USING DECIMALS

4% of 200 is what number?

$$0.04 \times 200 = \square$$

Translate to a multiplication statement and convert the percent to a decimal.

$$4\% = 4.\% = 4. \div 100$$
$$= 0.04$$

We are missing the answer to the multiplication statement, so multiply 0.04 by 200 to find the missing number.

$$\begin{array}{r} 200 \\ \times\ 0.04 \\ \hline 8.00 \end{array} = 8$$

 **Try These Problems**

Solve these problems in two ways **a.** *by using* decimals, *and* **b.** *by using* fractions.

3. 250% of 16 is what number?
4. What number is 8% of 650?

USING FRACTIONS

4% of 200 is what number?

$$\frac{4}{100} \times 200 = \boxed{}$$

Translate to a multiplication statement and convert the percent to a fraction.

$$4\% = 4 \div 100 = \frac{4}{100}$$

We are missing the answer to a multiplication statement, so multiply $\frac{4}{100}$ by 200 to find the missing number.

$$\frac{4}{100} \times 200$$

$$= \frac{4}{\underset{1}{\cancel{100}}} \times \frac{\overset{2}{\cancel{200}}}{1}$$

$$= 8$$

This answer is 8. ■

 Try Problems 3 and 4.

EXAMPLE 3   520 is what percent of 650?

SOLUTION   This problem can be solved by using either decimals or fractions. We show the problems done both ways.

USING DECIMALS

520 is what percent of 650?

$$520 = \boxed{\%} \times 650$$

Translate to a multiplication statement.

A multiplier is missing, so divide 520 by 650 to find the missing multiplier.

$$\begin{array}{r} .8 \\ 650 \overline{)520.0} \\ \underline{520} \end{array}$$

The answer to the multiplication statement goes inside the division symbol.

The missing multiplier is the decimal 0.8, but the question is *what percent,* so convert 0.8 to a percent.

$$0.8 = 0.8 \times 100\% = 80.\% = 80\%$$

USING FRACTIONS

520 is what percent of 650?

$$520 = \boxed{\%} \times 650$$

Translate to a multiplication statement.

▲ Try These Problems

Solve these problems in two ways **a.** *by using* decimals, *and* **b.** *by using* fractions.

5. 21 is what percent of 350?

6. What percent of 8 is 10?

A multiplier is missing, so divide 520 by 650 to find the missing multiplier.

$$520 \div 650 = \frac{520}{650}$$ The answer to the multiplication statement is written first.

$$= \frac{\cancel{10} \times 52}{\cancel{10} \times 65}$$

$$= \frac{52}{65}$$

$$= \frac{4 \times \cancel{13}}{5 \times \cancel{13}}$$ Reduce the fraction to lowest terms.

$$= \frac{4}{5}$$

The missing multiplier is the fraction $\frac{4}{5}$, but the question is *what percent*, so convert the fraction to a percent.

$$\frac{4}{5} = \frac{4}{5} \times 100\% = \frac{4}{\cancel{5}_1} \times \frac{\cancel{100}^{20}}{1}\% = 80\%$$

The answer is 80%. ∎

▲ Try Problems 5 and 6.

2 Choosing Decimals over Fractions to Solve Problems

The problems we have solved so far were easy to solve by using either fractions or decimals. Sometimes one of the methods is a lot easier to use than the other. You want to learn to choose the more convenient method. Next we look at examples where it is more convenient to use decimals.

EXAMPLE 4 170% of what number is 81.6?

SOLUTION We choose *decimals* to solve this problem because the number 81.6 is already in decimal form, and the percent 170% is easy to convert to a decimal.

170%   of   what number   is   81.6?
  ↓   ↓
 1.7   ×   ☐   =   81.6   Translate to a multiplication statement and convert the percent to a decimal.
                            170%
                          = 170. ÷ 100
                          = 1.70 = 1.7

9.4 The Missing Number in a Percent Statement

▼ **Try These Problems**

Use decimals to solve these problems.

7. What number is 8.9% of 46,000?
8. 33 is 0.6% of what number?
9. 34.2% of what number is 513?
10. 125% of 10.2 is what number?

A multiplier is missing, so divide 81.6 by 1.7 to find the missing multiplier.

$$1.7 \overline{)81.6} = 48$$

The answer to the multiplication statement goes inside the division symbol.

The answer is 48. ∎

EXAMPLE 5 What number is 0.8% of 4.5?

SOLUTION We choose *decimals* to solve this problem because 0.8% is easy to convert to a decimal, and the number 4.5 is already written in decimal form.

What number is 0.8% of 4.5?
  ☐   = 0.008 × 4.5

Translate to a multiplication statement and convert the percent to a decimal.
0.8%
= 0.8 ÷ 100
= 0.008

The answer to the multiplication problem is missing, so multiply 0.008 by 4.5 to find the missing number.

$$\begin{array}{r} 4.5 \\ \times\ 0.008 \\ \hline .0360 \end{array} = 0.036$$

The answer is 0.036. ∎

▲ **Try Problems 7 through 10.**

EXAMPLE 6 42.9 is what percent of 520?

SOLUTION We choose *decimals* to solve this problem because 42.9 is already in decimal form, and 520 can easily be converted to a decimal.

42.9 is what percent of 520?
42.9 =   ☐ % × 520

Translate to a multiplication statement.

A multiplier is missing, so divide 42.9 by 520 to find the missing multiplier.

$$520 \overline{)42.9000} \quad .0825$$
$$\begin{array}{r} 4160 \\ \hline 1300 \\ 1040 \\ \hline 2600 \\ 2600 \end{array}$$

The answer to the multiplication problem goes inside the division symbol.

Try These Problems

Use decimals to solve these problems.
11. What percent of 200 is 3.2?
12. 45.3 is what percent of 75?

The missing multiplier is the decimal 0.0825. The question is *what percent*, so we must convert the decimal to percent form.

$$0.0825 = 0.0825 \times 100\%$$
$$= 008.25\%$$
$$= 8.25\%$$

The answer is 8.25%. ∎

 Try Problems 11 and 12.

3 Choosing Fractions over Decimals to Solve Problems

Whenever the number you are trying to work with is a repeating decimal, you must choose fractions to solve the probem. Also, if the numbers are already in fractional form, you may want to choose fractions to do the problem. Here we show some examples.

EXAMPLE 7 Find $3\frac{1}{2}\%$ of $8\frac{1}{3}$.

SOLUTION We must use *fractions* to solve this problem because $8\frac{1}{3}$ is the repeating decimal $8.\overline{3}$.

$$\text{Find} \quad 3\frac{1}{2}\% \text{ of } 8\frac{1}{3}.$$

$$\text{What is} \quad 3\frac{1}{2}\% \text{ of } 8\frac{1}{3}?$$

$$\Box = \frac{7}{200} \times 8\frac{1}{3}$$

Translate to a multiplication statement, and convert $3\frac{1}{2}\%$ to a fraction.

$$3\frac{1}{2}\% = 3\frac{1}{2} \div 100$$
$$= \frac{7}{2} \div \frac{100}{1}$$
$$= \frac{7}{2} \times \frac{1}{100} = \frac{7}{200}$$

The answer to the multiplication statement is missing, so multiply $\frac{7}{200}$ by $8\frac{1}{3}$ to find the missing number.

$$\frac{7}{200} \times 8\frac{1}{3}$$
$$= \frac{7}{200} \times \frac{25}{3}$$
$$= \frac{7}{25 \times 8} \times \frac{\overset{1}{\cancel{25}}}{3}$$
$$= \frac{7}{24}$$

The answer is $\frac{7}{24}$. ∎

9.4 The Missing Number in a Percent Statement

EXAMPLE 8 $66\frac{2}{3}\%$ of what number is 7?

SOLUTION   We must use *fractions* to solve this problem because $66\frac{2}{3}\%$ converts to the repeating decimal $0.\overline{6}$.

$$66\frac{2}{3}\% \text{ of what number is 7?}$$

$$\frac{2}{3} \times \boxed{} = 7$$

Translate to a multiplication statement, and convert $66\frac{2}{3}\%$ to a fraction.

$$66\frac{2}{3}\% = 66\frac{2}{3} \div 100$$
$$= \frac{\cancel{200}^{2}}{3} \times \frac{1}{\cancel{100}_{1}}$$
$$= \frac{2}{3}$$

A multiplier is missing, so divide 7 by $\frac{2}{3}$ to find the missing multiplier.

$$7 \div \frac{2}{3}$$ The answer to the multiplication problem is written first.

$$= \frac{7}{1} \div \frac{2}{3}$$

$$= \frac{7}{1} \times \frac{3}{2}$$ Dividing by $\frac{2}{3}$ is equivalent to multiplying by $\frac{3}{2}$.

$$= \frac{21}{2}$$

$$= 10\frac{1}{2} \text{ or } 10.5$$

The answer is $10\frac{1}{2}$ or 10.5. ∎

EXAMPLE 9 $83\frac{1}{3}$ is what percent of 250?

SOLUTION   We must use *fractions* to solve this problem because $83\frac{1}{3}$ converts to the repeating decimal $83.\overline{3}$.

$$83\frac{1}{3} = \text{ what percent of 250?}$$

$$83\frac{1}{3} = \boxed{\%} \times 250$$

Translate to a multiplication statement.

A multiplier is missing, so divide $83\frac{1}{3}$ by 250 to find the missing multiplier.

Try These Problems

Use fractions to solve these problems.

13. Find $33\frac{1}{3}\%$ of 120.
14. $\frac{5}{6}$ is what percent of $3\frac{1}{3}$?
15. $16\frac{2}{3}\%$ of what number is 400?
16. $10\frac{1}{6}$ is 200% of what number?
17. What percent of 350 is $23\frac{1}{3}$?
18. $\frac{2}{3}\%$ of 400 is what number?

$$83\frac{1}{3} \div 250$$

The answer to the multiplication problem is written first.

$$= \frac{\overset{1}{\cancel{250}}}{3} \times \frac{1}{\underset{1}{\cancel{250}}}$$

$$= \frac{1}{3}$$

The missing multiplier is the fraction $\frac{1}{3}$. The question is *what percent*, so we must convert $\frac{1}{3}$ to percent form.

$$\frac{1}{3} = \frac{1}{3} \times 100\% = \frac{100}{3}\% = 33\frac{1}{3}\%$$

The answer is $33\frac{1}{3}\%$. ■

EXAMPLE 10 What percent of 50 is $\frac{1}{8}$?

SOLUTION We choose *fractions* to solve this problem because the number $\frac{1}{8}$ is already written in fractional form.

What percent of 50 is $\frac{1}{8}$?

$$\boxed{\%} \times 50 = \frac{1}{8} \qquad \text{Translate to a multiplication statement.}$$

A multiplier is missing, so divide $\frac{1}{8}$ by 50 to find the missing multiplier.

$$\frac{1}{8} \div 50$$

The answer to the multiplication problem is written first.

$$= \frac{1}{8} \times \frac{1}{50}$$

$$= \frac{1}{400}$$

The missing multiplier is the fraction $\frac{1}{400}$. The question is *what percent*, so we must convert $\frac{1}{400}$ to a percent.

$$\frac{1}{400} = \frac{1}{400} \times 100\%$$

$$= \frac{1}{\underset{4}{\cancel{400}}} \times \frac{\overset{1}{\cancel{100}}}{1}\%$$

$$= \frac{1}{4}\% \text{ or } 0.25\%$$

The answer is $\frac{1}{4}\%$ or 0.25%. ■

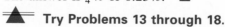 **Try Problems 13 through 18.**

Now we summarize the material presented in this section by giving a procedure for finding the missing number in a percent statement.

> ### Finding the Missing Number in a Percent Statement
> 1. If you are given the percent, convert it to a decimal or a fraction. Use fractions if repeating decimals are involved.
> 2. Translate the percent statement to a multiplication statement.
> 3. If you are missing the answer to the multiplication problem, multiply.
> 4. If you are missing one of the multipliers, divide. Be careful to divide in the correct order. Here is how it is done:
>
> known multiplier) answer to the multiplication problem
>
> or
>
> answer to the multiplication problem ÷ known multiplier
>
> 5. If the question asks *what percent*, convert the missing multiplier to percent form.

Answers to Try These Problems

1. 960 2. 200 3. 40 4. 52 5. 6% 6. 125%
7. 4094 8. 5500 9. 1500 10. 12.75 11. 1.6%
12. 60.4% 13. 40 14. 25% 15. 2400 16. $\frac{61}{12}$ or $5\frac{1}{12}$
17. $6\frac{2}{3}$% 18. $\frac{8}{3}$ or $2\frac{2}{3}$

EXERCISES 9.4

Solve these problems in two ways, by using decimals and by using fractions.

1. What number is 40% of 300?
2. 25% of 64 is what number?
3. 15 is what percent of 75?
4. What percent of 50 is 3?
5. 5% of what number is 8300?
6. 18 is 150% of what number?
7. What percent of 5000 is 5?
8. 2 is $\frac{1}{4}$% of what number?

Use decimals to solve these problems.

9. 228 is 240% of what number?
10. 3.2% of what number is 40?
11. What percent of 650 is 5.2?
12. 3600 is what percent of 2400?
13. 6.5% of 82.90 is what number?
14. What number is 0.3% of 97.8?
15. 0.5% of what number is 2.04?
16. 27.3 is what percent of 32.5?

Use fractions to solve these problems.

17. Find $16\frac{2}{3}$% of 126.
18. Find $33\frac{1}{3}$% of 1920.
19. $\frac{1}{3}$% of what number is 12?
20. $4\frac{1}{4}$ is 25% of what number?
21. $12\frac{2}{7}$ is what percent of $12\frac{2}{7}$?
22. $4\frac{5}{6}$ is what percent of $9\frac{2}{3}$?
23. What percent of 45 is 30?
24. What percent of 400 is $1\frac{1}{3}$?
25. $3\frac{1}{9}$% of $17\frac{1}{7}$ is what number?
26. 10 is 120% of what number?

You decide whether to use decimals or fractions to solve these problems.

27. $12\frac{1}{2}\%$ of 1244 is what number?   **28.** 105 is what percent of 30?
29. 0.2% of what number is 18?   **30.** What number is $66\frac{2}{3}\%$ of 630?
31. 5 is what percent of 15?   **32.** 100% of 45.96 is what number?
33. $17\frac{1}{3}$ is 80% of what number?   **34.** 0.21 is what percent of 1050?

9.5 SOLVING PERCENT APPLICATIONS BY TRANSLATING

OBJECTIVES
- Solve percent applications that contain the basic percent statement.
- Solve percent applications that contain the basic percent statement in an unclear form.

There are many situations in your everyday life where percents are used. At this time we introduce you to some of these situations, illustrating how you can use your knowledge of percents to solve problems related to these real-life situations. First we look at problems that contain the basic percent statement.

1 Problems Containing the Basic Percent Statement

If the problem contains a basic percent statement, you must recognize which sentence in the problem is the basic percent statement, then translate it to a multiplication statement. From this point you should know how to complete the problem by using what you have learned in the previous sections. Here we show some examples.

EXAMPLE 1 Two hundred fifty persons attended a convention. There were 100 women. What percent of those attending were women?

SOLUTION First, notice that the last sentence in this problem translates to a multiplication statement. The word *of* in this sentence translates to $\times$ and the word *were* translates to $=$.

What percent of those attending were women?
$$\boxed{\%} \quad \times \quad 250 \quad = \quad 100$$

After filling in the given information that 250 persons attended and 100 were women, we see that a multiplier is missing, so we divide 100 by 250 to find the missing multiplier.

$$250 \overline{)100.0} \quad \underline{100\ 0} \quad .4 = 0.4$$

The answer to the multiplication statement goes inside the division symbol.

The missing multiplier is 0.4, but because the question is *what percent*, we convert the decimal to percent form.

$$0.4 = 0.4 \times 100\% = 0\underset{\curvearrowright}{40}.\% = 40\%$$

40% of those attending were women. ∎

Try These Problems

Solve.

1. Sales tax in a certain city is 6.5% of the marked price. How much sales tax will Barbara pay when buying a dress marked $54?

2. A salesperson makes a $2100 bonus. The bonus is 14% of the base salary. What is the base salary?

3. A math teacher gave a test to 40 students. Five students received an A grade on the test. What percent of the class received an A grade?

4. The market value of a house in San Francisco is $400,000, while the replacement value is only $160,000. The market value is what percent of the replacement value?

5. Only 5% of the American troops serving in the Persian Gulf were women. If there were 26,500 women, how many American troops were there?

EXAMPLE 2 Ten years ago, Sally was a sales clerk in a major department store earning $15,000 per year. After completing her college degree and working for a few years, she now is a marketing manager for a major publisher. Her present annual salary is 420% of her salary ten years ago. What is her present annual salary?

SOLUTION The next to the last sentence in this problem is a basic percent statement, thus translates to a multiplication statement. The word *of* in this sentence translates to × and the word *is* translates to =.

Her present annual salary ($)	is	420%	of	her salary ten years ago ($)
↓		↓		↓
☐	=	420%	×	15,000
☐	=	4.2	×	15,000

Convert 420% to a decimal.
420% = 420. ÷ 100 = 4.2

The problem states that her salary 10 years ago was $15,000. After filling in this information, we see that her present salary can be obtained by finding the answer to the multiplication problem. After converting 420% to the decimal 4.2, we multiply 4.2 by 15,000 to obtain her present annual salary.

$$15,000 \times 4.2 = 63,000$$

Her present annual salary is $63,000. ■

EXAMPLE 3 A football team won 9 games, which were 60% of the total number of games played. How many games did the team play?

SOLUTION The first sentence is a basic percent statement, thus translates to a multiplication statement. The word *of* in the phrase *60% of the total* translates to ×, and the word *were* translates to =.

The 9 games won	were	60%	of	the total games played
	↓		↓	
9	=	60%	×	☐
9	=	0.6	×	☐

Convert 60% to a decimal.
60% = 60 ÷ 100
    = .60 = 0.6

A multiplier is missing, so divide 9 by 0.6 to find the missing multiplier. This missing multiplier is the total number of games played.

```
     1 5. = 15
0.6)9.0
    6
    3 0
    3 0
```

The answer to the multiplication statement goes inside the division symbol.

The team played a total of 15 games. ■

Try Problems 1 through 5.

Try These Problems

Solve.

6. The population of a small town decreased from 750 to 732. What percent of the original population was the decrease?

7. Mr. Ortez's hourly wage increased from $15 per hour to $20 per hour. The increase is what percent of the original?

EXAMPLE 4 The enrollment of a college in Pennsylvania increased from 25,000 to 28,000 students. The increase in enrollment is what percent of the original enrollment?

SOLUTION The last sentence is a basic percent statement, thus translates to a multiplication statement. The word *of* in the phrase *what percent of the original* translates to ×, and the word *is* translates to =.

$$\text{The increase} \;\; \underset{\downarrow}{is} \;\; \text{what percent} \;\; \underset{\downarrow}{of} \;\; \text{the original enrollment}$$

$$\text{The increase} = \boxed{\%} \times 25{,}000$$

The increase in enrollment is not given in the problem, but we can find it by subtracting 25,000 from 28,000.

$$\text{Increase in enrollment} = \text{larger enrollment} - \text{smaller enrollment}$$
$$\text{Increase} = 28{,}000 - 25{,}000$$
$$\text{Increase} = 3000$$

The increase in enrollment is 3000 students. Now we have the following.

$$\text{The increase} \;\; \underset{\downarrow}{is} \;\; \text{what percent} \;\; \underset{\downarrow}{of} \;\; \text{the original enrollment}$$

$$3000 = \boxed{\%} \times 25{,}000$$

A multiplier is missing, so divide 3000 by 25,000 to find the missing multiplier.

$$3000 \div 25{,}000 = \frac{3000}{25{,}000}$$
$$= \frac{3 \times \cancel{1000}}{25 \times \cancel{1000}}$$
$$= \frac{3}{25}$$

The answer to the multiplication problem is written first.

The missing multiplier is the fraction $\frac{3}{25}$, but the question is *what percent*, so we convert the fraction to a percent.

$$\frac{3}{25} = \frac{3}{25} \times 100\%$$
$$= \frac{3}{\underset{1}{\cancel{25}}} \times \frac{\overset{4}{\cancel{100}}}{1}\%$$
$$= 12\%$$

The increase is 12% of the original enrollment. ∎

 Try Problems 6 and 7.

2 Problems Containing Unclear Language

Suppose a truck contains 200 pieces of fruit, including 40 apples, 100 peaches, and 60 nectarines. It is easy to verify that 20% of the total are apples, 50% of the total are peaches, and 30% of the total are nectarines. In a situation like this, it is assumed that the parts are some percent *of the total,* so we often use language that is not as precise as the basic percent statement you are familiar with. Here are some examples of what might be said and what it really means.

Unclear Language	*Precise Translation*
20% are apples.	20% of the total are apples.
What percent are nectarines?	What percent of the total are nectarines?
The load is 50% peaches.	50% of the load is peaches.
The solution is 6% acid.	6% of the total solution is acid.

Observe that the unclear statements on the left do not contain the word *of* that translates to multiplication, so you must rephrase the statement into the precise form before translating it to a multiplication statement. It is important to remember that when language is used in this way, *the part is some percent of the total.*

EXAMPLE 5 Robyn took the percent test. She got 18 out of 20 questions correct. What percent did she get correct?

SOLUTION The question *what percent did she get correct* is really asking what percent *of the total* is the number she got correct? Now we translate this basic percent statement to a multiplication statement.

$$\underbrace{\text{What percent}}_{\boxed{\%}} \text{ of } \underbrace{\text{the total}}_{20} \text{ is } \underbrace{\text{the number she got correct?}}_{18}$$

$$\boxed{}\% \times 20 = 18$$

A multiplier is missing, so divide 18 by 20 to find the missing multiplier.

$$18 \div 20 = \frac{18}{20}$$ The answer to the multiplication probem is written first.

$$= \frac{18}{20} \times 100\%$$ Convert the fraction to a percent.

$$= \frac{18}{\underset{1}{\cancel{20}}} \times \frac{\overset{5}{\cancel{100}}}{1} \%$$

$$= 90\%$$

Robyn got 90% correct. ■

EXAMPLE 6   A parking lot contains $66\frac{2}{3}\%$ cars and the rest are trucks. If there are 480 cars, what is the total number of automobiles?

SOLUTION   The unclear phrase *a parking lot contains $66\frac{2}{3}\%$ cars* means that $66\frac{2}{3}\%$ *of the total* are cars. We have the following:

$$66\frac{2}{3}\% \text{ of the total are cars}$$

$$\frac{2}{3} \times \boxed{} = 480$$

Convert the percent to a fraction.

$$66\frac{2}{3}\% = \frac{200}{3} \div \frac{100}{1}$$

$$= \frac{\overset{2}{\cancel{200}}}{3} \times \frac{1}{\underset{1}{\cancel{100}}}$$

$$= \frac{2}{3}$$

A multiplier is missing so divide 480 by $\frac{2}{3}$ to find the missing multiplier.

$$480 \div \frac{2}{3} = \frac{\overset{240}{\cancel{480}}}{1} \times \frac{3}{\underset{1}{\cancel{2}}}$$

The answer to the multiplication problem is written first.

$$= 720$$

There are 720 automobiles. ■

EXAMPLE 7   A punch was made by mixing orange juice and champagne. The punch is 70% orange juice and 30% champagne. If the total amount of punch is 140 ounces, how many ounces of champagne are in the punch?

SOLUTION   The phrase *the punch is 30% champagne* means that 30% *of the total* is champagne. We have the following:

$$\begin{array}{cccc} 30\% & \text{of the total} & \text{is} & \text{champagne.} \\ \downarrow & (\text{oz}) \downarrow & & (\text{oz}) \\ 0.3 & \times \quad 140 & = & \boxed{} \end{array}$$

Convert 30% to a decimal.
$30\% = 30 \div 100 = 0.3$

The answer to the multiplication problem is missing, so multiply 0.3 by 140 to find the missing number.

$$0.3 \times 140 = 42$$

There are 42 ounces of champagne in the punch. ■

Try These Problems

8. A town in New Jersey has a population of 63,000. After a week of cold damp weather, 210 people in the town caught a virus. What percent caught the virus?

9. A box contains a total of 40 balls, including 20% red balls, 50% black balls, and 30% blue balls. How many blue balls are there?

10. As an inspector, you reject 7 out of 2000 products tested. What percent did you reject?

11. Chitat prepares a solution by mixing acid and water. The total volume of the solution is 390 milliliters. If the solution is $6\frac{2}{3}\%$ acid, what is the volume of acid in the solution?

12. A bottle of wine is 12% alcohol. If the alcohol volume is 90 milliliters, what is the volume of all the wine in the bottle?

EXAMPLE 8 A pharmacist mixes an alcohol solution that is 15% alcohol. If the solution contains 60 milliliters of alcohol, what is the total volume of the solution?

SOLUTION The phrase *alcohol solution that is 15% alcohol* means that 15% *of all the solution* is alcohol. We have the following:

15% of all the solution (mℓ) is alcohol (mℓ)

0.15 × ☐ = 60

Convert 15% to a decimal.
15% = 15 ÷ 100 = 0.15

A multiplier is missing, so divide 60 by 0.15 to find the missing multiplier.

$$0.15\overline{)60.00} = 400$$

The total volume of the alcohol solution is 400 milliliters.

Try Problems 8 through 12.

Answers to Try These Problems

1. $3.51   2. $15,000   3. 12.5% or $12\frac{1}{2}\%$ 4. 250%
5. 530,000 troops   6. 2.4% or $2\frac{2}{5}\%$ 7. $33\frac{1}{3}\%$ 8. $\frac{1}{3}\%$
9. 12 blue balls   10. 0.35% or $\frac{7}{20}\%$ 11. 26 mℓ
12. 750 mℓ

EXERCISES 9.5

Solve.

1. In a chemistry class, 9 students received grades of D or F. This represents 20% of the class. How many students are in the class?

2. Karl's regular salary is $12 per hour. His overtime pay is 150% of his regular pay. How much does Karl earn each hour when working overtime?

3. As an inspector in a factory, you reject 3 products in a sample of 1500 products. The number you rejected is what percent of the total sample?

4. The price of a portable radio decreased from $250 to $135. The lower price is what percent of the original price?

5. Sales tax in a town in Kentucky is 5.4% of the marked price. How much sales tax will George pay on a bedroom set marked $2800?

6. A basketball team won $66\frac{2}{3}\%$ of the total games played. If they played 126 games, how many games did they win?

7. Tom earns $32,000 per year as an auto mechanic and his girlfriend Gloria earns $48,000 per year as a teacher. Gloria's salary is what percent of Tom's salary?

8. A bowl of punch contains 6.6 ounces of pure alcohol, which is 2.4% of the total volume of the punch. What is the total volume?

9. The number of homeless people in a certain city increased from 2400 to 3600 in one year. The increase is what percent of the original number of homeless people?

10. The price of coffee decreased from $6.25 per pound to $4.25 per pound. What percent of the original price is the decrease in price?

11. A shipment of grapefruit consists of 100 crates of pink seedless grapefruit and 150 crates of white seedless grapefruit. What percent of the total are the white seedless grapefruit?

12. Joel sells men's colognes in a large department store. In addition to his regular monthly salary of $1500, he earns a commission that is $\frac{8}{10}$% of his total sales. If his total sales were $24,000 for the month of April, what were his total earnings for that month?

13. A can of concentrated soup contains 15% water. If the can contains 3 ounces of water, what is the total weight of the soup?

14. A bottle of gin contains 40% alcohol. What is the volume of alcohol in a 1000-milliliter bottle of gin?

15. A parking lot contains 720 automobiles. The lot includes 40% Hondas, 35% Toyotas, and 25% Buicks. How many Toyotas are there?

16. From 375 prospective jurors, only 15 were selected to serve on the jury. What percent were selected?

17. A biology class consists of 18 men and 12 women. What percent are women?

18. Manuel prepares a solution by mixing acid and water. If the solution is $7\frac{1}{3}$% acid and the volume of acid is 22 milliliters, what is the total volume of the solution?

19. You make a punch by mixing 16 ounces of orange juice, 14 ounces of lemon-lime soda, and 10 ounces of ginger ale. What percent is ginger ale?

20. A shipment of 700 crates of lettuce contains 15% romaine lettuce, 60% iceberg lettuce, and 25% butter lettuce. How many crates of romaine lettuce are there?

A statistician surveyed 1500 persons. She asked them to give their favorite color. She organized the results by making the circle graph shown here. Use the graph to answer Exercises 21 and 22.

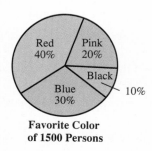

Favorite Color of 1500 Persons

21. How many persons chose red?
22. How many persons chose black?

9.6 APPLICATIONS: SALES TAX, COMMISSION, INTEREST, AND TIPS

OBJECTIVES
- Solve a percent application that involves sales tax.
- Solve a percent application that involves commission.
- Solve a percent application that involves simple interest earned.
- Solve a percent application that involves monthly interest paid on a charge card or account.
- Solve a percent application that involves restaurant tips.

In this section we look carefully at some particular situations where percents are used.

1 Problems Involving Sales Tax

Suppose you purchase a TV that is marked $350. You really pay more than $350 for this TV, because you are charged a sales tax in addition to the marked price. The **sales tax** is computed by taking a certain percent of the marked price. This percent varies depending on where you purchase the TV. Suppose you are in a city that computes the sales tax by taking 6% of the marked price.

$$\text{Sales tax} = 6\% \text{ of the marked price}$$
$$(\$) \qquad\qquad\qquad (\$)$$
$$= 0.06 \times 350$$
$$= 21$$

The tax charged for this TV is $21. The percent, 6%, is called the **sales tax rate.** Observe that the rate is multiplied by the marked price to obtain the tax. If you want to know the total cost of this TV, you must now add the $21 tax to the $350 marked price.

$$\text{Total cost} = \text{marked price} + \text{tax}$$
$$(\$) \qquad\qquad (\$) \qquad\quad (\$)$$
$$= \quad 350 \quad + \quad 21$$
$$= \quad 371$$

The total cost of the TV is $371.

The sales tax rate is always multiplied by the marked price to obtain the tax, and the tax is alway added to the marked price to obtain the total cost. You want to be familiar with these relationships since we encounter sales tax in our daily lives. Here we summarize the facts.

Rules Involving Sales Tax

Sales tax = sales tax rate (%) × marked price

Total cost = marked price + sales tax

Now we look at some examples where we use these rules.

Try These Problems

Solve.
1. The sales tax rate for a certain town in Texas is 5.8%. How much sales tax is paid on a $490 purchase?

2. The marked price of a car is $12,500. The sales tax amounts to $775. What is the sales tax rate? (Express the answer as a percent.)

3. Mr. Signorile buys a new pair of skis with bindings. If the sales tax is $37.80 and the tax rate is 7.5%, find the marked price of the skis.

4. Toni purchases some construction materials that are marked $6800. If the sales tax rate is 8.4%, what is the total cost of the materials?

5. Andrea buys 4 turtlenecks marked $12.50 each and 3 pairs of socks marked $2.80 each. If the sales tax rate is 7%, what is the total amount she pays? (Round off to the nearest cent.)

EXAMPLE 1 The marked price of a refrigerator was $760. A sales tax of $41.80 was added. What is the sales tax rate? Express the answer as a percent.

SOLUTION We have the following.

$$\underset{\text{tax}}{\text{Sales}} = \underset{\text{rate}}{\text{sales tax}} \times \underset{\text{price}}{\text{marked}}$$

$$41.80 = \boxed{\%} \times 760$$

A multiplier is missing, so divide 41.80 by 760 to find the missing multiplier.

$$\begin{array}{r} .055 = 5.5\% \\ 760\overline{)41.800} \\ \underline{38\ 00} \\ 3\ 800 \\ \underline{3\ 800} \end{array}$$

Convert the decimal to percent.

The sales tax rate is 5.5%. ■

 Try Problems 1 through 3.

EXAMPLE 2 Mr. Washington buys 3 pairs of shoes marked $140 each. If the sales tax rate is 7.2%, what is the total amount he pays?

SOLUTION First we find the total cost before the tax is added.

$$\underset{\text{before tax}}{\text{Total cost}} = 3 \times 140$$
$$(\$)$$
$$= 420$$

The marked price of the 3 pairs of shoes is $420. Next we compute the sales tax.

$$\underset{\substack{\text{tax} \\ (\$)}}{\text{Sales}} = \underset{\text{rate}}{\text{sales tax}} \times \underset{\substack{\text{price} \\ (\$)}}{\text{marked}}$$

$$\boxed{} = 7.2\% \times 420$$
$$\boxed{} = 0.072 \times 420$$

Convert the percent to a decimal

The answer to the multiplication problem is missing so multiply 0.072 by 420.

$$0.072 \times 420 = 30.24$$

The sales tax is $30.24. The question was to find the *total* amount he pays, so finally we add the tax to the marked price.

$$\underset{\text{after tax}}{\text{Total cost}} = \underset{\text{price}}{\text{marked}} + \underset{\text{tax}}{\text{sales}}$$
$$= 420 + 30.24$$
$$= 450.24$$

He pays a total of $450.24 for the 3 pairs of shoes. ■

 Try Problems 4 and 5.

2 Problems Involving Commission

Suppose you sell electronic goods for a large retail outlet. You earn $1500 per month, plus 1.2% of your total sales. Also assume that in July you sold $250,000 worth of merchandise. The $1500 is called your **base salary.** The additional amount earned that depends on your total sales is called the **commission.** The commission is computed by taking 1.2% of your total sales.

$$\begin{aligned}\text{Commission (\$)} &= 1.2\% \text{ of the total sales (\$)}\\ &= 0.012 \times 250{,}000\\ &= 3000\end{aligned}$$

The commission earned for the month of July is $3000. The percent 1.2% is called the **commission rate.** The commission rate varies widely depending on the particular situation. If you want to know the total salary for the month of July, you must add the commission to the base salary.

$$\begin{aligned}\text{Total salary (\$)} &= \text{base salary (\$)} + \text{commission (\$)}\\ &= 1500 + 3000\\ &= 4500\end{aligned}$$

You earned $4500 in July.

Often the commission is based on the total sales. When this is the case we have the following rules.

Rules Involving Commission Based on Total Sales

Commission = commission rate (%) × total sales

Total salary = base salary + commission

Now we look at some examples where we use these rules.

EXAMPLE 3 Mr. Nguyen is paid a commission for selling rugs at the Emporium that is $7\frac{1}{2}\%$ of his total sales. Last week he received a commission of $156. What were his total sales for the week?

SOLUTION We have the following:

$$\text{Commission (\$)} = 7\frac{1}{2}\% \times \text{total sales (\$)}$$
$$156 = 0.075 \times \boxed{} \quad \text{Convert the percent to a decimal.}$$

A multiplier is missing, so divide 156 by 0.075 to find the missing multiplier.

$$0.075\overline{)156.000} = 2080$$

He sold $2080 worth of rugs for the week.

Try These Problems

Solve.

6. When Ms. Leong sells a refrigerator for $560, she receives a $21 commission. If her commission is based on the total sales, what is her commission rate? (Express the answer as a percent.)

7. Mr. Praxmarer sells ski equipment and clothing to retail stores across the United States and Canada. He is paid a commission that is $\frac{2}{3}$% of the total sales. If he sells $75,000 worth of ski equipment to the Truckee Sport Shop, how much commission does he earn on this sale?

8. Don sells men's clothing in a large department store. In addition to his regular salary, he earns a commission that is 0.8% of his total sales. Last month he earned a $75 commission. What were his total sales for that month?

9. Ms. Alvarez sells appliances to department stores. She receives a base salary of $1850 per month, plus a commission that is 0.75% of her total sales. Last month she sold $95,800 worth of appliances. What were her total earnings for the month?

 Try Problems 6 through 8.

EXAMPLE 4 Ms. Lane sells computer software. She earns a base salary of $2700 per month, plus a commission that is 1.5% of her total sales. One month she sold $82,400 worth of software. What were her total earnings for that month?

SOLUTION First find her commission.

Commission ($) = 1.5% of total sales ($)

☐ = 0.015 × 82,400

The answer to the multiplication problem is missing, so multiply 0.015 by 82,400 to find the commission.

0.015 × 82,400 = 1236

The commission is $1236. Now we add the commission to the base salary to find the total earnings.

Total earnings ($) = base salary ($) + commission ($)
= 2700 + 1236
= 3936

She earned $3936 that month. ■

 Try Problem 9.

3 Problems Involving Interest Earned

Suppose you put $5000 in a savings account that pays an annual yield of 4%. If you make no further deposits and no withdrawals, at the end of one year you will have more than $5000 in that account, because the bank pays you additional money for leaving your money in the account. The amount of money that the account earns is called the **interest** or the **yield.** To compute the amount of interest earned, we take 4% of the **amount invested.**

Interest ($) = 4% × amount invested ($)
= 0.04 × 5000
= 200

This account earns $200 in one year. The 4% is called the **interest rate** or the **rate of return.** The interest rate varies depending on the bank and the type of account. To find out how much money is in the account at the end of one year, we add the interest to the original amount invested.

Final amount ($) = amount invested ($) + interest ($)
= 5000 + 200
= 5200

At the end of one year, the account contains $5200.

Try These Problems

Solve.

10. Melissa inherited some money from her grandmother. She invested all of it in an account that pays 12% interest. If she earned $1560 interest in one year, what was her inheritance?

11. Chuck invested $40,000 in an account and earned $3000 interest in one year. Find the interest rate. (Express the answer as a percent.)

12. Britt-Marie invested $12,900 in an account that earns $6\frac{2}{3}$% interest. How much interest did she earn in one year?

There is more than one type of interest that can be earned on an account. **Simple interest** is interest earned only on the amount invested, while **compound interest** is interest earned on both the amount invested and on the interest previously earned. In this text we assume that the interest earned is simple interest.

We also use the word *interest* to mean the earnings on any type of account, including savings accounts, money market funds, mutual fund accounts, and stock accounts. We use the phrase *interest rate* for the average annual rate of return on these accounts.

In general, if you invest a certain amount of money that earns simple interest, make no further deposits or withdrawals, and leave the money for one year, then the following rules apply.

Rules Involving Interest Earned

Interest (or yield) = interest rate (or rate of return) × amount invested

Final amount = amount invested + interest

Here are some examples where we use these rules.

EXAMPLE 5   Amy put $620 in a mutual fund account that earned a $58.90 yield in a year. Find the rate of return. (Express the answer as a percent.)

SOLUTION   We have the following.

$$\text{Yield} = \text{rate of return} \times \text{amount invested}$$
$$58.90 = \boxed{} \% \times 620$$

A multiplier is missing, so we divide 58.90 by 620 to find the missing multiplier.

$$620 \overline{)58.90}.095 = 9.5\%$$
$$55\,80$$
$$\overline{3\,100}$$
$$3\,100$$

The rate of return is 9.5%. ∎

Try Problems 10 through 12.

EXAMPLE 6   Mr. Kung made an investment of $7800 in a stock account that paid $8\frac{1}{3}$% interest for one year. How much money was in the account at the end of one year?

SOLUTION   First find the amount of interest that the account earned.

$$\begin{aligned}\text{Interest (\$)} &= \text{interest rate} \times \text{amount invested (\$)} \\ &= 8\frac{1}{3}\% \times 7800\end{aligned}$$

Try These Problems

Solve.

13. Ms. Allen put $860 in an account that earns interest at an annual rate of $5\frac{3}{4}$%. How much money is in the account at the end of one year?

$$= \frac{1}{12} \times \frac{7800}{1}$$

$$= \frac{1}{3 \times \cancel{4}} \times \frac{78 \times \cancel{100}^{25}}{1}$$

$$= \frac{78 \times 25}{3}$$

$$= \frac{\cancel{3} \times 26 \times 25}{\cancel{3}}$$

$$= 650$$

Convert $8\frac{1}{3}$% to a fraction.

$$8\frac{1}{3}\% = \frac{25}{3} \div 100$$

$$= \frac{25}{3} \times \frac{1}{\cancel{100}_{4}}$$

$$= \frac{1}{12}$$

The interest is $650. Now we add the interest to the amount invested to find the final amount in the account at the end of one year.

$$\begin{array}{c}\text{Final} \\ \text{amount} \\ (\$)\end{array} = \begin{array}{c}\text{amount} \\ \text{invested} \\ (\$)\end{array} + \begin{array}{c}\text{interest} \\ (\$)\end{array}$$

$$= 7800 + 650$$

$$= 8450$$

The amount in the account at the end of one year is $8450. ■

Try Problem 13.

4 Problems Involving Interest Paid

Suppose you purchase a sofa that costs $1500 by charging it on your credit card. When you receive your bill at the end of the month, assume you pay only $300 toward the $1500 bill. The amount that you did not pay, $1200, is called your **unpaid balance.** The next month when you receive your bill, the balance will be more than $1200, because the credit card company charges you **interest** for not paying the full bill by the end of last month. Assume your credit card charges 1.8% of the unpaid balance. The interest charged for one month is computed as follows.

$$\begin{array}{c}\text{Interest} \\ (\$)\end{array} = 1.8\% \text{ of } \begin{array}{c}\text{unpaid} \\ \text{balance} \\ (\$)\end{array}$$

$$= 0.018 \times 1200$$

$$= 21.60$$

The interest charged for the month is $21.60. Sometimes this interest is called a **service charge** or **finance charge.** The percent 1.8% is called the **monthly interest rate.** To compute the amount of the bill, we add the interest to the unpaid balance.

$$\begin{array}{c}\text{Total} \\ \text{bill} \\ (\$)\end{array} = \begin{array}{c}\text{unpaid} \\ \text{balance} \\ (\$)\end{array} + \begin{array}{c}\text{interest} \\ (\$)\end{array}$$

$$= 1200 + 21.60$$

$$= 1221.60$$

The balance is now $1221.60.

Try These Problems

Solve.

14. Mr. Hummer has a credit card that charges 1.9% interest on the unpaid balance each month. Last month Mr. Hummer's unpaid balance was $3700. How much interest did he pay for the month?

15. Pablo received his credit card bill. It showed an unpaid balance from the last month of $540 and a finance charge of $11.34. Find the monthly interest rate.

16. Sharon charged a grand piano on an account that charges a monthly finance charge that is $1\frac{2}{3}$% of the unpaid balance. If her unpaid balance last month was $4800, how much is her bill this month?

Here we summarize the facts discussed in the previous paragraph.

Rules Involving Interest Paid on an Unpaid Balance

Interest = monthly interest rate × unpaid balance

Total bill = unpaid balance + interest

Now we look at some examples where we use these rules.

EXAMPLE 7 Asher has a charge account at a department store that adds a 1.5% service charge to the unpaid balance at the end of each month. Last month his unpaid balance was $76.32. How much service charge did he pay? (Round off to the nearest cent.)

SOLUTION We have the following.

$$\begin{aligned}\text{Service charge (\$)} &= \text{interest rate} \times \text{unpaid balance (\$)} \\ &= 1.5\% \times 76.32 \\ &= 0.015 \times 76.32 \\ &= 1.1448 \\ &\approx 1.14\end{aligned}$$

Asher is charged a service charge of $1.14. ∎

 Try Problems 14 and 15.

EXAMPLE 8 Grace has a credit card that charges interest that is 2% of the unpaid balance. If her unpaid balance from last month was $640, and she makes no further pruchases, how much is her bill this month?

SOLUTION First we find the interest she was charged for leaving an unpaid balance of $640.

$$\begin{aligned}\text{Interest (\$)} &= \text{interest rate} \times \text{unpaid balance (\$)} \\ &= 2\% \times 640 \\ &= 0.02 \times 640 \\ &= 12.80\end{aligned}$$

Grace is charged $12.80 interest. Next we add the interest to the unpaid balance to find her total bill.

$$\begin{aligned}\text{Total bill (\$)} &= \text{unpaid balance (\$)} + \text{interest (\$)} \\ &= 640 + 12.80 \\ &= 652.80\end{aligned}$$

Her bill this month is $652.80. ∎

 Try Problem 16.

5 Solving Problems Involving Tips

Suppose you go out to eat at a restaurant where a person serves your table. It is customary to leave the server some money in addition to the dinner bill. This additional amount is called a **tip** or **gratuity.** If you were pleased with the service, it is common to leave a tip that is at least 15% of the dinner bill. The percent 15% is called the **rate.** Sometimes we take the percent of the dinner bill before tax is added. At other times we take the percent of the dinner bill after the tax is added. At the restaurant you get to choose how much money you want to leave for the tip. Here we summarize the rules about tips.

> *Rules Involving Tips*
> Tip = rate × dinner bill
> Total bill = dinner bill + tip

Here are some examples where we use these rules.

EXAMPLE 9 Mr. Castro was not pleased with the service at the seafood restaurant. In order to express his dissatisfaction, he left a $3.50 tip which was only 10% of the dinner bill. What was the dinner bill?

SOLUTION We have the following.

$$\text{Tip} = \text{rate} \times \text{dinner bill}$$
($)   ($)
3.50 = 10% × ☐
3.50 = 0.10 × ☐

A multiplier is missing, so divide 3.50 by 0.10 to find the dinner bill.

$$0.1\overline{)3.5} = 35$$

Note that we can divide by 0.1 instead of 0.10 because these decimals are equal.

The dinner bill was $35. ■

EXAMPLE 10 A dinner bill for three persons was $60. A $5 gratuity was left. The gratuity is what percent of the dinner bill? Is this an appropriate amount?

SOLUTION We have the following.

Gratuity is what percent of the dinner bill
5 = ☐% × 60

A multiplier is missing, so divide 5 by 60 to find the missing multiplier.

Try These Problems

Solve.

17. Sharmon took Ted out to eat at an exclusive French restaurant in New Orleans. She left a 15% tip that amounted to $9.30. What was the dinner bill?

18. Karla is pleased with the service at the Chinese restaurant. She wants to leave an 18% tip. If the dinner bill is $45, how much tip should she leave?

19. The dinner bill for a group of persons is $230. A $46 gratuity is left. The gratuity is what percent of the dinner bill?

$$5 \div 60 = \frac{5}{60}$$
$$= \frac{5}{60} \times \frac{100}{1}\%$$
$$= \frac{5}{6 \times \cancel{10}} = \frac{\cancel{10} \times 10}{1}\%$$
$$= \frac{50}{6}\%$$
$$= \frac{25}{3}\%$$
$$= 8\frac{1}{3}\%$$

Convert the fraction to a percent.

The gratuity is only $8\frac{1}{3}\%$ of the dinner bill. This is not an appropriate amount unless the server did a poor job. ∎

 Try Problems 17 through 19.

Here we summarize the information presented in this section.

Sales Tax

The sales tax is some percent of the marked price. The percent is called the sales tax rate.

Sales tax = sales tax rate (%) × marked price
Total cost = marked price + sales tax

Commission Based on Total Sales

The commission earned by a sales person is some percent of the total sales. The percent is called the commission rate.

Commission = commission rate (%) × total sales
Total salary = base salary + commission

Interest Earned

The interest earned on an investment is some percent of the original investment. The percent is called the monthly interest rate.

Interest = monthly interest rate (%) × amount invested
Final amount = amount invested + interest

Interest Paid

The interest paid on an unpaid balance is some percent of the unpaid balance. The percent is called the monthly interest rate.

Interest = monthly interest rate (%) × unpaid balance
Total bill = unpaid balance + interest

Tips

The tip (or gratuity) paid to a restaurant server is some percent of the dinner bill. The percent is usually about 15%.

Tip = rate (%) × dinner bill
Total bill = dinner bill + tip

Answers to Try These Problems

1. $28.42   2. 6.2%   3. $504   4. $7371.20   5. $62.49
6. $3\frac{3}{4}$% or 3.75%   7. $500   8. $9375   9. $2568.50
10. $13,000   11. 7.5% or $7\frac{1}{2}$%   12. $860   13. $909.45
14. $70.30   15. 2.1%   16. $4880   17. $62   18. $8.10
19. 20%

EXERCISES 9.6

Solve.

1. Gloria buys a bed that is marked $430. If the sales tax rate is 7.8%, how much sales tax does she pay?

2. The marked price of a radio is $135. A sales tax of $9.18 was added. What is the sales tax rate? (Express the answer as a percent.)

3. Larry bought a tool box and he was charged a sales tax of $1.88. If the tax rate is 8%, find the marked price of the tool box.

4. Leslie bought 4 sweaters at $42 each and 3 skirts at $69 each. If the sales tax rate is 5.4%, what is the total amount she paid for these items?

5. When Ms. Goldman sells a computer for $4000, she receives a $100 commission. If her commission is based on the total sales, what is her commission rate? (Express the answer as a percent.)

6. Mr. Lemus sells used cars. He is paid a commission that is 0.72% of the total sales. If he sells a car for $38,000, how much commission does he earn on this sale?

7. Paul sells commercial real estate. In addition to his base salary of $2000 per month, he receives a commission that is 0.38% of the total sales. Last month he sold $1,800,000 worth of real estate. What were his total earnings for that month?

8. Le Ha's commission rate for selling clothes at an exclusive boutique is 24%. For a three-day weekend, she earned a $480 commission. How many dollars worth of clothing did she sell during the three days?

9. Mr. Schectman invested some money in a stock account that earned $33\frac{1}{3}$% interest. If he earned $6400 interest after one year, how much did he invest?

10. Melissa earned $329 interest on an investment of $3500 in one year. Find the interest rate. (Express the answer as a percent.)

11. Ms. Jones invested $860 in an account that earns 7.6% annual interest. How much interest did she earn in one year?

12. Mr. DiGangi invested $20,000 in a growth stock account that earned 14% interest in one year. What was the total amount of money in this account at the end of one year?

13. Roberta received her credit card bill. It showed an unpaid balance from the last month of $630 and a finance charge of $11.97. Find the monthly interest rate. (Express the answer as a percent.)

14. Nell charged some vertical blinds on an account that charges a monthly finance charge that is $1\frac{5}{6}\%$ of the unpaid balance. If her unpaid balance last month was $714, how much interest is she charged on this month's bill?

15. Mr. Garcia charged a suit on an account that charges a monthly finance charge that is 1.88% of the unpaid balance. If his unpaid balance last month was $250, and he made no further charges, how much is his bill this month?

16. Susie charged a rug on an account that charges a monthly finance charge that is $1\frac{3}{4}\%$ of the unpaid balance. If her unpaid balance last month was $354 and she makes no additional purchases, how much is her bill this month? (Round up to the nearest cent.)

17. Edna, a lawyer, takes a client to lunch. She is very pleased with the service and food so she wants to leave a 20% tip. If the lunch bill is $38, how much tip should she leave?

18. Mr. Coyle is very displeased with the service he receives at a coffee shop. He left only a $0.75 tip for a bill that was $15. What percent tip did he leave?

19. Richard took his girlfriend Patricia out to eat at an expensive Italian restaurant in San Francisco. He left a $13.50 gratuity that was 15% of the dinner bill. What was the dinner bill?

20. Your dinner bill, not including tax, is $78.50. If the sales tax rate is 6%, and you leave a tip that is 15% of the dinner bill after the tax is added, what is your total bill? (Round off to the nearest cent.)

21. A student purchases 3 pens at $1.29 each and 5 notebooks at $4.69 each. If the sales tax rate is 5.2%, what is the total cost of these items? (Round off to the nearest cent.)

22. Delicia sells women's clothing for a department store. She makes a base salary of $1400 per month plus a commission that is $\frac{2}{3}\%$ of her total sales. If she sold $45,900 worth of clothing last month, what were her total earnings for the month?

23. Ms. Yee put $8000 of her royalty advances in a money market fund earning 8.8% interest annually. How much money is in the account at the end of one year?

24. David charged a television set on an account that charges a monthly service charge that is 1.5% of the unpaid balance. If his unpaid balance last month was $138 and he made no further purchases, how much is his bill this month?

25. Henry earns a weekly base salary of $420, plus 3.2% of all sales over $5000. If his total sales for one week were $6400, how much did he earn that week?

26. You plan to build a fence around a rectangular lot that is 15 yards wide and 20 yards long. If the fencing material costs $8.25 per yard and the sales tax rate is 5.6%, find the total cost of the fencing material.

A city contains 500,000 people. The bar graph below gives the age distribution of the population. Use the graph to answer Exercises 27 through 30.

27. What percent of the population is from 31 to 50 years old?
28. What percent of the population is 75 years old or older?
29. How many people in the city are 16 years old or younger?
30. How many people in the city are from 17 to 30 years old?

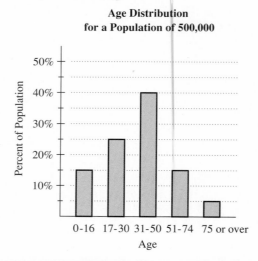

DEVELOPING NUMBER SENSE #8

FIGURING THE RESTAURANT TIP MENTALLY

Suppose you are at a restaurant and you want to leave the server a 15% tip. How can you quickly compute the tip without using paper and pencil or a calculator? There is more than one way to do this. We will discuss one method which involves finding 10% of the dinner bill, halving that amount to obtain 5% of the dinner bill, then adding those two quantities together. Here we look at an example.

Computing the tip for a $40 dinner bill.

- 10% of $40 = $4   The decimal point in 40 shifts 1 place to the left
- $4 ÷ 2 = $2   Dividing $4 in half is an easy way to obtain 5% of 40.
- Tip = $4 + $2 = $6   Add the 10% of 40 to 5% of 40 to obtain 15% of 40.

Finally, add $6 to $40 to obtain the total amount, $46.

Of course the dinner bill is not always going to be a nice number like 40. You can round off the bill to the nearest dollar before trying to do the mental work. Here is an example.

Computing the tip for a $24.27 dinner bill.

- 10% of $24 = $2.40   Round off the dinner bill to $24, then take 10% of 24. The decimal point in 24 shifts 1 place to the left.
- $2.40 ÷ 2 = $1.20   Divide $2.40 by 2 to obtain 5% of $24.
- Tip = $2.40 + $1.20 = $3.60   Add to obtain the tip.

Finally, add $3.60 to $24 to find the total amount, $27.60. (Of course you can also add the tip to the original dinner bill, $24.27, to obtain $27.87 for the total amount.)

Here we look at one more example.

Computing the tip for a $12.87 dinner bill.

- 10% of $13 = $1.30   Round off the dinner bill to $13, then take 10% of $13. The decimal point in 13 shifts 1 place to the left.
- $1.30 ÷ 2 ≈ $0.70   Divide $1.30 by 2 and round off to the nearest dime.
- Tip = $1.30 + $0.70 = $2.00   Add to obtain the tip

Finally, add the $2 tip to $13 to obtain $15 for the total amount. (Of course you can also add the tip to the original dinner bill $12.87 to obtain $14.87 for the total amount.)

Number Sense Problems

Mentally round off each to the nearest dollar and then take 10% of the result.

1. $8.67 **2.** $14.12 **3.** $25.06 **4.** $67.89

Mentally round off each to the nearest dollar, take 10% of the result, then divide that amount by 2. (Here we are obtaining an approximation for 5% of the given amount.)

5. $9.13 **6.** $18.64 **7.** $38.79 **8.** $66.34

Mentally compute a 15% tip for the following dinner bills. No rounding is necessary.

9. $10 **10.** $26 **11.** $60 **12.** $150

Mentally compute an approximate 15% tip for the following dinner bills. Begin by rounding off the dinner bill to the nearest dollar, then round off all other amounts to the nearest dime.

13. $6.78 **14.** $26.25 **15.** $48.33 **16.** $103.76

Mentally compute the total dinner bill after an approximate 15% tip has been added. Begin by rounding off the dinner bill to the nearest dollar (if necessary), then round off all other amounts to the nearest dime. Finally add the tip to the rounded dinner bill to find the total amount.

17. $16 **18.** $25 **19.** $27.78 **20.** $68.14

9.7 APPLICATIONS: PERCENT INCREASE, PERCENT DECREASE, AND MARKUP

OBJECTIVES
- Solve an application that involves percent increase.
- Solve an application that involves percent decrease.
- Solve an application that involves markup based on the cost.
- Solve an application that involves markup based on the selling price.

1 Problems Involving Percent Increase

Suppose you are earning $200 per week, and your boss informs you that you are to receive a 12% raise. The 12%, called the **percent increase**, is multiplied by the original salary to obtain the increase in salary.

$$\text{Increase in salary (\$)} = \text{percent increase} \times \text{original salary (\$)}$$
$$= 12\% \times 200$$
$$= 0.12 \times 200$$
$$= 24$$

The increase or raise is $24. To compute the final salary, add the increase to the original salary.

Try These Problems

Solve.
1. In one year, the price of gasoline has increased $16\frac{2}{3}\%$. If the original price was $1.20 per gallon, what is the price after the increase?

$$\begin{aligned}\text{Final salary}\ (\$) &= \text{original salary}\ (\$) + \text{increase in salary}\ (\$) \\ &= 200 + 24 \\ &= 224\end{aligned}$$

Your new salary will be $224 per week.

Also observe that the increase in salary is the difference between the final salary and the original salary.

$$\begin{aligned}\text{Increase in salary}\ (\$) &= \text{final salary}\ (\$) - \text{original salary}\ (\$) \\ &= 224 - 200 \\ &= 24\end{aligned}$$

There are many situations where we use percent increase to show how much a quantity has increased over its original value. The percent increase is taken of the original amount to obtain the increase. Here we give the rules that are relevant.

Rules Involving Increase and Percent Increase

Increase = percent increase × original amount

Final amount = original amount + increase

Increase = final amount − original amount

Here are some examples where we use these rules.

EXAMPLE 1 In Soraya's first year as president of the company, profits have increased 18%. If the original profit was $700,000, what is the profit after one year?

SOLUTION First we find the increase in the profit. We have the following:

$$\begin{aligned}\text{Increase in profit}\ (\$) &= \text{percent increase} \times \text{original profit}\ (\$) \\ &= 18\% \times 700{,}000 \\ &= 0.18 \times 700{,}000 \\ &= 126{,}000\end{aligned}$$

The increase in the profit is $126,000. To find the final profit, we add the increase to the original profit.

$$\begin{aligned}\text{Final profit}\ (\$) &= \text{original profit} + \text{increase in profit}\ (\$) \\ &= 700{,}000 + 126{,}000 \\ &= 826{,}000\end{aligned}$$

The profit after one year is $826,000. ∎

 Try Problem 1.

Try These Problems

Solve.

2. Dahlia receives a $1.44-per-hour raise, which is 4.5% over her original hourly wage. Find her original hourly wage.

3. The population of a small town in Wyoming increased by 20 persons. If the original population was 320, what is the percent increase?

4. A dog's weight increased from 10 pounds to 25 pounds. What percent increase is this?

EXAMPLE 2 Mr. Vetterli gained 9 pounds over the summer, which was a 6% increase. What was his original weight?

SOLUTION The percent increase is given as 6% and the increase in weight is given as 9 pounds. The percent increase is taken of the original weight to obtain the increase in weight, so we have the following.

$$\begin{array}{c}\text{Increase}\\ \text{in weight}\\ \text{(lb)}\end{array} = \begin{array}{c}\text{percent}\\ \text{increase}\end{array} \times \begin{array}{c}\text{original}\\ \text{weight}\\ \text{(lb)}\end{array}$$

$$9 = 6\% \times \boxed{}$$
$$9 = 0.06 \times \boxed{}$$

A multiplier is missing, so divide 9 by 0.06 to find the missing multiplier.

$$\begin{array}{r}150. = 150\\ 0.06\overline{)9.00}\\ \underline{6}\\ 30\\ \underline{30}\end{array}$$

His original weight was 150 pounds. ∎

 Try Problem 2.

EXAMPLE 3 Due to inflation, the price of a box of cereal increased from $2.50 to $3.00. What is the percent increase?

SOLUTION First we find the increase in price by subtracting the original price from the final price.

$$\begin{array}{c}\text{Increase}\\ \text{in price}\\ \text{(\$)}\end{array} = \begin{array}{c}\text{final}\\ \text{price}\\ \text{(\$)}\end{array} - \begin{array}{c}\text{original}\\ \text{price}\\ \text{(\$)}\end{array}$$
$$= 3.00 - 2.50$$
$$= 0.50$$

The increase in price is $0.50. This increase is a certain percent of the original price, so we have the following.

$$\begin{array}{c}\text{Increase}\\ \text{in price}\\ \text{(\$)}\end{array} = \begin{array}{c}\text{percent}\\ \text{increase}\end{array} \times \begin{array}{c}\text{original}\\ \text{price}\\ \text{(\$)}\end{array}$$
$$0.50 = \boxed{\%} \times 2.50$$

A multiplier is missing, so divide 0.50 by 2.50 to find the missing multiplier.

$$\begin{array}{r}.2 = 20\%\\ 2.5\overline{)0.50}\\ \underline{50}\end{array}$$

The percent increase is 20%. ∎

 Try Problems 3 and 4.

2 Problems Involving Percent Decrease

Suppose that the price of a comforter was $180, and you read in the newspaper that the store is having a 40%-off sale. The 40%, called the **percent decrease,** is multiplied by the original price to obtain the **decrease** (or **discount** or **savings**).

$$\begin{aligned}
\text{Decrease in price (\$)} &= \text{percent decrease} \times \text{original price (\$)} \\
&= 40\% \times \$180 \\
&= 0.40 \times \$180 \\
&= \$72
\end{aligned}$$

The decrease in price is $72. To compute the final price, subtract the decrease from the original price.

$$\begin{aligned}
\text{Final price (\$)} &= \text{original price (\$)} - \text{decrease in price (\$)} \\
&= 180 - 72 \\
&= 108
\end{aligned}$$

The final price without tax is $108.

There are many situations where we use percent decrease to show how much a quantity has decreased from its original value. The percent decrease is taken of the original amount to obtain the decrease. Here we give the rules that are relevant.

Rules Involving Decrease and Percent Decrease

Decrease = percent decrease × original amount

Final amount = original amount − decrease

Decrease = original amount − final amount

Here are some examples where we use these rules.

EXAMPLE 4 Due to financial problems, a company had to reduce the average hourly wage of its workers from $24 to $22.68. What percent decrease is this?

SOLUTION First, find the decrease in wage by subtracting the final wage from the original wage.

$$\begin{aligned}
\text{Decrease in wage (\$)} &= \text{original wage (\$)} - \text{final wage (\$)} \\
&= 24 - 22.68 \\
&= 1.32
\end{aligned}$$

9.7 Applications: Percent Increase, Percent Decrease, and Markup

 Try These Problems

Solve.

5. A television that was marked $350 is now on sale for 30% off. Find the savings.

6. While you were sick, your weight dropped 14 pounds. If this was an 8% decrease, what was your original weight?

7. The price of a scientific hand-held calculator decreased from $20 to $12. What percent decrease is this?

8. Due to less demand, the price of beef has decreased by $12\frac{1}{2}$% over the past few months. If the original price was $3.60 per pound, what is the current price?

The decrease is $1.32. This decrease in wage is a certain percent of the original wage, so we have the following.

$$\begin{array}{c}\text{Decrease}\\\text{in wage}\\(\$)\end{array} = \begin{array}{c}\text{percent}\\\text{decrease}\end{array} \times \begin{array}{c}\text{original}\\\text{wage}\\(\$)\end{array}$$

$$1.32 = \boxed{\%} \times 24$$

A multiplier is missing, so divide 1.32 by 24 to find the missing multiplier.

$$24\overline{)1.320} = 5.5\%$$
$$\underline{1\,20}$$
$$120$$
$$\underline{120}$$

The percent decrease is 5.5%. ■

EXAMPLE 5 The population of a mining town in Virginia decreased by $33\frac{1}{3}$%. If the original population was 24,000, what is the current population?

SOLUTION First we find the decrease in population. We have the following.

$$\begin{array}{c}\text{Decrease}\\\text{in population}\end{array} = \begin{array}{c}\text{percent}\\\text{decrease}\end{array} \times \begin{array}{c}\text{original}\\\text{population}\end{array}$$

$$= 33\frac{1}{3}\% \times 24{,}000$$

$$= \frac{1}{3} \times 24{,}000 \qquad \text{Convert } 33\frac{1}{3}\% \text{ to a fraction.}$$

$$= \frac{1}{\cancel{3}} \times \frac{\cancel{24{,}000}^{8000}}{1} \qquad 33\frac{1}{3}\% = 33\frac{1}{3} \div 100$$

$$= 8000 \qquad\qquad = \frac{\cancel{100}}{3} \times \frac{1}{\cancel{100}}$$

$$\qquad\qquad\qquad\qquad = \frac{1}{3}$$

The decrease in population is 8000. To compute the current population, subtract the decrease from the original population.

$$\begin{array}{c}\text{Final}\\\text{population}\end{array} = \begin{array}{c}\text{original}\\\text{population}\end{array} - \begin{array}{c}\text{decrease}\\\text{in population}\end{array}$$

$$= 24{,}000 - 8000$$

$$= 16{,}000$$

The current population is 16,000. ■

 Try Problems 5 through 8.

3 Problems Involving Markup Based on the Cost

A manufacturer encounters some expenses when producing a product. The expenses might include labor and materials. These direct expenses are called the **cost** in producing the product. The manufacturer then

▲ Try These Problems

Solve.

9. A jewelry manufacturer produces a ring at a cost of $200, and then marks up the ring 58%. Find the markup. (Assume the markup is based on the cost.)

10. A manufacturer produces a television at a cost of $174 and uses a percent markup of $33\frac{1}{3}\%$. Find the selling price of this television. (Assume the markup is based on the cost.)

---

sells the product to a retail business for a price that is higher than the cost. This price is called the **selling price.** The difference between the cost and the selling price is called the **markup.** Here are two statements that relate the cost, selling price, and markup.

$$\text{Selling price} = \text{cost} + \text{markup}$$
$$\text{Markup} = \text{selling price} - \text{cost}$$

The manufacturer usually computes the markup by taking a certain percent of the cost. When the markup is computed this way, we say the *markup is based on the cost.* This percent is called the **percent markup** or the **markup rate.** For markup based on the cost, we have the following.

> **Rule Involving Markup Based on the Cost**
> Markup = percent markup × cost

Markup is used at the retail business level as well as the manufacturer level. A retail business buys a product from the manufacturer, then sells the product to a customer at a higher price than was paid. The price paid for the product is called the **cost.** The price the product is sold to a customer is called the **selling price.** The difference between the cost and the selling price is called the **markup.** The relationship between the cost, selling price, and markup is the same as it was at the manufacturer level. Sometimes, but not usually, a retail business computes markup by taking a certain percent of the cost, thus bases the markup on cost. This percent is called the **percent markup** or **markup rate.**

Here are some examples involving markup based on cost.

EXAMPLE 6 A manufacturer of hardware determines that a markup rate of 40% is needed to make the desired profit. If it costs $21 to produce an electric drill, find the selling price of the drill. (Assume that the markup is based on the cost.)

SOLUTION First, we find the markup by taking 40% of the cost, $21.

$$\begin{aligned}\text{Markup} \atop (\$) &= {\text{markup} \atop \text{rate}} \times {\text{cost} \atop (\$)} \\ &= 40\% \times 21 \\ &= 0.40 \times 21 \\ &= 8.40\end{aligned}$$

The markup is $8.40. To obtain the selling price, we add the markup to the cost.

$$\begin{aligned}{\text{Selling} \atop \text{price}} &= \text{cost} + \text{markup} \\ &= 21 + 8.40 \\ &= 29.40\end{aligned}$$

The selling price for the electric drill is $29.40. ■

 Try Problems 9 and 10.

9.7 Applications: Percent Increase, Percent Decrease, and Markup

 Try These Problems

Solve.

11. A tennis shop buys a tennis racket for $120 and marks up the cost $54 before selling the racket. What is the markup rate? (Assume that the markup is based on the cost.)

12. A beauty supply store sells a hair brush for $11.56. If the store paid $8.50 for the brush, what is the markup rate? (Assume the markup is based on the cost.)

EXAMPLE 7 A college bookstore buys a textbook from the publisher for $30 and sells the textbook to the students for $37.50. What is the percent markup? (Assume that the markup is based on the cost.)

SOLUTION First we find the markup by subtracting the cost from the selling price.

$$\text{Markup (\$)} = \text{selling price (\$)} - \text{cost (\$)}$$
$$= 37.50 - 30$$
$$= 7.50$$

The markup is $7.50. Since the markup is based on the cost, we have the following.

$$\text{Markup} = \frac{\text{percent}}{\text{markup}} \times \text{cost}$$
$$7.50 = \boxed{\%} \times 30$$

A multiplier is missing, so divide 7.50 by 30 to find the percent markup.

$$30\overline{)7.50} \quad .25 = 25\%$$
$$\underline{6\,0}$$
$$1\,50$$
$$\underline{1\,50}$$

The percent markup is 25%. ■

 Try Problems 11 and 12.

Problems Involving Markup Based on the Selling Price

Although a manufacturer usually bases markup on the cost, a retail business usually does not base the markup on the cost. Instead, a retail business *bases the markup on the selling price;* that is, the markup is obtained by multiplying the percent markup by the selling price.

> ***Rule Involving Markup Based on the Selling Price***
> Markup = markup rate × selling price

Suppose a grocery store paid $4 for a product and sold the product for $5. The markup is $1. Here we show the markup rate computed in two ways: based on the cost and based on the selling price.

MARKUP BASED ON THE COST

$$\text{Markup} = \frac{\text{markup}}{\text{rate}} \times \text{cost}$$
$$1 = \boxed{\%} \times 4$$

The markup rate based on cost is $\frac{1}{4}$ or 25%.

Try These Problems

Solve.

13. A grocery store plans to sell a jar of mayonnaise for $4.50. If the markup rate is $33\frac{1}{3}\%$, what does the store pay the manufacturer for the jar of mayonnaise? (Assume the markup is based on the selling price.)

14. An art gallery pays an artist $150 for a painting. The gallery sells the painting for $600. What is the percent markup? (Assume the markup is based on the selling price.)

MARKUP BASED ON THE SELLING PRICE

$$\text{Markup} = \text{markup rate} \times \text{selling price}$$
$$1 = \boxed{}\% \times 5$$

The markup rate based on selling price is $\frac{1}{5}$ or 20%.

From the viewpoint of a mathematician or a manufacturer, it is more natural to base the markup on the cost; because, in this case, the percent markup is similar to percent increase. A quantity is increased over an original smaller amount. From the point of view of a retail business, however, it is more natural to base the markup on the selling price, because the selling price includes the business expenses, the cost of the product, and the profit. For the retail business, the selling price is the whole and the markup is part of the whole.

Now we look at an example that involves markup based on the selling price.

EXAMPLE 8   A shoe store plans to sell a pair of shoes for $64. If the markup rate is 38%, what does the shoe store pay the manufacturer for the pair of shoes? (Assume the markup is based on the selling price.)

SOLUTION   Because the markup is based on the selling price, we multiply the markup rate by the selling price to obtain the markup.

$$\underset{(\$)}{\text{Markup}} = \underset{}{\text{markup rate}} \times \underset{(\$)}{\text{selling price}}$$
$$= 38\% \times 64$$
$$= 0.38 \times 64$$
$$= 24.32$$

The markup is $24.32. To find the cost of the pair of shoes to the store, we subtract the markup from the selling price.

$$\underset{(\$)}{\text{Cost}} = \underset{(\$)}{\text{selling price}} - \underset{(\$)}{\text{markup}}$$
$$= 64 - 24.32$$
$$= 39.68$$

The store pays $39.68 for the pair of shoes.   ■

 Try Problems 13 and 14.

Here we summarize the information presented in this section.

> ### Percent Increase
> The percent increase is taken of the original amount to obtain the increase.
>
> $$\text{Increase} = \text{percent increase (\%)} \times \text{original amount}$$
> $$\text{Final amount} = \text{original amount} + \text{increase}$$
> $$\text{Increase} = \text{final amount} - \text{original amount}$$
>
> ### Percent Decrease
> The percent decrease is taken of the original amount to obtain the decrease.
>
> $$\text{Decrease} = \text{percent decrease (\%)} \times \text{original amount}$$
> $$\text{Final amount} = \text{original amount} - \text{decrease}$$
> $$\text{Decrease} = \text{original amount} - \text{final amount}$$
>
> ### Markup Based on the Cost
> The percent markup is taken of the cost to obtain the markup.
>
> $$\text{Markup} = \text{markup rate (\%)} \times \text{cost}$$
> $$\text{Selling price} = \text{cost} + \text{markup}$$
> $$\text{Markup} = \text{selling price} - \text{cost}$$
>
> ### Markup Based on the Selling Price
> The percent markup is taken of the selling price to obtain the markup.
>
> $$\text{Markup} = \text{markup rate (\%)} \times \text{selling price}$$
> $$\text{Cost} = \text{selling price} - \text{markup}$$
> $$\text{Markup} = \text{selling price} - \text{cost}$$

▲ **Answers to Try These Problems**

1. $1.40 **2.** $32 **3.** $6\frac{1}{4}\%$ or 6.25% **4.** 150% **5.** $105
6. 175 lb **7.** 40% **8.** $3.15 per lb **9.** $116 **10.** $232
11. 45% **12.** 36% **13.** $3 **14.** 75%

EXERCISES 9.7

Solve.

1. Ms. Garcia was earning $26.80 per hour. She received a 6.5% raise. What is her increase in salary? (Round off to the nearest cent.)
2. Your typing speed increased 8 words per minute, which was a $12\frac{1}{2}\%$ increase. What was your original typing speed?
3. The price of a stock increased from $35 per share to 41\frac{1}{8}$ per share. Find the percent increase.

4. A carpenter calculated that he needed 260 feet of lumber for a job. He wants to increase that amount by 12% to allow for waste. What is the total amount of lumber he should buy?
5. It is estimated that the value of a $5000 computer decreased by $1500 in the first year. What percent decrease is this?
6. By conserving water, the Jefferson family reduced their daily water usage from 450 gallons to 270 gallons. What percent decrease is this?
7. A hotel reduced the average amount of time a person spends waiting for the elevator by 1.8 minutes. If this represents a 72% decrease, what was the original average waiting time?
8. A bath towel that was selling for $14 is now on sale at 40% off. What is the sale price?
9. A manufacturer produces a pair of shoes at a cost of $48. A markup rate of 25% is used that is based on the cost. What is the selling price of the pair of shoes to the retailer?
10. A publisher produces a novel at a cost of $8.50. The publisher sells the novel to a bookstore for $10.03. If the publisher bases markup on cost, what is the percent markup?
11. A jewelry store buys a necklace from a manufacturer for $260. The store uses a markup rate of 320% that is based on the cost. Find the markup for this necklace.
12. A sporting goods store uses a percent markup of $33\frac{1}{3}$% that is based on the cost. If the markup for a pair of running shoes is $28, what does the store pay the manufacturer for the pair of running shoes?
13. A grocery store plans to sell a box of cereal for $2.60. If the markup rate, based on selling price, is 35%, what is the markup?
14. At an appliance store the markup on a refrigerator is $273.60. This is a 38% markup that is based on the selling price. Find the selling price of the refrigerator.
15. A department store buys a rug from the manufacturer for $440 and sells it for $800. If the markup is based on the selling price, what is the percent markup?
16. A fish market sells fresh trout at $5.40 per pound. The markup rate, based on the selling price, is 15%. How much does the market pay the fisherman for the trout?
17. Over a ten-year period, the cost of fuel increased from $4 a barrel to $40 a barrel. What percent increase is this?
18. Linda lost 6 pounds over the summer, which was a 5% decrease.
    a. What was Linda's original weight?
    b. What was Linda's weight at the end of the summer?
19. A retail store buys a sweater from the manufacturer for $54 and sells it to a customer for $72.
    a. Find the markup rate if the markup is based on the cost.
    b. Find the markup rate if the markup is based on the selling price.
20. A furniture store pays the manufacturer $62 for a table. The store uses a 40% markup based on cost. If the sales tax rate is 6.2%, what does the customer pay for this table? (Round off to the nearest cent.)

USING THE CALCULATOR #10

THE % KEY

Most calculators have a [%] key. On a scientific calculator this key may be a second function key so that you must enter the [2ndF], [INV], or [Shift] key before entering the [%] key.

The percent key can be used to find a percent of a number without first changing the percent to a fraction or decimal. Here we show an example.

To Compute	50% of 60
Enter	60 [×] 50 [%]
Result	30.

On most calculators you must enter the percent after the number you are taking the percent of. Also, if the percent key is a second function key on your calculator, you must enter the [2ndF], [INV], or [Shift] key before entering the [%] key.

The percent key can also be used to find the final amount after a quantity has been increased by a certain percent. The procedure is not the same on all calculators. Here we show how to compute 40 increased by 25% on two different calculators.

ON SOME CALCULATORS

To Compute	40 + (25% of 40)
Enter	40 [×] 25 [%] [+]
Result	50

ON OTHER CALCULATORS

To Compute	40 + (25% of 40)
Enter	40 [+] 25 [%]
Result	50

If the percent key is a second function key on your calculator, you must enter the [2ndF], [INV], or [Shift] key before entering the [%] key.

You can also use the percent key to find the final amount after a quantity has been decreased by a certain percent. The procedure is exactly like the above except enter the [−] key instead of the [+] key. Here we show how to compute 40 decreased by 25% on two different calculators.

ON SOME CALCULATORS

To Compute	40 − (25% of 40)
Enter	40 [×] 25 [%] [−]
Result	30

ON OTHER CALCULATORS

To Compute	40 − (25% of 40)
Enter	40 [−] 25 [%]
Result	30

If the percent key is a second function key on your calculator, you must enter the [2ndF], [INV], or [Shift] key before entering the [%] key.

The percent key applies to only very special situations. You will not be able to use it to solve all percent problems.

Calculator Problems

Evaluate each by using the [%] key.

1. $900 \times 40\%$
2. $75 \times 3.6\%$
3. 85% of 560
4. 125% of $9.80
5. 600, increased by 10%
6. 600, decreased by 10%
7. $45.76, increased by 75%
8. $45.76, decreased by 75%
9. 850, increased by 150%
10. 850, decreased by 0.06%

Evaluate each on the calculator, but without using the [%] key. If the decimal representation of the answer goes beyond 3 decimal places, then round off at 3 decimal places.

11. Find $83\frac{1}{3}\%$ of 600.
12. 81 is 15% of what number?
13. Find 90, increased by $3\frac{2}{3}\%$.
14. Find 120, decreased by $66\frac{2}{3}\%$.
15. What percent of 15.2 is 4.75?
16. $\frac{1}{4}\%$ of what number is 16?

9.8 USING PROPORTIONS TO SOLVE PERCENT PROBLEMS

OBJECTIVES
- Identify the amount, rate, and base in a basic percent statement.
- Write a proportion that corresponds to a basic percent statement.
- Find the missing number in a basic percent statement by using a proportion.
- Solve a percent application by using a proportion.

 Try These Problems

Identify the rate, base, and amount for each percent statement.
1. 40 is 50% of 80.
2. 6% of 550 is 33.
3. 200% of 45 is 90.

1 Identifying the Three Quantities in a Basic Percent Statement

Suppose that in an algebra class 5 students received As out of a total of 25 students. We say that $\frac{5}{25}$ or $\frac{1}{5}$ or 20% of the class received As. The basic percent statement that corresponds to this situation is as follows.

$$5 \text{ is } 20\% \text{ of } 25$$

or

$$20\% \text{ of } 25 \text{ is } 5$$

We give names to the three quantities in a percent statement. The percent 20% is called the **rate.** The quantity that follows the word *of* is called the **base,** thus 25 is the base. The other quantity, 5, is called the **amount.**

You know that the basic percent statement can be translated to a multiplication statement. For example,

$$5 \text{ is } 20\% \text{ of } 25$$

translates to

$$5 = 20\% \times 25$$

In general, it is true that the amount is the result of multiplying the rate by the base.

$$\textbf{Amount} = \textbf{rate} \times \textbf{base}$$

 Try Problems 1 through 3.

2 Writing a Proportion that Corresponds to a Basic Percent Statement

There is another way to view a basic percent statement. Let's look back at the example used previously. In an algebra class of 25 students, 5 of them got As. If we compare 5 to 25 by division, we obtain the fraction of the class who got As.

$$\frac{\text{Fraction}}{\text{who got As}} = \frac{\text{number of As}}{\text{total number}} = \frac{5}{25}$$

Converting this fraction to percent form gives us the percent who got As.

$$\frac{\text{Percent}}{\text{who got As}} = \frac{5}{25} = 20\%$$

Observe that the basic percent statement

5 is 20% of 25

can also be written in the form

$$20\% = \frac{5}{25}$$

We can view this as a **proportion** by converting the percent to a fraction.

$$\frac{20}{100} = \frac{5}{25}$$

We can check that these ratios are equal by looking at the **cross products.**

5 × 100 = 500

20 × 25 = 500

In general, the rate in a basic percent statement is always the ratio of the amount to the base. Remember that the base is the quantity that follows the word *of*.

$$\text{Rate} = \frac{\text{amount}}{\text{base}}$$

Here are some examples of translating a basic percent statement to a proportion.

EXAMPLE 1 Set up a proportion that corresponds to the statement *75% of 200 is 150*.

SOLUTION First, identify each of the three quantities.

75% of 200 is 150
Rate   Base   Amount

The rate is easy to identify because it is the percent. The base is the quantity that follows the word *of*. The amount is the other quantity mentioned.

Second, remember that the rate is always the ratio of the amount to the base. The base is always in the denominator.

$$\text{Rate} = \frac{\text{amount}}{\text{base}}$$

$$75\% = \frac{150}{200}$$

Last, convert the percent to a fraction form to obtain the proportion.

$$\frac{75}{100} = \frac{150}{200} \quad \blacksquare$$

436 Chapter 9 Percent

 Try These Problems

Set up a proportion that corresponds to each percent statement.

4. 16 is 25% of 64.
5. 200 is 150% of 80.
6. 0.8% of 2500 is 20.
7. $83\frac{1}{3}$% of 600 is 500.

EXAMPLE 2 Set up a proportion that corresponds to the statement *60 is 120% of 50*.

SOLUTION First, identify each of the three quantities.

$$\underset{\text{Amount}}{60} \text{ is } \underset{\text{Rate}}{120\%} \text{ of } \underset{\text{Base}}{50}$$

The rate is easy to identify because it is the percent. The base is the quantity that follows the word *of*. The amount is the other quantity mentioned.

Second, remember that the rate is always the ratio of the amount to the base. The base is always in the denominator.

$$\text{Rate} = \frac{\text{amount}}{\text{base}}$$

$$120\% = \frac{60}{50}$$

Last, convert the percent to a fraction form to obtain the proportion.

$$\frac{120}{100} = \frac{60}{50} \quad \blacksquare$$

 Try Problems 4 through 7.

3 Using Proportions to Solve Percent Problems

Three types of percent problems can be obtained by omitting either the rate, the base, or the amount in a basic percent statement. In Section 9.4, we solved these problems by translating the percent statement to a multiplication statement, then finding the missing number in the multiplication statement. At this time, we look at solving these problems by using ratios and proportions. The following examples illustrate the technique.

EXAMPLE 3 What number is 2% of 650?

SOLUTION First, identify each of the three quantities.

$$\underset{\text{Amount}}{\text{What number}} \text{ is } \underset{\text{Rate}}{2\%} \text{ of } \underset{\text{Base}}{650}?$$

Second, write the rate as a ratio of the amount to the base.

$$\text{Rate} = \frac{\text{amount}}{\text{base}}$$

$$2\% = \frac{\Box}{650}$$

$$\frac{2}{100} = \frac{\Box}{650} \qquad \text{Convert 2\% to } \tfrac{2}{100}.$$

Try These Problems

Set up a proportion and solve.
8. 77 is 35% of what number?
9. What number is 4% of 150?
10. 130% of what number is 104?

Finally, solve the proportion. That is, find the missing number that makes the two ratios equal.

$$\frac{2}{100} = \frac{\boxed{}}{650}$$

$2 \times 650 = 100 \times \boxed{}$ Set the cross products
$1300 = 100 \times \boxed{}$ equal to each other.

$1300 \div 100 = \boxed{}$ Divide 1300 by 100 to find the missing
$13 = \boxed{}$ multiplier

The missing amount is 13.

EXAMPLE 4 250% of what number is 400?

SOLUTION First, identify each of the three quantities.

250% of what number is 400?
  | | |
 Rate   Base   Amount

Second, write the rate as a ratio of the amount to the base.

$$\text{Rate} = \frac{\text{amount}}{\text{base}}$$

$$250\% = \frac{400}{\boxed{}}$$

$$\frac{250}{100} = \frac{400}{\boxed{}} \quad \text{Convert 250\% to } \tfrac{250}{100}.$$

Finally, solve the proportion for the missing number.

$$\frac{250}{100} = \frac{400}{\boxed{}}$$

$250 \times \boxed{} = 100 \times 400$ Set the cross products
$250 \times \boxed{} = 40{,}000$ equal to each other.
$\boxed{} = 40{,}000 \div 250$ Divide 40,000 by 250 to find the missing
$\boxed{} = 160$ multiplier.

The missing base is 160. ■

 Try Problems 8 through 10.

EXAMPLE 5 450 is what percent of 1500?

SOLUTION First, identify the three quantities.

450 is what percent of 1500?
 | | |
Amount   Rate   Base

Second, write the rate as the ratio of the amount to the base.

▲ Try These Problems

Set up a proportion and solve.
11. 6 is what percent of 24?
12. What number is 320% of 65?
13. 8% of what number is 192?
14. What percent of 160 is 168?

$$\text{Rate} = \frac{\text{amount}}{\text{base}}$$

$$\boxed{}\% = \frac{450}{1500}$$

$$\frac{\boxed{}}{100} = \frac{450}{1500}$$

Observe that we are viewing the rate as a missing number followed by the percent symbol. Percent means per 100 or divide by 100.

Next, solve the proportion for the missing number.

$$\frac{\boxed{}}{100} = \frac{450}{1500}$$

$$\boxed{} \times 1500 = 100 \times 450 \qquad \text{Set the cross products}$$
$$\boxed{} \times 1500 = 45{,}000 \qquad \text{equal to each other.}$$
$$\boxed{} = 45{,}000 \div 1500$$
$$\boxed{} = 30$$

Be careful! This missing number 30 is not the rate. The rate, which is what we are looking for, is this missing number 30 followed by a percent symbol.

$$\text{Rate} = \boxed{}\%$$
$$= 30\%$$

The missing rate is 30%. ∎

▲ Try Problems 11 through 14.

Next we look at some examples where fractions and decimals are involved.

EXAMPLE 6 0.01 is 0.4% of what number?

SOLUTION Identify the three quantities.

0.01 is 0.4% of what number?
| | |
Amount Rate   Base

Write the rate as the ratio of the amount to the base.

$$\text{Rate} = \frac{\text{amount}}{\text{base}}$$

$$0.4\% = \frac{0.01}{\boxed{}}$$

$$\frac{0.4}{100} = \frac{0.01}{\boxed{}} \qquad \text{Convert } 0.4\% \text{ to } \tfrac{0.4}{100}.$$

Now solve the proportion for the missing number.

$$\frac{0.4}{100} = \frac{0.01}{\boxed{}}$$

$$0.4 \times \boxed{} = 100 \times 0.01 \qquad \text{Set the cross products}$$
$$0.4 \times \boxed{} = 1 \qquad \text{equal to each other.}$$
$$\boxed{} = 1 \div 0.4 \qquad \text{Divide 1 by 0.4 to find}$$
$$\boxed{} = 2.5 \qquad \text{the missing multiplier.}$$

The missing base is 2.5. ∎

Try These Problems

Set up a proportion and solve.
15. 0.8% of 50 is what number?
16. 3.5% of what number is 3.36?
17. What percent of 186 is 9.3?
18. 224 is what percent of 1750?

EXAMPLE 7 64.4 is what percent of 36.8?

SOLUTION Identify the three quantities.

64.4 is what percent of 36.8?
| | | |
Amount   Rate   Base

Write the rate as the ratio of the amount to the base.

$$\text{Rate} = \frac{\text{amount}}{\text{base}}$$

$$\square\% = \frac{64.4}{36.8}$$ Observe that we are viewing the rate as a missing number followed by a percent symbol.

$$\frac{\square}{100} = \frac{64.4}{36.8}$$ Percent means per 100 or divide by 100.

$$\square \times 36.8 = 100 \times 64.4$$ Set the cross products equal to each other.

$$\square \times 36.8 = 6440$$

$$\square = 6440 \div 36.8$$ Divide 6440 by 36.8 to find the missing multiplier.

$$\square = 175$$

Be careful! The number 175 is not the rate. The rate is this missing number 175 followed by a percent symbol.

$$\text{Rate} = \square\%$$
$$= 175\%$$

The missing rate is 175%. ■

Try Problems 15 through 18.

EXAMPLE 8 $33\frac{1}{3}\%$ of 708 is what number?

SOLUTION Identify the three quantities.

$33\frac{1}{3}\%$ of 708 is what number?
| | | |
Rate   Base   Amount

Write the rate as the ratio of the amount to the base.

$$\text{Rate} = \frac{\text{amount}}{\text{base}}$$

$$33\frac{1}{3}\% = \frac{\square}{708}$$

$$\frac{33\frac{1}{3}}{100} = \frac{\square}{708}$$ Convert $33\frac{1}{3}\%$ to $\frac{33\frac{1}{3}}{100}$.

Solve the proportion for the missing number.

$$\frac{33\frac{1}{3}}{100} = \frac{\boxed{}}{708}$$

$33\frac{1}{3} \times 708 = 100 \times \boxed{}$ Set the cross products equal to each other.

$$\frac{100}{\cancel{3}_1} \times \frac{\cancel{708}^{236}}{1} = 100 \times \boxed{}$$

$23{,}600 = 100 \times \boxed{}$

$23{,}600 \div 100 = \boxed{}$ Divide 23,600 by 100 to find the missing multiplier.

$236 = \boxed{}$

The missing amount is 236. ■

EXAMPLE 9   What percent of 15.6 is 10.4?

SOLUTION   Identify the three quantities.

What percent of 15.6 is 10.4?
     Rate   Base   Amount

Write the rate as the ratio of the amount to the base.

$$\text{Rate} = \frac{\text{amount}}{\text{base}}$$

$\boxed{}\% = \dfrac{10.4}{15.6}$ Observe that we are viewing the rate as a missing number followed by the percent symbol.

$\dfrac{\boxed{}}{100} = \dfrac{10.4}{15.6}$ Percent means per 100 or divide by 100.

Solve the proportion for the missing number.

$$\frac{\boxed{}}{100} = \frac{10.4}{15.6}$$

$\boxed{} \times 15.6 = 100 \times 10.4$ Set the cross products equal to each other.

$\boxed{} \times 15.6 = 1040$

$\boxed{} = 1040 \div 15.6$ Divide 1040 by 15.6 to find the missing multiplier.

$\boxed{} = 66\frac{2}{3}$

Be careful! This missing number $66\frac{2}{3}$ is not the rate. The rate is this missing number followed by the percent symbol.

$\text{Rate} = \boxed{}\%$

$\phantom{\text{Rate}} = 66\frac{2}{3}\%$

The missing rate is $66\frac{2}{3}\%$. ■

Try These Problems

Set up a proportion and solve.

19. $12\frac{1}{2}\%$ of what number is 70?
20. $66\frac{2}{3}\%$ of 324 is what number?
21. 4 is what percent of 120?
22. What percent of 30.6 is 25.5?
23. 150% of what number is 6?

EXAMPLE 10 75 is $8\frac{1}{3}\%$ of what number?

SOLUTION Identify the three quantities.

75 is $8\frac{1}{3}\%$ of what number?
| | |
Amount Rate Base

Write the rate as the ratio of the amount to the base.

$$\text{Rate} = \frac{\text{amount}}{\text{base}}$$

$$8\frac{1}{3}\% = \frac{75}{\boxed{}}$$

$$\frac{8\frac{1}{3}}{100} = \frac{75}{\boxed{}} \qquad \text{Convert } 8\frac{1}{3}\% \text{ to } \frac{8\frac{1}{3}}{100}.$$

Solve the proportion for the missing number.

$$\frac{8\frac{1}{3}}{100} = \frac{75}{\boxed{}}$$

$$8\frac{1}{3} \times \boxed{} = 100 \times 75 \qquad \text{Set the cross products equal to each other.}$$

$$8\frac{1}{3} \times \boxed{} = 7500$$

$$\boxed{} = 7500 \div 8\frac{1}{3} \qquad \text{Divide 7500 by } 8\frac{1}{3} \text{ to find the missing multiplier.}$$

$$\boxed{} = 7500 \div \frac{25}{3}$$

$$\boxed{} = \frac{\overset{300}{\cancel{7500}}}{1} \times \frac{3}{\underset{1}{\cancel{25}}} \qquad \text{Dividing by } \frac{25}{3} \text{ is the same as multiplying by } \frac{3}{25}.$$

$$\boxed{} = 900$$

The missing base is 900. ∎

 Try Problems 19 through 23.

4 Solving Percent Applications by Using Proportions

In Sections 9.5, 9.6, and 9.7 we looked carefully at solving percent applications by finding the missing number in a multiplication statement. The multiplication statement came from a basic percent statement that was either stated in the problem or implied in the problem. Here we solve some of those same applications by translating the basic percent statement to a proportion then solving the proportion. The following examples illustrate the technique.

Try These Problems

Set up a proportion and solve.

24. A salesperson makes a $2100 bonus. The bonus is 14% of the base salary. What is the base salary?

25. A math teacher gave a test to 40 students. Six students received an A grade on the test. What percent of the class received an A grade?

26. Sales tax in a certain city is 6.5% of the marked price. How much sales tax will Lori pay when buying a dress marked $64?

EXAMPLE 11 A football team won 9 games, which were 60% of the total number of games played. How many games did the team play?

SOLUTION Observe that the first sentence is a basic percent statement and we use it to identify the amount, base, and rate.

The 9 games won (Amount) were 60% of (Rate) the total games played (Base)

Write the rate as the ratio of the amount to the base.

$$\text{Rate} = \frac{\text{amount}}{\text{base}}$$

$$60\% = \frac{9}{\boxed{}}$$

$$\frac{60}{100} = \frac{9}{\boxed{}} \quad \text{Convert 60\% to } \tfrac{60}{100}.$$

Solve the proportion for the missing number.

$$\frac{60}{100} = \frac{9}{\boxed{}}$$

$60 \times \boxed{} = 100 \times 9$ Set the cross products equal to each other.

$60 \times \boxed{} = 900$

$\boxed{} = 900 \div 60$ Divide 900 by 60 to find the missing multiplier.

$\boxed{} = 15$

The team played a total of 15 games. ■

 Try Problems 24 through 26.

EXAMPLE 12 Robyn took the percent test. She got 18 out of 20 questions correct. What percent did she get correct?

SOLUTION The last sentence in the problem is a basic percent statement that is written in an unclear form. The question *what percent did she get correct* is really asking what percent *of the total* is the amount she got correct. We use this basic percent statement to identify the amount, base, and rate.

What percent of the total is the number she got correct?

What percent (Rate) of 20 (Base) is 18 (Amount)?

Write the rate as the ratio of the amount to the base.

$$\text{Rate} = \frac{\text{amount}}{\text{base}}$$

$$\boxed{}\% = \frac{18}{20} \quad \begin{array}{l}\text{Observe that we are viewing the}\\\text{percent (or rate) as a missing number}\\\text{followed by the percent symbol.}\end{array}$$

$$\frac{\boxed{}}{100} = \frac{18}{20} \quad \begin{array}{l}\text{Percent means per 100}\\\text{or divide by 100.}\end{array}$$

Try These Problems

Set up a proportion and solve.

27. A box contains a total of 40 balls; including 20% red balls, 50% black balls, 30% blue balls. How many red balls are there?

28. As an inspector, you reject 9 out of 1500 products tested. What percent did you reject?

29. A bottle of wine is 12% alcohol. If the alcohol volume is 90 milliliters, what is the volume of all the wine in the bottle?

Solve the proportion for the missing number.

$$\frac{\Box}{100} = \frac{18}{20}$$

$\Box \times 20 = 100 \times 18$ Set the cross products equal to each other.
$\Box \times 20 = 1800$
$\Box = 1800 \div 20$ Divide 1800 by 20 to find the missing multiplier.
$\Box = 90$

Be careful! The number 90 is not the rate. The rate is the number 90 followed by a percent symbol.

$$\text{Percent she got correct} = \text{Rate} = \Box \%$$
$$= 90\%$$

Robyn got 90% of the problems correct. ∎

EXAMPLE 13 A punch was made by mixing orange juice and champagne. The punch is 70% orange juice and 30% champagne. If the total amount of punch is 140 ounces, how many ounces of champagne are in the punch?

SOLUTION The phrase *the punch is 30% champagne* means that *30% of the total* is champagne. We use this basic percent statement to identify the amount, base, and rate.

    30%   of   the total   is   champagne.

    30%   of   140   is   champagne.
    Rate   Base   Amount

Write the rate as the ratio of the amount to the base.

$$\text{Rate} = \frac{\text{amount}}{\text{base}}$$

$$30\% = \frac{\Box}{140}$$

$$\frac{30}{100} = \frac{\Box}{140} \quad \text{Convert 30\% to } \tfrac{30}{100}.$$

Solve the proportion for the missing number.

$$\frac{30}{100} = \frac{\Box}{140}$$

$30 \times 140 = 100 \times \Box$ Set the cross products equal to each other.
$4200 = 100 \times \Box$
$4200 \div 100 = \Box$ Divide 4200 by 100 to find the missing multiplier.
$42 = \Box$

There are 42 ounces of champagne in the punch. ∎

 Try Problems 27 through 29.

EXAMPLE 14 The marked price of a refrigerator was $760. A sales tax of $41.80 was added. What is the sales tax rate? Express the answer as a percent.

SOLUTION The sales tax is the result of taking a certain percent of the marked price. That percent is the sales tax rate. We can use this to identify the amount, base, and rate.

Sales tax is sales tax rate of marked price

$41.80 (Amount) is sales tax rate (Rate) of $760 (Base)

Write the rate as the ratio of the amount to the base.

$$\text{Rate} = \frac{\text{amount}}{\text{base}}$$

$$\boxed{}\% = \frac{41.8}{760}$$

Observe that we are viewing the percent (or rate) as a missing number followed by the percent symbol.

$$\frac{\boxed{}}{100} = \frac{41.8}{760}$$

Percent means per 100 or divide by 100.

$$\boxed{} \times 760 = 100 \times 41.8$$

Set the cross products equal to each other.

$$\boxed{} \times 760 = 4180$$

$$\boxed{} = 4180 \div 760$$

$$\boxed{} = 5.5$$

Divide 4180 by 760 to find the missing multiplier.

Be careful! The number 5.5 is not the sales tax rate. The rate is this number followed by a percent symbol.

$$\text{Sales tax rate} = \boxed{}\%$$

$$= 5.5\%$$

The sales tax rate is 5.5%. ■

EXAMPLE 15 Mr. Nguyen is paid a commission for selling rugs at the Emporium that is $7\frac{1}{2}\%$ of his total sales. Last week he received a commission of $156. What were his total sales for the week?

SOLUTION We have the following basic percent statement that we can use to identify the amount, base, and rate.

Commission is $7\frac{1}{2}\%$ of total sales.

$156 (Amount) is $7\frac{1}{2}\%$ (Rate) of total sales (Base).

Try These Problems

Set up a proportion and solve.

30. The marked price of a car is $12,500. The sales tax amounts to $825. What is the sales tax rate? (Express the answer as a percent.)

31. Mr. Gust purchases some construction materials that are marked $7800. If the sales tax rate is 7.4%, what is the total cost of the materials?

32. Ms. Wagner sells ski equipment and clothing to retail stores across the United States and Canada. She is paid a commission that is $\frac{2}{3}$% of the total sales. If she sells $45,000 worth of ski equipment to the Aspen Sport Shop, how much commission does she earn on this sale?

33. Mr. Cortez sells men's clothing in a large department store. In addition to his regular salary, he earns a commission that is 0.6% of his total sales. Last month he earned a $135 commission. What were his total sales for that month?

Write the rate as the ratio of the amount to the base.

$$\text{Rate} = \frac{\text{amount}}{\text{base}}$$

$$7\tfrac{1}{2}\% = \frac{156}{\boxed{}}$$

$$\frac{7\tfrac{1}{2}}{100} = \frac{156}{\boxed{}} \qquad \text{Convert } 7\tfrac{1}{2}\% \text{ to } \frac{7\tfrac{1}{2}}{100}.$$

Solve the proportion for the missing number.

$$\frac{7\tfrac{1}{2}}{100} = \frac{156}{\boxed{}}$$

$$7\tfrac{1}{2} \times \boxed{} = 100 \times 156 \qquad \text{Set the cross products equal to each other.}$$

$$7\tfrac{1}{2} \times \boxed{} = 15{,}600$$

$$\boxed{} = 15{,}600 \div 7\tfrac{1}{2} \qquad \text{Divide 15,600 by } 7\tfrac{1}{2} \text{ to find the missing multiplier.}$$

$$\boxed{} = 15{,}600 \div \frac{15}{2}$$

$$\boxed{} = \frac{\overset{1040}{\cancel{15{,}600}}}{1} \times \frac{2}{\underset{1}{\cancel{15}}} \qquad \text{Dividing by } \tfrac{15}{2} \text{ is the same as multiplying by } \tfrac{2}{15}.$$

$$\boxed{} = 2080$$

He sold $2080 worth of rugs for the week. ∎

 Try Problems 30 through 33.

EXAMPLE 16 Mr. Kung made an investment of $7800 in an account that pays $8\tfrac{1}{3}\%$ interest for one year. How much money was in the account at the end of one year?

SOLUTION The amount of interest earned is $8\tfrac{1}{3}\%$ of the amount invested. We use this basic percent statement to identify the amount, base, and rate.

Interest is interest rate of amount invested

$\underbrace{\text{Interest}}_{\text{Amount}}$ is $\underset{\text{Rate}}{8\tfrac{1}{3}\%}$ of $\underset{\text{Base}}{7800}$

Write the rate as the ratio of the amount to the base.

$$\text{Rate} = \frac{\text{amount}}{\text{base}}$$

$$8\tfrac{1}{3}\% = \frac{\boxed{}}{7800}$$

$$\frac{8\tfrac{1}{3}}{100} = \frac{\boxed{}}{7800} \qquad \text{Convert } 8\tfrac{1}{3}\% \text{ to } \frac{8\tfrac{1}{3}}{100}.$$

Solve the proportion for the missing number.

$$\frac{8\frac{1}{3}}{100} = \frac{\boxed{}}{7800}$$

$8\frac{1}{3} \times 7800 = 100 \times \boxed{}$ Set the cross products equal to each other.

$\frac{25}{3} \times \frac{7800}{1} = 100 \times \boxed{}$

$65{,}000 = 100 \times \boxed{}$

$65{,}000 \div 100 = \boxed{}$ Divide 65,000 by 100 to find the missing multiplier.

$650 = \boxed{}$

The amount of interest earned is $650. Now we add this to the amount invested to obtain the final amount at the end of one year.

$$\begin{array}{c}\text{Final}\\ \text{amount}\\ (\$)\end{array} = \begin{array}{c}\text{amount}\\ \text{invested}\\ (\$)\end{array} + \begin{array}{c}\text{interest}\\ (\$)\end{array}$$

$= 7800 + 650$

$= 8450$

The amount in the account at the end of one year is $8450. ∎

EXAMPLE 17 Due to inflation, the price of a box of cereal increased from $2.50 to $3.00. What is the percent increase?

SOLUTION First we find the increase in price by subtracting the original price from the final price.

$$\begin{array}{c}\text{Increase}\\ (\$)\end{array} = \begin{array}{c}\text{final}\\ \text{price}\\ (\$)\end{array} - \begin{array}{c}\text{original}\\ \text{price}\\ (\$)\end{array}$$

$= 3.00 - 2.50$

$= 0.50$

The increase in price is $0.50. This increase is a certain percent of the original price. This percent is the percent increase. We have the following.

Increase is percent increase of original price

$\underset{\text{Amount}}{0.50}$ is $\underset{\text{Rate}}{\boxed{}\%}$ of $\underset{\text{Base}}{2.50}$

Write the rate as the ratio of the amount to the base.

$$\text{Rate} = \frac{\text{amount}}{\text{base}}$$

$$\boxed{}\% = \frac{0.50}{2.50}$$

$$\frac{\boxed{}}{100} = \frac{0.50}{2.50}$$ Percent means per 100 or divide by 100.

Try These Problems

Set up a proportion and solve.

34. Ms. Youngman inherited some money from her grandmother. She invested all of it in an account that paid 9% interest. If she earned $7560 interest in one year, what was her inheritance?

35. Ms. Kelly put $960 in an account that earns interest at an annual rate of $4\frac{3}{4}$%. How much money is in the account at the end of one year?

36. The price of a scientific hand-held calculator decreased from $24 to $15. What percent decrease is this?

37. Mr. Welna receives a $1.77-per-hour raise, which is 7.5% over his original hourly wage. Find his original hourly wage.

Solve the proportion for the missing number.

$$\frac{\Box}{100} = \frac{0.50}{2.50}$$

$\Box \times 2.50 = 100 \times 0.50$ Set the cross products equal to each other.

$\Box \times 2.5 = 50$

$\Box = 50 \div 2.5$ Divide 50 by 2.5 to find the missing multiplier.

$\Box = 20$

Be careful! The number 20 is not the percent increase. The percent increase is the number 20 followed by a percent symbol.

Percent increase = Rate = $\Box$ %

= 20%

The percent increase is 20%. ■

Try Problems 34 through 37.

Answers to Try These Problems

1. Rate = 50%, Base = 80, Amount = 40
2. Rate = 6%, Base = 550, Amount = 33
3. Rate = 200%, Base = 45, Amount = 90
4. $\frac{25}{100} = \frac{16}{64}$ 5. $\frac{150}{100} = \frac{200}{80}$ 6. $\frac{0.8}{100} = \frac{20}{2500}$ 7. $\frac{83\frac{1}{3}}{100} = \frac{500}{600}$
8. 220   9. 6   10. 80   11. 25%   12. 208   13. 2400
14. 105%   15. 0.4   16. 96   17. 5%   18. 12.8%
19. 560   20. 216   21. $3\frac{1}{3}$%   22. $83\frac{1}{3}$%   23. 4
24. $15,000   25. 15%   26. $4.16   27. 8   28. 0.6%
29. 750 mℓ 30. 6.6%   31. $8377.20   32. $300
33. $22,500   34. $84,000   35. $1005.60   36. 37.5%
37. $23.60

EXERCISES 9.8

Set up a proportion and solve.

1. What number is 25% of 620?
2. 24 is 40% of what number?
3. 6% of what number is 222?
4. 8% of 850 is what number?
5. What percent of 45 is 9?
6. 75 is what percent of 600?
7. 120 is 150% of what number?
8. What number is 200% of 76?
9. 15 is what percent of 12?
10. What number is 0.6% of 800?
11. 0.15% of what number is 960?
12. What percent of 4500 is 1.8?
13. 34.2 is 3.6% of what number?
14. 33.2% of 5000 is what number?
15. $66\frac{2}{3}$% of 2400 is what number?
16. 14 is what percent of 42?
17. $\frac{1}{4}$% of what number is 68?
18. What percent of 6 is $2\frac{1}{2}$?
19. What number is $6\frac{1}{4}$% of 7500?
20. 250% of what number is $17\frac{1}{2}$?
21. A college has 650 chemistry students. If 16% of these students received a grade of A or B in chemistry, how many students received an A or B in chemistry?

22. On average, a ski resort in California receives a total of 180 inches of snow for the season. So far this year, the resort has received 117 inches of snow. What percent of the total have they already received?

23. The sales tax for a city is 7.8% of the marked price. Jean-Louis purchased stereo speakers and paid $28.08 in sales tax. What was the marked price of the speakers?

24. The price of a calculator decreased from $25 to $21. What percent of the original price is the decrease in price?

25. A neighborhood contains 760 houses, including 45% painted white, 35% painted grey, and 20% painted beige. How many houses are painted grey?

26. In a town of 5600 persons, 21 caught a virus. What percent caught the virus?

27. Jeff, a bartender, makes a drink by mixing vodka and orange juice. The drink is 80% orange juice. If the drink contains 12 ounces of orange juice, what is the total volume of the drink?

28. A parking lot contains 75% cars and the rest trucks. If the lot contains 180 cars, how many trucks are there?

29. Ms. Combs bought a coat and was charged a sales tax of $8.24. If the sales tax rate is 8%, find the marked price of the coat.

30. A real estate agent receives a 4.5% commission for selling a house. How much commission did the agent earn for selling a house for $65,000?

31. Ms. Tan earned $817 interest on an investment of $8600. What was the interest rate? (Express the answer as a percent.)

32. Mr. Austin invested some money in a money market account that earns $5\frac{1}{4}\%$ interest. If he earned $252 interest in one year, how much was his investment?

33. Ms. Wilson, a company executive, took two of her office staff members to lunch. Because she was very pleased with the food and service at the restaurant, she left an $11.25 tip for a bill that amounted to $62.50. What percent tip did she leave?

34. Karin's dinner bill, not including tax, is $32.60. If the sales tax rate is 5.5%, and she leaves a tip that is 20% of the dinner bill before the tax is added, what is her total bill? (Round off to the nearest cent.)

35. Mr. Valiente received his credit card bill. It showed an unpaid balance from the last month of $700 and a finance charge of $14.70. Find the monthly interest rate. (Express the answer as a percent.)

36. Mr. McAllister charged a suit on an account that charges a monthly finance charge that is 1.5% of the unpaid balance. If his unpaid balance last month was $204 and he made no further purchases, how much is his bill this month?

37. Ricor calculated that he needed 850 square feet of tile to cover a floor. He wants to increase that amount by 12% to allow for waste. What is the total amount of tile he should buy?

38. The sales of a certain company increased from 15 million dollars to 37.5 million dollars. What percent increase is this?

39. A store is selling answering machines at 40% off. If the savings on an answering machine is $28, what is the original price of the machine?

40. Since Peter got a new girlfriend, his average study time per week has decreased by 6 hours. If he used to study 24 hours per week, what percent decrease is this?

CHAPTER 9 SUMMARY

KEY WORDS AND PHRASES

percent [9.1]
percent of a number [9.2]
marked price [9.6]
sales tax [9.6]
sales tax rate [9.6]
base salary [9.6]
commission [9.6]
commission rate [9.6]
interest (yield) [9.6]
amount invested [9.6]
interest rate (rate of return) [9.6]
simple interest [9.6]
compound interest [9.6]
service charge (finance charge) [9.6]
unpaid balance [9.6]
monthly interest rate [9.6]
tip (gratuity) [9.6]
percent increase [9.7]
percent decrease [9.7]
cost [9.7]
selling price [9.7]
markup [9.7]
markup based on cost [9.7]
percent markup (markup rate) [9.7]
markup based on selling price [9.7]
rate [9.6, 9.8]
base [9.8]
amount [9.8]
proportion [9.8]

SYMBOLS

$0.1\overline{6}$ means $0.16666666666\ldots$
$\approx$ means *approximately equal to*
30% means $\frac{30}{100} = 0.30$

IMPORTANT RULES

Comparing a Percent to the Number 1 [9.1]

The percent, 100%, is equal to the number 1. A percent that is less than 100% represents a number that is less than 1. A percent that is more than 100% represents a number that is more than 1.

Converting a Percent to a Fraction or Decimal [9.1]

To convert a percent to a fraction or decimal, drop the percent symbol, %, and divide by 100.

Converting a Fraction or Decimal to a Percent [9.3]

To convert a fraction or decimal to a percent, multiply by 100 and attach the percent symbol, %.

Finding a Percent of a Number [9.2]

- Convert the percent to a fraction or decimal. Use fractions if repeating decimals are involved.
- Multiply the number by the converted percent.

Finding the Missing Number in a Basic Percent Statement [9.4, 9.8]

METHOD 1 [9.4]
- Translate the basic percent statement to a multiplication statement.
- Find the missing number in the multiplication statement.
- If the percent is given, be sure to convert it to a fraction or decimal before beginning. Use fractions if repeating decimals are involved.
- If the question is *what percent* be sure to convert the answer to percent form.

METHOD 2 [9.8]
- Identify the rate, base, and amount in the basic percent statement.
- Write the proportion that corresponds to the basic percent statement.
  Remember that $\text{Rate} = \frac{\text{amount}}{\text{base}}$.
- Solve the proportion for the missing number.
- If the rate is missing, and you have let the rate be a missing number followed by the percent symbol, be sure to attach the percent symbol to the missing number found.

Solving Problems Involving Sales Tax, Commission, Interest Earned, Interest Paid, and Tips [9.6, 9.8]
A summary of the information needed to solve these types of problems is on page 419.

Solving Problems Involving Percent Increase, Percent Decrease, and Markup [9.7, 9.8]
A summary of the information needed to solve these types of problems is on page 431.

CHAPTER 9 REVIEW EXERCISES

Indicate whether each percent is less than, equal to, or more than the number 1.
1. 100%
2. 85%
3. 4%
4. 230%

Convert each percent to a decimal.
5. 7%
6. 68%
7. 120%
8. 6.7%
9. 52.08%
10. 0.5%
11. 0.018%
12. $\frac{1}{4}$%
13. $7\frac{1}{2}$%
14. $13\frac{2}{5}$%
15. 300%
16. 460%

Convert each percent to a fraction.
17. 5%
18. 80%
19. 64%
20. 150%
21. 200%
22. $\frac{3}{10}$%
23. $\frac{5}{6}$%
24. $8\frac{3}{4}$%
25. $66\frac{2}{3}$%
26. $6\frac{1}{4}$%
27. 12.5%
28. 3.2%
29. 0.9%
30. 0.06%
31. 50.25%
32. 0.001%

Solve.
33. 7% of 120
34. 9.4% of 25.6
35. $\frac{1}{2}$% of 8000
36. $3\frac{1}{3}$% of 990
37. 45.2% of 8.5
38. 125% of $8\frac{4}{7}$

Convert each number to a percent.
39. 0.16
40. 0.046
41. 8.7
42. 0.001
43. 0.4
44. 4.18
45. 0.502
46. 0.0037
47. $\frac{67}{100}$
48. $\frac{3}{5}$
49. $\frac{7}{20}$
50. $\frac{1}{8}$
51. 8
52. $\frac{5}{2}$
53. $\frac{5}{6}$
54. $5\frac{1}{3}$

Complete the following chart. Simplify the fractions when possible.

	Fraction	Decimal	Percent
55.			125%
56.		0.75	
57.	$\frac{1}{3}$		
58.			2.5%
59.		0.002	
60.	$\frac{7}{8}$		

Solve.
61. 30% of what number is 411?
62. What percent of 72 is 60?
63. Find 350% of 20.
64. 0.03% of what number is 5.1?

65. 4.2 is what percent of 175?

66. 0.5% of 80.6 is what number?

67. What number is $12\frac{1}{2}$% of $\frac{8}{25}$?

68. 7 is what percent of $2\frac{1}{3}$?

69. What percent of 600 is 20?

70. $\frac{1}{9}$ is $\frac{2}{3}$% of what number?

Solve.

71. Frank works in the quality control department of a textbook bindery company. He rejected 3 textbooks, which was $2\frac{1}{2}$% of the sample he checked. How many textbooks did he check?

72. Only 15% of the population of a small town are smokers. If the town has 460 people, how many are smokers?

73. At a convention there were 208 men and 442 women. What percent were men?

74. In a chemistry lab, Pancho mixed a solution that contained salt and water. He put in 27.3 grams of salt and the rest was water. If the solution is 12% salt, what is the total weight of the solution?

75. A certain type of beer is 4% alcohol. If you drink 20 ounces of beer, how much alcohol have you consumed?

76. Three-fourths percent of the population of a town caught a certain virus. If the town has a population of 24,000, how many caught the virus?

77. Sharon purchased a new dress that was marked $140. If the sales tax rate is 5.8%, how much did she pay for the dress?

78. Mr. Bragg paid $9.61 in sales tax for tires that were marked $155. What is the sales tax rate? (Express the answer as a percent.)

79. A real estate agent receives 5.5% commission for selling property. If the agent earns an $11,440 commission that is based on the total sales, what was the total value of the sale?

80. Bill invests $4800 in an account that earns $13\frac{1}{3}$% interest. How much money does he have in the account at the end of one year?

81. Ms. Wong received her credit card bill. It showed an unpaid balance from last month of $840 and a finance charge of $14.70. Find the monthly interest rate. (Express the answer as a percent.)

82. You are pleased with the service and food at an Italian restaurant, so you want to leave a 15% tip. How much tip do you leave if the dinner bill is $38?

83. Chuck took his office staff out to eat at a seafood restaurant. He left an 18% tip which amounted to $54.90. What was the dinner bill?

84. Cynthia's monthly salary increased from $800 to $840. What percent increase in this?

85. Mr. Fredericks researches the stock market for his large corporate clients. The price of a stock he was observing decreased from $43 to 37\frac{5}{8}$ yesterday. What percent decrease is this?

86. Just prior to the wedding of a very famous couple, the London police were granted a 17% pay raise. A policewoman who was making the equivalent of $2400 a month is now making how much a month?

87. After Mr. Reid became vice-president in charge of finance, the company's profits decreased by 2.4 million dollars, which was a 12% decrease from last year. What were the profits last year?

88. A manufacturer of heavy farming equipment produces a tractor at a cost of $18,500. If the markup rate based on cost is 28%, what is the markup?

89. An exclusive men's clothing store buys a suit from the manufacturer for $240 and sells the suit to a customer for $540.
    a. Find the markup rate if the markup is based on the cost.
    b. Find the markup rate if the markup is based on the selling price.
90. Mr. Huynh plans to build a fence around his rectangular vegetable garden with fencing that costs $6.80 per foot. The garden is 64 feet long and 31 feet wide. If he increases the amount of fencing needed by 8% to allow for waste and the sales tax rate is 5%, what is the total amount he pays for the fencing? (Round off to the nearest cent.)

In Your Own Words

Write complete sentences to discuss each of the following. Support your comments with examples or pictures, if appropriate.

91. Discuss the basic meaning of the percent symbol and why you think the symbol was introduced.
92. Discuss how you can decide whether a percent is less than 1, equal to 1, or more than 1.
93. Discuss how the following two problems differ.
    a. 2.5% of what number is 80?
    b. What number is 2.5% of 80?
94. A business student wants to compute $66\frac{2}{3}\%$ of 900. The student calculates 0.66 times 900. Discuss why the student's result is not exactly correct, and how the correct result would be obtained.
95. Discuss a real-life situation that involves a percent that is more than 100%.

CHAPTER 9 PRACTICE TEST

Convert each percent to a decimal
1. 26.4%   2. 370%   3. 0.5%   4. $6\frac{1}{4}\%$

Convert each percent to a fraction. Simplify if possible.
5. 45%   6. 350%   7. $\frac{4}{5}\%$ 8. $6\frac{2}{3}\%$

Convert each number to a percent.
9. 4   10. 0.057   11. $\frac{9}{10}$ 12. $2\frac{1}{3}$

Solve.

13. Find 2.8% of 75
14. What percent of 85 is 34?
15. 220% of what number is $8\frac{1}{4}$?
16. 4.7 is what percent of 9400?
17. What number is 7% of 33.5?
18. $66\frac{2}{3}\%$ of what number is 144?
19. Due to inflation the price of gasoline has increased from $0.50 per gallon to $1.40 per gallon over the past few years.
    a. The new price is what percent of the original price?
    b. What percent increase is this?

20. Last year the Giants baseball team won 21 out of 50 games. What percent did they win?
21. Jim and Nell bought a new dishwasher for their home in Little Rock. The marked price of the dishwasher was $330. If the sales tax rate is 4%, how much did they pay for the dishwasher?
22. Mr. Dittmer sells real estate in Boston. His commission rate based on the total sales is 5%. On a recent sale of an apartment building he earned an $18,000 commission. What was the selling price of the building?
23. Ms. Kim enjoyed her meal at the Japanese restaurant. She wants to leave an 18% tip. How much tip will she leave if the dinner bill is $35?
24. A company survey revealed that 20.6% of the employees preferred increased dental coverage to a pay increase. Five hundred fifteen employees wanted increased dental coverage. How many employees took part in the survey?
25. A variety store uses a $33\frac{1}{3}\%$ markup that is based on cost. The markup on an umbrella is $4.05.
    a. How much did the store pay the manufacturer for the umbrella?
    b. How much does the customer pay for this umbrella?

CUMULATIVE REVIEW EXERCISES: CHAPTERS 1–9

Evaluate. Answer with a fraction.

1. $\frac{5}{36} + \frac{11}{42}$
2. $12\frac{5}{12} - 8\frac{22}{27}$
3. $\frac{450}{100} \times \frac{86}{21}$
4. $5\frac{1}{4} \div 4\frac{9}{10}$

Evaluate. Answer with a decimal.

5. $34 - 7.89$
6. $67 + 5.6 + 18.56$
7. $0.078 \times 43{,}000$
8. $40 \div 0.125$
9. $5.46 \div 1500$
10. $4.275 \div 9.5$

Solve.

11. Convert 0.308 to a fraction reduced to lowest terms.
12. Convert $\frac{27}{1500}$ to a decimal.
13. Convert $\frac{7}{11}$ to a repeating decimal.
14. Find the ratio of 1.5 to 60. Express the answer as **a.** a fraction, **b.** a decimal, and **c.** a percent.
15. Find the average of $5\frac{1}{3}$, $6\frac{2}{3}$, and $7\frac{5}{6}$. Answer with a fraction.
16. A pharmacist divides 2.4 liters of solution into 8 equal parts. How much is in each part?
17. If 4.5 centigrams of medicine are given for each 40 pounds of body weight, how much of this medicine should be given for a person who weighs 175 pounds? (Round off the answer to the nearest tenth of a centigram.)
18. Leon completed $\frac{1}{8}$ of the job on Monday and Tom completed $\frac{1}{4}$ of the job on Tuesday.
    a. What fraction of the job has been completed?
    b. What fraction of the job has not been completed?
    c. If the job is to write a 200-page report, how many pages did Leon write on Monday?
    d. What percent of the job did Tom complete on Tuesday?

19. Sue's hourly wage went from $32.60 per hour to $33.93 per hour.
    a. What is her increase in pay per hour?
    b. What is the percent increase? (Round off to the nearest whole percent.)
    c. The new salary is what percent of the old salary? (Round off to the nearest whole percent.)
    d. How much more will she make now than before for working an 8-hour day?
20. Mr. Tsu purchased 6 items at $4.69 each, 4 items at $9.29 each, and 2 items at $13.49 each. The sales tax rate is 7.4% on all of the items.
    a. What is the total cost of these items? (Round up the answer to the nearest cent.)
    b. How much change does Mr. Tsu receive from a $100 bill?

CHAPTER 10

Measurement: English and Metric Systems

10.1 LENGTH

OBJECTIVES

- Measure the length of a line segment by using a ruler that measures inches.
- Measure the length of a line segment by using a ruler that measures centimeters.
- Change units of length by multiplying or dividing two equal lengths by the same nonzero number.
- Change units of length by multiplying by ratios that equal one.
- Solve an application that involves converting units of length.

1 Measuring Units of Length—English System

A measurement includes both a number and a **unit of measure.** For example, 4 inches is a measurement.

The unit of measure, **inch,** is one of the standard units of measurement used to measure **length** or **distance.** A length that measures 1 inch is shown here.

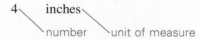

To measure the length of an object in inches, we can use a measuring stick that has markings at every inch and at fractional parts of an inch.

Try These Problems

Measure the length of each line segment to the nearest $\frac{1}{8}$ inch.

1. _____

2. _____

For example, here we use a ruler to measure the length of a line.

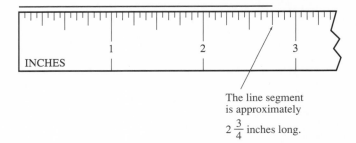

The line segment is approximately $2\frac{3}{4}$ inches long.

The line is approximately $2\frac{3}{4}$ inches long.

 Try Problems 1 and 2.

The unit of measure, **inch,** is one of several standard units of length in the **English measurement system.** English measurements are those whose units came from the English-speaking countries. Additional standard units of length in the English system are **foot, yard,** and **mile.**

A 1-foot length is the same distance as 12 inches, so we say that the measurements 1 foot and 12 inches are *equal* or *equivalent*. We use the symbol = to denote this equivalence.

$$1 \text{ foot} = 12 \text{ inches}$$

Observe that this equality statement makes no sense without the units attached to the numbers because the number 1 does not equal the number 12.

Here we list more equivalent units of length in the English measurement system.

Units of Length—English System

1 foot (ft) = 12 inches (in)
1 yard (yd) = 36 inches
1 yard   = 3 feet (ft)
1 mile (mi) = 5280 feet
1 mile   = 1760 yards (yd)

2 Measuring Units of Lengths—Metric System

About 200 years ago, another system of measurement was developed called the **metric system.** This system was invented to take advantage of the base ten place value system, thus is more convenient to use. Almost every country in the world, except the United States, has adopted this system. All scientists, even those in the United States, use the metric system.

The basic unit of length in the metric system is a **meter.** One meter is slightly more than the distance from a doorknob to the floor. One meter is slightly more than 1 yard.

Try These Problems

Measure the distance from X to Y to the nearest tenth of a centimeter.

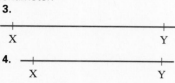

A **centimeter** is one-hundredth ($\frac{1}{100}$ or 0.01) of a meter. The prefix *centi-* means one-hundredth ($\frac{1}{100}$ or 0.01). A length that measures 1 centimeter is shown here.

To measure a distance in centimeters we can use a measuring stick that has markings at every centimeter and at fractional parts of a centimeter. For example, here we use a ruler to measure the distance from point A to point B.

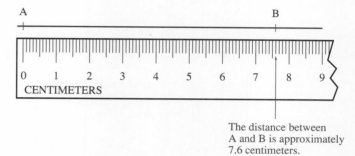

The distance between A and B is approximately 7.6 centimeters.

The distance from A to B is approximately 7.6 centimeters.

 Try Problems 3 and 4.

Here we list more equivalent units of length in the metric measurement system.

Units of Length—Metric System

1 millimeter (mm) = $\frac{1}{1000}$ or 0.001 meter (m)

1 centimeter (cm) = $\frac{1}{100}$ or 0.01 meter

1 decimeter (dm) = $\frac{1}{10}$ or 0.1 meter

1 dekameter (dam) = 10 meters (m)
1 hectometer (hm) = 100 meters
1 kilometer (km) = 1000 meters

The metric system of measurement is very logical. The prefix that precedes the basic unit *meter* corresponds to a specific numerical value. For example,

*kilo*meter   means   1000 meters

*milli*meter   means $\frac{1}{1000}$ or 0.001 meter

Here we list each prefix and the corresponding numerical value.

Prefix	Numerical Value
milli-	$\frac{1}{1000}$ or 0.001
centi-	$\frac{1}{100}$ or 0.01
deci-	$\frac{1}{10}$ or 0.1
deka-	10
hecto-	100
kilo-	1000

These prefixes are used not only for metric units of length but also in other metric units of measurement such as volume and weight. This makes the metric system very convenient to work with.

Study the following pictures to get some feel for the actual distances the units of length represent in both the English and metric systems.

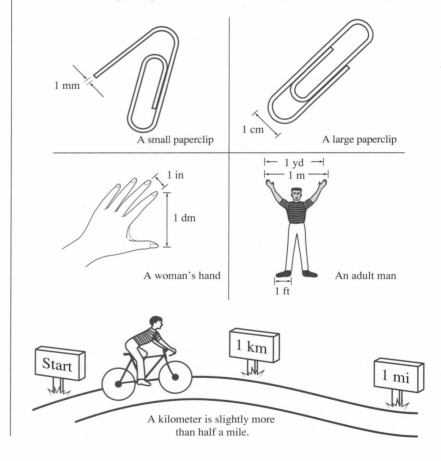

A kilometer is slightly more than half a mile.

3 Changing Units of Length by Multiplying or Dividing Two Equal Lengths by the Same Nonzero Number

If two measurements are equal, we can multiply each of them by the same nonzero number to produce two more equal measurements. Here we look at an example. We begin with an equivalence from a table.

$$1 \text{ foot} = 12 \text{ inches}$$

Observe that if we multiply both measurements by 3, we do not upset the equality.

$$1 \text{ foot} = 12 \text{ inches}$$
$$3 \times 1 \text{ foot} = 3 \times 12 \text{ inches} \quad \text{Multiply by 3}$$
$$3 \text{ feet} = 36 \text{ inches}$$

We have obtained that 3 feet is the same distance as 36 inches. We can also multiply both measurements by a fraction or decimal. For example, here we multiply by 0.5 or $\frac{1}{2}$.

$$1 \text{ foot} = 12 \text{ inches}$$
$$0.5 \times 1 \text{ foot} = 0.5 \times 12 \text{ inches} \quad \text{Multiply by 0.5}$$
$$0.5 \text{ foot} = 6 \text{ inches}$$

Here we see that half of a foot is the same distance as 6 inches.

You can also divide two equivalent measurements by the same nonzero number to produce two equivalent measurements. For example, here we divide by 5.

$$1 \text{ foot} = 12 \text{ inches}$$
$$\frac{1 \text{ foot}}{5} = \frac{12 \text{ inches}}{5} \quad \text{Divide by 5}$$
$$\frac{1}{5} \text{ foot} = 2\frac{2}{5} \text{ inches} \quad \text{Fractional form}$$
$$0.2 \text{ foot} = 2.4 \text{ inches} \quad \text{Decimal form}$$

A distance that measures 0.2 foot is the same as 2.4 inches.

We now state the multiplication and division rules for equivalent measurements.

Multiplication and Division Rules for Equivalent Measurements

1. Multiplying two equal measurements by the same nonzero number produces two equal measurements.
2. Dividing two equal measurements by the same nonzero number produces two equal measurements.

This rule can be used to convert units of measurement. Here are some examples.

Try These Problems

Convert the following measurements as indicated.

5. 15 yd = _?_ ft
6. 7.25 km = _?_ m
7. 1 ft = _?_ yd
8. 1 m = _?_ km

EXAMPLE 1   4 mi = _?_ ft

SOLUTION   Look in the table that gives equivalences between units of length in the English system. Choose the equivalence that relates miles to feet.

$$1 \text{ mi} = 5280 \text{ ft} \quad \text{From the table}$$

Because we want to convert this to an equivalence involving 4 miles, we multiply each measurement by 4.

$$1 \text{ mi} = 5280 \text{ ft}$$
$$4 \times 1 \text{ mi} = 4 \times 5280 \text{ ft} \quad \text{Multiply by 4}$$
$$4 \text{ mi} = 21{,}120 \text{ ft}$$

There are 21,120 feet in 4 miles. ∎

EXAMPLE 2   14.2 km = _?_ m

SOLUTION   Look in the table that gives equivalences between units of length in the metric system. Choose the equivalence that relates kilometers to meters.

$$1 \text{ km} = 1000 \text{ m}$$

Because we want to convert this to an equivalence that involves 14.2 kilometers, we multiply each measurement by 14.2.

$$1 \text{ km} = 1000 \text{ m}$$
$$14.2 \times 1 \text{ km} = 14.2 \times 1000 \text{ m} \quad \text{Multiply by 14.2}$$
$$14.2 \text{ km} = 14{,}200 \text{ m} \quad \text{Multiplying by 1000 moves the decimal point 3 places to the right.}$$

$$14.2 \times 1000 = 14200. = 14{,}200 \quad ∎$$

Try Problems 5 and 6.

EXAMPLE 3   1 in = _?_ yd

SOLUTION   Look in the table that gives equivalences between units of length in the English system. Choose the equivalence that relates inches to yards.

$$1 \text{ yd} = 36 \text{ in} \quad \text{From the table}$$

If we divide 36 by 36, we get 1. Therefore, divide each measurement by 36, and we get an equivalence that involves 1 inch.

$$1 \text{ yd} = 36 \text{ in}$$
$$\frac{1 \text{ yd}}{36} = \frac{36 \text{ in}}{36}$$
$$\frac{1}{36} \text{ yd} = 1 \text{ in} \quad \text{Because the fraction } \tfrac{1}{36} \text{ converts to a repeating decimal, we leave the answer in fractional form.}$$

The distance 1 inch is the same as $\tfrac{1}{36}$ of a yard. ∎

Try Problems 7 and 8.

Try These Problems

Convert the following measurements as indicated.
9. 4 in = _____?_____ yd
10. 55 dm = _____?_____ m

EXAMPLE 4 230 m = _____?_____ km

SOLUTION Look in the table that gives equivalences between units of length in the metric system. Choose the equivalence that relates meters to kilometers.

$$1 \text{ km} = 1000 \text{ m} \quad \text{From the table}$$

First find out how many kilometers are in 1 meter. To produce an equivalence involving one meter, divide each measurement by 1000.

$$1 \text{ km} = 1000 \text{ m}$$
$$\frac{1 \text{ km}}{1000} = \frac{1000 \text{ m}}{1000} \quad \text{Divide by 1000}$$
$$0.001 \text{ km} = 1 \text{ m}$$

A distance of 1 meter is the same as 0.001 kilometer. Now we can produce an equivalence involving 230 meters by multiplying each measurement by 230.

$$0.001 \text{ km} = 1 \text{ m}$$
$$230 \times 0.001 \text{ km} = 230 \times 1 \text{ m} \quad \text{Multiply by 230}$$
$$0.23 \text{ km} = 230 \text{ m}$$

A distance of 230 meters is the same as 0.23 kilometers. ■

 Try Problems 9 and 10.

Converting units of measurement by multiplying or dividing equivalent measurements by the same nonzero number is good to know because it enables you to do some easy conversions mentally. But when the situation becomes more complicated, this method is not convenient to use.

Suppose you want to convert 3 miles to inches. In the table, there is no equivalence that relates miles to inches, so it is not easy to see how to make the conversion. Now we introduce a conversion method that is easy to use no matter how complicated the situation.

4 Changing Units of Length by Multiplying by Ratios that Equal One

Consider the following equivalence from the table.

$$1 \text{ mile} = 5280 \text{ feet}$$

We can form a ratio by putting one of the measurements in the numerator and the other one in the denominator. Because both measurements are equal, the resulting ratio is equal to the number 1. Any nonzero quantity divided by itself equals the number 1.

$$\frac{1 \text{ mi}}{5280 \text{ ft}} = 1 \quad \text{and} \quad \frac{5280 \text{ ft}}{1 \text{ mi}} = 1$$

Suppose we want to know how many feet are in 3 miles. We can use one of these ratios to make this conversion. We choose the ratio that has miles in the denominator so that the miles units cancel, and we obtain the distance in feet.

Try These Problems

Convert the following measurements as indicated.

11. 10 ft = ____?____ yd
12. 8.3 m = ____?____ cm
13. 234 in = ____?____ yd
14. 934 mm = ____?____ m

$$3 \text{ mi} = \frac{3 \text{ mi}}{1} \times \boxed{\frac{5280 \text{ ft}}{1 \text{ mi}}}$$

— A ratio equal to 1

Put miles in the denominator so that the mile units cancel.

$$= 3 \times 5280 \text{ ft}$$
$$= 15{,}840 \text{ ft}$$

We have multiplied the measurement, 3 miles, by a ratio that is equal to 1, so the resulting measurement is the same distance as 3 miles. When a quantity is multiplied by 1, the value of that quantity does not change.

Now, if we want to know how many inches are in 3 miles, we can proceed in the same manner except we will need to use two different ratios that are equal to 1.

$$1 \text{ mi} = 5280 \text{ ft} \rightarrow \frac{1 \text{ mi}}{5280 \text{ ft}} \quad \text{or} \quad \frac{5280 \text{ ft}}{1 \text{ mi}}$$

$$1 \text{ ft} = 12 \text{ in} \rightarrow \frac{1 \text{ ft}}{12 \text{ in}} \quad \text{or} \quad \frac{12 \text{ in}}{1 \text{ ft}}$$

Be sure to choose the ratios so that the units cancel appropriately. For a unit to cancel, one of them must be in a numerator and another like it in a denominator.

$$3 \text{ mi} = \frac{3 \text{ mi}}{1} \times \boxed{\frac{5280 \text{ ft}}{1 \text{ mi}}} \times \boxed{\frac{12 \text{ in}}{1 \text{ ft}}}$$

Ratios equal to 1

Put miles in the denominator so that the miles units cancel.

Put feet in the denominator so that the feet units cancel.

$$= 3 \times 5280 \times 12 \text{ in}$$
$$= 190{,}080 \text{ in}$$

A distance of 3 miles is the same as 190,080 inches.

Now we look at some more examples.

EXAMPLE 5 $1\frac{1}{4}$ mi = ____?____ yd

SOLUTION   In the table for English units of length there is an equivalence that relates miles to yards, so we choose this equivalence and use it to write a ratio that equals the number 1.

$$1 \text{ mi} = 1760 \text{ yd} \rightarrow \frac{1 \text{ mi}}{1760 \text{ yd}} \quad \text{or} \quad \frac{1760 \text{ yd}}{1 \text{ mi}}$$

Now multiply the measurement $1\frac{1}{4}$ miles by the appropriate ratio, so that the miles units cancel, and the yards unit is left in the numerator.

$$1\frac{1}{4} \text{ mi}$$

— A ratio equal to 1

$$= \frac{1\frac{1}{4} \text{ mi}}{1} \times \boxed{\frac{1760 \text{ yd}}{1 \text{ mi}}}$$

Put yards in the numerator so that the answer is in yards.

$$= \frac{\frac{5}{4} \times 1760}{1} \text{ yd}$$

Put miles in the denominator so that the miles units cancel.

$$= 2200 \text{ yd}$$

The distance $1\frac{1}{4}$ miles is the same as 2200 yards. ■

Try Problems 11 through 14.

Try These Problems

Convert the following measurements as indicated.

15. 2 mi = _____?_____ in

16. 215 cm = _____?_____ dm

EXAMPLE 6   2030 mm = _____?_____ cm

SOLUTION   Choose the two equivalences from the table for metric units of length that involve millimeters and centimeters. Use these equivalences to write ratios that equal the number 1.

$$1 \text{ mm} = 0.001 \text{ m} \rightarrow \frac{1 \text{ mm}}{0.001 \text{ m}} \quad \text{or} \quad \frac{0.001 \text{ m}}{1 \text{ mm}}$$

$$1 \text{ cm} = 0.01 \text{ m} \rightarrow \frac{1 \text{ cm}}{0.01 \text{ m}} \quad \text{or} \quad \frac{0.01 \text{ m}}{1 \text{ cm}}$$

Now multiply the measurement, 2030 millimeters, by the appropriate ratios so that all units of measurement cancel except centimeters.

$$= \frac{2030 \text{ mm}}{1}$$
$$= \frac{2030 \text{ mm}}{1} \times \frac{0.001 \text{ m}}{1 \text{ mm}} \times \frac{1 \text{ cm}}{0.01 \text{ m}}$$
$$= \frac{2030 \times 0.001}{0.01} \text{ cm}$$
$$= \frac{2.03}{0.01} \text{ cm}$$
$$= 203 \text{ cm}$$

Ratios equal to 1

Put centimeter in the numerator so that the answer is in centimeters.

Put millimeters in the denominator so that the millimeters units cancel.

The distance 2030 millimeters is the same as 203 centimeters. ■

 Try Problems 15 and 16.

5 Solving Applications that Require Measurement Conversions

We look for length or distance to answer such questions as,

   How tall is she?

   How far did he run?

   What is the width of the window?

   What is the perimeter of the rectangle?

When solving problems that involve measurements, you must pay close attention to the units of measurement. It may be necessary for you to convert the units in one or more of the measurements before solving the problem. Here are some examples.

EXAMPLE 7   A rectangular window is 2.8 meters wide and 63 centimeters high. Find the perimeter of the window in centimeters.

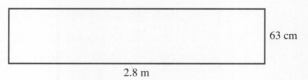

SOLUTION   The perimeter of the window is the distance all the way around the window, so we must add the lengths of the four sides.

$$\text{Perimeter} = 2.8 \text{ m} + 2.8 \text{ m} + 63 \text{ cm} + 63 \text{ cm}$$

Try These Problems

Solve.

17. Find the perimeter of this triangle. (Express the answer in meters.)

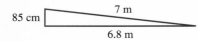

18. A sidewalk is $3\frac{1}{2}$ feet wide and 80 yards long. What is the perimeter of the sidewalk? (Express the answer in feet.)

19. Find the missing dimension in inches.

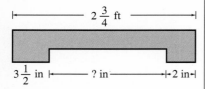

20. Wire sells for $4.25 per foot. What is the cost of 63 inches of this wire? (Round up to the nearest cent.)

21. Lumber sells for $1.29 per foot. What is the cost of 18 yards of this lumber?

Before adding these lengths, we must write them with the same unit of length. Because we want the answer in centimeters, we change the 2.8 meters to centimeters.

$$2.8 \text{ m} = \frac{2.8 \text{ m}}{1} \times \frac{1 \text{ cm}}{0.01 \text{ m}} \quad \text{A ratio equal to 1}$$

$$= \frac{2.8}{0.01} \text{ cm}$$

$$= 280 \text{ cm}$$

The width of the window is 280 centimeters.

Now we add up the lengths of the four sides to find the perimeter of the window.

$$\text{Perimeter} = 280 \text{ cm} + 280 \text{ cm} + 63 \text{ cm} + 63 \text{ cm}$$
$$= 686 \text{ cm}$$

The perimeter of the window is 686 centimeters. ■

 Try Problems 17 through 19.

EXAMPLE 8   Barnaby bought 6 feet of ribbon that sells for $1.29 per yard. What is the total cost?

SOLUTION   We have the following.

$$\frac{\text{Total}}{\text{cost}} = \frac{\text{cost}}{\text{per yard}} \times \frac{\text{how many}}{\text{yards}}$$

$$= \$1.29 \times \boxed{}$$

Because the cost is given per yard, we need to find the length of the ribbon, 6 feet, in yards.

$$6 \text{ ft} = \frac{6 \text{ ft}}{1} \times \frac{1 \text{ yd}}{3 \text{ ft}} \quad \text{Ratio equal to 1}$$

$$= \frac{6}{3} \text{ yd}$$

$$= 2 \text{ yd}$$

The length, 6 feet, is the same as 2 yards. Now we can find the total cost.

$$\frac{\text{Total}}{\text{cost}} = \frac{\text{cost}}{\text{per yard}} \times \frac{\text{how many}}{\text{yards}}$$

$$= \frac{\$1.29}{1 \text{ yd}} \times \frac{2 \text{ yd}}{1}$$

$$= \$2.58$$

The total cost of the ribbon is $2.58. ■

 Try Problems 20 and 21.

 Answers to Try These Problems

1. $1\frac{1}{2}$ in   2. $1\frac{3}{8}$ in   3. 4 cm   4. 3.4 cm   5. 45 ft   6. 7250 m
7. $\frac{1}{3}$ 8. 0.001 km   9. $\frac{1}{9}$ yd   10. 5.5 m   11. $\frac{10}{3}$ or $3\frac{1}{3}$ yd
12. 830 cm   13. $6\frac{1}{2}$ or 6.5 yd   14. 0.934 m   15. 126,720 in
16. 21.5 dm   17. 14.65 m   18. 487 ft   19. $27\frac{1}{2}$ or 27.5 in
20. $22.32   21. $69.66

EXERCISES 10.1

Measure the length of each line segment to the nearest $\frac{1}{8}$ inch.

1. ───────────────── 2. ──────────────────

Measure the distance from A to B to the nearest tenth of a centimeter.

3. A ──────────────────────────── B

4. A ──────────── B

5. Measure the length of each side of this triangle to the nearest $\frac{1}{4}$ inch. Use these measurements to find the perimeter of the triangle.

6. Measure the length and width of this rectangle to the nearest tenth of a centimeter. Use these measurements to find the perimeter of the rectangle.

Convert the following measurements by multiplying or dividing two equivalent measurements by the same nonzero number.

7. 5 yd = ___?___ ft
8. 3.5 ft = ___?___ in
9. 1 in = ___?___ ft
10. 1 in = ___?___ yd
11. 8 km = ___?___ m
12. 120 cm = ___?___ m
13. 1 m = ___?___ dam
14. 1 m = ___?___ cm
15. 48 in = ___?___ ft
16. 54 ft = ___?___ yd
17. 9.4 m = ___?___ mm
18. 250 m ___?___ hm

Convert the following measurements by using ratios that equal the number 1.

19. 8 in = ___?___ ft
20. 1.2 mi = ___?___ ft
21. 2200 yd = ___?___ mi
22. $2\frac{1}{3}$ yd = ___?___ ft
23. 0.2 mi = ___?___ in
24. 39,600 in = ___?___ mi
25. 62.5 m = ___?___ cm
26. 830 mm = ___?___ m
27. 95 dm = ___?___ m
28. 2468 m = ___?___ km
29. 8 mm = ___?___ cm
30. 0.45 dm ___?___ cm
31. 75 cm = ___?___ mm
32. 36.8 hm = ___?___ km

Solve.

33. Find the missing dimension in yards.

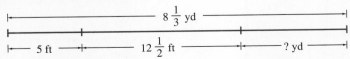

34. Find the total length of this shaft in centimeters.

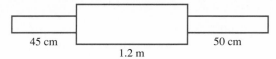

35. Find the perimeter of this triangle in feet.

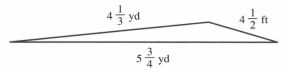

36. A rectangular strip of metal is 3.5 meters long and 4 centimeters wide. Find the perimeter of the strip in meters.
37. Stephanie purchases 13 feet of fabric that sells for $8.70 per yard. What is the total cost?
38. Rope sells for $3.25 per meter. What is the cost of 380 centimeters of this rope?
39. Anton wants to put weatherstripping all the way around a rectangular window that is 80 centimeters wide and 120 centimeters long. If the weatherstripping costs $0.70 per meter, what is the total cost of the stripping?
40. Farmer Anderson wants to put a fence around three sides of a rectangular region that is adjacent to a river. If the fencing material costs $10.80 per foot, find the total cost of the fence.

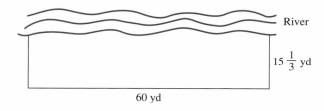

DEVELOPING NUMBER SENSE #9

CONVERTING METRIC UNITS BY USING A HORIZONTAL DISPLAY OF THE UNITS

When converting between units of length in the metric system, the decimal point moves either to the right or to the left. The digits in the measurement do not change. For example,

560 meters = 0.56 kilometers

0.245 decimeters = 2.45 centimeters

You can quickly see which direction to move the decimal and how many places to move the decimal by viewing the metric units of length from largest to smallest in the following horizontal display.

For example, to convert 560 meters to kilometers, move the decimal point 3 places to the left because kilometers is 3 places to the left of meters on the horizontal display.

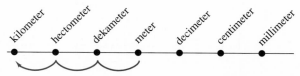

Cont. on page 467

Therefore, 560 meters
= .560 kilometer   The decimal point moves 3 places to the left.
= 0.56 kilometer

The horizontal display can be simplified by writing the basic unit of length, meter, then write only the prefix for the other units of length. Here we show the simplified display.

Now we look at another example. To convert 0.245 decimeter to centimeters, move the decimal point 1 place to the right since centimeters is 1 place to the right of decimeters on the horizontal display.

Therefore, 0.245 decimeter
= 02.45 centimeters   The decimal point moves 1 place to the right.
= 2.45 centimeters

Before using this horizontal display to convert units of length in the metric system, please make sure the horizontal display has the following two properties. First, the larger units are on the left and the smaller units on the right. For example, 1 kilometer is a greater distance than 1 millimeter. Second, list *all* metric units of length from kilometers to millimeters. As you move from right to left, each unit of length must be 10 times the previous unit of length.

Because the same prefixes are used for volume and mass (weight) as are used for length, this method of conversion can also be used to convert units of volume such as liters, centiliters, and milliliters, where *liters* is the basic volume unit, and units of mass (weight) such as grams, decigrams, and hectograms, where *grams* is the basic unit of mass (weight). Units of volume and mass (weight) are covered in Sections 10.2 and 10.3, respectively. For converting length, volume, or mass (weight), use the following horizontal display.

Number Sense Problems

Use the above horizontal display of units to convert the following units of length.

1. 3 meters = ___?___ millimeters
2. 490 kilometers = ___?___ centimeters
3. 0.4 decimeters = ___?___ meter
4. 30 centimeters = ___?___ dekameter

Use the above horizontal display of units to convert the following units of volume.

5. 675 centiliters = ___?___ liters
6. 4.5 liters = ___?___ milliliters
7. 0.08 kiloliters = ___?___ dekaliters
8. 240,000 centiliters = ___?___ hectoliters

Use the above horizontal display of units to convert the following units of mass (weight).

9. 74 grams = ___?___ centigrams
10. 9.8 kilograms = ___?___ decigrams
11. 6 milligrams = ___?___ grams
12. 4.5 dekagrams = ___?___ kilogram

10.2 AREA AND VOLUME

OBJECTIVES
- Find the area of a rectangular region that involves converting units of length.
- Solve a problem that involves finding the area of a composite of rectangular regions.
- Change units of area and volume.
- Solve an application that involves converting units of area or volume.
- Identify whether a unit of measurement is length, area, or volume.

1 Solving Area Problems that Require Unit Conversions

Area is a measure of the extent of a region. We look for area to answer such questions as,

How much wall surface do we paint?

What amount of carpeting do we need?

How much land did he buy?

Here we show a square that measures 1 centimeter on each side. We say that the area of this square is 1 *square centimeter*.

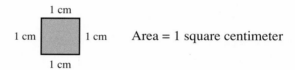

A rectangle that is 4 centimeters long and 3 centimeters wide has an area of 12 square centimeters, because 12 of these 1-square-centimeter squares fit inside.

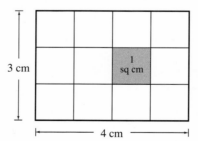

Observe that the area of this rectangle can be obtained by multiplying the length by the width.

$$\begin{aligned}\text{Area} &= \text{length} \times \text{width} \\ &= 4 \text{ cm} \times 3 \text{ cm} \\ &= 12 \text{ sq cm}\end{aligned}$$

Try These Problems

Solve.
1. A rectangular strip of metal measures 10 feet by 4 inches. What is the area of the strip in square feet?
2. A rectangle is 1.2 kilometers long and 500 meters wide. What is the area of the rectangle in square meters?

In general, the area of any rectangle can be computed by multiplying the length by the width.

Area of a Rectangle
The area of a rectangle is the length times the width.
$$\text{Area} = \text{length} \times \text{width}$$

When computing the area of a rectangle, make sure the length and the width are expressed with the same units of length. Here is how it works.

$$\text{Feet} \times \text{feet} = \text{square feet}$$
$$\text{Yards} \times \text{yards} = \text{square yards}$$
$$\text{Meters} \times \text{meters} = \text{square meters}$$

Units like square feet, square yards, and square meters are called **square units.** Area is often expressed in square units.

EXAMPLE 1 A remnant of carpeting measures 4 yards long and 5 feet wide. What is the area of the remnant in square feet?

SOLUTION We assume that the remnant is a rectangle. We have the following.

$$\text{Area} = \text{length} \times \text{width}$$
$$\boxed{} = 4 \text{ yd} \times 5 \text{ ft}$$

Before multiplying, write both dimensions with the same unit of length. Because we want the answer in square feet, we change 4 yards to feet.

$$4 \text{ yd} = \frac{4 \text{ yd}}{1} \times \boxed{\frac{3 \text{ ft}}{1 \text{ yd}}} \quad \text{A ratio equal to 1.}$$
$$= 12 \text{ ft}$$

The length, 4 yards, is the same as 12 feet. Now we can compute the area.

$$\text{Area} = \text{length} \times \text{width}$$
$$= 12 \text{ ft} \times 5 \text{ ft}$$
$$= 60 \text{ sq ft}$$

The area of the remnant is 60 square feet. ■

 Try Problems 1 and 2.

The following table gives equivalent units of area in both the English and metric systems of measurement.

> ### Units of Area
>
> **English System**
>
> 1 square foot (sq ft) = 144 square inches (sq in)
> 1 square yard (sq yd) = 9 square feet
> 1 acre   = 4840 square yards
> 1 square mile (sq mi) = 640 acres
>
> **Metric System**
>
> 1 square meter (sq m)   = 10,000 square centimeters (sq cm)
> 1 are   = 1 square dekameter (sq dam)
> 1 are   = 100 square meters
> 1 hectare (ha)   = 1 square hectometer (sq hm)
> 1 hectare   = 10,000 square meters
> 1 square kilometer (sq km) = 1,000,000 square meters

Do not confuse the English unit *acre* with the metric unit *are*. An *acre* is about three-fourths of the size of a football field. An *are* is approximately the floor space of a small two-bedroom apartment.

From the pictures below notice that 1 square meter is slightly more than 1 square yard.

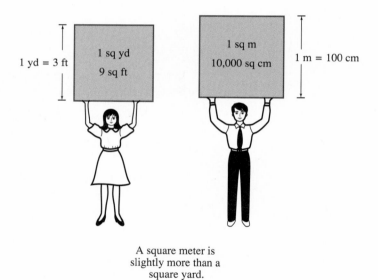

A square meter is slightly more than a square yard.

We can use the equivalent table for units of area to help us convert units of area. The process used is similar to that used to convert units of length. Here we look at some examples.

Try These Problems

Convert the following measurements as indicated.

3. 2.5 acres = ____?____ sq yd
4. 1250 sq cm = ____?____ sq m
5. 0.02 sq km = ____?____ ha

Solve.

6. Carpeting sells for $14.40 per square yard. What does it cost to carpet 250 square feet of floor surface?
7. Chicago covers 225.12 square miles and New Orleans covers 199.4 square miles. How many more acres of land does Chicago have than New Orleans?

EXAMPLE 2 60 sq m = ____?____ are

SOLUTION

$$60 \text{ sq m} = \frac{60 \text{ sq m}}{1} \times \frac{1 \text{ are}}{100 \text{ sq m}} \quad \text{A ratio equal to 1.}$$

$$= \frac{60}{100} \text{ are}$$

$$= 0.60 \text{ are}$$

$$= 0.6 \text{ are}$$

The area 60 square meters is the same as 0.6 are. ■

EXAMPLE 3 0.5 acre = ____?____ sq ft

SOLUTION

$$0.5 \text{ acre} = \frac{0.5 \text{ acre}}{1} \times \frac{4840 \text{ sq yd}}{1 \text{ acre}} \times \frac{9 \text{ sq ft}}{1 \text{ sq yd}} \quad \text{Ratios equal to 1.}$$

$$= \frac{0.5 \times 4840 \times 9}{1} \text{ sq ft}$$

$$= 21,780 \text{ sq ft}$$

The area 0.5 acre is the same as 21,780 square feet. ■

Try Problems 3 through 5.

EXAMPLE 4 A kitchen floor contains 27 square yards, and you want to cover the floor with tile. If the tile sells for $12.60 per square foot, what is the cost of tiling this floor?

SOLUTION Ignoring waste and taxes, we have the following.

$$\begin{aligned}\text{Total cost} &= \frac{\text{cost per}}{\text{square foot}} \times \frac{\text{how many}}{\text{square feet}} \\ &= \$12.60 \times \boxed{}\end{aligned}$$

Because the cost is given per square foot, we must get the area of the floor, 27 square yards, in square feet.

$$27 \text{ sq yd} = \frac{27 \text{ sq yd}}{1} \times \frac{9 \text{ sq ft}}{1 \text{ sq yd}} \quad \text{Ratio equal to 1}$$

$$= 27 \times 9 \text{ sq ft}$$

$$= 243 \text{ sq ft}$$

The area, 27 square yards, is the same as 243 square feet. Now we can compute the total cost of the tile.

$$\begin{aligned}\text{Total cost} &= \frac{\text{cost per}}{\text{square foot}} \times \frac{\text{how many}}{\text{square feet}} \\ &= \frac{\$12.60}{1 \text{ sq ft}} \times \frac{243 \text{ sq ft}}{1} \\ &= \$3061.80\end{aligned}$$

The total cost is $3061.80. ■

Try Problems 6 and 7.

EXAMPLE 5 Carpeting sells for $14.95 per square yard. What does it cost to carpet this floor surface? (Round off to the nearest cent.)

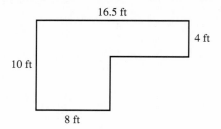

SOLUTION Ignoring waste and taxes, we have the following.

$$\text{Total cost} = \text{cost per square yard} \times \text{how many square yards}$$
$$= \$14.95 \times \boxed{}$$

Because the cost is given per square yard, we must eventually get the area of the floor in square yards.

To find the area of the floor, we view the floor as two rectangular regions. Because the dimensions are given in feet, we first find the area of the floor in square feet.

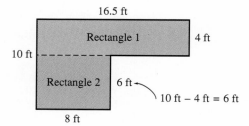

Area of rectangle 1 = 16.5 ft × 4 ft
    = 66 sq ft
Area of rectangle 2 = 8 ft × 6 ft
    = 48 sq ft

$$\text{Total area} = \text{area of rectangle 1} + \text{area of rectangle 2}$$
$$= 66 \text{ sq ft} + 48 \text{ sq ft}$$
$$= 114 \text{ sq ft}$$

The area of the floor surface is 114 square feet. Before finding the total cost of the carpeting, we must convert the area, 114 square feet, to square yards.

$$114 \text{ sq ft} = \frac{114 \cancel{\text{ sq ft}}}{1} \times \boxed{\frac{1 \text{ sq yd}}{9 \cancel{\text{ sq ft}}}} \quad \text{A ratio equal to 1.}$$
$$= \frac{114}{9} \text{ sq yd}$$
$$= \frac{\cancel{3} \times 38}{\cancel{3} \times 3} \text{ sq yd}$$
$$= \frac{38}{3} \text{ sq yd}$$

▲ Try These Problems

Solve.

8. Tile costs $3.19 per square foot. What is the cost of tiling this floor surface?

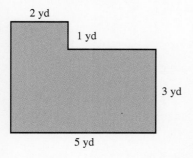

The area of the floor surface is $\frac{38}{3}$ square yards. Now we compute the total cost.

$$\text{Total cost} = \frac{\text{cost per}}{\text{square yard}} \times \frac{\text{how many}}{\text{square yards}}$$

$$= \frac{\$14.95}{1 \text{ sq yd}} \times \frac{38}{3} \text{ sq yd}$$

$$= \frac{\$14.95}{1} \times \frac{38}{3}$$

$$= \frac{\$568.10}{3}$$

$$\approx \$189.37 \qquad \text{Wait until the end to round off.}$$

The total cost of the carpeting is $189.37. ∎

▲ Try Problem 8.

2 Solving Volume Problems that Require Unit Conversions

Volume measures capacity, which is the space inside a solid. We look for volume to answer such questions as,

  How much water does the swimming pool hold?
  What is the amount of air inside the balloon?
  What is the capacity of this box?

Here we show a box that is 1 centimeter wide, 1 centimeter long, and 1 centimeter high. We say that the volume of this box is 1 cubic centimeter.

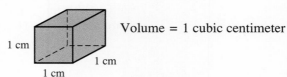

Volume = 1 cubic centimeter

A **rectangular solid** is like a box in that it has length, width, and height.

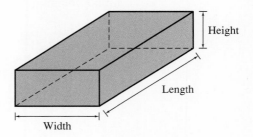

A rectangular solid that is 3 centimeters wide, 4 centimeters long, and 2 centimeters high contains 24 of the 1-cubic-centimeter boxes, so has a volume of 24 cubic centimeters.

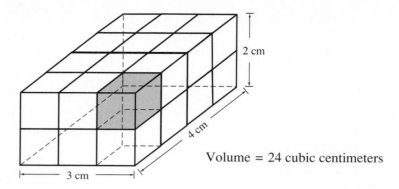

Volume = 24 cubic centimeters

Observe that the volume of this rectangular solid can be found by multiplying the length by the width by the height.

$$\begin{aligned}\text{Volume} &= \text{length} \times \text{width} \times \text{height} \\ &= 4 \text{ cm} \times 3 \text{ cm} \times 2 \text{ cm} \\ &= 24 \text{ cu cm}\end{aligned}$$

In general, the volume of a rectangular solid can be found by multiplying the length by the width by the height.

Volume of a Rectangular Solid

The volume of a rectangular solid is the length times the width times the height.

$$\text{Volume of a rectangular solid} = \text{length} \times \text{width} \times \text{height}$$

When computing the volume of a rectangular solid, be sure that the length, width, and height are all expressed with the same units of length before multiplying. Here is how it works.

$$\begin{aligned}\text{Feet} \times \text{feet} \times \text{feet} &= \text{cubic feet} \\ \text{Yards} \times \text{yards} \times \text{yards} &= \text{cubic yards} \\ \text{Meters} \times \text{meters} \times \text{meters} &= \text{cubic meters}\end{aligned}$$

Units like cubic feet, cubic yards, and cubic meters are called **cubic units.** Volume is often expressed in cubic units.

EXAMPLE 6 A swimming pool is in the shape of a rectangular solid. Find the volume of the pool in cubic feet if it is 2 yards wide, 5 yards long, and 4 feet deep.

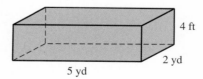

Try These Problems

Solve.

9. Find the capacity of a box that is $2\frac{1}{3}$ feet long, $1\frac{1}{4}$ feet wide, and $1\frac{1}{2}$ feet high.

10. A piece of ice is in the shape of a rectangular solid. It has dimensions 3 centimeters by 2 centimeters by 8 millimeters. What is the volume of this ice cube in cubic centimeters?

11. A cabinet is 5 feet wide, 4 feet high, and 16 inches deep. What is the capacity of the cabinet in cubic feet?

SOLUTION We have the following.

$$\text{Volume} = \text{length} \times \text{width} \times \text{height}$$
$$= 5 \text{ yd} \times 2 \text{ yd} \times 4 \text{ ft}$$

Before multiplying, write all three dimensions with the same unit of length. Because we want the answer in cubic feet, we convert the length, 5 yards, and the width, 2 yards, to feet.

$$5 \text{ yd} = \frac{5 \text{ yd}}{1} \times \boxed{\frac{3 \text{ ft}}{1 \text{ yd}}} = 15 \text{ ft}$$

Ratios equal to 1

$$2 \text{ yd} = \frac{2 \text{ yd}}{1} \times \boxed{\frac{3 \text{ ft}}{1 \text{ yd}}} = 6 \text{ ft}$$

The length, 5 yards, is the same as 15 feet, and the width, 2 yards, is the same as 6 feet. Now we compute the volume of the pool.

$$\text{Volume} = \text{length} \times \text{width} \times \text{height}$$
$$= 15 \text{ ft} \times 6 \text{ ft} \times 4 \text{ ft}$$
$$= 360 \text{ cu ft}$$

The volume of the pool is 360 cubic feet. ■

 Try Problems 9 through 11.

The following table gives equivalent units of volume in both the English and metric systems of measurement.

Units of Volume	
English System	
1 cubic yard (cu yd)	= 27 cubic feet (cu ft)
1 cubic foot	= 1728 cubic inches (cu in)
1 cubic foot	= 7.48 gallons (gal)
1 gallon	= 231 cubic inches (cu in)
1 gallon	= 4 quarts (qt)
1 quart	= 2 pints (pt)
1 pint	= 2 cups (c)
1 cup	= 8 fluid ounces (fl oz)
1 fluid ounce	= 2 tablespoons (T)
1 tablespoon	= 3 teaspoons (t)
Metric System	
1 liter (ℓ)	= 1 cubic decimeter (cu dm)
1 liter	= 1000 cubic centimeters (cu cm)
1 deciliter (dℓ)	= 0.1 liter
1 centiliter (cℓ)	= 0.01 liter
1 milliliter (mℓ)	= 0.001 liter
1 milliliter	= 1 cubic centimeter

Note from the previous table that the standard unit of volume in the metric system is the **liter**. The prefixes such as *centi-* and *milli-* have the

Try These Problems

Convert the following measurements as indicated.
12. 680 dℓ = ____?____ mℓ
13. 15 t = ____?____ fl oz

same meaning as they did for the metric units of length. Study the pictures that follow to get a feel for the actual size of some of these volume measurements.

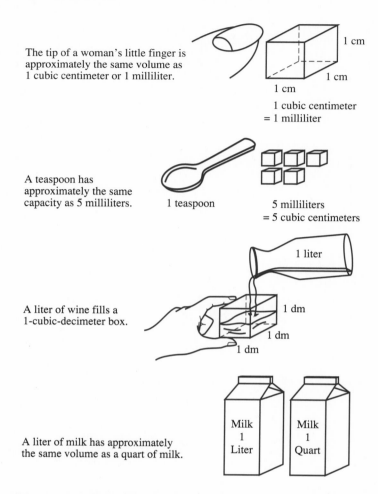

The tip of a woman's little finger is approximately the same volume as 1 cubic centimeter or 1 milliliter.

1 cubic centimeter = 1 milliliter

A teaspoon has approximately the same capacity as 5 milliliters.

1 teaspoon

5 milliliters = 5 cubic centimeters

A liter of wine fills a 1-cubic-decimeter box.

A liter of milk has approximately the same volume as a quart of milk.

We can use the equivalence table for units of volume to help us convert units of volume. The process used is similar to that used to convert units of length and units of area as shown in the examples that follow.

EXAMPLE 7   450 cℓ = ____?____ cu cm

SOLUTION

$$450 \text{ cℓ} = \frac{450 \text{ cℓ}}{1} \times \frac{0.01 \text{ ℓ}}{1 \text{ cℓ}} \times \frac{1000 \text{ cu cm}}{1 \text{ ℓ}} \quad \text{Ratios equal to 1}$$

$$= 450 \times 0.01 \times 1000 \text{ cu cm}$$

$$= 4500 \text{ cu cm}$$

The volume, 450 centiliters, is the same as 4500 cubic centimeters. ∎

 Try Problems 12 and 13.

Try These Problems

Convert the following measurements as indicated.
14. 16 pt = _____?_____ T
15. 64 fl oz = _____?_____ gal
16. 2 cu ft = _____?_____ c

EXAMPLE 8 $\frac{3}{4}$ gal = _____?_____ c

SOLUTION

$$\frac{3}{4} \text{ gal} = \frac{3 \text{ gal}}{4} \times \frac{4 \text{ qt}}{1 \text{ gal}} \times \frac{2 \text{ pt}}{1 \text{ qt}} \times \frac{2 \text{ c}}{1 \text{ pt}} \quad \text{Ratios equal to 1}$$

$$= \frac{3 \times 4 \times 2 \times 2}{4} \text{ c}$$

$$= 12 \text{ c}$$

A volume of $\frac{3}{4}$ gallon is the same as 12 cups. ■

EXAMPLE 9 20 fl oz = _____?_____ cu in

SOLUTION

20 fl oz

$$= \frac{20 \text{ fl oz}}{1} \times \frac{1 \text{ c}}{8 \text{ fl oz}} \times \frac{1 \text{ pt}}{2 \text{ c}} \times \frac{1 \text{ qt}}{2 \text{ pt}} \times \frac{1 \text{ gal}}{4 \text{ qt}} \times \frac{231 \text{ cu in}}{1 \text{ gal}}$$

Ratios equal to 1

$$= \frac{20 \times 231}{8 \times 2 \times 2 \times 4} \text{ cu in}$$

$$= \frac{4 \times 5 \times 231}{8 \times 2 \times 2 \times 4} \text{ cu in}$$

$$= \frac{1155}{32} \text{ cu in}$$

$$= 36\frac{3}{32} \text{ cu in}$$

A volume of 20 fluid ounces is the same as $36\frac{3}{32}$ cubic inches. ■

 Try Problems 14 through 16.

EXAMPLE 10 A recipe calls for 4 teaspoons of salt. Chef Pierre is preparing 15 times the recipe. How many cups of salt should he use?

SOLUTION First we find how many teaspoons of salt Chef Pierre needs.

$$\text{Total amount} = 15 \times \frac{\text{amount for}}{1 \text{ recipe}}$$

$$= 15 \times 4 \text{ t}$$

$$= 60 \text{ t}$$

Chef Pierre needs 60 teaspoons of salt. Now convert 60 teaspoons to cups.

$$60 \text{ t} = \frac{60 \text{ t}}{1} \times \frac{1 \text{ T}}{3 \text{ t}} \times \frac{1 \text{ fl oz}}{2 \text{ T}} \times \frac{1 \text{ c}}{8 \text{ fl oz}} \quad \text{Ratios equal to 1}$$

$$= \frac{60}{3 \times 2 \times 8} \text{ c}$$

$$= \frac{2 \times 3 \times 2 \times 5}{3 \times 2 \times 2 \times 4} \text{ c}$$

$$= \frac{5}{4} \text{ c} \quad \text{or} \quad 1\frac{1}{4} \text{ c}$$

Chef Pierre should use $1\frac{1}{4}$ cups of salt. ■

Try These Problems

Solve.

17. One can is $\frac{1}{5}$ the size of another can. The larger can holds 2.5 liters. How many milliliters does the smaller can hold?

18. A gallon of milk is divided equally among 8 children. How many fluid ounces of milk does each child receive?

19. How many 11.5-centiliter doses can a pharmacist make from 2.3 liters of medicine?

20. Robin made 5 gallons of punch by mixing 4 cups of wine with some orange juice. What fraction of the punch is wine?

 Try Problems 17 and 18.

EXAMPLE 11 How many 125-milliliter doses can a pharmacist make from 3 liters of medicine?

SOLUTION In this problem, the total volume 3 liters is being split into a number of equal parts. Each part has volume 125 milliliters. We have the following.

$$\text{Size of each part} \times \text{number of equal parts} = \text{total quantity}$$
$$125 \text{ m}\ell \times \boxed{} = 3 \ell$$

Before dividing, we must express both quantities using the same unit of volume. We choose to convert 3 liters to milliliters.

$$3 \ell = \frac{3 \ell}{1} \times \frac{1 \text{ m}\ell}{0.001 \ell} = \frac{3}{0.001} \text{ m}\ell = 3000 \text{ m}\ell$$

The total volume 3 liters is the same as 3000 milliliters. Now we have the following.

$$\text{Size of each part} \times \text{number of equal parts} = \text{total quantity}$$
$$125 \text{ m}\ell \times \boxed{} = 3000 \text{ m}\ell$$

To find the missing multiplier, divide 3000 by 125.

$$3000 \text{ m}\ell \div 125 \text{ m}\ell = \frac{3000 \text{ m}\ell}{125 \text{ m}\ell} = 24$$

The pharmacist can make 24 doses. ■

 Try Problems 19 and 20.

3 Recognizing Units of Length, Area, and Volume

In Sections 10.1 and 10.2 we looked carefully at three types of measurement: length, area, and volume. If you are given a measurement that is length, area, or volume, you can recognize its type by looking at the units of measurement. Here we list some of the units of measurement for each type.

Try These Problems

Identify each of these measurements as length, area, or volume.
21. 7.8 cubic inches
22. 56 inches
23. 16.2 square inches
24. 8 liters
25. 4.5 square kilometers
26. 18 hectometers

Length	English System	feet, inches, yards, and miles
	Metric System	meters, millimeters, centimeters, decimeters, dekameters, hectometers, kilometers
Area	English System	square feet, square inches, square yards, square miles, acres
	Metric System	square meters, square centimeters, square dekameters, square hectometers, square kilometers, ares, hectares
Volume	English System	cubic yards, cubic feet, cubic inches, gallons, quarts, pints, cups, fluid ounces, tablespoons, teaspoons
	Metric System	liter, deciliter, centiliter, milliliter, cubic decimeter, cubic centimeter

Observe that the word *square* followed by a unit of length always means area, while the word *cubic* followed by a unit of length means volume.

 Try Problems 21 through 26.

 Answers to Try These Problems

1. $\frac{10}{3}$ or $3\frac{1}{3}$ sq ft 2. 600,000 sq m 3. 12,100 sq yd
4. 0.125 sq m 5. 2 ha 6. $400 7. 16,460.8 acres
8. $488.07 9. $\frac{35}{8}$ or $4\frac{3}{8}$ cu ft 10. 4.8 cu cm
11. $\frac{80}{3}$ or $26\frac{2}{3}$ cu ft 12. 68,000 mℓ 13. $2\frac{1}{2}$ or 2.5 fl oz
14. 512 T 15. $\frac{1}{2}$ or 0.5 gal 16. 239.36 c 17. 500 mℓ
18. 16 fl oz 19. 20 doses 20. $\frac{1}{20}$ 21. volume 22. length
23. area 24. volume 25. area 26. length

EXERCISE 10.2

Identify each of these measurements as length, area, or volume.
1. 3.2 feet
2. 86 cubic feet
3. 200 square feet
4. 8.5 square kilometers
5. 150 kilometers
6. 5.42 cubic kilometers
7. 3 pints
8. 2 tablespoons
9. 5.25 acres
10. 0.65 hectare
11. 680 liters
12. 3.8 centiliters

Convert the following measurements as indicated.
13. 360 sq ft = _____?_____ sq yd
14. 6912 acres = _____?_____ sq mi
15. 2500 sq m = _____?_____ ha
16. 3.5 sq km = _____?_____ ares
17. 30 fl oz = _____?_____ qt
18. 24 t = _____?_____ c
19. 50 cu ft = _____?_____ pt
20. 12 c = _____?_____ gal
21. 32.8 mℓ = _____?_____ dℓ
22. 127.8 cu cm = _____?_____ ℓ

Solve.

23. A desk top measures 1.5 meters by 78 centimeters. What is the area of the desk top in square centimeters?
24. A piece of fabric is 54 inches wide and $6\frac{1}{3}$ yards long. What is the area of the fabric in square yards?
25. Carpeting sells for $13.59 per square yard. What does it cost to carpet 200 square feet of floor space?
26. Tile costs $4.28 per square foot. What is the cost of tiling this floor surface?

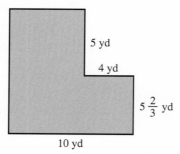

27. Jimmy bought 15.5 acres of land in Texas. He used 2 acres for a clubhouse, swimming pool, and tennis courts. He divided the remaining land into 75 lots of equal area. What is the area of each lot in square yards?
28. Which of these rectangles has the larger area and by how much? (Express the answer in square millimeters.)

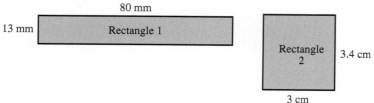

29. A cabinet is 50 centimeters wide, 65 centimeters high, and 24 centimeters deep. What is the capacity of this cabinet?
30. A book is $4\frac{1}{4}$ inches wide, 8 inches long, and $1\frac{1}{8}$ inches thick. How many cubic inches of space does this book occupy?
31. A box is 8.5 centimeters wide, 3.2 decimeters long, and 300 millimeters high. Find the volume of this box in cubic centimeters.
32. A box is 15 inches wide, 20 inches long, and 7 inches high. How many gallons of fluid does the box hold? (Round off the answer to the nearest hundredth of a gallon.)
33. A pharmacist combines 50 milliliters of water with 3.58 liters of water. What is the total volume of the water in milliliters?
34. How many 4-ounce glasses can a chef pour from 20 cups of orange juice?
35. A large tank is leaking a liter of water each hour. How many cubic centimeters of water have been lost at the end of $3\frac{1}{2}$ hours?
36. A recipe calls for 2 tablespoons of sugar. You are preparing 12 times the recipe. How many cups of sugar do you need?

37. A chemist mixes 5.7 liters of water with 30 centiliters of acid to make an acid solution. What is the ratio of the acid volume to the total volume?

38. A 5-quart bottle of liquid insecticide contains 1.4% active ingredient. How many fluid ounces of active ingredient are in the bottle?

10.3 WEIGHT, MASS, AND TIME

OBJECTIVES
- Change units of weight or mass.
- Change units of time.
- Solve an application that involves converting units of weight, mass, or time.

1 Converting Units of Weight and Mass

The **weight** of an object is a measure of how much the earth is pulling on the object. That is, the weight of an object measures how heavy it is. The pull that the earth has on an object becomes less as the object is moved away from the surface of the earth. Therefore, the weight of an object that is not on the surface of the earth is less than the weight of the same object on the surface of the earth. In this text, when we speak of the weight of an object, we will always mean the weight on the surface of the earth.

The standard units of weight in the English system are **pound, ounce, ton,** and **grain.** Four sticks of butter weigh 1 pound. A four-door automobile weighs about 1 ton. The following table gives equivalent units of weight in the English system.

Units of Weight
English System

1 ton = 2000 pounds (lb)
1 pound = 16 ounces (oz)
1 pound = 7000 grains (gr)
1 ounce = 437.5 grains

The **mass** of an object is a measure of the quantity of material that makes up the object. Mass and weight are closely related, but scientists prefer to work with mass, because the mass of an object does not change as it is moved away from the earth's surface.

The standard unit of mass in the metric system is the **gram.** A paper clip has a mass of about 1 gram, and the mass of a nickel is about 5 grams. The mass of an object in metric units can be thought of as its weight on the surface of the earth. However, the weight of an object in English units is not the mass of the object.

The following table gives equivalent units of mass in the metric system.

> **Units of Mass (Weight)**
> **Metric System**
>
> 1 milligram (mg) = $\frac{1}{1000}$ or 0.001 gram (g)
>
> 1 centigram (cg) = $\frac{1}{100}$ or 0.01 gram
>
> 1 decigram (dg) = $\frac{1}{10}$ or 0.1 gram
>
> 1 dekagram (dag) = 10 grams
> 1 hectogram (hg) = 100 grams
> 1 kilogram (kg) = 1000 grams

Observe that the prefixes *milli-*, *centi-*, *deci-*, *deka-*, *hecto-*, and *kilo-* have the same meaning as when they were used with length, area, and volume units of measurement.

Study the following pictures to get a feel for the heaviness of some of these weight measurements.

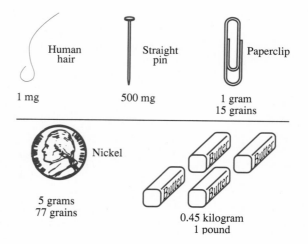

In both the metric and English systems there is an important relationship between the volume of water and the weight of water. One cubic centimeter (1 milliliter) of water weighs approximately 1 gram, and 1 cubic decimeter (1 liter) of water, weighs approximately 1 kilogram. Also, 1 fluid ounce of water weighs approximately 1 ounce, so 16 fluid ounces of water weighs approximately 16 ounces, which is 1 pound. Here we picture the situations.

Try These Problems

Convert the following measurements as indicated.

1. $3\frac{1}{8}$ lb = ____?____ oz
2. 12.3 dg = ____?____ cg

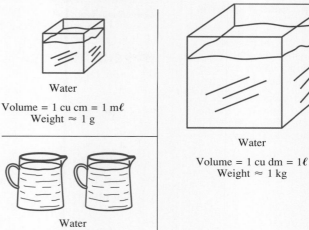

Water
Volume = 1 cu cm = 1 mℓ
Weight ≈ 1 g

Water
Volume = 1 cu dm = 1ℓ
Weight ≈ 1 kg

Water
Volume = 16 fl oz = 2 c
Weight ≈ 16 oz = 1 pound

The equivalence table for units of weight can be used to change units of weight. The procedure is similar to that used to change units of length, area, and volume as shown in the examples that follow.

EXAMPLE 1 3500 lb = ____?____ tons

SOLUTION

$$3500 \text{ lb} = \frac{3500 \text{ lb}}{1} \times \boxed{\frac{1 \text{ ton}}{2000 \text{ lb}}} \quad \text{A fraction equal to 1}$$

$$= \frac{3500}{2000} \text{ tons}$$

$$= \frac{35 \times 100}{20 \times 100} \text{ tons}$$

$$= \frac{5 \times 7}{4 \times 5} \text{ tons}$$

$$= \frac{7}{4} \text{ tons or } 1\frac{3}{4} \text{ tons or } 1.75 \text{ tons}$$

The weight 3500 pounds is the same as 1.75 tons. ■

EXAMPLE 2 500 kg = ____?____ cg

SOLUTION

$$500 \text{ kg} = \frac{500 \text{ kg}}{1} \times \frac{1000 \text{ g}}{1 \text{ kg}} \times \frac{1 \text{ cg}}{0.01 \text{ g}}$$

$$= \frac{500 \times 1000}{0.01} \text{ cg}$$

$$= \frac{500,000}{0.01} \text{ cg}$$

$$= 50,000,000 \text{ cg}$$

The mass or weight 500 kilograms is the same as 50,000,000 centigrams. ■

 Try Problems 1 and 2.

Try These Problems

Solve.
3. A wire cable weighs 2.4 kilograms per centimeter of length. What is the weight in grams of 40 centimeters of this cable?

4. Fresh mushrooms sell for $2.29 per pound. What is the cost of 10 ounces of mushrooms? (Round up the answer to the nearest cent.)

EXAMPLE 3   A manufacturer sells canned apricots for 5.6¢ per ounce. How much does he charge for a 1.2-pound can of apricots? (Round up the answer to the nearest cent.)

SOLUTION   Because the apricots cost 5.6¢ per ounce, we have the following.

$$\text{Total cost} = \frac{\text{cost}}{\text{per ounce}} \times \frac{\text{how many}}{\text{ounces}}$$
$$= \frac{5.6¢}{1 \text{ oz}} \times \boxed{}$$

We need to find the weight of the can in ounces, so we convert 1.2 pounds to ounces.

$$1.2 \text{ lb} = \frac{1.2 \text{ lb}}{1} \times \frac{16 \text{ oz}}{1 \text{ lb}} \quad \text{A ratio equal to 1}$$
$$= 19.2 \text{ oz}$$

The weight 1.2 pounds is the same as 19.2 ounces. Now we can find the total cost.

$$\text{Total cost} = \frac{\text{cost}}{\text{per ounce}} \times \frac{\text{how many}}{\text{ounces}}$$
$$= \frac{5.6¢}{1 \text{ oz}} \times \frac{19.2 \text{ oz}}{1}$$
$$= 5.6 \times 19.2¢$$
$$= 107.52¢$$
$$= \$1.0752 \quad \text{Convert } cents \text{ to } dollars.$$
$$\approx \$1.08 \quad \text{Round up to the nearest cent.}$$

The manufacturer charges $1.08 for the can of apricots. ∎

 Try Problems 3 and 4.

EXAMPLE 4   How many 5-gram weights does it take to make 2.3 kilograms? The total weight 2.3 kilograms is being split into equal parts with each part weighing 5 grams. We have the following.

$$\frac{\text{Weight of}}{\text{each part}} \times \frac{\text{how many}}{\text{parts}} = \frac{\text{total}}{\text{weight}}$$
$$5 \text{ g} \times \boxed{} = 2.3 \text{ kg}$$

Before dividing the total weight by the weight of each part, we must express both weights using the same unit of weight. We choose to convert 2.3 kilograms to grams.

$$2.3 \text{ kg} = \frac{2.3 \text{ kg}}{1} \times \frac{1000 \text{ g}}{1 \text{ kg}} \quad \text{A ratio equal to 1}$$
$$= 2300 \text{ g}$$

Try These Problems

Solve.

5. How many 12-ounce meat servings can a chef get from 60 pounds of meat?

6. Mr. Cortez transports 650 kilograms of dirt in loads of 1300 grams each. How many loads is this?

7. One-fourth ton of fertilizer contains 0.3% active ingredient. How many ounces of active ingredient are in the fertilizer?

The weight 2.3 kilograms is the same as 2300 grams. Now we divide the total weight by the weight of each part.

$$2300 \text{ g} \div 5 \text{ g} = \frac{2300 \text{ g}}{5 \text{ g}} = 460$$

There are 460 5-gram weights in 2.3 kilograms. ■

 Try Problems 5 and 6.

EXAMPLE 5 Twenty pounds of fertilizer contains 3.5% active ingredient. How many ounces of active ingredient are in the fertilizer?

SOLUTION The fertilizer contains 3.5% active ingredient, which means that 3.5% *of the total weight* is the weight of the active ingredient. We have the following.

$$\begin{aligned}\text{Weight of the active ingredient} &= 3.5\% \times \text{total weight} \\ &= 0.035 \times 20 \text{ lb} \\ &= 0.7 \text{ lb}\end{aligned}$$

The weight of the active ingredient is 0.7 pound, but the question is to find this weight in ounces, so we convert 0.7 pounds to ounces.

$$0.7 \text{ lb} = \frac{0.7 \text{ lb}}{1} \times \boxed{\frac{16 \text{ oz}}{1 \text{ lb}}} \quad \text{A ratio equal to 1}$$

$$= 11.2 \text{ oz}$$

The weight of the active ingredient is 11.2 ounces. ■

 Try Problem 7.

 Converting Units of Time

We all encounter the measurement of time in our daily lives. Some standard units of time are the **hour, minute, second, day, week, month,** and **year.** The following table gives equivalent units of time.

Units of Time

1 calendar year (yr)	= 365 days (da)
1 year	= 12 months (mo)
1 year	≈ 52 weeks (wk)
1 week	= 7 days
1 day	= 24 hours (hr)
1 hour	= 60 minutes (min)
1 minute	= 60 seconds (sec)

The equivalence table for units of time can be used to convert units of time. The procedure is similar to that used for changing units of length, area, volume, and weight. Here we look at some examples.

Try These Problems

Convert the following measurements as indicated.

8. 6.9 min = _____?_____ hr
9. $\frac{3}{4}$ da = _____?_____ min

EXAMPLE 6 3 wk = _____?_____ hr

SOLUTION

$$3 \text{ wk} = \frac{3 \text{ wk}}{1} \times \boxed{\frac{7 \text{ da}}{1 \text{ wk}}} \times \boxed{\frac{24 \text{ hr}}{1 \text{ da}}} \quad \text{Ratios equal to 1}$$

$$= \frac{3 \times 7 \times 24}{1} \text{ hr}$$

$$= 504 \text{ hr}$$

There are 504 hours in 3 weeks. ∎

 Try Problems 8 and 9.

EXAMPLE 7 A typist can type 65 words each minute. How many words can the typist type in $1\frac{1}{2}$ hours?

SOLUTION Because the typist types 65 words per minute, we have the following.

$$\begin{array}{c}\text{Total number}\\\text{of words}\end{array} = \begin{array}{c}\text{number of words}\\\text{each minute}\end{array} \times \begin{array}{c}\text{how many}\\\text{minutes}\end{array}$$

$$= \frac{65 \text{ words}}{1 \text{ min}} \times \text{? min}$$

We need to find the time in minutes, so we convert $1\frac{1}{2}$ hours to minutes.

$$1\frac{1}{2} \text{ hr} = \frac{3}{2} \text{ hr}$$

$$= \frac{3 \text{ hr}}{2} \times \boxed{\frac{60 \text{ min}}{1 \text{ hr}}} \quad \text{A ratio equal to 1}$$

$$= \frac{3 \times \overset{30}{\cancel{60}}}{\underset{1}{\cancel{2}}} \text{ min}$$

$$= 90 \text{ min}$$

Now we can find the total number of words the typist can type in $1\frac{1}{2}$ hours or 90 minutes.

$$\begin{array}{c}\text{Total number}\\\text{of words}\end{array} = \begin{array}{c}\text{number of words}\\\text{each minute}\end{array} \times \begin{array}{c}\text{how many}\\\text{minutes}\end{array}$$

$$= \frac{65 \text{ words}}{1 \text{ min}} \times \frac{90 \text{ min}}{1}$$

$$= 5850 \text{ words}$$

The typist can type 5850 words in $1\frac{1}{2}$ hours. ∎

EXAMPLE 8 How many 100-minute movies fit on a 6-hour videotape? (Truncate the answer to the nearest whole number.)

SOLUTION We are separating the total quantity of time, 6 hours, into a number of equal parts. We have the following.

$$\begin{array}{c}\text{Total}\\\text{time}\end{array} = \begin{array}{c}\text{amount of time}\\\text{in each part}\end{array} \times \begin{array}{c}\text{number of}\\\text{parts}\end{array}$$

$$6 \text{ hr} = 100 \text{ min} \times \boxed{}$$

Try These Problems

Solve.

10. How many 3-minute songs fit on a $1\frac{1}{2}$-hour tape cassette?

11. Moorgate Equipment Company charges $20 per minute of flying time to rent an airplane. What is the charge to rent a plane for the round trip between Los Angeles and New York if the flying time is $5\frac{1}{3}$ hours each way?

Before dividing, we must express both time periods with the same unit of time. We choose to convert 6 hours to minutes.

$$6 \text{ hr} = \frac{6 \text{ hr}}{1} \times \frac{60 \text{ min}}{1 \text{ hr}} = 360 \text{ min} \qquad \text{A ratio equal to 1}$$

There are 360 minutes in a 6-hour period. Now we can find how many 100-minute movies fit on the videotape.

$$6 \text{ hr} \div 100 \text{ min}$$
$$= 360 \text{ min} \div 100 \text{ min}$$
$$= \frac{360 \text{ min}}{100 \text{ min}}$$
$$= 3.6$$
$$\approx 3 \qquad \text{Truncate to the nearest whole number.}$$

Three of the 100-minute movies will fit on the videotape. ■

Try Problems 10 and 11.

Answers to Try These Problems

1. 50 oz 2. 123 cg 3. 96,000 g 4. $1.44 5. 80 servings
6. 500 loads 7. 24 oz 8. $\frac{23}{200}$ or 0.115 hr 9. 1080 min
10. 30 songs 11. $12,800

EXERCISES 10.3

Convert the following measurements as indicated.

1. 0.3 ton = ___?___ lb
2. 56 oz = ___?___ lb
3. 140,000 gr = ___?___ ton
4. 0.06 ton = ___?___ oz
5. 2.34 dag = ___?___ g
6. 0.065 g = ___?___ cg
7. 2.09 kg = ___?___ hg
8. 380 mg = ___?___ dg
9. $\frac{1}{4}$ da = ___?___ hr
10. 64 mo = ___?___ yr
11. $2\frac{1}{3}$ hr = ___?___ sec
12. 420 hr = ___?___ wk

Solve.

13. Canned chili sells for 6.8¢ per ounce. What is the cost of a 1.4-pound can of this chili? (Round up the answer to the nearest cent.)

14. A rope weighs 240 grams per meter of length. What is the weight in centigrams of 5.2 meters of this rope?

15. How many 8-ounce potato servings can a chef get from 12 pounds of potatoes?

16. A child transports 50 pounds of sand in loads of 80 ounces each. How many loads is this?

17. A chemist combines salt and water to make a salt solution. If the total mass of the solution is 6 kilograms and 5% of the solution is salt, what is the mass of the salt in grams?

18. A fertilizer that weighs 48 pounds contains 0.4% active ingredient. How many ounces of active ingredient are in the fertilizer?

19. A manufacturer spends 8 hours producing a certain product. During one-tenth of this production time, the product is in the fabricating department. How many minutes does the product spend in the fabricating department?

20. How many 4-minute speeches fit into a period of $1\frac{3}{4}$ hours? (Truncate the answer to the nearest whole number.)

21. A downtown parking garage charges as follows.

    One hour or less $1.25
    Each additional 15 minutes $0.75

    How much do you pay for parking in the garage for $2\frac{1}{4}$ hours?

22. Mr. Austin spent 9 days in Paris and 3 weeks in London. What fraction of the time was spent in Paris?

10.4 RATES

OBJECTIVES
- Solve a problem that involves interpreting a rate as a ratio.
- Solve a problem that involves multiplying a rate by a quantity so that units cancel.
- Change units of rate.
- Solve an application that involves changing units of rate.

Try These Problems

Find the missing number so that the two rates are equal.

1. $\frac{4 \text{ lb}}{5 \text{ ft}} = \underline{\quad?\quad}$ lb per ft

2. $\frac{1\frac{3}{4} \text{ mi}}{14 \text{ min}} = \underline{\quad?\quad}$ mi per min

1 Interpreting a Rate as a Ratio

Recall from Sections 4.5, 7.5, and 8.4 that a **rate** is a comparison by division. That is, a rate is the result of dividing one measurement by another. Some examples are given here.

$$55 \text{ miles per hour} \rightarrow \frac{55 \text{ miles}}{1 \text{ hour}}$$

$$300 \text{ words in 5 minutes} \rightarrow \frac{300 \text{ words}}{5 \text{ minutes}} = 60 \text{ words per minute}$$

$$12 \text{ pounds every 30 feet} \rightarrow \frac{12 \text{ pounds}}{30 \text{ feet}} = 0.4 \text{ pound per foot}$$

Observe that words like *per, in,* and *every* that are used to indicate rates translate to the fraction bar or division.

Try Problems 1 and 2.

EXAMPLE 1 A tank is leaking 2.4 gallons every 15 minutes. How many gallons per minute is this?

SOLUTION Write the rate *2.4 gallons every 15 minutes* as a ratio. Put minutes in the denominator since we want gallons per minute.

$$\begin{aligned} & 2.4 \text{ gal every 15 min} \\ &= \frac{2.4 \text{ gal}}{15 \text{ min}} \\ &= 0.16 \text{ gal per min} \end{aligned}$$

To obtain gallons per minute put minutes in the denominator. Divide 2.4 by 15.

The tank is leaking 0.16 gallon per minute. ∎

Try These Problems

Solve.

3. A 15-ounce can of tomatoes costs 81¢. What is the cost per ounce?
4. A faucet is leaking 5 centiliters of water every 20 minutes. How many centiliters per minute is this?
5. On a map, $\frac{3}{4}$ inch represents 3.6 miles. How many miles does 1 inch represent?
6. Peter earns $77 in 5 hours. How much does he earn in 1 hour?

Multiply as indicated. Be sure to give the unit of measurement with your answer.

7. $\dfrac{4.5¢}{1 \text{ oz}} \times \dfrac{15 \text{ oz}}{1}$
8. $\dfrac{2\frac{1}{2} \text{ t}}{5 \text{ persons}} \times \dfrac{30 \text{ persons}}{1}$

EXAMPLE 2 Cindy can jog 3.5 miles in 28 minutes. How long does it take her to go 1 mile?

SOLUTION To obtain how long it takes her to jog 1 mile, we want to find how many minutes per mile. Write the rate *3.5 miles in 28 minutes* as a ratio. Put miles in the denominator because we want to know minutes per mile.

$$3.5 \text{ mi in } 28 \text{ min}$$
$$= \dfrac{28 \text{ min}}{3.5 \text{ mi}} \qquad \text{To obtain minutes per mile put miles in the denominator.}$$
$$= 8 \text{ min per mi} \qquad \text{Divide 28 by 3.5.}$$

It takes Cindy 8 minutes to jog 1 mile. ∎

 Try Problems 3 through 6.

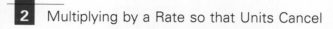 Multiplying by a Rate so that Units Cancel

If you drive 50 miles per hour, then you can drive 200 miles in 4 hours. We have the following.

$$\dfrac{50 \text{ mi}}{1 \text{ hr}} \times \dfrac{4 \text{ hr}}{1} = 200 \text{ mi}$$

Observe that if you multiply the rate $\frac{50 \text{ mi}}{1 \text{ hr}}$ by hours so that the hour units cancel, the resulting units must be miles. Here are more examples of how units cancel when multiplying by a rate.

$$\dfrac{\overset{90}{\cancel{270} \text{ yd}}}{\underset{1}{\cancel{3} \text{ sec}}} \times \dfrac{7 \text{ sec}}{1} = 90 \times 7 \text{ yd}$$
$$= 630 \text{ yd}$$

$$\dfrac{\overset{5}{\cancel{15} \text{ oz}}}{\underset{2}{\cancel{6} \text{ ft}}} \times \dfrac{8.5 \text{ ft}}{1} = \dfrac{5 \times 8.5}{2} \text{ oz}$$
$$= \dfrac{42.5}{2} \text{ oz}$$
$$= 21.25 \text{ oz}$$

 Try Problems 7 and 8.

EXAMPLE 3 A car can travel 700 miles on 20 gallons of gasoline. How far can the car travel on 8 gallons of gasoline?

SOLUTION The phrase *700 miles on 20 gallons* is a rate and can be written as a ratio.

$$700 \text{ mi on } 20 \text{ gal} \rightarrow \dfrac{700 \text{ mi}}{20 \text{ gal}} = \dfrac{35 \text{ mi}}{1 \text{ gal}}$$

The car travels 35 miles per gallon of gas. Because gallons is in the denominator, we can set up a multiplication statement so that the gallons units cancel.

Try These Problems

Solve.

9. A cable weighs 2 pounds every 5 feet. What is the weight of 16 feet of this cable?

10. Canned peaches cost 63¢ for 15 ounces. At this rate, what is the cost of 25 ounces?

11. A car can go 160 miles on 4 gallons of gas. At this rate, how many gallons of gas does it take for the car to go 960 miles?

12. Dave typed 280 words in 3.5 minutes. How long will it take him to type 440 words?

$$\frac{\text{miles}}{\cancel{\text{gallon}}} \times \frac{\cancel{\text{gallons}}}{1} = \text{miles}$$

$$\frac{35 \text{ mi}}{1 \cancel{\text{gal}}} \times \frac{8 \cancel{\text{gal}}}{1} = \underline{\quad ? \quad} \text{ mi}$$

The answer to the multiplication problem is missing, so multiply to find the missing number.

$$\frac{35 \text{ mi}}{1 \cancel{\text{gal}}} \times \frac{8 \cancel{\text{gal}}}{1} = 35 \times 8 \text{ mi}$$

$$= 280 \text{ mi}$$

The car can travel 280 miles on 8 gallons of gasoline. ■

EXAMPLE 4 A particle travels 2.6 meters every 6.5 seconds. How long does it take the particle to travel 30 meters?

SOLUTION The phrase *2.6 meters every 6.5 seconds* is a rate and can be written as a ratio.

$$2.6 \text{ m every } 6.5 \text{ sec}$$

$$= \frac{2.6 \text{ m}}{6.5 \text{ sec}}$$

$$= \frac{0.4 \text{ m}}{1 \text{ sec}} \qquad \text{Divide 2.6 by 6.5 to obtain 0.4.}$$

The particle travels 0.4 meter every second. Since seconds is in the denominator, we can set up a multiplication statement so that the seconds units cancel.

$$\frac{\text{meters}}{\cancel{\text{second}}} \times \frac{\cancel{\text{seconds}}}{1} = \text{meters}$$

$$\frac{0.4 \text{ m}}{1 \cancel{\text{sec}}} \times \underline{\quad ? \quad} \cancel{\text{sec}} = 30 \text{ m}$$

A multiplier is missing, so divide 30 by 0.4 to find the missing multiplier.

$$30 \div 0.4 = 75$$

It takes the particle 75 seconds to travel 30 meters. ■

 Try Problems 9 through 12.

3 Changing Units of Rate

Suppose that a jogger can run 1 mile in 8 minutes. How fast is this in miles per hour? To answer this question we need to convert the rate, *1 mile in 8 minutes,* to an equivalent rate where the units are *miles per hour.* We use the same procedure that we used to convert units of length, area, volume, weight, and time. Here we illustrate the procedure.

10.4 Rates 491

 Try These Problems

Find the missing number so that the two rates are equal.

13. 54 mi per hr = _____?_____ mi per min

14. 0.8 gal per hr = _____?_____ qt per hr

15. $4\frac{1}{2}$ in per da = _____?_____ ft per wk

16. 270 lb per cu ft = _____?_____ oz per cu in

$$= \frac{1 \text{ mi}}{8 \text{ min}}$$

$$= \frac{1 \text{ mi}}{8 \text{ min}} \times \frac{60 \text{ min}}{1 \text{ hr}} \quad \text{A ratio equal to 1}$$

$$= \frac{60 \text{ mi}}{8 \text{ hr}} \quad \text{Because we want the answer in miles per hour, we must have miles left in the numerator and hours left in the denominator.}$$

$$= 7.5 \text{ mi per hr}$$

The jogger runs 7.5 miles per hour. ■

Here are more examples of changing units of rate.

EXAMPLE 5 3.2 tons per truckload = _____?_____ lb per truckload

SOLUTION

3.2 tons per truckload

$$= \frac{3.2 \text{ tons}}{1 \text{ truckload}}$$

$$= \frac{3.2 \text{ tons}}{1 \text{ truckload}} \times \frac{2000 \text{ lb}}{1 \text{ ton}} \quad \text{A ratio equal to 1}$$

$$= \frac{6400 \text{ lb}}{1 \text{ truckload}}$$

$$= 6400 \text{ lb per truckload}$$

The rate 6400 pounds per truckload is the same as 3.2 tons per truckload. ■

 Try Problems 13 and 14.

EXAMPLE 6 150 m per sec = _____?_____ km per hr

SOLUTION

150 m per sec

$$= \frac{150 \text{ m}}{1 \text{ sec}}$$

$$= \frac{150 \text{ m}}{1 \text{ sec}} \times \frac{1 \text{ km}}{1000 \text{ m}} \times \frac{60 \text{ sec}}{1 \text{ min}} \times \frac{60 \text{ min}}{1 \text{ hr}} \quad \text{Ratios equal to 1}$$

$$= \frac{150 \times 60 \times 60 \text{ km}}{1000 \text{ hr}}$$

$$= \frac{15 \times \cancel{10} \times 6 \times \cancel{10} \times 6 \times \cancel{10} \text{ km}}{\cancel{10} \times \cancel{10} \times \cancel{10} \text{ hr}}$$

$$= \frac{540 \text{ km}}{1 \text{ hr}}$$

$$= 540 \text{ km per hr}$$

The rate 540 kilometers per hour is the same as 150 meters per second. ■

Try Problems 15 and 16.

Try These Problems

Solve.

17. Canned juice costs $1.68 for $1\frac{1}{4}$ quarts. What is the cost in cents per fluid ounce?

18. Water is entering a lake at the rate of 50 gallons every 3 hours. How many fluid ounces per second is this? (Round off at 1 decimal place.)

EXAMPLE 7 A runner can run 800 yards in 3.5 minutes. Find the runner's speed in miles per hour. (Round off the answer at one decimal place.)

SOLUTION

800 yd in 3.5 min

$$= \frac{800 \text{ yd}}{3.5 \text{ min}}$$

$$= \frac{800 \text{ yd}}{3.5 \text{ min}} \times \frac{60 \text{ min}}{1 \text{ hr}} \times \frac{3 \text{ ft}}{1 \text{ yd}} \times \frac{1 \text{ mi}}{5280 \text{ ft}} \quad \text{Ratios equal to 1}$$

$$= \frac{800 \times 60 \times 3 \text{ mi}}{3.5 \times 5280 \text{ hr}}$$

$$= \frac{144,000 \text{ mi}}{18,480 \text{ hr}}$$

$$\approx 7.8 \text{ mi per hr}$$

The rate 7.8 miles per hour is about the same as 800 yards in 3.5 minutes. ■

Try Problems 17 and 18.

Answers to Try These Problems

1. $\frac{4}{5}$ or 0.8 lb per ft   2. $\frac{1}{8}$ or 0.125 mi per min   3. 5.4¢ per oz
4. $\frac{1}{4}$ or 0.25 cℓ per min   5. 4.8 mi   6. $15.40   7. 67.5¢
8. 15 t   9. $6\frac{2}{5}$ or 6.4 lb   10. 105¢ or $1.05   11. 24 gal
12. 5.5 min   13. 0.9 mi per min   14. 3.2 qt per hr
15. $2\frac{5}{8}$ or 2.625 ft per wk   16. $2\frac{1}{2}$ or 2.5 oz per cu in
17. $4\frac{1}{5}$ or 4.2¢ per fl oz   18. 0.6 fl oz per sec

EXERCISES 10.4

Find the missing number so that the two rates are equal.

1. $\frac{\$48}{3 \text{ shirts}} = \$ \underline{\quad?\quad}$ per shirt   2. $\frac{\frac{3}{4} \text{ ton}}{5 \text{ loads}} = \underline{\quad?\quad}$ ton per load

3. $\frac{13\frac{1}{2} \text{ lb}}{3 \text{ cu ft}} = \underline{\quad?\quad}$ lb per cu ft   4. $\frac{20.8 \text{ g}}{0.25 \text{ m}} = \underline{\quad?\quad}$ g per m

5. 200 yd per sec = $\underline{\quad?\quad}$ yd per min
6. 4 mi per min = $\underline{\quad?\quad}$ ft per min   7. 8.2 kg per m = $\underline{\quad?\quad}$ g per m
8. 2.3 ¢ per cm = $\underline{\quad?\quad}$ ¢ per m   9. 5 lb per wk = $\underline{\quad?\quad}$ oz per da
10. 60 yd per min = $\underline{\quad?\quad}$ ft per sec   11. 3300 ft per sec = $\underline{\quad?\quad}$ mi per hr
12. 30 gal per min = $\underline{\quad?\quad}$ fl oz per hr

Multiply as indicated. Be sure to give the unit of measurement with your answer.

13. $\frac{\$35.40}{1 \text{ hr}} \times \frac{8 \text{ hr}}{1}$ 14. $\frac{4.5¢}{1 \text{ oz}} \times \frac{16 \text{ oz}}{1}$

15. $\frac{3\frac{1}{2} \text{ c}}{6 \text{ persons}} \times \frac{24 \text{ persons}}{1}$ 16. $\frac{90 \ \ell}{18 \text{ persons}} \times \frac{50 \text{ persons}}{1}$

Solve.

17. A 320-milliliter bottle of olive oil costs $5.69. What is the cost per milliliter? (Round off the answer to the nearest tenth of a cent.)
18. A certain fertilizer weighs 2 pounds for every 8 cubic inches. How much does 1 cubic inch of this fertilizer weigh?
19. On a map, $1\frac{1}{2}$ inches represents 30 miles. How many miles does 1 inch represent?
20. A car can go 488 miles on 16 gallons of gas. How many miles per gallon is this?
21. Darlene hikes $3\frac{1}{3}$ miles in 1 hour. How far can she hike in 9 hours?
22. Billy types 378 words in 6 minutes. How many words can he type in 30 minutes?
23. Mr. Nelson completed $\frac{1}{3}$ of the job in 2 hours. How long will it take him to complete $\frac{3}{4}$ of the job.
24. Five pounds of mushrooms cost $16. How many pounds can you buy for $11.20?
25. Don lost 42 pounds in 8 weeks. How many pounds per day did he lose?
26. A string weighs 4.2 grams every 2 meters of length. How many grams per centimeter is this?
27. A runner runs 880 yards in 4 minutes. What is the runner's speed in miles per hour?
28. One-half gallon of milk costs $2.60. What is the cost in cents per fluid ounce? (Round off the answer to the nearest tenth of a cent.)
29. Tom practices his guitar 45 minutes each day. How many hours does he practice over a 4-week period?
30. A typist charges $9 an hour and types 70 words each minute. What does it cost to have a 1400-word paper typed?

10.5 MEASUREMENTS INVOLVING MORE THAN ONE UNIT

OBJECTIVES
- Convert a measurement involving two or more units to an equivalent measurement involving only one unit.
- Convert a measurement involving only one unit to a measurement involving two or more units.
- Add measurements that involve two or more units.
- Subtract measurements that involve two or more units.
- Multiply a measurement involving two or more units by a pure number.
- Divide a measurement involving two or more units by a pure number.
- Solve an application that involves measurements expressed in two or more units.

The height of the woman shown here is $5\frac{1}{4}$ feet or 63 inches. Another way to express her height is to say that she is 5 feet 3 inches, where we are using two units of length to express a distance.

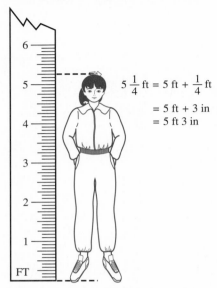

$$5\frac{1}{4} \text{ ft} = 5 \text{ ft} + \frac{1}{4} \text{ ft}$$
$$= 5 \text{ ft} + 3 \text{ in}$$
$$= 5 \text{ ft } 3 \text{ in}$$

This method of using more than one unit of measurement to express a single measurement is not used as much today as it was in previous times. There are a couple of reasons for this. This method is not used with metric units of measurement and the metric system is gradually becoming accepted in the United States. Also, when using a calculator or computer, it is easier to do calculations when a measurement is expressed using only one unit of measurement.

1 Converting a Measurement Involving Two or More Units to an Equivalent Measurement Involving Only One Unit

The examples that follow illustrate how to convert a measurement involving two or more units to an equivalent measurement involving only one unit.

EXAMPLE 1 5 ft 3 in = _____?_____ in

SOLUTION

$$5 \text{ ft } 3 \text{ in} = 5 \text{ ft} + 3 \text{ in}$$
$$= 60 \text{ in} + 3 \text{ in}$$
$$= 63 \text{ in}$$

Convert 5 ft to inches before adding.

$$5 \text{ ft} = \frac{5 \cancel{\text{ft}}}{1} \times \frac{12 \text{ in}}{1 \cancel{\text{ft}}} = 60 \text{ in}$$

The answer is 63 inches. ∎

10.5 Measurements Involving More Than One Unit

 Try These Problems

Find the missing number(s) so that the two measurements are equal.
1. 6 ft 8 in = _____?_____ in
2. 18 pt 7 fl oz = _____?_____ fl oz
3. 36 min 12 sec = _____?_____ min
4. 5 yd 2 ft = _____?_____ yd
5. 7 hr 40 min 13 sec = _____?_____ sec
6. 3 qt 1 pt 4 fl oz = _____?_____ pt

EXAMPLE 2 7 lb 10 oz = _____?_____ lb

SOLUTION

$$7 \text{ lb } 10 \text{ oz} = 7 \text{ lb} + 10 \text{ oz}$$
$$= 7 \text{ lb} + \frac{10}{16} \text{ lb} \quad \text{Convert 10 oz to pounds before adding.}$$
$$= 7\frac{10}{16} \text{ lb} \quad\quad 10 \text{ oz} = \frac{10 \cancel{\text{oz}}}{1} \times \frac{1 \text{ lb}}{16 \cancel{\text{oz}}}$$
$$= 7\frac{5}{8} \text{ lb} \quad\quad m = \frac{10}{16} \text{ lb}$$

The answer is $7\frac{5}{8}$ pounds. ∎

 Try Problems 1 through 4.

EXAMPLE 3 9 hr 45 min 20 sec = _____?_____ hr

SOLUTION We need to convert 45 min to hours, and 20 sec to hours. First we convert 45 min to hours.

$$45 \text{ min} = \frac{45 \cancel{\text{min}}}{1} \times \frac{1 \text{ hr}}{60 \cancel{\text{min}}}$$
$$= \frac{45}{60} \text{ hr}$$

Next we convert 20 sec to hours.

$$20 \text{ sec} = \frac{\overset{1}{\cancel{20 \text{ sec}}}}{1} \times \frac{1 \cancel{\text{min}}}{\underset{3}{\cancel{60 \text{ sec}}}} \times \frac{1 \text{ hr}}{60 \cancel{\text{min}}}$$
$$= \frac{1}{180} \text{ hr}$$

Finally, we convert 9 hours 45 minutes 20 seconds to hours.

$$9 \text{ hr } 45 \text{ min } 20 \text{ sec} = 9 \text{ hr} + 45 \text{ min} + 20 \text{ sec}$$
$$= 9 \text{ hr} + \frac{45}{60} \text{ hr} + \frac{1}{180} \text{ hr}$$
$$= 9 + \frac{45}{60} + \frac{1}{180} \text{ hr}$$
$$= 9 + \frac{45 \times 3}{60 \times 3} + \frac{1}{180} \text{ hr} \quad \text{Convert } \frac{45}{60} \text{ to a fraction with denominator 180.}$$
$$= 9 + \frac{135}{180} + \frac{1}{180} \text{ hr}$$
$$= 9 + \frac{136}{180} \text{ hr}$$
$$= 9\frac{34}{45} \text{ hr} \quad \text{Reduce } \frac{136}{180} \text{ to lowest terms.}$$

The answer is $9\frac{34}{45}$ hours. ∎

Try Problems 5 and 6.

496 Chapter 10 Measurement: English and Metric Systems

 Try These Problems

Find the missing number(s) so that the two measurements are equal.

7. 2117 ft = ___?___ yd
   ___?___ ft

8. 552 min = ___?___ hr
   ___?___ min

9. $8\frac{1}{2}$ hr = ___?___ hr
   ___?___ min

10. $9\frac{3}{4}$ lb = ___?___ lb
    ___?___ oz

2 Converting a Measurement Involving Only One Unit to an Equivalent Measurement Involving Two or More Units

Now we look at converting a measurement involving only one unit to an equivalent measurement involving two or more units. The procedures are illustrated in the following examples.

EXAMPLE 4   76 in = ___?___ ft ___?___ in

SOLUTION   We want to know how many whole feet are in 76 inches, and how many inches are left over. Since each foot contains 12 inches, divide 76 by 12.

$$\begin{array}{r} 6 \\ 12\overline{)76} \\ \underline{72} \\ 4 \end{array}$$

— The number of feet
— The number of inches left over

The answer is 6 ft 4 in. ∎

 Try Problems 7 and 8.

EXAMPLE 5 $4\frac{3}{4}$ hr = ___?___ hr ___?___ min

SOLUTION   It is clear that there are 4 whole hours in $4\frac{3}{4}$ hours. We only need to convert $\frac{3}{4}$ hours to minutes.

$$\frac{3}{4} \text{ hr} = \frac{3 \text{ hr}}{4} \times \frac{60 \text{ min}}{1 \text{ hr}} = 45 \text{ min} \quad \text{Ratio equal to 1}$$

Now we have the following:

$$4\frac{3}{4} \text{ hr} = 4 \text{ hr} + \frac{3}{4} \text{ hr}$$
$$= 4 \text{ hr} + 45 \text{ min}$$
$$= 4 \text{ hr } 45 \text{ min}$$

The answer is 4 hr 45 min. ∎

 Try Problems 9 and 10.

EXAMPLE 6   20,000 sec = ___?___ hr ___?___ min ___?___ sec

SOLUTION   First, we find how many whole minutes are in 20,000 seconds, and how many seconds are left over. Since each second contains 60 minutes, divide 20,000 by 60.

$$\begin{array}{r} 333 \\ 60\overline{)20000} \\ \underline{180} \\ 200 \\ \underline{180} \\ 200 \\ \underline{180} \\ 20 \end{array}$$

— The number of minutes

— The number of seconds left over

Now we know the following.

20,000 sec = 333 min 20 sec

10.5 Measurements Involving More Than One Unit

 Try These Problems

Find the missing number(s) so that the two measurements are equal.

11. 272 in = ____?____ yd ____?____ ft ____?____ in

12. 27,903 sec = ____?____ hr ____?____ min ____?____ sec

13. 53 pt = ____?____ gal ____?____ qt ____?____ pt

Perform the indicated operations and simplify.

14. 9 ft 6 in
      + 4 ft 4 in

15. 9 hr 25 min
      + 7 hr 56 min

16. 8 gal 3 qt
      + 3 gal 3 qt

Finally, we need to find how many whole hours are in 333 minutes, and how many minutes are left over. Since each hour contains 60 minutes, divide 333 by 60.

5 —— The number of hours
60)333
   300
    33 —— The number of minutes left over

The answer is 5 hr 33 min 20 sec. ∎

 Try Problems 11 through 13.

3 Adding Measurements Involving Two or More Units

Although it is easier to do calculations with measurements when they are expressed with only one unit, we can perform some arithmetic operations on measurements involving two or more units. Now we illustrate how this is done.

We can add measurements with *like* units. For example,

6 ft + 3 ft = 9 ft

9 hr + 5 hr = 14 hr

4 oz + 7 oz = 11 oz

To add measurements involving two or more units, we group together the like units. Here are some examples.

EXAMPLE 7 Add: 3 ft 7 in + 5 ft 3 in

SOLUTION For convenience, arrange the measurements vertically so that like units form a column.

  3 ft 7 in
+ 5 ft 3 in
  ─────────
  8 ft 10 in

The answer is 8 ft 10 in. ∎

EXAMPLE 8 Add: 9 hr 45 min + 12 hr 35 min

SOLUTION For convenience, arrange the measurements vertically so that like units form a column.

   9 hr 45 min
+ 12 hr 35 min
  ────────────
  21 hr 80 min

Because 80 minutes is more than 1 hour, the measurement 21 hr 80 min is not in simplified form. Now we simplify it.

21 hr 80 min = 21 hr + 80 min
             = 21 hr + 1 hr 20 min   Substitute 1hr 20 min for 80 min.
             = 22 hr 20 min

The answer is 22 hr 20 min. ∎

 Try Problems 14 through 16.

Try These Problems

Perform the indicated operations and simplify.

17. 25 lb 9 oz
   − 8 lb 3 oz

18. 7 hr 20 min
   − 4 hr 30 min

19. 18 gal 2 qt
   − 5 gal 3 qt

20. 15 ft 8 in
   − 9 ft 10 in

21. 14 hr 20 min 36 sec
   − 5 hr 35 min 18 sec

22. 31 yd 1 ft 5 in
   − 23 yd 2 ft 10 in

4 Subtracting Measurements Involving Two or More Units

To subtract measurements involving two or more units, we subtract like units. Sometimes the technique of **borrowing** is needed. The following examples illustrate the procedure.

EXAMPLE 9   Subtract: 12 ft 8 in − 5 ft 3 in

SOLUTION   For convenience, arrange the measurements vertically so that like units form a column.

   12 ft 8 in   Because 3 is less than 8 and 5 is less
−   5 ft 3 in   than 12, no borrowing is necessary.
   ────────
    7 ft 5 in

The answer is 7 ft 5 in. ■

EXAMPLE 10   Subtract: 14 hr 25 min − 6 hr 50 min

SOLUTION   For convenience, arrange the measurements vertically so that like units form a column.

   13   85
   1̶4̶ hr 2̶5̶ min   Because 50 min is more than
−   6 hr 50 min   25 min, we borrow. Borrow 1 hr
   ─────────────   from 14 hr. Mark out 14 and
    7 hr 35 min   write 13. The 1 hr we borrow is
                       60 min, so add 60 min to 25 min
                       to obtain 85 min.

The answer is 7 hr 35 min. ■

 Try Problems 17 through 20.

EXAMPLE 11   Subtract: 12 yd 2 ft 7 in − 3 yd 2 ft 10 in

SOLUTION   For convenience, arrange the measurements vertically so that like units form a column.

   11   1̶4̶   19
   1̶2̶ yd 2̶ ft 7̶ in   Because 10 in is more than 7 in,
−   3 yd 2 ft 10 in   we borrow. Borrow 1 ft from
   ──────────────   2 ft. Mark out 2 and write 1. The
    8 yd 2 ft 9 in   1 ft we borrow is 12 in, so add
                       12 in to 7 in to obtain 19 in.

                       Because 2 ft is more than 1 ft,
                       we borrow again. Borrow 1 yd
                       from 12 yd. Mark out 12 and
                       write 11. The 1 yd we borrow is
                       3 ft, so add 3 ft to 1 ft to obtain
                       4 ft.

The answer is 8 yd 2 ft 9 in. ■

 Try Problems 21 and 22.

10.5 Measurements Involving More Than One Unit

 Try These Problems

Perform the indicated operations and simplify.

23.   9 yd 1 ft 2 in
  ×   2

24.   7 yd 2 ft
  ×   9

25.   12 wk 5 da 17 hr
  ×   5

5 Multiplying a Measurement Involving Two or More Units by a Pure Number

A number without a unit of measurement attached to it is called a **pure number.** Now we look at multiplying a measurement involving two or more units by a pure number. First we discover how the multiplication is done.

Suppose you want to multiply 4 ft 2 in by 3. We should be able to obtain the correct answer by performing the following addition problem.

$$3 \times 4 \text{ ft } 2 \text{ in} \quad \text{means} \quad \begin{array}{r} 4 \text{ ft } 2 \text{ in} \\ 4 \text{ ft } 2 \text{ in} \\ + \ 4 \text{ ft } 2 \text{ in} \\ \hline 12 \text{ ft } 6 \text{ in} \end{array}$$

Observe that the correct answer can be obtained by multiplying each part of the measurement 4 ft 2 in by 3.

$$\begin{array}{r} 4 \text{ ft } 2 \text{ in} \\ \times \ \ \ \ \ 3 \\ \hline 12 \text{ ft } 6 \text{ in} \end{array}$$

In general, to multiply a measurement involving two or more units by a pure number, multiply the number by each part of the measurement. Now we look at some examples.

EXAMPLE 12   Multiply: 4×10 hr 5 min 3 sec

SOLUTION   The number 4 must be multiplied by all three parts of the measurement.

$$\begin{array}{r} 10 \text{ hr } \ \ 5 \text{ min } \ \ 3 \text{ sec} \\ \times \ 4 \\ \hline 40 \text{ hr } 20 \text{ min } 12 \text{ sec} \end{array}$$

Because 20 min is less than 1 hr and 12 sec is less than 1 min, the answer is in simplified form.

The answer is 40 hr 20 min 12 sec. ∎

EXAMPLE 13   Multiply: 7×1 lb 4 oz

SOLUTION   The number 7 must be multiplied by both parts of the measurement.

$$\begin{array}{r} 1 \text{ lb } \ \ 4 \text{ oz} \\ \times \ \ \ \ \ \ \ \ \ 7 \\ \hline 7 \text{ lb } 28 \text{ oz} \end{array}$$

Because 28 ounces is more than 1 pound, the measurement 7 lb 28 oz is not in simplified form. Now we simplify it.

$$\begin{aligned} 7 \text{ lb } 28 \text{ oz} &= 7 \text{ lb} + \ \ \ 28 \text{ oz} \\ &= \underbrace{7 \text{ lb} + 1 \text{ lb}}\ 12 \text{ oz} \\ &= 8 \text{ lb } 12 \text{ oz} \end{aligned}$$

Substitute 1 lb 12 oz for 28 oz.

The answer is 8 lb 12 oz. ∎

 Try Problems 23 through 25.

Try These Problems

Perform the indicated operations and simplify.

26. 8)16 hr 32 min
27. 4)9 yd 1 ft
28. 5)17 lb 13 oz
29. 7)28 lb 10 oz
30. 3)44 yd 2 ft

6 Dividing a Measurement Involving Two or More Units by a Pure Number

To divide a measurement by a pure number, we divide each part by the number. The following examples illustrate the procedure.

EXAMPLE 14 Divide 15 yd 3 ft 9 in by 3.

SOLUTION All three parts of the measurement must be divided by 3.

$$\begin{array}{r} 5 \text{ yd } 1 \text{ ft } 3 \text{ in} \\ 3\overline{)15 \text{ yd } 3 \text{ ft } 9 \text{ in}} \end{array}$$

The answer is 5 yd 1 ft 3 in. ∎

EXAMPLE 15 Divide 9 lb 12 oz by 4.

SOLUTION For convenience, begin with the larger unit and move to the right.

```
     2 lb        7 oz       Divide 9 lb by 4. The result is
  4)9 lb        12 oz ⎤     2 lb with 1 lb left over.
    8 lb                 ⎬— Convert the leftover 1 lb to
    1 lb  =    16 oz ⎦     16 oz, then add it to 12 oz,
               28 oz        before dividing by 4.
               28 oz        Finally, divide 28 oz by 4 to
                            obtain 7 oz with no oz left over.
```

The answer is 2 lb 7 oz. ∎

 Try Problems 26 through 28.

EXAMPLE 16 Divide 40 hr 16 min by 5.

SOLUTION For convenience, begin with the larger unit and move to the right.

```
     8 hr    3⅕ min      Divide 40 hr by 5 to obtain 8 hr.
  5)40 hr    16 min      Divide 16 min by 5 to obtain
    40 hr                3 min with 1 min left over.
             15 min
              1 min      When the last division does not
                         come out evenly, write the
                         remainder in fractional form.
```

The answer is 8 hr $3\frac{1}{5}$ min. ∎

 Try Problems 29 and 30.

Try These Problems

Perform the indicated operations and simplify.

31. $6\overline{)20 \text{ hr } 15 \text{ min } 36 \text{ sec}}$

32. $6\overline{)4 \text{ ft } 6 \text{ in}}$

33. $8\overline{)24 \text{ hr } 5 \text{ min } 20 \text{ sec}}$

34. $9\overline{)84 \text{ gal } 2 \text{ qt } 1 \text{ pt}}$

Solve.

35. Dede planted a tree in her yard that was 3 feet 4 inches tall. Two years later, it was 5 feet 2 inches tall. How much had the tree grown?

36. Sue practiced her violin for 1 hour 15 minutes on Monday, 50 minutes on Tuesday, and 3 hours on Wednesday. How long did she practice over the 3-day period?

EXAMPLE 17 Divide 3 wk 5 da 15 hr by 7.

SOLUTION For convenience, begin with the larger unit and move to the right.

$$
\begin{array}{r}
0 \text{ wk} \quad\quad 3 \text{ da} \\
7)\overline{3 \text{ wk} \quad\quad 5 \text{ da} \quad\quad 15 \text{ hr}} \\
\underline{0 \text{ wk}} \\
3 \text{ wk} = \quad 21 \text{ da} \\
\quad\quad\quad \overline{26 \text{ da}} \\
\quad\quad\quad \underline{21 \text{ da}} \\
\quad\quad\quad 5 \text{ da} = \quad 120 \text{ hr} \\
\quad\quad\quad\quad\quad\quad \overline{135 \text{ hr}}
\end{array}
$$

Because we will need to use long division to divide 135 by 7, we show this division separately.

$$
\begin{array}{r}
19\tfrac{2}{7} \text{ hr} \\
7)\overline{135 \text{ hr}} \\
\underline{7} \\
65 \\
\underline{63} \\
2
\end{array}
$$

When the last division does not come out evenly, write the remainder in fractional form.

The answer is 3 da $19\tfrac{2}{7}$ hr. ∎

Try Problems 31 through 34.

7 Solving Applications that Contain Measurements Expressed in Two or More Units

Now we look at some applications that involve performing arithmetic operations on measurements that have two or more units.

EXAMPLE 18 David studied for 3 hours 15 minutes yesterday and 4 hours 45 minutes today. What is the total time David studied in the two days?

SOLUTION We add to find the total time.

$$
\begin{array}{r}
3 \text{ hr } 15 \text{ min} \\
+ \; 4 \text{ hr } 45 \text{ min} \\
\hline
7 \text{ hr } 60 \text{ min } = 8 \text{ hr}
\end{array}
$$

David studied a total of 8 hours. ∎

EXAMPLE 19 Cresson has a piece of string that is 9 yards long. He cuts off a piece that is 5 yards 2 feet 8 inches long. How much string is left?

SOLUTION To find how much string is left, we subtract 5 yd 2 ft 8 in from 9 yd.

$$
\begin{array}{r}
\overset{8}{\cancel{9}} \text{ yd } \overset{2}{\cancel{0}} \text{ ft } \overset{12}{\cancel{0}} \text{ in} \\
- \; 5 \text{ yd } 2 \text{ ft } 8 \text{ in} \\
\hline
3 \text{ yd } 0 \text{ ft } 4 \text{ in}
\end{array}
\qquad
\begin{array}{l}
9 \text{ yd} = 8 \text{ yd} + 3 \text{ ft} \\
\phantom{9 \text{ yd}} = 8 \text{ yd} + 2 \text{ ft} + 12 \text{ in} \\
\phantom{9 \text{ yd}} = 8 \text{ yd } 2 \text{ ft } 12 \text{ in}
\end{array}
$$

The length of the string that is left is 3 yd 4 in. ∎

Try Problems 35 and 36.

Try These Problems

Solve.

37. A time period of 9 hours 50 minutes is divided into 6 equal periods. How long is each period?
38. A crew of 5 persons each work 13 hours 35 minutes. What is the total time they worked?

EXAMPLE 20 Nell has 7 cans of soup. Each can weighs 1 pound 4 ounces. What is the total weight?

SOLUTION We multiply the weight of each can by 7 to find the total weight.

$$\begin{array}{r} 1 \text{ lb } 4 \text{ oz} \\ \times \phantom{1 \text{ lb }} 7 \\ \hline 7 \text{ lb } 28 \text{ oz} \end{array} = 7 \text{ lb} + 28 \text{ oz}$$
$$= \underbrace{7 \text{ lb} + 1 \text{ lb}}\ 12 \text{ oz} \quad \text{Convert 28 oz to 1 lb 12 oz.}$$
$$= 8 \text{ lb } 12 \text{ oz}$$

The total weight is 8 lb 12 oz. ∎

EXAMPLE 21 A chef prepares 13 gallons 3 quarts of punch. She distributes the punch equally into 5 punch bowls. How much punch does each bowl contain?

SOLUTION The total quantity, 13 gallons 3 quarts, is being divided into 5 equal parts, so we divide 13 gal 3 qt by 5.

$$\begin{array}{r} 2 \text{ gal} \quad 3 \text{ qt} \\ 5\overline{)13 \text{ gal} \quad 3 \text{ qt}} \\ \underline{10 \text{ gal}} \\ 3 \text{ gal} = 12 \text{ qt} \\ \hline 15 \text{ qt} \\ \underline{15 \text{ qt}} \end{array} \Big\} \text{Add.}$$

Each bowl contains 2 gal 3 qt of punch. ∎

Try Problems 37 and 38.

Answers to Try These Problems

1. 80 in   2. 295 fl oz   3. $36\frac{1}{5}$ or 36.2 min   4. $5\frac{2}{3}$ yd
5. 27,613 sec   6. $7\frac{1}{4}$ or 7.25 pt   7. 705 yd 2 ft
8. 9 hr 12 min   9. 8 hr 30 min   10. 9 lb 12 oz
11. 7 yd 1 ft 8 in   12. 7 hr 45 min 3 sec
13. 6 gal 2 qt 1 pt   14. 13 ft 10 in   15. 17 hr 21 min
16. 12 gal 2 qt   17. 17 lb 6 oz   18. 2 hr 50 min
19. 12 gal 3 qt   20. 5 ft 10 in   21. 8 hr 45 min 18 sec
22. 7 yd 1 ft 7 in   23. 18 yd 2 ft 4 in   24. 69 yd
25. 64 wk 13 hr   26. 2 hr 4 min   27. 2 yd 1 ft
28. 3 lb 9 oz   29. $4 \text{ lb } 1\frac{3}{7}$ oz   30. $14 \text{ yd } 2\frac{2}{3}$ ft
31. $10 \text{ hr } 7 \text{ min } 46\frac{1}{2}$ sec   32. 9 in   33. 3 hr 40 sec
34. $9 \text{ gal } 1 \text{ qt } 1\frac{2}{9}$ pt   35. 1 ft 10 in   36. 5 hr 5 min
37. $1 \text{ hr } 38\frac{1}{3}$ min   38. 67 hr 55 min

EXERCISES 10.5

Find the missing number so that the two measurements are equal.

1. 8 lb 3 oz = __?__ oz
2. 15 gal 2 qt = __?__ gal
3. 5 hr 12 min = __?__ hr
4. 8 ft 9 in = __?__ in
5. 4 hr 45 min 20 sec = __?__ sec
6. 9 yd 2 ft 8 in = __?__ in
7. 12 gal 3 qt 1 pt = __?__ gal
8. 1 wk 5 da 6 hr = __?__ da

9. 80 ft = ___?___ yd ___?___ ft
10. 450 min = ___?___ hr ___?___ min
11. $4\frac{1}{2}$ mi = ___?___ mi ___?___ ft
12. 3.6 tons = ___?___ tons ___?___ lb
13. 3700 sec = ___?___ hr ___?___ min ___?___ sec
14. 905 in = ___?___ yd ___?___ ft ___?___ in
15. $17\frac{1}{3}$ da = ___?___ wk ___?___ da ___?___ hr
16. 9.5 qt = ___?___ gal ___?___ qt ___?___ pt

Perform the indicated operations and simplify.

17.   18 lb 4 oz
  +   3 lb 14 oz

18.   5 yd 2 ft
  + 2 yd 2 ft

19.   9 hr 34 min 35 sec
  + 5 hr 45 min 46 sec

20.   12 gal 3 qt 1 pt
  + 4 gal 2 qt 1 pt

21.   13 ft 10 in
  − 7 ft 6 in

22.   25 hr 40 min
  − 8 hr 29 min

23.   9 wk 4 da
  − 2 wk 6 da

24.   13 ft 7 in
  − 8 ft 11 in

25.   19 hr 42 min
  − 12 hr 50 min

26.   30 lb 4 oz
  − 18 lb 12 oz

27.   15 hr 24 min 54 sec
  − 8 hr 30 min 45 sec

28.   30 yd 2 ft 8 in
  − 12 yd 2 ft 10 in

29.   5 lb 12 oz
  ×   6

30.   8 wk 4 da
  ×   8

31.   7 wk 3 da 18 hr
  ×   9

32.   5 yd 2 ft 9 in
  ×   7

33. 7)59 gal 2 qt
34. 4)15 hr 20 min
35. 3)19 yd 2 ft
36. 5)20 lb 8 oz
37. 9)4 hr 7 min 39 sec
38. 2)8 wk 3 da 6 hr
39. 4)74 yd 1 ft 9 in
40. 6)34 gal 3 qt 1 pt

Solve.

41. Doug spent 5 weeks 3 days in Chicago and 3 weeks 5 days in Detroit. How much longer did Doug spend in Chicago than in Detroit?

42. Chef Theresa started with 10 pounds of butter. She used 1 pound 3 ounces on Monday, 2 pounds 4 ounces on Tuesday, and 6 ounces on Wednesday. How much butter is left?

43. The government is distributing surplus cheese to 6 families that were victims of a mining tragedy. The total weight of the cheese is 38 pounds 10 ounces. If the cheese is distributed equally, how much does each family receive?

44. The height of the sail is what fraction of the total height of this sailboat?

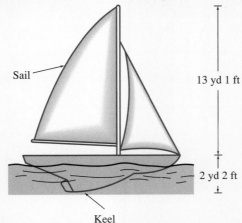

Sail — 13 yd 1 ft — 2 yd 2 ft — Keel

45. Mr. Garcia is painting the interior of a two-story townhouse. It takes 3 gallons and 2 quarts of paint to do the walls upstairs. The walls downstairs take twice as much paint as the walls upstairs. What is the total amount of paint he needs?

46. Diane practiced her piano as follows:
- Monday   1 hr 20 min
- Tuesday   35 min
- Wednesday   1 hr 15 min

On the average, how long did she practice each day?

CHAPTER 10 SUMMARY

KEY WORDS AND PHRASES

unit of measure [10.1]	area [10.2]	liter [10.2]
length (distance) [10.1]	area of a rectangle [10.2]	weight [10.3]
English measurement system [10.1]	square units [10.2]	mass [10.3]
metric system [10.1]	volume [10.2]	gram [10.3]
meter [10.1]	volume of a rectangular solid [10.2]	time [10.3]
perimeter [10.1]	cubic units [10.2]	rate [10.4]
		pure number [10.5]

SYMBOLS
$45.8 \approx 46$ means 45.8 is approximately equal to 46.

IMPORTANT RULES

Multiplication and Division Rules for Equivalent Measurements [10.1]
- Multiplying two equal measurements by the same nonzero number produces two equal measurements.
- Dividing two equal measurements by the same nonzero number produces two equal measurements.

Area of a Rectangle [10.2]
The area of a rectangle is the length times the width.

$$\text{Area of a rectangle} = \text{length} \times \text{width}$$

Volume of a Rectangular Solid [10.2]
The volume of a rectangular solid is the length times the width times the height.

$$\text{Volume of a rectangular solid} = \text{length} \times \text{width} \times \text{height}$$

CHAPTER 10 REVIEW EXERCISES

1. Measure the length of this line segment to the nearest $\frac{1}{8}$ of an inch.

2. Find the distance between points P and Q to the nearest centimeter by measuring.

3. Measure the length and width of this rectangle to the nearest $\frac{1}{4}$ inch. Use your measurements to find the approximate area of the rectangle.

Identify each of these measurements as length, area, volume, or weight.
4. 5.6 quarts
5. 85 meters
6. 4 square centimeters
7. 19 cubic inches
8. 5 tons
9. 200 liters
10. 53 miles
11. 3 tablespoons
12. 120 kilometers
13. 5 acres
14. 5.4 milligrams
15. 25.8 square feet

Find the missing number so that the two measurements are equal.
16. 15 ft = ___?___ yd
17. 5 dm = ___?___ m
18. 4 sq yd = ___?___ sq ft
19. 0.04 acre = ___?___ sq yd
20. 8 qt = ___?___ c
21. 82 mℓ = ___?___ ℓ
22. 56.7 g = ___?___ cg
23. 0.002 ton = ___?___ oz
24. $\frac{2}{3}$ wk = ___?___ min
25. $4\frac{1}{3}$ gal = ___?___ cu in
26. 23 gal 3 qt = ___?___ qt
27. 8 wk 3 da = ___?___ wk
28. 8 hr 16 min 40 sec = ___?___ min
29. 100 ft = ___?___ yd ___?___ ft
30. 8.25 tons = ___?___ tons ___?___ lb
31. 180,455 sec = ___?___ hr ___?___ min ___?___ sec

Find the missing number so that the two rates are equal.
32. $\dfrac{2\frac{1}{2} \text{ T}}{6 \text{ persons}}$ = ___?___ T per person
33. $\dfrac{18.3 \text{ g}}{15 \text{ m}}$ = ___?___ g per m
34. 0.2 ton per load = ___?___ lb per load
35. 4.5 kg per da = ___?___ kg per wk
36. 84 yd per min = ___?___ ft per sec
37. 3¢ per oz = $ ___?___ per lb
38. 540 mℓ per min = ___?___ cℓ per hr
39. $6\frac{3}{4}$ mi per hr = ___?___ ft per sec

Perform the indicated operations and simplify.
40. 24 yd 2 ft
    + 9 yd 2 ft

41. 65 lb 8 oz
    − 46 lb 14 oz

42. 10 wk 5 da 5 hr
    − 2 wk 6 da 9 hr

43. 9 lb 7 oz
    ×   5

44. 8 yd 1 ft 9 in
    ×   7

45. 6)45 hr 6 min

46. 3)10 yd 2 ft 3 in

47. 5)15 wk 4 da 5 hr

Solve.

48. Carol jogs 6.2 miles in one hour. How far does she jog in $3\frac{1}{2}$ hours?
49. Wire sells for $8.49 per meter. What is the cost of 40 centimeters of wire? (Round up to the nearest cent.)
50. A piece of fabric is 45 inches wide and $7\frac{1}{4}$ yards long. What is the area of the fabric in square yards?
51. Tile costs $3.60 per square foot. What is the cost of tiling this floor surface?

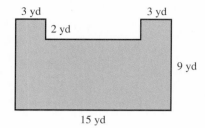

52. How many 150-milliliter doses can a pharmacist make from 4.2 liters of medicine?
53. A box is 15 centimeters wide, 20 centimeters long, and 4 centimeters high. How many liters of fluid does this box hold?
54. A chemist mixes alcohol with water to make 8 liters of a 12% alcohol solution. She then pours in an additional 750 milliliters of pure alcohol. How many milliliters of pure alcohol are in the final solution?
55. How many 7.5-gram weights does it take to make 3.9 kilograms?
56. Fresh mushrooms sell for $1.29 per $\frac{1}{2}$ pound. What is the cost of 12 ounces of these mushrooms? (Round up to the nearest cent.)
57. A manufacturer spends 20 hours producing a certain product. The product is in the fabricating department $\frac{1}{30}$ of the production time. How many minutes is the product in the fabricating department?
58. Mr. Henry mixes 12 ounces of decaffeinated coffee with 3 pounds of regular coffee. What percent of the total mixture is the regular coffee?
59. A time period of $3\frac{1}{4}$ hours is divided into 25 equal periods. How many minutes long is each of the shorter periods?
60. A rope that is 3 feet long weighs 8.4 ounces. What is the weight of 32 feet of this rope?
61. A particle travels 5 meters in 37 seconds. How long does it take the particle to travel 1 meter?
62. Sound travels at about 750 miles per hour. What is the speed of sound in feet per second?
63. Chuck completed $\frac{2}{5}$ of the job in 6 hours. What fraction of the job can he do in 30 minutes?
64. Bonnie is writing a report. She spent 1 hour 25 minutes working on the report yesterday. Today she spent 3 times as long as yesterday. What is the total time spent on the report in the two days?
65. A steak weighs 2 pounds 5 ounces. If you cut off 10 ounces of fat from the steak, what is the final weight?

66. Ms. Kawasaki has 3 puppies weighing 4 pounds 8 ounces, 4 pounds 10 ounces, and 5 pounds 3 ounces. What is the average weight of the 3 puppies?

67. A segment of fencing is 2 yards 1 foot in length. How many of these segments are needed to extend a total distance of 560 feet?

In Your Own Words

Write complete sentences to discuss each of the following. Support your comments with examples or pictures, if appropriate.

68. Discuss how you could help a friend to understand that there are 9 square feet in 1 square yard.

69. A rectangular sidewalk is 30 inches wide and 7.5 feet long. Without actually calculating, discuss the procedure you would use to find the area of this sidewalk.

70. Discuss how the following two measurements differ:
  a. 15 milliliters   **b.** 15 millimeters

71. Discuss how the following three measurements differ:
  a. 4 feet   **b.** 4 square feet   **c.** 4 cubic feet

72. Discuss how these two measurements are similar, and how they differ:
  a. 5 meters   **b.** 5 centimeters

CHAPTER 10 PRACTICE TEST

1. Measure the length of this line segment to the nearest $\frac{1}{4}$ inch.

2. Measure the length and width of this rectangle to the nearest tenth of a centimeter, and use your measurements to find the perimeter of the rectangle.

Find the missing number(s) so that the two measurements are equal.

3. 6 gal = ____?____ qt   **4.** $2\frac{3}{4}$ tons = ____?____ lb   **5.** 732 in = ____?____ yd

6. 26 cg = ____?____ g   **7.** 7.08 ℓ = ____?____ mℓ

8. 12,000 cm = ____?____ km   **9.** 157 oz = ____?____ lb ____?____ oz

10. 4 hr 12 min = ____?____ hr

Find the missing number so that the two rates are equal.

11. $\frac{126 \text{ mi}}{3 \text{ hr}}$ = ____?____ mi per hr   **12.** 3.2 ℓ per hr = ____?____ cℓ per hr

13. 30 mi per hr = ____?____ ft per sec   **14.** 0.9 g per km = ____?____ cg per m

Perform the indicated operations and simplify.

15.   3 wk 4 da 18 hr
       + 4 wk 6 da 6 hr

16.   20 gal 2 qt
       − 7 gal 3 qt

17.   3 yd 2 ft 7 in
       ×   6

18. 7)20 lb 2 oz

Solve.

19. A rectangular strip of metal is 22 centimeters wide and 8.4 meters long. What is the area in square centimeters?
20. Mary bicycles 8.5 miles in 1 hour. At this rate, how far can she go in $4\frac{1}{2}$ hours?
21. A piece of wire 3.5 yards long is cut into pieces that are each 4.2 inches long. How many of the smaller pieces can be cut?
22. Clara planted a rose bush in her yard that was 14 inches tall. Five years later, the bush was 2 feet 8 inches tall. How much had the bush grown?
23. A $14\frac{1}{2}$-ounce can of beef broth sells for 87¢. What is the cost per ounce?
24. Tommy bought a 90-minute tape. He recorded 7 songs that are each 4 minutes 15 seconds long. After each song, he left a 5-second pause. How much time is left on the tape after the last 5-second pause?
25. A recipe calls for $1\frac{1}{4}$ teaspoons of vanilla flavoring. A chef is preparing 80 times the recipe. How many cups of vanilla flavoring are needed?

CHAPTER 11

Introduction to Statistics

11.1 TABLES, BAR GRAPHS, AND LINE GRAPHS

OBJECTIVES
- Read information from a table, a bar graph, or a line graph.
- Make a table, bar graph, or line graph that organizes data.

1 Defining Statistics

All of us are faced daily with making decisions. Some of these decisions are easy for us to handle using our built-in decision-making system. However, we also have built-in prejudices that affect many decisions. One advantage of statistical methods is that they can help us to make decisions without prejudice. We often are faced with not knowing which way to turn; in these instances statistics can be of help in making decisions.

A first step for statistical decision making is often the gathering of numerical information or **data**. This information must then be organized, analyzed, and interpreted. Procedures for evaluating the data make up the central topics in statistics. In short, we can say that **statistics** *is the study of how to collect, organize, analyze, and interpret numerical data.*

In this text we look at only a very brief introduction to statistics. We begin by looking at ways of organizing data.

2 Making a Table

A **table** is the most basic way to organize information. A table involves arranging the information to form rows and columns. Here we give an example.

A tennis club wanted to know which meal its members were most likely to eat at the club. A survey of 1256 members showed that 286 of them said they were most likely to eat breakfast, 406 of them lunch, 125 said dinner, and 391 said they would only eat a snack at the club. There were 48 who didn't know.

The last two sentences of the previous paragraph contain a lot of numbers, making the sentences difficult to read. It is important to be able to organize information so that it can be presented to other people. A table would be a far more effective way to present this information than a sentence full of numbers. The following table displays the results of the tennis club survey.

| WHICH MEAL WILL MEMBERS EAT AT THE TENNIS CLUB? ||
Meal Choice	Number of Members
Breakfast	286
Lunch	406
Dinner	125
Snack	391
Don't Know	48

Note that each column is given a **column heading** and the table is given a **caption**, "Which Meal Will Members Eat at the Tennis Club?"

In general, tables should be properly labeled so that the reader does not have to scan the text to see what the table is about. The table caption and the column headings help make the table a self-contained body of information.

Here we look at another example.

EXAMPLE 1 It is costly in both time and money to go to college. Does it pay off? The results of a survey show that the answer is yes. The average yearly income of a household headed by a person with the stated educational background is as follows: $20,000 if grade school is the top level of education received, $32,500 for high school graduates, $53,200 for college graduates, and $64,700 for completion of one or more years of postgraduate studies. Arrange this information in a table with an appropriate caption and column headings.

SOLUTION Because the information reveals that an increase in education corresponds to an increase in salary, we choose the caption, "More Education Yields Higher Salaries."

There will be two columns, one for the education level and one for the corresponding yearly salary. We choose to arrange these columns from the lowest level to the highest level of education so that the table clearly reveals that more education means higher salaries.

Finally, to avoid using large numbers and repeated use of the dollar sign, we list the salaries in thousands of dollars. For clarity we have included the units, thousands of dollars, in the column heading for the yearly salaries. Here we show the final table.

Try These Problems

1. The number of fish caught in Paradise Lake over a 5-month period is as follows: 500 in April, 850 in May, 1200 in June, 2000 in July, and 1500 in August. Arrange this information in a table that includes an appropriate caption and column headings.

MORE EDUCATION YIELDS HIGHER SALARIES	
Top Level of Education	**Yearly Salary (thousands of dollars)**
Grade school	20.0
High School Graduate	32.5
College Graduate	53.2
Some Postgraduate Work	64.7

 Try Problem 1.

 Reading and Making a Bar Graph

Newspapers and magazines often use graphs instead of tables to present information. Graphs make it easier for the reader to compare facts. Frequently used graphs are bar graphs, line graphs, pictographs, and circle graphs. Here we look at how to read and make a **bar graph.**

The total bankruptcy filings in northern California, including both personal and business, increased sharply from 1989 to 1992. In 1989, there were 18,000 filings; in 1990, there were 20,000 filings; in 1991, there were 26,000 filings; and in 1992, there were 30,000 filings. Here is a bar graph that displays the information in a way that it is easy to see the increase in bankruptcy filings over the four-year period.

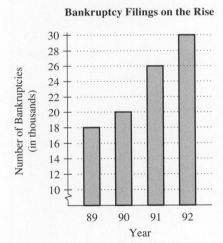

In this bar graph the years are displayed along the **horizontal axis** and the number of bankruptcy filings are displayed along the **vertical axis.** The squiggle marking, $\lessgtr$, at the bottom of the vertical axis indicates that the first increment has a different length than the remaining increments. In this case, the first increment has length 10 units, while the remaining increments have length 2 units. Each axis is labeled appropriately and the graph is given an appropriate **caption;** in this case, "Bankruptcy Filings on the Rise." In this particular graph, the length of the bars becomes longer as you move from left to right which illustrates the increase in filings over the four-year period.

Try These Problems

Use the horizontal bar graph entitled Bankruptcy Filings on the Rise *located on this page to answer Problems 2 through 4.*

2. How many bankruptcies were filed in the year 1991?

3. Which of the four years had the most bankruptcies, and how many bankruptcies were there for that year?

4. Which of the four years had fewer than 25,000 bankruptcies filed?

5. In 1987 the United States led the world in vehicle miles traveled. The following table gives the miles per person traveled in several countries. Make a bar graph that displays this information. Label both axes and give the graph a caption.

U.S. LEADS WORLD IN VEHICLE MILES TRAVELED	
Country	Miles per Person (in thousands)
United States	12
United Kingdom	6
Sweden	5.5
France	5.5
Germany	5
Italy	5
Japan	3

The previous bar graph has *vertical bars*. We could have made the graph with *horizontal bars*. To do this we display the years along the vertical axis and the number of bankruptcies along the horizontal axis. Here we show the graph made in this way.

Bankruptcy Filings on the Rise

[Horizontal bar graph showing years 89, 90, 91, 92 with bars extending to approximately 18, 20, 26, and 30 thousand bankruptcies]

In general, the bars of a bar graph can be vertical or horizontal, but they should have the same width and be spaced the same distance apart. Before making a bar graph, it may be helpful for you to first make a table that organizes the information, then make the bar graph.

 Try Problems 2 through 5.

4 Reading and Making a Line Graph

Sometimes **line graphs** are used instead of bar graphs to display information. A bar graph with vertical bars can be changed into a line graph by placing points at the top and in the middle of each bar. The bars are removed and the points are connected by straight lines. Here we show how a line graph is obtained from a bar graph.

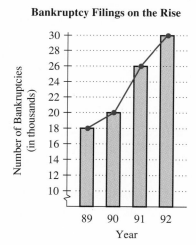
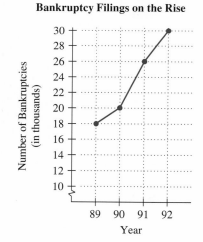

Line Graph on a Bar Graph   **Line Graph**

Observe that the line graph has some of the same features as a bar graph. Specifically, both the horizontal and the vertical axes are labeled

Try These Problems

Use the line graph entitled U.S. Gasoline Costs *located on this page to answer Problems 6 through 8.*

6. In what year was the price per gallon the highest?

7. What was the decrease in the price per gallon from 1982 to 1988?

8. What was the percent decrease in price per gallon from 1982 to 1988?

9. Use the information displayed in this table to make a line graph. Label both axes and give the graph a caption.

TUBERCULOSIS IN NEW YORK CITY	
Year	Number of Cases (in thousands)
1960	4.75
1970	2.5
1980	0.5
1990	3.5

appropriately and the graph is given a caption. Labeling the graph makes it a self-contained body of information.

A line graph can be drawn without first drawing a bar graph. Simply plot the point at the top of where the bar would have been drawn. Then connect the points with straight lines. Here is another example.

EXAMPLE 2 The average price of a gallon of gasoline, adjusted for inflation, fluctuated from 1974 to 1988. The following table gives the price for each year. Make a line graph that displays this information.

U.S. GASOLINE COSTS	
Year	Gas Price (in dollars per gallon)
1974	1.30
1976	1.25
1978	1.20
1980	1.75
1982	1.60
1984	1.40
1986	1.00
1988	0.80

SOLUTION We display the years along the horizontal axis and the gasoline prices along the vertical axis. We label each axis appropriately.

Above each year we plot a point whose height corresponds with the price of gas for that year. For example, above 1974 we plot a point that is 1.30 units high.

Finally, we give the graph the same caption as the table.

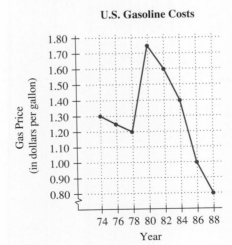

 Try Problems 6 through 9.

Answers to Try These Problems

1.

FISH CAUGHT IN PARADISE LAKE

Month	Number Caught
April	500
May	850
June	1200
July	2000
August	1500

2. 26,000   3. 1992; 30,000   4. 1989; 1990

5. **U.S. Leads World in Vehicle Miles Traveled**

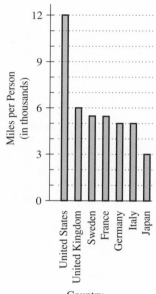

6. 1980   7. $0.80 per gal   8. 50%

9. **Tuberculosis in New York City**

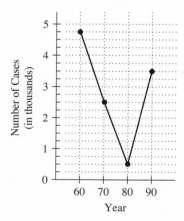

EXERCISES 11.1

The following table gives a prediction of what it will cost to attend an average university for four years. The expenses include room, board, tuition, and supplies. Use the table to answer Exercises 1 through 6.

WHAT YOU'LL PAY FOR FOUR YEARS AT A UNIVERSITY		
Entrance Year	**Cost of Public University (in thousands of dollars)**	**Cost of Private University (in thousands of dollars)**
1996	50	100
2000	60	125
2004	75	175
2008	100	225
2010	125	250

1. What will four years of college cost if you enter a public university in the year 2000?
2. What will four years of college cost if you enter a private university in the year 2010?
3. In 1996, what is the predicted cost per year of attending a private university?
4. In the year 2000, what is the predicted cost per year of attending a public university?
5. From 1996 to 2000, what is the percent increase in the cost of attending a public university?
6. From 2000 to 2004, what is the percent increase in the cost of attending a public university?

Car manufacturers are under tremendous pressure from state and federal governments to make cars that are more fuel efficient. The bar graph shows the average miles per gallon for cars in four past years and predictions of what the average will be for two future years. Use the graph to answer Exercises 7 through 12.

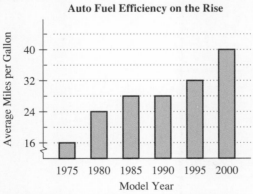

7. What is the average miles per gallon for a 1990 car?
8. What is the predicted average miles per gallon for a 2000 car?
9. On average, how many miles can a 1995 car go on 20 gallons of gasoline?
10. On average, how many miles can a 1980 car go on 20 gallons of gasoline?
11. On average, how many gallons of gasoline will be needed for a 1995 car to go a distance of 600 miles?
12. On average, how many gallons of gasoline are needed for a 1980 car to go a distance of 600 miles?

The line graph shows how the percent of Americans in poverty has fluctuated from 1974 to 1992. Use the graph to answer Exercises 13 through 18.

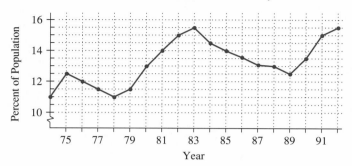

13. 14%   **14.** 11%
15. 1974, 1976, 1977, 1978, and 1979
16. 1982, 1983, 1991, and 1992
17. 14%
18. 14.125%
19. a.

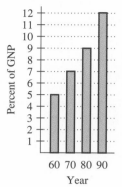

13. What percent of Americans were in poverty in 1981?
14. What percent of Americans were in poverty in 1978?
15. In what years was the poverty rate below 12.5%?
16. In what years was the poverty rate at least 15%?
17. Find the average poverty rate for the years 1980, 1981, and 1982.
18. Find the average poverty rate for the years 1989, 1990, 1991, and 1992.
19. Over the past 30 years, the cost of health care in the United States has risen dramatically. The table gives the amount spent on health care as a percent of the gross national product (GNP) for four different years. (The gross national product is the nation's total economic output.) Use the information in the table to make **a.** a bar graph and **b.** a line graph. For both graphs, label the axes appropriately and give the graph a caption.

20. a.

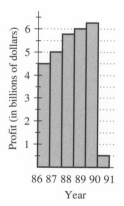

U.S. HEALTH CARE COST	
Year	Percent of GNP
1960	5
1970	7
1980	9
1990	12

20. The profits of the Macro Computer Company (MCC) fell drastically in 1991. The table below gives the annual net earnings for six successive years. Use the information in the table to make **a.** a bar graph and **b.** a line graph. For both graphs, label the axes appropriately and give the graph a caption.

21. b.

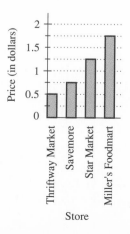

MCC ANNUAL NET EARNINGS	
Year	Profit (in billions of dollars)
1986	4.5
1987	5
1988	5.75
1989	6
1990	6.25
1991	0.5

21. Enrique was buying a turkey for his family before Thanksgiving. After checking the ads in several newspapers, he found that the price varied widely from store to store. Turkey costs $0.50 a pound at Thriftway Market, $0.75 a pound at Savemore, $1.25 a pound at Star Market, and $1.75 a pound at Miller's Foodmart. Make **a.** a table, **b.** a bar graph, and **c.** a line graph, displaying this information. Label appropriately.

22. A papillon puppy weighed 2 pounds at 2 months, 5.5 pounds at 4 months, 8.5 pounds at 6 months, and 9 pounds at 8 months. Make **a.** a table, **b.** a bar graph, and **c.** a line graph, displaying this information. Label appropriately.

22. b.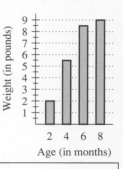

11.2 PICTOGRAPHS AND CIRCLE GRAPHS

OBJECTIVES
- Read information from a pictograph.
- Read information from a circle graph.
- Find the degree measure of the sector of a circle when given the fraction or percent of the circle that the sector occupies.
- Measure the central angle of a circle by using a protractor.
- Make a circle graph that organizes given data.

1 Interpreting a Pictograph

Another way to display information that involves data is to use a **pictograph.** The pictograph is similar to the bar graph, but symbols or pictures are used instead of bars. Like the bar graph, the pictograph can be arranged vertically or horizontally. Here is a pictograph that shows the number of houses sold by Four Star Realty over a four-month period.

 represents 25 houses

Several characteristics of this graph are typical of pictographs. First, notice at the bottom of the graph there is a statement of what each picture represents. In this case, each picture represents 25 houses. This is a necessary feature; however, the statement does not have to be placed at the bottom of the graph.

Try These Problems

The pictograph shows the profits of the Ferguson Hardware for each quarter of 1993. Use the graph to answer Problems 1 through 4.

Ferguson Hardware Profits

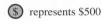

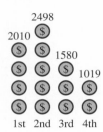

Quarters of 1993

1. What are the exact profits earned in the 3rd quarter of 1993?
2. The pictures give only the approximate profits. What are the approximate profits earned in the 1st quarter?
3. How much money does each individual picture represent?
4. Using the approximate profits, rather than the exact profits, compute the approximate percent decrease in profits from the 2nd quarter to the 3rd quarter.

The pictorial display in a pictograph is often showing only an approximation of the exact information. In this graph, the pictures indicate that approximately 100 houses were sold in September. The number, 108, placed at the end of the column, gives the more accurate data. Placing this number at the end of the column (or row) is not only helpful for improving accuracy but keeps the reader from having to count the number of pictures.

Finally, observe that the graph is given a caption, "Houses Sold by Four Star Realty." This is a common element of most displays of information.

 Try Problems 1 through 4.

 Interpreting a Circle Graph

Another popular pictorial representation of data is a **circle graph.** A circle graph is especially useful for showing the division of a quantity into its component parts. The entire circular region represents the total quantity and the wedge-shaped slices or **sectors of the circle** represent the component parts. Here is an example.

Shambala, a statistician for an insurance company, made a study to obtain the causes of death for 1000 persons. She made the following circle graph to display the results.

Cause of Death for 1000 Persons

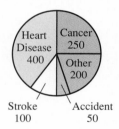

Note that this circle graph gives the actual number of people in each of the five categories. For example, 250 people out of the total 1000 died of cancer. Observe that the cancer-death sector is occupying exactly $\frac{1}{4}$ (or 25%) of the entire circular region; thus, 25% of the 1000 persons died of cancer.

Visually, we can also see that the heart-disease sector is occupying somewhat less than $\frac{1}{2}$ (or 50%) of the entire circular region, so we know that somewhat less than 50% of the persons died of heart disease. It is difficult to see visually the exact percent who died of heart disease, but this percent can be obtained as follows.

$$\text{Percent of persons who died of heart disease} = \frac{\text{number of heart-disease deaths}}{\text{total number of persons}}$$

$$= \frac{400}{1000}$$

$$= \frac{40}{100} \text{ or } \frac{2}{5}$$

$$= 40\%$$

Thus, exactly $\frac{2}{5}$ or 40% of the 1000 persons died of heart disease.

Try These Problems

A Phoenix newspaper surveyed 500 of its readers to find out whether they are very pleased, somewhat pleased, or not pleased with the economic policies of the federal government. The results are shown here in the circle graph. Use the graph to answer Problems 5 through 9.

Survey Response to "Are you pleased with the government's economic policies?"

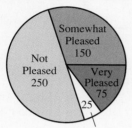

Total Number Surveyed = 500

5. How many people in the survey are not pleased with the government's economic policies?

6. How many people in the survey are very pleased with the government's economic policies?

7. What percent of those surveyed are not pleased with the government's economic policies?

8. What percent of those surveyed are very pleased with the government's economic policies?

9. Make a table that displays the information in the graph and include a column that gives the percent of the total for each category.

The following table shows how to compute what percent of the total is the number in each category.

CAUSE OF DEATH FOR 1000 PERSONS			
Cause	Number of Persons	Fractional Part of Total	Percent of Total
Cancer	250	$\frac{250}{1000} = \frac{1}{4} = 0.25$	25%
Heart Disease	400	$\frac{400}{1000} = \frac{2}{5} = 0.4$	40%
Stroke	100	$\frac{100}{1000} = \frac{1}{10} = 0.1$	10%
Accident	50	$\frac{50}{1000} = \frac{1}{20} = 0.05$	5%
Other	200	$\frac{200}{1000} = \frac{1}{5} = 0.2$	20%
Total	1000		100%

Instead of including the actual number of persons in each sector of the circle graph, Shambala could have included the percents of the total. Here we show the graph done in this way.

Cause of Death for 1000 Persons

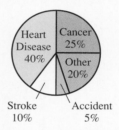

Observe that the percents add up to be 100%. This should come as no surprise because the entire circular region represents all 1000 persons, which is 100% of them.

 Try Problems 5 through 9.

3 Understanding Central Angle Measurement

The straight-line borders of a sector of a circle form an angle that is called a **central angle** of the circle. Here are several examples.

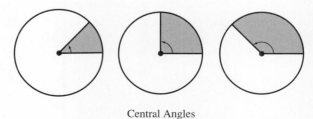

Central Angles

The straight-line pieces that form the angle are the **sides of the angle,** and the center point of the circle is called the **vertex of the angle.**

Angles are given a measurement. The measurement depends on the amount one side rotates about the vertex away from the other side. The unit of measurement often used is **degree.** Here we display the measure, in degrees, of several central angles. These **degree measures** are also referred to as the degree measures of the sectors.

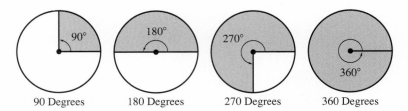

Observe that a complete rotation all the way around has measure 360°, and the other measures are fractional parts of the total 360°.

If we know what fractional part of the circle a sector is, we can compute the sector's degree measure by multiplying the fraction by 360. Here are some examples.

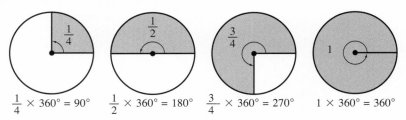

EXAMPLE 1 Here we show a circle with sectors named A, B, and C. The fractional portion of the circle that the sector occupies is given. Use this information to find the degree measure of each central angle.

SOLUTION To obtain the number of degrees in each sector, we must multiply the given fraction by 360°, the total degree measure for a complete rotation.

A. $\frac{1}{3} \times 360 = 120$

The central angle for sector A measures 120°.

Try These Problems

10. Here we show a circle with sectors A, B, and C. The fractional portion of the circle that the sector occupies is given. Use this information to find the degree measure of each central angle.

11. Show that the fractions $\frac{1}{6}$, $\frac{1}{8}$, and $\frac{17}{24}$ from Problem 10 add up to 1.

12. Here we show a circle with sectors A, B, and C. The percent of the circle that the sector occupies is given. Use this information to find the degree measure of each central angle.

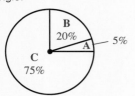

B. $\dfrac{5}{12} \times 360 = \dfrac{5}{\underset{1}{\cancel{12}}} \times \dfrac{\overset{30}{\cancel{360}}}{1}$

$= 150$

The central angle for sector B measures 150°

C. $\dfrac{1}{4} \times 360 = 90$

The central angle for sector C measures 90°. ■

Try Problems 10 and 11.

We often use percents, instead of fractions, when discussing parts of a circle. If we know what percent of the whole circle a sector is, we can obtain the degree measure by multiplying the percent by 360. Of course, the percent is changed to its fraction or decimal form before multiplying.

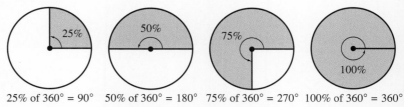

25% of 360° = 90°   50% of 360° = 180°   75% of 360° = 270°   100% of 360° = 360°

EXAMPLE 2 Here we show a circle with sectors named A and B. The percent of the circle that the sector occupies is given. Use this information to find the degree measure of the sector.

SOLUTION To find the degree measure of each sector, multiply each percent by 360°, the total degree measure for a complete revolution.

A. 30% of 360° = 0.3 × 360°
$\phantom{\text{A. 30\% of 360°}} = 108°$

The central angle for sector A measures 108°.

B. 70% of 360° = 0.7 × 360°
$\phantom{\text{B. 70\% of 360°}} = 252°$

The central angle for sector B measures 252°. ■

Try Problem 12.

Try These Problems

13. Use a protractor to measure each of the central angles, labeled A, B, C, D. It can be helpful to trace the angles on another piece of paper and extend the sides before measuring.

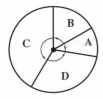

4 Using a Protractor to Measure Angles

An instrument used to measure angles in degrees is called a **protractor.** The protractor can measure angles with measurement from 0° to 180°. Here we show a protractor measuring two central angles.

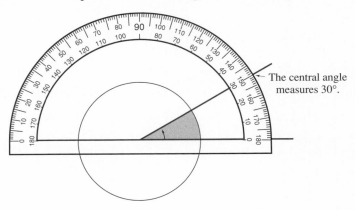

The central angle measures 30°.

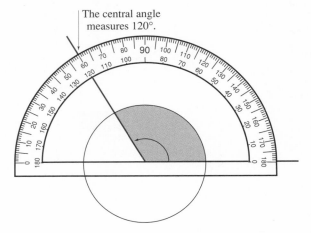

The central angle measures 120°.

Observe that the center point of the protractor, that is aligned with both the 0° (180°) marking and the 90° marking, must be placed at the vertex of the angle (center of the circle). One side of the angle must be aligned with the center point and the 0° (180°) marking. The other side will align with a marking that gives two different measurements, one less than 90° and the other more than 90°. Choose the measure that is appropriate for the angle you are measuring.

 Try Problem 13.

Because a protractor is limited to measuring angles with measures from 0° to 180°, how can we obtain the measure of central angles that measure more than 180°? The following example illustrates one procedure.

Try These Problems

14. Find the measure of both of these central angles. Use a protractor to help you.

EXAMPLE 3 Find the measure of both of these central angles. Use a protractor to help you.

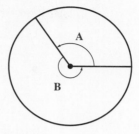

SOLUTION

A. The protractor can be used to measure angle A because it measures less than 180°.

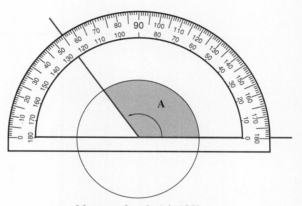

Measure of angle **A** is 125°

B. Because a complete rotation measures 360°, we can find the measure of angle B by subtracting the measure of angle A from 360°.

$$\text{Measure of angle B} = 360° - \text{measure of angle A}$$
$$= 360° - 125°$$
$$= 235°$$ ■

Try Problem 14.

3 Making a Circle Graph

Now we want to discuss how to make a circle graph. The following example illustrates the procedure.

EXAMPLE 4 Juan Loeza, a college student, has $600 a month to live on. Last month he spent $150 for food, $180 for rent, $60 for entertainment, $90 for transportation, and the rest for miscellaneous items.

a. Make a circle graph that gives the actual dollars spent in each category.

b. Make a circle graph that gives the percent of the total for each category.

SOLUTION First, organize the information in a table with four columns as shown here.

LOEZA'S MONTHLY BUDGET (TOTAL AMOUNT IS $600)

Item	Amount	Percent of Total	Degree Measure
Food	$150	$\frac{150}{600} = \frac{1}{4} = 25\%$	$\frac{1}{4} \times 360° = 90°$
Rent	$180	$\frac{180}{600} = \frac{3}{10} = 30\%$	$\frac{3}{10} \times 360° = 108°$
Entertainment	$60	$\frac{60}{600} = \frac{1}{10} = 10\%$	$\frac{1}{10} \times 360° = 36°$
Transportation	$90	$\frac{90}{600} = \frac{3}{20} = 15\%$	$\frac{3}{20} \times 360° = 54°$
Miscellaneous	$120	$\frac{120}{600} = \frac{1}{5} = 20\%$	$\frac{1}{5} \times 360° = 72°$
Totals	$600	100%	360°

The second column contains the amount spent in each category. The $120 spent on miscellaneous items was found by adding all of the other amounts and subtracting from the total $600.

The third column contains the percent of the total for each category. For example, the percent spent on food is computed as follows.

$$\begin{aligned}\text{Percent spent on food} &= \frac{\text{amount spent on food}}{\text{total amount}} \\ &= \frac{150}{600} \\ &= \frac{1}{4} \\ &= 25\%\end{aligned}$$

Finally, the fourth column contains the degree measure of each sector. These angle measurements are needed before making the circle graph. For example, the degree measure of the food sector is computed as follows.

$$\begin{aligned}\text{Degree measure of food sector} &= 25\% \text{ of } 360° \\ &= \frac{1}{4} \times 360° \\ &= 90°\end{aligned}$$

After you have the degree measure of each sector, you are ready to make the circle graph.

Draw a circle and locate the center point. Make the circle large enough that some labeling will fit inside. A circle can be drawn in several ways. You can trace a circle from this text or you can use a plastic template that has large circular regions cut out of it. There is also an instrument, called a **compass,** that can be used to draw a circle. (See Section 15.4 for more information on how to use a compass.) A template or a compass can be purchased in the college bookstore or in a stationery store.

Try These Problems

15. The total student enrollment for a college in the state of Washington is 8000. The ethnic breakdown is as follows: 3600, Caucasian; 2000, Hispanic; 1000, Asian-American; 800, African-American; and 600, other.
a. Make a circle graph that gives the actual number of students in each category.
b. Make a circle graph that gives the percent of the total for each category.

After drawing the circle and locating the center, draw one radius from the center out to the circle. This will be one side of a central angle. Now use your protractor to make central angles (or sectors) that measure 90°, 108°, 36°, 54°, and 72°.

Now we can make the two types of circle graphs.

a. Here we label each sector with the actual amount spent in each category.

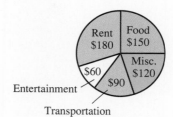

Loeza's Monthly Budget
(Total Amount Is $600)

b. Here we label each sector with the percent each category is of the total.

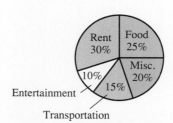

Loeza's Monthly Budget
(Total Amount Is $600)

 Try Problem 15.

 Answers to Try These Problems

1. $1580 **2.** $2000 **3.** $500 **4.** 40% **5.** 250 **6.** 75
7. 50% **8.** 15%
9.

SURVEY RESPONSE OF 500 PERSONS "ARE YOU PLEASED WITH THE GOVERNMENT'S ECONOMIC POLICIES?"		
Response	Number of Persons	Percent of Total
Somewhat pleased	150	$\frac{150}{500} = \frac{3}{10} = 30\%$
Not pleased	250	$\frac{250}{500} = \frac{1}{2} = 50\%$
Very pleased	75	$\frac{75}{500} = \frac{3}{20} = 15\%$
No opinion	25	$\frac{25}{500} = \frac{1}{20} = 5\%$
Total	500	100%

10. A. 60° B. 45° C. 255°
11. $\frac{1}{6} + \frac{1}{8} + \frac{17}{24} = \frac{4}{24} + \frac{3}{24} + \frac{17}{24} = \frac{24}{24} = 1$
12. A. 18° B. 72° C. 270° 13. A. 40° B. 60° C. 150° D. 110°
14. A. 120° B. 240°

15. a. **Ethnic Breakdown for a College of 8000 Students**

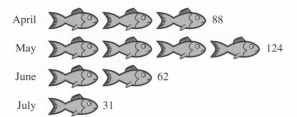

15. b. **Ethnic Breakdown for a College of 8000 Students**

EXERCISES 11.2

The pictograph shows the number of fish caught in Lake Willis for each of four months. Use the graph to answer Exercises 1 through 6.

Number of Fish Caught in Lake Willis

 = 30 fish

April 🐟🐟🐟 88
May   🐟🐟🐟🐟 124
June   🐟🐟 62
July   🐟 31

1. What is the exact number of fish caught during the month of May?
2. How many fish does each individual picture represent?
3. The pictures give only the approximate number of fish caught. What was the approximate number of fish caught in April?
4. Using the pictures, what was the approximate number of fish caught in May?

5. Using the approximate values, rather than the exact values, compute the percent increase in the number of fish caught from April to May.

6. Using the exact values, rather than the approximate values, compute the average number of fish caught per month for the four-month period.

An advertising agency was creating an advertisement to be shown during the television show "A Thrilling Event." Before deciding on a theme for the advertisement, research was done to find out the age distribution of the viewers of this program. The circle graph summarizes the results of the research that was done on 1200 randomly selected viewers. Use the graph to answer Exercises 7 through 12.

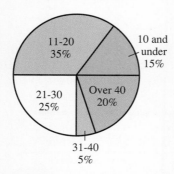

7. What percent of the 1200 viewers are over 40 years old?
8. What percent of the 1200 viewers are from 21 to 30 years old?
9. How many of the 1200 viewers are over 40 years old?
10. How many of the 1200 viewers are from 21 to 30 years old?
11. What is the degree measure of the sector that contains the 11- to 20-year-olds? (Do not use a protractor.)
12. What is the degree measure of the sector that contains the 31- to 40-year-olds? (Do not use a protractor.)
13. Here we show a circle with sectors A, B and C. The percent of the circle that the sector occupies is given. Use this information to find the degree measure of each central angle. (Do not use a protractor.)

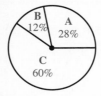

14. Here we show a circle with sectors A, B and C. The percent of the circle that the sector occupies is given. Use this information to find the degree measure of each central angle. (Do not use a protractor.)

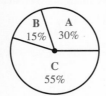

In Exercises 15 through 18, find the measure of each central angle. Use a protractor to help you. It can be helpful to trace the angles on another piece of paper and extend the sides before measuring.

15.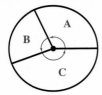

16.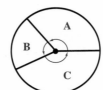

17.

18.

19. A sociology statistician wanted to know how important friendships are to adults. One of the questions she asked in her survey of 800 adults was, "How many close friends would you say you have?" The results are given in the table.

800 ADULTS ANSWER THE QUESTION "HOW MANY CLOSE FRIENDS WOULD YOU SAY YOU HAVE?"	
Number of Close Friends	**Number of Responses**
0–2	40
3–5	320
6–8	200
9 or more	240

a. Make a circle graph that gives the actual number of responses in each category.
b. Make a circle graph that gives the percent of the total for each category.

20. A newspaper conducted a survey to find out from its readers what was important to them if they had an opportunity to build a dream house. One of the questions asked was, "What is the most important room in your dream house?" The results from 1500 readers are given in the table.

1500 PERSONS ANSWER THE QUESTION "WHAT IS THE MOST IMPORTANT ROOM IN YOUR DREAM HOUSE?"	
Most Important Room	**Number of Responses**
Kitchen	600
Family room	525
Master suite	180
Living room	150
Other	45

a. Make a circle graph that gives the actual number of responses in each category.
b. Make a circle graph that gives the percent of the total for each category.

11.3 MEASURES OF CENTRAL TENDENCY: MEAN, MEDIAN, AND MODE

OBJECTIVES
- To find the mean (or average) of a collection of data.
- To find the median of a collection of data.
- To find the mode or modes of a collection of data, or recognize that the mode does not exist.

Here are some statements that might appear in a newspaper or magazine.

1. The average age of the students attending Yakima Community College in the state of Washington is 24 years old.
2. The automobile averages 35 miles per gallon.
3. On a flight from Nashville to Dallas the plane averaged 648 miles per hour.
4. The median income for a family of four in a region of Indiana is $38,000.
5. At the Baker Street Bistro most customers have two cups of coffee with their meal.

In each of the above statements one number is being used to represent a group of numbers. Numbers used in this way are **measures of average** or **measures of central tendency.** Loosely stated, an average means the center of the distribution or the most typical case. There are several different methods for obtaining an average. In this section we look at the three most commonly used measures of central tendency: the **mean,** the **median,** and the **mode.**

1 Finding the Mean

The **mean,** often referred to as the **average,** has been covered several times throughout this text. We looked at this type of averaging in Sections 2.5, 5.7, 7.5, and 8.5. The mean is found by adding the values of the data and dividing by the total number of values. Here are some examples.

EXAMPLE 1 Juan receives scores of 77, 93, 85, and 81 on four biology exams. What is the mean score?

SOLUTION To calculate the mean, add up the four scores and divide by 4.

$$\text{Mean} = \frac{77 + 93 + 85 + 81}{4}$$
$$= \frac{336}{4}$$
$$= 84$$

The mean score is 84.

Try These Problems

1. The annual salaries of six auto-industry employees are as follows.

   $44,000 $20,500 $35,000
   $28,920 $44,600 $32,780

   What is the mean salary of these six employees?

2. The annual salaries of seven computer-industry employees are as follows.

   $48,000 $52,000 $65,000
   $32,000 $74,000 $600,000
   $1,425,000

   What is the mean salary of these seven employees?

EXAMPLE 2   Because Carolyn has been out of school for 15 years, she has a little trouble adjusting to college. Consequently, she fails her first algebra exam. Her scores for the semester are 25, 86, 92, 85, and 97 on five algebra exams. What is the mean score?

SOLUTION   To calculate the mean, add up the five scores and divide by 5.

$$\text{Mean} = \frac{25 + 86 + 92 + 85 + 97}{5}$$
$$= \frac{385}{5}$$
$$= 77$$

The mean score is 77. ■

Let's look back at Examples 1 and 2 to see if the mean is an appropriate average for each of these sets of data. Juan's average of 84 in Example 1 is an appropriate measure of how well he did in the biology course. If scores of 80–89 are given a B, Juan will receive a B in this class. The mean worked well here because Juan made no extreme scores. His scores were all relatively near the computed mean.

However, Carolyn's average of 77 in Example 2 is not a good indicator of how well she did in the algebra course. The extremely low score of 25 is making the mean be an unfair measure of her performance, especially if this was her first exam score, rather than a later exam score. A fair-minded instructor would not use the mean score to assign Carolyn a grade. If scores of 70–79 are given a C and scores of 80–89 are given a B, a fair-minded instructor would give Carolyn a B, not a C.

Here we give a rule for computing the mean.

To Compute the Mean
1. Add up all of the data values.
2. Divide the sum by the number of values.

 Try Problems 1 and 2.

2   Finding the Median

As we discussed previously, the mean of a set of data is not always a good indicator of the center of the data. Another type of average is the **median.** If we order the data from smallest to largest, the median will have just as many values above it as below it. The median is the middle value.

EXAMPLE 3   Carolyn receives scores of 25, 86, 92, 85, and 97 on five algebra exams. What is the median score?

11.3 Measures of Central Tendency: Mean, Median, and Mode

▲ **Try These Problems**

3. The weight in pounds of five football players is as follows.

    184   210   220   195   200

    Find the median weight of the five players.

4. The annual salaries of seven computer-industry employees are as follows.

    $48,000 $52,000 $65,000
    $32,000 $74,000 $600,000
    $1,425,000

    What is the median salary of these seven employees?

SOLUTION First, list the scores from smallest to largest.

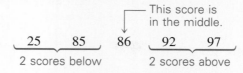

The score 86 is in the middle. There are two scores below 86 and two scores above 86. The median score is 86. ■

▲ **Try Problems 3 and 4.**

In the previous example there were 5 scores, an *odd* number of scores. Whenever there is an odd number of data values, there will always be a middle value. However, when there are an *even* number of data values, there is not a middle value. Instead, there are two middle values. In this case, we average the two middle values to obtain the median. That is, we add the two middle values and divide by 2. Here is an example.

EXAMPLE 4 Juan receives scores of 77, 93, 85, and 81 on four biology exams. What is the median score?

SOLUTION First, arrange the scores from smallest to largest.

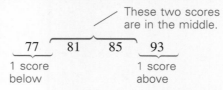

The scores 81 and 85 are the two middle scores. There is one score below them and one score above them. Average these two scores to find the median.

$$\text{Median} = \frac{81 + 85}{2}$$

$$= \frac{166}{2}$$

$$= 83$$

The median score is 83. ■

EXAMPLE 5 Thaddeus called six different stores to check the price of a certain printing calculator. Here are the six prices he obtained.

$40 $44 $35 $52 $48 $60

What is the median price?

SOLUTION First, arrange the prices from lowest to highest.

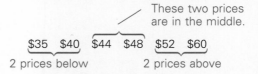

Try These Problems

5. George bowled four games with the following scores.

   140 132 108 135

   What is his median score?

6. The annual salaries of six auto-industry employees are as follows.

   $44,000 $20,500 $35,000
   $28,940 $44,600 $32,780

   What is the median salary of these six employees?

The prices $44 and $48 are the two middle prices. There are two prices below them and two prices above them. Average these two prices to obtain the median price.

$$\text{Median} = \frac{44 + 48}{2}$$
$$= \frac{92}{2}$$
$$= 46$$

The median price is $46. ■

In Example 5 note that the median price would remain $46, no matter what the lowest two prices were, as long as they were less than or equal to $44. Unlike the mean, the median does not use the *specific value* of each data entry; instead, it uses the *position* of the data.

Here we give a rule for finding the median.

To Find the Median

1. Arrange the data values from smallest to largest.
2. The median is the middle value. It has just as many data values below it as above it.
3. When there is an even number of data values, there are two middle values. In this case, add the two middle values and divide by 2.

 Try Problems 5 and 6.

3 Finding the Mode

A third way of averaging is to look at the data value that occurs most frequently. This value is called the **mode.** We prefer the mode when we want the most typical case.

EXAMPLE 6 The owner of the Baker Street Bistro recorded the number of cups of coffee each of 10 customers had with their dinner. The data are shown here.

Number of Cups of Coffee
(for 10 customers)

1 0 2 1 2
2 3 0 2 1

Find the mode.

SOLUTION First, list the data values from smallest to largest so that it is easy to see the frequency of each data value.

0 0 1 1 1 2 2 2 2 3

Observe that the data value 2 cups of coffee occurs 4 times, more than any other data value. Therefore, the mode is 2 cups of coffee. ■

EXAMPLE 7 A Pennsylvania highway patrol officer clocked the speeds, in miles per hour, of 15 automobiles on Interstate 95 outside of

Try These Problems

7. On a day during finals, eight students at Greenville Technical College in South Carolina were asked how many hours sleep they had the night before. Here are the student responses in hours.

   4 5 5 6 6 6 7 8

   Find the mode.

8. Mr. Lopez, a technical assistant in the computer lab at a community college, wanted to know how long each student was spending at the computer terminals in a single sitting. He recorded the time in minutes for 12 students. Here is the data.

   Time Spent at Computer Terminals (in Minutes)

   15 20 35 15 40 35
   45 35 20 35 15 25

   What is the mode?

9. Hillary checked the fuel efficiency for eight different 1993-model cars. Here are the results in miles per gallon.

   36 24 28 40
   27 38 32 35

   Find **a.** the mean, **b.** the median, and **c.** the mode.

Philadelphia. The data are shown here.

Speed of 15 Autos on I-95
(in miles per hour)

65 50 58 75 75
60 65 45 80 65
65 58 65 38 65

Find the mode.

SOLUTION First, arrrange the data from smallest to largest and obtain the frequency of each data value. When there are lots of data, a tally can help you.

Data Values (mi per hr)	38	45	50	58	60	65	75	80
Tally	|	|	|	||	|	ⅣⅠ	||	|
Frequency (number of autos)	1	1	1	2	1	6	2	1

The data value 65 miles per hour occurs 6 times, more than any other data value. The mode is 65 miles per hour. ∎

 Try Problems 7 and 8.

A set of data always has exactly one mean and exactly one median. However, a data set can have more than one mode or no mode at all. These points are illustrated in the following examples.

EXAMPLE 8 The heights of seven basketball players were measured in inches and found to be as follows.

77 80 73 74 70 79 72

Find **a.** the mean, **b.** the median, and **c.** the mode.

SOLUTION

a. To find the mean we add up all of the data values and divide by 7, the number of data values.

$$\text{Mean} = \frac{77 + 80 + 73 + 74 + 70 + 79 + 72}{7}$$

$$= \frac{525}{7}$$

$$= 75$$

The mean height is 75 inches.

b. To find the median we list the data values from smallest to largest and choose the middle value.

⎯The middle height is 74.

$\underbrace{70 \ \ 72 \ \ 73}_{\text{3 heights below}} \ \ 74 \ \ \underbrace{77 \ \ 79 \ \ 80}_{\text{3 heights above}}$

The median height is 74 inches.

c. Because no data value occurs more than once, we say there is no mode. ∎

 Try Problem 9.

Try These Problems

10. Ms. Combs checked the cost of a certain computer printer at seven different stores. Here are the results in dollars.

    2400 2000 2500 2400
    2600 2200 2600

Find **a.** the mean, **b.** the median, and **c.** the mode. (Round off the mean to the nearest ten dollars.)

EXAMPLE 9 Six students were asked how many brothers and sisters they have. Here are the results.

$$2 \quad 1 \quad 3 \quad 2 \quad 3 \quad 0$$

Find **a.** the mean, **b.** the median, and **c.** the mode. (Round off the mean to 1 decimal place.)

SOLUTION **a.** To find the mean we add up all of the data values and divide by 6, the number of data values.

$$\text{Mean} = \frac{2 + 1 + 3 + 2 + 3 + 0}{6}$$
$$= \frac{11}{6}$$
$$= 1.8\overline{3}$$
$$\approx 1.8 \qquad \text{Here we round off to 1 decimal place.}$$

The mean is 1.8 brothers and sisters.

b. To find the median we arrange the data values from smallest to largest and locate the middle value.

                      There are two middle values.

$$0 \quad 1 \quad \overbrace{2 \quad 2} \quad 3 \quad 3$$

There are two middle values, 2 and 2. The average of these two values is the median.

$$\text{Median} = \frac{2 + 2}{2}$$
$$= \frac{4}{2}$$
$$= 2$$

The median is 2 brothers and sisters.

c. The mode is the data value that occurs most frequently. Observe that the data values 2 and 3 both occur twice and this is the most any data value occurs. Therefore, we say there are two modes, 2 and 3. ■

 Try Problem 10.

EXAMPLE 10 Ten persons were asked the question, "How many pets do you have?" Here are the results.

    0   0   1   2   0
    5   1   0   2   0

Find **a.** the mean, **b.** the median, and **c.** the mode.

SOLUTION

a. To find the mean we add up all of the data values and divide by 10, the number of data values.

11.3 Measures of Central Tendency: Mean, Median, and Mode

 Try These Problems

11. Alice asked ten persons the question, "How many cigarettes do you smoke each day?" Here are the responses.

Number of Cigarettes Smoked Per Day By Each of 10 Persons

0	4	12	8	0
0	2	0	4	0

Find **a.** the mean, **b.** the median, and **c.** the mode.

$$\text{Mean} = \frac{0+0+1+2+0+5+1+0+2+0}{10}$$

$$= \frac{11}{10}$$

$$= 1.1$$

The mean is 1.1 pets.

b. To find the median, first list the data from smallest to largest.

The two middle values are 0 and 1.

0 0 0 0 0 1 1 2 2 5

4 values below   4 values above

There are two middle values, 0 and 1. The median is the average of these two values.

$$\text{Median} = \frac{0+1}{2}$$

$$= \frac{1}{2}$$

$$= 0.5$$

The median is 0.5 pet.

c. The mode is the data value appearing most frequently. Observe that the data value 0 pets appears five times, more than any other data value. Therefore, the mode is 0 pets. ∎

 Try Problem 11.

Observe that in Example 8 there was no mode because no two basketball players were the same height, while in Example 10 the mode was 0 because most of the persons surveyed had no pets. Be careful not to say the mode is 0 when there is no mode because, as you see in Example 10, it is possible to have 0 as a data value that occurs most frequently.

To Find the Mode

1. List the data values from smallest to largest.
2. The data value that occurs most frequently is the mode.
3. When no data value occurs more than once, there is no mode.
4. When several data values have the largest frequency, then all of these data are modes.

 Answers to Try These Problems

1. $34,300   2. $328,000   3. 200 lb   4. $65,000
5. 133.5   6. $33,890   7. 6 hr   8. 35 min
9. **a.** 32.5 mi per gal **b.** 33.5 mi per gal **c.** There is no mode.
10. **a.** $2390 **b.** $2400 **c.** $2400 and $2600
11. **a.** 3 cigarettes **b.** 1 cigarette **c.** 0 cigarettes

EXERCISES 11.3

Find **a.** *the mean,* **b.** *the median, and* **c.** *the mode (or modes) for each of the given sets of data.*

1. 20 20 10 12 16
2. 0 4 4 8 0 0 5
3. 270 350 220 380
4. 1 1 0 4 3 4 1 4
5. 8 14 0 9 15
   18 24 0 18 18
6. 10 80 10 0
   30 0 10 30
   20 20 90 70
7. 3.5 4.0 3.5 9.2 7.3
8. 0.6 1.2 0.5 1.2
9. $13\frac{1}{2}$ $7\frac{2}{3}$ $8\frac{1}{3}$ $10\frac{1}{2}$
10. $14\frac{5}{8}$ $8\frac{3}{4}$ $6\frac{1}{8}$

For each situation find **a.** *the mean,* **b.** *the median, and* **c.** *the mode (or modes).*

11. Five secretaries were given a typing test, and the times (in minutes) to complete it were as follows.

    12 9 7 9 10

12. Response times for seven emergency police calls in Houston were measured to the nearest minute and found to be as follows.

    8 6 10 12 15 12 12

    Round off the mean to one decimal place.

13. Eight shoppers were asked about how much money they had spent on clothing in the past month. Here are the results in dollars.

    50 200 0 85
    150 0 35 50

14. Twelve women are participating in a weight loss program. For the first five weeks the following weight losses (in pounds) were recorded.

    8 5 0 12 8 0
    5 8 2 15 12 6

15. An architectural firm looked at the annual salaries of a beginning architect at six other firms. Here are the data rounded to the nearest hundred dollars.

    30,000 28,800 32,000 48,000 35,700 32,500

16. The vice-president of Great Lakes Bank looked at the checking account balance for ten of his customers. Here are the amounts rounded to the nearest ten dollars.

    620 800 1500 50 750
    20 160 300 480 1200

17. A scholarship committee at Forrest Hill Community College in St. Louis looked at the grade point average of fifteen students. The data are listed here.

    3.2 3.5 2.9 4.0 3.6
    3.0 2.9 3.4 3.2 3.0
    3.4 3.6 3.2 3.8 4.0

18. A mail carrier recorded the weights of nine packages. Here are the weights rounded to the nearest tenth of a pound.

    5.1 12.3 8.2 10.1 4.8
    3.9 14.5 12.8 8.4

19. The closing price (in dollars) of a certain stock for the last five days was as follows.

$$21\tfrac{5}{8} \quad 20\tfrac{1}{8} \quad 22\tfrac{1}{4} \quad 23 \quad 22\tfrac{3}{4}$$

20. Lynnette wanted to know how much dry dog food she should feed her 30-week-old puppy that weighs 8 pounds. The amount of food recommended varies depending on the brand of dog food. Lynnette recorded the amount for six different brands. Here are the results in cups per day.

$$1\tfrac{1}{2} \quad 1\tfrac{3}{4} \quad 2 \quad 1\tfrac{3}{4} \quad 2\tfrac{1}{2} \quad 2\tfrac{1}{4}$$

CHAPTER SUMMARY

KEY WORDS AND PHRASES

statistics [11.1]
data [11.1]
table [11.1]
column heading {11.1}
caption [11.1]
bar graph [11.1]
horizontal axis [11.1]
vertical axis [11.1]

vertical bars [11.1]
horizontal bars [11.1]
line graph [11.1]
pictograph [11.2]
circle graph [11.2]
sector of a circle [11.2]
central angle [11.2]
side of an angle [11.2]

vertex of an angle [11.2]
degree [11.2]
degree measurement [11.2]
protractor [11.2]
compass [11.2]
mean (average) [11.3]
median [11.3]
mode [11.3]

IMPORTANT RULES

Finding the Degree Measure of the Sector of a Circle [11.2]

Multiply the percent (or fraction) the sector is of the circle times 360°.

Computing the Mean of a Set of Data [11.3]

1. Add up all of the data values.
2. Divide the sum by the number of values.

Finding the Median of a Set of Data [11.3]

1. List the data from smallest to largest.
2. The median is the middle value. It has just as many data values below it as above it.
3. When there is an even number of data values, there are two middle values. In this case, add the two values and divide by 2 to obtain the median.

Finding the Mode of a Set of Data [11.3]

1. List the data from smallest to largest.
2. The data value that occurs most frequently is the mode.
3. When no data value occurs more than once, there is no mode.
4. When several data values have the largest frequency, then all of these data are modes.

CHAPTER 11 REVIEW EXERCISES

Rita, a newspaper reporter, was attempting to show that teacher's incomes are much lower than other professional incomes even though comparable education is required. She found the monthly incomes of several individual professionals in the San Francisco Bay Area to be as shown in the bar graph. Use the graph to answer Exercises 1 through 5.

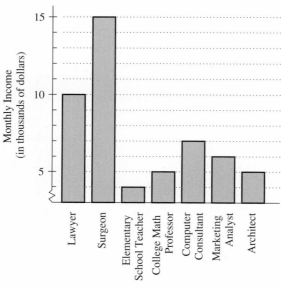

1. What is the monthly income of the marketing analyst?
2. What is the yearly income of the surgeon?
3. How much more per year does the surgeon make than the elementary school teacher?
4. What is the ratio of the lawyer's income to the college professor's income?
5. The bar that represents the college professor's income is twice as high as the bar that represents the elementary school teacher's income. Does this mean the college professor's income is twice the elementary school teacher's income? Explain.

The Computerworld transportation department ships personal computer products from the warehouse to franchise locations and to individuals. The line graph gives the pounds shipped per month for each of five months. Use the graph to answer Exercises 6 through 10.

6. How many pounds of computer products did Computerworld ship in February?
7. How many more pounds were shipped in March than in February?
8. From January to February did the weight of the shipments increase or decrease and by how much?
9. What is the average (mean) pounds per month for the five-month period?

---

1. $6000
2. $180,000
3. $132,000
4. $\frac{2}{1}$
5. The professor's income is $5000 per month and the elementary teacher's income is $4000 per month. Therefore, the professor does not make twice as much as the elementary teacher. Since the squiggle ($\lessgtr$) is used on the vertical axis, the visual heights of the bars cannot be used to compare the incomes by ratios.
6. 0.8 million lb
7. 0.4 million lb
8. decreased by 0.2 million lb
9. 1.16 million lb per mo
10. $33\frac{1}{3}$%
11.

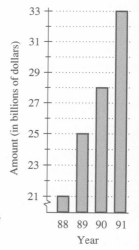

10. From April to May what is the percent increase in the number of pounds shipped?

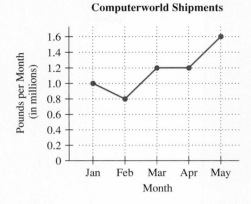

11. In only three years, the flow of products from the United States to Mexico has increased dramatically. The table gives the United States exports to Mexico in each of four years. Make a bar graph that displays this information.

U.S. EXPORTS TO MEXICO	
Year	Amount (in billions of dollars)
1988	21
1989	25
1990	28
1991	33

12. To determine how affordable homes are in a certain area, one can look at the share of income that goes to the mortgage payment. The table gives the percent of income for five different cities in 1992. Make a line graph that displays this information.

HOUSING MARKET AFFORDABILITY COMPARISON	
City	Ratio of Mortgage Payment to Income
Miami, Fl	17%
Atlanta, GA	15%
Washington, D.C.	18%
San Antonio, TX	10%
Mobile, AL	13%
West Palm Beach, FL	19%

13. The number of households of unmarried couples is more than five times what it was in 1970. In 1970 there were 0.5 million such households, 1.25 million in 1975, 1.5 million in 1980, 2.25 million in 1985, and 2.75 million in 1990. Make **a.** a table, **b.** a bar graph, and **c.** a line graph displaying this information. Be sure to label appropriately and give each graph a caption.

12.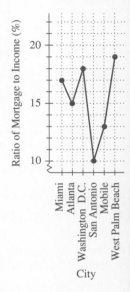

13. b.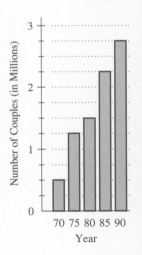

The pictograph shows the average household income in 1992 for several cities. Use the graph to answer Exercises 14 through 17.

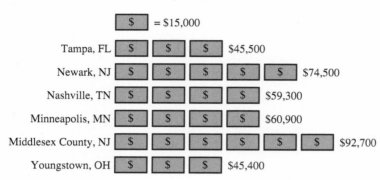

14. How much money does each separate picture represent?
15. Using the pictures, what is the approximate average household income in Newark?
16. Using the pictures, how much more is the average household income in Middlesex County than in Newark?
17. Which of the six cities has the least average household income and what is the income?

Ross wanted to attend a university that is strong in both engineering and business since he was interested in both of these fields. He found a university with 25% of its student body majoring in engineering and 30% majoring in business. This university has a total of 8500 students and the breakdown by majors is given in the circle graph. Use the graph to answer Exercises 18 through 22.

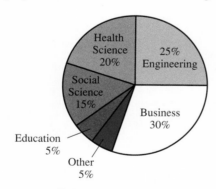

University Student Body Majors
(8500 Total Students)

18. What percent of the students major in health science?
19. What percent of the students major in education?
20. How many students major in engineering?
21. How many students major in business?
22. What is the degree measure of the sector that contains the social science majors? (Do not use a protractor.)

A personal computer company shipped 2 million pounds of computer products during October 1993. There are three ways the products are shipped: by truck, by next-day air freight, or by second-day air freight. The breakdown for the month is given in the circle graph. Use the graph to answer Exercises 23 through 27.

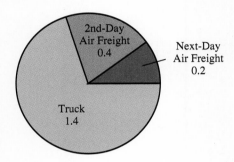

October 1993 Shipment Weights in Millions of Pounds
(Total Weight = 2 Million Pounds)

23. How many pounds were shipped by truck in October 1993?
24. How many pounds were shipped by next-day air freight in October 1993?
25. What percent of the total was shipped by truck?
26. What percent of the total was shipped by next-day air freight?
27. If the company projects a 15% increase in total shipment weight for the coming year, what is the projected total shipment weight for October 1994?

In Exercises 28 and 29, we show a circle with sectors A, B, and C. The percent of the circle that the sector occupies is given. Use this information to find the degree measure of each central angle. (Do not use a protractor.)

28. 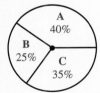 29.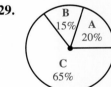

30. Use a protractor to help find the measures of the sectors A, B, and C. It is helpful to trace the figure on a separate piece of paper and extend the sides of the sectors before measuring.

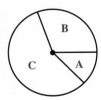

31. Draw a circle with sectors measuring 60°, 145°, and 155°.

For the year 1993 a city had a total budget of $2.5 billion. The table below gives how much money was spent in various categories. Use the table to answer Exercises 32 through 36.

1993 CITY BUDGET OF $2.5 BILLION	
Category	Amount (in millions of dollars)
Public Protection	375
Public Services	750
Health and Welfare	625
Education	500
Recreation and Culture	125
Administration	125
Total	2500

32. What percent of the total budget was spent on health and welfare?
33. What percent of the total budget was spent on education?
34. If a circle graph is made to display this information, what would be the degree measure of the sector containing health and welfare?
35. If a circle graph is made to display this information, what would be the degree measure of the sector containing education?
36. Make a circle graph that displays this information. Label the sectors with the percent of the budget spent in each category.
37. Martha wanted to look carefully at how much wine she was drinking so she decided to record the number of glasses of wine she drank for a 30-day period. Here are the original data.

    Use this data to find **a.** the mean, **b.** the median, and **c.** the mode for the number of glasses of wine consumed. Round off the mean to the nearest tenth of a glass.

GLASSES OF WINE CONSUMED FOR EACH OF 30 DAYS					
0	1	2	1	2	0
3	4	3	2	1	3
0	0	4	2	1	5
0	0	3	4	2	2
1	3	1	0	0	3

38. Mr. Sigman compared the acceleration capability of six different automobiles. He looked at the time it takes for the auto to go from 0 to 60 miles per hour. Here are the times in seconds.

    4.5   6.9   6.6   9.2   7.5   9.1

    Find **a.** the mean, **b.** the median, and **c.** the mode.

In Your Own Words

Write complete sentences to discuss each of the following. Support your comments with examples or pictures, if appropriate.

39. Alonzo wondered how many cigarettes he smoked each day. He recorded the number of cigarettes he smoked for a 14-day period. Here are the original data.

    8   10   7   6   8   12   14
    6   7   8   8   10   8   9

    Discuss the procedure you would use to find the mean of this data.

40. Discuss the procedure you would use to find the median of the data in Exercise 39.
41. Discuss the procedure you would use to find the mode of the data in Exercise 39.
42. Suppose you wanted to make a circle graph with a sector that corresponds to 15% of the circular region. Discuss the procedure you would use to find the measurement of the central angle for this sector.
43. A statistical company took a survey of 800 voters immediately after the 1992 presidential election. The survey asked, "Which one or two issues mattered most in deciding how you voted?" The following bar graph summarizes partial results of the survey. Make several statements about information you can obtain from the graph.

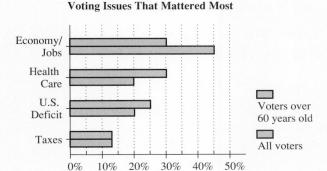

CHAPTER 11 PRACTICE TEST

Chuck was on a diet to lose weight. He recorded his weight on the first day of the month for six consecutive months. The results are given in the line graph. Use the graph to answer Exercises 1 through 4.

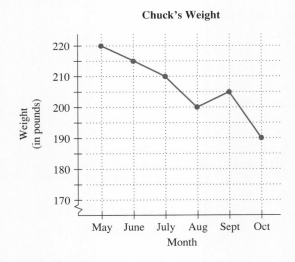

1. How much did Chuck weigh on the first of July?
2. During what month did Chuck gain weight and how much did he gain?
3. How much weight did Chuck lose over the 5-month period?
4. What is the average (mean) of the 6 weights? Round off to the nearest pound.

A newspaper asked 800 of its readers if they approve of the way Congress is handling the economy. The circle graph gives the results of the survey. Use the graph to answer Exercises 5 through 8.

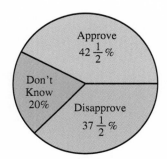

800 Persons Respond to
"Do you approve of how Congress
is handling the economy?"

5. What percent of the persons surveyed disapprove of the way Congress is handling the economy?
6. How many persons answered, "Don't know"?
7. How many persons approve of the way Congress is handling the economy?
8. What is the degree measure of the sector that contains those that approve of the way Congress is handling the economy? Do not use a protractor.
9. A taxi driver looked at how many miles she drove her taxi each day for a week. The results were: Monday, 175 miles; Tuesday, 180 miles; Wednesday, 200 miles; Thursday, 170 miles; Friday, 190 miles; Saturday, 150 miles; and Sunday, 120 miles. Construct a bar graph that displays this information. Label the axes appropriately and give the graph a caption.
10. Ten college students were asked how many hours per day they watch television. The results to the nearest hour were as follows.

$$\begin{array}{ccccc} 1 & 2 & 3 & 4 & 2 \\ 1 & 3 & 1 & 0 & 1 \end{array}$$

Find **a.** the mean, **b.** the median, and **c.** the mode.

11. Due to bad weather most flights were delayed leaving Boston International Airport one morning. An airline employee recorded the delay time for 5 flights. Here are the times to the nearest tenth of an hour.

$$1.8 \quad 0.5 \quad 2.0 \quad 1.2 \quad 1.5$$

Find **a.** the mean, **b.** the median, and **c.** the mode.

CUMULATIVE REVIEW EXERCISES: CHAPTERS 1–11

Evaluate. Answer with a fraction or whole numeral.

1. $4\frac{2}{15} - \frac{8}{9}$
2. $\frac{150}{600} \div 2\frac{1}{12}$
3. $66\frac{2}{3}\%$ of 2100
4. $\frac{3}{4}$ of 900
5. $\frac{\frac{5}{18}}{8\frac{1}{3}}$
6. $14\frac{9}{38} + 27\frac{17}{57}$

Solve.

7. Add 23.8, 78.54, and 460, then divide the sum by 0.04.
8. Multiply 34.08 by 1500, then decrease the result by 82.96.
9. Reduce 18.75 by 2.8%.
10. Increase 380 by 125%.
11. Find the sum of 2 feet and 4 inches.
    a. Express the answer in feet.
    b. Express the answer in inches.
12. Find the difference between $2\frac{1}{4}$ hours and 30 minutes.
    a. Express the answer in minutes.
    b. Expess the answer in hours.
13. Find **a.** the area and **b.** the perimeter of this rectangle.

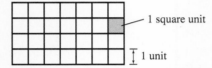

14. Aracely was walking on a Trackmaster machine at the athletic club. The electronic display showed that she had walked 0.6 miles in 12 minutes. How many miles per hour is this?

15. Stan wanted to barbecue a London Broil steak on his charcoal grill. He was having 12 guests for dinner and wanted to serve each person 0.4 pound of steak.
    a. How much London Broil does he need?
    b. If the London Broil costs $3.87 per pound, what is the total cost without tax? (Round up to the nearest cent.)
    c. If the steak costs $2.98 per pound and Stan lives in a state that charges a 6.2% tax on food, what is the total cost of the steak? (Round up to the nearest cent.)

16. A state allocates $1,500,000 for salaries for teaching aides for summer school. If 400 aides, each earning $1200, have already been hired, how many aides earning $850 each can be hired?

17. How many 6.2-centigram doses of medicine can a pharmacist make from 40.3 grams of medicine?

18. On a trip, Lucinda spent 3 days in Indianapolis, 4 days in Seattle, 1 week in Boston, and 2 weeks in Atlanta.
    a. What was the total length of her trip in weeks?
    b. What was the total length of her trip in days?
    c. What fraction of her trip was spent in Indianapolis?
    d. How much longer did she stay in Atlanta than in Seattle?

19. Sixteen adults were asked the question, "How many times have you been married?" Here are the data collected.

0	1	0	2	3	0	1	0
0	1	1	2	4	2	1	2

    a. What is the mean of these data?
    b. What is the median of these data?
    c. What is the mode of these data?

20. Mr. Azcarate recorded the total time he spent on the stationary bicycle each week for a 5-week period. Here are the data collected.

$$\begin{array}{ll} \text{1st week} & 1\tfrac{1}{2} \text{ hours} \\ \text{2nd week} & 2 \text{ hours} \\ \text{3rd week} & 3\tfrac{1}{4} \text{ hours} \\ \text{4th week} & 2\tfrac{3}{4} \text{ hours} \\ \text{5th week} & 3\tfrac{1}{2} \text{ hours} \end{array}$$

a. Construct a bar graph that displays these data. Be sure to label the axes and give the graph a caption.
b. How many hours per week did he average over the 5-week period?
c. What was the increase in time from the fourth week to the fifth week?
d. What was the percent increase in time from the fourth week to the fifth week? (Round off to the nearest whole percent.)

CHAPTER 12

Order of Operation, Signed Numbers, and Variables

12.1 ORDER OF OPERATION

OBJECTIVES
- Understand the different ways to represent multiplication.
- Understand the use of grouping symbols.
- Use the rules for the order to operate to evaluate a numerical expression containing more than one operation.

1 Representing Multiplication in New Ways

For the remainder of the text we use a raised dot, $\cdot$, instead of the multiplication symbol, $\times$, to indicate the operation of multiplication. We change the symbol because letters of the alphabet are used to represent numbers in algebra. To avoid confusion, we do not want to use a symbol for multiplication that resembles a letter of the alphabet. For example,

$$5 \cdot 7 \quad \text{means} \quad 5 \times 7$$

Be sure to raise the dot high enough so that it does not look like a decimal point.

Parentheses are often used to hold a quantity. If we omit the operation between two quantities, when one or both of them is wrapped in parentheses, the operation is understood to be multiplication. For example,

$$(3)(8) \quad \text{means} \quad 3 \times 8$$
$$3(8) \quad \text{means} \quad 3 \times 8$$
$$(3)8 \quad \text{means} \quad 3 \times 8$$

Be careful! It is *not* the parentheses that mean multiplication, it is the absence of an operation symbol that implies multiplication. Here we

Try These Problems

Evaluate.
1. $8 \cdot 7$
2. $4 \cdot 5 \cdot 6$
3. $(4) + (25)$
4. $(7)6$
5. $(3)(6)(2)$
6. $[300] - 19$

show parentheses used in two multiplication problems, as well as an addition problem and a subtraction problem.

$$5(7) = 5 \times 7 = 35$$
$$(12)(9) = 12 \times 9 = 108$$
$$5 + (7) = 5 + 7 = 12$$
$$(12) - (9) = 12 - 9 = 3$$

Brackets, [ ], and braces, { }, can be used in place of parentheses.

 Try Problems 1 through 6.

2 Interpreting Grouping Symbols

We have studied four different operations: addition, subtraction, multiplication, and division. We now look at **numerical expressions** that contain more than one of these operations. For example, consider carefully this numerical expression:

$$3 + 4 \cdot 2$$

Observe that there are two operations: addition and multiplication. If we add first, then multiply, the expression equals 14.

However, if we multiply first, then add, the expression equals 11.

$$3 + 4 \cdot 2 = 3 + 8$$
$$= 11$$

Which number does this numerical expression represent, 14 or 11? The correct answer is 11, and by the end of this section you will understand why.

To avoid ambiguity of this type we establish rules for the order to operate. One way we define the order is to use **grouping symbols:** parentheses, ( ); brackets, [ ]; and braces, { }. We agree that the operation inside the grouping symbol is done first. Here are some examples.

EXAMPLE 1 Evaluate: $9 \cdot (8 + 2)$

SOLUTION There are two operations to perform: addition and multiplication. Do the addition first because the addition problem is inside the parentheses.

$9 \cdot (8 + 2)$
$= 9 \cdot (10)$ Operate inside the grouping
$= 9 \cdot 10$ symbol first.
$= 90$ ■

Try These Problems

Evaluate.
7. $6 \cdot (25 - 13)$
8. $10(5 + 18)$
9. $30 + (5 \cdot 8)$
10. $45 - (35 - 16)$
11. $(6 + 7)8$
12. $10 - [2 \cdot 3]$
13. $(9 + 3)(3 + 7)$
14. $(8 \cdot 9) - (33 - 14)$
15. $[4 + 3][3 \cdot 4]$

EXAMPLE 2 Evaluate: $5(14 - 7)$

SOLUTION There are two operations to perform: subtraction and multiplication. Do the subtraction first because it is inside the parentheses.

$$5(14 - 7)$$
$$= 5(7)$$
$$= 5 \cdot 7$$
$$= 35$$

Operate inside the grouping symbol first.

If you drop the parentheses, be sure to insert a multiplication symbol, because 5(7) means 5 times 7, not 57. ■

EXAMPLE 3 Evaluate: $20 - (8 + 5)$

SOLUTION There are two operations to perform: subtraction and addition. Do the addition first because it is inside the parentheses.

$$20 - (8 + 5)$$
$$= 20 - (13)$$
$$= 20 - 13$$
$$= 7$$

Operate inside the grouping symbol first.

Try Problems 7 through 12.

EXAMPLE 4 Evaluate: $(24 + 6)(9 - 5)$

SOLUTION There are three operations to perform: addition, subtraction, and multiplication. Do the addition and subtraction first because each of them is inside a parentheses.

$$(24 + 6)(9 - 5)$$
$$= (30)(4)$$
$$= 30 \cdot 4$$
$$= 120$$

Operate inside both pairs of parentheses first.

The expression (30)(4) means 30 times 4.

Try Problems 13 through 15.

3 Interpreting the Fraction Bar as a Grouping Symbol

The fraction bar, which indicates division, is understood to be a grouping symbol. For example,

$$\frac{23 - 8}{5} \quad \text{means} \quad \frac{(23 - 8)}{5}$$

Therefore, the subtraction in the numerator is done first, followed by the division. Here we show the process.

$$\frac{23 - 8}{5} = \frac{15}{5}$$
$$= 3$$

Operate within the numerator first.

The fraction bar, as a grouping symbol, means to operate within the numerator and denominator before dividing. Here we show more examples.

Try These Problems

Evaluate.

16. $\dfrac{14}{2+5}$

17. $\dfrac{56-16}{8}$

18. $\dfrac{8+12}{50-10}$

19. $\dfrac{18-3}{3 \cdot 6}$

EXAMPLE 5 Evaluate: $\dfrac{6 \cdot 10}{4+6}$

SOLUTION There are three operations to perform: multiplication, addition, and division. Operate within the numerator and denominator before dividing.

$\dfrac{6 \cdot 10}{4+6}$ Observe that you cannot cancel the 6s here because 6 is not a factor of the denominator, it is a term.

$= \dfrac{60}{10}$

$= 6$ ∎

EXAMPLE 6 Evaluate: $\dfrac{48-12}{12+8}$

SOLUTION There are three operations: subtraction, addition and division. Operate within the numerator and denominator before dividing.

$\dfrac{48-12}{12+8}$ Don't cancel these 12s. They are terms, not factors.

$= \dfrac{36}{20}$ When the division does not come out evenly, leave the answer as a fraction reduced to lowest terms or as a decimal.

$= \dfrac{\cancel{4} \cdot 9}{\cancel{4} \cdot 5}$ Cancel the common factor 4.

$= \dfrac{9}{5}$ or $1\dfrac{4}{5}$ or 1.8 ∎

 Try Problems 16 through 19.

4 Defining the Order of Operation

If we used only grouping symbols to define the order to operate, the number of grouping symbols in an expression would become overwhelming, causing numerical expressions to be awkward to work with. For this reason, we make the following rules to define the **order of operation.**

> *Order of Operation*
> 1. Operate inside grouping symbols, beginning with the innermost.
> 2. Multiply and divide, from left to right.
> 3. Add and subtract, from left to right.

At this time we investigate several aspects of this rule carefully.

12.1 Order of Operation 551

 Try These Problems

Evaluate.
20. $6 + 8 \cdot 4$
21. $7 \cdot 35 - 5 \cdot 8$
22. $18 - 15 \div 3$
23. $9 \cdot 24 - 4$
24. $(7)(3) + (12)(5)$
25. $3 \div 4 + 2 \div 3$

5 Multiplying and Dividing Before Adding and Subtracting

Observe that the rule for the order to operate states that multiplication and division are done before addition and subtraction, unless grouping symbols indicate otherwise. For example, the expression

$$3 + 4 \cdot 2 \quad \text{means} \quad 3 + (4 \cdot 2)$$

The multiplication is performed first, with or without the presence of the parentheses.

$$3 + 4 \cdot 2 = 3 + 8$$
$$= 11$$

Here are more examples.

EXAMPLE 7 Evaluate: $66 - 6 \cdot 9$
SOLUTION There are two operations: subtraction and multiplication. Do the multiplication first.

$$66 - 6 \cdot 9$$
$$= 66 - 54 \qquad \text{Do the multiplication first.}$$
$$= 12 \quad \blacksquare$$

EXAMPLE 8 Evaluate: $20 \cdot 6 + 4 \cdot 12$
SOLUTION There are two multiplications and an addition. Do the multiplications first.

$$20 \cdot 6 + 4 \cdot 12$$
$$= 120 + 48 \qquad \text{Do the multiplications first.}$$
$$= 168 \quad \blacksquare$$

 Try Problems 20 through 23.

EXAMPLE 9 Evaluate: $(4)(3) - 1 \div 2$
SOLUTION There are three operations: multiplication, subtraction, and division. Do the multiplication and division first.

$$(4)(3) - 1 \div 2$$
$$= 12 - \frac{1}{2} \qquad \text{Do the multiplication and division first.}$$
$$= \frac{12}{1} - \frac{1}{2} \qquad \text{Before subtracting write the fractions with a common denominator.}$$
$$= \frac{12 \times 2}{1 \times 2} - \frac{1}{2}$$
$$= \frac{24}{2} - \frac{1}{2}$$
$$= \frac{23}{2} \quad \text{or} \quad 11\frac{1}{2} \quad \text{or} \quad 11.5 \quad \blacksquare$$

 Try Problems 24 and 25.

Try These Problems

Evaluate.
26. $7 + 4(8 - 2)$
27. $9(14) - 6(10 - 8)$
28. $98 - 10(15 - 3 \cdot 2)$

EXAMPLE 10 Evaluate: $95 - 5(1 + 4 \cdot 3)$

SOLUTION How many operations do you count in this expression? There are four: subtraction, addition, and *two* multiplications. Observe that one of the multiplications is between the number 5 and the quantity inside the parentheses.

$95 - 5(1 + 4 \cdot 3)$	This multiplication must be done before the subtraction, so the subtraction must be done last.
$= 95 - 5(1 + 12)$	Begin by operating inside the grouping symbol. Multiply first.
$= 95 - 5(13)$	Add inside the grouping symbol.
$= 95 - 5 \cdot 13$	If you drop the parentheses, be sure to insert the multiplication dot between the numbers 5 and 13.
$= 95 - 65$	Multiply before subtracting.
$= 30$	Finally, we can do the subtraction. ∎

Try Problems 26 through 28.

6 Multiplying and Dividing from Left to Right (→)

When an expression has a mixture of multiplication and division, the rule does not state that one of these operations is done before the other. It depends on where the operation appears in the expression. We operate a mixture of multiplication and division from left to right (→) like you would read a book. For example, in the expression

$$36 \div 6 \cdot 2$$

the division is done first because it appears first as we move from left to right (→). However, in the expression

$$8 \cdot 4 \div 2$$

the multiplication is done first. Here we evaluate each expression.

$36 \div 6 \cdot 2 = 6 \cdot 2 = 12$	Divide first here because the division appears first.
$8 \cdot 4 \div 2 = 32 \div 2 = 16$	Multiply first here because the multiplication appears first.

Also, repeated division is evaluated by moving left to right (→), unless grouping symbols indicate otherwise. Here we show some examples.

EXAMPLE 11 Evaluate: $12 \div 4 \div 4$

SOLUTION The expression contains repeated division, so calculate from left to right (→) as you read a book.

$12 \div 4 \div 4$	
$= 3 \div 4$	First, divide 12 by 4.
$= \dfrac{3}{4}$ or 0.75 ∎	

12.1 Order of Operation 553

 Try These Problems

Evaluate.
29. $40 \div 5 \cdot 2$
30. $100 \div 20 \div 5$
31. $3 \div 8 \cdot 2$
32. $12 \div 4 \cdot 10 \div 2$
33. $7(5 \div 6) \div 15$
34. $(2)(4)(6) \div (8 \cdot 3)(5)$

EXAMPLE 12 Evaluate: $6 \cdot 15 \div 3 \cdot 2$

SOLUTION The expression contains a mixture of multiplication and division, so calculate from left to right ($\rightarrow$).

$$6 \cdot 15 \div 3 \cdot 2$$
$$= 90 \div 3 \cdot 2 \quad \text{First, multiply 6 by 15 to obtain 90.}$$
$$= 30 \cdot 2 \quad \text{Next, divide 90 by 3 to obtain 30.}$$
$$= 60 \quad \blacksquare$$

EXAMPLE 13 Evaluate: $4 \div 8(2 \div 5)$

SOLUTION The expression contains three operations: two divisions and one multiplication. We operate inside the grouping symbol first.

$$4 \div 8(2 \div 5)$$
$$= 4 \div 8\left(\frac{2}{5}\right) \quad \text{First divide 2 by 5 to obtain } \tfrac{2}{5}.$$
$$= 4 \div 8 \cdot \frac{2}{5} \quad \text{If you drop the parentheses, be sure to insert the multiplication dot between the numbers 8 and } \tfrac{2}{5}.$$
$$= \frac{4}{8} \cdot \frac{2}{5} \quad \text{Next divide 4 by 8 because it appears first.}$$
$$= \frac{\overset{1}{\cancel{4}}}{\underset{1}{\cancel{4}} \cdot \cancel{2}} \cdot \frac{\overset{1}{\cancel{2}}}{5} \quad \text{Cancel the common factors 4 and 2.}$$
$$= \frac{1}{5} \quad \blacksquare$$

Although a mixture of multiplication and division or repeated division must be evaluated from left to right ($\rightarrow$), you have more flexibility when calculating repeated multiplication. Recall from Sections 2.1, 4.1, and 7.1 that **repeated multiplication can be done in any order.** For example, here we show an expression of this type evaluated in several different ways.

$$(\underbrace{5)(3}_{})(2)(9) = (15)(2)(9) \quad \text{Multiply 5 by 3.}$$
$$= (30)(9) \quad \text{Multiply 15 by 2.}$$
$$= 270$$

$$(\underbrace{5)(3}_{})(\underbrace{2)(9}_{}) = (15)(18) \quad \text{Multiply 5 by 3 and 2 by 9.}$$
$$= 270$$

$$(\underbrace{5)(3)(2}_{})(9) = (10)(27) \quad \text{Multiply 5 by 2 and 3 by 9.}$$
$$= 270$$

Observe that the last method is the easiest because the calculation can be done mentally.

 Try Problems 29 through 34.

7 Adding and Subtracting from Left to Right (→)

A mixture of addition and subtraction (with no grouping symbols) is evaluated in the same way as a mixture of multiplication and division, that is, from left to right (→) as you would read a book. For example, in the expression

$$25 - 8 + 2$$

the subtraction is done first because it appears first as we move from left to right (→). In the expression

$$3 + 27 - 9$$

the addition is done first because it appears on the left. Here we evaluate each expression.

$25 - 8 + 2 = 17 + 2 = 19$ Subtract first here because the subtraction appears first.

$3 + 27 - 9 = 30 - 9 = 21$ Add first here because the addition appears first.

Also, repeated subtraction is evaluated by moving left to right (→), unless grouping symbols indicate otherwise. Here we show some examples.

EXAMPLE 14   Evaluate: $10 - 7 + 2$

SOLUTION   The expression contains a mixture of addition and subtraction, so calculate from left to right (→) as you read a book.

$10 - 7 + 2$ First, subtract 7 from 10.
$= 3 + 2$
$= 5$ ∎

EXAMPLE 15   Evaluate: $20 - 3 - 4 - 2$

SOLUTION   The expression contains repeated subtraction, so calculate from left to right (→).

$20 - 3 - 4 - 2$
$= 17 - 4 - 2$ First, subtract 3 from 20.
$= 13 - 2$ Next, subtract 4 from 17.
$= 11$ ∎

EXAMPLE 16   Evaluate: $54 - (35 - 17 + 2)$

SOLUTION   The expression contains three operations: two subtractions and one addition. We operate inside the grouping symbol first.

$54 - (35 - 17 + 2)$ Inside the grouping symbol, the subtraction must be done first because it appears first.
$= 54 - (18 + 2)$ First, subtract 17 from 35.
$= 54 - (20)$ Next, add 18 and 2.
$= 34$ ∎

Try These Problems

Evaluate.
35. 31 − 7 + 4
36. 14 − 6 − 2
37. 82 − 10 − 5 − 4
38. 61 − 15 + 8 + 9
39. 40 − 12 + (8 − 2 − 1)
40. (30 − 7) − (11 + 7) + (7 − 5)
41. 95 + 4[20 − 3(6 − 4)]
42. 280 − 2[5(14 − 2 · 3) + 8]
43. 74[37 + 3(8 − 5 + 2)]

Although a mixture of addition and subtraction or repeated subtraction must be evaluated from left to right (→), you have more flexibility when calculating repeated addition. Recall from Section 1.2 that **repeated addition can be done in any order.** For example, here we show an expression of this type evaluated in several different ways.

$$7 + 5 + 8 + 3 = 12 + 8 + 3 \quad \text{Add 7 and 5.}$$
$$= 20 + 3 \quad \text{Add 12 and 8.}$$
$$= 23$$

$$7 + 5 + 8 + 3 = 10 + 13 \quad \text{Add 7 to 3 and 5 to 8.}$$
$$= 23$$

$$7 + 5 + 8 + 3 = 12 + 11 \quad \text{Add 7 to 5 and 8 to 3.}$$
$$= 23 \ \blacksquare$$

 Try Problems 35 through 40.

8 Working with Grouping Symbols Within Grouping Symbols

Sometimes we encounter expressions that contain grouping symbols within grouping symbols. In this case, we operate inside the innermost grouping symbol first. Here we look at some examples.

EXAMPLE 17 Evaluate: $84 - 5[15 + 3(7 - 3)]$

SOLUTION Observe that there are two pairs of grouping symbols one within the other. We begin by operating inside the innermost.

$$184 - 5[15 + 3(7 - 3)]$$
$$= 184 - 5[15 + 3(4)] \quad \text{Subtract 3 from 7.}$$
$$= 184 - 5[15 + 12] \quad \text{Multiply 3 by 4.}$$
$$= 184 - 5[27] \quad \text{Add 15 and 12.}$$
$$= 184 - 135 \quad \text{Multiply 5 by 27.}$$
$$= 49 \ \blacksquare$$

EXAMPLE 18 Evaluate: $60 - [38 - (13 - 8)]$

SOLUTION Notice the parentheses that are within the brackets. Operate inside the parentheses first because they are innermost.

$$60 - [38 - (13 - 8)]$$
$$= 60 - [38 - 5] \quad \text{Subtract 8 from 13.}$$
$$= 60 - 33 \quad \text{Subtract 5 from 38.}$$
$$= 27 \ \blacksquare$$

 Try Problems 41 through 43.

9 Using the Order of Operation

At this time we restate the rule for the order to operate and look at some miscellaneous examples.

> ### Order of Operation
> 1. Operate inside grouping symbols beginning with the innermost.
> 2. Multiply and divide, from left to right.
> 3. Add and subtract, from left to right.

EXAMPLE 19 Evaluate: $(6)(8) - (2)(1) + (5)(3)(2)$

SOLUTION

$(6)(8) - (2)(1) + (5)(3)(2)$
$= 48 - 2 + 30$ Multiply first.
$= 46 + 30$ Subtract next.
$= 76$ ∎

EXAMPLE 20 Evaluate: $78 - 2(7 \cdot 3 + 6 \cdot 2)$

SOLUTION Be careful! Outside the parentheses there are two operations: subtraction and multiplication. Because multiplication is done before subtraction, this subtraction must be done last.

$78 - 2(7 \cdot 3 + 6 \cdot 2)$ This multiplication must be done before the subtraction, so save the subtraction until last.
$= 78 - 2(21 + 12)$ Multiply inside the grouping symbol first.
$= 78 - 2(33)$ Next, add inside the grouping symbol.
$= 78 - 66$ Multiply before subtracting.
$= 12$ ∎

EXAMPLE 21 Evaluate: $23 + 7[15 - 3(6 - 3 + 2)]$

SOLUTION Notice that there are parentheses within brackets. Begin operating inside the parentheses because they are innermost.

$23 + 7[15 - 3(6 - 3 + 2)]$
$= 23 + 7[15 - 3(3 + 2)]$ Subtract 3 from 6.
$= 23 + 7[15 - 3(5)]$ Add 3 and 2.
$= 23 + 7[15 - 15]$ Multiply 3 by 5.
$= 23 + 7[0]$ Subtract 15 from 15.
$= 23 + 0$
$= 23$ ∎

Try These Problems

Evaluate.

44. $(4)(12)(5) - (3)(6) - (2)(4)$
45. $5(8 - 2)(10 - 5 + 2)$
46. $80 + 4(45 - 15 \cdot 2)$
47. $7 \cdot 3 - 2(13 - 5 - 3)$
48. $\dfrac{2(5 - 3)}{4 - 2 + 2(3)}$
49. $68 - 2[4 \cdot 7 - 5(12 - 2 \cdot 4)]$

EXAMPLE 22 Evaluate: $\dfrac{14 + 4 \cdot 4}{5(15 - 2 + 7)}$

SOLUTION

$\dfrac{14 + 4 \cdot 4}{5(15 - 2 + 7)}$

$= \dfrac{14 + 16}{5(13 + 7)}$ Multiply first in the numerator. Subtract first in the denominator.

$= \dfrac{30}{5(20)}$

$= \dfrac{30}{100}$

$= \dfrac{3 \cdot \cancel{10}}{10 \cdot \cancel{10}}$ Cancel the common factor 10.

$= \dfrac{3}{10}$ ∎

Try Problems 44 through 49.

Answers to Try These Problems

1. 56 2. 120 3. 29 4. 42 5. 36 6. 281 7. 72 8. 230
9. 70 10. 26 11. 104 12. 4 13. 120 14. 53 15. 84
16. 2 17. 5 18. $\frac{1}{2}$ 19. $\frac{5}{6}$ 20. 38 21. 205 22. 13 23. 212
24. 81 25. $\frac{17}{12}$ or $1\frac{5}{12}$ 26. 31 27. 114 28. 8 29. 16 30. 1
31. $\frac{3}{4}$ 32. 15 33. $\frac{7}{18}$ 34. 10 35. 28 36. 6 37. 63 38. 63
39. 33 40. 7 41. 151 42. 184 43. 3848 44. 214 45. 210
46. 140 47. 11 48. $\frac{1}{2}$ 49. 52

EXERCISES 12.1

Evaluate.

1. $6 \cdot 8$
2. $4 \cdot 2 \cdot 3$
3. $(9)5$
4. $(9) + 5$
5. $(8) + (3) + (5)$
6. $(8)(3)(5)$
7. $7(5 + 4)$
8. $[15 - 11] - [13 - 9]$
9. $1 - (1 \div 3)$
10. $(4 + 13)(10 - 4)$
11. $\dfrac{8 + 4}{8 - 3}$
12. $\dfrac{3 \cdot 6 \cdot 5}{12 - 7}$
13. $12 - 5 \cdot 2$
14. $25 + 5 \div 5$
15. $8 \cdot 3 + 7 \cdot 4$
16. $(20)(4) - (2)(6)(4)$
17. $1 \div 2 + 4 \div 5$
18. $3 \cdot 4 - 2 \div 3$
19. $48 + 2(13 - 5)$
20. $43 - 3(2 + 3 \cdot 4)$
21. $[2 + 4 \cdot 5][15 - 3 \cdot 5]$
22. $[16 - 6 + 2][4 + 2 \cdot 3]$
23. $80 \div 5 \cdot 2$
24. $360 \div 12 \div 6$

25. $63 - 7 + 3$
26. $35 - 8 - 5$
27. $240 \div 24 \cdot 2 \div 4$
28. $56 \div 2 \div 2 \div 7$
29. $75 - 18 - 13 - 5$
30. $60 - 14 + 16 - 4 + 3$
31. $32 \div 8(3 \div 5)$
32. $6 \cdot (8 \div 3) \cdot 4$
33. $32 - (4 + 7) - 3$
34. $(9 - 5 - 4) + (18 - 6) - (16 - 9)$
35. $(8)(4) - (3)(2) + (2)(7)$
36. $(5)(9) - (3)(3) - (2)(5)(3)$
37. $76 + 5(14 - 5 \cdot 2)$
38. $300 - 5(4 + 5 \cdot 6)$
39. $64 - 4(3)(2)$
40. $81 - 4(5)(2)$
41. $2(6 - 3) + 3(9 + 4) - 7$
42. $25(16 - 7) - 10(4 + 6 + 1)$
43. $(19 - 5)(8 + 2)(13 - 5)$
44. $8(9 - 2 + 1)(16 - 4 - 4) + 12$
45. $200 - 30(15 - 6 - 3)$
46. $4(2)(6) - 8(16 - 3 \cdot 5) + 6(4 \cdot 5)$
47. $\dfrac{4 \cdot 5 - 6}{3 - 2 + 4}$
48. $\dfrac{(6)(8)(6)}{6 + 8 + 6}$
49. $\dfrac{22 - 6 - 2}{5(7)}$
50. $\dfrac{8(4 + 5)}{8 + 2 \cdot 7}$
51. $3 \div 4 + 5 \div 8$
52. $6 \cdot 2 \div 5 - 1 + 5$
53. $6(1 \div 2) + (4 \div 3)$
54. $4 + 5(7 \div 8) - 1 \div 3$
55. $(4)(6) \div 9(6) - 3$
56. $45 \div 9(5) \div 100 \cdot 12$
57. $450 - 4(9)(30 - 14 - 5)$
58. $(15 - 8) - (9 + 5) \cdot (2 \div 7)$
59. $(1 \div 3) \cdot 5(9 - 2 - 1) + 50$
60. $(4 + 2) \div (5 + 4) \div (7 - 1)$
61. $150 - 4[32 - 3(5 + 2)]$
62. $97 - [28 - (16 - 9)]$
63. $95 + 5[15 + (13 - 5 + 2)]$
64. $8(6)[15 - 2(14 - 8 - 5) - 1]$

USING THE CALCULATOR #11

CALCULATORS AND THE ORDER OF OPERATION

To compute the expression $5 \cdot 6 + 3 \cdot 8$, we multiply first, then add. The result is 54. On a scientific calculator you can enter the numbers and operations as they appear from left to right because the scientific calculator is programmed to use the rules for the order to operate. Here we show how this expression is computed on a scientific calculator.

ON A SCIENTIFIC CALCULATOR

```
To Compute    5 · 6 + 3 · 8
Enter         5 [×] 6 [+] 3 [×] 8 [=]
Result        54.
```

If you enter this same sequence of key strokes on a basic calculator, the result is 264, which is not the value of the expression. A basic calculator is not programmed to use the order to operate. It performs the operations as they are entered. To obtain the result 264, the basic calculator adds the number 3 to 30, then multiplies 33 by 8. To compute this expression correctly on a basic calculator you can use the memory keys. Here we show one way to do this. Clear the memory before beginning.

```
To Compute    5 · 6 + 3 · 8
Enter         5 [×] 6 [M+] 3 [×] 8
              [M+] [MR]
Result        54.
```

Cont. page 559

Calculator Problems

Compute the following. Answer with a decimal. If the decimal representation of the answer goes more than 3 places past the decimal point, then round off at 3 decimal places. If you are using a basic calculator, you may need to use the memory keys to avoid writing down intermediate steps.

1. $45 + 8(15)$
2. $800 - 5(19)$
3. $36(7) + 235$
4. $7(14) - 8(56)$
5. $94(87) - 35(74)$
6. $\frac{14}{2031} + \frac{15}{570}$
7. $\frac{9}{19} - \frac{3}{70}$
8. $9(45) - 3(54) + 6(28)$
9. $4.5(3) + 0.6(94) - 3(5.6)$
10. $20 - \frac{12}{17}$
11. $\frac{3}{7} \cdot \frac{18}{46}$
12. $\frac{11}{5} \div \frac{2}{3}$

USING THE CALCULATOR #12

THE PARENTHESES KEYS

Most scientific calculators have parentheses keys like $[($ and $)]$. These keys can be helpful when computing expressions that involve grouping symbols. Here we show an example.

To Compute $64 - (50 - 8)$
Enter $64\ [-]\ [(\ 50\ [-]\ 8\ )]\ [=]$
Result   22.

The parentheses keys can also be used to compute $\frac{715}{40 + 25}$. Since the fraction bar serves as a grouping symbol, the addition in the denominator is done first. We enter the appropriate parentheses keys before entering 40 and after entering 25. Here we show the procedure.

To Compute $\frac{715}{40 + 25}$
Enter $715\ [\div]\ [(\ 40\ [+]\ 25\ )]\ [=]$
Result   11.

Basic calculators do not have parentheses keys. However, the above expressions can be computed on a basic calculator by using the memory keys. Refer to Using the Calculator #4, #5, #9, and #11 for instruction on how to use the memory keys.

Calculator Problems

Compute each of the following by using the parentheses keys or the memory keys. Answer with a decimal. If the decimal representation of the answer goes beyond 3 decimal places, then round off at 3 decimal places.

1. $40 - (34 - 6)$
2. $150 - 18(4)$
3. $475 - 0.3(475)$
4. $56.8 - (46 + 1.65)$
5. $\frac{876}{65 + 8}$
6. $\frac{1500}{(40)(75)}$
7. $\frac{15 - 3.6}{30 - 1.8}$
8. $\frac{7.5 + 48}{50.1 - 35.98}$
9. Find $750 reduced by 15%.
10. Find $25.50 decreased by 8%.

12.2 AN INTRODUCTION TO SIGNED NUMBERS

OBJECTIVES
- Write a signed number for a given situation.
- Give the integers that are associated with given points on a number line.
- Compare integers.
- Evaluate numerical expressions that involve the opposite of a number and the absolute value of a number.

1 Explaining the Need for Signed Numbers

One use of signed numbers is to measure temperature. Here we show a Fahrenheit scale and a Celsius scale with some familiar temperatures.

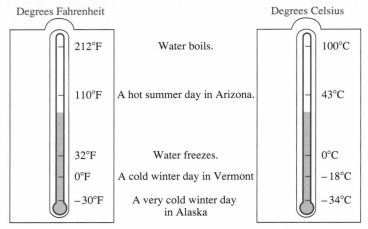

The temperature is much colder at $-30°F$ than at $30°F$.

$-30°F$ means 30 degrees *below* zero on the Fahrenheit scale
$30°F$ means 30 degrees *above* zero on the Fahrenheit scale

The numbers 30 and -30 are two different **signed numbers.**

$+30$ or 30 is read   "positive thirty" or "thirty"
-30 is read   "negative thirty"

The positive sign, $+$, is usually omitted.

In general, signed numbers are needed to measure quantities that involve not only magnitude, but two different directions. Using a signed number allows one to express both the magnitude and the direction with one single number. Here are some more situations that correspond to signed numbers.

Situation	*Signed Number*
You receive $20.	$+20$ or 20
You pay $20.	-20
A gain of 15 yards.	$+15$ or 15
A loss of 15 yards.	-15
700 feet above sea level.	$+700$ or 700
700 feet below sea level.	-700

Try These Problems

Write a signed number for each.
1. You deposit $55.
2. 8° below zero on the Celsius scale.
3. 40 feet below sea level.
4. A gain of 22 yards.

 Try Problems 1 through 4.

2 Integers on the Number Line

Numbers like $+4$, $+36$, 70, $\frac{3}{4}$, and 0.54 are called **positive numbers.** Numbers like -6, -47, -820, $-4\frac{1}{2}$, and -8.3 are called **negative numbers.**

The set of numbers

$$\ldots, -6, -5, -4, -3, -2, -1, 0, 1, 2, 3, 4, 5, 6, \ldots$$

is called the **integers.** Numbers like 1, 19, 50, and 200 are called **positive integers.** Numbers like -1, -7, -25, and -300 are called **negative integers.** The number 0 is considered neither positive nor negative.

We picture signed numbers on a number line. Here we show a portion of a number line with some integers labeled.

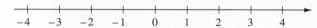

The positive integers are placed to the right of the number 0, and the negative integers to the left of 0. For example, the number 3 is placed 3 units to the right of 0 while -3 is placed 3 units to the left of 0.

EXAMPLE 1 Give the integers associated with points A, B, and C.

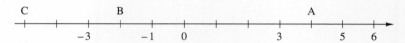

SOLUTION First observe that the distance between consecutive tick marks is 1 unit.

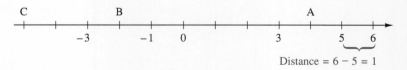

Now we fill in the missing numbers. The positive numbers are placed to the right of 0 and the negative numbers are placed to the left of 0.

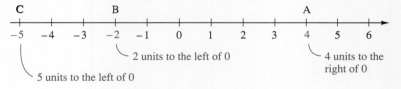

A. 4   B. -2 C. -5 ∎

Try These Problems

Give the numbers associated with points A and B.

5.
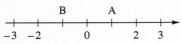

6.
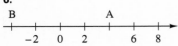

7.

EXAMPLE 2 Give the integers associated with points A, B, and C.

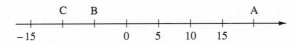

SOLUTION First observe that the distance between consecutive tick marks is 5 units.

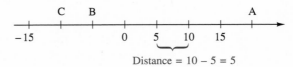

Now we fill in the missing numbers. The positive numbers are placed to the right of 0 and the negative numbers are placed to the left of 0.

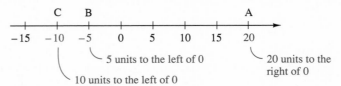

A. 20 B. −5 C. −10. ■

Try Problems 5 through 7.

3 Comparing Numbers

An advantage to picturing numbers on the number line is that it helps to compare them. The arrow, located on the right end of the number line and pointing to the right (→), indicates that the numbers get larger as you move to the right.

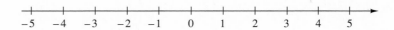

For example, 4 is more than 1 (4 > 1) because 4 is to the right of 1 on the number line. Also, −3 is less than 2 (−3 < 2) because −3 is to the left of 2 on the number line.

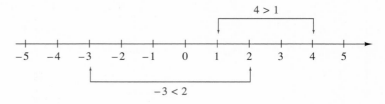

Two equal numbers occupy the *same* position on the number line.

Try These Problems

8. Which is more, −5 or −8?
9. Which is less, −30 or 0?

In Problems 10 through 12, use the symbol <, =, or > to compare each pair of numbers.
10. 12, 9
11. 15, −18
12. −65, −34
13. List from smallest to largest: 4, −7, −15, 0

Give the value of each.
14. |12|
15. |−3|
16. |0|
17. |−25|

For example, −1.5 and −1½ have the same position on the number line, halfway between −1 and −2.

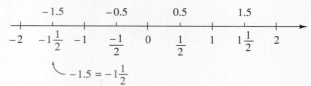

 Try Problems 8 through 12.

EXAMPLE 3 List from smallest to largest: −2, 5, 0, −6

SOLUTION First we picture the numbers on the number line.

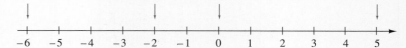

The numbers get larger as we move from left to right (→). Listed from smallest to largest, we have −6, −2, 0, and 5. ■

 Try Problem 13.

4 Determining the Absolute Value of a Number

The numbers 4 and −4 are different numbers, but they have something in common. Both of them are a distance of 4 units from 0 on the number line.

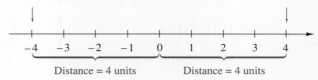

We say 4 and −4 have the same **absolute value** (or **magnitude**). The absolute value of each number is 4. In general, *the absolute value of a number is the distance that number is from 0 on the number line.* Here are more examples of taking the absolute value of a number.

The absolute value of 8 is 8.

The absolute value of −8 is 8.

The absolute value of 0 is 0.

Observe that *the absolute value of a number is never negative.*

The symbol, | |, is used to mean *the absolute value of.* The number that we are taking the absolute value of is put between the vertical bars. For example,

|8| = the absolute value of 8 = 8

|−8| = the absolute value of −8 = 8

|0| = the absolute value of 0 = 0

 Try Problems 14 through 17.

Try These Problems

Give the value of each.

18. $-(4)$
19. $-(-9)$
20. $-(0)$
21. $-(-34)$
22. $-(-(7))$
23. $-(-(-16))$
24. $-|-6|$
25. $-|3|$

5 Determining the Opposite of a Number

The numbers 4 and -4 are the same distance from 0 on the number line, but on opposite sides.

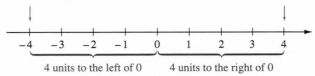

4 units to the left of 0   4 units to the right of 0

We say that 4 is the **opposite of** -4, and -4 is the opposite of 4. In general, *to take the opposite of a number means to change its sign.* Here are more examples of taking the opposite of a number.

The opposite of 8 is -8.

The opposite of -8 is 8.

The opposite of 0 is 0.

Observe that 0 is the only number that is the opposite of itself.

The negative sign, $-$, is used to mean *the opposite of*. For example,

$-(8) =$ the opposite of $8 = -8$

$-(-8) =$ the opposite of $-8 = 8$

$-(0) =$ the opposite of $0 = 0$

Try Problems 18 through 21.

EXAMPLE 4   Evaluate: $-(-(-12))$

SOLUTION   Begin from within and work out.

$-(-(\underline{-12}))$

$= -(12)$ First, take the opposite of -12 to obtain 12.

$= -12$ The opposite of 12 is -12. ■

EXAMPLE 5   Evaluate: $-|-18|$

SOLUTION   First, apply the absolute value to the number inside the vertical bars.

$-|\underline{-18}|$

$= -(18)$ Take the absolute value of -18 to obtain 18.

$= -18$ The opposite of 18 is -18. ■

Try Problems 22 through 25.

Answers to Try These Problems

1. 55   2. -8 3. -40 4. 22   5. A. 1   B. -1
6. A. 4   B. -4 7. A. 10   B. -10 8. -5
9. -30 10. $>$ 11. $>$ 12. $<$ 13. $-15, -7, 0, 4$
14. 12   15. 3   16. 0   17. 25   18. -4 19. 9   20. 0
21. 34   22. 7   23. -16 24. -6 25. -3

EXERCISES 12.2

Write a signed number for each.
1. You pay $120.
2. You receive $120.
3. 400 feet above sea level.
4. 400 feet below sea level.
5. 25°F below zero.
6. 25°F above zero.
7. A gain of 8 yards.
8. A loss of 8 yards.
9. The absolute value of −6.
10. The opposite of −6.
11. The opposite of 32.
12. The absolute value of 32.

Give the integers associated with points B and C.

13.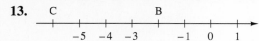

14.

15.

16.

Use the symbol <, =, or > to compare each pair of numbers.
17. 5, 13
18. 24, −30
19. −17, −75
20. 0, 26
21. −5, 0
22. −27, 3
23. Which is more, 16 or −24?
24. Which is less, −35 or −13?
25. List from smallest to largest: −13, −67, 0, −36
26. List from smallest to largest: 19, −14, −24, 0
27. Which has larger absolute value, −17 or −8?
28. Which has larger absolute value, 68 or −85?

Give the value of each.
29. $|25|$
30. $-(25)$
31. $-(-14)$
32. $|-14|$
33. $|-29|$
34. $-(-29)$
35. $-(62)$
36. $|62|$
37. $|0|$
38. $-(0)$
39. $-|17|$
40. $-|-17|$
41. $-(-(-18))$
42. $-(-(-(-45)))$
43. $-|-(-5)|$
44. $-(-|-13|)$

12.3 ADDITION OF SIGNED NUMBERS

OBJECTIVES
- Add two or more signed numbers.
- Solve an application that involves adding integers.

1 Adding Two Signed Numbers

Recall that to add means to total or combine. When adding signed numbers, we want to preserve this meaning of addition. We begin by putting an interpretation on the signed numbers so that you understand

Try These Problems

Add.
1. $-9 + (-6)$
2. $-13 + (-4)$
3. $24 + 36$
4. $-24 + (-36)$

why we add them the way we do. A positive number corresponds to winning money, a negative number corresponds to losing money, and the number 0 corresponds to neither winning nor losing. For example,

8   You win $8.
−8   You lose $8.
0   You win nothing and you lose nothing.

Here we show six different types of addition problems that one can encounter and an explanation of how to obtain each answer.

$8 + 5 = 13$ You win $8. You win $5. Total amount *won* is $13.

$-8 + (-5) = -13$ You lose $8. You lose $5. Total amount *lost* is $13.

$8 + (-5) = 3$ You win $8. You lose $5. The combined result is that you *won* $3.

$-8 + 5 = -3$ You lose $8. You win $5. The combined result is that you *lost* $3.

$8 + 0 = 8$ You win $8. You win nothing and lose nothing. Total *won* is $8.

$-8 + 0 = -8$ You lose $8. You win nothing and lose nothing. Total *lost* is $8.

If you study these six different examples carefully, you see that we can group them into only three different types of problems: adding numbers with like signs, adding numbers with unlike signs, and adding a number to zero. Here we look at each type carefully so that you can discover a procedure for adding two signed numbers that is not dependent on putting an interpretation on the numbers.

Type #1—Adding numbers with *like* signs.

$$8 + 5 = 13$$
$$-8 + (-5) = -13$$

Observe that in both cases the 13 can be obtained by adding 8 and 5, the absolute values of the two numbers. The sign of the answer is the same as the sign of the numbers being added. Here are more examples.

$7 + 12 = 19$ Add 7 and 12, then attach the like sign.

$-7 + (-12) = -19$ Add 7 and 12, then attach the like sign.

$-20 + (-30) = -50$ Add 20 and 30, then attach the like sign.

 Try Problems 1 through 4.

Try These Problems

Add.
5. $10 + (-3)$
6. $-10 + 3$
7. $-50 + 16$
8. $50 + (-16)$
9. $33 + (-14)$
10. $-33 + 14$
11. $0 + (-9)$
12. $0 + 9$

Type #2—Adding numbers with *unlike* signs.
$$8 + (-5) = 3$$
$$-8 + 5 = -3$$

Observe that in both cases the 3 can be obtained by subtracting 5 from 8, the smaller absolute value from the larger absolute value. The sign of the answer is the same as the sign of the number with the larger absolute value. Here are more examples.

$7 + (-12) = -5$ Subtract 7 from 12 to obtain 5, then attach the sign of -12 because the absolute value of -12 is more than the absolute value of 7.

$-7 + 12 = 5$ Subtract 7 from 12 to obtain 5, then attach the sign of 12 because the absolute value of 12 is more than the absolute value of -7.

$20 + (-30) = -10$ Subtract 20 from 30 to obtain 10, then attach the sign of -30 because the absolute value of -30 is more than the absolute value of 20.

$-20 + 30 = 10$ Subtract 20 from 30 to obtain 10, then attach the sign of 30 because the absolute value of 30 is more than the absolute value of -20.

When adding two numbers with unlike signs, we use subtraction to obtain the answer, but it is important for you to realize that the operation being performed is *addition*, not subtraction.

 Try Problems 5 through 10.

Type #3—Adding a number to zero.
$$8 + 0 = 8$$
$$-8 + 0 = -8$$

Observe that any number added to zero is that number.

 Try Problems 11 and 12.

Now we summarize the previous discussion by giving a rule for adding two signed numbers.

To Add Two Signed Numbers
1. **With Like Signs:** Add the absolute values. Attach the like sign.
2. **With Unlike Signs:** Subtract the smaller absolute value from the larger absolute value. Attach the sign of the number with the larger absolute value.
3. **With Zero:** Any number added to zero is that number.

Try These Problems

Add.
13. $-3 + (-2)$
14. $-5 + 7$
15. $8 + (-15)$
16. $-27 + 0$
17. $-62 + 25$
18. Add in *three* different ways:
$-5 + (-9) + 12 + (-24)$

Here are some more examples.

EXAMPLE 1 Add: $-45 + 18$

SOLUTION Observe that we are adding numbers with *unlike* signs. The first number is *negative* while the second number is *positive*.

$$-45 + 18$$
$$= -27$$

Subtract 18 from 45 to obtain 27. Attach the sign of -45 because -45 has larger absolute value than 18. ∎

EXAMPLE 2 Add: $-26 + (-14)$

SOLUTION Observe that we are adding numbers with *like* signs. Both of the numbers are negative.

$$-26 + (-14)$$
$$= -40$$

Add 26 to 14 to obtain 40. Attach the negative sign, the like sign. ∎

 Try Problems 13 through 17.

2 Adding More Than Two Numbers

You know that repeated addition of *positive numbers* can be done in any order. Can we evaluate repeated addition of *signed numbers* in any order? Here we do a problem in three ways.

1. $-6 + 14 + (-9) + 3 = 8 + (-9) + 3$ Here we add from left to right.
$= -1 + 3$
$= 2$

2. $-6 + 14 + (-9) + 3 = -15 + 17$ Here we add numbers with like signs first.
$= 2$

3. $-6 + 14 + (-9) + 3 = 8 + (-6)$ Here we add the first two numbers, the last two numbers, then add these results.
$= 2$

Observe that the answer is 2 no matter in what order we do the addition. In general, **repeated addition of signed numbers can be done in any order.**

 Try Problem 18.

Because it is easier to add numbers with like signs than to add numbers with unlike signs, it is often convenient to begin a repeated addition problem by adding numbers with like signs. Here we show an example.

EXAMPLE 3 Add: $34 + (-16) + 13 + (-5)$

SOLUTION This addition problem can be done in any order, but for convenience we add the two positive numbers and the two negative numbers first.

$$34 + (-16) + 13 + (-5)$$
$$= 47 + (-21)$$ Add like signs first.
$$= 26$$ ∎

Try These Problems

Add.
19. 13 + (−15) + (−4)
20. −16 + 23 + (−4) + (−12)
21. 27 + (−35) + (−8) + (−27)
22. 13 + (−24) + 13 + 24 + (−15)

Solve by using addition of signed numbers.
23. George's bank balance was −$35. He deposited $120, withdrew $43, then withdrew another $67. What is his balance now?

 Try Problems 19 and 20.

Here are some examples of adding a number to its opposite.

$$-8 + 8 = 0$$
$$17 + (-17) = 0$$
$$-67 + 67 = 0$$

Observe that **any number added to its opposite yields zero.** We can take advantage of this fact when doing a repeated addition problem that involves adding opposites. Here we show an example.

EXAMPLE 4 Add: $38 + (-36) + (-13) + 36 + (-28)$

SOLUTION Observe that the opposites −36 and 36 are in the list of numbers to be added, so we add these first.

$$38 + (-36) + (-13) + 36 + (-28)$$
$$= 38 + (-13) + (-28) \quad \text{Begin by adding } -36 \text{ and 36 to obtain 0.}$$
$$= 38 + (-41) \quad \text{Next, add the two negative numbers.}$$
$$= -3 \quad \blacksquare$$

 Try Problems 21 and 22.

3 Solving Applications by Using Addition of Signed Numbers

Now we look at applying the addition of signed numbers to real-life situations.

EXAMPLE 5 The temperature was −25°F early one winter morning in Anchorage, Alaska. During the day, the temperature increased 8°, decreased 12°, then increased 19°. What was the temperature after these changes?

SOLUTION We can interpret a change in temperature as addition of a signed number as follows.

increase of 8°   corresponds to   adding 8
decrease of 12°   corresponds to   adding −12
increase of 19°   corresponds to   adding 19

$$\text{Final temperature} = \text{original temperature} + 8° \text{ increase} + 12° \text{ decrease} + 19° \text{ increase}$$
$$= -25 + 8 + (-12) + 19$$
$$= -37 + 27 \quad \text{Add } -25 \text{ and } -12 \text{ to obtain } -37. \text{ Add 8 and 19 to obtain 27.}$$
$$= -10$$

The final temperature is −10°F. ∎

 Try Problem 23.

Try These Problems

Solve by using addition of signed numbers.

24. You are walking along a number line and your starting position is 6. You walk 12 units to the left, 7 units to the right, then 18 units to the left. What is your final position?

EXAMPLE 6 You are walking along a number line. Your position is the number that you are standing on. You begin at the position -4, then walk 5 units to the left and 8 units to the right. What is your position after these movements?

SOLUTION First, we view the situation on a number line.

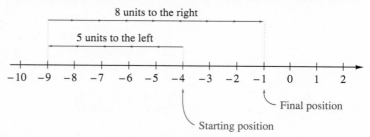

Second, we solve the problem by using addition of signed numbers. A change in position on the number line can be interpreted as addition of a signed number as follows.

Moving 5 units to the left corresponds to adding -5

Moving 8 units to the right corresponds to adding 8

$$\begin{aligned}\text{Final position} &= \text{starting position} + \text{5 units to the left} + \text{8 units to the right} \\ &= -4 + (-5) + 8 \\ &= -9 + 8 \\ &= -1\end{aligned}$$

Your final position is -1. ∎

 Try Problem 24.

 Answers to Try These Problems

1. -15 2. -17 3. 60 4. -60 5. 7 6. -7 7. -34
8. 34 9. 19 10. -19 11. -9 12. 9 13. -5 14. 2
15. -7 16. -27 17. -37 18. -26 19. -6 20. -9
21. -43 22. 11 23. $-\$25$ 24. -17

EXERCISES 12.3

Add.

1. $-15 + 8$
2. $17 + (-4)$
3. $-5 + (-7)$
4. $-3 + (-4)$
5. $14 + (-9)$
6. $-18 + 8$
7. $-10 + (-15)$
8. $-18 + (-5)$
9. $12 + (-12)$
10. $-26 + 26$
11. $0 + (-34)$
12. $35 + 0$
13. $32 + (-24)$
14. $(-65) + (-19)$
15. $-75 + 38$
16. $-45 + (-17)$
17. $200 + (-145)$
18. $-111 + 85$
19. $-458 + (-235)$
20. $-763 + (-383)$
21. $-5 + (-6) + 8$
22. $3 + (-4) + 7$

23. $16 + (-9) + (-6) + 5$
24. $-7 + 23 + (-4) + (-15)$
25. $-43 + 17 + (-16) + (-7)$
26. $-18 + 14 + 54 + (-32)$
27. $19 + (-54) + (-19) + 48 + 7$
28. $-23 + 67 + (-64) + 23 + (-5)$
29. $175 + (-96) + (-150)$
30. $-260 + (-65) + 85$

31.
$$\begin{array}{r} -24 \\ -5 \\ 10 \\ +8 \\ \hline \end{array}$$

32.
$$\begin{array}{r} -24 \\ -36 \\ 30 \\ +26 \\ \hline \end{array}$$

33.
$$\begin{array}{r} 36 \\ -54 \\ -50 \\ +16 \\ \hline \end{array}$$

34.
$$\begin{array}{r} 18 \\ -35 \\ 35 \\ +-84 \\ \hline \end{array}$$

Solve by using addition of signed numbers.

35. The temperature was $-15°F$ one winter evening in Jackson Hole, Wyoming. During the next 24 hours, the temperature increased 6°, decreased 9°, then decreased 13°. What was the temperature after these changes?

36. Nicole's bank balance was $58. She deposited $45, withdrew $38, then withdrew $75. What is her bank balance now?

37. You are walking along a number line and your starting position is -7. You walk 4 units to the left, 8 units to the right, then 15 units to the left. What is your final position?

38. On four successive football plays, the Chicago Bears lost 14 yards, gained 8 yards, lost 6 yards, then gained 16 yards. Represent the outcome of these four plays with a signed number. Interpret the answer.

12.4 SUBTRACTION OF SIGNED NUMBERS

OBJECTIVES
- Subtract two signed numbers.
- Translate an English phrase involving subtraction to math symbols.
- Find the distance between two integers on the number line by using subtraction of integers.
- Solve an application that involves subtraction of integers.
- Evaluate a numerical expression that involves repeated subtraction or subtraction mixed with addition.

1 Subtracting Two Signed Numbers

Subtraction is the reverse of addition. For example,

$$8 - 5 = 3 \quad \text{because} \quad 3 + 5 = 8$$

These two numbers must add to give 8.

The answer to a subtraction problem added to the number being subtracted is equal to the number we are subtracting from. When subtracting signed numbers, we want to preserve this property that subtraction is the reverse of addition.

Here we look at two more subtraction problems. At this time, we use the idea that subtraction is the reverse of addition to obtain the answers. After we see what the answers must be, we will discover a more efficient method for subtracting signed numbers.

1. $5 - 8 = ?$ The problem is 5 minus 8.
The answer must add to 8 to give 5.

$$? + 8 = 5$$

By trial and error, we see that -3 works.

$$-3 + 8 = 5$$

$5 - 8 = -3$

2. $8 - (-5) = ?$ The problem is 8 minus -5.
The answer must add to -5 to give 8.

$$? + (-5) = 8$$

By trial and error, we see that 13 works.

$$13 + (-5) = 8$$

$8 - (-5) = 13$

Observe that the symbol, $-$, is being used in two different ways: to indicate the operation of subtraction and to indicate a negative number (or taking the opposite of a number). It is important that you recognize which role the symbol is playing when you look at an expression. For the symbol to indicate subtraction, there must be a number immediately in front of the symbol as well as following it. However, when the symbol is indicating a negative number or taking the opposite of a number, the symbol is only followed by a number.

The above process for subtracting signed numbers gives us the correct results because it preserves the notion that subtraction is the reverse of addition. Although this process works well for the problem $8 - 5$, it is awkward to use on a problem like $8 - (-5)$. Now we discover a more straightforward approach to subtracting signed numbers.

Look again at the problem $8 - 5 = 3$. Notice that *subtracting 5 is the same as adding -5*.

$$\begin{array}{c} 8 - 5 \\ \downarrow \downarrow \\ = 8 + (-5) \\ = 3 \end{array}$$

Here we change the subtraction operation to addition, and change the sign of the number being subtracted.

Next, we look again at the problem $5 - 8 = -3$. Notice that *subtracting 8 is the same as adding -8*.

$$\begin{array}{c} 5 - 8 \\ \downarrow \downarrow \\ = 5 + (-8) \\ = -3 \end{array}$$

Here we change the subtraction operation to addition, and change the sign of the number being subtracted.

Finally, we look again at the problem $8 - (-5) = 13$. Notice that *subtracting -5 is the same as adding 5*.

$$\begin{array}{c} 8 - (-5) \\ \downarrow \downarrow \\ = 8 + 5 \\ = 13 \end{array}$$

Here we change the subtraction operation to addition, and change the sign of the number being subtracted.

12.4 Subtraction of Signed Numbers

▲ Try These Problems

Subtract by first converting to an addition problem.
1. $14 - 20$
2. $10 - (-3)$
3. $-5 - 12$
4. $-15 - (-3)$
5. $-6 - (-34)$
6. $-9 - 16$

In each of the above problems observe that *subtracting a number is the same as adding its opposite*. In general, this is always true. This gives us a straightforward approach to subtracting signed numbers. Here we state a rule.

> **To Subtract Two Signed Numbers**
> 1. Change the subtraction operation to addition.
> 2. Change the sign of the number that is being subtracted (the second number).
> 3. Follow the rules for adding signed numbers.
>
> Note: Do *not* alter the number that is being subtracted from, that is, the first number.

Here are some more examples.

EXAMPLE 1 Subtract: $7 - 16$

SOLUTION Here 16 is being subtracted from 7.

$$7 - 16$$
$$= 7 + (-16) \quad \text{Change the subtraction to addition, and change 16 to } -16.$$
$$= -9 \quad \text{Add 7 and } -16. \ ∎$$

EXAMPLE 2 Subtract: $8 - (-2)$

SOLUTION Here -2 is being subtracted from 8.

$$8 - (-2)$$
$$= 8 + 2 \quad \text{Change the subtraction to addition, and change } -2 \text{ to } 2.$$
$$= 10 \quad \text{Add 8 and 2.} \ ∎$$

EXAMPLE 3 Subtract: $-4 - 25$

SOLUTION Here 25 is being subtracted from -4.

$$-4 - 25$$
$$= -4 + (-25) \quad \text{Change the subtraction to addition, and change 25 to } -25.$$
$$= -29 \quad \text{Add } -4 \text{ and } -25. \ ∎$$

EXAMPLE 4 Subtract: $-18 - (-12)$

SOLUTION Here -12 is being subtracted from -18.

$$-18 - (-12)$$
$$= -18 + 12 \quad \text{Change the subtraction to addition, and change } -12 \text{ to } 12.$$
$$= -6 \ ∎$$

▲ **Try Problems 1 through 6.**

Try These Problems

Subtract by first converting to an addition problem.

7. $32 - 0$
8. $-9 - 0$
9. $0 - 16$
10. $0 - (-35)$
11. $0 - 75$
12. $\quad 32$
    $-\ -7$
13. $\quad -48$
    $-\ \ 13$
14. $\quad -12$
    $-\ -5$

Now we look at some subtraction problems that involve the number zero. The rule stated above applies in these problems, that is, *subtracting a number is the same as adding its opposite*. Recall that the opposite of 0 is 0.

EXAMPLE 5 Subtract: $-13 - 0$

SOLUTION Here 0 is being subtracted from -13.

$$-13 - 0$$
$$\downarrow$$
$$= -13 + 0 \quad \text{Change the subtraction to addition, and since 0 is the opposite of itself, it does not change.}$$
$$= -13 \quad \text{Add } -13 \text{ and } 0.$$

EXAMPLE 6 Subtract: $0 - 7$

SOLUTION Here 7 is being subtracted from 0.

$$0 - 7$$
$$\downarrow \ \downarrow$$
$$= 0 + (-7) \quad \text{Change the subtraction to addition, and change 7 to } -7.$$
$$= -7 \quad \text{Add 0 and } -7.$$

EXAMPLE 7 Subtract: $0 - (-28)$

SOLUTION Here -28 is being subtracted from 0.

$$0 - (-28)$$
$$\downarrow \ \ \downarrow$$
$$= 0 + 28 \quad \text{Change the subtraction to addition, and change } -28 \text{ to } 28.$$
$$= 28 \quad \text{Add 0 and 28.}$$

 Try Problems 7 through 11.

Sometimes we encounter subtraction problems written in vertical form. For example,

$$\begin{array}{r} -15 \\ -\ -6 \\ \hline \end{array} \quad \text{means} \quad \begin{array}{r} -15 \\ +\ \ 6 \\ \hline \end{array}$$

We rewrite the problem as addition by using the fact that subtracting a number is the same as adding its opposite.

 Try Problems 12 through 14.

2 Reversing the Order of Subtraction

In an addition problem it makes no difference what order the numbers are written. For example, here we look at the sum of -8 and 2 written in two different ways.

$$-8 + 2 = -6 \quad \text{and} \quad 2 + (-8) = -6$$

Try These Problems

Translate each statement to a subtraction statement using math symbols.

15. 18 minus 27 equals −9.

16. 17 less than −30 is −47.

17. Negative five subtracted from negative twenty equals negative fifteen.

However, in a subtraction problem, it does make a difference what order the numbers are written. For example,

$$7 - 2 = 5 \quad \text{but} \quad 2 - 7 = -5$$

Because switching the order of subtraction changes the result, we must be very careful to use the appropriate language when stating a subtraction problem. The following chart gives several correct ways to read a subtraction problem that involves signed numbers.

Math Symbols	English
$-4 - 5 = -9$	Negative 4 minus 5 equals negative 9.
	−4 minus 5 is −9.
	5 subtracted from −4 equals −9.
	5 less than −4 is −9.

First, observe that the word *minus* is used to read the operation of subtraction, while the word *negative* is used to read a negative number. Second, observe that when the phrases *subtracted from* and *less than* are used, the numbers are said in the reverse order than they appear in the symbolic statement.

Here are some examples.

EXAMPLE 8 Translate *−8 subtracted from 5 is 13* to a subtraction statement using math symbols.

SOLUTION Because the phrase *subtracted from* is used, we *reverse* the order of the −8 and 5 when writing the symbolic statement.

$$\text{−8 subtracted from 5 is 13}$$
$$5 - (-8) = 13$$

The symbolic statement is $5 - (-8) = 13$. ∎

Try Problems 15 through 17.

EXAMPLE 9 What number is −14 minus −12?

SOLUTION Because the word *minus* is used, we do *not* reverse the order of the numbers when writing the symbolic statement.

What number is −14 minus −12?

$$\boxed{} = -14 - (-12)$$

Now we perform the subtraction.

$$-14 - (-12)$$
$$= -14 + 12 \quad \text{Subtracting −12 is the same as adding 12.}$$
$$= -2 \quad \text{Add −14 and 12.} \quad \blacksquare$$

Try These Problems

Solve.

18. What number is 17 less than 12?
19. What number is −14 subtracted from 18?
20. Fifty minus negative seventeen is what number?
21. Negative nine subtracted from negative twelve equals what number?

EXAMPLE 10   Sixteen less than negative thirty is what number?

SOLUTION   Because the phrase *less than* is used, we *reverse* the order of the numbers 16 and −30 when writing the symbolic statement.

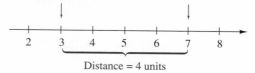

Now we perform the subtraction.

$$-30 - 16$$
$$= -30 + (-16) \quad \text{Subtracting 16 is the same as adding } -16.$$
$$= -46 \quad \text{Add } -30 \text{ and } -16.$$

 Try Problems 18 through 21.

3 Finding the Distance Between Two Numbers on the Number Line

A very important application of subtraction of signed numbers is finding how far apart two numbers are, that is, finding the **distance between two numbers.** Here we investigate some situations. We begin by looking at the numbers 7 and 3 on the number line.

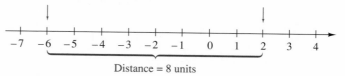

Distance = 4 units

Observe that there are 4 units between 7 and 3 on the number line, so the distance between 7 and 3 is 4. To obtain this distance by subtraction, we subtract 3 from 7, the smaller number from the larger number.

$$\text{Distance between 3 and 7} = \text{larger number} - \text{smaller number}$$
$$= 7 - 3$$
$$= 4$$

In general the distance between any two positive numbers can be found by subtracting the smaller number from the larger number. Will this same method work when one or both of the numbers are negative? Here we look at the numbers −6 and 2 on the number line.

Distance = 8 units

There are 8 units between −6 and 2, so the distance between −6 and 2 is 8. To obtain this distance by subtraction, observe that we can subtract −6 from 2, the smaller number from the larger number.

Try These Problems

Find the distance between the given pair of numbers by using subtraction.
22. 9 and 2
23. −4 and 10
24. −5 and −15

$$\begin{aligned}\text{Distance between } -6 \text{ and } 2 &= \text{larger number} - \text{smaller number} \\ &= 2 - (-6) \\ &= 2 + 6 \quad \text{Subtracting } -6 \text{ is the same as adding 6.} \\ &= 8\end{aligned}$$

Here we see that when one number is positive and the other is negative, we can still subtract the smaller number from the larger number to obtain the distance between them. Will this same method work when both numbers are negative? Here we look at the numbers −8 and −3 on the number line.

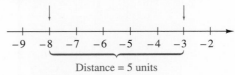

There are 5 units between −8 and −3, so the distance between −8 and −3 is 5. To obtain this distance by subtraction, observe that we can subtract −8 from −3, the smaller number from the larger number.

$$\begin{aligned}\text{Distance between } -8 \text{ and } -3 &= \text{larger number} - \text{smaller number} \\ &= -3 - (-8) \\ &= -3 + 8 \quad \text{Subtracting } -8 \text{ is the same as adding 8.} \\ &= 5\end{aligned}$$

In general, the distance between two unequal numbers can always be found by subtracting the smaller number from the larger number. Here we state a rule.

Distance Between Two Numbers

The distance between two unequal numbers is the smaller number subtracted from the larger number. The distance between two equal numbers is zero.

$$\text{Distance between two unequal numbers} = \text{larger number} - \text{smaller number}$$

The distance between two numbers is always positive or zero.

 Try Problems 22 through 24.

EXAMPLE 11 A place in the Sierra Mountains is 9250 feet above sea level while a place in the Pacific Ocean is 4500 feet below sea level. What is the difference in the two altitudes?

SOLUTION First, these altitudes can be represented by using signed numbers. We assume that sea level corresponds to an altitude of 0.

9,250 ft *above* sea level → altitude of 9250 ft

4500 ft *below* sea level → altitude of −4500 ft

Try These Problems

Solve by using subtraction of signed numbers.

25. A submarine in the ocean is 4600 feet below sea level. Another submarine directly above it is 1400 feet below sea level. How far apart are the two submarines?

26. The temperature in Moscow is $-14°C$ while the temperature in London is $12°C$. What is the difference in these two temperatures?

Second, the word *difference* used in real-life situations usually indicates to find how far apart the numbers are as a *positive number*. This is equivalent to finding the distance between the two numbers.

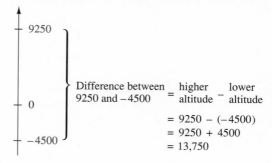

The difference in the two altitudes is 13,750 feet. ∎

 Try Problems 25 and 26.

4 Evaluating Repeated Subtraction and a Mixture of Addition and Subtraction

We know that repeated addition can be done in any order; however, repeated subtraction *cannot* be done in any order. In Section 12.1 the rule for the order to operate states that repeated subtraction is done by operating from left to right. Here we show an example:

$$9 - 4 - 3 - 7 = 5 - 3 - 7$$
$$= 2 - 7$$
$$= -5$$

Here we operate from left to right.

The correct answer to this repeated subtraction problem is -5.

Now that we have introduced signed numbers there is a better way to think about a repeated subtraction problem. Because subtracting a number is the same as adding its opposite, we can think of each subtraction as addition, and then the resulting addition problem can be done in any order. Here we show the above problem done in this way.

$$9 - 4 - 3 - 7$$
$$= 9 + (-4) + (-3) + (-7)$$
$$= 9 + (-14)$$
$$= -5$$

Subtracting 4 is the same as adding -4.
Subtracting 3 is the same as adding -3.
Subtracting 7 is the same as adding -7.

Now add the numbers in any order.
We add the negative numbers first.

Observe that we obtain the same result as when we operated from left to right.

Try These Problems

Evaluate in two ways.
27. $12 - 5 - 4 - 6$
28. $7 + 13 - 8 - 3 + 2$

Try Problem 27.

Similarly, a mixture of addition and subtraction can be done from left to right. However, now that we have signed numbers, it is better to convert the subtraction to addition and then add in any order. Here we evaluate an expression both ways.

1. $2 - 8 - 4 + 2 = -6 - 4 + 2$ Here we operate from left to right.
$= -6 + (-4) + 2$
$= -10 + 2$
$= -8$

2. $2 - 8 - 4 + 2$
$= 2 + (-8) + (-4) + 2$ Subtracting 8 is the same as adding -8.
Subtracting 4 is the same as adding -4.
$= 4 + (-12)$ Now add the numbers in any order. We add like signs first.
$= -8$

Try Problem 28.

Here we summarize the above discussion.

Evaluating Repeated Subtraction and a Mixture of Addition and Subtraction

1. Convert each subtraction to addition by using the fact that subtracting a number is the same as adding its opposite.
2. Add in any order.

(The correct answer can also be obtained by operating from left to right.)

Here we look at more examples.

EXAMPLE 12   Evaluate: $-10 - (-15) - 14$

SOLUTION   Observe that there are two subtractions. We convert each of these subtractions to addition before doing any calculations.

$-10 - (-15) - 14$
$= -10 + 15 + (-14)$ Subtracting -15 is the same as adding 15. Subtracting 14 is the same as adding -14.
$= -10 + 1$ Now add in any order. We add 15 and -14 first to obtain 1. ■
$= -9$

Try These Problems

Evaluate.
29. $5 - 10 - (-8) - 12$
30. $-15 - (-4) - (-6) - 11$
31. $81 + (-13) - 7 + 3$
32. $-17 - (-5) + (-15) - 40 + 8$
33. $7 - 8 - (-9 + 5 - 9) + 6$
34. $-17 + (-5) - (16 - 9 - 1) + 20$

EXAMPLE 13 Evaluate: $25 - (-28) - 8 + 12$

SOLUTION There are two subtractions. Convert each of these subtractions to addition, then add in any order. Be careful! You cannot begin by adding the 8 and 12.

$$25 - (-28) - 8 + 12$$
$$= 25 + 28 + (-8) + 12$$

Subtracting -28 is the same as adding 28. Subtracting 8 is the same as adding -8.

$$= 65 + (-8)$$
$$= 57$$

Add in any order. We add the three positive numbers first.

Try Problems 29 through 32.

EXAMPLE 14 Evaluate: $6 - 8 - (2 - 3 + 5)$

SOLUTION First, we perform the two operations inside the parentheses.

$$6 - 8 - (2 - 3 + 5)$$
$$= 6 - 8 - (2 + (-3) + 5)$$

Operate inside the parentheses first. Subtracting 3 is the same as adding -3.

$$= 6 - 8 - (7 + (-3))$$

Add inside the parentheses in any order. We add the two positive numbers first.

$$= 6 - 8 - 4$$
$$= 6 + (-8) + (-4)$$

Subtracting 8 is the same as adding -8. Subtracting 4 is the same as adding -4.

$$= 6 + (-12)$$
$$= -6$$

Try Problems 33 and 34.

Answers to Try These Problems

1. $14 + (-20) = -6$ 2. $10 + 3 = 13$ 3. $-5 + (-12) = -17$
4. $-15 + 3 = -12$ 5. $-6 + 34 = 28$ 6. $-9 + (-16) = -25$
7. $32 + 0 = 32$ 8. $-9 + 0 = -9$ 9. $0 + (-16) = -16$
10. $0 + 35 = 35$ 11. $0 + (-75) = -75$ 12. $32 + 7 = 39$
13. $-48 + (-13) = -61$ 14. $-12 + 5 = -7$ 15. $18 - 27 = -9$
16. $-30 - 17 = -47$ 17. $-20 - (-5) = -15$ 18. -5 19. 32
20. 67 21. -3 22. $9 - 2 = 7$ 23. $10 - (-4) = 14$
24. $-5 - (-15) = 10$ 25. $-1400 - (-4600) = 3200$; 3200 ft
26. $12 - (-14) = 26$; 26°C 27. -3 28. 11 29. -9
30. -16 31. 64 32. -59 33. 18 34. -8

EXERCISES 12.4

Subtract.

1. $8 - 13$
2. $13 - 8$
3. $11 - 5$
4. $5 - 11$
5. $-3 - (-5)$
6. $-24 - (-7)$
7. $-4 - 6$
8. $-5 - 14$
9. $50 - (-25)$
10. $45 - (-60)$
11. $36 - 42$
12. $-19 - 16$
13. $-72 - 18$
14. $67 - 100$
15. $-16 - (-12)$
16. $34 - (-47)$
17. $53 - (-17)$
18. $-76 - (-51)$
19. $28 - (-28)$
20. $-28 - (-28)$
21. $-4 - 0$
22. $0 - (-16)$
23. $0 - 48$
24. $77 - 0$
25. -16
    $\underline{-8}$
26. 49
    $\underline{-\,-16}$
27. 68
    $\underline{-\,80}$
28. -35
    $\underline{-\,-83}$

Translate each statement to a subtraction statement using math symbols.

29. 4 subtracted from 3 equals -1.
30. 9 less than -5 is -14.
31. -35 minus 18 equals -53.
32. -15 subtracted from -40 is -25.
33. Eighty minus negative seventy-five equals one hundred fifty-five.
34. Nineteen less than eleven equals negative eight.

Solve by using subtraction of signed numbers.

35. What number is -4 minus -15?
36. What number is 8 subtracted from -24?
37. Eighteen less than negative eleven is what number?
38. Negative twenty minus negative sixty-three is what number?
39. What is the distance between -5 and -39?
40. What is the distance between 48 and -23?
41. Point A has a position of -37 on the number line, while point B has a position of 25. How far is point A from point B?
42. Point P has a position of -16 on the number line, while point Q has a position of -59. What is the distance between points P and Q?
43. A sunken ship is 2700 feet below sea level. An airplane flying directly overhead is 6800 feet above sea level. How far is the ship from the airplane?
44. Blue Mountain in Oregon has an elevation of 7420 feet. A place in the Atlantic Ocean has an elevation of -750 feet. What is the difference in these elevations?
45. One evening in Boston, the temperature was $-15°F$. The next day, the temperature reached $23°F$. What is the difference in these temperatures?
46. A football player ran from 12 yards behind the line of scrimmage to 26 yards beyond the line of scrimmage. How far did he run?

Evaluate.

47. $12 - 15 - 4 - 6$
48. $7 - 17 - 9 - 15 - 4$
49. $-8 - 5 + 9 - 11 + 4$
50. $-7 + 3 - 5 - 13 + 6$
51. $9 + (-6) - (-8) - 7 + 6$
52. $12 - 24 + (-7) - (-34) + 8$
53. $-11 - 5 + (-15) - 8 - 36$
54. $-26 + 7 - (-18) - 17 + (-9)$
55. $29 - (16 + (-8) - 5) + 3$
56. $-38 + (10 - 9 - (-5)) - 7$
57. $-64 - 25 - (9 - 13 - 13)$
58. $-95 - 76 + (8 + (-16) - 3)$
59. $35 - [16 - (5 - 7) + 45]$
60. $-80 + [15 - 17 - (-1 + 4)]$

12.5 MULTIPLICATION AND DIVISION OF SIGNED NUMBERS

OBJECTIVES
- Multiply two or more signed numbers.
- Evaluate a numerical expression that involves several operations.
- Divide two signed numbers.
- Translate an English phrase involving division to math symbols.
- Recognize when two signed fractions are equal.
- Convert a signed fraction to decimal form.

 Try These Problems

Multiply.
1. $9 \cdot (-3)$
2. $-4 \cdot 7$
3. $5(-13)$
4. $(-6)30$

1 Multiplying Two Signed Numbers

First we look at what it means to multiply a positive number by a positive number. For example,

$$3 \cdot 5 = 15$$

can be interpreted as repeated addition as follows.

$$3 \cdot 5 = 5 + 5 + 5 = 15$$

or

$$3 \cdot 5 = 3 + 3 + 3 + 3 + 3 = 15$$

We can use this concept to discover how to multiply a positive number by a negative number. For example,

$$3 \cdot (-5) = (-5) + (-5) + (-5) = -15$$

and

$$-3 \cdot 5 = (-3) + (-3) + (-3) + (-3) + (-3)$$
$$= -15$$

Observe that the answer can be obtained by multiplying 3 and 5, the absolute values of the numbers, then attaching the negative sign. Here are more examples.

$$6 \cdot (-7) = -42$$
$$-8 \cdot 4 = -32$$
$$-5(7) = -35$$
$$8(-6) = -48$$

Here we state a rule for multiplying a positive number by a negative number.

> *To Multiply a Positive and a Negative Number*
> 1. Multiply the absolute values of the numbers to obtain the absolute value of the answer.
> 2. Attach the negative sign.

 Try Problems 1 through 4.

Try These Problems

Multiply.
5. 0(−8)
6. 0(0)
7. 35 · 0

Next we look at multiplying a signed number by zero. Recall that zero multiplied by any positive number or zero gives zero. For example,

$$17 \cdot 0 = 0 \quad \text{or} \quad 0 \cdot 17 = 0$$

and

$$0 \cdot 0 = 0$$

Also, any negative number multiplied by zero is zero. For example,

$$-17 \cdot 0 = 0 \quad \text{or} \quad 0 \cdot (-17) = 0$$

Here we state a rule for multiplying by zero.

Multiplying by Zero
Any number multiplied by zero is zero.

Try Problems 5 through 7.

The only type of multiplication problem we have left to discuss is multiplying a negative number by a negative number. It is not so easy to see intuitively what the answer should be to this type of problem. To help you discover how to multiply a negative number by a negative number, we look at some multiples of −4.

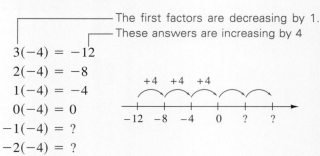

If we continue the established pattern, we see the results of multiplying two negative numbers.

$$2(-4) = -8$$
$$1(-4) = -4$$
$$0(-4) = 0$$
$$-1(-4) = 4$$
$$-2(-4) = 8$$
$$-3(-4) = 12$$

The above discussion should convince you that a negative number times a negative number is a positive number. Here are more examples.

$$-9(-4) = 36$$
$$-45(-2) = 90$$
$$(-14)(-6) = 84$$
$$-20 \cdot (-50) = 1000$$

Try These Problems

Multiply.
8. $(-3)(-5)$
9. $-7(-3)$
10. $-60(-24)$
11. $-16 \cdot (-8)$
12. $7(-5)$
13. $-6(9)$
14. $-7(-5)$
15. $-6(-9)$
16. $-15(0)$
17. $-35(240)$

Multiply in three different ways.
18. $7 \cdot (-3) \cdot (-4) \cdot 2$
19. $(-9)(-1)(2)(5)(-3)$

Now we state a rule for multiplying two negative numbers.

Multiplying a Negative Number by a Negative Number.
Multiply the absolute values of the numbers to obtain the answer. The result is a positive number.

 Try Problems 8 through 11.

We summarize the previous discussion by writing a rule for multiplying any two signed numbers.

To Multiply Two Signed Numbers
1. Multiply the absolute values of the two numbers.
2. Attach the appropriate sign by using these sign rules.

$$(+)(+) = (+)$$
$$(+)(-) = (-)$$
$$(-)(+) = (-)$$
$$(-)(-) = (+)$$

3. Zero multiplied by any number is zero.

 Try Problems 12 through 17.

2 Multiplying More Than Two Numbers

You know that repeated multiplication of *positive* numbers can be done in any order. Can we evaluate repeated multiplication of *signed* numbers in any order? Here we do a problem in three ways.

1. $-5 \cdot 4 \cdot (-6) \cdot (-2) = -20 \cdot (-6) \cdot (-2)$ Here we multiply
   $= 120 \cdot (-2)$ from left to right.
   $= -240$

2. $-5 \cdot 4 \cdot (-6) \cdot (-2) = -20 \cdot 12$ Here we multiply the first
   $= -240$ two numbers, the last two numbers, then multiply these results.

3. $-5 \cdot 4 \cdot (-6) \cdot (-2) = (-5)(-6)(-2)4$ Here we rearrange
   $= 30(-2)4$ the order of the factors and multiply
   $= -60(4)$ the negative
   $= -240$ numbers first.

Observe that the answer is -240 no matter in what order we do the multiplication. In general, **repeated multiplication of signed numbers can be done in any order.**

 Try Problems 18 and 19.

Try These Problems

Multiply.
20. $(-1)(-4)(-2)(-5)(-4)$
21. $(-3)(9)(0)(5)(-3)$
22. $6(-4)(-1)(-2)(8)(-5)$
23. $11(-6)(-1)(3)(-20)$

If one of the factors in a repeated multiplication problem is zero, then the result is zero. For example,

$$(-5)(-15)(0)(7)(19) = 0$$

One factor is 0.

If there are several negative factors and no zero factors, it is convenient to decide the sign of the answer before performing any multiplication. This can be done by grouping the negative factors in pairs. Here is an example.

$$(-20)(-1)(8)(-3)(6)(-5) = \text{a positive number}$$

The product of each pair of negative numbers is a positive number. In this problem, after pairing the negatives, we are left with no negatives, so the result is positive. To evaluate, we simply need to multiply the absolute values of the numbers in any order.

$$\begin{aligned}(-20)(-1)(8)(-3)(6)(-5) &= 20(1)(8)(3)(6)(5) \\ &= 20 \quad (24) \quad (30) \\ &= 600(24) \\ &= 14{,}400\end{aligned}$$

In general, when we pair the negative factors and no negative factors are left, this means there are an *even* number of negatives. In the previous problem, there were 4 negative factors, an even number. We conclude that *an even number of negatives multiply to give a positive.*

Here is another example.

$$2(-4)(5)(-2)(-1)(10)(-8)(-3) = \text{a negative number}$$

One negative factor left after pairing all other negative factors.

The product of each pair of negatives is positive. In this problem, after pairing the negatives, we are left with one negative, so the result is negative. To evaluate, we simply need to multiply the absolute values in any order and attach the negative sign.

$$\begin{aligned}2(-4)(5)(-2)(-1)(10)(-8)(-3) &= -(2)(4)(5)(2)(1)(10)(8)(3) \\ &= -(8)\ (10)\ (10)\ (24) \\ &= -(192)(100) \\ &= -19{,}200\end{aligned}$$

In general, when we pair the negative factors and one negative factor is left, this means there are an *odd* number of negatives. In the previous problem, there were 5 negative factors, an odd number. We conclude that *an odd number of negatives multiply to give a negative.*

 Try Problems 20 through 23.

Try These Problems

Evaluate.
24. $8 - 10(-8 + 4)$
25. $(-5 + 10)(-7 + (-3))$
26. $|5(-3) - 7|$
27. $|5 + (-9) - 4|$

3 Mixing the Operations

In Section 12.1 we worked with the order of operation for positive numbers. These same rules also apply to signed numbers. In Section 12.2 we introduced two new operations: taking the opposite of a number and taking the absolute value of a number. These new operations were needed to help explain how to add, subtract, and multiply signed numbers. At this time we incorporate these two new operations into the rules for the order of operation.

> *Order of Operation*
> 1. Operate inside grouping symbols, beginning with the innermost.
> 2. Take absolute values.
> 3. Take opposites, multiply, and divide, from left to right. (Repeated multiplication can be done in any order.)
> 4. Add and subtract, from left to right. (Repeated addition can be done in any order.)

Here are some examples of using these rules.

EXAMPLE 1 Evaluate: $-12 - 4(-1 - 2)$

SOLUTION Begin by operating inside the grouping symbol.

$$-12 - 4(-1 - 2)$$
$$= -12 - 4(-1 + (-2))$$ Subtracting 2 is the same as adding -2.
$$= -12 - 4(-3)$$ Add -1 and -2 inside the parentheses.
$$= -12 - (-12)$$ Multiply 4 by -3 to obtain -12.
$$= -12 + 12$$ Subtracting -12 is the same as adding 12.
$$= 0 \blacksquare$$

Try Problems 24 and 25.

EXAMPLE 2 Evaluate: $|-6 - 8 + 2|$

SOLUTION Evaluate inside the absolute value symbol before taking the absolute value because *the absolute value symbol is understood to be a grouping symbol.*

$$|-6 - 8 + 2|$$
$$= |-6 + (-8) + 2|$$ Subtracting 8 is the same as adding -8.
$$= |-14 + 2|$$ Add the numbers inside in any order.
$$= |-12|$$
$$= 12$$ The absolute value of -12 is 12. $\blacksquare$

Try Problems 26 and 27.

Try These Problems

Divide.
28. $15 \div (-3)$
29. $-35 \div (-5)$
30. $-24 \div 4$
31. $-63 \div 1$
32. $\frac{-8}{2}$
33. $\frac{27}{-9}$
34. $\frac{-250}{-50}$
35. $\frac{-13}{-13}$

4 Dividing Two Signed Numbers

Division is the reverse of multiplication. For example,
$$30 \div 5 = 6 \quad \text{because} \quad 5 \cdot 6 = 30$$
These two numbers must multiply to give 30.

We want to divide signed numbers so that this property of division is preserved. For example,
$$30 \div (-5) = -6 \quad \text{because} \quad -6 \cdot (-5) = 30$$
$$-30 \div 5 = -6 \quad \text{because} \quad -6 \cdot 5 = -30$$
$$-30 \div (-5) = 6 \quad \text{because} \quad 6 \cdot (-5) = -30$$

In each of the above problems observe that to divide two signed numbers, we can divide the absolute values, then attach the appropriate sign by using sign rules similar to the ones used for multiplication. Here we summarize the sign rules for division of signed numbers.

$$(+) \div (+) = (+)$$
$$(+) \div (-) = (-)$$
$$(-) \div (+) = (-)$$
$$(-) \div (-) = (+)$$

Here are more examples:
$$28 \div (-4) = -7$$
$$-42 \div (-7) = 6$$
$$-75 \div (-75) = 1$$
$$45 \div (-1) = -45$$

Try Problems 28 through 31.

We also use the fraction bar to indicate division. Here we show some examples.

$$\frac{20}{4} = 5$$
$$\frac{-21}{-3} = 7$$
$$\frac{60}{-6} = -10$$
$$\frac{-88}{-4} = 22$$

Try Problems 32 through 35.

5 Division Involving Zero

First we look at dividing zero by a nonzero number. For example,
$$0 \div 6 = 0 \quad \text{because} \quad 0 \cdot 6 = 0$$
or
$$\frac{0}{6} = 0 \quad \text{because} \quad 0 \cdot 6 = 0$$

Observe that *zero divided by a nonzero number equals zero.*

Try These Problems

Divide, if possible. If not possible, write no answer.

36. $19 \div 0$
37. $0 \div (-5)$
38. $\frac{0}{48}$
39. $\frac{0}{0}$
40. $-45 \div 9$
41. $342 \div (-6)$
42. $-40 \div (-8)$
43. $0 \div 0$
44. $\frac{14}{-1}$
45. $\frac{16}{0}$
46. $\frac{-90}{-15}$
47. $\frac{-46}{-46}$

Now we try to divide a nonzero number by zero.

$$6 \div 0 = ? \quad \text{or} \quad \frac{6}{0} = ?$$

The answer must multiply by 0 to give 6. That is, $0 \cdot ? = 6$. There is no number that multiplies by 0 to give 6 because 0 multiplied by any number gives 0. Therefore, the problem has no answer.

$$6 \div 0 = \text{no answer} \quad \text{or} \quad \frac{6}{0} = \text{no answer}$$

What happens if we try to divide zero by zero?

$$0 \div 0 = ? \quad \text{or} \quad \frac{0}{0} = ?$$

The answer must multiply by 0 to give 0. That is, $0 \cdot ? = 0$. Any number multiplied by zero yields zero, so any number will work! We do not want a division problem to have many answers, so we agree that this problem has no answer.

$$0 \div 0 = \text{no answer} \quad \text{or} \quad \frac{0}{0} = \text{no answer}$$

In general, *division by zero is undefined.*

 Try Problems 36 through 39.

We summarize the above discussion by writing a rule for dividing any two signed numbers.

To Divide Two Signed Numbers

1. Divide the absolute values of the two numbers.
2. Attach the appropriate sign by using these sign rules.

$$(+) \div (+) = (+)$$
$$(+) \div (-) = (-)$$
$$(-) \div (+) = (-)$$
$$(-) \div (-) = (+)$$

3. Zero divided by a nonzero number equals zero.

$$\frac{0}{\text{nonzero number}}$$
$$= 0 \div \text{nonzero number}$$
$$= 0$$

4. Division by zero is undefined.

$$\frac{\text{any number}}{0}$$
$$= \text{any number} \div 0$$
$$= \text{no answer}$$

 Try Problems 40 through 47.

6 Reversing the Order of Division

In a multiplication problem it makes no difference in what order the numbers are written. For example, here we look at the product of 9 and -3 written in two different ways.

$$9(-3) = -27 \quad \text{and} \quad -3(9) = -27$$

However, in a division problem, it does make a difference in what order the numbers are written. For example,

$$50 \div (-5) = \frac{50}{-5}$$
$$= -10 \quad \text{This number is less than } -1.$$

but

$$-5 \div 50 = \frac{-5}{50}$$
$$= \frac{-1}{10} \quad \text{The number is more than } -1.$$

Because switching the order of division changes the result, we must be very careful to use the appropriate language when stating a division problem. Study the following chart to see how to say a division statement correctly.

Math Symbols	*English*
$18 \div 6 = \frac{18}{6} = 3$	18 *divided by* 6 is 3.
	6 *divided into* 18 is 3.
$6 \div 18 = \frac{6}{18} = \frac{1}{3}$	6 *divided by* 18 is $\frac{1}{3}$.
	18 *divided into* 6 is $\frac{1}{3}$.

Observe that when the phrase *divided by* is used, the numbers are said in the *same order* as they are written in the symbolic statement. However, when the phrase *divided into* is used, the numbers are said in the *reverse order* from the way they are written in the symbolic statement.

Here are some examples.

EXAMPLE 1 Evaluate each.

a. 450 divided by -9. **b.** 450 divided into -9.

SOLUTION

a. The phrase *divided by* is used, so we do *not* reverse the order of the numbers when translating to the symbolic statement.

Try These Problems

Evaluate.
48. 30 divided by −5
49. 30 divided into −5
50. −14 divided into −21
51. −14 divided by −21

$$450 \text{ divided by } -9$$
$$= 450 \div (-9)$$ Keep the numbers in the same order.
$$= \frac{450}{-9}$$
$$= -50$$

b. The phrase *divided into* is used, so we *reverse* the order of the numbers when translating to the symbolic statement.

$$450 \text{ divided into } -9$$
$$= -9 \div 450$$ Reverse the order of the numbers.
$$= \frac{-9}{450}$$
$$= \frac{-1}{50}$$ ■

EXAMPLE 2 Evaluate each.

a. −12 divided into −18. **b.** −12 divided by −18

SOLUTION

a. The phrase *divided into* is used, so we *reverse* the order of the numbers when translating to the symbolic statement.

$$-12 \text{ divided into } -18$$
$$= -18 \div (-12)$$ Reverse the order of the numbers.
$$= \frac{-18}{-12}$$
$$= \frac{\cancel{6} \cdot 3}{\cancel{6} \cdot 2}$$ Because a negative divided by a negative is positive we drop the negative signs.
$$= \frac{3}{2} \text{ or } 1\frac{1}{2}$$ Simplify the fraction.

b. The phrase *divided by* is used, so we do *not* reverse the order of the numbers when translating to the symbolic statement.

$$-12 \text{ divided by } -18$$
$$= -12 \div (-18)$$ Keep the numbers in the same order.
$$= \frac{-12}{-18}$$
$$= \frac{\cancel{6} \cdot 2}{\cancel{6} \cdot 3}$$ Because a negative divided by a negative is positive we drop the negative signs.
$$= \frac{2}{3}$$ Simplify the fraction. ■

 Try Problems 48 through 51.

Try These Problems

52. Which of these numbers are equal to the fraction, $\frac{7}{-8}$?
    a. $\frac{-7}{8}$
    b. $\frac{-7}{-8}$
    c. $-1\frac{1}{8}$
    d. -0.875
    e. $-\frac{7}{8}$

53. Which of these numbers are equal to the fraction, $-\frac{1}{9}$?
    a. $\frac{-1}{-9}$
    b. -9
    c. $\frac{-1}{9}$
    d. $\frac{1}{-9}$
    e. $0.\overline{1}$

Divide and simplify.

54. $5 \div (-40)$
55. Divide -15 into 27.
56. Divide 56 by -64.

Convert each fraction to decimal form.

57. $\frac{-3}{8}$
58. $\frac{-11}{9}$

7 Expressing a Negative Fraction in Several Ways

Here we view some negative fractions on the number line.

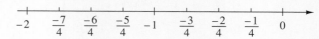

A negative fraction can be written in several ways. For example, the fraction *negative three-fourths* is shown here written in four different ways.

$$\text{negative three-fourths} = -\left(\frac{3}{4}\right) = -\frac{3}{4} = \frac{3}{-4} = \frac{-3}{4}$$

⎣— The preferred way to write negative three-fourths.

The negative sign may be written in front of the entire fraction with the fraction wrapped or not wrapped in parentheses, written in the denominator, or written in the numerator. Of these four methods, the last one is usually preferred. That is, *when a fraction is negative, we prefer to associate the negative sign with the numerator.*

▲ **Try Problems 52 through 56.**

Of course the number negative three-fourths can also be written in decimal form, generating the decimal negative seventy-five hundredths.

$$\frac{-3}{4} = -0.75 \qquad \text{Divide 3 by 4, and attach the negative sign.}$$

We often prefer to leave fractions in fractional form because many fractions convert to decimals that do not terminate. For example,

$$\frac{-2}{3} = -0.6666666\ldots = -0.\overline{6} \qquad \text{Divide 2 by 3, and attach the negative sign.}$$

The fractional form $\frac{-2}{3}$ is easier to work with than the decimal form $-0.\overline{6}$, since the decimal is nonterminating.

▲ **Try Problems 57 and 58.**

▲ **Answers to Try These Problems**

1. -27 2. -28 3. -65 4. -180 5. 0 6. 0 7. 0 8. 15
9. 21 10. 1440 11. 128 12. -35 13. -54 14. 35
15. 54 16. 0 17. -8400 18. 168 19. -270 20. -160
21. 0 22. 1920 23. -3960 24. 48 25. -50 26. 22 27. 8
28. -5 29. 7 30. -6 31. -63 32. -4 33. -3 34. 5
35. 1 36. no answer 37. 0 38. 0 39. no answer 40. -5
41. -57 42. 5 43. no answer 44. -14 45. no answer
46. 6 47. 1 48. -6 49. $\frac{-1}{6}$ 50. $\frac{3}{2}$ or $1\frac{1}{2}$ 51. $\frac{2}{3}$
52. $\frac{-7}{8}, -0.875, -\frac{7}{8}$ 53. $\frac{-1}{9}, \frac{1}{-9}$ 54. $\frac{-1}{8}$ 55. $\frac{-9}{5}$ or $-1\frac{4}{5}$
56. $\frac{-7}{8}$ 57. -0.375 58. $-1.\overline{2}$

EXERCISES 12.5

Multiply.

1. $(-4)8$
2. $7(-5)$
3. $-9(-4)$
4. $(-6)(-8)$
5. $-7(-12)$
6. $8(-25)$
7. $(32)(-4)$
8. $0 \cdot 0$
9. $-5 \cdot 0$
10. $25 \cdot 19$
11. $-65(-40)$
12. $75(-88)$
13. $-200(90)$
14. $-78(-3000)$
15. $-704(-603)$
16. $670(-3002)$
17. $4 \cdot (-5) \cdot 3$
18. $-2 \cdot 6 \cdot (-2)$
19. $5(-7)(-8)$
20. $(-4)(-6)(-9)$
21. $-2(-3)(5)(-4)$
22. $3(-6)(-3)(10)$
23. $-9(-7)(-4)(0)(-12)$
24. $7(-15)(-8)(-15)(0)$
25. $-10(-5)(-1)(5)(-6)$
26. $-1(-3)(-15)(-4)(-8)$

Evaluate.

27. $-9 + (-5)(-6)$
28. $4 - 7(-3)$
29. $(-5 + 7)(-3 - 8)$
30. $25 - 4(-3)(-2)$
31. $-12 + 6(-1 - 5 + 3)$
32. $7 - 8(9 - 3 \cdot 6)$
33. $|-12 - 7 - 1|$
34. $7|2 - 9|$
35. $|(-3)(6) + (-5)(2)|$
36. $|-5 + 3| - |-4 - (-2)|$

Divide, if possible. If not possible, write no answer.

37. $35 \div (-7)$
38. $-36 \div (6)$
39. $-81 \div (-9)$
40. $-42 \div (-6)$
41. $-75 \div 5$
42. $-84 \div (-6)$
43. $-200 \div (-4)$
44. $-416 \div 4$
45. $-36 \div 36$
46. $-15 \div (-1)$
47. $0 \div (-18)$
48. $14 \div 0$
49. $\frac{-48}{-8}$
50. $\frac{18}{-9}$
51. $\frac{300}{-5}$
52. $\frac{-450}{-6}$
53. $\frac{87}{-1}$
54. $\frac{-16}{-16}$
55. $\frac{0}{0}$
56. $\frac{0}{-19}$

57. Which of these numbers are equal to the fraction $-\frac{15}{2}$?
    a. $\frac{-15}{2}$
    b. $\frac{-2}{-15}$
    c. -7.5
    d. $\frac{15}{-2}$
    e. $-\left(\frac{15}{2}\right)$

58. Which of these numbers are equal to the fraction $\frac{-2}{3}$?
    a. $-1\frac{1}{2}$
    b. $-0.\overline{6}$
    c. $\frac{-2}{-3}$
    d. $\frac{2}{3}$
    e. $\frac{2}{-3}$

Evaluate. Write the answer as a fraction reduced to lowest terms.

59. $6 \div (-2)$
60. $-2 \div 6$
61. $-3 \div (-27)$
62. $-27 \div (-3)$
63. $21 \div 35$
64. $35 \div 21$
65. $-24 \div (-15)$
66. $-15 \div (-24)$
67. -24 divided by -3.
68. -9 divided into 2.
69. -6 divided into 40.
70. -15 divided by -20.
71. Divide -480 by 12.
72. Divide -760 into -200.

Convert each fraction to decimal form.

73. $\frac{5}{8}$
74. $\frac{-7}{8}$
75. $\frac{5}{-6}$
76. $\frac{-1}{-6}$
77. $\frac{-4}{3}$
78. $\frac{-8}{-3}$
79. $-4\frac{1}{4}$
80. $-8\frac{3}{4}$

12.6 SIGNED FRACTIONS AND DECIMALS

OBJECTIVES
- Name signed fractions and decimals associated with points on a number line.
- Add, subtract, multiply, or divide two signed fractions or decimals.
- Evaluate numerical expressions that involve signed fractions and decimals.

1 Signed Fractions on the Number Line

Here is a number line with some fractions with denominator 4.

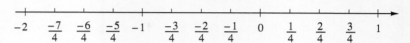

The fractions get larger as we move in the direction of the arrow. On this number line, the numbers are increasing by $\frac{1}{4}$ unit as we move from left to right, so that the length of each smaller segment is $\frac{1}{4}$ unit. Recall that the distance between any two points can be found by subtracting the smaller number from the larger number.

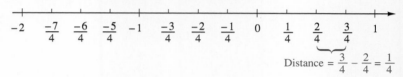

EXAMPLE 1 Give the fractions associated with points A, B, and C.

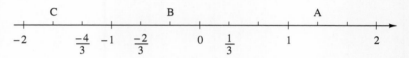

SOLUTION We can see that each of the smaller segments has length $\frac{1}{3}$ by subtracting any two numbers that are associated with adjacent tick marks. For example, the distance between 0 and $\frac{1}{3}$ is $\frac{1}{3}$.

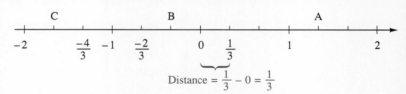

Now fill in the missing numbers by adding $\frac{1}{3}$ as you move from left to right ($\rightarrow$) from one marking to the next marking or by adding $\frac{-1}{3}$ as you move from right to left ($\leftarrow$).

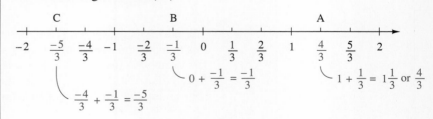

A. $\frac{4}{3}$ or $1\frac{1}{3}$ B. $\frac{-1}{3}$ C. $\frac{-5}{3}$ or $-1\frac{2}{3}$ ∎

594 Chapter 12 Order of Operation, Signed Numbers, and Variables

 Try These Problems

Give the fractions associated with points B and C.

1.
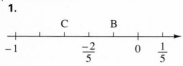

2.

3.

EXAMPLE 2 Give the fractions associated with points A, B, and C.

SOLUTION First, find the distance between two given numbers. For example, the distance between 0 and 1 is 1.

Second, count the number of segments between 0 and 1. There are 5 segments between 0 and 1. If a 1-unit length is divided into 5 equal parts then each part has length $\frac{1}{5}$ unit. Thus each of the smaller segments has length $\frac{1}{5}$ unit.

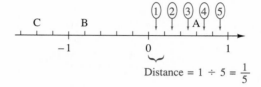

Finally, fill in the missing numbers by adding $\frac{1}{5}$ as you move from left to right ($\rightarrow$) or by adding $\frac{-1}{5}$ as you move from right to left ($\leftarrow$).

C B A
$\frac{-8}{5}$ $\frac{-7}{5}$ $\frac{-6}{5}$ -1 $\frac{-4}{5}$ $\frac{-3}{5}$ $\frac{-2}{5}$ $\frac{-1}{5}$ 0 $\frac{1}{5}$ $\frac{2}{5}$ $\frac{3}{5}$ $\frac{4}{5}$ 1

A. $\frac{3}{5}$ B. $\frac{-4}{5}$ C. $\frac{-7}{5}$ or $-1\frac{2}{5}$ ∎

 Try Problems 1 through 3.

2 Signed Decimals on the Number Line

Here is a number line with some decimals between -2 and 1.

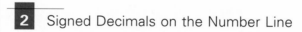

The decimals get larger as we move in the direction of the arrow. On this number line the numbers are increasing by 0.25 as we move from left to right so that the length of each of the smaller segments is 0.25. Recall that the distance between any two numbers can be found by subtracting the smaller number from the larger number.

12.6 Signed Fractions and Decimals   595

 Try These Problems

Give the decimals associated with points A and B.

4.
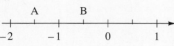

5.
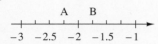

6.

EXAMPLE 3 Give the decimals associated with points A and B.

SOLUTION We can see that each of the smaller segments has length 0.2 by subtracting any two numbers that are associated with adjacent tick marks. For example, the distance between 0.6 and 0.8 is 0.2.

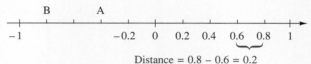

Now fill in the missing numbers by adding 0.2 as you move from left to right (→) or by adding −0.2 as you move from right to left (←).

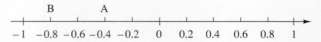

A. −0.4   B. −0.8 ■

EXAMPLE 4 Give the decimals associated with points A and B.

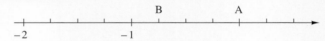

SOLUTION First, find the distance between two given numbers. For example, the distance between −2 and −1 is 1.

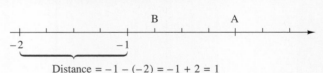

Second, count how many segments are between −2 and −1. There are 4 segments between −2 and −1. If a 1-unit segment is divided into 4 equal parts then each part has length $\frac{1}{4}$ or 0.25 unit.

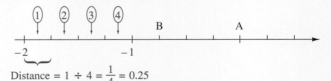

Now fill in the missing numbers by adding 0.25 as you move from left to right (→) or by adding −0.25 as you move from right to left (←).

A. 0   B. −0.75 ■

 Try Problems 4 through 6.

Try These Problems

Evaluate.
7. $\frac{1}{4} + \frac{-2}{3}$
8. $\frac{-1}{5} + \frac{-3}{10}$
9. $-8 - \frac{-1}{3}$
10. $\frac{1}{2} - \frac{5}{6} + 1$

3 Operating with Signed Fractions

Next we look at adding, subtracting, multiplying, and dividing signed fractions. We do not need any new rules. We simply use a combination of rules that have already been introduced. We will need the rules for operating with positive fractions together with the rules for operating with signed numbers. Here are some examples.

EXAMPLE 5 Evaluate: $\frac{-1}{2} - \frac{-3}{4}$

SOLUTION Because the operation is subtraction, we need a common denominator. The least common multiple (LCM) of 2 and 4 is 4.

$$\frac{-1}{2} - \frac{-3}{4}$$

$$= \frac{-1}{2} + \frac{3}{4} \qquad \text{Subtracting } \frac{-3}{4} \text{ is the same as adding } \frac{3}{4}.$$

$$= \frac{-1 \cdot 2}{2 \cdot 2} + \frac{3}{4} \qquad \text{Convert the fraction } \frac{-1}{2} \text{ to a fraction with denominator 4 by multiplying numerator and denominator by 2.}$$

$$= \frac{-2}{4} + \frac{3}{4}$$

$$= \frac{1}{4} \qquad \text{Add } -2 \text{ and } 3 \text{ to obtain } 1. \text{ Keep the same denominator.}$$

EXAMPLE 6 Evaluate: $2 + \frac{-4}{5} - \frac{2}{3}$

SOLUTION Before adding and subtracting fractions, we need a common denominator. The least common multiple (LCM) of 3 and 5 is 15.

$$2 + \frac{-4}{5} - \frac{2}{3}$$

$$= \frac{2}{1} + \frac{-4}{5} - \frac{2}{3} \qquad \text{The whole number 2 can be written as the fraction } \frac{2}{1}.$$

$$= \frac{2 \cdot 15}{1 \cdot 15} + \frac{-4 \cdot 3}{5 \cdot 3} - \frac{2 \cdot 5}{3 \cdot 5} \qquad \text{Convert each of the three numbers to a fraction with denominator 15.}$$

$$= \frac{30}{15} + \frac{-12}{15} - \frac{10}{15}$$

$$= \frac{30}{15} + \frac{-12}{15} + \frac{-10}{15} \qquad \text{Subtracting } \frac{10}{15} \text{ is the same as adding } \frac{-10}{15}.$$

$$= \frac{30 + (-12) + (-10)}{15} \qquad \text{Add the numerators. Keep the same denominator.}$$

$$= \frac{30 + (-22)}{15}$$

$$= \frac{8}{15}$$

 Try Problems 7 through 10.

Try These Problems

Evaluate.
11. $\frac{-5}{6} \cdot \frac{-4}{3}$
12. $\frac{1}{2} \cdot (-7)$
13. $\frac{-4}{15} \div \frac{14}{5}$
14. $\frac{-2}{3} \div (-12)$

EXAMPLE 7 Evaluate: $\frac{5}{8} \cdot \frac{-2}{7}$

SOLUTION Observe that a *positive* fraction is being multiplied by a *negative* fraction, so the result should be a *negative* fraction.

$$\frac{5}{8} \cdot \frac{-2}{7}$$

$$= \frac{5}{2 \cdot 4} \cdot \frac{\overset{1}{-\cancel{2}}}{7} \quad \text{Cancel the common factor 2 from the numerator and denominator.}$$

$$= \frac{-5}{28} \quad \begin{array}{l} \text{Multiply 5 by } -1 \text{ to obtain } -5. \\ \text{Multiply 4 by 7 to obtain 28.} \end{array}$$

EXAMPLE 8 Evaluate: $-6 \div \frac{-9}{4}$

SOLUTION Observe that a *negative* number is being divided by a *negative* number, so the result should be a *positive* number.

$$-6 \div \frac{-9}{4}$$

$$= \frac{-6}{1} \cdot \frac{4}{-9} \quad \begin{array}{l} \text{Change the division} \\ \text{to multiplication and} \\ \text{invert the divisor.} \end{array}$$

$$= \frac{-2 \cdot \cancel{3}}{1} \cdot \frac{4}{-3 \cdot \cancel{3}} \quad \begin{array}{l} \text{Cancel the common factor} \\ \text{3 from the numerator} \\ \text{and denominator.} \end{array}$$

$$= \frac{-8}{-3} \quad \begin{array}{l} \text{Multiply } -2 \text{ by 4 to obtain } -8. \\ \text{Multiply 1 by } -3 \text{ to obtain } -3. \end{array}$$

$$= \frac{8}{3} \text{ or } 2\frac{2}{3} \quad \begin{array}{l} \text{A negative divided by a} \\ \text{negative is positive.} \end{array}$$

Try Problems 11 through 14.

4 Operating with Signed Decimals

To operate with signed decimals we use the rules for operating with positive decimals together with the rules for operating with signed numbers. Here are some examples.

EXAMPLE 9 Evaluate: $-12.2 - 0.75$

SOLUTION First, we convert the subtraction to addition by using the fact that subtracting a number is the same as adding its opposite.

$$-12.2 - 0.75$$

$$= -12.2 + (-0.75) \quad \begin{array}{l} \text{Subtracting 0.75 is the same} \\ \text{as adding } -0.75. \end{array}$$

$$= -12.95 \quad \begin{array}{l} \text{Add the decimals 12.2 and} \\ \text{0.75 and attach the like sign.} \end{array}$$

$$\begin{array}{r} 12.20 \\ +\ 0.75 \\ \hline 12.95 \end{array} \quad \begin{array}{l} \text{Align the decimal} \\ \text{points.} \end{array}$$

The answer is -12.95.

Try These Problems

Evaluate.
15. $5.7 - 8.3$
16. $-6.23 - 4$
17. $0.92 + (-6.4)$
18. $23.1 - (-4.5)$
19. $-0.75(500)$
20. $-0.3(-1.24)$
21. $-4.01 \div (-0.5)$
22. $\frac{-104}{0.16}$

EXAMPLE 10 Evaluate: $0.86 + (-5)$

SOLUTION Because we are adding unlike signs, and -5 has the larger absolute value, the answer is negative. To find the absolute value of the answer, we subtract 0.86 from 5.

$$0.86 + (-5)$$
$$= -4.14$$

Subtract 0.86 from 5 and attach the sign of -5.

$$\begin{array}{r} 49 \\ \cancel{5}.\cancel{0}{}^{1}0 \\ -\ 0.8\ 6 \\ \hline 4.1\ 4 \end{array}$$

Align the decimal points.

The answer is -4.14. ■

Try Problems 15 through 18.

EXAMPLE 11 Evaluate: $-0.04(-6.7)$

SOLUTION A negative number is multiplied by a negative number so the result is positive.

$$-0.04(-6.7)$$
$$= 0.268$$

Multiply 0.04 by 6.7 and the result is positive.

$$\begin{array}{r} 6.7 \\ \times\ 0.0\ 4 \\ \hline 0.2\ 6\ 8 \end{array}$$

1 decimal place
2 decimal places
3 decimal places

The answer is 0.268. ■

EXAMPLE 12 Evaluate: $\dfrac{108}{-0.12}$

SOLUTION A positive number is divided by a negative number so the result is negative.

$$\frac{108}{-0.12}$$
$$= 108 \div (-0.12)$$
$$= -900$$

Divide 108 by 0.12 and attach the negative sign.

$$\begin{array}{r} 9\ 0\ 0. \\ 0.1\ 2\overline{)1\ 0\ 8.0\ 0} \\ \underline{1\ 0\ 8} \\ 0 \end{array}$$

The result is -900. ■

Try Problems 19 through 22.

5 Mixing the Operations

Now we look at evaluating expressions with fractions and decimals that contain more than one operation. In solving these problems, we use not only the rules for signed numbers and the rules for fractions and decimals, but also the rules for the order to operate. Here we review the order of operation rules.

12.6 Signed Fractions and Decimals

> *Order of Operation*
> 1. Operate inside grouping symbols, beginning with the innermost.
> 2. Take absolute values.
> 3. Take opposites, multiply, and divide, from left to right. (Repeated multiplication can be done in any order.)
> 4. Add and subtract, from left to right. (Repeated addition can be done in any order.)

EXAMPLE 13 Evaluate: $\frac{1}{2}(9.6)(8 - 3.5)$

SOLUTION Begin by subtracting inside the grouping symbol.

$\frac{1}{2}(9.6)(8 - 3.5)$ Subtract 3.5 from 8.

$= \frac{1}{2}(9.6)(4.5)$
$$\begin{array}{r} \overset{7}{\cancel{8}}.\overset{1}{0} \\ -\ 3.5 \\ \hline 4.5 \end{array}$$
Align decimal points.

$= \frac{1}{2}(43.20)$ Multiply the numbers in any order. We multiply 9.6 by 4.5.

$= \frac{1}{2} \cdot \frac{43.2}{1}$
$$\begin{array}{r} 9.6 \\ \times\ \ 4.5 \\ \hline 4\ 8\ 0 \\ 38\ 4\ \ \ \\ \hline 43.2\ 0 \end{array}$$
1 decimal place
1 decimal place

2 decimal places

$= \frac{43.2}{2}$ Divide 43.2 by 2.

$= 21.6$ ∎

EXAMPLE 14 Evaluate: $4\left(\frac{-2}{3}\right) - \frac{1}{2}\left(\frac{-4}{5}\right) + 2$

SOLUTION Begin by performing the two multiplications.

$4\left(\dfrac{-2}{3}\right) - \dfrac{1}{2}\left(\dfrac{-4}{5}\right) + 2$

$= \dfrac{4}{1} \cdot \dfrac{-2}{3} - \dfrac{1}{\underset{1}{\cancel{2}}} \cdot \dfrac{\overset{2}{\cancel{-4}}}{5} + 2$

$= \dfrac{-8}{3} - \dfrac{-2}{5} + 2$ Multiply first.

$= \dfrac{-8}{3} + \dfrac{2}{5} + \dfrac{2}{1}$ Subtracting $\frac{-2}{5}$ is the same as adding $\frac{2}{5}$.

$= \dfrac{-8 \cdot 5}{3 \cdot 5} + \dfrac{2 \cdot 3}{5 \cdot 3} + \dfrac{2 \cdot 15}{1 \cdot 15}$ Convert each fraction to an equivalent fraction with denominator 15.

$= \dfrac{-40}{15} + \dfrac{6}{15} + \dfrac{30}{15}$

Try These Problems

Evaluate.
23. $\frac{3}{4}(6.4)(5 - 2.75)$
24. $\frac{-1}{3}\left(\frac{-2}{3}\right) - 5\left(\frac{1}{4}\right) - 1$

$$= \frac{-40 + 6 + 30}{15}$$

$$= \frac{-40 + 36}{15}$$

$$= \frac{-4}{15} \quad \blacksquare$$

Add the numerators. Keep the same denominator.

Try Problems 23 and 24.

Answers to Try These Problems

1. B. $\frac{-1}{5}$ C. $\frac{-3}{5}$ 2. B. $\frac{-2}{3}$ C. $\frac{-7}{3}$ or $-2\frac{1}{3}$ 3. B. $-4\frac{1}{2}$ C. $-3\frac{1}{2}$
4. A. -1.5 B. -0.5 5. A. -2.25 B. -1.75
6. A. -0.8 B. -0.2 7. $\frac{-5}{12}$ 8. $\frac{-1}{2}$ 9. $\frac{-23}{3}$ or $-7\frac{2}{3}$ 10. $\frac{2}{3}$
11. $\frac{10}{9}$ or $1\frac{1}{9}$ 12. $\frac{-7}{2}$ or $-3\frac{1}{2}$ 13. $\frac{-2}{21}$ 14. $\frac{1}{18}$ 15. -2.6
16. -10.23 17. -5.48 18. 27.6 19. -375 20. 0.372
21. 8.02 22. -650 23. 10.8 24. $\frac{-73}{36}$ or $-2\frac{1}{36}$

EXERCISES 12.6

Give the fractions associated with points A, B, and C.

1.

2.

3.

4.

Give the decimals associated with points A and B.

5.

6.

7.

8.

Evaluate.

9. $\frac{-2}{5} + \frac{-3}{10}$ 10. $\frac{-7}{8} + \frac{1}{6}$ 11. $\frac{3}{4} - \frac{-3}{5}$ 12. $\frac{-3}{7} - \frac{11}{14}$

13. $\frac{1}{3} - 2 + \frac{-5}{2}$ 14. $\frac{3}{8} - \frac{1}{2} + 3$ 15. $\frac{5}{9} \cdot \frac{-3}{7}$ 16. $(\frac{-15}{4})(\frac{-3}{25})$

17. $-5(\frac{2}{3})$ 18. $8 \cdot \frac{-3}{16}$ 19. $\frac{-8}{9} \div \frac{-4}{5}$ 20. $\frac{12}{5} \div \frac{-4}{15}$

21. $-35 \div \frac{7}{5}$ 22. $\frac{-18}{4} \div (-6)$ 23. $-0.67 + 2.4$

24. $-5.46 + (-3.8)$ 25. $6 - (-3.4)$ 26. $0.08 - 5$

27. $-9.04 - 12.2$ 28. $56.8 + (-0.65) - 38$ 29. $-7.8 - 9 + 8.25$

30. $-2.5(-30.8)$ 31. $0.86(-700)$ 32. $-5.2(-6)(-4.5)$

33. $-4(9.12)(-3.5)$ 34. $60 \div (-0.4)$ 35. $-7.8 \div (-1.3)$

36. $\frac{-376.5}{7.5}$ 37. $\frac{-20}{-0.025}$ 38. $-2.4(9 - 4.8)$

39. $(-5.8)(-0.4) - 2.3$ 40. $\frac{5(-1.2) - 1}{-3.2 + 0.4}$ 41. $\frac{1}{2}(8.6)(8.2 - 6.78)$

42. $7 - 3.5(4 - 5.2)$ 43. $\frac{-1}{5} + 8(\frac{-3}{10})$ 44. $\frac{-3}{4}(\frac{2}{3} - \frac{-5}{9})$

45. $-4(\frac{-7}{8}) - 8(\frac{5}{12}) + \frac{9}{2}$ 46. $\frac{7}{4} + \frac{3}{8}(\frac{1}{4} - \frac{9}{4})$ 47. $\frac{2}{3}(0.6) + 4.2$

48. $\frac{4}{5}(20.5 + 15)$ 49. $\frac{-3}{4}(6.8)(0.12)$ 50. $\frac{1}{4}(8.24) - \frac{2}{5}(10.2)$

USING THE CALCULATOR #13

THE +/− KEY

Most all calculators have a plus/minus key, +/−. This key can be used to change the sign of an entry from positive to negative or from negative to positive. In this feature we show some examples of using this key.

To Compute   −97 + 89
Enter   97 +/− + 89 =
Result   −8.

A scientific calculator displays the negative sign in front of the number 8 as shown here. However, on a basic calculator the negative sign may not be displayed in front of the number 8. Some basic calculators will display the negative to the right of the number 8 while others put the negative in the right or left corner of the display screen away from the number 8. Examine your calculator to see how it displays a negative number.

Here we look at another example.

To Compute |17 − 28| + 54
Enter   17 − 28 = +/− + 54 =
Result   65.

The result of subtracting 28 from 17 is a negative number, −11, so we change the sign since we want to add the absolute value of −11 to 54.

Calculator Problems

Use the plus/minus key to compute each of the following. If the decimal representation of the answer goes beyond 3 decimal places, then round off at 3 decimal places.

1. $-5 + (-6)$ 2. $-30 - 16$ 3. $-8(1.75)$ 4. $-19(0.5) + 7$

5. $-6.6(-45)(15)$ 6. $18|-3 - 6|$ 7. $|14 - 90| - 130$ 8. $\frac{-2.5}{-0.004}$

Use the plus/minus key to compute each of the following. If you are working on a basic calculator you may also need to use the memory keys. If you are working on a scientific calculator, you will not need the parentheses or memory keys because a scientific calculator uses the order of operation.

9. $-50 - (-9)(8)$ 10. $450 + 75(-36)$

11. $-6(-7.4) - 8(19)$ 12. $\frac{-4}{5} + \frac{1}{8}$

12.7 VARIABLES, SIMPLE ALGEBRAIC EXPRESSIONS, AND FORMULAS

OBJECTIVES
- Translate an English phrase to an algebraic expression.
- Evaluate an expression for the given values of the variables.
- Evaluate a formula for given values of the variables.

 Try These Problems

Assume n represents some unknown number. Translate each phrase to an algebraic expression.
1. The sum of n and 8.
2. The product of n and 8.
3. The ratio of 8 to n.
4. 8 less than n.
5. 8 minus n.
6. 7 subtracted from 5 times n.

1 Working with Variables and Algebraic Expressions

In algebra we often work with unknown quantities. We represent an unknown quantity with a letter of the alphabet. For example, we could let T represent the amount of taxes you will pay next year, we could let n represent the number of years you will go to school, or we could let x represent the smallest of the five numbers that will be drawn next week in the New York lottery. Letters used in this way are called **variables.**

An **algebraic expression** is a combination of numbers, variables, and operations put together in a meaningful way so that the expression represents a number when the variables are assigned values. For example, suppose we let x represent some unknown number. Here are some algebraic expressions and how to say them using English.

Algebraic Expression	English
$x + 3$	x plus 3 x increased by 3 The sum of x and 3
$x - 3$	x minus 3 x take away 3 x decreased by 3 3 subtracted from x 3 less than x
$3x$	3 times x 3 multiplied by x The product of 3 and x
$\dfrac{x}{3}$	x divided by 3 3 divided into x The ratio of x to 3
$2x - 5$	2 times x, minus 5 Twice x, decreased by 5 5 less than two times x 5 subtracted from twice x

 Try Problems 1 through 6.

2 Evaluating Algebraic Expressions for Given Values of the Variables

An algebraic expression takes on a specific value when the variable is given a value. For example, the expression $2x - 5$ can be evaluated for $x = 8$.

Try These Problems

Evaluate for the given values of the variables.

7. $8x + 9$ for $x = 2$
8. $15 - c$ for $c = 20$
9. $6m - n$ for $n = -4$ and $m = 10$
10. $-xy + 3y$ for $x = 8$ and $y = -2$
11. $8(12 - x)$ for $x = 5$
12. $-n(9 + n)$ for $n = -6$
13. $(x + y)(x - y)$ for $x = 2$ and $y = -5$
14. $(4m - 10)(5m + 7)$ for $m = 2$

$$2x - 5$$
$$= 2(8) - 5 \quad \text{Substitute 8 for } x. \text{ Be sure to wrap 8 in parentheses to indicate 2 times 8.}$$
$$= 16 - 5 \quad \text{Multiply first, then subtract.}$$
$$= 11$$

Here are more examples of evaluating algebraic expressions for given values of the variables.

EXAMPLE 1 Evaluate $w - y$ for $w = 9$ and $y = -5$.

SOLUTION Carefully substitute in the given values before making any calculations.

$$w - y$$
$$= 9 - (-5) \quad \text{Substitute 9 for } w \text{ and } -5 \text{ for } y.$$
$$\quad \text{Wrap } -5 \text{ in parentheses.}$$
$$= 9 + 5 \quad \text{Subtracting } -5 \text{ is the same as adding 5.}$$
$$= 14 \quad ∎$$

EXAMPLE 2 Evaluate $-n + 4m$ for $n = -6$ and $m = -7$.

SOLUTION After substituting in the given values, follow the rules for the order to operate.

$$-n + 4m$$
$$= -(-6) + 4(-7) \quad \text{Substitute } -6 \text{ for } n \text{ and } -7 \text{ for } m.$$
$$= 6 + (-28) \quad \text{The opposite of } -6 \text{ is 6.}$$
$$\quad 4 \text{ times } -7 \text{ is } -28.$$
$$= -22 \quad \text{Add 6 and } -28 \text{ to obtain } -22. \quad ∎$$

 Try Problems 7 through 10.

EXAMPLE 3 Evaluate $-20(5 + a)$ for $a = -11$.

SOLUTION After substituting in the given values, follow the rules for the order to operate.

$$-20(5 + a)$$
$$= -20(5 + (-11)) \quad \text{Substitute } -11 \text{ for } a.$$
$$\quad \text{Wrap } -11 \text{ in parentheses.}$$
$$= -20(-6) \quad \text{Add 5 and } -11 \text{ to obtain } -6.$$
$$= 120 \quad \text{Multiply } -20 \text{ by } -6 \text{ to obtain } 120. \quad ∎$$

EXAMPLE 4 Evaluate $(4y - 6)(8 + y)$ for $y = 3$.

SOLUTION After substituting in the given values, follow the rules for the order to operate.

$$(4y - 6)(8 + y)$$
$$= (4(3) - 6)(8 + 3) \quad \text{Substitute 3 for } y.$$
$$= (12 - 6)(11) \quad \text{Multiply 4 by 3 to obtain 12.}$$
$$\quad \text{Add 8 and 3 to obtain 11.}$$
$$= (6)(11) \quad \text{Subtract 6 from 12 to obtain 6.}$$
$$= 66 \quad ∎$$

 Try Problems 11 through 14.

Try These Problems

Evaluate for the given values of the variables.

15. $-m + n$ for $m = \frac{1}{2}$ and $n = \frac{5}{8}$
16. $8ab$ for $a = \frac{1}{4}$ and $b = \frac{-7}{8}$
17. $a - c$ for $a = 8.1$ and $c = 12$.
18. $xy + 9$ for $x = 0.52$ and $y = 3.5$

EXAMPLE 5 Evaluate $-4xy$ for $x = 6$ and $y = \frac{2}{3}$.

SOLUTION The expression $-4xy$ means -4 times x times y, so be sure to indicate the multiplication after substituting in the values for x and y.

$$-4xy$$
$$= -4 \cdot 6 \cdot \frac{2}{3} \quad \text{Substitute 6 for } x \text{ and } \frac{2}{3} \text{ for } y.$$
$$= -24 \cdot \frac{2}{3} \quad \text{Multiply the three numbers in any order. We multiply } -4 \text{ by 6 first.}$$
$$= \frac{-\overset{8}{\cancel{24}}}{1} \cdot \frac{2}{\underset{1}{\cancel{3}}} \quad \text{Cancel the common factor 3.}$$
$$= -16 \quad \blacksquare$$

 Try Problems 15 and 16.

EXAMPLE 6 Evaluate $3.4 - 5w$ for $w = 1.2$.

SOLUTION Substitute 1.2 for w.

$$3.4 - 5w$$
$$= 3.4 - 5(1.2)$$
$$= 3.4 - 6 \quad \text{Multiply 5 by 1.2 to obtain 6.}$$
$$= 3.4 + (-6) \quad \text{Subtracting 6 is the same as adding } -6.$$
$$= -2.6 \quad \text{Subtract 3.4 from 6 to obtain 2.6. Attach the negative sign.} \quad \blacksquare$$

 Try Problems 17 and 18.

3 Working with Formulas

Recall from Sections 2.5, 4.5, 7.5, 8.4, and 10.2, that the *area* of a rectangle can be found by multiplying the length by the width. Suppose we let A represent the area of a rectangle, L represent the length, and W represent the width. Then we have the following:

$$\text{Area} = \text{length} \times \text{width}$$
$$A = L \cdot W$$
$$A = LW$$

The statement, $A = LW$, is an example of a **formula**. *A formula is a statement containing at least two variables that indicates that two algebraic expressions are equal.* The formula gives a relationship between two or more unknown quantities.

We can evaluate formulas for given values of the variables. For example, we can use the above formula to find the area of a rectangle, if we are given the length and width.

12.7 Variables, Simple Algebraic Expressions, and Formulas

 Try These Problems

19. Evaluate $A = LW$ for $L = 24$ inches and $W = 20\frac{1}{3}$ inches.

20. Evaluate $P = 2L + 2W$ for $L = 60$ meters and $W = 45.5$ meters.

EXAMPLE 7 Find the area of a rectangle with width 6.85 feet and length 7.2 feet.

SOLUTION We use the formula, $A = LW$.

$A = LW$
$A = (7.2)(6.85)$ — Substitute 7.2 for L and 6.85 for W.
$A = 49.32$

The area is 49.32 square feet. ■

Recall from Sections 1.4, 5.7, 6.7, 8.4, and 10.1 that the *perimeter* of a rectangle is the distance all the way around. Therefore, the perimeter is the sum of the lengths of the four sides. Suppose we let P represent the perimeter of a rectangle, L represent the length, and W represent the width. We have the following:

$\text{Perimeter} = L + L + W + W$
$P = L + L + W + W$
$P = 2L + 2W$

Because $L + L$ is the same as 2 times L and $W + W$ is the same as 2 times W, we simplify the formula.

This formula for the perimeter of a rectangle is one you should be familiar with. Here we look at an example.

EXAMPLE 8 Find the perimeter of a rectangle with length $7\frac{1}{2}$ inches and width $5\frac{2}{3}$ inches.

SOLUTION We use the formula $P = 2L + 2W$.

$P = 2L + 2W$

$= 2\left(7\frac{1}{2}\right) + 2\left(5\frac{2}{3}\right)$ — Substitute $7\frac{1}{2}$ for L and $5\frac{2}{3}$ for W.

$= \dfrac{\overset{1}{\cancel{2}}}{1} \cdot \dfrac{15}{\underset{1}{\cancel{2}}} + \dfrac{2}{1} \cdot \dfrac{17}{3}$ — Convert the two mixed numerals to improper fractions.

$= \dfrac{15}{1} + \dfrac{34}{3}$

$= \dfrac{15 \cdot 3}{1 \cdot 3} + \dfrac{34}{3}$ — Convert $\dfrac{15}{1}$ to an equivalent fraction with denominator 3.

$= \dfrac{45}{3} + \dfrac{34}{3}$

$= \dfrac{79}{3}$ or $26\dfrac{1}{3}$

The perimeter is $26\frac{1}{3}$ inches. ■

 Try Problems 19 and 20.

Formulas are used in many situations. You should know how to evaluate a formula for given values of the variables. Here is an example that involves the formula that relates Celsius temperature to Fahrenheit temperature.

Try These Problems

21. Evaluate $C = \frac{5}{9}(F - 32)$ for $F = 14°$.
22. Evaluate $F = \frac{9}{5}C + 32$ for $C = 10°$.

EXAMPLE 9

The formula
$$C = \frac{5}{9}(F - 32)$$
where C is the Celsius temperature and F is the Fahrenheit temperature, gives the relationship between degrees Celsius and degrees Fahrenheit. If the temperature in Cleveland is 23° Fahrenheit, what is the temperature on the Celsius scale?

SOLUTION We substitute 23 for F in the given formula.

$$C = \frac{5}{9}(F - 32)$$
$$= \frac{5}{9}(23 - 32) \quad \text{Substitute 23 for } F.$$
$$= \frac{5}{9}(23 + (-32)) \quad \text{Subtracting 32 is the same as adding } -32.$$
$$= \frac{5}{9}(-9) \quad \text{Add 23 and } -32 \text{ to obtain } -9.$$
$$= \frac{5}{\cancel{9}_1} \cdot \frac{\cancel{-9}^1}{1}$$
$$= -5$$

The temperature is $-5°$ Celsius. ∎

Try Problems 21 and 22.

Answers to Try These Problems

1. $n + 8$ or $8 + n$ 2. $8n$ 3. $\frac{8}{n}$ 4. $n - 8$ 5. $8 - n$
6. $5n - 7$ 7. 25 8. -5 9. 64 10. 10 11. 56
12. 18 13. -21 14. -34 15. $\frac{1}{8}$ 16. $\frac{-7}{4}$ or $-1\frac{3}{4}$ 17. -3.9
18. 10.82 19. 488 sq in 20. 211 m 21. $-10°C$ 22. $50°F$

EXERCISES 12.7

Assume x and y represent unknown numbers. Translate each phrase to an algebraic expression.

1. x multiplied by y.
2. x plus y.
3. The ratio of y to x.
4. x subtracted from y.
5. x minus y.
6. The sum of x and 4.
7. 4 less than y.
8. 4 divided into y.
9. 6 minus twice y.
10. 3 times x, increased by 12.

Evaluate for the given values of the variables.

11. $5x + 1$ for $x = 6$
12. $7xy$ for $x = 3$ and $y = -2$
13. $4x - 6y$ for $x = -3$ and $y = 7$
14. $-5n + m$ for $n = 4$ and $m = 8$
15. $n(9 - m)$ for $n = 7$ and $m = -16$

16. $6a(a - b)$ for $a = -1$ and $b = 10$
17. $(x + 2y)(x - y)$ for $x = 3$ and $y = 6$
18. $(3m - 1)(m + 5)$ for $m = -4$
19. $5w - 6x + 8$ for $w = -5$ and $x = 0$
20. $12 - 3m + 4x$ for $m = -7$ and $x = -1$
21. $6x$ for $x = \frac{1}{3}$
22. $6 - x$ for $x = \frac{1}{3}$
23. $m + n$ for $m = \frac{3}{8}$ and $n = \frac{1}{4}$
24. $-3mn$ for $m = \frac{5}{6}$ and $n = 12$
25. $30 - 4m$ for $m = 6.25$
26. $\frac{m}{5}$ for $m = 0.45$
27. $20xy$ for $x = 0.8$ and $y = -1.4$
28. $35x + 50$ for $x = 1.6$

Use the formula $A = LW$ to find the area of the rectangle with the given length and width.

29. $L = 84$ inches, $W = 20$ inches
30. $L = 40.8$ meters, $W = 25$ meters
31. $L = 0.8$ mile, $W = 0.5$ mile
32. $L = 12\frac{2}{3}$ yards, $W = 9$ yards

Use the formula $P = 2L + 2W$ to find the perimeter of the rectangle with the given length and width.

33. $L = 15$ feet, $W = 12$ feet
34. $L = 230$ centimeters, $W = 85$ centimeters
35. $L = 2\frac{1}{4}$ miles, $W = 1\frac{1}{2}$ miles
36. $L = 24.6$ meters, $W = 15.36$ meters

Use the temperature conversion formula $F = \frac{9}{5}C + 32$ to find the Fahrenheit temperature that corresponds to the given Celsius temperature.

37. $C = 100°$
38. $C = 35°$
39. $C = -15°$
40. $C = -30°$

Use the temperature conversion formula $C = \frac{5}{9}(F - 32)$ to find the Celsius temperature that corresponds to the given Fahrenheit temperature.

41. $F = 50°$
42. $F = 14°$
43. $F = -4°$
44. $F = -13°$

Evaluate each formula for the given values of the variables.

45. $D = 25 - 4ac$ for $a = 5$ and $c = 3$
46. $d = |x - y|$ for $x = 8$ and $y = 2$
47. $y = \frac{3}{4}a - b$ for $a = -12$ and $b = -4$
48. $A = \frac{1}{2}bh$ for $b = 36$ and $h = 9$

CHAPTER 12 SUMMARY

KEY WORDS AND PHRASES

numerical expression [12.1]
grouping symbol [12.1]
order of operation [12.1, 12.5, 12.6]
signed number [12.2]
positive numbers [12.2]
negative numbers [12.2]
integer [12.2]
positive integer [12.2]
negative integer [12.2]
absolute value (magnitude) of a number [12.2]
opposite of a number [12.2]
distance between two numbers [12.4]
variable [12.7]
algebraic expression [12.7]
formula [12.7]

SYMBOLS

$4 \cdot 5$ or $4(5)$ or $(4)(5)$ means 4 multiplied by 5.

The grouping symbols, parentheses, ( ), brackets, [ ], and braces, { }, are used to help define the order to operate. We operate inside the grouping symbol first, beginning with the innermost.

-8 means negative eight or the opposite of eight.

$+8$ means positive eight or eight.

$|x|$ means the absolute value of x.

$\frac{-1}{2}$ or $\frac{1}{-2}$ or $-\frac{1}{2}$ or -0.5 or $-\left(\frac{1}{2}\right)$ means negative one-half.

Letters like x, n, y, W, and B are used to represent unknown numbers.

$5c$ means 5 multiplied by c.

xy means x multiplied by y.

IMPORTANT RULES

Order of Operation [12.1, 12.5, 12.6]

1. Operate inside grouping symbols, beginning with the innermost.
2. Take absolute values.
3. Take opposites, multiply, and divide, from left to right. (Repeated multiplication can be done in any order.)
4. Add and subtract, from left to right. (Repeated addition can be done in any order.)

How to Add Two Signed Numbers [12.3]

1. **With Like Signs:** Add the absolute values. Attach the like sign.
2. **With Unlike Signs:** Subtract the smaller absolute value from the larger absolute value. Attach the sign of the number with the larger absolute value.
3. **With Zero:** Any number added to zero is that number.

How to Subtract Two Signed Numbers [12.4]

1. Change the subtraction operation to addition.
2. Change the sign of the number that is being subtracted (the second number.)
3. Follow the rules for adding signed numbers.

Note: Do *not* alter the number that is being subtracted from, that is, the first number.

How to Multiply Two Signed Numbers [12.5]

1. Multiply the absolute values of the two numbers.
2. Attach the appropriate sign by using these sign rules.

$$(+)(+) = (+)$$
$$(+)(-) = (-)$$
$$(-)(+) = (-)$$
$$(-)(-) = (+)$$

How to Divide Two Signed Numbers [12.5]

1. Divide the absolute values of the two numbers.
2. Attach the appropriate sign by using these sign rules.

$$(+) \div (+) = (+) \qquad (-) \div (-) = (+)$$
$$(+) \div (-) = (-) \qquad (-) \div (+) = (-)$$

3. Zero divided by a nonzero number equals zero.

$$\frac{0}{\text{nonzero number}} = 0 \div \text{nonzero number} = 0$$

4. Division by zero is undefined.

$$\frac{\text{any number}}{0} = \text{any number} \div 0 = \text{no answer}$$

CHAPTER 12 REVIEW EXERCISES

Evaluate.

1. $30 - 8(4)$
2. $-6 + 12 + (-7)$
3. $-6(12)(-7)$
4. $4(-7) + (-5)(-3)$
5. $8(10 - 7 + 1)$
6. $6 - 14(5 - 2 \cdot 4)$
7. $-|-17|$
8. $|-6| - |19|$
9. $(7 - 15)(4 - (-3))$
10. $160 - 6[25 - 7(3 + 1)]$
11. $15 - 3[10 - 2(3 - 7) + 15]$
12. $-5\{|-7 - (-6) - 8| - |-9 - 6|\}$

Give the numbers associated with points A, B, and C.

13. (number line with C, B near 0–20 range and A near 40)

14. (number line with C near -10, B between -10 and 0, A near or past 0 toward 10)

Use the symbol $<$, $=$, or $>$ to compare each pair of numbers.

15. $-75, 20$
16. $-18, -27$

Solve.

17. List from smallest to largest: $0, -8, 12, -15$.
18. Which has the larger absolute value, -32 or 14?
19. Subtract -40 from 8.
20. Find the product of -5, 8, and 3.
21. What number is 6 less than -5?
22. Divide -64 by 8.
23. Divide -64 into 8.
24. Find the ratio of -240 to -6.

In Exercises 25 and 26, solve by using addition of signed numbers.

25. Monica's bank balance was $110. She withdrew $75, withdrew $100, then deposited $50. What is her balance now? Interpret the answer.
26. The temperature in a refrigerator vault was $-6°C$. Over a ten-hour period the temperature increased 4°, decreased 18°, then increased 12°. What was the final temperature after these changes?

In Exercises 27 and 28, solve by using subtraction of signed numbers.

27. A football quarterback ran from 13 yards behind the line of scrimmage to 34 yards beyond the line of scrimmage. How far did he run?
28. A sunken treasure is 6500 feet below sea level. A submarine directly above the treasure is 3300 feet below sea level. How far is the submarine from the sunken treasure?
29. Which of these numbers are equal to $\frac{-3}{-8}$?
    a. $\frac{3}{8}$ b. $\frac{-3}{8}$ c. -0.38 d. 0.375 e. $2\frac{2}{3}$
30. Which of these numbers are equal to $-\frac{6}{5}$?
    a. $\frac{-6}{-5}$ b. $\frac{-6}{5}$ c. $\frac{6}{-5}$ d. $1\frac{1}{5}$ e. -1.2

Convert each of these fractions to decimal form.

31. $\frac{-17}{3}$
32. $\frac{-11}{-20}$

Give the fractions associated with points A, B, and C.

33. C B A on number line from −2 to 2

34. C B A on number line from −3 to −1

Give the decimals associated with points A, B, and C.

35. C B A on number line with marks at −27, −26, −25

36. C B A on number line from −3 to 3

Evaluate.

37. $\frac{-4}{5} + \frac{3}{10} - \frac{-1}{5}$
38. $\frac{-25}{27} \div \frac{15}{2}$
39. $3(-4.5) - 8.65$
40. $\frac{20.5}{-0.4}$
41. $\frac{0.7(-8.6) + 4}{-6.7 + 1.7}$
42. $\frac{7}{9} \cdot \frac{3}{2} + 5 \cdot \frac{2}{3}$

Assume w represents an unknown number. Translate each phrase to an algebraic expression.

43. The product of 5 and w.
44. The sum of 5 and w.
45. 5 less than w.
46. The ratio of 5 to w.

Evaluate for the given values of the variables.

47. $3y - 7$ for $y = 6$
48. $-5m + 4x$ for $m = -6$ and $x = -7$
49. $4mn - 6n$ for $m = -3$ and $n = 10$
50. $(x - y)(3x + y)$ for $x = -5$ and $y = 13$
51. $x + y - 6$ for $x = \frac{5}{6}$ and $y = 2$
52. $-80cb$ for $c = 4.8$ and $b = 0.09$

Use the formula $A = LW$ or $P = 2L + 2W$ to find **a.** the area and **b.** the perimeter of the rectangle with the given length and width.

53. $L = 60$ inches, $W = 45$ inches
54. $L = 13.9$ meters, $W = 8.7$ meters

Use the temperature conversion formula $C = \frac{5}{9}(F - 32)$ to find the Celsius temperature that corresponds to the given Fahrenheit temperature.

55. $F = -40°$
56. $F = 11°$

Evaluate each formula for the given values of the variables.

57. $A = \frac{1}{2}bh$ for $b = 30$ and $h = 34.6$

58. $d = |x - y|$ for $x = -16$ and $y = -8$

In Your Own Words

Write complete sentences to discuss each of the following. Support your comments with examples or pictures, if appropriate.

59. Discuss some real-life situations where signed numbers can be used.
60. Discuss what happens if you reverse the order of subtraction.
61. If the sum of two numbers is 0, what can you say about the two numbers?

62. Discuss what happens when a negative number is added to a positive number.

63. Let the expression $-N$ represent a signed number. Discuss whether this expression represents a positive number or a negative number.

CHAPTER 12 PRACTICE TEST

Evaluate.

1. $12 + 8 \cdot 4$ **2.** $-12 - (-5 - 7)$ **3.** $|(-5)(-4) + 8(-3)|$

4. $50 - 5(8 - 4 + 9)$ **5.** $-24(\frac{1}{3} + \frac{3}{4})$ **6.** $36 - 5(0.05)(3.6)$

Evaluate for the given values of the variable.

7. $15R - 40$ for $R = -6$ **8.** $(w + 8)(2w - 4)$ for $w = -1$

9. $3y + 5(y - 8)$ for $y = 3$ **10.** $9ab - 6b + a$ for $a = -2$ and $b = 5$

11. $6y + 4x$ for $x = \frac{5}{2}$ and $y = \frac{4}{3}$

12. $m - n + 2.6$ for $m = 0.54$ and $n = 4.38$

Give the numbers associated with points A, B and C.

13. C B   A
       —+—+—+—+—+—→
         −7 −6

14.   C   B   A
       —+—+—+—+—+—
         −2 −1 0 1

15.   A   C   B
       —+—+—+—+—+—
         −6 −5 −4

16. Which of these numbers are equal to $-\frac{4}{5}$?
   a. $\frac{-4}{-5}$ **b.** $\frac{-4}{5}$ **c.** $\frac{4}{-5}$ **d.** -0.45 **e.** -0.8

17. List from smallest to largest: $45, -24, 8, -15, 0$

Assume n represents an unknown number. Translate each phrase to an algebraic expression.

18. The sum of n and 20. **19.** 20 less than n.

Solve.

20. Find the product of -9 and 20. **21.** Subtract -6 from -11.

22. You are walking along a number line and your starting position is -27. You walk 9 units to the left, 20 units to the right, then 35 units to the left. What is your final position?

23. A place in the Rocky Mountains is 10,800 feet above sea level while a place in the Atlantic Ocean is 3600 feet below sea level. What is the difference in the two altitudes?

24. The temperature at a ski resort in Argentina was reported to be $-10°C$. Use the formula

$$F = \frac{9}{5}C + 32$$

to find the corresponding Fahrenheit temperature.

25. Evaluate $d = |x - y|$ for $x = -18$ and $y = -6$.

Exponents and Square Roots

13.1 POSITIVE INTEGER EXPONENTS

OBJECTIVES
- Write an expression in its exponential form.
- Evaluate a number raised to a positive exponent.
- Translate an English phrase to an exponential expression.
- Evaluate an expression that involves positive exponents mixed with other operations.

1 The Meaning of a Positive Integer Exponent

A shortcut notation is often used when a number is multiplied by itself repeatedly. For example,

$$4^3 \quad \text{means} \quad 4 \cdot 4 \cdot 4 = 64$$

The **exponent** 3, written to the right of the **base** 4 and raised up slightly, tells how many times to use 4 as a factor. For the expression 4^3 we say, "four cubed" or "four to the third power." This chart gives several examples.

Exponential Form	Meaning and Value	English
3^2	$3 \cdot 3 = 9$	Three squared. Three raised to the second power. The second power of three.
5^3	$5 \cdot 5 \cdot 5 = 125$	Five cubed. The cube of five. Five raised to the third power.
10^4	$10 \cdot 10 \cdot 10 \cdot 10$ $= 10,000$	Ten to the fourth power. The fourth power of ten.
8^1	8	Eight to the first power.

Try These Problems

Write the exponential form for each. Do not evaluate.
1. $9 \cdot 9 \cdot 9$
2. $(7)(7)(7)(7)(7)$
3. Thirty squared.

Evaluate.
4. 6^2
5. 3^4
6. Eight cubed.
7. $4(2)^3 - 15$
8. $2 \cdot 3^4 + 50$
9. $(15 - 7)^2$

Try Problems 1 through 6.

2 Mixing Exponentiation with Other Operations

Raising a number to an exponent is an operation called **exponentiation.** Here we look at how exponentiation fits into the order to operate.

> *Order of Operation*
> 1. Operate inside grouping symbols, beginning with innnermost.
> 2. Apply exponents and take absolute values.
> 3. Take opposites, multiply, and divide, from left to right. (Repeated multiplication can be done in any order.)
> 4. Add and subtract, from left to right. (Repeated addition can be done in any order.)

Note that, after working inside grouping symbols, we apply exponents along with taking absolute values. Exponents have priority over the opposite of, multiplication, division, addition, and subtraction. Here are some examples where we evaluate numerical expressions with exponentiation mixed with other operations.

EXAMPLE 1 Evaluate: $3(5)^2 + 60$

SOLUTION Begin by applying the exponent 2 to the number 5.

$3(5)^2 + 60$
$= 3(25) + 60$ Square 5 to obtain 25.
$= 75 + 60$ Multiply 3 by 25 to obtain 75.
$= 135$ ∎

EXAMPLE 2 Evaluate: $(3 + 4)^3$

SOLUTION Because the addition is inside the parentheses, begin by adding.

$(3 + 4)^3$
$= 7^3$ Add 3 and 4 to obtain 7.
$= 7 \cdot 7 \cdot 7$
$= 7 \cdot 49$
$= 343$ ∎

Try Problems 7 through 9.

Consider the expression

-4^2

Does this expression equal 16 or -16? Notice that there are *two* operations: the opposite of and exponentiation. Remember that exponents are applied before the opposite of, so this exponent 2 applies only to 4, not -4. Therefore, here is the meaning of the expression.

$-4^2 = -(4)(4) = -16$ The opposite of the square of 4.

Try These Problems

Evaluate.
10. $(-2)^3$
11. $(-5)^2$
12. -6^2
13. -3^4
14. $(1-5)^2$
15. $(-7)^2 + (-3)^3$
16. $-3^2 + 4(-1)^2$
17. $-2^4 + (-2)^3 - 4(-2)$

The value is -16, not 16. If you want to indicate that -4 is squared, you must wrap -4 in a grouping symbol and put the exponent outside the grouping symbol. For example,

$$(-4)^2 = (-4)(-4) = 16 \qquad \text{The square of } -4.$$

 Try Problems 10 through 13.

EXAMPLE 3 Evaluate: $(2-7)^3$

SOLUTION Subtract inside the parentheses first.

$$(2-7)^3$$
$$= (-5)^3$$
$$= (-5)(-5)(-5)$$
$$= -5(25)$$
$$= -125 \ \blacksquare$$

EXAMPLE 4 Evaluate: $-12^2 - 3(12) + 20$

SOLUTION

$$-12^2 - 3(12) + 20$$
$$= -(12)(12) - 3(12) + 20 \qquad \text{The exponent 2 applies only to 12, not to } -12.$$
$$= -144 - 36 + 20$$
$$= -144 + (-36) + 20 \qquad \text{Subtracting 36 is the same as adding } -36.$$
$$= -180 + 20 \qquad \text{Add the three numbers in any order.}$$
$$= -160 \ \blacksquare$$

 Try Problems 14 through 17.

EXAMPLE 5 Evaluate $x^2 + y^2$ for $x = -3$ and $y = 7$.

SOLUTION

$$x^2 + y^2$$
$$= (-3)^2 + 7^2 \qquad \text{Wrap } -3 \text{ in parentheses and place the exponent outside the parentheses.}$$
$$= 9 + 49 \qquad \text{Apply exponents first.}$$
$$= 58 \ \blacksquare$$

EXAMPLE 6 Evaluate $(x+y)^2$ for $x = 6$ and $y = -10$.

SOLUTION

$$(x+y)^2$$
$$= (6 + (-10))^2 \qquad \text{Add inside the grouping symbol before applying the exponent 2.}$$
$$= (-4)^2$$
$$= 16 \ \blacksquare$$

Try These Problems

Evaluate for the given values of the variables.

18. $m^2 - n^2$ for $m = 5$ and $n = -4$
19. $5x^2 - 7x + 4$ for $x = 10$
20. $W = -6x^3 + 2x^2 - x$ for $x = -3$
21. $5(m - n)^2$ for $m = -6$ and $n = -12$
22. $-w^2$ for $w = 7$
23. $A = -y^2 - 9y + 16$ for $y = 6$

EXAMPLE 7 Evaluate $P = 4x^3 - 3x + 2$ for $x = -2$.

SOLUTION

$P = 4x^3 - 3x + 2$
$= 4(-2)^3 - 3(-2) + 2$
$= 4(-2)(-2)(-2) - 3(-2) + 2$
$= 4(-8) \quad - 3(-2) + 2$ The cube of -2 is -8.
$= -32 \quad - (-6) + 2$ Multiply 4 by -8 to obtain -32.
     Multiply 3 by -2 to obtain -6.
$= -32 + 6 + 2$ Subtracting -6 is the same as adding 6.
$= -32 + 8$ Add the three numbers in any order.
$= -24$ ■

 Try Problems 18 through 21.

EXAMPLE 8 Evaluate $-y^2$ for $y = 8$.

SOLUTION The exponent 2 applies only to the variable y, not to $-y$.

$-y^2$
$= -(8)^2$ Substitute 8 for y.
$= -(8)(8)$ The exponent 2 applies only to 8, the value of y.
$= -64$ ■

EXAMPLE 9 Evaluate $C = -m^2 + 8m + 12$ for $m = -5$.

SOLUTION The exponent 2 applies only to the variable m, not to $-m$.

$C = -m^2 + 8m + 12$
$= -(-5)^2 + 8(-5) + 12$
$= -(25) + 8(-5) + 12$ The square of -5 is 25.
$= -25 + (-40) + 12$ Multiply 8 by -5 to obtain -40.
$= -65 + 12$ Add the three numbers in any order.
$= -53$ ■

 Try Problems 22 and 23.

Answers to Try These Problems

1. 9^3 2. 7^5 3. 30^2 4. 36   5. 81   6. 512   7. 17   8. 212
9. 64   10. -8 11. 25   12. -36 13. -81 14. 16   15. 22
16. -5 17. -16 18. 9   19. 434   20. $W = 183$ 21. 180
22. -49 23. $A = -74$

EXERCISES 13.1

Write the exponential form for each. Do not evaluate.

1. $79 \cdot 79$
2. $8 \cdot 8 \cdot 8 \cdot 8$
3. $(19)(19)(19)(19)(19)(19)$
4. $(25)(25)(25)$
5. The square of 13
6. 27 raised to the fifth power
7. The square of -9
8. The fourth power of -5

Evaluate.

9. 9^2
10. 2^6
11. 4^3
12. 3^4
13. $(-7)^2$
14. $(-6)^3$
15. -8^2
16. -2^4
17. $-(-4)^2$
18. $-(-3)^2$
19. $4^2 + 5^2$
20. $3^2 - 6^2$
21. $(7 - 2)^2$
22. $(4 - 5)^3$
23. $-7(3)^2 + (-8)$
24. $(-2)^3 - (8)(-4)$
25. $(-6 + 8)^3 - (3 - 6)^2$
26. $(30(-5) + 100)^2(28 - 30)^3$

Evaluate for the given values of the variables.

27. $4m^2$ for $m = 20$
28. $-20x^3$ for $x = 3$
29. $-y^2$ for $y = 6$
30. $-y^2$ for $y = -6$
31. $-R^2$ for $R = -8$
32. $-R^2$ for $R = 8$
33. $4d^3 + 10$ for $d = 20$
34. $50x^2 - 100$ for $x = -4$
35. $5y^2 + 6y - 40$ for $y = 3$
36. $-12w^3 - 5w + 30$ for $w = -1$
37. $x^2 + y^2$ for $x = 4$ and $y = 8$
38. $a^2 - b^2$ for $a = 6$ and $b = 7$
39. $(a - b)^2$ for $a = 1$ and $b = 8$
40. $(x + y)^2$ for $x = 8$ and $y = 3$
41. $(a + b)(a - b)^3$ for $a = -4$ and $b = 1$
42. $-16w^3(3w^2 - 40)$ for $w = -2$

Evaluate $D = b^2 - 4ac$ for the given values of the variables.

43. $a = 1, b = 3, c = 5$
44. $a = 2, b = 4, c = -6$
45. $a = 3, b = -2, c = -1$
46. $a = 4, b = -1, c = 8$

Solve.

47. The volume V of a box, with all edges the same length, is given by the formula $V = s^3$ where s is the length of each edge. Find the volume of a box with each edge measuring 5 centimeters.

48. Ms. Herrera owns a small business. A consultant informed her that her profit P (in dollars) for selling x products is approximated by the formula $P = -x^2 + 500x$. Find the profit when 100 products are sold.

USING THE CALCULATOR #14

THE x^y KEY

One way to evaluate an exponential expression like 4^3 on the calculator is to multiply the 4s repeatedly.

To compute 4^3

Enter   4 ⊠ 4 ⊠ 4 ⊟

Result   64

If you enter another ⊠ instead of entering ⊟, you will also obtain the result 64.

If you are working on a basic calculator, the above procedure will have to be used to compute exponential expressions. This procedure is very convenient for exponents 2 and 3.

A scientific calculator has a key that looks like $\boxed{x^y}$. On some calculators this key may be a second function (inverse or shift) key. The key $\boxed{x^y}$ can be used to compute 4^3 without repeatedly multiplying by 4.

To compute	4^3
Enter	4 $\boxed{x^y}$ 3 $\boxed{=}$
Result	64

In case the key $\boxed{x^y}$ is a second function (inverse or shift) key on your calculator, you will need to enter $\boxed{\text{2nd}}$ or $\boxed{\text{INV}}$ or $\boxed{\text{Shift}}$ before entering $\boxed{x^y}$.

Calculator Problems

Evaluate. If the result is not a whole number, round off to 3 decimal places.

1. 3^4
2. 2^5
3. 15^6
4. 3^{12}
5. $(4.6)^3$
6. $(0.72)^4$
7. $(0.8)^9$
8. $(4.6)^5$

13.2 ZERO AND NEGATIVE EXPONENTS

OBJECTIVES
- Evaluate a number raised to an integer exponent.
- Evaluate an expression involving integer exponents mixed with other operations.
- Solve an application that involves evaluating a formula that contains a negative exponent.

1 Understanding the Meaning of a Zero and Negative Exponent

In Section 13.1 you became familiar with positive integer exponents. For example,

$$5^3 = (5)(5)(5) = 125$$

The exponent 3 tells how many times to use the base 5 as a factor.

Now we look at what is meant by a zero exponent and negative exponents. Here we observe some powers of 3 to discover what the meaning of zero and negative exponents must be.

Note that these results are decreasing by a factor of 3.

$$3^4 = (3)(3)(3)(3) = 81$$
$$3^3 = (3)(3)(3) = 27 \quad 81 \div 3 = 27$$
$$3^2 = (3)(3) = 9 \quad 27 \div 3 = 9$$
$$3^1 = 3 = 3 \quad 9 \div 3 = 3$$
$$3^0 = ?$$
$$3^{-1} = ?$$
$$3^{-2} = ?$$

Observe that the base 3 is not changing, but the exponent is decreasing by 1 at each step. Also, observe that reducing the exponent by 1 is

Try These Problems

1. Evaluate each of these powers of 2.
   a. 2^5
   b. 2^4
   c. 2^3
   d. 2^2
   e. 2^1
   f. 2^0
   g. 2^{-1}
   h. 2^{-2}
   i. 2^{-3}
   j. 2^{-4}

2. Evaluate each of these powers of 10.
   a. 10^3
   b. 10^2
   c. 10^1
   d. 10^0
   e. 10^{-1}
   f. 10^{-2}
   g. 10^{-3}
   h. 10^{-4}

Evaluate.

3. 5^0
4. 43^0
5. 18^0
6. 5^{-1}
7. 12^{-1}
8. 40^{-1}
9. 4^{-2}
10. 5^{-2}
11. 9^{-2}

equivalent to dividing by the base 3. This pattern should continue to give the following.

Each result is the previous result divided by the base 3.

$3^4 = (3)(3)(3)(3) = 81$
$3^3 = (3)(3)(3) = 27$ $81 \div 3 = 27$
$3^2 = (3)(3) = 9$ $27 \div 3 = 9$
$3^1 = 3 = 3$ $9 \div 3 = 3$
$3^0 = 1 = 1$ The result should be $3 \div 3 = 1$.
$3^{-1} = \frac{1}{3} = \frac{1}{3}$ The result should be $1 \div 3 = \frac{1}{3}$.
$3^{-2} = \frac{1}{3^2} = \frac{1}{9}$ The result should be $\frac{1}{3} \div 3 = \frac{1}{3} \cdot \frac{1}{3} = \frac{1}{3^2}$ or $\frac{1}{9}$.

 Try Problems 1 and 2.

From the previous display of powers of 3 you see that

$$3^0 = 1$$

Therefore, the exponent 0 means to not use the factor 3 any number of times and that leaves a factor of 1, because the factor 1 is always understood to be present even when not written. In general, **any nonzero number raised to the zero power is 1.** Here are some more examples.

$2^0 = 1 \quad 4^0 = 1 \quad 10^0 = 1 \quad 75^0 = 1$

 Try Problems 3 through 5.

From the previous display of some powers of 3, you see that

$$3^{-1} = \frac{1}{3}$$

Therefore, the exponent -1 applied to 3 means to take the **reciprocal** of 3, which is $\frac{1}{3}$. In general, **if the exponent -1 is applied to any nonzero number, it means to take the reciprocal of the number.** Here are more examples.

$2^{-1} = \frac{1}{2} \quad 4^{-1} = \frac{1}{4} \quad 10^{-1} = \frac{1}{10} \quad 75^{-1} = \frac{1}{75}$

 Try Problems 6 through 8.

From the previous display of some powers of 3, you see that

$$3^{-2} = \frac{1}{3^2} = \frac{1}{9}$$

Therefore, the exponent -2 applied to 3 means to divide 1 by the square of 3. In general, **if the exponent -2 is applied to any nonzero number, it means to divide 1 by the square of the number.** Here are more examples.

$2^{-2} = \frac{1}{2^2} = \frac{1}{4} \quad 10^{-2} = \frac{1}{10^2} = \frac{1}{100} \quad 6^{-2} = \frac{1}{6^2} = \frac{1}{36}$

 Try Problems 9 through 11.

Try These Problems

Evaluate.
12. 2^{-3}
13. 4^{-4}
14. 10^{-3}
15. 2^{-5}
16. 6^{-1}
17. 8^0
18. 7^{-2}
19. 4^{-3}
20. 2^{-4}
21. $5(3^{-1}) - 1$
22. $20(4^{-2}) + 5$

From the previous display of some powers of 3, you see that:

$$3^0 = 1$$
$$3^{-1} = \frac{1}{3}$$
$$3^{-2} = \frac{1}{3^2} = \frac{1}{9}$$

This pattern continues. That is, we have the following:

$$3^{-3} = \frac{1}{3^3} = \frac{1}{27}$$
$$3^{-4} = \frac{1}{3^4} = \frac{1}{81}$$
$$3^{-5} = \frac{1}{3^5} = \frac{1}{243}$$

Try Problems 12 through 15.

Now we state a definition that gives the meaning of any integer exponent.

Definition of an Integer Exponent

1. For n a positive integer,

$$x^n = xxxx \cdots x \quad (n \text{ factors of } x)$$

$$x^{-n} = \frac{1}{x^n} \quad (n \text{ factors of } x \text{ in the denominator})$$

2. $x^0 = 1$ for nonzero x.

Try Problems 16 through 20.

2 Mixing Exponentiation with Other Operations

Now we look again at evaluating numerical expressions that involve exponentiation mixed with other operations. If you need to review the rules for the order to operate, they are listed in Section 13.1.

EXAMPLE 1   Evaluate: $3(4^{-1}) + 2$

SOLUTION

$3(4^{-1}) + 2$

$= \frac{3}{1} \cdot \frac{1}{4} + 2$ Apply the exponent -1 to 4 to obtain the fraction $\frac{1}{4}$.

$= \frac{3}{4} + 2$ Multiply 3 by $\frac{1}{4}$ to obtain $\frac{3}{4}$.

$= \frac{3}{4} + \frac{2 \cdot 4}{1 \cdot 4}$ Write 2 as a fraction with denominator 4.

$= \frac{3}{4} + \frac{8}{4}$

$= \frac{11}{4}$ or $2\frac{3}{4}$ ∎

Try Problems 21 and 22.

Try These Problems

Evaluate.
23. $5^{-1} + 4^{-1}$
24. $(5 + 4)^{-1}$
25. $3^{-2} + 4^{-2}$
26. $(3 + 4)^{-2}$

Evaluate for the given values of the variables.
27. $P = 32n^{-2} + 4n - 10$ for $n = 4$
28. $A = 400x^{-3} + 80x^2 - 100$ for $x = 2$

EXAMPLE 2 Evaluate: **a.** $3^{-1} + 2^{-1}$ **b.** $(3 + 2)^{-1}$

SOLUTION

a.
$$3^{-1} + 2^{-1}$$
$$= \frac{1}{3} + \frac{1}{2}$$ Apply the exponents first.
$$= \frac{1 \cdot 2}{3 \cdot 2} + \frac{1 \cdot 3}{2 \cdot 3}$$ Write each fraction with denominator 6.
$$= \frac{2}{6} + \frac{3}{6}$$
$$= \frac{5}{6}$$

b.
$$(3 + 2)^{-1}$$
$$= 5^{-1}$$ Add inside the grouping symbol first.
$$= \frac{1}{5}$$ Apply the exponent. ∎

Try Problems 23 through 26.

EXAMPLE 3 Evaluate $200x^{-3} + 5x^4 - 40$ for $x = 2$.

SOLUTION

$$200x^{-3} + 5x^4 - 40$$
$$= 200(2)^{-3} + 5(2)^4 - 40$$
$$= 200 \cdot \frac{1}{2^3} + 5(16) - 40$$ Apply exponents first.
$$= \frac{200}{1} \cdot \frac{1}{8} + 80 - 40$$
$$= 25 + 80 - 40$$ Multiply 200 by $\frac{1}{8}$.
$$= 105 - 40$$
$$= 65$$ ∎

Try Problems 27 and 28.

Answers to Try These Problems

1. a. 32 b. 16 c. 8 d. 4 e. 2 f. 1 g. $\frac{1}{2}$ h. $\frac{1}{4}$ i. $\frac{1}{8}$ j. $\frac{1}{16}$
2. a. 1000 b. 100 c. 10 d. 1 e. $\frac{1}{10}$ f. $\frac{1}{100}$ g. $\frac{1}{1000}$ h. $\frac{1}{10000}$
3. 1 4. 1 5. 1 6. $\frac{1}{5}$ 7. $\frac{1}{12}$ 8. $\frac{1}{40}$ 9. $\frac{1}{16}$ 10. $\frac{1}{25}$
11. $\frac{1}{81}$ 12. $\frac{1}{8}$ 13. $\frac{1}{256}$ 14. $\frac{1}{1000}$ 15. $\frac{1}{32}$ 16. $\frac{1}{6}$
17. 1 18. $\frac{1}{49}$ 19. $\frac{1}{64}$ 20. $\frac{1}{16}$ 21. $\frac{2}{3}$ 22. $\frac{25}{4}$
23. $\frac{9}{20}$ 24. $\frac{1}{9}$ 25. $\frac{25}{144}$ 26. $\frac{1}{49}$ 27. 8 28. 270

EXERCISES 13.2

Evaluate.

1. 5^{-1}
2. 3^{-1}
3. 6^{-2}
4. 10^{-2}
5. 8^{-3}
6. 4^{-3}
7. 3^{-4}
8. 10^{-4}
9. 2^{-5}
10. 3^{-5}
11. 7^0
12. 8^0

13. 1^{-4} 14. 1^{-5} 15. $5 \cdot 4^{-1}$ 16. $18 \cdot 2^{-3}$ 17. $10(5^{-2})$
18. $20(8^{-1})$ 19. $4 \cdot 2^{-2} + 3$ 20. $24(3^{-1}) - 8$
21. $4^{-1} - 3^{-1}$ 22. $3^{-2} + 2^{-2}$ 23. $(4 - 3)^{-1}$
24. $(3 + 2)^{-2}$ 25. $2 \cdot 5^{-2} + 2 \cdot 5^{-1} + 1$
26. $40(4^{-1}) - 8(4^{-2}) + 20$

Evaluate for the given values of the variables.

27. n^{-1} for $n = 6$
28. $3n^{-2}$ for $n = 9$
29. $5x^{-3} + 2$ for $x = 2$
30. $80m^{-2} + 6$ for $m = 4$
31. $5x^{-1} - x^2 + 6$ for $x = 2$
32. $36y^{-2} + 2y^3 - 5$ for $y = 3$
33. $(x + y)^{-2}$ for $x = 10$ and $y = 5$
34. $(m - n)^{-1}$ for $m = 20$ and $n = 4$
35. $x^{-2} + y^{-2}$ for $x = 10$ and $y = 5$
36. $m^{-1} - n^{-1}$ for $m = 20$ and $n = 4$
37. $R = 200n^{-2} - 8n$ for $n = 10$
38. $A = 40m^{-1} + 15$ for $m = 5$

Solve.

39. An advertising company promotes the sale of a certain product. The company estimates that, at first, the promotion will bring in extra sales and, after the promotion is over, the extra sales S (in dollars) will decrease as indicated by the formula $S = 4000 \cdot 2^{-t}$ where t represents the number of months after the promotion.
    a. Find the amount of extra sales 1 month after the promotion. (Find S when $t = 1$.)
    b. Find the amount of extra sales 3 months after the promotion. (Find S when $t = 3$.)
    c. Find the amount of extra sales just after the promotion, before any months have gone by. (Find S when $t = 0$.)

40. The mass M (in grams) of a certain radioactive substance is given by the formula $M = 800 \cdot 4^{-x}$ where x represents the number of minutes after 3 PM.
    a. What is the mass of the substance at 3:02 PM? (Find M when $x = 2$.)
    b. What is the mass of the substance at 3:03 PM? (Find M when $x = 3$.)
    c. What is the mass of the substance at 3 PM? (Find M when $x = 0$.)

13.3 LAWS OF EXPONENTS

OBJECTIVES

- Use the first law of exponents to multiply exponential expressions with a common base.
- Use the second law of exponents to divide exponential expressions with a common base.
- Use the third law of exponents to apply an exponent to an exponential expression.

In this section, we discover some properties of exponents. These properties are useful when multiplying or dividing exponential expressions.

1 Introducing the First Law of Exponents

Consider this numerical expression.

$$2^3 \cdot 2^4$$

Suppose we are not interested in evaluating this, but we want to see if it can be written in a simpler exponential form. Notice that if we apply the definition of exponents, we see that there are 7 factors of 2.

$$2^3 \cdot 2^4$$
$$= (2)(2)(2) \cdot (2)(2)(2)(2) \qquad \text{There are 3 factors of 2 multiplied by 4 factors of 2.}$$
$$= 2^7 \qquad \text{There are a total of 7 factors of 2.}$$

The expression $2^3 \cdot 2^4$ can be written in the simpler exponential form 2^7. This product can be simplified because both factors have a common base 2. The exponent 7 can be obtained by adding the exponents 3 and 4.

$$2^3 \cdot 2^4$$
$$= 2^{3+4} \qquad \text{When multiplying with a common base,}$$
$$= 2^7 \qquad \text{add exponents.}$$

Here we look at another product of exponential expressions with a common base.

$$3^2 \cdot 3^4$$
$$= (3)(3) \cdot (3)(3)(3)(3) \qquad \text{There are 2 factors of 3 multiplied by 4 factors of 3.}$$
$$= 3^6 \qquad \text{There are a total of 6 factors of 3.}$$

The expression $3^2 \cdot 3^4$ can be written in the simpler exponential form 3^6. This product can be simplified because both factors have a common base 3. Again observe that the exponent 6 can be obtained by adding the exponents 2 and 4.

$$3^2 \cdot 3^4$$
$$= 3^{2+4} \qquad \text{Add the exponents and apply the resulting}$$
$$= 3^6 \qquad \text{exponent to the common base.}$$

Here we look at a product of exponential expressions where the bases are different.

$$2^4 \cdot 3^2$$
$$= (2)(2)(2)(2) \cdot (3)(3) \qquad \text{There are 4 factors of 2 multiplied by 2 factors of 3.}$$

In this case, we cannot write the expression in a simpler exponential form as before.

In general, when multiplying exponential expressions with a common base, we add the exponents and apply the resulting exponent to the common base. Here are more examples.

$$8^3 \cdot 8^9 = 8^{3+9} = 8^{12}$$
$$3^7 \cdot 3^3 = 3^{7+3} = 3^{10}$$
$$10^{17} \cdot 10 = 10^{17} \cdot 10^1 \qquad \text{When no exponent is written,}$$
$$= 10^{17+1} \qquad \text{it is assumed to be 1.}$$
$$= 10^{18} \qquad 10 = 10^1.$$
$$m^2 \cdot m^3 = m^{2+3} = m^5$$

Try These Problems

Write each in a simpler exponential form. Do not evaluate.
1. $2^3 \cdot 2^2$
2. $4^5 \cdot 4^3$
3. $7^5 \cdot 7^9 \cdot 7$
4. $y^4 y^3$
5. $3^{-1} \cdot 3^3$
6. $10^4 \cdot 10^{-7}$
7. $5^{-2} \cdot 5^{-8} \cdot 5$
8. $x^{-3} x^{-5}$

The previous discussion suggests the first law of exponents that is stated here.

First Law of Exponents
When multiplying exponential expressions with a common base, add the exponents.

$$x^m \cdot x^n = x^{m+n}$$

 Try Problems 1 through 4.

The first law of exponents also holds true when negative exponents are involved. Here are some examples.

$$4^{-1} \cdot 4^5 = 4^{-1+5}$$
$$= 4^4$$
$$6^{-4} \cdot 6^{-7} = 6^{-4+(-7)}$$
$$= 6^{-11} \text{ or } \frac{1}{6^{11}}$$
$$12^3 \cdot 12^{-8} \cdot 12 = 12^3 \cdot 12^{-8} \cdot 12^1$$
$$= 12^{3+(-8)+1}$$
$$= 12^{-4} \text{ or } \frac{1}{12^4}$$
$$a^{-5} a^8 = a^{-5+8}$$
$$= a^3$$

 Try Problems 5 through 8.

2 Introducing the Second Law of Exponents

The first law of exponents deals with multiplying exponential expressions with the same base. Now we look at dividing exponential expressions with the same base.

To discover the second law of exponents, we look carefully at three expressions. Suppose we are not interested in evaluating the expressions but are interested in seeing if there is a simpler exponential form.

$$\frac{4^5}{4^3} = \frac{(4)(4)(4)(4)(4)}{(4)(4)(4)} = \frac{(4)(4)}{1} = 4^2$$

$$\frac{4^3}{4^5} = \frac{(4)(4)(4)}{(4)(4)(4)(4)(4)} = \frac{1}{(4)(4)} = \frac{1}{4^2} \text{ or } 4^{-2}$$

$$\frac{4^3}{4^3} = \frac{(4)(4)(4)}{(4)(4)(4)} = 1 = 4^0$$

Observe that in each case you can subtract the exponent in the denominator from the exponent in the numerator to obtain the final exponent that is applied to the common base 4.

Try These Problems

Write each in a simpler exponential form. Do not evaluate.

9. $\dfrac{3^6}{3^2}$

10. $\dfrac{3^2}{3^6}$

11. $\dfrac{10^{13}}{10^5}$

12. $\dfrac{a^3}{a^9}$

$$\dfrac{4^5}{4^3} = 4^{5-3} = 4^2$$

$$\dfrac{4^3}{4^5} = 4^{3-5} = 4^{-2} \text{ or } \dfrac{1}{4^2}$$

$$\dfrac{4^3}{4^3} = 4^{3-3} = 4^0 = 1$$

In general, when dividing exponential expressions with a common base, we subtract exponents and apply the resulting exponent to the common base. Be careful not to subtract backwards. The exponent in the denominator is subtracted from the exponent in the numerator. Here are more examples:

$$\dfrac{6^7}{6^{10}} = 6^{7-10} = 6^{-3}$$

$$\dfrac{10^{22}}{10^8} = 10^{22-8} = 10^{14}$$

$$\dfrac{x^4}{x^3} = x^{4-3} = x^1 = x$$

Here we state the second law of exponents.

Second Law of Exponents

When dividing exponential expressions with a common base, subtract the exponents. Subtract the exponent in the denominator from the exponent in the numerator, and apply the result to the common base.

$$\dfrac{x^m}{x^n} = x^{m-n}$$

 Try Problems 9 through 12.

The second law of exponents also holds true when negative exponents are involved. Here are some examples.

$$\dfrac{5^{-1}}{5^3} = 5^{-1-3}$$
$$= 5^{-1+(-3)}$$
$$= 5^{-4} \text{ or } \dfrac{1}{5^4}$$

Subtracting 3 is the same as adding −3.

$$\dfrac{10^{-5}}{10^{-2}} = 10^{-5-(-2)}$$
$$= 10^{-5+2}$$
$$= 10^{-3} \text{ or } \dfrac{1}{10^3}$$

Subtracting −2 is the same as adding 2.

$$\dfrac{a^4}{a^{-1}} = a^{4-(-1)}$$
$$= a^{4+1}$$
$$= a^5$$

Subtracting −1 is the same as adding 1.

Try These Problems

Write each in a simpler exponential form. Do not evaluate.

13. $\dfrac{12^{-2}}{12^{-6}}$

14. $\dfrac{12^{-6}}{12^{-2}}$

15. $\dfrac{8^{14}}{8^{-5}}$

16. $\dfrac{x^{-1}}{x^{3}}$

17. $(2^4)^3$
18. $(5^2)^3$
19. $(10^{-1})^{-4}$
20. $(n^8)^{-7}$

Try Problems 13 through 16.

3 Introducing the Third Law of Exponents

Now we look at applying an exponent to an exponential expression. Study the following examples carefully to see if you can discover another law of exponents.

$$(3^4)^2 = 3^4 \cdot 3^4$$
$$= 3^{4+4}$$
$$= 3^8$$
$$(2^5)^3 = 2^5 \cdot 2^5 \cdot 2^5$$
$$= 2^{5+5+5}$$
$$= 2^{15}$$

Observe that, in each case, the final exponent can be obtained by multiplying the original two exponents. For example,

$$(3^4)^2 = 3^{(4)(2)} = 3^8$$
$$(2^5)^3 = 2^{(5)(3)} = 2^{15}$$

In general, when applying an exponent to an exponential expression, we multiply exponents. This law applies to negative exponents as well as positive exponents. Here are more examples.

$$(10^4)^3 = 10^{(4)(3)} = 10^{12}$$
$$(8^{-3})^2 = 8^{(-3)(2)} = 8^{-6} \text{ or } \dfrac{1}{8^6}$$
$$(4^{-5})^{-4} = 4^{(-5)(-4)} = 4^{20}$$
$$(x^6)^{-3} = x^{6(-3)} = x^{-18}$$

Here we state the third law of exponents.

Third Law of Exponents

When applying an exponent to an exponential expression, multiply exponents.

$$(x^m)^n = x^{mn}$$

Try Problems 17 through 20.

Here we summarize the results in this section by stating the three laws of exponents that have been discussed.

Laws of Exponents

1. $x^m \cdot x^n = x^{m+n}$
2. $\dfrac{x^m}{x^n} = x^{m-n}$
3. $(x^m)^n = x^{mn}$

▲ **Answers to Try These Problems**

1. 2^5 2. 4^8 3. 7^{15} 4. y^7 5. 3^2 6. 10^{-3} or $\frac{1}{10^3}$
7. 5^{-9} or $\frac{1}{5^9}$ 8. x^{-8} or $\frac{1}{x^8}$ 9. 3^4 10. 3^{-4} or $\frac{1}{3^4}$ 11. 10^8
12. a^{-6} or $\frac{1}{a^6}$ 13. 12^4 14. 12^{-4} or $\frac{1}{12^4}$ 15. 8^{19} 16. x^{-4} or $\frac{1}{x^4}$
17. 2^{12} 18. 5^6 19. 10^4 20. n^{-56} or $\frac{1}{n^{56}}$

EXERCISES 13.3

1. Which of these expressions are equal to $2^3 \cdot 2^2$?
   a. 4^6 **b.** 4^5 **c.** 2^5 **d.** 2^6 **e.** 32
2. Which of these expressions are equal to $3^4 \cdot 3^2$?
   a. 729 **b.** 3^8 **c.** 3^6 **d.** 9^8 **e.** 9^6

Use the first law of exponents to write each expression in a simpler exponential form. Do not evaluate.

3. $5^2 \cdot 5^6$
4. $10^7 \cdot 10^3$
5. $2^{18} \cdot 2^5$
6. $7^8 \cdot 7^3$
7. $3^{-4} \cdot 3^{-6}$
8. $4^{-9} \cdot 4^{-6}$
9. $12^{-9} \cdot 12^{13}$
10. $8^{14} \cdot 8^{-20}$
11. $n^5 n^3$
12. $n^3 n^{12}$
13. $y^{-1} y^6$
14. $y^{-3} y^{-2}$

15. Which of these expressions are equal to $\frac{3^4}{3^6}$?
    a. $\frac{1}{6}$ **b.** -9 **c.** $\frac{1}{9}$ **d.** 3^{-2} **e.** 3^2
16. Which of these expressions are equal to $\frac{10^6}{10^2}$?
    a. 10^3 **b.** 10^4 **c.** 10^8 **d.** 1000 **e.** 10,000

Use the second law of exponents to write each expression in a simpler exponential form. Do not evaluate.

17. $\frac{7^5}{7^3}$
18. $\frac{8^{12}}{8^6}$
19. $\frac{10^{20}}{10^{22}}$
20. $\frac{4^8}{4^{11}}$
21. $\frac{3^9}{3^{-2}}$
22. $\frac{5^{-6}}{5^{-3}}$
23. $\frac{4^{-1}}{4^{-6}}$
24. $\frac{10^{-4}}{10^{15}}$
25. $\frac{x^7}{x^3}$
26. $\frac{x^{-8}}{x^{-4}}$
27. $\frac{n^{13}}{n^{-5}}$
28. $\frac{a^5}{a^{10}}$

29. Which of these expressions are equal to $(2^4)^2$?
    a. 64 **b.** 2^6 **c.** 256 **d.** 2^8 **e.** 2^{16}
30. Which of these expressions are equal to $(10^3)^2$?
    a. 10^5 **b.** 10^6 **c.** 10^9 **d.** 100,000 **e.** 1,000,000

Use the third law of exponents to write each expression in a simpler exponential form. Do not evaluate.

31. $(4^3)^6$
32. $(3^5)^4$
33. $(9^7)^2$
34. $(10^5)^3$
35. $(8^2)^{-3}$
36. $(6^{-5})^5$
37. $(10^{-8})^{-4}$
38. $(12^{-1})^{-6}$
39. $(a^6)^5$
40. $(x^3)^4$
41. $(n^{-6})^{-4}$
42. $(y^8)^{-7}$

*Evaluate **a.** $x^2 + y^2$ and **b.** $(x + y)^2$ for the given values of x and y.*
43. $x = 3, y = 4$
44. $x = 1, y = 5$
45. $x = -3, y = 6$
46. $x = 7, y = -2$
47. In general, is the expression $x^2 + y^2$ equal to or not equal to the expression $(x + y)^2$? Use the results of Exercises 43 through 46 to help you decide.

*Evaluate **a.** $x^3 + y^3$ and **b.** $(x + y)^3$ for the given values of x and y.*
48. $x = 1, y = 1$
49. $x = 2, y = 3$
50. $x = -2, y = 4$
51. $x = -1, y = 5$
52. In general, is the expression $x^3 + y^3$ equal to or not equal to the expression $(x + y)^3$? Use the results of Exercises 48 through 51 to help you decide.

*Evaluate **a.** $(xy)^3$ and **b.** $x^3 y^3$ for the given values of x and y.*
53. $x = 2, y = 3$
54. $x = 4, y = 5$
55. $x = -1, y = 3$
56. $x = -2, y = 5$
57. In general, is the expression $(xy)^3$ equal to or not equal to the expression $x^3 y^3$? Use the results of Exercises 53 through 56 to help you decide.

*Evaluate **a.** $(xy)^2$ and **b.** $x^2 y^2$ for the given values of x and y.*
58. $x = 3, y = 4$
59. $x = 6, y = 5$
60. $x = -2, y = 7$
61. $x = 5, y = -3$
62. In general, is the expression $(xy)^2$ equal to or not equal to the expression $x^2 y^2$? Use the results of Exercises 58 through 61 to help you decide.

13.4 POWERS OF TEN AND SCIENTIFIC NOTATION

OBJECTIVES
- Write a power of 10 as a decimal numeral.
- Multiply a decimal by a power of 10.
- Write a decimal numeral in scientific notation.
- Convert the product of a decimal and a power of 10 to scientific notation.
- Multiply and divide numbers using scientific notation.
- Solve an application that involves multiplication or division by using scientific notation.

1 Converting Powers of Ten to Decimal Numerals

When we apply an integer exponent to the number 10, we obtain a **power of 10.** Here are some powers of 10 and the corresponding decimal numerals.

$$10^{-4} = \frac{1}{10^4} = \frac{1}{10000} = 0.0001$$

$$10^{-3} = \frac{1}{10^3} = \frac{1}{1000} = 0.001$$

$$10^{-2} = \frac{1}{10^2} = \frac{1}{100} = 0.01$$

Try These Problems

Write as a decimal numeral.
1. 10^8
2. 10^9
3. 10^{11}
4. 10^{-8}
5. 10^{-9}
6. 10^{-11}
7. 6.18×10^5
8. 9×10^8
9. 0.017×10^{12}
10. 45.8×10^{-7}
11. 66×10^{-5}
12. 7.3×10^{-9}

$$10^{-1} = \frac{1}{10^1} = \frac{1}{10} = 0.1$$
$$10^0 = 1 = 1$$
$$10^1 = 10 = 10$$
$$10^2 = 10 \cdot 10 = 100$$
$$10^3 = 10 \cdot 10 \cdot 10 = 1000$$
$$10^4 = 10 \cdot 10 \cdot 10 \cdot 10 = 10{,}000$$

Observe that 10^4 is the numeral 10,000, which consists of the digit 1 followed by 4 zeros. The number of zeros is equal to the exponent applied to 10. This pattern continues for all of the positive powers of 10. For example,

$$10^7 = 10{,}000{,}000 \quad \text{There are 7 zeros.}$$
$$10^{12} = 1{,}000{,}000{,}000{,}000 \quad \text{There are 12 zeros.}$$

 Try Problems 1 through 3.

In the previous display, observe that 10^{-4} is the numeral 0.0001, which consists of a decimal point followed by 3 zeros and a 1. There are 4 decimal places. The number of decimal places agrees with the absolute value of the exponent. This pattern continues for the negative powers of 10. For example,

$$10^{-7} = 0.000\,0001 \quad \text{There are 7 decimal places.}$$
$$10^{-12} = 0.000\,000\,000\,001 \quad \text{There are 12 decimal places.}$$

 Try Problems 4 through 6.

2 Multiplying a Decimal by a Power of Ten

It is easy to multiply a decimal numeral by a power of 10. For example, consider

$$5.8 \times 10^6$$

The exponent 6 tells you that 5.8 is being multiplied by 10 six times, and each time you multiply by 10 the decimal point shifts one place to the right. Therefore, we have the following:

$$5.8 \times 10^6 = 5800000. \quad \text{The decimal point shifts 6 places to the right.}$$
$$= 5{,}800{,}000$$

 Try Problems 7 through 9.

Now we look at multiplying a decimal numeral by a negative power of 10. For example, consider

$$3 \times 10^{-5}$$

The exponent -5 tells you that 3 is being divided by 10 five times, and each time you divide by 10 the decimal point shifts 1 place to the left. Therefore, we have the following:

$$3 \times 10^{-5} = 0.00003 \quad \text{The decimal point shifts 5 places to the left.}$$
$$= 0.00003$$

 Try Problems 10 through 12.

Try These Problems

Write each in scientific notation.
13. 12,000,000
14. 7,000,000,000
15. 0.000 098
16. 0.000 000 000 0367

3 Converting Numbers to Scientific Notation

Scientists often work with very large and very small numbers. Television signals travel at 30,000,000,000 centimeters per second. The diameter of a red corpuscle is approximately 0.00008 centimeter. It is inconvenient to write these numbers as ordinary decimal numerals. Another notation, called **scientific notation,** is often used. Here are some examples of decimal numerals written in scientific notation.

Decimal Numeral	Scientific Notation
150,000	1.5×10^5
30,000,000,000	3×10^{10}
0.00583	5.83×10^{-3}
0.00008	8×10^{-5}
0.0001	1×10^{-4}

Writing a number in scientific notation involves writing it as the product of two numbers. The first factor is a decimal numeral and the second factor is a power of 10. Of course, these two factors must multiply to give the original decimal numeral. Notice that, in each of the examples above, the decimal factor has the decimal point placed after the first nonzero digit. This means that the decimal factor is a number more than or equal to 1 and less than 10.

Here are some examples of how to write a decimal numeral in scientific notation.

EXAMPLE 1 Write 45,000 in scientific notation.

SOLUTION

$45{,}000$

$= 4.5 \times 10^?$ The decimal factor is 4.5. The decimal is placed after the first nonzero digit to make a number between 1 and 10.

$= 4.5 \times 10^4$ Choose the appropriate power of 10 that makes the product equal the original decimal. The decimal in 4.5 must move 4 places to the right to give 45,000, so we choose the exponent 4. ∎

EXAMPLE 2 Write 0.000 000185 in scientific notation.

SOLUTION

$0.000\,000185$

$= 1.85 \times 10^?$ The decimal factor is 1.85. The decimal is placed after the first nonzero digit to make a number between 1 and 10.

$= 1.85 \times 10^{-7}$ Choose the appropriate power of 10 that makes the product equal the original decimal. The decimal in 1.85 must move 7 places to the left to give 0.000000185, so we choose the exponent −7. ∎

 Try Problems 13 through 16.

Try These Problems

Write each in scientific notation.
17. 1500×10^8
18. 0.056×10^5
19. 24×10^{-20}
20. 0.8×10^{-7}

Sometimes a number is written as the product of a decimal numeral and a power of ten but it is not in scientific notation because the decimal factor is not between 1 and 10. For example, the number

$$792 \times 10^5$$

is not in scientific notation because the factor 792 is not a number between 1 and 10. The factor should be 7.92. Here we show how to shift the number into scientific notation.

$$792 \times 10^5$$
$$= 7.92 \times 10^2 \times 10^5$$
$$= 7.92 \times 10^7$$

First, write the factor 792 in scientific notation.
$792 = 7.92 \times 10^2$

Finally, use the first law of exponents to multiply the powers of 10.
$10^2 \cdot 10^5 = 10^7$

Now the number is in scientific notation.
Here is another example.

EXAMPLE 3 Write 0.65×10^{-9} in scientific notation.

SOLUTION The number is not already in scientific notation because the decimal factor 0.65 is not a number between 1 and 10.

$$0.65 \times 10^{-9}$$
$$= 6.5 \times 10^{-1} \times 10^{-9}$$
$$= 6.5 \times 10^{-10}$$

First, write the factor 0.65 in scientific notation.
$0.65 = 6.5 \times 10^{-1}$

Finally, use the first law of exponents to multiply the powers of 10.
$10^{-1} \cdot 10^{-9} = 10^{-10}$ ∎

Try Problems 17 through 20.

4 Multiplying and Dividing Numbers in Scientific Notation

An Iowa farmer has estimated that it takes him about 50,000 pounds of water to grow one bushel of corn. How much water does he need to grow 300,000 bushels of corn? To solve the problem we need to multiply 50,000 by 300,000. This can be done using scientific notation.

$$\text{Total pounds of water} = \frac{\text{pounds per}}{\text{bushel}} \times \frac{\text{how many}}{\text{bushels}}$$
$$= 50{,}000 \times 300{,}000$$
$$= (5 \times 10^4)(3 \times 10^5)$$
$$= 5 \times 3 \times 10^4 \times 10^5$$
$$= 15 \times 10^9$$
$$= 15{,}000{,}000{,}000$$
$$\text{or } 1.5 \times 10^{10}$$

The four factors can be written in any order.

The farmer will need 15 billion pounds of water. It can be very helpful to use scientific notation when multiplying or dividing very large or very

Try These Problems

Evaluate each by using scientific notation. Express the answer in scientific notation and as an ordinary decimal numeral.

21. $(6 \times 10^{20})(4 \times 10^{-9})$
22. $(4.6 \times 10^8)(5.5 \times 10^6)$
23. $\dfrac{9 \times 10^{-8}}{3 \times 10^{15}}$
24. $\dfrac{7.8 \times 10^{-12}}{1.3 \times 10^{-7}}$

small numbers even when you have a calculator to help you. Here are more examples.

EXAMPLE 4 Divide 950,000 by 0.000 000 0019 by using scientific notation. Express the answer in scientific notation and as an ordinary decimal numeral.

SOLUTION

$$\frac{950{,}000}{0.000\,000\,0019}$$

$$= \frac{9.5 \times 10^5}{1.9 \times 10^{-9}} \qquad \text{Write each number in scientific notation.}$$

$$= \frac{9.5}{1.9} \times \frac{10^5}{10^{-9}}$$

$$= 5 \times 10^{5-(-9)} \qquad \text{Divide 9.5 by 1.9 to obtain 5. Divide the powers of 10 by subtracting exponents.}$$

$$= 5 \times 10^{14} \qquad \text{Scientific notation}$$

$$= 500{,}000{,}000{,}000{,}000 \qquad \text{Decimal numeral} \quad \blacksquare$$

EXAMPLE 5 Multiply 0.000 000 023 by 0.000 008 by using scientific notation. Express the answer in scientific notation and as an ordinary decimal numeral.

SOLUTION

$$0.000\,000\,023 \times 0.000008$$

$$= (2.3 \times 10^{-8})(8 \times 10^{-6}) \qquad \text{Write each number in scientific notation.}$$

$$= (2.3 \times 8) \times 10^{-8} \times 10^{-6} \qquad \text{Rearrange the order of the factors.}$$

$$= 18.4 \times 10^{-8+(-6)} \qquad \text{Multiply 2.3 by 8 to obtain 18.4.}$$

$$= 18.4 \times 10^{-14} \qquad \text{Multiply the powers of 10 by adding exponents.}$$

The result is not in scientific notation because the factor 18.4 is not a decimal between 1 and 10. Now we shift the result to scientific notation, then write the ordinary decimal numeral.

$$\overbrace{18.4 \times 10^{-14}}$$

$$= \underline{1.84 \times 10^1} \times 10^{-14} \qquad \text{First write 18.4 in scientific notation. } 18.4 = 1.84 \times 10^1$$

$$= 1.84 \times 10^{1+(-14)} \qquad \text{Multiply the powers of 10 by adding exponents.}$$

$$= 1.84 \times 10^{-13} \qquad \text{Scientific notation}$$

$$= 0.000\,000\,000\,000\,184 \qquad \text{Decimal numeral} \quad \blacksquare$$

Try Problems 21 through 24.

Try These Problems

Solve. Express the answer in scientific notation and as an ordinary decimal numeral.

25. The mass of one hydrogen atom is 9×10^{-28} gram. What is the mass of 500,000,000 hydrogen atoms?

26. One year the national debt was estimated at 10^{12} dollars. That same year, the population of the entire nation was estimated at 2.5×10^8 persons. What amount of money would be owed by each person if each took an equal share of the national debt?

5 Solving Applications by Using Scientific Notation

Now we look at some applications that involve multiplying and dividing numbers that are very large or very small.

EXAMPLE 6 Electricity travels in a computer circuit at a rate of about 200,000 miles per second. How far will electricity travel in 30 seconds? Write the answer as an ordinary decimal numeral and in scientific notation.

SOLUTION The phrase *200,000 miles per second* is a rate. We can set up a multiplication statement using this rate.

$$\text{Total miles} = \frac{\text{miles}}{\text{second}} \cdot \text{seconds}$$
$$= 200{,}000 \cdot 30$$
$$= (2 \times 10^5)(3 \times 10) \quad \text{Write each factor in scientific notation.}$$
$$= (2 \times 3) \times 10^5 \times 10^1$$
$$= 6 \times 10^6$$
$$= 6{,}000{,}000$$

Electricity will travel 6,000,000 or 6×10^6 miles in 30 seconds. ∎

EXAMPLE 7 The concentration of hydrogen ions in milk is 1.6×10^{-5} mole for every 1000 liters. What is the concentration in moles per liter? Write the answer as an ordinary decimal numeral and in scientific notation.

SOLUTION The phrase *1.6×10^{-5} mole for every 1000 liters* is a rate. We write this rate in its ratio form.

$$1.6 \times 10^{-5} \text{ mole for every 1000 liters}$$
$$= \frac{1.6 \times 10^{-5} \text{ mole}}{1000 \text{ liters}}$$

Because mole is in the numerator and liters is in the denominator, we can obtain the concentration in moles per liter by performing the indicated division.

$$\frac{1.6 \times 10^{-5}}{1000} = \frac{1.6 \times 10^{-5}}{10^3}$$
$$= 1.6 \times \frac{10^{-5}}{10^3}$$
$$= 1.6 \times 10^{-5-3}$$
$$= 1.6 \times 10^{-8}$$
$$= 0.000\,000\,016$$

The concentration of hydrogen ions in milk is 0.000 000 016 or 1.6×10^{-8} mole per liter. ∎

 Try Problems 25 and 26.

13.4 Powers of Ten and Scientific Notation

▲ **Answers to Try These Problems**

1. 100,000,000 2. 1,000,000,000 3. 100,000,000,000
4. 0.000 000 01 5. 0.000 000 001 6. 0.000 000 000 01
7. 618,000 8. 900,000,000 9. 17,000,000,000
10. 0.000 00458 11. 0.00066 12. 0.000 000 0073
13. 1.2×10^7 14. 7×10^9 15. 9.8×10^{-5} 16. 3.67×10^{-11}
17. 1.5×10^{11} 18. 5.6×10^3 19. 2.4×10^{-19} 20. 8×10^{-8}
21. 2.4×10^{12} or 2,400,000,000,000
22. 2.53×10^{15} or 2,530,000,000,000,000
23. 3×10^{-23} or 0.000 000 000 000 000 000 000 03
24. 6×10^{-5} or 0.000 06
25. 4.5×10^{-19} or 0.000 000 000 000 000 000 45 gram
26. 4×10^3 or 4000 dollars

EXERCISES 13.4

Write as a decimal numeral.

1. 10^5
2. 10^7
3. 10^{-6}
4. 10^{-9}
5. 7×10^8
6. 36.7×10^{12}
7. 0.8×10^{-10}
8. 125×10^{-13}

9. The planet Mars has a radius of 3.3×10^6 meters. Write the radius of Mars as an ordinary decimal numeral.

10. A certain computer can perform one addition problem in 4×10^{-8} second. Write this length of time as an ordinary decimal numeral.

Write each in scientific notation.

11. 4,980,000,000
12. 80,000,000,000,000
13. 32,000,000,000
14. 760,000,000
15. 0.000 000 0009
16. 0.000 000 000 00134
17. 0.000 004
18. 0.000 000 000 078

19. Water is entering a reservoir at the rate of 52,000 cubic feet per second. Write this rate in scientific notation.

20. The wavelength of a certain light is 0.000 000 000 54 centimeter. Express this length in scientific notation.

Evaluate each by using scientific notation. Express the answer in scientific notation and as an ordinary decimal numeral.

21. $(2 \times 10^5)(4 \times 10^8)$
22. $(1.5 \times 10^4)(5 \times 10^{-7})$
23. $(4 \times 10^{12})(3.25 \times 10^{-6})$
24. $(2 \times 10^{-5})(5.6 \times 10^{-6})$
25. $(25,300,000)(0.000\ 0006)$
26. $(4,500,000)(70,000)$
27. $\dfrac{1.2 \times 10^{16}}{3 \times 10^{12}}$
28. $\dfrac{4.5 \times 10^{-7}}{1.5 \times 10^3}$
29. $\dfrac{10^8}{4 \times 10^{13}}$
30. $\dfrac{6 \times 10^{-4}}{2 \times 10^{-8}}$
31. $\dfrac{600,000,000}{800,000}$
32. $\dfrac{7,200,000}{5,000,000,000,000}$

Solve. Express the answer in scientific notation and as an ordinary decimal numeral.

33. Television signals travel at 3×10^{10} centimeters per second. How far will the television signal travel in 200 seconds?

634 Chapter 13 Exponents and Square Roots

34. During the summer, water evaporates from a large lake at the rate of 400,000 gallons per day. How many gallons evaporate in 150 days?
35. The earth orbits around the sun at a speed of 3×10^4 meters per second. How long does it take the earth to orbit a distance of 600 meters?
36. The earth's mass is 6×10^{24} kilograms and the sun's mass is 2×10^{30} kilograms. What is the ratio of the earth's mass to the sun's mass?
37. It takes light 3.3×10^{-4} second to travel 1 kilometer. How long does it take light to travel 400 kilometers?
38. A rectangular piece of paper is 30 centimeters long, 20 centimeters wide, and 1.5×10^{-3} centimeters thick. What is the volume of the piece of paper?

USING THE CALCULATOR #15

HANDLING VERY LARGE AND VERY SMALL NUMBERS

Try evaluating 200,000 × 4,000,000 on your calculator. If you get an error message, then your calculator can not handle very large or very small numbers. You probably have a basic calculator, rather than a scientific calculator.

Most scientific calculators are limited to an 8-digit or a 10-digit display, but they can handle very large and very small numbers by using scientific notation. Here we compute 200,000 × 4,000,000 on a scientific calculator.

To Compute	200,000 × 4,000,000
Enter	200 000 [×] 4 000 000 [=]
Result	8.^{11}

The result, 8.^{11}, displayed by the calculator means 8×10^{11}, which equals 800,000,000,000.

If you want to enter the number 800,000,000,000 into a calculator that is limited to an 8- or 10-digit display, you must enter it in scientific notation. This can be done by using the [EXP] key. Here we show an example.

To Compute	5 × 800,000,000,000
Enter	5 [×] 8 [EXP] 11 [=]
Result	4.^{12}

The result, 4.^{12}, displayed by the calculator means 4×10^{12}, which is equal to 4,000,000,000,000.

A scientific calculator can also handle very small numbers. Here we show an example.

To Compute	0.000 02 × 0.000 004
Enter	0.000 02 [×] 0.000 004 [=]
Result	8.$^{-11}$

The result, 8.$^{-11}$, displayed by the calculator means 8×10^{-11}, which equals 0.000 000 000 08.

If you want to enter the number 0.000 000 000 08 into the calculator, you must enter it in scientific notation, because it contains more than 10 digits. This can be done by using the [EXP] key. Here we show an example.

To Compute	0.000 000 000 08 ÷ 4
Enter	8 [EXP] 11 [±] [÷] 4 [=]
Result	2.$^{-11}$

Observe that the [±] key, pressed after entering 11, changes the exponent 11 to −11. The result, 2.$^{-11}$, displayed by the calculator means 2×10^{-11}, which equals 0.000 000 000 02.

Calculator Problems

Evaluate. Express the answer in scientific notation.

1. 500,000 × 800,000
2. 980,000 × 76,000,000
3. 60,000,000 ÷ 0.000 005
4. 420,000 ÷ 0.000 006
5. 0.000 005 ÷ 800,000
6. 0.00082 ÷ 4,100,000
7. 560,000,000,000 × 940
8. 0.6 ÷ 0.000 000 000 15

13.5 SQUARE ROOTS

OBJECTIVES
- Know the difference between finding the square of a number and the square roots of a number.
- Find the negative or positive square root of a number that is a perfect square.
- Use a calculator to approximate the positive square root of a positive number.
- Make a number line that locates the approximate positions of numbers expressed as square roots.
- Compare numbers expressed as square roots to other numbers.
- Evaluate an expression that involves square roots mixed with other operations.
- Solve an application that involves evaluating a formula that contains a square root.

 Try These Problems

Solve.
1. What is the square of 8?
2. What are the two square roots of 25?
3. What is the positive square root of 16?
4. What is the negative square root of 16?

1 Defining Square Root

To reverse the effect of addition, we subtract. To reverse the effect of multiplication, we divide. Now we introduce an operation that reverses the effect of squaring a number.

$$6^2 = 6 \cdot 6 = 36$$

The number 36 is called the **square** of 6, while the number 6 is called a **square root** of 36. A square root of 36 is a number that multiplies by itself to give 36. Observe that the number -6 also multiplies by itself to give 36.

$$(-6)^2 = (-6)(-6) = 36$$

Therefore, -6 is also a square root of 36. That is, the two square roots of 36 are 6 and -6. We call 6 the *positive square root* of 36 and -6 the *negative square root* of 36.

 Try Problems 1 through 4.

We write

$$\sqrt{36} = 6 \quad \text{and} \quad -\sqrt{36} = -6$$

That is, $\sqrt{36}$ means the *positive* square root of 36, while $-\sqrt{36}$ means the *negative* square root of 36. The symbol $\sqrt{}$ is often called a **radical.** Here are more examples.

$$\sqrt{49} = 7 \quad \text{because} \quad 7^2 = 7 \cdot 7 = 49$$
$$\sqrt{100} = 10 \quad \text{because} \quad 10^2 = 10 \cdot 10 = 100$$
$$-\sqrt{64} = -8 \quad \text{because} \quad (-8)^2 = (-8)(-8) = 64$$

Chapter 13 Exponents and Square Roots

 Try These Problems

Find the indicated square root.

5. $\sqrt{9}$
6. $\sqrt{144}$
7. $\sqrt{81}$
8. $-\sqrt{49}$

It is easy to find the square root of numbers like 0, 1, 4, 9, 16, 25, and so on. These numbers are squares of whole numbers. They are called **perfect squares.** Here we list some perfect squares. Becoming familiar with these can help you with square roots.

Some Perfect Squares			
$0^2 = 0$	$1^2 = 1$	$2^2 = 4$	$3^2 = 9$
$4^2 = 16$	$5^2 = 25$	$6^2 = 36$	$7^2 = 49$
$8^2 = 64$	$9^2 = 81$	$10^2 = 100$	$11^2 = 121$
$12^2 = 144$	$14^2 = 196$	$15^2 = 225$	$16^2 = 256$

 Try Problems 5 through 8.

Suppose you wanted to find the positive square root of 6.

$$\sqrt{6} = ?$$

The number $\sqrt{6}$ is a positive number whose square is 6. That is, the number $\sqrt{6}$ must multiply by itself to give 6. There is no whole number that when multiplied by itself equals 6. However, we do know the following:

$$\sqrt{4} < \sqrt{6} < \sqrt{9}$$

This inequality statement says that $\sqrt{6}$ is between $\sqrt{4}$ and $\sqrt{9}$. Since $\sqrt{4} = 2$ and $\sqrt{9} = 3$, we have the following:

$$2 < \sqrt{6} < 3$$

This inequality statement says that the number $\sqrt{6}$ is between 2 and 3. Another way to say this is that the number $\sqrt{6}$ is more than 2, but less than 3. Therefore, we have roughly approximated $\sqrt{6}$ as a number between 2 and 3. For more on approximating square roots without using a calculator and for related exercises, see Developing Number Sense #10 on page 638.

 Approximating Square Roots by Using a Calculator

There are many square roots that cannot be expressed as a terminating or repeating decimal but can be approximated with a decimal numeral. We can use a calculator to find the approximate value. For example, to obtain the approximate value of the number $\sqrt{6}$ on a calculator, first enter 6, then enter $\boxed{\sqrt{}}$. Here we show what the result should be.

$$\sqrt{6} \approx 2.4494897$$

This approximation is accurate to 7 decimal places.

Try These Problems

Approximate the indicated square root to 3 decimal places by using a calculator.

9. $\sqrt{2}$
10. $\sqrt{21}$
11. $\sqrt{150}$
12. $\sqrt{678}$

13. Make a number line that shows the numbers 6, 7, and 8, and the approximate positions of the numbers $\sqrt{39}$ and $\sqrt{60}$.

14. Make a number line that shows the numbers 0, 1, and 2, and the approximate positions of the numbers $\sqrt{0.5}$ and $\sqrt{0.1}$.

15. List from smallest to largest: 12, 11, $\sqrt{131}$, 10.5, $\sqrt{120}$

16. List from smallest to largest: $\sqrt{650}$, $\sqrt{600}$, 24, 25, 24.2

Here are more examples. We round off each of the following results to 3 decimal places.

$$\sqrt{3} \approx 1.7320508 \approx 1.732$$
$$\sqrt{75} \approx 8.660254 \approx 8.660$$
$$\sqrt{581} \approx 24.103942 \approx 24.104$$

Using a calculator, enter 3, then enter $\boxed{\sqrt{}}$.

 Try Problems 9 through 12.

3 Approximating the Positions of Square Roots on the Number Line

Numbers like $\sqrt{6}$, $\sqrt{3}$, and $\sqrt{75}$ have positions on the number line. Even though we do not know the exact location of these numbers on a number line, we can approximate the position. Here we show the approximate positions of $\sqrt{6}$, $\sqrt{3}$, and $\sqrt{75}$ on a number line.

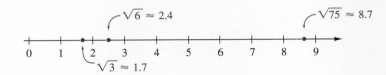

Observe that before approximating the position of the square root on the number line, we need the decimal approximation of the square root.

 Try Problems 13 and 14.

One of the reasons we like to view numbers on a number line is that it helps us to compare them. Here is an example.

EXAMPLE 1 List from smallest to largest: $\sqrt{28}$, 5, 5.5, 5.25

SOLUTION First, approximate the value of the number $\sqrt{28}$.

$$\sqrt{28} \approx 5.2915026$$
$$\approx 5.29$$

Next, picture the numbers on the number line. Numbers on the number line get larger as you move to the right.

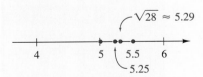

From smallest to largest, we have 5, 5.25, $\sqrt{28}$, and 5.5. ■

 Try Problems 15 and 16.

DEVELOPING NUMBER SENSE #10

ROUGHLY APPROXIMATING A SQUARE ROOT WITHOUT USING A CALCULATOR

You should be able to obtain a rough approximation of a number like $\sqrt{40}$ without using a calculator. How can this be done? You know that $\sqrt{40}$ is between $\sqrt{36} = 6$ and $\sqrt{49} = 7$, therefore $\sqrt{40}$ is between 6 and 7.

$$\sqrt{36} < \sqrt{40} < \sqrt{49}$$
$$6 < \sqrt{40} < 7$$

The number $\sqrt{40}$ is more than 6, but less than 7.

In general, we should be able to obtain consecutive whole numbers that a square root lies between.

What about $\sqrt{475}$? Not many people are familiar with the perfect squares near 475, so we can begin to create some perfect squares until we find two on each side of 475. Here we look at a few.

$20^2 = 20(20) = 400$

$21^2 = 21(21) = 441$

$22^2 = 22(22) = 484$

These calculations would be done by hand. Remember, we are working without a calculator.

Because 475 lies between 441 and 484, we do not need to look at any more perfect squares. We have the following:

$$\sqrt{441} < \sqrt{475} < \sqrt{484}$$
$$21 < \sqrt{475} < 22$$

Therefore, $\sqrt{475}$ is more than 21, but less than 22.

Number Sense Problems

Name the two consecutive whole numbers that each of these square roots lies between. Do not use a calculator.

1. $\sqrt{7}$ 2. $\sqrt{56}$ 3. $\sqrt{120}$ 4. $\sqrt{230}$ 5. $\sqrt{579}$ 6. $\sqrt{974}$

 Try These Problems

Evaluate.

17. $\sqrt{5^2 - 3^2}$
18. $\sqrt{6^2 + 8^2}$

4 Mixing Square Roots with Other Operations

Now we look at evaluating expressions that involve the square root operation mixed with other operations.

The square root symbol is understood to be a grouping symbol, so we operate under the radical before taking the square root. Here are some examples.

EXAMPLE 2 Evaluate: $\sqrt{3^2 + 4^2}$

SOLUTION Perform the operations under the radical before taking the square root.

$$\sqrt{3^2 + 4^2} = \sqrt{9 + 16}$$
$$= \sqrt{25}$$
$$= 5 \blacksquare$$

 Try Problems 17 and 18.

Try These Problems

Evaluate.
19. $6\sqrt{49} + 4\sqrt{25}$
20. $15 - 5\sqrt{100 - 19}$

In Problems 21 and 22, evaluate for the given values of the variables.
21. $\sqrt{b^2 - 4ac}$ for $a = 2$, $b = 5$, and $c = -3$
22. $\sqrt{a^2 + b^2}$ for $a = 5$ and $b = 12$
23. If an object is dropped, the velocity (in feet per second) is given by the formula $v = 8\sqrt{d}$, where d represents the distance (in feet) that the object has fallen. Find the velocity when the object has fallen 49 feet.

EXAMPLE 3 Evaluate: $3\sqrt{25 - 16}$

SOLUTION Begin by performing the subtraction under the radical.

$$3\sqrt{25 - 16}$$
$$= 3\sqrt{9} \qquad \text{Subtract 16 from 25 to obtain 9.}$$
$$= 3(3) \qquad \text{Take the positive square root of 9 to obtain 3.}$$
$$= 9 \qquad \text{Multiply.} \quad \blacksquare$$

Try Problems 19 and 20.

EXAMPLE 4 Evaluate $\sqrt{b^2 - 4ac}$ for $a = 5$, $b = 3$, and $c = -2$.

SOLUTION First we substitute in the given values of a, b, and c.

$$\sqrt{b^2 - 4ac}$$
$$= \sqrt{3^2 - 4(5)(-2)} \qquad \text{Substitute 5 for } a, \text{ 3 for } b, \text{ and } -2 \text{ for } c.$$
$$= \sqrt{9 - (-40)} \qquad \text{Square 3 to obtain 9. Multiply 4, 5, and } -2 \text{ to obtain } -40.$$
$$= \sqrt{9 + 40} \qquad \text{Subtracting } -40 \text{ is the same as adding 40.}$$
$$= \sqrt{49} \qquad \text{Add under the radical.}$$
$$= 7 \qquad \text{Finally, take the indicated square root.} \quad \blacksquare$$

Try Problems 21 and 22.

EXAMPLE 5 If an object is dropped, the time t (in seconds) that it takes the object to fall a distance d (in feet) is given by the formula

$$t = \sqrt{\frac{d}{16}}$$

How long does it take an object to fall a distance of 1024 feet?

SOLUTION We want to find the value of t when d is equal to 1024.

$$t = \sqrt{\frac{d}{16}}$$
$$= \sqrt{\frac{1024}{16}} \qquad \text{Substitute 1024 for } d.$$
$$= \sqrt{64} \qquad \text{Divide 1024 by 16 to obtain 64.}$$
$$= 8 \qquad \text{Take the positive square root of 64 to obtain 8.}$$

It takes the object 8 seconds to fall 1024 feet. $\blacksquare$

Try Problem 23.

Answers to Try These Problems

1. 64 2. 5, −5 3. 4 4. −4 5. 3 6. 12 7. 9 8. −7
9. 1.414 10. 4.583 11. 12.247 12. 26.038
13. $\sqrt{39} \approx 6.24$, $\sqrt{60} \approx 7.75$ (on number line between 6, 7, 8)
14. $\sqrt{0.5} \approx 0.71$, $\sqrt{0.1} \approx 0.32$ (on number line between 0, 1, 2)
15. 10.5, $\sqrt{120}$, 11, $\sqrt{131}$, 12   16. 24, 24.2, $\sqrt{600}$, 25, $\sqrt{650}$
17. 4 18. 10 19. 62 20. −30 21. 7 22. 13
23. 56 ft per sec

USING THE CALCULATOR #16

THE $\sqrt{}$ KEY

Both basic and scientific calculators have a key that looks like $\boxed{\sqrt{}}$. This key can be used to evaluate square roots. Here we show how to evaluate an expression like $\sqrt{36}$.

To Compute	$\sqrt{36}$
Enter	36 $\boxed{\sqrt{}}$
Result	6

The result came out to be a whole number because 36 is a perfect square.

When the number under the square root symbol is not a perfect square, the result is not a whole number. Here we evaluate $\sqrt{20}$.

To Compute	$\sqrt{20}$
Enter	20 $\boxed{\sqrt{}}$
Result	4.472136

The decimal 4.472136 is not the exact square root of 20 but it is an approximation of $\sqrt{20}$ to 6 decimal places.

Calculator Problems

Evaluate. If the result is not a whole number, round off to 3 decimal places.

1. $\sqrt{81}$
2. $\sqrt{2809}$
3. $\sqrt{320}$
4. $\sqrt{82.6}$
5. $\sqrt{0.08}$
6. $\sqrt{\frac{3}{4}}$

EXERCISES 13.5

Solve.

1. What is the square of 12?
2. What is the square of 6?
3. What are the two square roots of 49?
4. What are the two square roots of 100?
5. What is the positive square root of 64?
6. What is the negative square root of 64?

Find the indicated square root without the use of a calculator.

7. $\sqrt{25}$
8. $\sqrt{144}$
9. $\sqrt{400}$
10. $\sqrt{625}$
11. $-\sqrt{81}$
12. $-\sqrt{169}$

Approximate the indicated square root to 3 decimal places by using a calculator.

13. $\sqrt{6}$
14. $\sqrt{28}$
15. $\sqrt{76}$
16. $\sqrt{286}$
17. $\sqrt{45.3}$
18. $\sqrt{0.07}$

Make a number line that shows the positions of the following numbers. Use a calculator to help you.

19. 13, 14, 15, $\sqrt{175}$, $\sqrt{200}$
20. 4, 4.5, 5, $\sqrt{18}$, $\sqrt{24}$
21. -1, 0, 1, $-\sqrt{0.5}$, $\sqrt{0.6}$
22. 0, 0.5, 1, $\sqrt{0.8}$, $\sqrt{0.1}$

List each group of numbers from smallest to largest. Use a calculator to help you.

23. 8.5, 8.25, $\sqrt{68}$, $\sqrt{83}$, 9
24. 20.3, 20.5, $\sqrt{420}$, $\sqrt{405}$, 20

Evaluate.

25. $\sqrt{3^2 + 4^2}$
26. $\sqrt{5^2 + 12^2}$
27. $\sqrt{10^2 - 6^2}$
28. $\sqrt{13^2 - 5^2}$
29. $5\sqrt{36} + 10\sqrt{4}$
30. $7\sqrt{81} - 8\sqrt{9}$
31. $25 + 6\sqrt{40 + 9}$
32. $30 - 8\sqrt{4 + 60}$

Evaluate for the given values of the variable.

33. $\sqrt{x^2 - y^2}$ for $x = 10$ and $y = 8$
34. $3\sqrt{4a + 9b}$ for $a = 7$ and $b = 8$
35. $\sqrt{b^2 - 4ac}$ for $a = 3$, $b = 7$, and $c = 2$
36. $-b + \sqrt{b^2 - 4ac}$ for $a = 6$, $b = -1$, and $c = -1$

Solve.

37. A square floor for a dance hall is to be designed. The length s (in feet) of each side of the floor is given by the formula $s = \sqrt{A}$ where A represents the area of the floor. Find the length of each side if the area is 3600 square feet.
38. The length c (in centimeters) of the longest side of a right triangle is given by the formula $c = \sqrt{a^2 + b^2}$ where a and b are the lengths of the shorter sides. Find the length of the longest side of a right triangle if the shorter sides have lengths 12 centimeters and 16 centimeters.

CHAPTER 13 SUMMARY

KEY WORDS AND PHRASES
positive exponent [13.1]
base [13.1]
exponentiation [13.1]
zero exponent [13.2]

negative exponent [13.2]
reciprocal [13.2]
power of ten [13.4]
scientific notation [13.4]

square [13.5]
square root [13.5]
radical [13.5]
perfect squares [13.5]

SYMBOLS
$\sqrt{x}$ means the *positive* square root of x. [13.5]
$-\sqrt{x}$ means the *negative* square root of x. [13.5]

IMPORTANT RULES

Meaning of Integer Exponents [13.1, 13.2]
For n a positive integer,
x^n means $xxxx \cdots x$ (n factors of x)
x^{-n} means $\dfrac{1}{x^n}$ for nonzero x
x^0 means 1 for nonzero x

How Exponents Fit into the Order to Operate [13.1, 13.2]
After grouping symbols, exponents have the highest priority. Exponents are done before the opposite of, multiplication, division, addition, and subtraction.

Three Laws of Exponents [13.3]
$x^n \cdot x^m = x^{n+m}$
$\dfrac{x^n}{x^m} = x^{n-m}$
$(x^n)^m = x^{nm}$

Meaning of Scientific Notation [13.4]
A decimal numeral is in scientific notation if it is written as a power of ten or as the product of a decimal between 1 and 10 and a power of ten.

How Square Roots Fit into the Order to Operate [13.5]
The square root symbol is understood to be a grouping symbol, so operate under the radical before taking the square root. Taking square roots has the same priority as applying exponents and taking absolute values.

CHAPTER 13 REVIEW EXERCISES

Solve.

1. Find the square of 36.
2. Find the two square roots of 36.
3. Find the cube of 5.
4. What is the fifth power of -2?
5. What is the positive square root of 625?
6. What is the negative square root of 400?

Evaluate.

7. 4^3
8. $(-3)^6$
9. 10^{-4}
10. 32^{-1}
11. $\sqrt{81}$
12. $\sqrt{900}$
13. $(2 - 17)^2$
14. $4 \cdot 8^{-2}$
15. $14 - 2\sqrt{64}$
16. $6^3 + 4^3$
17. $5^{-1} + 4^{-2}$
18. $\sqrt{20^2 - 12^2}$
19. $3^2 - 5[24 + (-1)^3]$
20. $6^2(3(5)^2 - 4(15))$
21. $38 + \sqrt{13 + 3(4)}$
22. $-5 + \sqrt{(-5)^2 - 4(3)(2)}$

Evaluate for the given values of the variables.

23. $4h^2 + 8$ for $h = -6$
24. $5x^2 - 8x + 3$ for $x = 4$
25. $-c^2 + 3c - 8$ for $c = 3$
26. $(2y^3 + 4y)(x^2 - x)$ for $x = 5$ and $y = 2$
27. $15a^{-1}$ for $a = 9$
28. $45b + 20b^{-2}$ for $b = 2$
29. $(a + b)^{-2}$ for $a = 3$ and $b = 4$
30. $a^{-2} + b^{-2}$ for $a = 3$ and $b = 4$
31. $4\sqrt{3x + 4y}$ for $x = 20$ and $y = 10$
32. $\sqrt{b^2 - 4ac}$ for $a = 8$, $b = 10$, and $c = -3$

Use an appropriate law of exponents to write each expression in a simpler exponential form. Do not evaluate.

33. $12^3 \cdot 12^8$
34. $10^{-6} \cdot 10^{20}$
35. $\dfrac{12^3}{12^8}$
36. $\dfrac{10^{-6}}{10^{20}}$
37. $(8^{-5})^{-4}$
38. $(4^7)^3$
39. $x^{-5} x^9$
40. $\dfrac{x^{-5}}{x^9}$
41. $(x^{-5})^9$
42. $\dfrac{a^{-3}}{a^{-7}}$

Write a decimal numeral for each.

43. 10^5
44. 10^{-9}
45. 5.9×10^{-12}
46. 63×10^7

Write each in scientific notation.

47. 1,750,000
48. 300,000,000,000,000
49. 0.0073
50. 0.000 000 000 472

Solve. Express the answer in scientific notation and as an ordinary decimal numeral.

51. During the summer, water evaporates from a large lake at the rate of 1.2×10^7 gallons per day. How many gallons per hour is this?
52. A rectangular sheet of foil is 0.05 meter wide, 2.5×10^2 meters long, and 4×10^{-6} meter thick. What is the volume of this sheet of foil?
53. In water, light travels about 6×10^{13} meters in 5000 minutes. How many meters per minute is this?
54. The time it takes light to travel 1 kilometer is 3.3×10^{-4} second. How long does it take light to travel 8×10^7 kilometers?

Solve.

55. Ms. Gomez is a consultant for a small business. She determined that the profit P (in dollars) is approximated by the formula $P = -x^2 + 500x$, where x is the number of products sold. What is the profit when 25 products are sold?
56. Due to pollution and overuse, the population P of rainbow trout in a lake declined as given by the formula $P = 400 \cdot 2^{-N}$, where N represents the number of months after April 1, 1992.
    a. How many trout were in the lake on April 1?
    b. How many trout were in the lake 1 month after April 1?
    c. How many trout were in the lake 3 months after April 1?
57. If an object is dropped, the velocity (in feet per second) is given by the formula $v = 8\sqrt{d}$, where d represents the distance (in feet) that the object has fallen. Find the velocity when the object has fallen 81 feet.

Make a number line that shows the positions of the following numbers.

58. $11.5, 12.5, \sqrt{150}, \sqrt{125}, 12$
59. $-1, 0, 1, \sqrt{0.5}, -\sqrt{0.2}$

List each group of numbers from smallest to largest.

60. $\sqrt{45}$, 7, 6.5, $\sqrt{50}$, 7.5
61. 2, 2.5, $\sqrt{5}$, $\sqrt{6}$, 2.3

Use a calculator to evaluate each to 3 decimal places.

62. $\sqrt{58}$
63. $\sqrt{5.98}$
64. $\sqrt{0.75}$

In Your Own Words

Write complete sentences to discuss each of the following. Support your comments with examples or pictures, if appropriate.

65. Look at values of the expression x^2 for some values of x between 0 and 1. Discuss how x^2 compares with x for these values of x.
66. Look at values of the expression $\sqrt{w}$ for some values of w that are between 0 and 1. Discuss how $\sqrt{w}$ compares with w for these values of w.
67. Look at values of the expression $\frac{1}{x}$ for values of x that are very large. Discuss what you observe about the value of $\frac{1}{x}$ as x gets larger and larger.
68. Look at the values of the expression $\frac{1}{x}$ for $x = 0.5$, $x = 0.2$, $x = 0.1$, $x = 0.01$, $x = 0.001$, and $x = 0.0001$. Discuss what you observe about the value of $\frac{1}{x}$ as x gets closer and closer to 0.
69. Compare the value of $m^2 + n^2$ to the value of $(m + n)^2$ for various values of n and m. Discuss what you observe.

CHAPTER 13 PRACTICE TEST

Evaluate.

1. 2^6
2. 3^{-4}
3. $\sqrt{121}$
4. $(5 - 9)^3$
5. $2^{-1} + 5^{-1}$
6. $\sqrt{5^2 + 12^2}$

Evaluate for the given values of the variables.

7. $4c^2 - 5c + 6$ for $c = 5$
8. $-x^2 + 50x$ for $x = 20$
9. $3y^{-2}$ for $y = 6$
10. $(a + b)^{-1}$ for $a = 6$ and $b = 8$
11. $14 + x\sqrt{18x}$ for $x = 2$
12. $\sqrt{b^2 - 4ac}$ for $a = 4$, $b = -5$, and $c = -6$
13. The volume of a box, with all edges the same length, is given by the formula $V = s^3$, where s represents the length of each edge. Find the volume of a box with each edge measuring 8 feet.

Use an appropriate law of exponents to write each expression in a simpler exponential form. Do not evaluate.

14. $\dfrac{5^8}{5^{-3}}$
15. $9^7 \cdot 9^8$
16. $y^{-3}y^{-6}$
17. $(a^{-1})^{-7}$

Write a decimal numeral for each.

18. 7×10^8
19. 3.96×10^{-13}

Write each in scientific notation.

20. 0.000 000 000 045

21. 940,000

22. Make a number line that shows the approximate positions of the numbers 4.5, 5.2, $\sqrt{23}$, $\sqrt{20}$, and 5.5.

23. List these numbers from smallest to largest: 8, $\sqrt{62}$, $\sqrt{67}$, 8.5

Solve. Express the answer in scientific notation and as an ordinary decimal numeral.

24. One year the government of a country was estimated to be in debt 5.4×10^{13} dollars. The population of the country that year was about 6×10^8. If each person paid an equal share of the debt, how much would each person owe?

25. Water is entering a tank at the rate of 7×10^4 gallons per second. How much water has entered the tank after 10^{-2} second?

CUMULATIVE REVIEW EXERCISES: CHAPTERS 1–13

Evaluate.

1. $\frac{7}{15} + \frac{2}{3} + \frac{4}{9}$

2. $200.3 - 76.875$

3. $28 - 6(7 - 4)$

4. $-8(-5) + 9(-5) - 12$

5. $0.8 \div 25$

6. $15\frac{5}{12} - 9\frac{2}{3}$

7. 6^3

8. 3^{-2}

9. $\sqrt{2500}$

Evaluate each by using a calculator. Round off the answer to 3 decimal places.

10. $(0.85)^8$

11. $\sqrt{133} + \sqrt{82}$

12. $50\sqrt{\frac{187}{65.8}}$

13. $\sqrt{\frac{5.6(9.2)}{197}}$

Evaluate for the given values of the variables.

14. $5c - 9$ for $c = 4$

15. $7xy + 6y$ for $x = -3$ and $y = 4$

16. $4n^2 + 16$ for $n = 8$

17. $200m^{-1}$ for $m = 5$

Solve.

18. Find the sum of 0.083 and 7.36.

19. What is the ratio of 0.05 to 2.5?

20. What number is 5 less than 1?

21. Three-fourths of what number is 87?

22. Walter owns $\frac{7}{8}$ of a building and Thelma owns the rest. What fraction of the building does Thelma own?

23. In the last seven days Scott has jogged the following distances in miles.

3.5   4.0   3.0   3.0   4.5   5.0   3.0

Find **a.** the mean, **b.** the median, and **c.** the mode. (Round off to the nearest tenth of a mile.)

24. How many 1.8-meter pieces of wire can be cut from a wire that is 34.2 meters long?

25. Ms. Martinez wants to have a fence built around a rectangular piece of land that is 9.5 feet wide and 22.4 feet long. The fencing material costs $8.50 per foot. What is the total cost of the fence? (Round up the answer to the nearest dollar.)

26. A bath towel that was selling for $18 is now on sale at 40% off. What is the sale price?

27. A history class consists of 16 men and 24 women. What percent are women?

28. String sells for $1.39 per meter. What is the cost of 30 centimeters of this string? (Round up to the nearest cent.)

29. Ms. Tokerud is driving from Miami to Jacksonville. She begins with a full tank of gas. When she has gone $\frac{2}{3}$ of the distance from Miami to Jacksonville, she stops to fill her tank with gas. The gasoline costs $1.20 per gallon and she spent a total of $10.50 for the gas. Her car averages 32 miles per gallon.
   a. How many gallons of gasoline did she buy with $10.50?
   b. When she stopped for gas, how many miles had she traveled?
   c. Using the information in this problem, find the distance from Miami to Jacksonville.

The bar graph shown here gives the rate of return earned (or lost) on a mutual fund account for each of five years. Use the graph to answer Exercises 30 through 33.

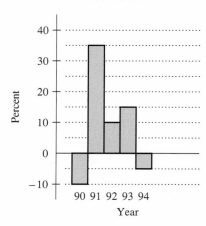

30. What was the return rate for 1991? Interpret the answer.
31. What was the return rate for 1990? Interpret the answer.
32. If you had an investment of $860 in the fund for the entire year of 1993, how much money did you earn (or lose)?
33. If you had $1500 in the fund for the entire year of 1994, how much money did you earn (or lose)?

CHAPTER 14

Introduction to Algebra

14.1 SIMPLIFYING BASIC ALGEBRAIC EXPRESSIONS

OBJECTIVES

- Simplify an algebraic expression by rearranging the order of the terms.
- Simplify an algebraic expression by rearranging the order of the factors.
- Simplify an algebraic expression by using the first law of exponents.

Recall from Section 12.7 that an **algebraic expression** is a combination of variables, numbers, and operations put together in a meaningful way so that the expression represents a number.

We simplify algebraic expressions to make them easier to work with. In general, to simplify the expression means to write it with as few symbols as possible. In simplifying an algebraic expression, we change the form of the expression but not the value it represents. Since the variables represent numbers, we must follow the rules for operating with numbers when we rearrange the way the expression is written. Here are several rules that we have already studied that are used repeatedly when simplifying algebraic expressions.

Some Number Properties Needed to Simplify Algebraic Expressions

1. Repeated addition can be done in any order.
2. Repeated multiplication can be done in any order.
3. Subtracting a number is the same as adding its opposite.
4. The first law of exponents: $x^n x^m = x^{n+m}$

648 Chapter 14 Introduction to Algebra

 Try These Problems

Simplify.
1. $a + b + 5 + 7$
2. $y + 8 + u + 6$
3. $(b + 7) + (d + 9)$
4. $(w + 3) + (6 + r) + 9$

1 Rearranging the Order of the Terms

Here we simplify some basic algebraic expressions by using the fact that addition can be done in any order. We look at expressions with several **terms.** Recall from Section 1.2 that terms are quantities that are added.

EXAMPLE 1 Simplify: $6 + x + 5 + y$

SOLUTION The only operation in this expression is addition. Because addition can be done in any order, we rearrange the order of the terms so we can add the numbers 6 and 5.

$$6 + x + 5 + y$$
$$= 6 + 5 + x + y \quad \text{Rearrange the order of the addition.}$$
$$= 11 + x + y \quad \text{Add 6 and 5 to obtain 11.}$$

The remaining three terms, 11, x, and y, can be written in any order. For example, the answer can also be written $x + y + 11$. ∎

EXAMPLE 2 Simplify: $(w + 4) + (a + 9) + 8$

SOLUTION Because the only operation in this expression is addition and addition can be done in any order, the parentheses serve no purpose. Therefore, we drop the parentheses and rearrange the order of the addition so that we can add together the numerical terms.

$$(w + 4) + (a + 9) + 8$$
$$= w + 4 + a + 9 + 8 \quad \text{Drop the parentheses.}$$
$$= w + a + 4 + 9 + 8 \quad \text{Rearrange the order of the terms.}$$
$$= w + a + 21 \quad \text{Add 4, 9, and 8 to obtain 21.}$$

The remaining 3 terms can be written in any order. For example, the answer can also be written $w + 21 + a$. ∎

Try Problems 1 through 4.

EXAMPLE 3 Simplify: $c - 4 + 6 + a - 9$

SOLUTION This expression contains a mixture of addition and subtraction. Subtracting a number is the same as adding its opposite, so we can view this expression as repeated addition and are able to rearrange the order of the terms.

$$c - 4 + 6 + a - 9$$
$$= c + (-4) + 6 + a + (-9) \quad \text{Subtracting 4 is the same as adding } -4. \text{ Subtracting 9 is the same as adding } -9.$$
$$= c + a + (-4) + 6 + (-9) \quad \text{Rearrange the order of the terms.}$$
$$= c + a + (-7) \quad \text{Add } -4, 6, \text{ and } -9 \text{ to obtain } -7.$$
$$= c + a - 7 \quad \text{Adding } -7 \text{ is the same as subtracting 7.}$$

Try These Problems

Simplify.
5. $3 - 5 + x + y - 16$
6. $4 - d - 13 + k + 7$
7. $(8 - r) + (x - 6) - 1$
8. $(13 - c) + (a + 5) + (x - 14)$

The remaining 3 terms can be written in any order as long as the minus symbol stays associated with the 7. For example, the answer can also be written $c - 7 + a$. ■

In Example 3, notice that we could have rearranged the order of the terms without actually changing the subtractions to additions. Here we do the problem in this way.

$$c - 4 + a + 6 - 9$$
$$= c + a - 4 + 6 - 9 \qquad \text{The terms can be written in any order as long as the minus symbols stay with the 4 and 9.}$$
$$= c + a - 7 \qquad \text{Add } -4, 6, \text{ and } -9 \text{ to obtain } -7.$$

EXAMPLE 4 Simplify: $(15 - x) + (y - 8) + 4$

SOLUTION Removing the parentheses does not change the meaning of the expression since repeated addition can be done in any order.

$$(15 - x) + (y - 8) + 4$$
$$= 15 - x + y - 8 + 4 \qquad \text{Drop the parentheses.}$$
$$= 15 - 8 + 4 - x + y \qquad \text{Rearrange the order of the terms so that the numerical terms are together. Keep the minus symbols with the 8 and } x.$$
$$= 11 - x + y \qquad \text{Add } 15, -8, \text{ and } 4 \text{ to obtain } 11.$$

The answer can be written in several other ways. Two examples are $y - x + 11$ or $-x + y + 11$. ■

 Try Problems 5 through 8.

2 Rearranging the Order of the Factors

Here we simplify some basic algebraic expressions by using the fact that repeated multiplication can be done in any order. We look at expressions with several **factors.** Recall that factors are quantities that are multiplied.

EXAMPLE 5 Simplify: $4x(5y)$

SOLUTION The only operation in this expression is multiplication. Since multiplication can be done in any order, we drop the parentheses and rearrange the order of the factors so that the numerical factors are written first.

$$4x(5y)$$
$$= 4x \cdot 5y \qquad \text{Drop the parentheses.}$$
$$= 4(5)xy \qquad \text{Rearrange the order of the factors so that the numerical factors are written first.}$$
$$= 20xy \qquad \text{Multiply 4 by 5 to obtain 20.}$$

Try These Problems

Simplify.

9. $-6w(3a)$
10. $(5r)(4w)(2w)$
11. $(-4wc)(-7wc)$
12. $8vm(3m)(vm)$
13. $-xy(6y)(5xy)$
14. $(-ac)(-ab)(7ab)$

Although the expression represents the same value no matter what order the factors are written, we prefer to write the numerical factor first. The variable factors can be written in any order. For example, the answer can also be written $20yx$. ■

EXAMPLE 6 Simplify: $2x(-7w)(5x)$

SOLUTION Because this expression is repeated multiplication, we can rearrange the order of the factors, then simplify.

$$2x(-7w)(5x)$$
$$= 2(-7)(5)xxw$$ Drop the original parentheses and rearrange the order of the factors so that the numerical factors are first and the x factors are together.
$$= -70x^2w$$ Multiply 2, -7, and 5 to obtain -70. Write xx as x^2.

This answer can also be written $-70wx^2$. ■

Try Problems 9 through 12.

Observe that taking the opposite of a number is the same as multiplying the number by -1. Here we look at some examples.

$$-(20) = -1(20)$$
$$-(-16) = -1(-16)$$
$$-x = -1x$$ The opposite of x is the same as -1 times x.
$$-ab = -1ab$$

Therefore, if a negative sign precedes a variable or several variable factors, the numerical factor is really -1.

EXAMPLE 7 Simplify: $4b(-ab)(-3ab)$

SOLUTION Because the expression is repeated multiplication, we can rearrange the order of the factors.

$$4b(-ab)(-3ab)$$
$$= 4b(-1ab)(-3ab)$$ Observe that $-ab$ means $-1ab$. Rearrange the order of the factors.
$$= 4(-1)(-3)aabbb$$
$$= 12a^2b^3$$ Multiply 4, -1, and -3 to obtain 12. Write aa as a^2. Write bbb as b^3. ■

Try Problems 13 and 14.

3 Applying the First Law of Exponents

Recall that the first law of exponents was introduced in Section 13.3. Here are two examples of using this law.

$$4^2 \cdot 4^3 = 4^{2+3} = 4^5$$
$$m^4m^5 = m^{4+5} = m^9$$

Try These Problems

Simplify.
15. $9x^2(-2x)$
16. $12y^3(8y^3)$
17. $7x^2y(2xy^3)$
18. $xw(-xw^3)(4x^2w)$

This law indicates to add exponents when multiplying exponential expressions with a common base.

Here are some examples using the first law of exponents to simplify some algebraic expressions.

EXAMPLE 8 Simplify: $(5x^2)(7x^3)$

SOLUTION Because multiplication can be done in any order, we can drop the parentheses and rearrange the order of the factors.

$(5x^2)(7x^3)$
$= 5x^2 \cdot 7x^3$ Drop the parentheses.
$= 5(7)x^2x^3$ Rearrange the order of the factors so that the numerical factors are first.
$= 35x^5$ Multiply 5 by 7 to obtain 35.
Multiply x^2 by x^3 to obtain x^5 by using the first law of exponents.

EXAMPLE 9 Simplify: $4xy^2(-6x^3y)(-yw)$

SOLUTION Because multiplication can be done in any order, we can drop the parentheses and rearrange the order of the factors.

$4xy^2(-6x^3y)(-yw)$
$= 4xy^2(-6x^3y)(-1yw)$ Note that $-yw$ means $-1yw$.
$= (4)(-6)(-1)xx^3y^2yyw$ Rearrange the order of the factors, listing the numerical factors first, then group the xs together and the ys together.
$= 24x^4y^4w$ Multiply 4, -6, and -1 to obtain 24.
Multiply x by x^3 to obtain x^4.
Multiply y^2, y, and y to obtain y^4.

Try Problems 15 through 18.

Answers to Try These Problems

1. $a + b + 12$ 2. $y + 14 + u$ 3. $b + d + 16$
4. $w + 18 + r$ 5. $x + y - 18$ 6. $k - d - 2$ 7. $1 - r + x$
8. $4 - c + a + x$ 9. $-18wa$ 10. $40rw^2$ 11. $28w^2c^2$
12. $24v^2m^3$ 13. $-30x^2y^3$ 14. $7a^3b^2c$ 15. $-18x^3$ 16. $96y^6$
17. $14x^3y^4$ 18. $-4x^4w^5$

EXERCISES 14.1

Simplify.
1. $16 + a + 8$
2. $30 + x + 5$
3. $x - 12 + 7$
4. $-13 - y + 6$
5. $8 + a + 15 + c$
6. $c + 35 + d + 9$
7. $15 - r + 12 - 5$
8. $-16 + x - y - 4$
9. $7 + (w + 3)$

10. $(18 + x) + 9$ 11. $(w - 5) - 14$ 12. $32 + (c - 17)$
13. $(5 + x) + (10 - y)$ 14. $y - 3 + (7 - c + 8)$ 15. $-(-4x)$
16. $5x(9y)$ 17. $-16a(-6b)$ 18. $(15c)(-20a)$
19. $8x(5x)(-x)$ 20. $5xy(8xy)(4y)$ 21. $w(-5w)(-xw)$
22. $-ab(8ab)(5b)$ 23. $x^2(3x^3)$ 24. $-5y^3(4y^3)$
25. $13x(x)(7x^2)$ 26. $45y^3(2y)(3y)$ 27. $(-6xy^2)(5x^2y^2)$
28. $5xy^3(15x^3y^2)$ 29. $(-7ab)(5a^2b^3)(3ab^2)$ 30. $50ab^3(-a^3b^3)(-4ab)$

14.2 COMBINING LIKE TERMS

OBJECTIVES
- Know the difference between terms and factors.
- Multiply by using the distributive law.
- Take out factors common to all terms.
- Simplify an algebraic expression by combining like terms.
- Simplify an algebraic expression by removing parentheses and combining like terms.

 Try These Problems

Answer true or false.
1. The number 4 is a factor of the expression $4 + y + x$.
2. The number 4 is a factor of the expression $4yx$.
3. The variable y is a term of the expression $4yx$.
4. The variable y is a term of the expression $4 + y + x$.
5. The number 7 is a term of the expression $7m + 3n$.
6. The number 7 is a factor of the expression $7m + 3n$.

1 Defining Necessary Terminology

In this section we introduce a new property that relates addition and multiplication; we then use this property to simplify some algebraic expressions.

When working with algebraic expressions, you must be able to distinguish terms from factors. **Terms** are quantities that are added together, while **factors** are quantities that are multiplied. For example, in the expression

$$n + m + 7$$

there are three *terms*, n, m, and 7. In the expression

$$7mn$$

there are three *factors*, 7, m, and n. In the expression

$$5x + 3y$$

there are two *terms*, $5x$ and $3y$. The number 5 is a factor of only the first term, but 5 is not a factor of the entire expression. Likewise, y is a factor of only the second term, but y is not a factor of the entire expression. In the expression

$$5(x + 2)$$

there are two *factors*, 5 and the quantity $x + 2$. Here 5 is a factor of the entire expression.

 Try Problems 1 through 6.

Try These Problems

Using the expression $7ac - 9a - 8$, answer Problems 7 through 9.

7. List the terms.
8. What is the numerical coefficient of the first term?
9. What is the coefficient of the second term?

When an expression has several terms, we think of these terms as being added, even though the subtraction symbol may appear. For example, the expression

$$4x^2 - 5x + 2$$

can be written as

$$4x^2 + (-5x) + 2$$

so there are three terms: $4x^2$, $-5x$, and 2. When the expression is viewed as addition of terms, the numerical factor of each term is called the **numerical coefficient** (or **coefficient**) of that term. In this example, the numerical coefficients of the three terms are 4, -5, and 2, respectively.

Try Problems 7 through 9.

2 Using the Distributive Law of Multiplication over Addition

Consider the numerical expression

$$5(4 + 8)$$

If we use the rules for the order of operation to evaluate the expression, we see that it equals 60.

$$5(4 + 8) = 5(12)$$
$$= 60$$

Add inside the parentheses first, then multiply.

There is another way to evaluate this expression to obtain the same result.

$$5(4 + 8) = 5 \cdot 4 + 5 \cdot 8$$
$$= 20 + 40$$
$$= 60$$

Multiply each term inside the parentheses by 5, then add.

Here is a diagram that illustrates why $5(4 + 8) = 5 \cdot 4 + 5 \cdot 8$. In this diagram we count the total number of dots in two ways as illustrated. Note that the number of dots in a rectangular array can be counted by multiplying the number of rows by the number of columns.

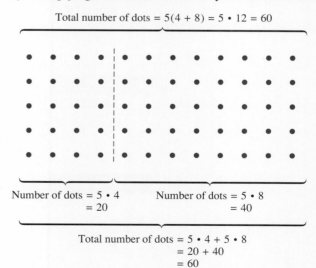

Try These Problems

Evaluate each in two ways: **a.** *by using the order to operate and* **b.** *by using the distributive law.*

10. 8(12 + 18)

11. 4(3 + 9 + 7)

The above discussion suggests a very important relationship between multiplication and addition. We say that **multiplication distributes over addition.** Here is a general statement of this law.

Distributive Law of Multiplication over Addition
For numbers a, b, and c,
$$a(b + c) = ab + ac$$

Try Problems 10 and 11.

The distributive law is not really needed to evaluate *numerical* expressions because using the rule for the order to operate is usually an easier way to evaluate an expression like

$$5(4 + 8)$$

However, the distributive law of multiplication over addition is necessary when working with *algebraic* expressions, because it gives us an alternative way of looking at the expression without changing the value the expression represents. For example,

$$6(x + 9) = 6 \cdot x + 6 \cdot 9$$
$$= 6x + 54$$

When viewing the expression as $6(x + 9)$, we see the *factors* of the expression so we call this the *factored form*. When viewing the expression as $6x + 54$, we see the *terms* of the expression, and we say the expression is *multiplied out* or *written without parentheses*. After more experience with algebra, you will see that sometimes we prefer the factored form, and at other times, we prefer the multiplied out form. Both forms of the expression are considered simplified.

Here are some examples of using the distributive law to multiply a quantity by a sum.

EXAMPLE 1 Multiply: $4(a + 3)$

SOLUTION

$$4(a + 3) = 4 \cdot a + 4 \cdot 3$$
$$= 4a + 12$$

Multiply each term in the parentheses by 4.

EXAMPLE 2 Multiply: $x(y + 5 + a)$

SOLUTION

$$x(y + 5 + a) = xy + 5x + xa$$

Multiply each term in the parentheses by x.

EXAMPLE 3 Multiply: $4w(w + 7)$

SOLUTION

$$4w(w + 7) = 4w \cdot w + 4w \cdot 7$$
$$= 4w^2 + 28w$$

Multiply each term in the parentheses by $4w$.

Try These Problems

Multiply, by using the distributive law.

12. $2(7 + c)$
13. $w(a + b + 8)$
14. $3c(c + 4a)$
15. $6w(w^2 + 5w + 9)$
16. $-7(3 + 5x)$
17. $-8w(w + 6y)$
18. $-3a(5a^2 + 2a + 12)$

Try Problems 12 through 15.

The previous three examples involved a factor outside the parentheses that had a positive sign associated with it. Now we look at situations where the factor outside the parentheses has a negative sign associated with it.

EXAMPLE 4 Multiply: $-6(2x + 4)$

SOLUTION

$$-6(2x + 4) = (-6)2x + (-6)4$$
$$= -12x + (-24)$$
$$= -12x - 24$$

Multiply -6 by each term inside the parentheses.

Adding -24 is the same as subtracting 24. ∎

EXAMPLE 5 Multiply: $-8c(2c^2 + 3c + 1)$

SOLUTION

$$-8c(2c^2 + 3c + 1)$$
$$= (-8c)2c^2 + (-8c)3c + (-8c)1$$
$$= -16c^3 + (-24c^2) + (-8c)$$
$$= -16c^3 - 24c^2 - 8c$$

Multiply $-8c$ by each term inside the parentheses.

Adding $-24c^2$ is the same as subtracting $24c^2$.

Adding $-8c$ is the same as subtracting $8c$. ∎

Try Problems 16 through 18.

In the expression

$$x(x - 6)$$

the operation inside the parentheses is subtraction, and because subtracting a number is the same as adding its opposite, the distributive law of multiplication over addition applies. Here we look at the situation in detail.

$$x(x - 6)$$
$$= x(x + (-6))$$ Subtracting 6 is the same as adding -6.
$$= xx + x(-6)$$ Multiply each term inside the parentheses by x.
$$= x^2 + (-6x)$$
$$= x^2 - 6x$$ Adding $-6x$ is the same as subtracting $6x$.

In the previous example, we can think subtracting 6 is the same as adding −6 without actually writing it down. Here we look at the problem done in this way.

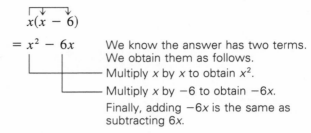

$= x^2 - 6x$ We know the answer has two terms. We obtain them as follows.
- Multiply x by x to obtain x^2.
- Multiply x by -6 to obtain $-6x$.

Finally, adding $-6x$ is the same as subtracting $6x$.

Although the second procedure requires fewer steps to be written down, this shortcut should not be used until you fully understand what is really happening.

Here are more examples.

EXAMPLE 6   Multiply: $4(2w - 10)$

SOLUTION
LONG METHOD

$4(2w - 10)$
$= 4(2w + (-10))$ Subtracting 10 is the same as adding −10.
$= 4 \cdot 2w + 4(-10)$ Multiply each term inside the parentheses by 4.
$= 8w + (-40)$
$= 8w - 40$ Adding −40 is the same as subtracting 40.

SHORTCUT

$4(2w - 10)$
$= 8w - 40$ We know the answer has two terms. We obtain them as follows.
- Multiply 4 by $2w$ to obtain $8w$.
- Multiply 4 by -10 to obtain -40.

Adding -40 is the same as subtracting 40. ∎

EXAMPLE 7   Multiply: $-3n(n^2 - 7n + 2)$

SOLUTION
LONG METHOD

$-3n(n^2 - 7n + 2)$
$= -3n(n^2 + (-7n) + 2)$ Subtracting $7n$ is the same as adding $-7n$.
$= -3n \cdot n^2 + (-3n)(-7n) + (-3n)2$ Multiply $-3n$ by each term inside the parentheses.
$= -3n^3 + 21n^2 + (-6n)$
$= -3n^3 + 21n^2 - 6n$ Adding $-6n$ is the same as subtracting $6n$.

14.2 Combining Like Terms

Try These Problems

Multiply, by using the distributive law.
19. $-5(8 - a)$
20. $-b(a - 1)$
21. $-3w(4x + 5y - 9)$
22. $-30c(4c^2 - c - 2)$

SHORTCUT

$$-3n(n^2 - 7n + 2)$$
$$= -3n^3 + 21n^2 - 6n$$

We obtain the three terms as follows.
- Multiply $-3n$ by n^2 to obtain $-3n^3$.
- Multiply $-3n$ by $-7n$ to obtain $21n^2$.
- Multiply $-3n$ by 2 to obtain $-6n$. ∎

▲ **Try Problems 19 through 22.**

2 Taking Out All Factors Common to All Terms

The expression

$$4x + 4y$$

has two terms: $4x$ and $4y$. Note that 4 is a factor of each term. The distributive law of multiplication over addition can be used to write the expression in its factored form by taking out the factor common to both terms.

$$4x + 4y = 4(x + y)$$

Observe that if you multiply out the expression $4(x + y)$, you obtain the original expression $4x + 4y$. The expression $4x + 4y$ is written without parentheses, while the expression $4(x + y)$ is written in factored form. Both forms of the expression are considered simplified.

Here we look at more examples of *taking out factors common to all terms* by using the distributive law. This process is often referred to as **factoring**.

EXAMPLE 8 Take out all factors common to all terms: $18 + 24m$

SOLUTION Observe that 6 is a factor of each term.

$$18 + 24m$$
$$= 6 \cdot 3 + 6 \cdot 4m$$ The number 6 is a factor of both terms.
$$= 6(3 + 4m)$$

- This term should multiply by 6 to give back 18.
- This term should multiply by 6 to give back $24m$.

CHECK

$$6(3 + 4m) = 6 \cdot 3 + 6 \cdot 4m$$
$$= 18 + 24m$$

The answer is $6(3 + 4m)$. ∎

Try These Problems

Take out all factors common to all terms.

23. $8m + 8n$
24. $15x + 35$
25. $9x - 18y$
26. $20 - 30ac$

EXAMPLE 9 Take out all factors common to all terms: $12x - 4$

SOLUTION Observe that 4 is a factor of each term.

$$12x - 4$$
$$= 4 \cdot 3x - 4 \cdot 1 \quad \text{The number 4 is a factor of each term.}$$
$$= 4(3x - 1)$$

This term should multiply by 4 to give back $12x$.
This term should multiply by 4 to give back -4.

CHECK
$$4(3x - 1) = 12x - 4$$

The answer is $4(3x - 1)$. ■

 Try Problems 23 through 26.

EXAMPLE 10 Take out all factors common to all terms: $3xy + 2x + ax$

SOLUTION Observe that x is a factor of all three terms.

$$3xy + 2x + ax$$
$$= 3y \cdot x + 2 \cdot x + a \cdot x \quad \text{The variable } x \text{ is a factor of all three terms.}$$
$$= x(3y + 2 + a)$$

This term should multiply by x to give back $3xy$.
This term should multiply by x to give back $2x$.
This term should multiply by x to give back ax.

CHECK
$$x(3y + 2 + a) = 3xy + 2x + ax$$

The answer is $x(3y + 2 + a)$. ■

EXAMPLE 11 Take out all factors common to all terms: $15y^2 - y$

SOLUTION Observe that there is one factor of y common to both terms.

$$15y^2 - y$$
$$= 15y \cdot y - 1 \cdot y \quad \text{The variable } y \text{ is a factor of each term.}$$
$$= y(15y - 1)$$

This term must multiply by y to give back $15y^2$.
This term must multiply by y to give back $-y$.

CHECK
$$y(15y - 1) = 15y^2 - y$$

The answer is $y(15y - 1)$. ■

Try These Problems

Take out all factors common to all terms.
27. $6y + xy$
28. $3w^2 + 5w$
29. $wc - c + 8xc^2$
30. $7m^3 - 9m^2 + m$
31. $9x + 6ax$
32. $24w^2 + 8w$
33. $45ac - 5c$
34. $27mn - 18xm$

Try Problems 27 through 30.

EXAMPLE 12 Take out all factors common to all terms: $4x^2 + 6x$

SOLUTION Observe that 2 and x are factors of both terms.

$4x^2 + 6x$
$= 2 \cdot 2 \cdot x \cdot x + 2 \cdot 3 \cdot x$ 2 and x are factors of both terms.
$= 2x(2x + 3)$ Take out the common factor $2x$.
     This term must multiply by $2x$ to give back $4x^2$.
     This term must multiply by $2x$ to give back $6x$.

CHECK

$$2x(2x + 3) = 4x^2 + 6x$$

The answer is $2x(2x + 3)$. ∎

EXAMPLE 13 Take out all factors common to all terms: $14ax - 7x$

SOLUTION Observe that 7 and x are factors of both terms.

$14ax - 7x$
$= 7x(2a - 1)$
     This term must multiply by $7x$ to give back $14ax$.
     This term must multiply by $7x$ to give back $-7x$.

CHECK

$$7x(2a - 1) = 14ax - 7x$$

The answer is $7x(2a - 1)$. ∎

 Try Problems 31 through 34.

3 Combining Like Terms

Consider the expression

$$2x + 3x$$

where the variable x is a common factor of each term. The expression can be written as follows by factoring out the common factor x.

$$2x + 3x = x(2 + 3)$$
$$= x(5)$$
$$= 5x$$

Note that the terms add to give only one term because the variable factors in the original two terms are exactly alike: both are x.

Here is another example.

$$2ac - 9ac = ac(2 - 9)$$
$$= ac(-7)$$
$$= -7ac$$

The terms combine into a single term because each of the original terms has exactly the same variable factors, *ac*. Terms of this type are called **like terms.** The process of adding (or subtracting) like terms is often referred to as *combining like terms.*

In the expression

$$4y + 3x$$

the two terms have different variable factors: the first term has the factor *y* and the second term has the factor *x*. Therefore, these two terms are not like terms and cannot be combined to form a single term.

In the expression

$$8x^2 - 5x$$

the two terms have different variable factors: the first term has the factor x^2, while the second term has the factor *x*. Therefore, these two terms are not like terms and cannot be combined to form a single term. Because *x* is a factor common to both terms, this expression can be written in a factored form. For example,

$$8x^2 - 5x = x(8x - 5)$$

Both forms of this expression are considered simplified.

Combining like terms helps in simplifying algebraic expressions. Here are more examples.

EXAMPLE 14 Combine like terms: $5y + 4y$

SOLUTION The two terms are like terms because they have the same variable factor, *y*.

$$5y + 4y$$
$$= (5 + 4)y$$
$$= 9y \blacksquare$$

EXAMPLE 15 Combine like terms: $12w - 17w + 3w$

SOLUTION All three terms are like terms because they have the same variable factor, *w*.

$$12w - 17w + 3w$$
$$= (12 - 17 + 3)w$$
$$= -2w \qquad \text{Add 12, } -17, \text{ and 3 to obtain } -2. \blacksquare$$

EXAMPLE 16 Combine like terms: $20m^2 - 5m^2$

SOLUTION The two terms are like terms because they have the same variable factor, m^2.

$$20m^2 - 5m^2$$
$$= (20 - 5)m^2$$
$$= 15m^2 \blacksquare$$

14.2 Combining Like Terms   **661**

 Try These Problems

Simplify by combining like terms.
35. $7x + 3x$
36. $16c + 5c - 3c$
37. $2ax - 6ax - 8ax$
38. $3y^2 + 4y^2$
39. $13a^2 - 22a^2$
40. $9x + x$
41. $12ab - ab - 8ab$
42. $-5c + c - 6c$

 Try Problems 35 through 39.

EXAMPLE 17   Combine like terms: $8x + x$

SOLUTION   The two terms are like terms because they have the same variable factor, x.

$$8x + x$$
$$= 8x + 1x$$
$$= (8 + 1)x$$
$$= 9x$$

You can think of the numerical factor of the second term as being 1 because $x = 1x$.

EXAMPLE 18   Combine like terms: $15w - w - 20w$

SOLUTION   All three terms have the same variable factor, so they can be combined to form a single term.

$$15w - w - 20w$$
$$= 15w - 1w - 20w$$
$$= (15 - 1 - 20)w$$
$$= -6w$$

You can think of the numerical factor of the second term as being -1 because $-w = -1w$.

Add 15, -1, and -20 to obtain -6.

Try Problems 40 through 42.

Before simplifying more complex expressions, we want to look at how to combine like terms without writing down the factoring step. For example,

$$5b + 8b = (5 + 8)b$$
$$= 13b$$

and we can omit the step with the parentheses if we observe that the numerical factor, 13, can be obtained by adding the numerical coefficients of the two like terms. The variable factor, b, is the same as the common factor in each term.

Here we look at another example.

$$3x^2 - x^2 - 7x^2$$
$$= -5x^2$$

Add the coefficients, 3, -1, and -7 to obtain the numerical factor -5. Keep the common variable factor, x^2.

Here are more examples.

EXAMPLE 19   Combine like terms: $16c - 8 - 10c$

SOLUTION   Only the first and last terms are like terms, so those two terms can be combined to form a single term.

$$16c - 8 - 10c$$
$$= 16c - 10c - 8$$
$$= 6c - 8$$

Rearrange the order of the terms so that like terms are grouped together.

Add 16 and -10 to obtain 6, the coefficient of the c term.

Chapter 14 Introduction to Algebra

 Try These Problems

Simplify by combining like terms.
43. $3x + 5y - 9x$
44. $n^2 - 4n + 13n^2 - 5n$
45. $-8m - 4mn + 7 + 7mn - 8$
46. $5ac + 14a - 12ac - a$

In the previous example, the expression cannot be written with fewer than two terms because the terms are not like terms.

EXAMPLE 20 Combine like terms: $7y^2 - 15y + 5y^2 + 8y$

SOLUTION

$$7y^2 - 15y + 5y^2 + 8y$$
$$= 7y^2 + 5y^2 - 15y + 8y$$
$$= 12y^2 - 7y$$

Rearrange the order of the terms so that like terms are grouped together.
Add 7 and 5 to obtain 12. Keep the common variable factor, y^2.
Add -15 and 8 to obtain -7. Keep the common variable factor, y. ∎

EXAMPLE 21 Combine like terms: $-9 + 5xy - 4 - xy + 7$

SOLUTION

$$-9 + 5xy - 4 - xy + 7$$
$$= -9 - 4 + 7 + 5xy - xy$$
$$= -6 + 4xy$$

Rearrange the order of the terms so that like terms are grouped together.
Add -9, -4, and 7 to obtain -6.
Add 5 and -1 to obtain 4. Keep the common variable factor, xy. ∎

 Try Problems 43 through 46.

EXAMPLE 22 Remove parentheses and simplify: $(2x - 9) + (6x + 8)$

SOLUTION Because addition of the terms can be done in any order, drop the parentheses and rearrange the order of the terms so that like terms are grouped together.

$$(2x - 9) + (6x + 8)$$
$$= 2x - 9 + 6x + 8$$
$$= 2x + 6x - 9 + 8$$
$$= 8x - 1$$

Group the x terms together and the numerical terms together.
Add $2x$ and $6x$ to obtain $8x$.
Add -9 and 8 to obtain -1. ∎

Try These Problems

Remove parentheses and combine like terms.
47. $(x + 5y) + (2x - 12y)$
48. $(2x^2 - x) + (9 - x - 5x^2)$
49. $6(3a - 1) + 4(a + 3)$
50. $-4(5 - 6y) + 12y(1 + y)$

EXAMPLE 23 Remove parentheses and simplify:
$4(8 - 2w) + 2(3w + 1)$

SOLUTION The parentheses can be removed by using the distributive law; we can then combine like terms.

$4(8 - 2w) + 2(3w + 1)$
$= 32 - 8w + 6w + 2$ — Multiply 4 by each term inside the first parentheses.
— Multiply 2 by each term inside the second parentheses.
$= 32 + 2 - 8w + 6w$
$= 34 - 2w$ Combine like terms. ■

 Try Problems 47 through 50.

Answers to Try These Problems

1. false 2. true 3. false 4. true 5. false 6. false
7. $7ac$, $-9a$, and -8 8. 7 9. -9 10. 240 11. 76
12. $14 + 2c$ 13. $aw + bw + 8w$ 14. $3c^2 + 12ac$
15. $6w^3 + 30w^2 + 54w$ 16. $-21 - 35x$ 17. $-8w^2 - 48wy$
18. $-15a^3 - 6a^2 - 36a$ 19. $-40 + 5a$ 20. $-ab + b$
21. $-12wx - 15wy + 27w$ 22. $-120c^3 + 30c^2 + 60c$
23. $8(m + n)$ 24. $5(3x + 7)$ 25. $9(x - 2y)$ 26. $10(2 - 3ac)$
27. $y(6 + x)$ 28. $w(3w + 5)$ 29. $c(w - 1 + 8xc)$
30. $m(7m^2 - 9m + 1)$ 31. $3x(3 + 2a)$ 32. $8w(3w + 1)$
33. $5c(9a - 1)$ 34. $9m(3n - 2x)$ 35. $10x$ 36. $18c$
37. $-12ax$ 38. $7y^2$ 39. $-9a^2$ 40. $10x$ 41. $3ab$ 42. $-10c$
43. $5y - 6x$ 44. $14n^2 - 9n$ 45. $-8m + 3mn - 1$
46. $-7ac + 13a$ 47. $3x - 7y$ 48. $-3x^2 - 2x + 9$
49. $22a + 6$ 50. $-20 + 36y + 12y^2$

EXERCISES 14.2

Multiply, by using the distributive law.

1. $6(3w + 8)$
2. $5(2a + 1)$
3. $a(a + y)$
4. $w(y + 2x)$
5. $3b(a + b + 7)$
6. $3m(m^2 + m + 12)$
7. $-6(x + 5y)$
8. $-n(m + 6n)$
9. $3x(5 - x + y)$
10. $7m(2n + m - 5)$
11. $-15(x - 3)$
12. $-8(mn - 5n)$
13. $-7x(2x - 5y + 1)$
14. $-6c(-2a + 3 - 4b)$

Take out all factors common to all terms.

15. $6x + 6y$
16. $4a + 4c$
17. $4m + 6n$
18. $20mn - 45x$
19. $5x - 6xy$
20. $4m + mn$
21. $ac - ab + 8a$
22. $x^2 + 9x - ax$
23. $10ac + 8ab$
24. $15x^2 - 9xy$
25. $21xy + 14y^2$
26. $24yw - 18xy$
27. $5 - 5a$
28. $9wc + w$
29. $x^3 - 3x^2 - x$
30. $27a^2 + 18a + 9$
31. $7c - 14ac + 7cx$
32. $5xy + 5x + 15x^2$

Simplify, by combining like terms.

33. $8y + 2y$
34. $5x^2 + 6x^2$
35. $6xy + 4xy + xy$
36. $8ac + ac + 5ac$
37. $7y^2 - 2y^2$
38. $12x - 7x$
39. $5ac - 13ac$
40. $3xy - 9xy$
41. $3c - 4c + 7c$
42. $2x^2 + 5x^2 - 3x^2$
43. $-12ac + 8ac - ac$
44. $-xy - 14xy + 5xy$
45. $8n - 5 + 4n - 6$
46. $18 - 6x - 9 + 13x$
47. $20c^2 - 8c + 3c^2 - 12c$
48. $-13x - 4x^2 + 5x - 5x^2$
49. $4ac - 5c + 5a - 8c + 6ac$
50. $-12 + 14xy - 7y + 18 - 20xy$
51. $x^2 - 3x + 8 - 4x^2 + 9x$
52. $8 - 7a + 9a^2 - 17a - 11$
53. $3 - 2x + 7y - 7y - 2x - 3$
54. $-7 - 12x + 6y - 7 + 12x + 6y$
55. $-8c + 13w - 8c - 13w - 5 + 5$
56. $15 - 14m - 15 + 14m - 30n + 30n$

Remove parentheses and combine like terms.

57. $(3x + 5) + (5x + 1)$
58. $(4w + 3x) + (9w + 8x)$
59. $(5y - 2x) + (6y - 12x)$
60. $(10 - 6a) + (7a + 12)$
61. $5(3x - 5) + 3(2 + x)$
62. $4(3x + 5y) + 7(3y - 8x)$
63. $-6(6c - 1) + 12(c + 5)$
64. $-3(20w + 15) + 15(-w + 3)$
65. $w(w + 5) + 2w(3w - 6)$
66. $y(7y - 1) + 4y(2y - 3)$
67. $-8c(c + 7x) + 4c(2c - x)$
68. $-5x(3x - 6) + x(12x - 8)$

Remove parentheses, when present, and simplify.

69. $5c(6c)$
70. $5c + 6c$
71. $5c(6 + c)$
72. $(5 + c) + (6 + c)$
73. $8x + 7x$
74. $8x(7x)$
75. $(8 + x) + (7 + x)$
76. $8x(7 + x)$
77. $-4(7x)(2x)$
78. $-4(7x + 2x)$
79. $(-4 + 7x) + 2x$
80. $(-4 + 7x)2x$
81. $-4(7 + x) + (2 + x)$
82. $-4(7 + x) + 2x$

14.3 SOLVING EQUATIONS WITH THE VARIABLE APPEARING ONLY ON ONE SIDE

OBJECTIVES
- Test a number to see if it is the solution to a given equation.
- Solve an equation of the form $x + a = b$.
- Solve an equation of the form $ax = b$.
- Solve an equation of the form $ax + b = c$.
- Solve an equation that, when simplified, is of the form $ax + b = c$.

Try These Problems

Determine whether the given value is a solution to the given equation. Answer yes or no.

1. $5y + 20 = 60$; $y = 8$
2. $3(1 - w) = 9$; $w = 4$
3. $15 - x = -3$; $x = 18$
4. $35 = 7m$; $m = 5$

1 Introducing Equations

An **equation** is a statement that two expressions are equal. For example,

$$4x - 2 = 10$$

is an equation. The expression to the left of the symbol $=$ is referred to as the *left side of the equation*, and the expression to the right of the symbol $=$ is referred to as the *right side of the equation*.

$$\underbrace{4x - 2}_{\text{left side}} = \underbrace{10}_{\text{right side}}$$

A value for x that makes the equation true is called a **solution** to the equation. For example, if we substitute 3 for x in the equation $4x - 2 = 10$, a true statement results.

Let $x = 3$,
$4x - 2 = 10$
$4(3) - 2 \stackrel{?}{=} 10$ Substitute 3 for x.
$12 - 2 \stackrel{?}{=} 10$
$10 = 10$ The statement is true.

Because the equation is true when $x = 3$, the number 3 is a solution to the equation.

In the equation $4x - 2 = 10$, if we let x be any number other than 3, the equation is not true. For example, here we let x be 2.

Let $x = 2$,
$4x - 2 = 10$
$4(2) - 2 \stackrel{?}{=} 10$ Substitute 2 for x.
$8 - 2 \stackrel{?}{=} 10$
$6 \neq 10$ The left side does not equal 10.

In this case, we say that the number 2 is not a solution to the equation. This equation has only one solution, the number 3.

An equation can have one solution, two solutions, many solutions, or no solutions. In this text, however, we limit our discussion to equations with only one solution.

 Try Problems 1 through 4.

The process of searching for the solution to an equation is called **solving the equation.** Here we look at a straightforward procedure for solving equations.

An equation is a statement that two quantities are equal; take care not to upset the equality. The value of one side of the equation can be changed as long as the value of the other side is changed in exactly the same way. To solve an equation, we change the equation to a simpler equation that has the same solution. Two equations with the same solution are called **equivalent equations.** Here is a list of what can be done to an equation without changing the solution of the equation, thus producing an equivalent equation.

666 Chapter 14 Introduction to Algebra

> *To Solve an Equation*
>
> These operations can be done without changing the solution of the equation.
>
> 1. Add the same number to both sides.
> 2. Subtract the same number from both sides.
> 3. Multiply both sides by the same nonzero number.
> 4. Divide both sides by the same nonzero number.
>
> Choose the operations so that eventually the variable is isolated on one side of the equation by itself, and a number is left on the other side. That number is the solution.

2 Solving an Equation by Adding or Subtracting the Same Number to Both Sides

Addition and subtraction are reverse operations of each other, that is, addition undoes subtraction and subtraction undoes addition. We can use this notion to help in solving equations that involve addition and/or subtraction. The following examples illustrate the procedure.

EXAMPLE 1   Solve for x: $x + 8 = 12$

  SOLUTION   We want to leave x on the left side by itself, so we need to undo the addition of 8. Subtracting 8 will undo the addition of 8, so we subtract 8 from both sides.

$$x + 8 = 12$$
$$x + 8 - 8 = 12 - 8 \quad \text{Subtract 8 from both sides.}$$
$$x = 4$$

Instead of subtracting 8 from both sides, we could have added -8 to both sides.
CHECK

$$x + 8 = 12$$
$$4 + 8 \stackrel{?}{=} 12 \quad \text{Substitute 4 for } x.$$
$$12 = 12$$

The solution is $x = 4$. ∎

EXAMPLE 2   Solve for w: $w - 4 = 7$

  SOLUTION   We want to leave w on the left side by itself so we need to undo the subtraction of 4. Adding 4 will undo the subtraction of 4, so we add 4 to both sides.

$$w - 4 = 7$$
$$w - 4 + 4 = 7 + 4 \quad \text{Add 4 to both sides.}$$
$$w = 11$$

Try These Problems

Solve for the variable and check the solution.

5. $9 + m = 12$
6. $y - 6 = 4$
7. $m + 5 = 1$
8. $y - 12 = -9$
9. $15 = -7 + n$
10. $-40 = -20 + x$

CHECK

$$w - 4 = 7$$
$$11 - 4 \stackrel{?}{=} 7 \qquad \text{Substitute 11 for } w.$$
$$7 = 7$$

The solution is $w = 11$. ■

 Try Problems 5 and 6.

EXAMPLE 3 Solve for c: $\quad 3 = 15 + c$

SOLUTION We want to leave c on the right side by itself so we need to undo the addition of 15. Subtracting 15 will undo the addition of 15, so we subtract 15 from both sides.

$$3 = 15 + c$$
$$3 - 15 = 15 - 15 + c \qquad \text{Subtract 15 from both sides.}$$
$$-12 = c$$

Instead of subtracting 15 from both sides, we could have added -15 to both sides.

CHECK

$$3 = 15 + c$$
$$3 \stackrel{?}{=} 15 + (-12) \qquad \text{Substitute } -12 \text{ for } c.$$
$$3 = 3$$

The solution is $c = -12$. ■

EXAMPLE 4 Solve for y: $\quad -8 = -6 + y$

SOLUTION We want to leave y on the right side by itself so we need to undo the addition of -6. Adding 6 will undo the addition of -6, so we add 6 to both sides.

$$-8 = -6 + y$$
$$-8 + 6 = -6 + 6 + y \qquad \text{Add 6 to both sides.}$$
$$-2 = y \qquad \text{Note that } -6 \text{ plus 6 gives 0, and 0 plus } y \text{ is } y.$$

CHECK

$$-8 = -6 + y$$
$$-8 \stackrel{?}{=} -6 + (-2) \qquad \text{Substitute } -2 \text{ for } y.$$
$$-8 = -8$$

The solution is $y = -2$. ■

 Try Problems 7 through 10.

3 Solving Equations by Multiplying or Dividing Both Sides by the Same Nonzero Number

Multiplication and division are reverse operations of each other, that is, multiplication undoes division and division undoes multiplication. We can use this notion to help in solving equations that involve multiplication and/or division. The following examples illustrate the procedure.

Try These Problems

Solve for the variable and check the solution.
11. $3m = 21$
12. $-5x = 40$

EXAMPLE 5 Solve for x: $4x = 24$

SOLUTION We want to leave x on the left side by itself so we need to undo the multiplication by 4. Dividing by 4 will undo the multiplication by 4, so we divide both sides by 4.

$$4x = 24$$
$$\frac{4x}{4} = \frac{24}{4} \quad \text{Divide both sides by 4.}$$
$$x = 6$$

CHECK
$$4x = 24$$
$$4(6) \stackrel{?}{=} 24 \quad \text{Substitute 6 for } x.$$
$$24 = 24$$

The solution is $x = 6$. ∎

EXAMPLE 6 Solve for n: $60 = -5n$

SOLUTION We want to leave n on the right side by itself so we need to undo the multiplication by -5. Dividing by -5 will undo the multiplication by -5, so we divide both sides by -5.

$$60 = -5n$$
$$\frac{60}{-5} = \frac{-5n}{-5} \quad \text{Divide both sides by } -5.$$
$$-12 = n$$

CHECK
$$60 = -5n$$
$$60 \stackrel{?}{=} -5(-12) \quad \text{Substitute } -12 \text{ for } n.$$
$$60 = 60$$

The solution is $n = -12$. ∎

 Try Problems 11 and 12.

EXAMPLE 7 Solve for y: $4 = 8y$

SOLUTION We want to leave y on the right side by itself so we need to undo the multiplication by 8. Dividing by 8 will undo the multiplication by 8, so we divide both sides by 8.

$$4 = 8y$$
$$\frac{4}{8} = \frac{8y}{8} \quad \text{Divide both sides by 8.}$$
$$\frac{1}{2} = y$$

Try These Problems

Solve for the variable and check the solution.
13. $9 = 27m$
14. $4a = -6$
15. $\dfrac{m}{8} = 3$
16. $-7 = \dfrac{w}{4}$

CHECK

$$4 = 8y$$
$$4 \stackrel{?}{=} 8 \cdot \frac{1}{2} \qquad \text{Substitute } \tfrac{1}{2} \text{ for } y.$$
$$4 \stackrel{?}{=} \frac{\overset{4}{\cancel{8}}}{1} \cdot \frac{1}{\underset{1}{\cancel{2}}}$$
$$4 = 4$$

The solution is $y = \tfrac{1}{2}$. ■

 Try Problems 13 and 14.

EXAMPLE 8 Solve for w: $\dfrac{w}{6} = 30$

SOLUTION We want to leave w on the left side by itself so we need to undo the division by 6. Multiplying by 6 will undo the division by 6, so we multiply both sides by 6.

$$\frac{w}{6} = 30$$
$$\frac{6}{1} \cdot \frac{w}{6} = 6 \cdot 30 \qquad \text{Multiply both sides by 6.}$$
$$w = 180$$

CHECK

$$\frac{w}{6} = 30$$
$$\frac{180}{6} \stackrel{?}{=} 30 \qquad \text{Substitute 180 for } w.$$
$$30 = 30$$

The solution is $w = 180$. ■

 Try Problems 15 and 16.

EXAMPLE 9 Solve for y: $-y = 5$

SOLUTION We want the left side of the equation to be y, not $-y$. Therefore, we multiply both sides by -1 to change the sign.

$$-y = 5$$
$$-1y = 5 \qquad \text{The expression } -y \text{ means } -1y.$$
$$(-1)(-y) = (-1)5 \qquad \text{Multiply both sides by } -1.$$
$$y = -5$$

Note that dividing both sides by -1 will produce the same result.

Try These Problems

Solve for the variable and check the solution.
17. $-m = 15$
18. $-x = -5$

CHECK

$$-y = 5$$
$$-(-5) \stackrel{?}{=} 5 \quad \text{Substitute } -5 \text{ for } y.$$
$$5 = 5$$

The solution is $y = -5$. ■

 Try Problems 17 and 18.

4 Solving Equations of the Form $ax + b = c$.

Now we look at equations that require more than one step to solve. For example, consider the equation

$$3x - 5 = 16$$

We want to leave x on the left side by itself so we need to undo the multiplication by 3 and the subtraction of 5. Which operation should we undo first, the multiplication or the subtraction? It is easier to undo the subtraction first because it is the last operation that would be performed if we were evaluating the left side for a given value of x. Here we show the procedure.

$$3x - 5 = 16$$
$$3x - 5 + 5 = 16 + 5 \quad \text{Add 5 to both sides.}$$
$$3x = 21$$
$$\frac{3x}{3} = \frac{21}{3} \quad \text{Divide both sides by 3.}$$
$$x = 7$$

The solution to the equation $3x - 5 = 16$ is $x = 7$.
Here are more examples.

EXAMPLE 10 Solve for y: $5y + 2 = 17$

SOLUTION Begin by undoing the addition of 2.

$$5y + 2 = 17$$
$$5y + 2 - 2 = 17 - 2 \quad \text{Subtract 2 from both sides.}$$
$$5y = 15$$
$$\frac{5y}{5} = \frac{15}{5} \quad \text{Divide both sides by 5.}$$
$$y = 3$$

CHECK

$$5y + 2 = 17$$
$$5(3) + 2 \stackrel{?}{=} 17 \quad \text{Substitute 3 for } y.$$
$$15 + 2 \stackrel{?}{=} 17$$
$$17 = 17$$

The solution is $y = 3$. ■

Try These Problems

Solve for the variable and check the solution.
19. $2m + 3 = 13$
20. $4a - 7 = 17$
21. $-6x + 2 = -22$
22. $3y - 17 = -8$
23. $28 = -4 - 8w$
24. $17 = -3 - 4c$

Try Problems 19 and 20.

EXAMPLE 11 Solve for w: $-25 = 9w - 7$

SOLUTION Begin by undoing the subtraction of 7.

$$-25 = 9w - 7$$
$$-25 + 7 = 9w - 7 + 7 \quad \text{Add 7 to both sides.}$$
$$-18 = 9w$$
$$\frac{-18}{9} = \frac{9w}{9} \quad \text{Divide both sides by 9.}$$
$$-2 = w$$

CHECK

$$-25 = 9w - 7$$
$$-25 \stackrel{?}{=} 9(-2) - 7 \quad \text{Substitute } -2 \text{ for } w.$$
$$-25 \stackrel{?}{=} -18 - 7$$
$$-25 = -25$$

The solution is $w = -2$. ∎

EXAMPLE 12 Solve for a: $6 - 4a = 42$

SOLUTION The left side of the equation is really $6 + (-4a)$. We begin by undoing the addition of 6.

$$6 - 4a = 42$$
$$6 + (-4a) = 42 \quad \text{Subtracting } 4a \text{ is the same as adding } -4a.$$
$$6 - 6 + (-4a) = 42 - 6 \quad \text{Subtract 6 from both sides.}$$
$$-4a = 36$$
$$\frac{-4a}{-4} = \frac{36}{-4} \quad \text{Divide both sides by } -4.$$
$$a = -9$$

CHECK

$$6 - 4a = 42$$
$$6 - 4(-9) \stackrel{?}{=} 42 \quad \text{Substitute } -9 \text{ for } a.$$
$$6 - (-36) \stackrel{?}{=} 42$$
$$6 + 36 \stackrel{?}{=} 42$$
$$42 = 42$$

The solution is $a = -9$. ∎

Try Problems 21 through 24.

Try These Problems

Solve for the variable and check the solution.
25. $45 - y = 5$
26. $-18 = -6 - w$

EXAMPLE 13 Solve for n: $\quad -16 = 8 - n$

SOLUTION The right side of the equation is really $8 + (-n)$. We begin by undoing the addition of 8.

$$-16 = 8 - n$$
$$-16 = 8 + (-n) \qquad \text{Subtracting } n \text{ is the same as adding } -n.$$
$$-16 - 8 = 8 - 8 + (-n) \qquad \text{Subtract 8 from both sides.}$$
$$-24 = -n$$
$$-1(-24) = -1(-n) \qquad \text{Multiply both sides by } -1.$$
$$24 = n$$

CHECK

$$-16 = 8 - n$$
$$-16 \stackrel{?}{=} 8 - 24$$
$$-16 = -16$$

The solution is $n = 24$. ∎

 Try Problems 25 and 26.

EXAMPLE 14 Solve for x: $\quad 8x - 7 = 3$

SOLUTION Begin by undoing the subtraction of 7.

$$8x - 7 = 3$$
$$8x - 7 + 7 = 3 + 7 \qquad \text{Add 7 to both sides.}$$
$$8x = 10$$
$$\frac{8x}{8} = \frac{10}{8} \qquad \text{Divide both sides by 8.}$$
$$x = \frac{5}{4}$$

CHECK

$$8x - 7 = 3$$
$$8\left(\frac{5}{4}\right) - 7 \stackrel{?}{=} 3 \qquad \text{Substitute } \tfrac{5}{4} \text{ for } x.$$
$$\frac{\overset{2}{\cancel{8}}}{1} \cdot \frac{5}{\underset{1}{\cancel{4}}} - 7 \stackrel{?}{=} 3$$
$$10 - 7 \stackrel{?}{=} 3$$
$$3 = 3$$

The solution is $x = \tfrac{5}{4}$. ∎

Try These Problems

Solve for the variable and check the solution.

27. $2m + 4 = 7$
28. $5 - 6m = 1$
29. $-12 = 5x - 8$
30. $-3 = -5 - 6y$
31. $5c - 16 + 4 - 3c = 14$
32. $(45 - 6w) + (11w - 15) = 115$

EXAMPLE 15 Solve for m: $25 = 10 - 20m$

SOLUTION The right side of the equation is really $10 + (-20m)$. We begin by undoing the addition of 10.

$$25 = 10 - 20m$$
$$25 = 10 + (-20m)$$ Subtracting $20m$ is the same as adding $-20m$.
$$25 - 10 = 10 - 10 + (-20m)$$ Subtract 10 from both sides.
$$15 = -20m$$
$$\frac{15}{-20} = \frac{-20m}{-20}$$ Divide both sides by -20.
$$\frac{-3}{4} = m$$

CHECK

$$25 = 10 - 20m$$
$$25 \stackrel{?}{=} 10 - 20\left(\frac{-3}{4}\right)$$ Substitute $\frac{-3}{4}$ for m.
$$25 \stackrel{?}{=} 10 - \frac{\overset{5}{\cancel{20}}}{1} \cdot \frac{-3}{\underset{1}{\cancel{4}}}$$
$$25 \stackrel{?}{=} 10 - (-15)$$
$$25 = 25$$

The solution is $m = \frac{-3}{4}$. ∎

Try Problems 27 through 30.

Sometimes one side or both sides of the equation are not in simplified form. When this is the case, it is best to simplify before beginning to operate on both sides. Here we look at some examples.

EXAMPLE 16 Solve for x: $(x - 30) + (2x + 40) = 70$

SOLUTION Begin by simplifying the left side of the equation.

$$(x - 30) + (2x + 40) = 70$$
$$x - 30 + 2x + 40 = 70$$
$$3x + 10 = 70$$ Combine like terms on the left side.
$$3x + 10 - 10 = 70 - 10$$ Subtract 10 from both sides.
$$3x = 60$$
$$\frac{3x}{3} = \frac{60}{3}$$ Divide both sides by 3.
$$x = 20$$ ∎

Try Problems 31 and 32.

Try These Problems

Solve for the variable and check the solution.

33. $4(5y - 1) = 16$

34. $5(x + 3) + 10(2x - 4) = 250$

EXAMPLE 17 Solve for c: $3(c - 2) + 6(3c + 5) = 213$

SOLUTION Begin by simplifying the left side of the equation.

$$3(c - 2) + 6(3c + 5) = 213$$
$$3c - 6 + 18c + 30 = 213 \quad \text{Remove the parentheses by using the distributive law.}$$
$$21c + 24 = 213$$
$$21c + 24 - 24 = 213 - 24 \quad \text{Subtract 24 from both sides.}$$
$$21c = 189$$
$$\frac{21c}{21} = \frac{189}{21} \quad \text{Divide both sides by 21.}$$
$$c = 9 \quad \blacksquare$$

Try Problems 33 and 34.

Answers to Try These Problems

1. yes 2. no 3. yes 4. yes 5. $m = 3$ 6. $y = 10$
7. $m = -4$ 8. $y = 3$ 9. $n = 22$ 10. $x = -20$ 11. $m = 7$
12. $x = -8$ 13. $m = \frac{1}{3}$ 14. $a = \frac{-3}{2}$ 15. $m = 24$
16. $w = -28$ 17. $m = -15$ 18. $x = 5$ 19. $m = 5$
20. $a = 6$ 21. $x = 4$ 22. $y = 3$ 23. $w = -4$ 24. $c = -5$
25. $y = 40$ 26. $w = 12$ 27. $m = \frac{3}{2}$ 28. $m = \frac{2}{3}$ 29. $x = \frac{-4}{5}$
30. $y = \frac{-1}{3}$ 31. $c = 13$ 32. $w = 17$ 33. $y = 1$ 34. $x = 11$

EXERCISES 14.3

Solve for the variable and check the solution.

1. $8 + x = 20$
2. $x - 5 = 4$
3. $9 = x - 12$
4. $7 = 5 + x$
5. $-13 = y - 4$
6. $-15 + y = 3$
7. $3y = 18$
8. $36 = 4y$
9. $40 = -8w$
10. $42 = -6w$
11. $5w = 2$
12. $3 = 4w$
13. $50 = 35v$
14. $45v = 18$
15. $-6v = 75$
16. $-4 = -20v$
17. $4v = 0$
18. $0 = -6v$
19. $\frac{w}{7} = 8$
20. $14 = \frac{w}{3}$
21. $6 = \frac{c}{-8}$
22. $\frac{c}{15} = -2$
23. $-c = -7$
24. $-c = 6$
25. $16 = -a$
26. $-50 = -a$
27. $3a + 11 = 32$
28. $5a - 7 = 13$
29. $42 = 6m - 6$
30. $56 = 4m + 12$
31. $6m + 7 = -5$
32. $-8 + 7m = -29$
33. $-8n - 45 = 3$
34. $-51 = -9n + 3$
35. $16 - 12n = 100$
36. $20 - 25n = -180$

37. $3x - 4 = 1$
38. $7x + 2 = 6$
39. $15 = 13 + 4x$
40. $-56 = 14 + 21x$
41. $8 - 6x = 12$
42. $17 - 8x = -11$
43. $7x - 8 = 0$
44. $16x + 4 = 0$
45. $2x + 9x - 6 = 27$
46. $4x + 5 - 8 = 45$
47. $5x - 7 - 2x + 15 = 47$
48. $8x + 10 + 4x - 22 = 36$
49. $4y - 7 - 8y + 2 = 1$
50. $-14 - 8y - 3 + 33y = 18$
51. $3m - 8 + m + 10 = 0$
52. $-16m + 32 - 2m - 5 = 0$
53. $14n + 17 - 18n - 3 = 14$
54. $-25 = 20n - 16 + 8n - 9$
55. $(9x - 1) + (5x - 17) = 24$
56. $(-5x + 3) + (2x - 7) = 47$
57. $(6x - 8) + (14x - 7) = 73$
58. $(7x + 15) + (4 - 49x) = 12$
59. $2(a + 9) + 3(2a - 1) = 87$
60. $-4(a - 5) + 2(8a - 3) = 74$
61. $44 = 15(a - 3) + 10(2a + 4)$
62. $20 = 4(3 - 2a) + 5(a + 2)$

14.4 APPLICATIONS

OBJECTIVES
- Solve an application when given a formula.
- Translate an English sentence to an equation.
- Translate an English sentence to an equation when there is more than one unknown quantity.
- Translate an English sentence to an equation when the quantities are from real-life situations.
- Solve an application by using an equation.

Try These Problems

Solve by using the given formula.

1. The perimeter of a rectangle is 118 meters and its width is 24 meters. Find the length of the rectangle. (Use the formula $P = 2L + 2W$.)

1 Solving Applications When Given a Formula

Recall from Section 12.7 that the perimeter P of any rectangle is given by the **formula**

$$P = 2L + 2W$$

where L represents the length and W represents the width of the rectangle. If given the values of any two of these three variables, we can use this formula to find the value of the remaining variable.

For example, suppose that a rectangle, with length 90 inches, has a perimeter of 340 inches. We can use the formula, $P = 2L + 2W$, to find the width of the rectangle.

$$P = 2L + 2W$$
$$340 = 2(90) + 2W \qquad \text{Substitute 340 for } P \text{ and 90 for } L.$$
$$340 = 180 + 2W \qquad \text{Solve this equation for } W.$$
$$340 - 180 = 180 - 180 + 2W \qquad \text{Subtract 180 from both sides.}$$
$$160 = 2W$$
$$\frac{160}{2} = \frac{2W}{2} \qquad \text{Divide both sides by 2.}$$
$$80 = W$$

The width of the rectangle is 80 inches.

▲ Try Problem 1.

Try These Problems

Solve by using the given formula.

2. The smaller of two numbers is −19 and the distance between the two numbers is 31. What is the larger number? (Use the formula $D = L - S$, where D is the distance between the larger number L and the smaller number S.)

3. The number of products in demand D of a certain product is given by the formula $D = 1200 - 6p$ where p is the price (in dollars) per product. Find the price p that causes the demand to be 810.

Here are more examples of solving applications when given a formula.

EXAMPLE 1 The distance D between two numbers on the number line is given by the formula

$$D = L - S$$

where L represents the larger number and S represents the smaller number. If the larger number is 13 and the distance between the numbers is 27, find the smaller number.

SOLUTION We want to find S when $L = 13$ and $D = 27$.

$$D = L - S$$
$$27 = 13 - S \quad \text{Substitute 27 for } D \text{ and 13 for } L. \text{ Solve this equation for } S.$$
$$27 - 13 = 13 - 13 - S \quad \text{Subtract 13 from both sides.}$$
$$14 = -S$$
$$-1(14) = -1(-S) \quad \text{Multiply both sides by } -1.$$
$$-14 = S$$

The smaller number is −14. ■

 Try Problem 2.

EXAMPLE 2 Tom's monthly cost C (in dollars) at the tennis club is given by the formula

$$C = 50 + 8x$$

where x represents the number of hours of court time. If Tom's total bill in January was $178, how many hours of court time did he purchase that month?

SOLUTION We want to find x when C equals $178.

$$C = 50 + 8x$$
$$178 = 50 + 8x \quad \text{Substitute 178 for } C. \text{ Solve this equation for } x.$$
$$178 - 50 = 50 - 50 + 8x \quad \text{Subtract 50 from both sides.}$$
$$128 = 8x$$
$$\frac{128}{8} = \frac{8x}{8} \quad \text{Divide both sides by 8.}$$
$$16 = x$$

Tom purchased 16 hours of court time during the month of January. ■

 Try Problem 3.

Try These Problems

Solve by using the given formula.

4. Find the base of a triangle that has area 150 square yards and an altitude of 30 yards. (Use the formula $A = \frac{1}{2}bh$.)

EXAMPLE 3 The area A of a triangle is given by the formula

$$A = \frac{1}{2}bh$$

where b is the base and h is the altitude. If a triangle, with base 6 feet, has an area of 15 square feet, find the altitude of the triangle.

SOLUTION We want to find h when $A = 15$ square feet and $b = 6$ feet.

$$A = \frac{1}{2}bh$$

$15 = \frac{1}{2} \cdot 6h$ Substitute 15 for A and 6 for b.

$15 = \frac{1}{2} \cdot \frac{6}{1}h$

$15 = 3h$ Multiply $\frac{1}{2}$ by 6 to obtain 3.

$\frac{15}{3} = \frac{3h}{3}$ Divide both sides by 3.

$5 = h$

The altitude of the triangle is 5 feet. ■

Try Problem 4.

2 Translating English Sentences to Equations

In the previous application problems, we were given a formula (or equation) to work with, but some application problems require that you create an equation. Before solving this type of application we look at translating sentences to equations.

It is a good idea at this time for you to review the language chart in Section 12.7 that translates English phrases to algebraic expressions.

Here we list the words and phrases that often translate to the equality symbol, =.

Words or Phrases that Often Translate to =

is equal to	represents
equals	will be
is the same as	was
is the result of	makes
is	leaves
are	yields
were	

Try These Problems

In Problems 5 through 7, let N represent the unknown number and translate each sentence to an equation. Do not solve the equation.

5. A number minus 15 equals 40.
6. The product of a number and 9 yields 189.
7. Sixty-seven represents a number increased by 13.

In Problems 8 and 9, let x represent the unknown number and translate each sentence to an equation. Do not solve the equation.

8. Thirty more than 5 times a number is 85.
9. One hundred twenty represents 6 times the sum of 12 and a number.

EXAMPLE 4 Let N represent the unknown number. Translate to an equation, "the sum of a number and 24 is 80." Do not solve the equation.

SOLUTION Begin by stating clearly what the variable represents.

Let $N =$ the unknown number.

$$\underbrace{\text{The sum of a number and 24}}_{N + 24} \underbrace{\text{is}}_{=} \; 80$$

The equation is $N + 24 = 80$. ■

EXAMPLE 5 Let c represent the unknown number. Translate to an equation, "8 less than a number yields 45." Do not solve the equation.

SOLUTION Begin by stating what the variable represents.
Also, recall that the phrase *less than* means the same as *subtracted from*, so that the numbers in the algebraic expression appear in the reverse order than they do in the English phrase.

Let $c =$ the unknown number.

$$\underbrace{\text{8 less than a number}}_{c - 8} \underbrace{\text{yields}}_{=} \; 45$$

The equation is $c - 8 = 45$. ■

Try Problems 5 through 7.

EXAMPLE 6 Let y represent the unknown number. Translate to an equation, "75 represents 130 decreased by twice a number." Do not solve the equation.

SOLUTION Begin by stating what the variable represents.
Let $y =$ the unknown number.

$$\underbrace{75}_{75} \underbrace{\text{represents}}_{=} \underbrace{130}_{130} \underbrace{\text{decreased by}}_{-} \underbrace{\text{twice a number}}_{2 \cdot y}$$

The equation is $75 = 130 - 2y$. ■

Try Problems 8 and 9.

3 Translating with More Than One Unknown Quantity

In each of the previous examples, there is only one unknown quantity. The given variable represents this unknown number. Now we look at situations that involve more than one unknown quantity. Be careful to note which of the unknowns the variable represents. Before translating the designated sentence to an equation, write an algebraic expression, using the given variable, that represents each of the other unknown quantities. Here are some examples.

Try These Problems

In Problems 10 through 13, write expressions that represent each of the unknown quantities before translating the given sentence to an equation. Use only one variable, the given variable. Do not solve the equation.

10. One number is 8 more than another number. Let S represent the smaller of the two numbers. Translate to an equation, "the sum of the two numbers is 76." Do not solve the equation.

11. One number is 15 less than another number. Let L represent the larger of the two numbers. Translate to an equation, "five times the smaller number, decreased by the larger number, yields 9." Do not solve the equation.

12. A larger number is twice a smaller number. Let n represent the smaller number. Translate to an equation, "five times the smaller number, increased by the larger number, yields 119." Do not solve the equation.

13. A larger number is 4 times a smaller number. Let y represent the smaller number. Translate to an equation, "the difference between the two numbers is 33." Do not solve the equation.

EXAMPLE 7 One number is 38 less than another number. Let L represent the larger of the two numbers. Translate to an equation, "the larger number, increased by 4 times the smaller number, equals 128." Do not solve the equation.

SOLUTION In this problem there are two unknown numbers. Begin by writing expressions, using the given variable, that represent each of the unknown numbers.

Let L = the larger number.

$L - 38$ = the smaller number.

The second sentence states to let L be the larger number. The first sentence in the problem has information that allows us to represent the smaller number.

The larger number **increased by** 4 **times** the smaller number **equals** 128

$L \quad + \quad 4 \quad \cdot \quad (L - 38) \quad = \quad 128$

The equation is $L + 4(L - 38) = 128$ or $5L - 152 = 128$. ■

 Try Problems 10 and 11.

EXAMPLE 8 A larger number is 3 times a smaller number. Let x represent the smaller number. Translate to an equation, "twice the larger number, decreased by the smaller number, is 150." Do not solve the equation.

SOLUTION

Let x = the smaller number.

$3x$ = the larger number.

The second sentence states to let x be the smaller number. The first sentence in the problem has information that allows us to represent the larger number.

Twice the larger number **decreased by** the smaller number **is** 150

$2 \cdot \quad (3x) \quad - \quad x \quad = \quad 150$

The equation is $2(3x) - x = 150$ or $5x = 150$. ■

 Try Problems 12 and 13.

4 Translating with Quantities from Real-Life Situations

Now we look at translating sentences to equations where the sentence is making a statement about quantities from a real-life situation.

Try These Problems

14. The weights of four objects are 7 pounds, 12 pounds, 13 pounds, and 9 pounds. Let w represent the weight of a fifth object. Translate to an equation, "the average (mean) weight of all 5 objects is 11 pounds." Do not solve the equation.

In Problem 15, write expressions that represent each of the unknown quantities before translating the given sentence to an equation. Use only one variable, the given variable. Do not solve the equation.

15. Fred belongs to a health club where he pays $15 per month, plus $4 for each hour of racquetball. Let N represent how many hours of racquetball he played last month. Translate to an equation, "his total bill for last month was $63." Do not solve the equation.

EXAMPLE 9 Paz scored 80, 76, and 83 on the first three chemistry exams. Let x represent her score on the fourth exam. Translate to an equation, "the average (mean) of the four scores is 78." Do not solve the equation.

SOLUTION

Let x = the 4th score.   The second sentence states to let x be the fourth score.

$$\underbrace{\frac{80 + 76 + 83 + x}{4}}_{\text{average of the four scores}} \overset{\text{is}}{=} 78$$

The equation is $\dfrac{80 + 76 + 83 + x}{4} = 78$ or $\dfrac{239 + x}{4} = 78$. ∎

 Try Problem 14.

EXAMPLE 10 A salesperson receives a base salary of $1200 per month, plus $84 for each product sold. Let N represent the number of products the person sold in July. Translate to an equation, "the total amount earned in July was $3300." Do not solve the equation.

SOLUTION

Let N = number of products sold.   The second sentence of the problem says to let N be the number of products sold in July.

$84N$ = additional earnings for selling the N products ($).   The first sentence states that the person receives $84 for each product sold.

$$\underbrace{\underbrace{1200}_{\text{Base salary (\$)}} + \underbrace{84N}_{\text{additional earnings (\$)}}}_{\text{The total amount earned in July}} \overset{\text{was}}{=} 3300$$

The equation is $1200 + 84N = 3300$. ∎

Try Problem 15.

EXAMPLE 11 Let L be the length (in meters) of a rectangle. The width is 20 meters less than twice the length. Translate to an equation, "the perimeter of the rectangle is 50 meters." Do not solve the equation.

Try These Problems

In Problems 16 through 18, write expressions that represent each of the unknown quantities before translating the given sentence to an equation. Use only one variable, the given variable. Do not solve the equation.

16. Let W be the width (in feet) of a rectangle. The length is 3 times the width. Translate to an equation, "the perimeter of the rectangle is 104 feet." Do not solve the equation.

17. Let L be the length (in inches) of a rectangle. The width is 14 inches less than the length. Translate to an equation, "the area of the rectangle is 3920 square inches." Do not solve the equation.

18. A carpenter earns $25 per hour and an engineer earns $75 per hour. Let x represent the number of hours the engineer worked. Assume that the carpenter worked 15 more hours than the engineer. Translate to an equation, "the total amount they earned was $2875." Do not solve the equation.

SOLUTION Begin by writing expressions to represent the unknown quantities.

Let L = length of the rectangle (m). *The first sentence says to let L be the length.*

$2L - 20$ = width of the rectangle (m). *The second sentence gives information that allows us to write an expression that represents the width of the rectangle.*

Now translate the designated sentence to an equation.

Perimeter is 50 meters

Twice the length (m) + twice the width (m) = 50 meters

$2L + 2(2L - 20) = 50$

The equation is $2L + 2(2L - 20) = 50$ or $6L - 40 = 50$. ∎

Try Problems 16 and 17.

EXAMPLE 12 Jan and Denny were selling auction tickets to raise money for charity. Jan sold N tickets at $8 each. Denny sold 25 more tickets than Jan at the same price. Translate to an equation, "the total value of all the tickets is $1000."

SOLUTION Begin by writing expressions to represent the unknown quantities.

Let N = the number of tickets Jan sold. *The second sentence says Jan sold N tickets.*

$N + 25$ = the number of tickets Denny sold. *The third sentence says Denny sold 25 more tickets than Jan.*

$8N$ = amount collected for the sale of Jan's tickets ($). *The second sentence says Jan's tickets sold for $8 each.*

$8(N + 25)$ = amount collected for the sale of Denny's tickets ($). *The third sentence says that Denny's tickets also sold for $8 each.*

Now translate the designated sentence to an equation.

Total value of all the tickets is $1000

Value of Jan's tickets ($) + value of Denny's tickets ($) = $1000

$8N + 8(N + 25) = 1000$

The equation is $8N + 8(N + 25) = 1000$ or $16N + 200 = 1000$. ∎

 Try Problem 18.

Try These Problems

Solve by using an equation.

19. Ten times the sum of a number and 18 yields 150. Find the number.
20. Sixteen represents 8 more than twice a number. Find the number.

5 Solving Applications

Now we look at solving application problems by using an equation. Here is the general procedure that will be used.

> *Solving an Application by Using an Equation*
> 1. Choose a letter of the alphabet to be a variable that represents an unknown quantity. Write an English phrase that clearly states what the variable represents.
> 2. If there are other unknown quantities, write an algebraic expression that represents each of the other unknowns using the variable introduced in step 1. Write an English phrase that clearly states what each algebraic expression represents.
> 3. Write an equation that states a relationship between the known and unknown quantities.
> 4. Solve the equation for the variable.
> 5. Reread the question and answer it precisely. Write an English sentence that answers the question.
> 6. Finally, check the answer.

Many students who have trouble solving application problems do not spend enough time thinking about steps 1 and 2. They tend to rush into step 3, writing the equation. Steps 1 and 2 should not only be thought about, they should be written down. Here are some examples.

EXAMPLE 13 Five times the sum of a number and 25 equals 160. Find the number.

SOLUTION This problem has only one unknown quantity, the number mentioned. We choose a variable to represent this number.

Let N = the unknown number.

Now we translate the first sentence to an equation.

Five times the sum of a number and 25 equals 160.
$$5 \cdot (N + 25) = 160$$

Next we solve the equation for N.

$$5(N + 25) = 160$$
$$5N + 125 = 160$$
$$5N + 125 - 125 = 160 - 125$$
$$5N = 35$$
$$\frac{5N}{5} = \frac{35}{5}$$
$$N = 7$$

The number is 7. ■

 Try Problems 19 and 20.

Try These Problems

Solve by using an equation.

21. A larger number is 12 times a smaller number. The sum of the two numbers is 143. Find the two numbers.

22. A larger number is 15 more than a smaller number. Twice the larger number decreased by the smaller number yields 47. Find the two numbers.

EXAMPLE 14 A larger number is 8 times a smaller number. The larger number decreased by 4 times the smaller number yields 3. Find the two numbers.

SOLUTION This problem has two unknown quantities, the two numbers. We let a variable represent one of the numbers, then write an algebraic expression that represents the other number.

We use only the first sentence in the problem to do this preliminary work. Because the larger number is described in terms of the smaller number, we let the variable be the smaller number.

$$\text{Let } x = \text{the smaller number.}$$
$$8x = \text{the larger number.}$$

Now we use the second sentence to write an equation.

$$\underbrace{\text{The larger number}}_{8x} \quad \underbrace{\text{decreased by}}_{-} \quad 4 \quad \text{times} \quad \underbrace{\text{the smaller number}}_{x} \quad \text{yields} \quad 3$$
$$8x \quad - \quad 4 \quad \cdot \quad x \quad = \quad 3$$

Next we solve the equation for x.

$$8x - 4x = 3$$
$$4x = 3$$
$$\frac{4x}{4} = \frac{3}{4}$$
$$x = \frac{3}{4} \quad \text{The smaller number}$$

Recall that x represents the smaller number, so we have found that the smaller number is $\frac{3}{4}$.

The question is to find both numbers. Now we find the larger number which was represented by the expression $8x$.

$$\text{Larger number} = 8x$$
$$= 8\left(\frac{3}{4}\right) \quad \text{Substitute } \tfrac{3}{4} \text{ for } x.$$
$$= \frac{\overset{2}{\cancel{8}}}{1} \cdot \frac{3}{\underset{1}{\cancel{4}}}$$
$$= 6$$

The numbers are $\frac{3}{4}$ and 6. ∎

 Try Problems 21 and 22.

Try These Problems

Solve by using an equation.

23. Arthur is thinking of three numbers. The second number is 3 more than twice the first number. The third number is 7 more than the second number. The sum of the second and third numbers, decreased by 6 times the first number yields 12. Find the three numbers.

EXAMPLE 15 Kimberly is thinking of three numbers. The second number is 11 more than the first number. The third number is twice the second number. The sum of all three numbers is 13. Find the three numbers.

SOLUTION There are three unknown quantities in this problem, the three numbers. We let a variable represent one of the numbers, then write algebraic expressions to represent the other two numbers.

We use the second and third sentences of the problem to do this preliminary work. Because the second number is expressed in terms of the first number, it is best to let the variable be the first number.

$$\text{Let } n = \text{the first number.}$$
$$n + 11 = \text{the second number.}$$
$$2(n + 11) = \text{the third number.}$$

Now we translate the fourth sentence to an equation.

$$\underbrace{\text{first number}}_{n} + \underbrace{\text{second number}}_{(n+11)} + \underbrace{\text{third number}}_{2(n+11)} = 13$$

Next solve this equation for n.

$$n + (n + 11) + 2(n + 11) = 13$$
$$n + n + 11 + 2n + 22 = 13$$
$$4n + 33 = 13$$
$$4n + 33 - 33 = 13 - 33$$
$$4n = -20$$
$$\frac{4n}{4} = \frac{-20}{4}$$
$$n = -5 \quad \text{The first number}$$

Recall that n represents the first number, so we have found that the first number is -5.

The question is to find all three numbers. We now find the other two numbers by using the expressions that represent them.

$$\text{Second number} = n + 11$$
$$= -5 + 11 \quad \text{Substitute } -5 \text{ for } n.$$
$$= 6$$
$$\text{Third number} = 2(n + 11)$$
$$= 2(-5 + 11) \quad \text{Substitute } -5 \text{ for } n.$$
$$= 2(6)$$
$$= 12$$

The three numbers are -5, 6, and 12. ∎

 Try Problem 23.

Try These Problems

Solve by using an equation.

24. On a 7-mile trip across town, Annette jogged part of the distance and walked the rest. She jogged 2 fewer miles than she walked. Find both her jogging distance and her walking distance.

25. Last month Mr. Herrera earned 4 times as much as his wife, Ms. Herrera. Together they earned $6250. How much did each earn?

EXAMPLE 16 A board is 85 centimeters long. Toni cuts it into two pieces such that one piece is 56 centimeters longer than the other piece. Find the length of the two pieces.

SOLUTION There are two unknown quantities, the lengths of the two pieces. We let a variable be the length of one piece, then write an expression to represent the length of the other piece.

We use the second sentence to do the preliminary work. Because the second sentence gives the length of the larger piece in terms of the length of the shorter piece, we let the variable represent the length of the shorter piece.

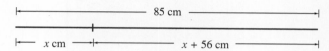

Let x = length of the shorter piece (cm).
$x + 56$ = length of the longer piece (cm).

Now we use the fact that the total length is 85 centimeters to write an equation.

$$\underbrace{\underbrace{x}_{\substack{\text{Length of} \\ \text{shorter piece} \\ \text{(cm)}}} + \underbrace{(x + 56)}_{\substack{\text{length of} \\ \text{longer piece} \\ \text{(cm)}}}}_{\text{Total length (cm)}} = 85$$

Next solve the equation for x.

$$x + (x + 56) = 85$$
$$x + x + 56 = 85$$
$$2x + 56 = 85$$
$$2x + 56 - 56 = 85 - 56$$
$$2x = 29$$
$$\frac{2x}{2} = \frac{29}{2}$$
$$x = \frac{29}{2} \text{ or } 14\frac{1}{2} \quad \text{Length of the smaller piece in centimeters}$$

Now we use the expression $x + 56$ to find the length of the longer piece.

$$\text{Length of longer piece in centimeters} = x + 56$$
$$= 14\frac{1}{2} + 56$$
$$= 70\frac{1}{2}$$

The lengths of the two pieces are $14\frac{1}{2}$ and $70\frac{1}{2}$ centimeters. ■

 Try Problems 24 and 25.

Try These Problems

Solve by using an equation.

26. Ms. Cabrillo earns a base salary of $880 per month, plus $32 for each product she sells. During the month of December she earned a total of $5680. How many products did she sell?

EXAMPLE 17 George belongs to a tennis club where he pays $40 per month, plus $5 per hour for court time. His bill for the month of August was $125. How many hours of court time did he use that month?

SOLUTION There are two unknown quantities in this problem. We do not know how many hours of court time he used and we also do not know the total amount he paid for this court time.

We let the variable n be the number of hours of court time; because he pays $5 each hour, we can use the expression $5n$ to represent the total amount he paid for the court time.

Let n = The number of hours of court time.
$5n$ = total amount he paid for the court time ($).

Now we use the fact that the total bill was $125 to write an equation.

$$\underbrace{\$40 + \underbrace{\text{charge for court time (\$)}}} = \$125$$

$$40 + 5n = 125$$

Next solve this equation for n.

$$40 + 5n = 125$$
$$40 - 40 + 5n = 125 - 40$$
$$5n = 85$$
$$\frac{5n}{5} = \frac{85}{5}$$
$$n = 17$$

George used 17 hours of court time in August. ∎

Try Problem 26.

EXAMPLE 18 The length of a rectangle is 36 inches more than the width. The perimeter is 120 inches. Find the length and width of the rectangle.

SOLUTION This problem has two unknown quantities, the length and the width of the rectangle. Because the first sentence gives the length in terms of the width, we let the variable be the width of the rectangle.

Let w = width of the rectangle (in).
$w + 36$ = length of the rectangle (in).

We use the fact that the perimeter is 120 inches to write an equation.

$$\underbrace{\text{twice the length (in)}} + \underbrace{\text{twice the width (in)}} = 120 \text{ in}$$

$$2(w + 36) + 2w = 120$$

Try These Problems

Solve by using an equation.
27. The width of a rectangle is 6 feet less than the length. The perimeter is 100 feet. Find the length and width of the rectangle.
28. The length of a rectangle is 4 times the width. The perimeter is 1050 meters. Find the length and the width.

Next we solve the equation for w.

$$2(w + 36) + 2w = 120$$
$$2w + 72 + 2w = 120$$
$$4w + 72 = 120$$
$$4w + 72 - 72 = 120 - 72$$
$$4w = 48$$
$$\frac{4w}{4} = \frac{48}{4}$$
$$w = 12 \quad \text{The width of the rectangle in inches}$$

The question is to find both the length and the width, so we use the expression $w + 36$ to find the length.

$$\text{Length in inches} = w + 36$$
$$= 12 + 36$$
$$= 48$$

The length is 48 inches and the width is 12 inches. ■

 Try Problems 27 and 28.

EXAMPLE 19 Juan sold a total of 80 tickets to the school football game. Some of the tickets sold for $5 each and the rest sold for $8 each. If he collected a total of $499 on the sale of the tickets, how many of each type of ticket did he sell?

   SOLUTION There are four unknown quantities in this problem. We do not know the number of $5 tickets sold, the number of $8 tickets sold, the total amount of money collected for the $5 tickets, and the total amount collected for the $8 tickets.

   If we let the variable be the number of $5 tickets, then we can use the first sentence to write an expression that represents the number of $8 tickets.

$$\text{Let } x = \text{ the number of \$5 tickets.}$$

The first sentence says that the total number of tickets is 80. So we have the following.

$$\underbrace{\text{number of \$5 tickets}}_{x} + \text{number of \$8 tickets} = 80$$
$$x + \text{number of \$8 tickets} = 80$$

Subtracting x from both sides of this equation, we see that the number of $8 tickets can be represented by the expression $80 - x$.

$$80 - x = \text{ the number of \$8 tickets.}$$
$$5x = \text{ the total amount of money collected for the \$5 tickets (\$).}$$
$$8(80 - x) = \text{ total amount collected for the \$8 tickets (\$).}$$

Try These Problems

Solve by using an equation.

29. Barry and Steve sold tickets to an auction to raise money for the homeless. Steve sold 20 more tickets than Barry. If the tickets sold for $2 each and they collected a total of $340, how many tickets did each sell?

30. Hans and Sally worked a total of 70 hours last week. Hans, a mason, earns $25 per hour, while Sally, a lawyer, earns $80 per hour. If their combined earnings for the week were $3950, how many hours did each person work?

Now we can write an equation using the fact that the total amount collected for all the tickets was $499.

$$\underbrace{\underbrace{5x}_{\text{Amount collected for \$5 tickets (\$)}} + \underbrace{8(80 - x)}_{\text{amount collected for \$8 tickets (\$)}}}_{\text{Total amount collected}} = \overset{\text{was \$499}}{499}$$

Next we solve the equation for x.

$$5x + 8(80 - x) = 499$$
$$5x + 640 - 8x = 499$$
$$-3x + 640 = 499$$
$$-3x + 640 - 640 = 499 - 640$$
$$-3x = -141$$
$$\frac{-3x}{-3} = \frac{-141}{-3}$$
$$x = 47 \quad \text{The number of \$5 tickets}$$

The question is to find how many of each type of ticket he sold. We can find the number of $8 tickets by using the expression $80 - x$.

$$\text{Number of \$8 tickets} = 80 - x$$
$$= 80 - 47 \quad \text{Substitute 47 for } x.$$
$$= 33$$

Juan sold 47 $5 tickets and 33 $8 tickets. ∎

Try Problems 29 and 30.

Answers to Try These Problems

1. 35 *m* 2. 12 3. $65 4. 10 yd 5. $N - 15 = 40$
6. $9N = 189$ 7. $67 = N + 13$ 8. $30 + 5x = 85$
9. $120 = 6(12 + x)$ 10. $S + (S + 8) = 76$ or $2S + 8 = 76$
11. $5(L - 15) - L = 9$ or $4L - 75 = 9$
12. $5n + 2n = 119$ or $7n = 119$ 13. $4y - y = 33$ or $3y = 33$
14. $\frac{7 + 12 + 13 + 9 + w}{5} = 11$ or $\frac{41 + w}{5} = 11$ 15. $15 + 4N = 63$
16. $2W + 2(3W) = 104$ or $8W = 104$
17. $L(L - 14) = 3920$ or $L^2 - 14L = 3920$
18. $25(x + 15) + 75x = 2875$ or $100x + 375 = 2875$
19. -3 20. 4 21. 11; 132 22. 17; 32 23. $\frac{1}{2}$, 4, 11
24. walking distance, $4\frac{1}{2}$ mi; jogging distance, $2\frac{1}{2}$ mi
25. Ms. Herrera, $1250; Mr. Herrera, $5000 26. 150 products
27. length, 28 ft; width, 22 ft 28. width, 105 m; length, 420 m
29. Barry, 75 tickets; Steve, 95 tickets
30. Hans, 30 hr; Sally, 40 hr

EXERCISES 14.4

Solve by using the given formula.

1. The perimeter of a rectangle is 340 feet and its length is 95 feet. Find the width of the rectangle. (Use the formula $P = 2L + 2W$.)
2. The perimeter of a rectangle is 1260 inches and its width is 300 inches. Find the length of the rectangle. (Use the formula $P = 2L + 2W$.)
3. The larger of two numbers is -4 and the distance between the two numbers is 25. Find the smaller number. (Use the formula $D = L - S$, where D is the distance between the larger number L and the smaller number S.)
4. The smaller of two numbers is -12 and the distance between the two numbers is 59. Find the larger number. (Use the formula $D = L - S$, where D is the distance between the larger number L and the smaller number S.)
5. The number of products in demand D of a certain product is given by the formula $D = 2000 - 4p$, where p is the price (in dollars) per product. Find the price p that causes the demand to be 920.
6. The cost C (in dollars) for parking in a downtown garage is given by the formula $C = 1 + 2x$, where x is the total number of hours parked. How many hours can you park for $15?
7. Find the altitude of a triangle that has area 120 square meters and a base of 20 meters. (Use the formula $A = \frac{1}{2}bh$.)
8. Find the base of a triangle that has area 200 square feet and an altitude of 50 feet. (Use the formula $A = \frac{1}{2}bh$.)

In Exercises 9 through 16, let N represent the unknown number and translate each sentence to an equation. Do not solve the equation.

9. A number subtracted from 50 equals 18.
10. Sixty represents the product of a number and 8.
11. The ratio of a number to 3 is 36.
12. Seventeen increased by a number is 46.
13. Seventy-five minus twice a number yields 31.
14. Four times a number, increased by 7, yields 39.
15. Five times the sum of a number and 13 equals 75.
16. Six times the product of a number and 8 yields 240.

In Exercises 17 through 22, write expressions that represent each of the unknown quantities before translating the given sentence to an equation. Use only one variable, the given variable. Do not solve the equation.

17. One number is 15 less than another number. Let L be the larger of the two numbers. Translate to an equation, "the sum of the two numbers is 155." Do not solve the equation.
18. One number is 34 more than another number. Let S be the smaller of the two numbers. Translate to an equation, "five times the smaller number, increased by the larger number, yields 430." Do not solve the equation.
19. A larger number is 8 times a smaller number. Let x represent the smaller number. Translate to an equation "the difference in the two numbers is 49." Do not solve the equation.

20. A larger number is 6 times a smaller number. Let x represent the smaller number. Translate to an equation, "three times the larger number, decreased by 8 times the smaller number, yields 30." Do not solve the equation.

21. The sum of two numbers is 50. Let N be the smaller number. Translate to an equation, "twice the smaller number, decreased by the larger number, yields 1." Do not solve the equation.

22. The sum of two numbers is 98. Let L be the larger number. Translate to an equation, "three times the larger number, increased by the smaller number, gives 206." Do not solve the equation.

In Exercises 23 through 28, write expressions that represent each of the unknown quantities before translating the given sentence to an equation. Use only one variable, the given variable. Do not solve the equation.

23. Wade scored 86 and 78 on the first two history exams. Let x represent his score on the third exam. Translate to an equation, "the average (mean) of the three scores is 80." Do not solve the equation.

24. Ms. Wong belongs to a tennis club where she pays $60 per month, plus $8 for each hour of court time. Let N represent how many hours of court time she used last month. Translate to an equation, "the total bill for last month was $156." Do not solve the equation.

25. Let W be the width (in meters) of a rectangle. The length is 5 meters more than the width. Translate to an equation, "the area of the rectangle is 204 square meters." Do not solve the equation.

26. Let L be the length (in centimeters) of a rectangle. The width is 100 centimeters less than 4 times the length. Translate to an equation, "the perimeter of the rectangle is 100 centimeters." Do not solve the equation.

27. Brenda earns $18 per hour and her older brother, John, earns $38 per hour. In one week Brenda worked 10 more hours than John. Let J represent the number of hours John worked. Translate to an equation, "the total amount of money they earned for the week is $1860." Do not solve the equation.

28. Mr. Jackson purchased some shirts and sweaters. The shirts cost $35 each and the sweaters cost $45 each. Let x represent how many shirts he bought and assume that he bought 3 fewer sweaters than shirts. Translate to an equation, "the total cost is $425." Do not solve the equation.

Solve by using an equation.

29. Fifteen minus twice a number yields 27. Find the number.

30. Eight times the sum of a number and 3 is 152. Find the number.

31. A larger number is 2 more than a smaller number. Six times the larger number decreased by 4 times the smaller number equals 15. Find the two numbers.

32. One number is 10 times another number. The sum of the two numbers increased by 7 equals 40. Find the two numbers.

33. Marjorie is thinking of three numbers. The first number is 1 less than the second number. The third number is 1 more than the second number. The sum of the first two numbers increased by 4 times the third number is 6. Find the three numbers.

34. Harold is thinking of three numbers. The second number is 15 more than the first number. The third number is 3 more than the second number. The sum of the three numbers is 3. Find the three numbers.

35. You buy a car by making a $2000 down payment in addition to 36 equal monthly payments. If you paid a total of $11,000 for the car, how much was the monthly payment?
36. Cheryl was charged $610 for having her car serviced. The cost included $358 for materials and $42 per hour for labor. For how many hours of labor did she get charged?
37. Jose received a bonus of $4800 at the end of the year. He invested part of it in stocks and spent the rest. If he invested 3 times as much as he spent, how much did he invest and how much did he spend?
38. Maria has a ribbon that is 27 inches long. She cuts it into two pieces such that one piece is 2 inches longer than the other piece. Find the length of each piece.
39. The length of a rectangle is twice its width. The perimeter of the rectangle is 222 feet. Find the length and the width.
40. The width of a rectangle is 28 yards less than 3 times the length. The perimeter of the rectangle is 48 yards. Find the area of the rectangle.
41. A theater sold 550 tickets to a play. Some of the tickets sold for $24 each while the rest sold at $10 each. If the total amount received for the tickets was $8440, how many of each type of ticket were sold?
42. Betty and Roger were selling boxes of candy at $5 each to raise money for the school. Roger sold 28 more boxes than Betty. If they brought in a total of $360, how many boxes did each sell?

14.5 SOLVING EQUATIONS WITH THE VARIABLE APPEARING ON BOTH SIDES

OBJECTIVES
- Solve an equation with the variable appearing on both sides.
- Solve an application by using an equation.

1 Solving Equations with the Variable Appearing on Both Sides

In Section 14.3 we solved equations that had the variable appearing on only one side. Now we look at solving equations with the variable on both sides. For example, consider the equation

$$4x + 7 = 2x - 5$$

Observe that the variable, x, appears on both the left side and the right side. To solve the equation, we eventually want the variable, x, isolated by itself on one side and a number on the other side. We begin by getting all of the x terms on one side of the equation. In this case, that can be done by subtracting the term, $2x$, from both sides.

$$4x + 7 = 2x - 5$$
$$4x - 2x + 7 = 2x - 2x - 5$$ Subtract $2x$ from both sides.
$$2x + 7 = -5$$

▲ Try These Problems

Solve for the variable.
1. $5y + 3 = 27 - 3y$
2. $14c - 7 = 4c - 37$

Instead of subtracting $2x$, we could have added $-2x$. Now the variable, x, appears on only one side, so we continue as we did in Section 14.3.

$$2x + 7 = -5$$
$$2x + 7 - 7 = -5 - 7 \qquad \text{Subtract 7 from both sides.}$$
$$2x = -12$$
$$\frac{2x}{2} = \frac{-12}{2} \qquad \text{Divide both sides by 2.}$$
$$x = -6$$

The solution is -6.

The previous problem could have been solved in several other ways. For example, we could have begun by subtracting $4x$ from both sides, thus resulting in an equation with the variable appearing only on the right side. Here we show the equation solved in this way.

$$4x + 7 = 2x - 5$$
$$4x - 4x + 7 = 2x - 4x - 5 \qquad \text{Subtract } 4x \text{ from both sides.}$$
$$7 = -2x - 5$$
$$7 + 5 = -2x - 5 + 5 \qquad \text{Add 5 to both sides.}$$
$$12 = -2x$$
$$\frac{12}{-2} = \frac{-2x}{-2} \qquad \text{Divide both sides by } -2.$$
$$-6 = x$$

Of course we obtain the same solution, -6.

Here are more examples.

EXAMPLE 1 Solve for w: $10w - 20 = 8 - 4w$

SOLUTION Begin by getting all of the w terms on one side of the equation. To undo the subtraction of $4w$, we add $4w$. Therefore, we add the term, $4w$, to both sides.

$$10w - 20 = 8 - 4w$$
$$10w + 4w - 20 = 8 - 4w + 4w \qquad \text{Add } 4w \text{ to both sides.}$$
$$14w - 20 = 8$$
$$14w - 20 + 20 = 8 + 20 \qquad \text{Add 20 to both sides.}$$
$$14w = 28$$
$$\frac{14w}{14} = \frac{28}{14} \qquad \text{Divide both sides by 14.}$$
$$w = 2 \qquad ■$$

▲ Try Problems 1 and 2.

Try These Problems

Solve for the variable.
3. $3a + 4 = 7a - 10$
4. $-5c - 1 = 4 - 3c$
5. $15x - 7 + 3x = 12x - 1 - 12$
6. $13 - 7x - 7 = 2x + 14 - 12x$

EXAMPLE 2 Solve for y: $y + 8 = 3y + 7$

SOLUTION Begin by getting all of the y terms on one side of the equation. To undo the addition of y, we subtract y. Therefore, we subtract y from both sides of the equation.

$$y + 8 = 3y + 7$$
$$y - y + 8 = 3y - y + 7 \quad \text{Subtract } y \text{ from both sides.}$$
$$8 = 2y + 7$$
$$8 - 7 = 2y + 7 - 7 \quad \text{Subtract 7 from both sides.}$$
$$1 = 2y$$
$$\frac{1}{2} = \frac{2y}{2} \quad \text{Divide both sides by 2.}$$
$$\frac{1}{2} = y \quad \blacksquare$$

EXAMPLE 3 Solve for x: $-2x - 12 = 4x + 9$

SOLUTION Begin by getting all of the x terms on one side of the equation. To undo the addition of $-2x$, we add $2x$. Therefore, we add $2x$ to both sides of the equation.

$$-2x - 12 = 4x + 9$$
$$-2x + 2x - 12 = 4x + 2x + 9 \quad \text{Add } 2x \text{ to both sides.}$$
$$-12 = 6x + 9$$
$$-12 - 9 = 6x + 9 - 9 \quad \text{Subtract 9 from both sides.}$$
$$-21 = 6x$$
$$\frac{-21}{6} = \frac{6x}{6} \quad \text{Divide both sides by 6.}$$
$$\frac{-7}{2} = x \quad \blacksquare$$

 Try Problems 3 and 4.

EXAMPLE 4 Solve for c: $4c + 18 - 5c = -8 + 8c - 1$

SOLUTION We simplify each side before beginning to operate on both sides.

$$4c + 18 - 5c = -8 + 8c - 1$$
$$-c + 18 = -9 + 8c \quad \text{Combine like terms on both sides.}$$
$$-c + c + 18 = -9 + 8c + c \quad \text{Add } c \text{ to both sides.}$$
$$18 = -9 + 9c$$
$$18 + 9 = -9 + 9 + 9c \quad \text{Add 9 to both sides.}$$
$$27 = 9c$$
$$\frac{27}{9} = \frac{9c}{9} \quad \text{Divide both sides by 9.}$$
$$3 = c \quad \blacksquare$$

 Try Problems 5 and 6.

Try These Problems

Solve for the variable.
7. $4 + 2(7y - 1) = 3y - 20$
8. $5(5w - 4) - 7w = 8(2 + 2w)$

Solve by using an equation.
9. Seven more than 5 times a number is 12 decreased by 10 times the number. Find the number.

EXAMPLE 5 Solve for v: $40v + 8(20v - 5) = 5(10v + 40)$

SOLUTION We begin by using the distributive law to remove the parentheses. Also, we simplify each side before beginning to operate on both sides.

$40v + 8(20v - 5) = 5(10v + 40)$	
$40v + 160v - 40 = 50v + 200$	Remove the parentheses.
$200v - 40 = 50v + 200$	Combine like terms.
$200v - 50v - 40 = 50v - 50v + 200$	Subtract $50v$ from both sides.
$150v - 40 = 200$	
$150v - 40 + 40 = 200 + 40$	Add 40 to both sides.
$150v = 240$	
$\dfrac{150v}{150} = \dfrac{240}{150}$	Divide both sides by 150.
$v = \dfrac{8}{5}$ ∎	

 Try Problems 7 and 8.

2 Solving Applications

Now we look at solving some applications that involve setting up an equation that has the variable appearing on both sides.

EXAMPLE 6 Twice a number, decreased by 5, equals 13 less than 3 times the number. Find the number.

SOLUTION This problem has only one unknown quantity, the missing number. We choose a letter of the alphabet to represent this number.

Let N = the unknown number.

Now we translate the given sentence to an equation.

Twice   a number   decreased by   5   equals   13 less than   3 times the number
  2 · N $-$ 5   = $3N$ $-$ 13

Next solve the equation for N.

$2N - 5 = 3N - 13$	
$2N - 2N - 5 = 3N - 2N - 13$	Subtract $2N$ from both sides.
$-5 = N - 13$	
$-5 + 13 = N - 13 + 13$	Add 13 to both sides.
$8 = N$	

The number is 8. ∎

 Try Problem 9.

Try These Problems

Solve by using an equation.

10. A larger number is 5 more than a smaller number. Four times the smaller number is equal to 5 more than 3 times the larger number. Find the two numbers.

EXAMPLE 7 A larger number is 3 times a smaller number. Nine times the smaller number is equal to 4 times the larger number, decreased by 36. Find the two numbers.

SOLUTION This problem has two unknown quantities, the two numbers. We use the first sentence of the problem to write expressions to represent each number.

Let x = the smaller number.
$3x$ = the larger number.

Now we translate the second sentence to an equation.

9 times   the smaller number   is equal to   4 times   the larger number   decreased by   36
9 · x = 4 · $(3x)$ − 36

Now we solve the equation for x.

$$9x = 4(3x) - 36$$
$$9x = 12x - 36 \quad \text{Remove the parentheses.}$$
$$9x - 12x = 12x - 12x - 36 \quad \text{Subtract } 12x \text{ from both sides.}$$
$$-3x = -36$$
$$\frac{-3x}{-3} = \frac{-36}{-3} \quad \text{Divide both sides by } -3.$$
$$x = 12 \quad \text{The smaller number}$$

The question was to find both numbers. Therefore, we use the expression $3x$ to find the larger number.

$$\text{Larger number} = 3x = 3(12) = 36$$

The two numbers are 12 and 36. ■

Try Problem 10.

EXAMPLE 8 A gardener and his assistant worked a total of 75 hours last week. The gardener is paid $25 per hour while the assistant is paid only $10 per hour. If the gardener's pay for that week was 5 times the assistant's pay, how many hours did each work that week?

SOLUTION Observe that there are several unknown quantities in this problem. We do not know the number of hours the gardener worked, the number of hours his assistant worked, nor the total earned by each for the week.

Begin by letting a variable be the number of hours the gardener worked. Let G = the number of hours the gardener worked. The first sentence says that they worked a total of 75 hours. Therefore we have the following:

Number of hours the gardener worked + number of hours the assistant worked = 75

G + number of hours the assistant worked = 75

Subtracting G from both sides of this equation, we see that the expression, $75 - G$, represents the number of hours the assistant worked.

Try These Problems

Solve by using an equation.

11. A theater sold a total of 220 tickets. Some of them sold for $6 each and the rest sold for $3 each. The total amount collected for the $6 tickets is $30 less than 3 times the amount collected for the $3 tickets. How many of each type of ticket were sold?

G = number of hours the gardener worked.

$75 - G$ = number of hours the assistant worked.

$25G$ = amount the gardener earned for the week ($).

$10(75 - G)$ = amount the assistant earned for the week ($).

Now we translate the last sentence to an equation.

Gardener's pay ($) was 5 times the assistant's pay ($)

$$25G = 5 \cdot 10(75 - G)$$

Next we solve the equation for G.

$$25G = 5 \cdot 10(75 - G)$$
$$25G = 50(75 - G)$$
$$25G = 3750 - 50G$$
$$25G + 50G = 3750 - 50G + 50G$$
$$75G = 3750$$
$$\frac{75G}{75} = \frac{3750}{75}$$
$$G = 50 \quad \text{The number of hours the gardener worked}$$

The question is to find the number of hours each worked. We use the expression $75 - G$ to find the number of hours the assistant worked.

$$\text{Number of hours the assistant worked} = 75 - G$$
$$= 75 - 50$$
$$= 25$$

The gardener worked 50 hours and the assistant worked 25 hours. ∎

Try Problem 11.

Answers to Try These Problems

1. $y = 3$ 2. $c = -3$ 3. $a = \frac{7}{2}$ 4. $c = \frac{-5}{2}$ 5. $x = -1$
6. $x = \frac{8}{3}$ 7. $y = -2$ 8. $w = 18$ 9. $\frac{1}{3}$ 10. 20, 25
11. 130 $6 tickets; 90 $3 tickets

EXERCISES 14.5

Solve for the variable.

1. $8x + 6 = 2x + 30$
2. $9x - 5 = 4x + 10$
3. $3y - 12 = 7y + 8$
4. $2y + 12 = 10y + 44$
5. $5v - 17 = 18 - 2v$
6. $12v + 14 = 6 - 8v$
7. $16a + 3 = -4 - 5a$
8. $7a - 15 = -3 - 5a$
9. $6 - w = 14 + 5w$
10. $20 - 3w = 8 + w$
11. $-45 - 7w = 15 + 3w$
12. $23 - 8w = -16 + 5w$
13. $1 - 4x = 7 - 5x$
14. $3 - 6x = 17 - 8x$
15. $3 - 10y = 7 - 4y$
16. $60 - 15y = 10 - 5y$
17. $-n + 3 = 5n - 7$
18. $-4n - 35 = 10 - 7n$
19. $-9 + 16n = -14n + 11$
20. $-17 - 8n = -12n - 3$
21. $-3c + 7 - 5c = 24 + 12c - 2$
22. $15 - 9c + 3 = -8c - 6 + 2c$
23. $14 - 17c + 5c - 24 = 8c$
24. $45c - 20 + 8 = c - 12 + 4c$
25. $30w = 5(w - 100)$

26. $4(2w - 1) = w + 24$ **27.** $8w + 6(3w + 3) = 2w + 108$
28. $84 - 20w = 30 + 5(4w - 2)$ **29.** $-8(5c + 1) - 12 = 1 + (2c + 7)$
30. $-4(8c - 2) + 3(c + 6) = c - 4$

Solve by using an equation.

31. Ten minus 6 times a number equals 24 more than the number. Find the number.

32. Five times a number, decreased by 7, is equal to 4 times the number, increased by 3. Find the number.

33. One number is 8 more than another number. Twice the smaller number, plus 103, gives 5 times the larger number. Find the two numbers.

34. One number is 7 less than another number. Three times the larger number is equal to 1 more than, 5 times the smaller number. Find the two numbers.

35. The sum of two numbers is 50. Three times the smaller number is equal to 2 less than the larger number. Find the two numbers.

36. The sum of two numbers is 84. Eight more than twice the smaller number equals 26 more than the larger number. Find the two numbers.

37. Victoria is thinking of three numbers. The second number is 9 more than the first number. The third number is 10 less than 5 times the second number. Seven times the sum of the first and second numbers equals 1 more than the third number. Find the three numbers.

38. Edward is thinking of three numbers. The second number is 2 more than the first number. The third number is 27 times the second number. Six times the sum of the first and second numbers, increased by 12, equals 5 less than the third number. Find the three numbers.

39. Bernice was selling tickets to a spaghetti supper to raise money for her church. Regular tickets were selling for $12 each and tickets for children or senior citizens were selling for $8 each. She sold 16 more $8 tickets than $12 tickets. If she collected $52 more for the sale of the $12 tickets than for the $8 tickets, how many of each type of ticket did she sell?

40. Mr. Huynh has two jobs. His construction job pays $32 per hour, and he charges $15 per hour for gardening. Last week he worked a total of 65 hours. If he made $35 more gardening than on the construction job, how many hours did he work on each job?

14.6 BASIC ALGEBRAIC FRACTIONS

OBJECTIVES
- Evaluate an algebraic fraction for given values of the variable.
- Simplify an algebraic fraction.
- Multiply and divide algebraic fractions.
- Add and subtract algebraic fractions.
- Combine like terms with fractional and decimal coefficients.

1 Evaluating Algebraic Fractions for Given Values of the Variable

In Section 12.6 you worked with signed fractions such as $\frac{2}{3}$ and $\frac{-7}{8}$. Now we look at fractions that contain a variable in the numerator or the

Try These Problems

Evaluate for the given values of the variable.

1. $\dfrac{y}{5}$ for $y = 35$
2. $\dfrac{n}{-12}$ for $n = -6$
3. $\dfrac{45}{m}$ for $m = -30$

denominator. Fractions of this type are called **algebraic fractions**. For example, consider the fraction

$$\frac{x}{3}$$

We do not know the value of this fraction unless we are given the value of the variable x. Here we evaluate the fraction for various values of x.

For $x = 1$, $\dfrac{x}{3} = \dfrac{1}{3}$

For $x = -8$, $\dfrac{x}{3} = \dfrac{-8}{3}$ or $-2\dfrac{2}{3}$

For $x = 12$, $\dfrac{x}{3} = \dfrac{12}{3} = 4$

Here are more examples of evaluating algebraic fractions for given values of the variable.

EXAMPLE 1 Evaluate $\dfrac{20}{m}$ for $m = -25$.

SOLUTION

$\dfrac{20}{m} = \dfrac{20}{-25}$ Substitute -25 for m.

$= \dfrac{4 \cdot \cancel{5}}{-5 \cdot \cancel{5}}$ Observe the common factor 5 in the numerator and denominator.

$= \dfrac{4}{-5}$ Reduce the fraction to lowest terms.

$= \dfrac{-4}{5}$ When the fraction is negative, we prefer to write the negative sign in the numerator. ■

EXAMPLE 2 Evaluate $\dfrac{-18}{c}$ for $c = -24$.

SOLUTION

$\dfrac{-18}{c} = \dfrac{-18}{-24}$

$= \dfrac{18}{24}$ The fraction is positive because a negative number divided by a negative number is a positive number.

$= \dfrac{3 \cdot \cancel{6}}{4 \cdot \cancel{6}}$ Observe the common factor 6 in the numerator and the denominator.

$= \dfrac{3}{4}$ ■

 Try Problems 1 through 3.

Try These Problems

Evaluate for the given values of the variable.

4. $\dfrac{w - 6}{2}$ for $w = 1$

5. $\dfrac{y - x}{-3}$ for $y = 4$ and $x = 6$

6. $\dfrac{8 + 5x}{4x}$ for $x = 4$

7. $\dfrac{w + 7}{6 - w}$ for $w = -2$

Now we look at fractions where there are operations to be performed in the numerator or denominator. These operations should be performed before doing the division that is indicated by the fraction bar. Here are some examples.

EXAMPLE 3 Evaluate $\dfrac{x + y}{5}$ for $x = 10$ and $y = -2$.

SOLUTION

$\dfrac{x + y}{5} = \dfrac{10 + (-2)}{5}$ Substitute 10 for x and -2 for y.

$= \dfrac{8}{5}$ The addition in the numerator is done first because the fraction bar is an understood grouping symbol. ∎

EXAMPLE 4 Evaluate $\dfrac{5w - 6}{2w}$ for $w = 4$.

SOLUTION

$\dfrac{5w - 6}{2w} = \dfrac{5(4) - 6}{2(4)}$ Substitute 4 for w.

$= \dfrac{20 - 6}{8}$ Operate within the numerator and denominator first.

$= \dfrac{14}{8}$

$= \dfrac{\cancel{2} \cdot 7}{\cancel{2} \cdot 4}$ Reduce the fraction to lowest terms.

$= \dfrac{7}{4}$ ∎

 Try Problems 4 through 7.

2 Simplifying Algebraic Fractions

When we simplify a numerical fraction, we cancel common *factors* from the numerator and denominator. For example, the fraction

$$\dfrac{35}{42}$$

can be simplified by canceling the common factor 7 from the numerator and denominator.

$$\dfrac{35}{42} = \dfrac{\cancel{7} \cdot 5}{\cancel{7} \cdot 6} = \dfrac{5}{6}$$

Canceling a common *factor* from the numerator and denominator of a fraction does not change the value of the fraction. A *factor* of the numerator is a quantity that is *multiplied* by another quantity to create the numerator. Likewise, a *factor* of the denominator is a quantity that is *multiplied* by another quantity to create the denominator.

Try These Problems

Simplify.

8. $\dfrac{63}{72}$

9. $\dfrac{15y^2}{2y}$

10. $\dfrac{24ac}{21c}$

11. $\dfrac{50w^2m}{45wm}$

Now we look at simplifying algebraic fractions. For example, consider the fraction

$$\frac{12xy}{9x}$$

In the numerator and denominator there are the common factors 3 and x. We cancel these common factors to reduce the fraction to lowest terms.

$$\frac{12xy}{9x} = \frac{\cancel{3} \cdot 4\cancel{x}y}{\cancel{3} \cdot 3\cancel{x}}$$

$$= \frac{4y}{3}$$

Here is another example of reducing an algebraic fraction to lowest terms.

EXAMPLE 5   Simplify: $\dfrac{36x^2}{20xw}$

SOLUTION

$$\frac{36x^2}{20xw} = \frac{\cancel{4} \cdot 9\cancel{x}x}{\cancel{4} \cdot 5\cancel{x}w}$$

$$= \frac{9x}{5w} \blacksquare$$

 Try Problems 8 through 11.

Consider the fraction

$$\frac{4+3}{4+6}$$

If we perform each of the additions, we see that the fraction is equal to $\frac{7}{10}$.

$$\frac{4+3}{4+6} = \frac{7}{10}$$

Observe that you cannot cancel the 4s and obtain an equal fraction. The number 4 is not a *factor* of the numerator because it is not *multiplied* by 3 to create the numerator, it is added to 3. Likewise, the number 4 is not a factor of the denominator. This illustrates that you cannot cancel common *terms* from the numerator and denominator of a fraction.

When working with algebraic fractions you must be careful not to cancel common *terms* from the numerator and denominator. For example, the fraction

$$\frac{x+5}{5}$$

cannot be reduced. The number 5 is not a factor of the numerator, because it is added to x, not multiplied by x, to create the numerator. Therefore, these 5s cannot be canceled.

14.6 Basic Algebraic Fractions 701

 Try These Problems

Simplify.

12. $\dfrac{3(2 + c)}{6}$

13. $\dfrac{8(10 - x)}{4x}$

Here are more examples.

EXAMPLE 6 Simplify: $\dfrac{2(y - 6)}{12}$

SOLUTION Observe that 2 is a *factor* of the numerator because 2 is multiplied by the quantity, $y - 6$, to create the numerator. Of course, 2 is also a factor of the denominator.

$\dfrac{2(y - 6)}{12} = \dfrac{\cancel{2}(y - 6)}{\cancel{2} \cdot 6}$ Cancel the common factor 2.

$= \dfrac{y - 6}{6}$ These 6s cannot be canceled because 6 is not a factor of the numerator. ■

 Try Problems 12 and 13.

Consider the fraction

$$\dfrac{2(3) + 1}{2(4)}$$

If we evaluate the numerator and denominator, we see that the fraction is equal to $\tfrac{7}{8}$.

$\dfrac{2(3) + 1}{2(4)} = \dfrac{6 + 1}{8}$

$= \dfrac{7}{8}$

Observe that you cannot cancel the 2s in the original fraction and obtain an equal fraction. The number 2 cannot be canceled because 2 is not a *factor of the entire numerator,* it is only a factor of the first term of the numerator.

The algebraic fraction

$$\dfrac{3x + 4}{15}$$

cannot be reduced. The number 3 is a factor of the denominator because $15 = (3)(5)$. However, 3 is not a factor of the numerator; 3 is only a factor of the first term of the numerator. You cannot cancel 3s in this fraction.

EXAMPLE 7 Simplify: $\dfrac{4(7x + 2)}{28}$

SOLUTION

$\dfrac{4(7x + 2)}{28} = \dfrac{\cancel{4}(7x + 2)}{\cancel{4}(7)}$ Cancel the common factor 4.

$= \dfrac{7x + 2}{7}$ These 7s cannot be canceled. ■

Try These Problems

Simplify.

14. $\dfrac{5(3y + 2)}{15y}$

15. $\dfrac{30(8w + 1)}{24}$

16. $\dfrac{y(4y - 7)}{8y}$

17. $\dfrac{4x - 24}{2x}$

18. $\dfrac{c^2 + 3c}{6c}$

19. $\dfrac{15y^2 + 10y}{25}$

20. $\dfrac{7w^2 - 21w}{14w}$

EXAMPLE 8 Simplify: $\dfrac{18w(5 - 2w)}{12w}$

SOLUTION

$\dfrac{18w(5 - 2w)}{12w} = \dfrac{3(6w)(5 - 2w)}{2(6w)}$ Cancel the common factor $6w$.

$= \dfrac{3(5 - 2w)}{2}$ These 2s cannot be canceled.

or $\dfrac{15 - 6w}{2}$ The numerator can be left in factored form or multiplied out. ∎

 Try Problems 14 through 16.

EXAMPLE 9 Simplify: $\dfrac{5w + 10}{20}$

SOLUTION Notice that the number 5 is a factor of *both terms* of the numerator. Therefore, we can factor out 5 in the numerator by using the distributive law.

$\dfrac{5w + 10}{20} = \dfrac{5(w + 2)}{20}$ Write the numerator in factored form by factoring out 5.

$= \dfrac{\cancel{5}(w + 2)}{\cancel{5}(4)}$ Cancel the common factor 5.

$= \dfrac{w + 2}{4}$ Do not cancel 2s here. This fraction is simplified. ∎

EXAMPLE 10 Simplify: $\dfrac{6y^2 + 6y}{3y^2}$

SOLUTION Do not attempt to reduce the fraction until the numerator is written in a *factored form*. Observe that $6y$ is a factor of both terms in the numerator, so we use the distributive law to factor out $6y$.

$\dfrac{6y^2 + 6y}{3y^2} = \dfrac{6y(y + 1)}{3y^2}$ Write the numerator in factored form by factoring out $6y$.

$= \dfrac{2(\cancel{3})\cancel{y}(y + 1)}{\cancel{3}\,\cancel{y}\,y}$ Cancel the common factor $3y$.

$= \dfrac{2(y + 1)}{y}$

or $\dfrac{2y + 2}{y}$ The numerator can be left in factored form or multiplied out. ∎

 Try Problems 17 through 20.

14.6 Basic Algebraic Fractions

 Try These Problems

Multiply or divide as indicated and simplify.

21. $\dfrac{6cw}{5} \cdot \dfrac{4c}{3w}$

22. $\dfrac{3w}{2a} \cdot \dfrac{5a^2}{10}$

23. $\dfrac{30xw}{40y} \div \dfrac{7w}{4x}$

24. $\dfrac{14x^2}{35} \div \dfrac{2x}{3y}$

3 Multiplying and Dividing Algebraic Fractions

To multiply and divide algebraic fractions we use exactly the same rules as we used to multiply and divide numerical fractions. Here we illustrate the procedure with some examples.

EXAMPLE 11 Multiply: $\dfrac{4y}{w} \cdot \dfrac{5w}{8}$

SOLUTION

$\dfrac{4y}{w} \cdot \dfrac{5w}{8} = \dfrac{\cancel{4}y}{\cancel{w}} \cdot \dfrac{5\cancel{w}}{\cancel{4}(2)}$ Cancel the common factors 4 and w from the numerator and denominator.

$= \dfrac{5y}{2}$ Multiply the numerators to obtain the numerator of the answer. Multiply the denominators to obtain the denominator of the answer. ∎

EXAMPLE 12 Divide: $\dfrac{12c^2}{9c} \div \dfrac{7a}{15}$

SOLUTION

$\dfrac{12c^2}{9c} \div \dfrac{7a}{15} = \dfrac{12c^2}{9c} \cdot \dfrac{15}{7a}$ Change the division to multiplication and invert the divisor.

$= \dfrac{\cancel{3}(4)\cancel{c}c}{\cancel{3}(\cancel{3})\cancel{c}} \cdot \dfrac{\cancel{3}(5)}{7a}$ Cancel the common factors 3, 3, and c from the numerator and denominator.

$= \dfrac{20c}{7a}$ Multiply the numerators to obtain the numerator of the answer. Multiply the denominators to obtain the denominator of the answer. ∎

 Try Problems 21 through 24.

When the numerator or denominator of a fraction contains more than one term, you must be careful with canceling. We cancel common *factors* of the numerator and denominator, not common *terms*. Here we look at some examples.

EXAMPLE 13 Multiply: $\dfrac{4}{3a} \cdot \dfrac{a+3}{16}$

SOLUTION Observe that the second numerator has two terms. The only factor of that numerator is the entire quantity $a + 3$. The variable a is not a factor of the second numerator and the number 3 is also not a factor of that numerator.

$\dfrac{4}{3a} \cdot \dfrac{(a+3)}{16}$ Wrap $a+3$ in parentheses to remind you that the whole quantity is the factor for that numerator.

$= \dfrac{\cancel{4}}{3a} \cdot \dfrac{(a+3)}{\cancel{4}(4)}$ Cancel the common factor 4 from the numerator and denominator.

$= \dfrac{a+3}{12a}$ ∎

Try These Problems

Multiply or divide as indicated and simplify.

25. $\dfrac{w+8}{5} \cdot \dfrac{15w}{4w}$

26. $\dfrac{4}{a^2} \cdot \dfrac{a(2a-1)}{8}$

27. $\dfrac{36}{12w} \div \dfrac{3}{w-3}$

28. $\dfrac{20(y+4)}{10} \div \dfrac{yw}{3w^2}$

EXAMPLE 14 Divide: $\dfrac{6(x+7)}{5} \div 14x$

SOLUTION Observe that the factors of the first numerator are 2, 3 and the quantity $x + 7$. The variable x is not a factor of the first numerator and the number 7 is also not a factor of that numerator.

$\dfrac{6(x+7)}{5} \div 14x$

$= \dfrac{6(x+7)}{5} \div \dfrac{14x}{1}$ The expression $14x$ is the same as $\dfrac{14x}{1}$.

$= \dfrac{6(x+7)}{5} \cdot \dfrac{1}{14x}$ Change the division to multiplication and invert the divisor.

$= \dfrac{\cancel{2}(3)(x+7)}{5} \cdot \dfrac{1}{\cancel{2}(7)x}$ Cancel the common factor 2 from the numerator and the denominator.

$= \dfrac{3(x+7)}{35x}$ The final answer may be left with the numerator in factored form or multiplied out.

or $\dfrac{3x+21}{35x}$ ∎

 Try Problems 25 through 28.

EXAMPLE 15 Multiply: $\dfrac{9}{m} \cdot \dfrac{m^2 + 4m}{12}$

SOLUTION Observe that the variable m is a factor of both terms of the second numerator, therefore we can factor an m out of that numerator.

$\dfrac{9}{m} \cdot \dfrac{m^2 + 4m}{12}$

$= \dfrac{9}{m} \cdot \dfrac{m(m+4)}{12}$ Write the second numerator in factored form.

$= \dfrac{\cancel{3}(3)}{\cancel{m}} \cdot \dfrac{\cancel{m}(m+4)}{\cancel{3}(4)}$ Cancel the common factors 3 and m from the numerator and denominator.

$= \dfrac{3(m+4)}{4}$ The final numerator may be left in factored form or multiplied out.

or $\dfrac{3m+12}{4}$ ∎

Try These Problems

Multiply or divide as indicated and simplify.

29. $\dfrac{4c}{9c^2} \cdot \dfrac{6c-3}{7}$

30. $\dfrac{6}{m} \cdot \dfrac{m^2+2m}{2m}$

31. $\dfrac{8w^2}{15} \div \dfrac{4w}{5w-20}$

32. $\dfrac{3n^2+9n}{n^2} \div \dfrac{24}{5n}$

Add or subtract as indicated and simplify.

33. $\dfrac{c}{6} + \dfrac{3}{5}$

34. $\dfrac{R}{4} - \dfrac{5}{12}$

EXAMPLE 16 Divide: $\dfrac{12v-18}{4v} \div \dfrac{3}{5v^2}$

SOLUTION Observe that the number 6 is a factor of both terms of the first numerator; therefore, we can factor out a 6 from that numerator.

$$\dfrac{12v-18}{4v} \div \dfrac{3}{5v^2}$$

$$= \dfrac{12v-18}{4v} \cdot \dfrac{5v^2}{3} \quad \text{Change the division to multiplication and invert the divisor.}$$

$$= \dfrac{6(2v-3)}{4v} \cdot \dfrac{5v^2}{3} \quad \text{Write the first numerator in factored form.}$$

$$= \dfrac{2(3)(2v-3)}{2(2)v} \cdot \dfrac{5v\!\!\!/v}{\cancel{3}} \quad \text{Cancel the common factors 2, 3, and } v \text{ from the numerator and denominator.}$$

$$= \dfrac{5v(2v-3)}{2}$$

or $\dfrac{10v^2-15v}{2}$ ∎

 Try Problems 29 through 32.

4 Adding and Subtracting Algebraic Fractions

To add or subtract numerical fractions, we must first write the fractions with a common denominator. The same procedure is used to add or subtract algebraic fractions. Here we illustrate the procedure with some examples.

EXAMPLE 17 Add: $\dfrac{x}{4} + \dfrac{3}{8}$

SOLUTION The least common multiple of 4 and 8 is 8. Therefore, we write the first fraction with denominator 8.

$$\dfrac{x}{4} + \dfrac{3}{8}$$

$$= \dfrac{2x}{2(4)} + \dfrac{3}{8} \quad \text{Multiply the numerator and the denominator of the first fraction by 2 so that it has a denominator of 8. This does not change the value of that fraction.}$$

$$= \dfrac{2x}{8} + \dfrac{3}{8}$$

$$= \dfrac{2x+3}{8} \quad \text{Add the numerators to obtain the numerator of the answer. Keep the common denominator.} \quad ∎$$

 Try Problems 33 and 34.

Try These Problems

Add or subtract as indicated and simplify.

35. $\dfrac{y}{4} + \dfrac{y}{6}$

36. $\dfrac{5w}{2} - \dfrac{2w}{5} + \dfrac{3}{10}$

37. $\dfrac{5c}{8} + \dfrac{1}{6} + \dfrac{5c}{24}$

EXAMPLE 18 Write as a single fraction and simplify: $\dfrac{y}{3} - \dfrac{1}{2} + \dfrac{5y}{3}$

SOLUTION The least common multiple of 2 and 3 is 6. Therefore, we write each fraction with a denominator of 6.

$$\dfrac{y}{3} - \dfrac{1}{2} + \dfrac{5y}{3}$$

$$= \dfrac{2y}{2(3)} - \dfrac{3(1)}{3(2)} + \dfrac{2(5y)}{2(3)}$$

— Multiply numerator and denominator by 2.
— Multiply numerator and denominator by 3.
— Multiply numerator and denominator by 2.

$$= \dfrac{2y}{6} - \dfrac{3}{6} + \dfrac{10y}{6}$$

$$= \dfrac{2y - 3 + 10y}{6}$$
Add the numerators to obtain the numerator of the answer. Keep the common denominator.

$$= \dfrac{12y - 3}{6}$$
Combine like terms in the numerator.

$$= \dfrac{3(4y - 1)}{6}$$
Factor the numerator.

$$= \dfrac{\cancel{3}(4y - 1)}{2(\cancel{3})}$$
Cancel the common factor 3 from the numerator and denominator.

$$= \dfrac{4y - 1}{2} \blacksquare$$

 Try Problems 35 through 37.

EXAMPLE 19 Subtract: $\dfrac{2}{x} - \dfrac{3}{4}$

SOLUTION The least common multiple of 4 and x is $4x$. Notice that the variable x is a factor in the common denominator. We must write each fraction with the denominator $4x$.

$$\dfrac{2}{x} - \dfrac{3}{4}$$

$$= \dfrac{4(2)}{4x} - \dfrac{3x}{4x}$$
Write each fraction with denominator $4x$.
— Multiply numerator and denominator by 4.
— Multiply numerator and denominator by x.

$$= \dfrac{8}{4x} - \dfrac{3x}{4x}$$

$$= \dfrac{8 - 3x}{4x}$$
Subtract the numerators. Keep the same denominator. $\blacksquare$

Try These Problems

Add or subtract as indicated and simplify.

38. $\dfrac{3}{y} + \dfrac{x}{5}$

39. $\dfrac{4}{3x} - \dfrac{2}{y}$

40. $\dfrac{3}{20} + \dfrac{1}{4m} - \dfrac{3}{5m}$

EXAMPLE 20 Add: $\dfrac{5}{6c} + \dfrac{4}{9c} + \dfrac{1}{3}$

SOLUTION The least common multiple (LCM) of $6c$, $9c$, and 3 is $18c$. We need to put only one factor of c in the LCM because only one c appears as a factor in either one of the denominators.

$$\dfrac{5}{6c} + \dfrac{4}{9c} + \dfrac{1}{3}$$

$$= \dfrac{3(5)}{3(6c)} + \dfrac{2(4)}{2(9c)} + \dfrac{6c(1)}{6c(3)} \quad \text{Write each fraction with denominator } 18c.$$

Multiply numerator and denominator by 3.
Multiply numerator and denominator by 2.
Multiply numerator and denominator by $6c$.

$$= \dfrac{15}{18c} + \dfrac{8}{18c} + \dfrac{6c}{18c}$$

$$= \dfrac{15 + 8 + 6c}{18c}$$

$$= \dfrac{23 + 6c}{18c} \quad \blacksquare$$

Try Problems 38 through 40.

EXAMPLE 21 Write as a single fraction and simplify: $2x + \dfrac{7x}{6}$

SOLUTION Before we can add $2x$ to the fraction $\dfrac{7x}{6}$, we must write $2x$ as a fraction with denominator 6.

$$2x + \dfrac{7x}{6}$$

$$= \dfrac{2x}{1} + \dfrac{7x}{6} \quad 2x \text{ is the same as } \dfrac{2x}{1}.$$

$$= \dfrac{6(2x)}{6(1)} + \dfrac{7x}{6} \quad \text{Multiply the numerator and denominator of the first term by 6.}$$

$$= \dfrac{12x}{6} + \dfrac{7x}{6}$$

$$= \dfrac{12x + 7x}{6} \quad \text{Add the numerators. Keep the same denominator.}$$

$$= \dfrac{19x}{6} \quad \text{Combine like terms in the numerator.} \quad \blacksquare$$

Try These Problems

Add or subtract as indicated and simplify.

41. $w - \dfrac{2}{3}$

42. $\dfrac{7}{5c} + 4$

EXAMPLE 22 Write as a single fraction: $\dfrac{x}{y} - 3$

SOLUTION Before we can perform this subtraction, we must write the number 3 as a fraction with denominator y.

$$\dfrac{x}{y} - 3$$

$$= \dfrac{x}{y} - \dfrac{3}{1} \qquad \text{3 is equal to } \tfrac{3}{1}.$$

$$= \dfrac{x}{y} - \dfrac{3y}{1y} \qquad \text{Multiply the numerator and denominator of the second term by the variable } y.$$

$$= \dfrac{x}{y} - \dfrac{3y}{y}$$

$$= \dfrac{x - 3y}{y} \qquad \text{Subtract the numerators. Keep the same denominator.} \blacksquare$$

 Try Problems 41 and 42.

5 Combining Like Terms with Fractional Coefficients

Recall that when two or more terms have exactly the same variable factor, we can combine the terms to make a single term. For example,

$$5M + 4M = (5 + 4)M$$
$$= 9M$$

It is the distributive law of multiplication over addition that allows us to factor out the common factor M.

Now we look at a situation where the numerical coefficients are fractions.

$$\dfrac{1}{2}M + \dfrac{3}{4}M = \left(\dfrac{1}{2} + \dfrac{3}{4}\right)M \qquad \text{The common factor } M \text{ can be factored out.}$$

$$= \left(\dfrac{2}{4} + \dfrac{3}{4}\right)M \qquad \text{Write the fraction } \tfrac{1}{2} \text{ with denominator 4 so that it can be added to the fraction } \tfrac{3}{4}.$$

$$= \dfrac{5}{4}M \quad \text{or} \quad \dfrac{5M}{4} \qquad \text{The expression } \tfrac{5}{4}M \text{ can be written as } \tfrac{5M}{4} \text{ because } \tfrac{5}{4}M = \tfrac{5}{4} \cdot \tfrac{M}{1} = \tfrac{5M}{4}.$$

Do not write the expression $\dfrac{5}{4}M$ as $1\dfrac{1}{4}M$. In algebra, we do not usually write mixed numerals in algebraic expressions. In arithmetic,

$$1\dfrac{1}{4} \quad \text{means} \quad 1 + \dfrac{1}{4}$$

Try These Problems

Combine like terms.

43. $\frac{3}{5}y + \frac{2}{5}y$

44. $\frac{1}{2}B + \frac{3}{8}B$

45. $5w - \frac{3}{4}w$

but, in algebra, when we omit the operation symbol, we mean multiplication. For example,

$$1\left(\frac{1}{4}\right) \quad \text{means} \quad 1 \cdot \frac{1}{4}$$

To avoid any confusion, we do not use the mixed numeral notation in algebraic expressions.

Here are more examples of combining like terms with fractional coefficients.

EXAMPLE 23 Combine like terms: $4c - \frac{2}{3}c$

SOLUTION Because each term has the same variable factor, c, we can factor out c.

$$\begin{aligned}
4c - \frac{2}{3}c &= \left(4 - \frac{2}{3}\right)c \\
&= \left(\frac{4 \cdot 3}{1 \cdot 3} - \frac{2}{3}\right)c &&\text{Write the number 4 with denominator 3.} \\
&= \left(\frac{12}{3} - \frac{2}{3}\right)c \\
&= \frac{10}{3}c \quad \text{or} \quad \frac{10c}{3} &&\text{The expression } \frac{10}{3}c \text{ can be written as } \frac{10c}{3} \text{ because } \frac{10}{3}c = \frac{10}{3} \cdot \frac{c}{1} = \frac{10c}{3}.
\end{aligned}$$

 Try Problems 43 through 45.

EXAMPLE 24 Combine like terms: $w - \frac{2}{5}w$

SOLUTION Because both terms have the same variable factor, w, we can factor out w.

$$\begin{aligned}
w - \frac{2}{5}w &= 1w - \frac{2}{5}w &&\text{The expression } w \text{ equals } 1w. \\
&= \left(1 - \frac{2}{5}\right)w &&\text{Factor out the common factor } w. \\
&= \left(\frac{5}{5} - \frac{2}{5}\right)w \\
&= \frac{3}{5}w \quad \text{or} \quad \frac{3w}{5} &&\text{Subtract the fractions inside the parentheses.}
\end{aligned}$$

Chapter 14 Introduction to Algebra

▲ **Try These Problems**

Combine like terms.

46. $x + \dfrac{4}{5}x$

47. $2.5y - 0.75y$

48. $m - 0.8m$

EXAMPLE 25 Combine like terms: $n + 0.35n$

SOLUTION Because both terms have the same variable factor, n, we can factor out n.

$n + 0.35n = 1n + 0.35n$ The expression n equals $1n$.

$= (1 + 0.35)n$ Factor out the common factor n.

$= 1.35n$ ∎

▲ Try Problems 46 through 48.

▲ **Answers to Try These Problems**

1. 7   2. $\dfrac{1}{2}$ 3. $\dfrac{-3}{2}$ 4. $\dfrac{-5}{2}$ 5. $\dfrac{2}{3}$ 6. $\dfrac{7}{4}$ 7. $\dfrac{5}{8}$ 8. $\dfrac{7}{8}$ 9. $\dfrac{15y}{2}$

10. $\dfrac{8a}{7}$ 11. $\dfrac{10w}{9}$ 12. $\dfrac{2+c}{2}$ 13. $\dfrac{2(10-x)}{x}$ or $\dfrac{20-2x}{x}$

14. $\dfrac{3y+2}{3y}$ 15. $\dfrac{5(8w+1)}{4}$ or $\dfrac{40w+5}{4}$ 16. $\dfrac{4y-7}{8}$

17. $\dfrac{2(x-6)}{x}$ or $\dfrac{2x-12}{x}$ 18. $\dfrac{c+3}{6}$

19. $\dfrac{y(3y+2)}{5}$ or $\dfrac{3y^2+2y}{5}$ 20. $\dfrac{w-3}{2}$ 21. $\dfrac{8c^2}{5}$

22. $\dfrac{3aw}{4}$ 23. $\dfrac{3x^2}{7y}$ 24. $\dfrac{3xy}{5}$ 25. $\dfrac{3(w+8)}{4}$ or $\dfrac{3w+24}{4}$

26. $\dfrac{2a-1}{2a}$ 27. $\dfrac{w-3}{3}$ 28. $\dfrac{6w(y+4)}{y}$ or $\dfrac{6wy+24w}{y}$

29. $\dfrac{4(2c-1)}{21c}$ or $\dfrac{8c-4}{21c}$ 30. $\dfrac{3(m+2)}{m}$ or $\dfrac{3m+6}{m}$

31. $\dfrac{2w(w-4)}{3}$ or $\dfrac{2w^2-8w}{3}$ 32. $\dfrac{5(n+3)}{8}$ or $\dfrac{5n+15}{8}$

33. $\dfrac{5c+18}{30}$ 34. $\dfrac{3R-5}{12}$ 35. $\dfrac{5y}{12}$

36. $\dfrac{21w+3}{10}$ or $\dfrac{3(7w+1)}{10}$ 37. $\dfrac{5c+1}{6}$ 38. $\dfrac{15+xy}{5y}$

39. $\dfrac{4y-6x}{3xy}$ or $\dfrac{2(2y-3x)}{3xy}$ 40. $\dfrac{3m-7}{20m}$ 41. $\dfrac{3w-2}{3}$

42. $\dfrac{7+20c}{5c}$ 43. y 44. $\dfrac{7}{8}B$ or $\dfrac{7B}{8}$ 45. $\dfrac{17}{4}w$ or $\dfrac{17w}{4}$

46. $\dfrac{9}{5}x$ or $\dfrac{9x}{5}$ 47. $1.75y$ 48. $0.2m$

EXERCISES 14.6

Evaluate for the given values of the variable.

1. $\dfrac{40}{w}$ for $w = 8$

2. $\dfrac{c}{120}$ for $c = -30$

3. $\dfrac{-75}{3x}$ for $x = 50$

4. $\dfrac{5y}{-45}$ for $y = -24$

5. $\dfrac{w+5}{10}$ for $w=2$ **6.** $\dfrac{y-6}{y}$ for $y=4$

7. $\dfrac{6c-5}{3c}$ for $c=-2$ **8.** $\dfrac{5(m+6)}{36}$ for $m=21$

Simplify.

9. $\dfrac{84}{60}$ **10.** $\dfrac{56}{104}$ **11.** $\dfrac{18x}{27}$ **12.** $\dfrac{30}{24w}$

13. $\dfrac{15ac}{4a}$ **14.** $\dfrac{7v^2}{8v}$ **15.** $\dfrac{48mc}{52c^2}$ **16.** $\dfrac{63m}{14mn}$

17. $\dfrac{5(x+1)}{15}$ **18.** $\dfrac{x(2x-5)}{2x}$ **19.** $\dfrac{4m(m+8)}{8m}$ **20.** $\dfrac{14v(3v-1)}{21v^2}$

21. $\dfrac{75(3-c)}{9c}$ **22.** $\dfrac{8mn(7+4m)}{32}$ **23.** $\dfrac{5c-10}{10c}$ **24.** $\dfrac{35-14m}{14}$

25. $\dfrac{6c-24}{9}$ **26.** $\dfrac{3xy+6}{3y}$ **27.** $\dfrac{x^2+x}{8x}$ **28.** $\dfrac{8xy-y}{2xy}$

29. $\dfrac{15ac-5c}{15ac}$ **30.** $\dfrac{12m^2+8m}{2mn}$

Multiply or divide as indicated and simplify.

31. $\dfrac{6}{8} \cdot \dfrac{3}{9}$ **32.** $\dfrac{24}{18} \cdot \dfrac{15}{25}$ **33.** $\dfrac{-8}{9} \div \dfrac{16}{21}$ **34.** $\dfrac{-60}{85} \div \dfrac{-12}{34}$

35. $\dfrac{5}{n} \cdot \dfrac{m}{4}$ **36.** $\dfrac{x}{y} \cdot \dfrac{2x}{m}$ **37.** $\dfrac{7n}{2} \div \dfrac{5m}{3}$ **38.** $\dfrac{3m}{n} \div \dfrac{1}{2m}$

39. $\dfrac{7m}{14} \cdot \dfrac{m}{3}$ **40.** $\dfrac{15}{x} \cdot \dfrac{2y}{3}$ **41.** $\dfrac{1}{5a} \cdot \dfrac{3ac}{10}$ **42.** $\dfrac{3}{7} \cdot \dfrac{8m}{m^2}$

43. $\dfrac{30}{11x} \div \dfrac{12y}{5x}$ **44.** $\dfrac{n}{21} \div \dfrac{m^2}{14mn}$ **45.** $\dfrac{xy}{45x} \div \dfrac{y^2}{36}$

46. $\dfrac{20c}{8} \div \dfrac{25c}{m}$ **47.** $\dfrac{27m}{3n} \cdot 20mn$ **48.** $50x^2 \cdot \dfrac{9}{75xy}$

49. $\dfrac{32ab}{28a} \div b^2$ **50.** $\dfrac{200m}{40m^2} \div 7mx$ **51.** $\dfrac{x+4}{4} \cdot \dfrac{1}{x}$

52. $\dfrac{5}{y} \cdot \dfrac{3y-1}{6}$ **53.** $\dfrac{w(w+4)}{8w} \cdot \dfrac{6}{5}$ **54.** $\dfrac{12(n-2)}{20n} \cdot \dfrac{n^2}{9}$

55. $\dfrac{4a-4}{24} \cdot \dfrac{3}{a}$ **56.** $\dfrac{8x}{3} \cdot \dfrac{2xy-3y}{4y}$ **57.** $\dfrac{49}{2m} \div \dfrac{21}{m^2-m}$

58. $\dfrac{20a+50}{5a} \div \dfrac{8}{2a}$ **59.** $\dfrac{2nm-4m}{6n} \div \dfrac{2mn}{9}$ **60.** $\dfrac{6}{y^2} \div \dfrac{6x^2+3x}{4xy}$

Add or subtract as indicated and simplify.

61. $\dfrac{7}{30} + \dfrac{11}{60}$ **62.** $\dfrac{17}{45} - \dfrac{8}{9}$ **63.** $\dfrac{13}{20} - \dfrac{7}{16}$ **64.** $\dfrac{13}{27} + \dfrac{7}{18}$

65. $\dfrac{2x}{5} + \dfrac{x}{5}$ **66.** $\dfrac{4y}{9} - \dfrac{5}{9}$ **67.** $\dfrac{2}{3} - \dfrac{a}{9}$ **68.** $\dfrac{y}{8} + \dfrac{5}{2}$

69. $\dfrac{y}{12} - \dfrac{y}{20}$ **70.** $\dfrac{m}{15} + \dfrac{2m}{9}$ **71.** $\dfrac{x}{5} + \dfrac{1}{3} - \dfrac{2x}{15}$

72. $\dfrac{3y}{7} - \dfrac{9}{2} + \dfrac{y}{14}$
73. $\dfrac{3}{x} - \dfrac{5}{x}$
74. $\dfrac{y}{w} + \dfrac{3}{w}$

75. $\dfrac{4}{a} + \dfrac{2}{3}$
76. $\dfrac{7}{c} - \dfrac{2}{5}$
77. $\dfrac{1}{5y} - \dfrac{5}{4}$

78. $\dfrac{3}{10} + \dfrac{5}{2w}$
79. $\dfrac{3}{x} + \dfrac{2}{y}$
80. $\dfrac{5}{w} - \dfrac{1}{3x}$

81. $\dfrac{m}{2x} + \dfrac{2}{3} + \dfrac{m}{6x}$
82. $\dfrac{2}{5} - \dfrac{3}{2w} + \dfrac{1}{10}$
83. $\dfrac{3w}{5} - 2$

84. $1 + \dfrac{y}{7}$
85. $\dfrac{8}{x} + 1$
86. $3 - \dfrac{2}{5x}$

87. $\dfrac{3x}{10} + \dfrac{x}{5} - 1$
88. $\dfrac{5}{2} + \dfrac{3}{2x} - 2$

Combine like terms.

89. $\dfrac{2}{3}x + \dfrac{4}{3}x$
90. $\dfrac{7}{2}y - \dfrac{6}{2}y$
91. $\dfrac{4}{5}m - \dfrac{1}{2}m$

92. $\dfrac{5}{6}n + \dfrac{1}{2}n$
93. $\dfrac{7}{8}v - 3v$
94. $4x + \dfrac{2}{5}x$

95. $a + \dfrac{5}{6}a$
96. $c + \dfrac{1}{8}c$
97. $y - \dfrac{4}{5}y$

98. $w - \dfrac{1}{3}w$
99. $3.2n + 4.1n$
100. $0.87x + 1.3x$

101. $3x - 0.45x$
102. $2.4n - 5n$
103. $y + 0.45y$

104. $w + 1.5w$
105. $w - 0.65w$
106. $y - 0.07y$

14.7 SOLVING PROPORTIONS AND OTHER EQUATIONS WITH FRACTIONS

OBJECTIVES
- Solve a proportion for the variable.
- Solve an application that involves a proportion.
- Solve an equation that involves fractions or decimals.
- Solve an application that involves an equation with fractions or decimals.

In this section we look at solving equations that involve fractions and decimals. We also solve applications that lead to solving these types of equations.

Try These Problems

Solve for the variable.

1. $\dfrac{x}{35} = \dfrac{40}{56}$
2. $\dfrac{y}{20} = \dfrac{21}{100}$
3. $\dfrac{1.5}{8} = \dfrac{w}{48}$
4. $\dfrac{2.4}{50} = \dfrac{c}{75}$

1 Solving Proportions

A **proportion** is a statement that two fractions, ratios, or rates are equal. For example, here we list some true proportions:

$$\frac{5}{10} = \frac{1}{2}$$

$$\frac{150 \text{ miles}}{3 \text{ hours}} = \frac{50 \text{ miles}}{1 \text{ hour}}$$

Not only does $\frac{150}{3} = \frac{50}{1}$, but also the rate unit, $\frac{\text{miles}}{\text{hour}}$, is the same on both sides.

$$\frac{\$2.50}{2 \text{ pounds}} = \frac{\$5}{4 \text{ pounds}}$$

Note that $\frac{2.50}{2} = \frac{5}{4}$ because both fractions equal 1.25. Also, the rate unit, $\frac{\$}{\text{pounds}}$, is the same on both sides.

If one of the quantities in a proportion is missing, we can solve for the unknown quantity by using the rules for solving an equation. The procedure is illustrated in the following examples.

EXAMPLE 1 Solve for N: $\dfrac{N}{15} = \dfrac{16}{40}$

SOLUTION We want to leave the variable N on the left side by itself, so we need to undo the division by 15. Multiplying by 15 will undo division by 15, so we multiply both sides by 15.

$$\frac{N}{15} = \frac{16}{40}$$

$$\frac{15}{1} \cdot \frac{N}{15} = \frac{16}{40} \cdot \frac{15}{1} \qquad \text{Multiply both sides by 15.}$$

$$N = \frac{(16)(15)}{40}$$

$$N = 6$$

Try Problems 1 and 2.

EXAMPLE 2 Solve for w: $\dfrac{3.5}{8} = \dfrac{w}{120}$

SOLUTION We want to leave the variable w on the right side by itself, so we need to undo the division by 120. Multiplying by 120 will undo division by 120, so we multiply both sides by 120.

$$\frac{3.5}{8} = \frac{w}{120}$$

$$\frac{120}{1} \cdot \frac{3.5}{8} = \frac{w}{120} \cdot \frac{120}{1} \qquad \text{Multiply both sides by 120.}$$

$$\frac{(120)(3.5)}{8} = w$$

$$52.5 = w$$

Try Problems 3 and 4.

▲ **Try These Problems**

Solve for the variable.

5. $\dfrac{6}{n} = \dfrac{15}{20}$

6. $\dfrac{1.02}{9} = \dfrac{5.1}{m}$

Consider the equal fractions $\tfrac{3}{4}$ and $\tfrac{6}{8}$.

$$\frac{3}{4} = \frac{6}{8}$$

If we invert each of the fractions, the resulting fractions are equal. For example,

$$\frac{4}{3} = \frac{8}{6}$$

In general, *if two fractions are equal, then their reciprocals are also equal.* We can use this to solve a proportion when the variable is in the denominator.

EXAMPLE 3   Solve for m: $\dfrac{6}{9} = \dfrac{14}{m}$

SOLUTION   Observe that the variable, m, is in the denominator on the right side of the equation. This equation would be easier to solve if the variable was in the numerator. The solution to the equation will not change if each fraction is inverted because if two fractions are equal, their reciprocals are also equal.

$$\frac{6}{9} = \frac{14}{m}$$

$$\frac{9}{6} = \frac{m}{14} \qquad \text{Invert each fraction so that the variable, } m, \text{ is in the numerator.}$$

$$\frac{14}{1} \cdot \frac{9}{6} = \frac{m}{14} \cdot \frac{14}{1} \qquad \text{Multiply both sides by 14.}$$

$$\frac{(14)(9)}{6} = m$$

$$21 = m \ \blacksquare$$

EXAMPLE 4   Solve for a: $\dfrac{8.25}{a} = \dfrac{7.2}{24}$

SOLUTION   Because the variable is in the denominator, we rewrite the proportion so that each fraction is inverted.

$$\frac{8.25}{a} = \frac{7.2}{24}$$

$$\frac{a}{8.25} = \frac{24}{7.2} \qquad \text{Invert each fraction so that the variable, } a, \text{ is in the numerator.}$$

$$\frac{8.25}{1} \cdot \frac{a}{8.25} = \frac{24}{7.2} \cdot \frac{8.25}{1} \qquad \text{Multiply both sides by 8.25.}$$

$$a = \frac{24(8.25)}{7.2}$$

$$a = 27.5 \ \blacksquare$$

▲ **Try Problems 5 and 6.**

14.7 Solving Proportions and Other Equations with Fractions

▲ Try These Problems

Solve for the variable.

7. $\dfrac{2\frac{1}{4}}{25} = \dfrac{B}{800}$

8. $\dfrac{c}{6} = \dfrac{3\frac{1}{2}}{7\frac{1}{2}}$

9. $\dfrac{40\frac{1}{2}}{c} = \dfrac{27}{\frac{1}{20}}$

10. $\dfrac{8\frac{2}{3}}{100} = \dfrac{6\frac{1}{2}}{c}$

EXAMPLE 5 Solve for c: $\dfrac{\frac{2}{3}}{80} = \dfrac{c}{400}$

SOLUTION We want to leave the variable, c, on the right side so we need to undo the division by 400. Multiplying by 400 will undo division by 400, so we multiply both sides by 400.

$$\dfrac{\frac{2}{3}}{80} = \dfrac{c}{400}$$

$$\dfrac{\overset{5}{\cancel{400}}}{1} \cdot \dfrac{\frac{2}{3}}{\underset{1}{\cancel{80}}} = \dfrac{c}{\underset{1}{\cancel{400}}} \cdot \dfrac{\overset{1}{\cancel{400}}}{1} \qquad \text{Multiply both sides by 400.}$$

$$5 \cdot \dfrac{2}{3} = c$$

$$c = \dfrac{10}{3} \quad \text{or} \quad 3\tfrac{1}{3} \quad \blacksquare$$

▲ Try Problems 7 and 8.

EXAMPLE 6 Solve for n: $\dfrac{2\frac{2}{3}}{n} = \dfrac{15}{1\frac{1}{4}}$

SOLUTION Because the variable, n, is in the denominator, we invert each fraction.

$$\dfrac{2\frac{2}{3}}{n} = \dfrac{15}{1\frac{1}{4}}$$

$$\dfrac{n}{2\frac{2}{3}} = \dfrac{1\frac{1}{4}}{15} \qquad \text{Invert each fraction so that the variable, } n, \text{ is in the numerator.}$$

$$\dfrac{\overset{1}{\cancel{2\frac{2}{3}}}}{1} \cdot \dfrac{n}{\underset{1}{\cancel{2\frac{2}{3}}}} = \dfrac{1\frac{1}{4}}{15} \cdot \dfrac{2\frac{2}{3}}{1} \qquad \text{Multiply both sides by } 2\tfrac{2}{3}.$$

$$n = \dfrac{\overset{5}{\cancel{1\tfrac{1}{4}}} \cdot \overset{2}{\cancel{\tfrac{8}{3}}}}{15}$$

$$n = \dfrac{\frac{10}{3}}{15}$$

$$n = \dfrac{10}{3} \div 15$$

$$n = \dfrac{\overset{2}{\cancel{10}}}{3} \cdot \dfrac{1}{\underset{3}{\cancel{15}}}$$

$$n = \dfrac{2}{9} \quad \blacksquare$$

▲ Try Problems 9 and 10.

2 Solving Applications by Using Proportions

Any situation that involves equal ratios or equivalent rates can be solved by setting up a proportion. Here are some examples.

Try These Problems

Set up a proportion and solve.

11. The sales tax on an $80 purchase is $6.40. At this rate, what is the sales tax on a $350 purchase?

12. A gardener fertilizes 200 square feet of lawn with 3 pounds of fertilizer. At this rate, how many square feet of lawn can be fertilized with 19.5 pounds of fertilizer?

EXAMPLE 7 The Smith family pays $300 in property tax on their home, which is worth $75,000. At this rate, what is the property tax on a home worth $120,000?

SOLUTION We want the rate

$$\frac{\text{property tax}}{\text{home value}}$$

to be the same in both cases.

Let n = the property tax for the home worth $120,000.

Set up a proportion that states the two rates are equal and solve for the variable.

$$\frac{n}{120{,}000} = \frac{300}{75{,}000} \quad \begin{array}{l}\leftarrow \text{Property tax in both numerators} \\ \leftarrow \text{Home value in both denominators}\end{array}$$

$$\frac{\cancel{120{,}000}^{1}}{1} \cdot \frac{n}{\cancel{120{,}000}_{1}} = \frac{300}{75{,}000} \cdot \frac{120{,}000}{1} \quad \text{Multiply both sides by 120,000.}$$

$$n = \frac{300(120{,}000)}{75{,}000}$$

$$n = 480$$

The property tax on a home worth $120,000 is $480. ∎

Try Problem 11.

EXAMPLE 8 The label on a bottle of concentrated rug shampoo says to mix 8 cups of water with 3 tablespoons of the concentrated rug shampoo. At this rate, how many tablespoons of concentrated rug shampoo should be mixed with 10 cups of water?

SOLUTION We want the rate

$$\frac{\text{tablespoons of shampoo}}{\text{cups of water}}$$

to be the same in both cases.

Let x = how many tablespoons of concentrated shampoo are needed for 10 cups of water.

Set up a proportion that states the two rates are equal and solve for the variable.

$$\frac{x}{10} = \frac{3}{8} \qquad \text{Here the proportion is written so that the tablespoons of shampoo are in the numerator and the cups of water are in the denominator.}$$

$$\frac{\cancel{10}^{1}}{1} \cdot \frac{x}{\cancel{10}_{1}} = \frac{3}{8} \cdot \frac{10}{1} \qquad \text{Multiply both sides by 10.}$$

$$x = \frac{30}{8}$$

$$x = \frac{15}{4} \quad \text{or} \quad 3\frac{3}{4} \quad \text{or} \quad 3.75$$

3.75 tablespoons of concentrated rug shampoo should be mixed with 10 cups of water. ∎

Try Problem 12.

EXAMPLE 9 Farmer Jones grows acres of cotton and acres of corn in the ratio 3 to 5. If he grows 75 acres of corn, how many acres of cotton does he grow?

Try These Problems

Set up a proportion and solve.

13. A truck load contains crates of apples and crates of oranges in the ratio 2 to 7. If there are 175 crates of oranges, how many crates of apples are there?

14. A cable television channel shows drama movies and comedy movies in the ratio 4 to 5. If 92 drama movies were shown last month, how many comedies were shown?

SOLUTION The first sentence says that the ratio of the acres of cotton to the acres of corn is 3 to 5. That is,

$$\frac{\text{acres of cotton}}{\text{acres of corn}} = \frac{3}{5}$$

The acres of cotton correspond to 3 and the acres of corn correspond to 5.

Let y = the number of acres of cotton.

75 = the number of acres of corn.

Set up a proportion that states the two ratios are equal and solve for the variable.

$$\frac{y}{75} = \frac{3}{5} \qquad \frac{\text{acres of cotton}}{\text{acres of corn}}$$

$$\frac{\cancel{75}^1}{1} \cdot \frac{y}{\cancel{75}_1} = \frac{3}{5} \cdot \frac{75}{1} \qquad \text{Multiply both sides by 75.}$$

$$y = \frac{3(75)}{5}$$

$$y = 45$$

He grows 45 acres of cotton. ∎

 Try Problem 13.

EXAMPLE 10 A disc jockey plays oldies-rock tunes and country-western tunes in the ratio 8 to 15. If the disc jockey played 104 oldies-rock tunes, how many country-western tunes did she play?

SOLUTION The first sentence says that the ratio of oldies-rock tunes to country-western tunes is 8 to 15. That is,

$$\frac{\text{oldies-rock tunes}}{\text{country-western tunes}} = \frac{8}{15}$$

The number of oldies-rock tunes corresponds to 8 and the number of country-western tunes corresponds to 15.

Let R = the number of country-western tunes.

104 = the number of oldies-rock tunes.

Set up a proportion that states the two ratios are equal and solve for the variable.

$$\frac{104}{R} = \frac{8}{15} \qquad \frac{\text{oldies-rock tunes}}{\text{country-western tunes}}$$

$$\frac{R}{104} = \frac{15}{8} \qquad \text{Invert each fraction so that the variable, } R, \text{ is in the numerator.}$$

$$\frac{\cancel{104}^1}{1} \cdot \frac{R}{\cancel{104}_1} = \frac{15}{8} \cdot \frac{104}{1} \qquad \text{Multiply both sides by 104.}$$

$$R = \frac{15(104)}{8}$$

$$R = 195$$

The disc jockey played 195 country-western tunes. ∎

 Try Problem 14.

Try These Problems

Set up a proportion and solve.

15. On a map, $\frac{3}{4}$ inch represents 125 miles. How many miles are represented by $3\frac{3}{4}$ inches?

16. A recipe calls for $3\frac{1}{2}$ cups of flour for 8 servings. How many cups of flour are needed for 12 servings?

Solve for the variable.

17. $\frac{y}{5} = 20$

18. $12\frac{1}{3} = \frac{x}{6}$

EXAMPLE 11 A rectangular photograph is to be enlarged so that the ratio of the width to the height remains the same. The photograph was $3\frac{1}{2}$ inches wide and 6 inches high. The height of the enlargement is 8 inches. Find the width of the enlargement.

SOLUTION We want the ratio of the width to the height to remain the same for the original photograph and the enlargement. That is,

$$\frac{\text{original width}}{\text{original height}} = \frac{\text{width of enlargement}}{\text{height of enlargement}}$$

Let w = width of the enlargement (in).

8 in = height of the enlargement.

Set up a proportion that states the two ratios are equal and solve for the variable.

$$\frac{3\frac{1}{2}}{6} = \frac{w}{8} \qquad \frac{\text{width}}{\text{height}}$$

$$\frac{8}{1} \cdot \frac{3\frac{1}{2}}{6} = \frac{w}{\overset{1}{\cancel{8}}} \cdot \frac{\overset{1}{\cancel{8}}}{1} \qquad \text{Multiply both sides by 8.}$$

$$\frac{\overset{4}{\cancel{8}}}{1} \cdot \frac{7}{\underset{1}{\cancel{2}}} = w \qquad \text{Convert } 3\frac{1}{2} \text{ to } \frac{7}{2}.$$

$$\frac{28}{6} = w$$

$$w = \frac{14}{3} \quad \text{or} \quad 4\frac{2}{3}$$

The width of the enlarged photograph is $4\frac{2}{3}$ inches.

Try Problems 15 and 16.

3 Solving Other Equations with Fractions and Decimals

A proportion is a very special type of equation: it simply states that two fractions are equal. Now we look at solving other equations that contain fractions or decimals. The following examples illustrate the procedure.

EXAMPLE 12 Solve for w: $\frac{w}{8} = 16$

SOLUTION We want to leave w on the left side by itself, so we need to undo the division by 8. Multiplying by 8 will undo the division by 8, so we multiply both sides by 8.

$$\frac{w}{8} = 16$$

$$\frac{\overset{1}{\cancel{8}}}{1} \cdot \frac{w}{\underset{1}{\cancel{8}}} = \frac{8}{1} \cdot \frac{16}{1} \qquad \text{Multiply both sides by 8.}$$

$$w = 128 \quad \blacksquare$$

Try Problems 17 and 18.

14.7 Solving Proportions and Other Equations with Fractions

 Try These Problems

Solve for the variable.

19. $\dfrac{4w}{7} = 24$

20. $90 = \dfrac{-3m}{4}$

EXAMPLE 13 Solve for x: $-75 = \dfrac{2x}{3}$

SOLUTION We want to leave x on the right side by itself, so we need to undo the division by 3 and the multiplication by 2. We begin by undoing the division by 3.

$$-75 = \dfrac{2x}{3}$$

$$3(-75) = \dfrac{\cancel{3}}{1} \cdot \dfrac{2x}{\cancel{3}} \quad \text{Multiply both sides by 3.}$$

$$-225 = 2x$$

$$\dfrac{-225}{2} = \dfrac{2x}{2} \quad \text{Divide both sides by 2.}$$

$$x = \dfrac{-225}{2} \quad \text{or} \quad -112\dfrac{1}{2} \quad \text{or} \quad -112.5 \quad \blacksquare$$

 Try Problems 19 and 20.

EXAMPLE 14 Solve for y: $6\dfrac{2}{3} = \dfrac{3}{5}y$

SOLUTION The right side of the equation, $\dfrac{3}{5}y$, is the same as $\dfrac{3y}{5}$ because

$$\dfrac{3}{5}y = \dfrac{3}{5} \cdot \dfrac{y}{1} = \dfrac{3y}{5}$$

Viewing the expression this way makes it easier to see that we need to undo the division by 5 and the multiplication by 3 to leave y on the right side by itself.

$$6\dfrac{2}{3} = \dfrac{3}{5}y$$

$$6\dfrac{2}{3} = \dfrac{3y}{5} \quad \text{The expression } \dfrac{3}{5}y \text{ equals } \dfrac{3y}{5}.$$

$$\dfrac{5}{1} \cdot 6\dfrac{2}{3} = \dfrac{\cancel{5}}{1} \cdot \dfrac{3y}{\cancel{5}} \quad \text{Multiply both sides by 5.}$$

$$\dfrac{5}{1} \cdot \dfrac{20}{3} = 3y$$

$$\dfrac{100}{3} = 3y$$

$$\dfrac{100}{3} \div 3 = \dfrac{3y}{3} \quad \text{Divide both sides by 3.}$$

$$\dfrac{100}{3} \cdot \dfrac{1}{3} = y \quad \text{Dividing by 3 is the same as multiplying by } \tfrac{1}{3}.$$

$$\dfrac{100}{9} = y$$

$$y = \dfrac{100}{9} \quad \text{or} \quad 11\dfrac{1}{9} \quad \blacksquare$$

Try These Problems

Solve for the variable.

21. $\frac{3}{4}m = 21$

22. $8\frac{1}{3} = \frac{5}{6}A$

23. $0.3R = 231$

24. $1.02 = 0.75u$

 Try Problems 21 and 22.

EXAMPLE 15 Solve for B: $0.8B = 96$

SOLUTION We show this equation solved in two ways.

METHOD 1 Here we undo the multiplication by 0.8 by dividing by 0.8. Of course, we must divide both sides by 0.8.

$$0.8B = 96$$
$$\frac{0.8B}{0.8} = \frac{96}{0.8} \quad \text{Divide both sides by 0.8.}$$
$$B = 120$$

METHOD 2 Here we begin by multiplying both sides by 10, because 10 times 0.8 is the whole number 8. Then, when we divide on both sides, we will be dividing by a whole number instead of a decimal. If you are working without a calculator, this procedure will be easier.

$$0.8B = 96$$
$$10(0.8B) = 10(96) \quad \text{Multiply both sides by 10 so that the equation has no decimals.}$$
$$8B = 960$$
$$\frac{8B}{8} = \frac{960}{8} \quad \text{Divide both sides by 8.}$$
$$B = 120 \quad \blacksquare$$

 Try Problems 23 and 24.

EXAMPLE 16 Solve for x: $\frac{x}{3} + 2 = \frac{8}{3}$

SOLUTION We solve this problem in two ways.

METHOD 1 We want the x term on the left side by itself, so we can begin by undoing the addition of 2. Therefore, we subtract 2 from both sides.

$$\frac{x}{3} + 2 = \frac{8}{3}$$
$$\frac{x}{3} + 2 - 2 = \frac{8}{3} - 2 \quad \text{Subtract 2 from both sides.}$$
$$\frac{x}{3} = \frac{8}{3} - \frac{2}{1}$$
$$\frac{x}{3} = \frac{8}{3} - \frac{2(3)}{1(3)} \quad \text{Write each number on the right side with a common denominator 3 before subtracting.}$$
$$\frac{x}{3} = \frac{8}{3} - \frac{6}{3}$$
$$\frac{x}{3} = \frac{2}{3}$$
$$\frac{3}{1} \cdot \frac{x}{3} = \frac{3}{1} \cdot \frac{2}{3} \quad \text{Multiply both sides by 3.}$$
$$x = 2$$

Try These Problems

Solve for the variable.

25. $\dfrac{y}{4} - 6 = 10$

26. $\dfrac{21}{2} = 8 + \dfrac{w}{2}$

METHOD 2 Another approach to solving this problem is to begin by undoing the division by 3 so that the resulting equation has no fractions. If this is done, we must multiply both sides by 3, which means that on the left side each term must be multiplied by 3.

$$\dfrac{x}{3} + 2 = \dfrac{8}{3}$$

$$3\left(\dfrac{x}{3} + 2\right) = \dfrac{3}{1} \cdot \dfrac{8}{3} \qquad \text{Multiply both sides by 3.}$$

$$\dfrac{3}{1} \cdot \dfrac{x}{3} + 3(2) = 8 \qquad \text{Each term on the left side must be multiplied by 3.}$$

$$x + 6 = 8 \qquad \text{Observe that multiplying both sides by 3 creates an equivalent equation without fractions.}$$

$$x + 6 - 6 = 8 - 6 \qquad \text{Subtract 6 from both sides.}$$

$$x = 2 \quad \blacksquare$$

Try Problems 25 and 26.

EXAMPLE 17 Solve for W: $\quad 1 = \dfrac{4}{15} + \dfrac{2}{5}W$

SOLUTION We show this problem done in two ways.

METHOD 1 If we want to leave the W term on the right side by itself, then we need to undo the addition of $\dfrac{4}{15}$. Therefore, we begin by subtracting $\dfrac{4}{15}$ from both sides.

$$1 = \dfrac{4}{15} + \dfrac{2}{5}W$$

$$1 - \dfrac{4}{15} = \dfrac{4}{15} - \dfrac{4}{15} + \dfrac{2}{5}W \qquad \text{Subtract } \dfrac{4}{15} \text{ from both sides.}$$

$$\dfrac{15}{15} - \dfrac{4}{15} = \dfrac{2}{5}W \qquad \text{Each number on the left side must be written with the common denominator 15 before subtracting. } 1 = \dfrac{15}{15}$$

$$\dfrac{11}{15} = \dfrac{2}{5}W$$

$$\dfrac{11}{15} = \dfrac{2W}{5} \qquad \text{The expression } \dfrac{2}{5}W \text{ equals } \dfrac{2W}{5}.$$

$$\dfrac{\cancel{5}^{1}}{1} \cdot \dfrac{11}{\cancel{15}_{3}} = \dfrac{\cancel{5}^{1}}{1} \cdot \dfrac{2W}{\cancel{5}_{1}} \qquad \text{Multiply both sides by 5.}$$

$$\dfrac{11}{3} = 2W$$

$$\dfrac{11}{3} \div 2 = \dfrac{2W}{2} \qquad \text{Divide both sides by 2.}$$

$$\dfrac{11}{3} \cdot \dfrac{1}{2} = W \qquad \text{Dividing by 2 is the same as multiplying by } \dfrac{1}{2}.$$

$$\dfrac{11}{6} = W$$

$$W = \dfrac{11}{6} \quad \text{or} \quad 1\dfrac{5}{6}$$

Try These Problems

Solve for the variable.

27. $\dfrac{3}{8}A + 1 = \dfrac{1}{4}$

28. $\dfrac{1}{3} + \dfrac{3}{2}B = 2$

METHOD 2 If we multiply both sides of the equation by the least common multiple of 5 and 15, which is 15, we will undo both of those divisions at once. The resulting equation will have no fractions.

$$1 = \frac{4}{15} + \frac{2}{5}W$$

$$1 = \frac{4}{15} + \frac{2W}{5} \qquad \text{The expression } \frac{2}{5}W \text{ equals } \frac{2W}{5}.$$

$$15(1) = \frac{15}{1} \cdot \frac{4}{15} + \frac{15}{1} \cdot \frac{2W}{5} \qquad \text{Multiply both sides by 15, the LCM of 5 and 15. Each term must be multiplied by 15.}$$

$$15 = 4 + 6W \qquad \text{Observe that after multiplying both sides by 15, the LCM of 5 and 15, we get an equation without fractions.}$$

$$15 - 4 = 4 - 4 + 6W \qquad \text{Subtract 4 from both sides.}$$

$$11 = 6W$$

$$\frac{11}{6} = \frac{6W}{6} \qquad \text{Divide both sides by 6.}$$

$$W = \frac{11}{6} \quad \text{or} \quad 1\frac{5}{6} \quad \blacksquare$$

 Try Problems 27 and 28.

EXAMPLE 18 Solve for L: $2L + \dfrac{2}{3}L = 120$

SOLUTION We show this equation solved in two ways.

METHOD 1 Because both terms on the left side of the equation have a common variable factor, L, we can combine these into a single term.

$$2L + \frac{2}{3}L = 120$$

$$\left(2 + \frac{2}{3}\right)L = 120 \qquad \text{Factor out the common factor } L.$$

$$\left(\frac{2 \cdot 3}{1 \cdot 3} + \frac{2}{3}\right)L = 120 \qquad \text{Write the number 2 as a fraction with denominator 3 so that we can add the fractions.}$$

$$\left(\frac{6}{3} + \frac{2}{3}\right)L = 120$$

$$\frac{8}{3}L = 120 \qquad \text{Add the fractions inside the parentheses.}$$

$$\frac{\cancel{3}^1}{1} \cdot \frac{8}{\cancel{3}_1}L = 3(120) \qquad \text{Multiply both sides by 3.}$$

$$8L = 360$$

$$\frac{8L}{8} = \frac{360}{8} \qquad \text{Divide both sides by 8.}$$

$$L = 45$$

Try These Problems

Solve for the variable.

29. $3x + \dfrac{3}{4}x = 45$

30. $N - \dfrac{5}{6}N = 330$

31. $40 + 0.8x = 56$

32. $0.12y + 3.3 = 6$

METHOD 2 We can also begin by multiplying both sides by 3. Because this undoes the division by 3, the resulting equation will not have fractions.

$$2L + \dfrac{2}{3}L = 120$$

$$3\left(2L + \dfrac{2}{3}L\right) = 3(120) \quad \text{Multiply both sides by 3.}$$

$$3(2L) + \dfrac{\cancel{3}^{1}}{1} \cdot \dfrac{2L}{\cancel{3}_{1}} = 360 \quad \text{Each term must be multiplied by 3.}$$

$$6L + 2L = 360 \quad \text{Observe that the result of multiplying both sides by 3 is an equation without fractions.}$$

$$8L = 360 \quad \text{Combine like terms on the left side.}$$

$$\dfrac{8L}{8} = \dfrac{360}{8} \quad \text{Divide both sides by 8.}$$

$$L = 45 \quad \blacksquare$$

Try Problems 29 and 30.

EXAMPLE 19 Solve for n: $0.25n - 3.5 = 15$

SOLUTION We show this problem solved in two ways.

METHOD 1 One way to begin is to add 3.5 to both sides.

$$0.25n - 3.5 = 15$$

$$0.25n - 3.5 + 3.5 = 15 + 3.5 \quad \text{Add 3.5 to both sides.}$$

$$0.25n = 18.5$$

$$\dfrac{0.25n}{0.25} = \dfrac{18.5}{0.25} \quad \text{Divide both sides by 0.25.}$$

$$n = 74$$

METHOD 2 Because 0.25 has 2 decimal places, we could begin by multiplying both sides by 100, which will give an equivalent equation without decimals.

$$0.25n - 3.5 = 15$$

$$100(0.25n - 3.5) = 100(15) \quad \text{Multiply both sides by 100.}$$

$$100(0.25n) - 100(3.5) = 1500 \quad \text{On the left side, each term must be multiplied by 100.}$$

$$25n - 350 = 1500 \quad \text{Observe that after multiplying both sides by 100, we get an equation without decimals.}$$

$$25n - 350 + 350 = 1500 + 350 \quad \text{Add 350 to both sides.}$$

$$25n = 1850$$

$$\dfrac{25n}{25} = \dfrac{1850}{25} \quad \text{Divide both sides by 25.}$$

$$n = 74 \quad \blacksquare$$

Try Problems 31 and 32.

 **Try These Problems**

Solve for the variable.
33. $1.4B - 0.2B = 51$
34. $C + 0.3C = 650$

EXAMPLE 20 Solve for A: $A + 0.25A = 400$

SOLUTION We show this equation solved in two ways.

METHOD 1 Because both terms on the left side have the same variable factor A, we can combine the two terms into a single term.

$$A + 0.25A = 400$$
$$1A + 0.25A = 400 \quad \text{The expression } A \text{ equals } 1A.$$
$$(1 + 0.25)A = 400 \quad \text{Factor out the common factor } A \text{ on the left side.}$$
$$1.25A = 400$$
$$\frac{1.25A}{1.25} = \frac{400}{1.25} \quad \text{Divide both sides by 1.25.}$$
$$A = 320$$

METHOD 2 There are 2 decimal places in 0.25, so if we multiply both sides by 100, the resulting equation will have no decimals.

$$A + 0.25A = 400$$
$$100(A + 0.25A) = 100(400) \quad \text{Multiply both sides by 100.}$$
$$100A + 100(0.25A) = 40{,}000 \quad \text{Each term must be multiplied by 100.}$$
$$100A + 25A = 40{,}000$$
$$125A = 40{,}000 \quad \text{Combine like terms on the left side.}$$
$$\frac{125A}{125} = \frac{40{,}000}{125} \quad \text{Divide both sides by 125.}$$
$$A = 320 \quad \blacksquare$$

 Try Problems 33 and 34.

4 Solving More Applications

Now we look at some applications that require setting up equations involving fractions or decimals.

EXAMPLE 21 Ms. Gamez pays 0.4 of her yearly salary in taxes. If she paid $20,800 in taxes last year, find her salary for that year.

SOLUTION There is only one unknown quantity in this problem, her salary for last year. Therefore, we introduce a variable to represent her salary last year.

Let $S = $ her salary last year ($).

The first sentence translates to a multiplication statement, therefore translates to an equation. It says the following:

$$0.4 \text{ of } \underbrace{\text{her salary (\$)}}_{S} \text{ is } \underbrace{\text{her taxes (\$)}}_{20{,}800}$$

$$0.4 \cdot S = 20{,}800$$

Try These Problems

Set up an equation and solve.

35. Sales tax in New York is 0.08 of the marked price. Mr. Ramirez purchased a car and paid $1000 in sales tax. Find the marked price of the car.

36. Eighteen hundred men attended a lawyer's convention. This represented $\frac{2}{3}$ of the total attendance. Find the total attendance.

37. A taxi charges $1.25, plus $0.45 per mile. How far did you travel in the taxi if your fare was $3.95?

38. Ms. Hamilton sells linens in a large department store. She earns $1200 each month, plus 0.01 of her total sales. If she earned a total of $1975 in July, what were her total sales for that month?

Now we solve this equation for S.

$$0.4S = 20{,}800$$
$$\frac{0.4S}{0.4} = \frac{20{,}800}{0.4} \quad \text{Divide both sides by 0.4.}$$
$$S = 52{,}000$$

Ms. Gamez earned $52,000 last year. ■

 Try Problems 35 and 36.

EXAMPLE 22 A taxi charges $0.95, plus $0.40 per mile. If Wade's taxi fare is $3.15, how far did he travel?

SOLUTION The only unknown quantity in this problem is how far Wade went in the taxi. Therefore, we introduce a variable to represent this distance.

Let d = how far Wade went in the taxi (mi).

The first sentence explains how the fare is computed. We use this to write an equation.

$$\text{Total fare(\$)} = \$0.95 + \$0.40 \cdot \underbrace{\text{how many miles Wade went}}_{d}$$

$$3.15 = 0.95 + 0.40 \cdot d$$
$$3.15 = 0.95 + 0.40d$$

Now we solve the equation for d.

$$3.15 = 0.95 + 0.40d$$
$$100(3.15) = 100(0.95 + 0.40d) \quad \text{Multiply both sides by 100.}$$
$$315 = 100(0.95) + 100(0.40d) \quad \text{Each term must be multiplied by 100.}$$
$$315 = 95 + 40d$$
$$315 - 95 = 95 - 95 + 40d \quad \text{Subtract 95 from both sides.}$$
$$220 = 40d$$
$$\frac{220}{40} = \frac{40d}{40} \quad \text{Divide both sides by 40.}$$
$$\frac{11}{2} = d$$
$$d = \frac{11}{2} \quad \text{or} \quad 5\frac{1}{2} \quad \text{or} \quad 5.5.$$

Wade went 5.5 miles in the taxi. ■

 Try Problems 37 and 38.

Try These Problems

Set up an equation and solve.

39. Cindy made 87, 73, 83, and 65 on the first four chemistry exams. What score does she need on the fifth exam to have an average (or mean) of 80?

EXAMPLE 23   Roger made 78, 88, and 96 on the first three math exams. What score does he need on the fourth exam to have an average (or mean) of 90?

SOLUTION   The only unknown quantity in this problem is Roger's score on his fourth math exam. We introduce a variable to be this fourth score.

Let x = his score on the fourth math exam.

To compute his average, we would add up all four scores and divide by how many scores there are. We use this to write an equation.

$$\text{Average} = \frac{\text{sum of all scores}}{\text{the number of scores}}$$

$$90 = \frac{78 + 88 + 96 + x}{4}$$

Now we solve this equation for x.

$$90 = \frac{78 + 88 + 96 + x}{4}$$

$$90 = \frac{262 + x}{4} \qquad \text{Add 78, 88, and 96 to obtain 262.}$$

$$4(90) = \frac{4}{1} \cdot \frac{262 + x}{4} \qquad \text{Multiply both sides by 4.}$$

$$360 = 262 + x$$

$$360 - 262 = 262 - 262 + x \qquad \text{Subtract 262 from both sides.}$$

$$98 = x$$

Roger needs a score of 98 on the fourth exam. ■

 Try Problem 39.

EXAMPLE 24   The Santos family installed fiberglass insulation in their attic that will reduce their monthly gas bill by 15%. If their monthly gas bill after the insulation was installed is $34, what was their original bill?

SOLUTION   First, you must realize what it means for the gas bill to be *reduced by 15%*. Here we explain.

$$\begin{array}{c}\text{Amount} \\ \text{of decrease}\end{array} = 15\% \text{ of } \begin{array}{c}\text{the original} \\ \text{bill}\end{array}$$

We do not know the amount of decrease, and we do not know the original bill. We should let a variable represent one of these unknown quantities. Because the amount of decrease is expressed in terms of the original bill, we should let the variable represent the original bill. Then we can write an algebraic expression to represent the amount of decrease using the variable we have introduced.

Let B = the original gas bill ($).

$0.15B$ = amount of decrease ($).

We can write an equation by using the following relationship.

$$\underbrace{\begin{array}{c}\text{Original} \\ \text{bill}\end{array}}_{B} - \underbrace{\text{decrease}}_{0.15B} = \underbrace{\begin{array}{c}\text{final} \\ \text{bill}\end{array}}_{34}$$

Try These Problems

Set up an equation and solve.

40. A suit sold for $147 after a 30% discount off the original marked price. Find the original price of the suit.

41. The price of chicken has increased 225% over the past few years. If the price now is $3.90 per pound, what was the price a few years ago?

Now we solve this equation for B.

$$B - 0.15B = 34$$
$$100(B - 0.15B) = 100(34)$$ Multiply both sides by 100.
$$100B - 100(0.15B) = 3400$$ Each term must be multiplied by 100.
$$100B - 15B = 3400$$
$$85B = 3400$$ Combine like terms on the left side.
$$\frac{85B}{85} = \frac{3400}{85}$$ Divide both sides by 85.
$$B = 40$$

The original bill was $40. ∎

 Try Problems 40 and 41.

 Answers to Try These Problems

1. $x = 25$ 2. $y = \frac{21}{5}$ or $4\frac{1}{5}$ or 4.2 3. $w = 9$
4. $c = \frac{18}{5}$ or $3\frac{3}{5}$ or 3.6 5. $n = 8$ 6. $m = 45$
7. $B = 72$ 8. $c = \frac{14}{5}$ or $2\frac{4}{5}$ or 2.8
9. $c = \frac{3}{40}$ or 0.075 10. $c = 75$ 11. $28   12. 1300 sq ft
13. 50 crates of apples   14. 115 comedies   15. 625 mi
16. $5\frac{1}{4}$ c   17. $y = 100$ 18. $x = 74$ 19. $w = 42$
20. $m = -120$ 21. $m = 28$ 22. $A = 10$ 23. 770
24. $u = 1.36$ 25. $y = 64$ 26. $w = 5$ 27. $A = -2$
28. $B = \frac{10}{9}$ or $1\frac{1}{9}$ 29. $x = 12$ 30. $N = 1980$ 31. $x = 20$
32. $y = 22.5$ 33. $B = 42.5$ 34. $C = 500$ 35. $12,500
36. 2700   37. 6 mi   38. $77,500   39. 92   40. $210
41. $1.20 per lb

EXERCISES 14.7

Solve for the variable.

1. $\dfrac{x}{52} = \dfrac{2}{13}$ 2. $\dfrac{x}{7} = \dfrac{270}{63}$ 3. $\dfrac{1}{12} = \dfrac{x}{8}$

4. $\dfrac{45}{6} = \dfrac{x}{5}$ 5. $\dfrac{2}{90} = \dfrac{52}{w}$ 6. $\dfrac{36}{m} = \dfrac{45}{6}$

7. $\dfrac{P}{160} = \dfrac{7.5}{20}$ 8. $\dfrac{1.24}{4} = \dfrac{P}{15}$ 9. $\dfrac{A}{0.002} = \dfrac{7000}{0.8}$

10. $\dfrac{54}{40.8} = \dfrac{A}{3.4}$ 11. $\dfrac{51}{w} = \dfrac{0.3}{0.05}$ 12. $\dfrac{2.5}{w} = \dfrac{0.2}{6}$

13. $\dfrac{\frac{4}{5}}{80} = \dfrac{c}{400}$ 14. $\dfrac{c}{38} = \dfrac{\frac{5}{16}}{95}$ 15. $\dfrac{c}{12} = \dfrac{7\frac{1}{2}}{2}$

16. $\dfrac{2\frac{1}{7}}{270} = \dfrac{P}{42}$ 17. $\dfrac{3}{12\frac{3}{5}} = \dfrac{1\frac{2}{3}}{w}$ 18. $\dfrac{10}{w} = \dfrac{2\frac{1}{5}}{7\frac{7}{10}}$

Set up a proportion and solve.

19. A cable weighs 5 pounds for every 12 feet. What is the weight of 9 feet of this cable?

20. A car weighing 3500 pounds is taxed $18. At this rate, how much should a 4200-pound car be taxed?

21. An Arkansas farmer plants cotton and rice in the ratio 7 to 10. If 91 acres of cotton are planted, how many acres of rice are planted?
22. A basketball team finished the season with wins to losses in the ratio of 8 to 5. If the team lost 45 games, how many games did they win?
23. The height of a building and the length of its shadow are in the ratio 50 to 7. If the shadow is 4.2 feet long, how high is the building?
24. Five pounds of apples cost $4.40. What is the cost of 3.5 pounds?
25. Your car burns 2.5 quarts of oil on a 1500-mile trip. At this rate, how many quarts will your car burn on a 375-mile trip?
26. Gerson paid 314 francs for dinner in Paris. If the exchange rate is 6.28 francs to 1 dollar, what was his dinner bill in United States currency?
27. A recipe calls for $2\frac{1}{4}$ cups of milk for 6 persons. How much milk is needed for 8 persons?
28. On a map $\frac{1}{2}$ inch represents 25 miles. How many miles are represented by $6\frac{1}{4}$ inches?
29. A string is cut into two pieces that are in the ratio 2 to 5. If the longer piece is $2\frac{2}{3}$ yards long, how long is the shorter piece?
30. The length and width of a rectangle are in the ratio 3 to 2. If the length is $8\frac{1}{3}$ feet, find the width.

Solve for the variable.

31. $\frac{x}{6} = 12$
32. $\frac{3}{5} = \frac{x}{8}$
33. $\frac{x}{7} = 2\frac{1}{3}$
34. $4.8 = \frac{x}{5}$
35. $\frac{5w}{36} = 25$
36. $42 = \frac{7w}{2}$
37. $\frac{2}{3} = 5w$
38. $\frac{5}{8}w = 100$
39. $24 = \frac{3}{10}m$
40. $\frac{3}{4}m = 5\frac{1}{2}$
41. $36.2 = 4m$
42. $3m = 0.045$
43. $0.6y = 150$
44. $36 = 2.4y$
45. $0.25y = 17.5$
46. $3.43 = 0.35y$
47. $\frac{c}{3} - 2 = 10$
48. $5 + \frac{c}{2} = 3$
49. $\frac{c}{5} + 1 = 2\frac{1}{5}$
50. $\frac{c}{4} - 3 = \frac{5}{4}$
51. $\frac{3}{8}a + 2 = \frac{1}{4}$
52. $\frac{2}{3} + \frac{1}{6}a = 5$
53. $\frac{1}{2} = \frac{2}{3}a - 4$
54. $\frac{5}{2} = \frac{4}{5}a + \frac{1}{2}$
55. $\frac{1}{2}B + 3B = 49$
56. $85 = 5B - \frac{3}{4}B$
57. $B - \frac{3}{5}B = 160$
58. $B + \frac{7}{8}B = 225$
59. $20 + 0.75v = 500$
60. $0.18v - 4 = 32$
61. $0.3v + 2.2 = 7$
62. $3.2 = 2.5v - 8$
63. $1.5R - 0.2R = 26$
64. $5R + 1.4R = 200$
65. $R + 0.85R = 259$
66. $R - 0.7R = 31.5$

Set up an equation and solve.

67. Mr. Johnson pays 0.24 of his yearly salary in federal taxes. If he paid $10,800 in federal taxes last year, find his salary for that year.
68. Courtney's soccer team won 80% of the games they played. If they won 28 games, how many games did they play?
69. A parking garage charges $1.50, plus $0.45 per hour. If your fee is $4.20, how long was the car parked in the garage?
70. David is a sales clerk in the electronics department of a store. He is paid $1800 each month, plus 0.015 of his total sales. During the month of December, he earned a total of $3600. What were his total sales?
71. Samantha scored 86 and 97 on her first two algebra exams. What score does she need on the third exam to have an average (or mean) of 94?
72. The Adams family have packed 3 suitcases weighing 36 pounds, 58 pounds, and 43 pounds. How much should a fourth suitcase weigh so that the average weight is 45 pounds?
73. The price of an automobile decreased 4.8%. If the price after the decrease was $8092, what was the original price of the automobile?
74. A compact disc sold for $9 after a 40% markdown. Find the original price of the compact disc.
75. A television set costs $798.75 after a 6.5% sales tax was added to the marked price. What was the marked price of the television set?
76. Gloria's gas and electric bill for the month of January was up 12% over last January's bill. If this January's bill is $89.60, what was last year's bill?

CHAPTER 14 SUMMARY

KEY WORDS AND PHRASES

algebraic expression [14.1]
term [14.1, 14.2]
factor [14.1, 14.2]
numerical coefficient (coefficient) [14.2]
distributive law [14.2]
factoring [14.2]
like terms [14.2]
combine like terms [14.2, 14.6]
equation [14.3]
solution [14.3]
solving equations [14.3, 14.5, 14.7]
equivalent equations [14.3]
formula [14.4]
algebraic fractions [14.6]
proportion [14.7]

SYMBOLS

x means $1 \cdot x$ or $1x$
$-x$ means $-1 \cdot x$ or $-1x$
$x - y$ means $x + (-y)$

IMPORTANT RULES

The First Law of Exponents [14.1]

For numbers x, n, and m,
$$x^n x^m = x^{n+m}$$

The Distributive Law of Multiplication over Addition [14.2]

For numbers a, b, and c,
$$a(b + c) = ab + ac$$

How to Solve an Equation [14.3, 14.5, 14.7]
These operations can be done without changing the solution of the equation.

1. Add the same number to both sides.
2. Subtract the same number from both sides.
3. Multiply both sides by the same nonzero number.
4. Divide both sides by the same nonzero number.

Choose the operations so that eventually the variable is isolated on one side of the equation by itself, and a number is left on the other side. That number is the solution.

How to Solve an Application Problem by Using an Equation [14.4, 14.5, 14.7]
The general procedure for solving applications by using an equation is given on page 682.

CHAPTER 14 REVIEW EXERCISES

Multiply, by using the distributive law.

1. $7(y + 4)$
2. $5m(4 - 6m)$
3. $-6v(2v^2 - 3v + 1)$

Take out all factors common to all terms.

4. $10x + 15$
5. $3ac - a$
6. $4y^3 + 2y^2 - 2y$

Remove parentheses, when present, and simplify.

7. $-10 + (9 - x)$
8. $(5 - 2x) + (6 + 5x)$
9. $3n(5n)(4n)$
10. $3n + 5n + 4n$
11. $3(n + 5) - 4n$
12. $5w^3(7w^2)$
13. $m^2(3my^3)(4y^2)$
14. $9u + 4u$
15. $5c^2 - c^2$
16. $6xy - 6y + 4xy - 7y$
17. $5w + 9 - 16w + 6$
18. $8x^2 - 3x + 7 + 5x^2 + x$
19. $(6d - 3) + (4d + 2)$
20. $8(a - 4) + 3(2a + 1)$
21. $-c(5c + 4) + c(6c - 7)$
22. $2y(3y) - 4(3y^2)$
23. $2(4v^2 - 2v + 5) - 6v$

Solve for the variable.

24. $x - 6 = 5$
25. $45 = 9v$
26. $\dfrac{c}{3} = 36$
27. $16 = w - 14$
28. $5x - 7 = 13$
29. $-3x + 8 = -1$
30. $9 = 2y + 6$
31. $3 = 8w - 7$
32. $18 = 10 - 6v$
33. $-14 - y = 32$
34. $7x + 5 - 2x + 2 = 32$
35. $(-8m - 3) + (-4m + 7) = 68$
36. $120 = 8(y - 5) + 5(2y + 20)$
37. $13x - 5 = 3x + 15$
38. $17x + 8 = 9 - 3x$
39. $2w - 3 + 5w = 8 - 4w$
40. $8(v - 7) - 6v = 12v$
41. $\dfrac{3}{4}m = 18$
42. $-8m = \dfrac{5}{6}$
43. $\dfrac{2}{7}v = \dfrac{1}{2}$
44. $6m = 4.5$
45. $7.2n = 288$
46. $.24 = 60B$
47. $\dfrac{4}{24} = \dfrac{x}{42}$
48. $\dfrac{24}{w} = \dfrac{15}{35}$
49. $\dfrac{1.2}{150} = \dfrac{w}{3.75}$
50. $\dfrac{3.12}{v} = \dfrac{1.04}{4}$
51. $\dfrac{3\frac{1}{3}}{2} = \dfrac{c}{12}$
52. $\dfrac{30}{\frac{3}{8}} = \dfrac{24}{c}$

53. $\dfrac{3}{4}v + 1 = \dfrac{1}{4}$ **54.** $\dfrac{3w}{2} - 5 = \dfrac{7}{3}$ **55.** $2A - \dfrac{3}{8}A = 26$

56. $A + \dfrac{1}{4}A = 200$ **57.** $0.4v + 8 = 9$ **58.** $1.4 = 3.2w - 5$

59. $3w - 2.4w = 9$ **60.** $m - 0.88m = 510$

Evaluate for the given value of the variable.

61. $\dfrac{60}{w}$ for $w = 5$ **62.** $\dfrac{-5x}{4}$ for $x = 14$ **63.** $\dfrac{20 - 3w}{10}$ for $w = 40$

Simplify.

64. $\dfrac{63w}{45w^2}$ **65.** $\dfrac{40ac}{25c}$ **66.** $\dfrac{8n - 4}{6}$

67. $\dfrac{4w(w + 3)}{8w^2}$ **68.** $\dfrac{6 - 2x}{12}$ **69.** $\dfrac{y^2 - 7y}{14y}$

Perform the indicated operations and simplify.

70. $\dfrac{16}{m} \cdot \dfrac{3m^2}{30}$ **71.** $\dfrac{5mn}{m^2} \div \dfrac{15n}{2}$ **72.** $\dfrac{4xy}{6y} \div 20x^3$

73. $\dfrac{18mn^2}{27} \cdot 4m$ **74.** $\dfrac{y}{5} \cdot \dfrac{y + 5}{2}$ **75.** $\dfrac{12}{8c} \div \dfrac{3}{4c(c - 3)}$

76. $\dfrac{6c^2 + 24c}{2c^2} \cdot \dfrac{5c}{4}$ **77.** $\dfrac{7v}{3} - \dfrac{2v}{3}$ **78.** $\dfrac{w}{6} + \dfrac{8}{2}$

79. $\dfrac{3x}{2} + \dfrac{3}{4} - \dfrac{x}{4}$ **80.** $\dfrac{9}{m} + \dfrac{2}{m}$ **81.** $\dfrac{3}{5} - \dfrac{x}{y}$

82. $\dfrac{6}{5n} - \dfrac{2}{3n}$ **83.** $8 - \dfrac{v}{3}$

84. $\dfrac{4v}{y} + 1$ **85.** $5 - \dfrac{7}{3} + \dfrac{6}{4m}$

Combine like terms.

86. $\dfrac{2}{5}B - \dfrac{1}{5}B$ **87.** $\dfrac{7}{3}B + \dfrac{5}{3}B$ **88.** $2B - \dfrac{4}{5}B$

89. $B + \dfrac{2}{3}B$ **90.** $1.4m - 0.9m$ **91.** $6m - 0.78m$

92. $6.5m + m$ **93.** $m - 0.3m$

Set up an equation and solve.

94. Ming-Lai pays 32% of her income on rent. If her monthly rent is $864, find her monthly income.

95. There were 480 women at a lawyer's convention. This represented $\dfrac{3}{8}$ of the total attendance. How many people attended the convention?

96. A gardener uses 3 pounds of fertilizer for every 100 square feet of lawn. At this rate, how many pounds of fertilizer are needed for 750 square feet of lawn?

97. To make a punch, Francisco mixes orange juice and ginger ale in the ratio 5 to 2. How much ginger ale should he mix with 35 ounces of orange juice?

98. The label on a box of cereal says that $\dfrac{2}{3}$ cup cereal contains 90 calories. How many calories are in $1\dfrac{1}{3}$ cups of this cereal?

99. Ms. Takeuchi paid $88.80 in sales tax for a stereo system marked $1200. At this rate, what is the sales tax on a pair of downhill skis marked $550?

100. A painter mixes red paint with blue paint in the ratio 1 to 6. How much red paint should be mixed with 0.75 gallon of blue paint?

101. A string is cut into two pieces that are in the ratio 4 to 5. If the length of the shorter piece is $8\frac{1}{2}$ feet, what is the length of the longer piece?

102. At the tennis club, Mr. Ramirez pays a monthly fee of $65, plus $17.50 per hour of court time. During the month of October, his bill was $380. How many hours of court time was he charged for that month?

103. For an end-of-the-year bonus Debra receives $500, plus $\frac{1}{300}$ of her total sales for the year. If her bonus this year is $1800, what were her total sales?

104. The rainfall in Seattle for a 3-day period was as follows:

    Monday   3.1 inches
    Tuesday   2.5 inches
    Wednesday   2.8 inches

    How much rainfall is needed on Thursday so that the average daily rainfall for the 4-day period is 3 inches?

105. The value of a stock increased 8%. If the value after the increase is $35.10, what was the original value?

106. A microwave oven is on sale at 40% off. If the sale price is $177, what was the original price?

In Your Own Words

Write complete sentences to discuss each of the following. Support your comments with examples or pictures, if appropriate.

107. Look at the expressions $A + A$ and AA for several values of A. Give the simplified form for each expression and discuss how they are different.

108. Suppose you have a child that does not understand how to solve equations. How would you explain to your child how to solve the equation $3B - 6 = 9$? Be sure to explain your overall strategy and give reasons for each of your steps.

109. You are helping a friend, Ted, who is having trouble with simplifying the fraction $\frac{4+5}{12}$. Here we show Ted's incorrect work.

$$\text{Ted's incorrect work} \rightarrow \frac{\cancel{4}^{1} + 5}{\cancel{12}_{3}} = \frac{6}{3} = 2$$

    Discuss how you can help Ted to understand his error.

110. Suppose that the variable y represents an odd whole number (a number like 1, 3, 5, 7, 9, . . .). Look at values of the expression $y + 2$ for various values of y, and discuss what type of number you observe $y + 2$ to be.

111. You know that multiplication distributes over addition. For example, $2(5 + 4) = 2(5) + 2(4) = 10 + 8 = 18$. Investigate whether exponents distribute over addition and state your conclusion.

CHAPTER 14 PRACTICE TEST

Multiply by using the distributive law.
1. $7(a + 3)$
2. $6x(3x - 2)$

Take out all factors common to all terms.
3. $5y + 10$
4. $16y^2 - 8y$

Combine like terms.
5. $5v - 7v + 3v$
6. $8c^2 + c^2$
7. $0.9x - 2x$
8. $\frac{2}{5}W + \frac{3}{4}W$
9. $5w - 4 + 8w + 2$

Remove parentheses, and simplify.
10. $4m(5m)$
11. $-15 + (8m - 7)$
12. $5(x + 1) + 2(3x + 8)$
13. $12x^2(-4x)$
14. $10(x^2 - 1) + 8x(x + 6)$

Solve for the variable.
15. $5y - 6 = 19$
16. $2w + 8 = 6w - 20$
17. $3x = \frac{5}{6}$
18. $\frac{30}{72} = \frac{c}{84}$
19. $\frac{6.5}{x} = \frac{26}{40}$
20. $\frac{2}{3}B + 5 = \frac{7}{3}$
21. $y - 0.02y = 735$
22. Evaluate $\frac{w - 6}{3w}$ for $w = 2$.

Simplify.
23. $\frac{25aw}{15w}$
24. $\frac{12c + 8}{4c}$

Perform the indicated operations and simplify.
25. $\frac{36x}{20c} \cdot \frac{10c^2}{7}$
26. $\frac{2(x - 8)}{4x} \div \frac{14y}{xy}$
27. $\frac{3w}{5} + \frac{1}{2}$
28. $\frac{7}{v} - \frac{2}{3}$
29. $2 + \frac{5}{x}$

Set up an equation and solve.
30. A researcher compared the salaries of assembly-line workers to top management in a major automobile industry. She found the ratio to be 1 to 35. If a top executive made $700,000, what was the assembly-line workers salary?
31. On a map, every $1\frac{1}{3}$ inches represents 25 miles. If two cities are 12 inches apart on the map, how many miles apart are they?
32. A taxi charges $2.50, plus $1.60 per mile. How far can you travel for $14.50?
33. Jack's average daily water usage is down 15% from last year. This year he uses 102 gallons per day. How many gallons per day did he use last year?

CHAPTER 15
Geometry

15.1 POINTS, LINES, AND ANGLES

OBJECTIVES
- Use and interpret math symbols that name points, lines, line segments, and angles.
- Use and interpret math symbols that state that two lines are parallel or perpendicular.
- Use and interpret markings on drawings that indicate equal line segments and equal angles.
- Use a ruler or protractor to draw a given geometric figure involving points, lines, line segments, or angles.
- Solve problems involving points, lines, line segments, angles, and perpendicular lines.
- Solve problems involving angles formed on one side of a line, angles formed around a point, and angles formed by intersecting lines.

Architects, engineers, artists, and many other professionals use geometric figures in their daily work. In this chapter we limit our discussion to plane geometric figures, that is, those figures that lie on a flat surface. The flat surface is called a **plane.**

15.1 Points, Lines, and Angles

 Try These Problems

Draw and properly label each figure that is described in Problems 1 through 3.

1. A slanted line, $\overleftrightarrow{l}$, with point A on $\overleftrightarrow{l}$.
2. A horizontal line, $\overleftrightarrow{CD}$, with point B that does not lie on $\overleftrightarrow{CD}$.
3. A vertical line, $\overleftrightarrow{AB}$, with points C and D on $\overleftrightarrow{AB}$ between A and B.

1 Introducing Points, Lines, and Line Segments

A **point** indicates position. It has no size. We represent a point with a dot and name it with an uppercase letter. Here we picture points A, B, C, and D.

Points A, B, C, and D

A **geometric figure** is any collection of points. Some special collections of points form lines. In ordinary conversation, one refers to different types of lines: straight lines, broken lines, and curved lines.

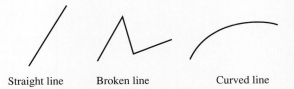

Straight line   Broken line   Curved line

However, in this text, when we use the word **line,** we mean a *straight line*. We assume the line extends indefinitely in two directions. It has no width or thickness. Here we picture three types of lines: **horizontal line, vertical line,** and **slanted (oblique) line.**

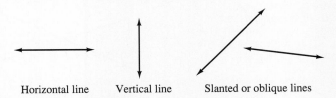

Horizontal line   Vertical line   Slanted or oblique lines

The arrows at each end are used to emphasize that the line extends beyond what we have drawn. Lines are named by using two points on the line or by using a single lowercase letter. Here we picture line AB, written $\overleftrightarrow{AB}$, and line m, written $\overleftrightarrow{m}$.

Line AB or $\overleftrightarrow{AB}$ Line m or $\overleftrightarrow{m}$

 Try Problems 1 through 3.

A **line segment** consists of two points and all points on the line that are between them. Here we picture line segment AB, written $\overline{AB}$.

Line segment AB or $\overline{AB}$

Try These Problems

4. Draw and properly label a line segment, $\overline{EF}$, that is $1\frac{1}{4}$ inches long.

5. In the given figure, use a centimeter ruler to measure the lengths of all the line segments.

6. Without measuring, find the distance from R to T.

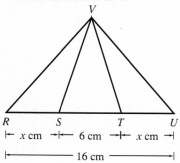

7. In traveling from point A to point B, which path is shorter, the path along the line segment, $\overline{AB}$, or the path along line segments, $\overline{AC}$ and $\overline{CB}$?

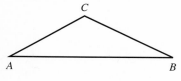

The points A and B are called **endpoints** of the line segment. A line segment has length, but no width. We write AB to represent the length of line segment AB, $\overline{AB}$.

To help indicate that line segments have equal length we use special markings. The markings in the figure below indicate that $AB = DC$ and $AD = BC$.

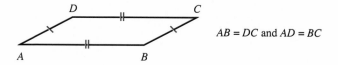

$AB = DC$ and $AD = BC$

One of the most basic properties of a line segment is that it is the *shortest path between two points*.

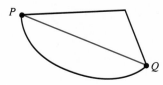

The shortest path from P to Q is the line segment PQ, $\overline{PQ}$.

 Try Problems 4 through 7.

When two lines cross, they **intersect** in exactly one point. Here we show two lines intersecting at point P.

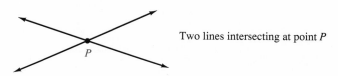

Two lines intersecting at point P

Actually, an infinite number of lines can pass through a single point. Here we picture eight lines intersecting at point T.

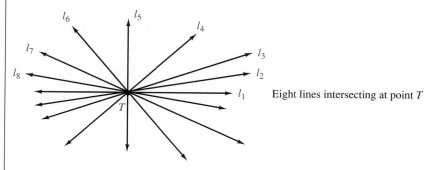

Eight lines intersecting at point T

Only one line can pass through two points. We say *two points determine a line.* Here we show the only line that can pass through the points P and Q.

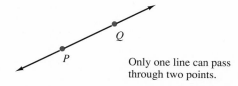

Only one line can pass through two points.

Try These Problems

Draw and properly label each figure that is described in Problems 8 through 11.

8. An oblique line, $\overleftrightarrow{m}$, and a vertical line, $\overleftrightarrow{n}$, intersecting at point H.

9. A line segment, $\overline{AB}$, and a horizontal line, $\overleftrightarrow{l}$, intersecting at point Q.

10. Two lines, $\overleftrightarrow{m}$, and $\overleftrightarrow{n}$, that never intersect.

11. Two line segments, $\overline{AB}$ and $\overline{BC}$, with a common endpoint B such that the points A, B, and C do not all lie on the same line.

12. How many lines are determined by the three points L, M, and N?

How many lines are determined by the four points E, F, G, and H?

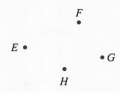

Because each pair of points determines a line, the four points determine six lines. Here we show them.

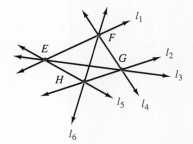

Four points determine six lines.

 Try Problems 8 through 12.

2 Viewing Angles as Sets of Points and Naming Angles

If we begin at a specific point on a line and extend in only one direction along the line, we form a **ray.** The beginning point is called the **endpoint of the ray.** Here we picture ray AB, written $\overrightarrow{AB}$.

Ray AB or $\overrightarrow{AB}$

Two rays with a common endpoint form an **angle.** The common endpoint is called the **vertex of the angle** and the two rays without the vertex point are called the **sides of the angle.** Here we picture two angles.

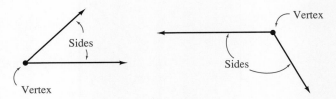

We name angles in several ways. One method is to name the angle by its vertex point. Here we picture angle A, written ∡A.

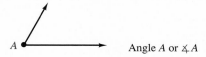

Angle A or ∡A

We also use numbers to name angles. Here we picture angle 1, written ∡1.

Angle 1 or ∡1

In some figures, we refer to an angle even though we do not indicate that the sides extend indefinitely. Here we picture angles 1, 2, and 3, written ∡1, ∡2, and ∡3, respectively.

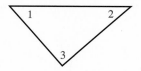

Sometimes we use an arc and a number to designate an angle. The arc is drawn from one side of the angle to the other side and the number is placed next to the arc. This can help in situations where several angles have the same vertex and at least one of them has the side of another angle running through its interior. Here we picture ∡1, ∡2, and ∡3 with a common vertex C. The arc is helping to designate ∡3.

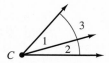

In the previous diagram, do not refer to ∡C because it would not be clear which of the three angles is being named.

Another way to name angles is to use three letters, one naming the vertex of the angle and the other two letters naming points on each side of the angle. Here we picture ∡ABC.

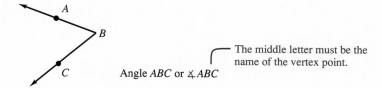

Angle ABC or ∡ABC

The middle letter must be the name of the vertex point.

It is important to observe that the middle letter must be the name of the vertex point.

Try These Problems

In Problems 13 and 14, draw and properly label each figure described.

13. An angle whose sides are CB and CA.

14. ∡1, ∡2, and ∡3 with a common vertex P.

15. Use the three-letter method to name ∡5, ∡6, and ∡7.

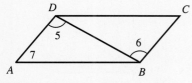

16. How many angles in the figure have vertex C? Name each of them by using the three-letter method.

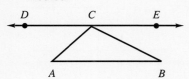

The three-letter method for naming angles allows us to name angles with a common vertex without the use of arcs. The following figure shows three different angles with common vertex B, ∡ABC, ∡CBD, and ∡ABD.

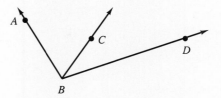

 Try Problems 13 through 16.

3 Measuring Angles

Angles have measurement. The measurement depends on the amount one side rotates about the vertex away from the other side. There are two types of units used to measure angles: degrees and radians. In this text, we look at only **degree measurement.**

When two sides have rotated so that they form a straight line, the angle measures 180 degrees, written 180°.

When two sides have not rotated away from each other at all, that is the two sides coincide, the angle has measure 0 degrees, written 0°.

Rotations between the two above extremes give us angles that measure between 0° and 180°. Here we picture a few.

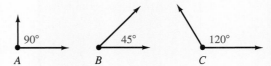

We write $m\angle A = 90°$, $m\angle B = 45°$, and $m\angle C = 120°$. The letter m is written in front of the angle name to make it clear that we mean the *measure* of the angle, not the angle itself, which is a collection of points.

It is possible for angles to measure more than 180°. Angles of this type were discussed in Section 11.2. It is also possible for angles to have negative measurement, but this is not discussed in this text. In this chapter, we limit the discussion to angles that measure from 0° to 180°.

740 Chapter 15 Geometry

 Try These Problems

In Problems 17 and 18, draw and properly label each figure described.

17. A right angle, ∡1, and an obtuse angle, ∡2, that have a common vertex C but do not have a common side.

18. A straight angle, ∡ABC, and an acute angle, ∡DBC, that have both a common vertex and a common side.

Use a protractor to measure each indicated angle. You may find it helpful to trace the angle on another piece of paper, extend the sides, then measure.

19.
20.
21.

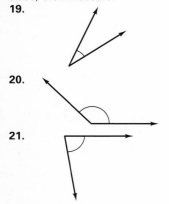

Use the protractor to draw each angle. Label the angle properly.

22. ∡R measuring 15°.
23. ∡CPT measuring 148°.

We classify angles according to their measure. The following chart illustrates four types.

Type of Angle	Measurement	Picture
Right angle	90°	
Acute angle	between 0° and 90°	
Obtuse angle	between 90° and 180°	
Straight angle	180°	

 Try Problems 17 and 18.

An instrument used to measure angles is called a **protractor.** Here we picture a protractor measuring an acute angle, ∡A, and an obtuse angle, ∡Q.

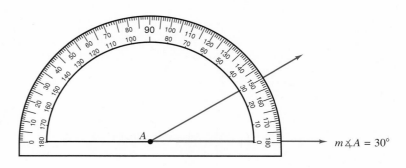

$m\angle A = 30°$

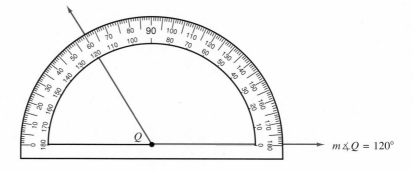

$m\angle Q = 120°$

 Try Problems 19 through 23.

15.1 Points, Lines, and Angles   **741**

▲ Try These Problems

24. Given that $m\angle MQN = 130°$, find $m\angle MQP$.

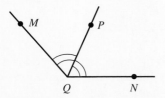

25. Given that $m\angle CBD = 42°$, find the measure of $m\angle ABC$.

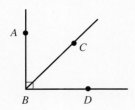

A special marking is used to indicate that an angle measures 90°. The figure shows two right angles, $\angle DBC$ and $\angle ADC$.

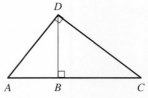

Right angles, $\angle ADC$ and $\angle DBC$

Special markings are also used to indicate that angles have equal measure. In the figure that follows, $m\angle A = m\angle B$, and $m\angle D = m\angle C$.

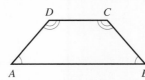

$m\angle A = m\angle B, m\angle D = m\angle C$

▲ Try Problems 24 and 25.

4 Some Interesting Observations

Here we picture four nonoverlapping angles with a common vertex on one side of a straight line.

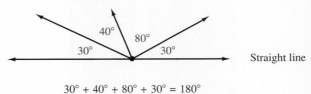

Straight line

$30° + 40° + 80° + 30° = 180°$

Observe that the measures add up to be 180°. This should not be surprising since they make up a straight angle. In general, the following is true.

Angles on One Side of a Line

Nonoverlapping angles with a common vertex formed on one side of a line have measures that total 180°.

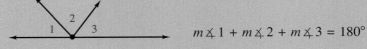

$m\angle 1 + m\angle 2 + m\angle 3 = 180°$

Try These Problems

26. Find the measure of ∡1.

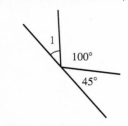

27. Assume $m\angle FBC = 160°$, $m\angle EBD = 100°$, and $m\angle FBE = m\angle CBD$. Find $m\angle FBE$ and $m\angle FBA$.

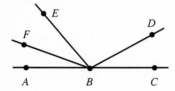

EXAMPLE 1 Find the measure of the obtuse angle.

SOLUTION The obtuse angle and the acute angle are nonoverlapping angles with a common vertex on one side of a straight line so their measures total 180°. Therefore, we have the following.

$$(5x - 6) + (x + 6) = 180$$

Solve this equation for x.

$$5x - 6 + x + 6 = 180$$
$$6x = 180$$
$$x = 30$$

Because the obtuse angle measures $5x - 6$ degrees, we evaluate this expression for $x = 30$.

$$5x - 6 = 5(30) - 6$$
$$= 150 - 6$$
$$= 144$$

The obtuse angle measures 144°. ∎

 Try Problems 26 and 27.

The previous observation can be extended. Here we picture four nonoverlapping angles with a common vertex formed around a point.

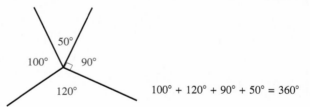

Observe that the measures of the angles add up to be 360°. In general, the following is true.

Angles Around a Point

Nonoverlapping angles with a common vertex formed around a point have measures that total 360°.

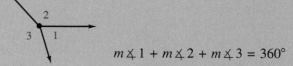

Try These Problems

28. ∡1, ∡2, ∡3, ∡4, ∡5, and ∡6 each have equal measure. What is the measure of each?

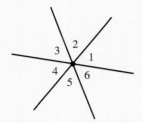

29. Find the measures of ∡SRT, ∡TRP, and ∡PRQ without the use of a protractor.

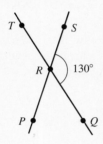

Try Problem 28.

Special pairs of angles are formed when two straight lines intersect. Here we look at two specific situations.

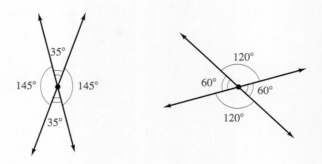

Observe that when two lines intersect, pairs of angles with equal measure are formed. Pairs of equal angles formed in this way are called **vertical angles**. In general, the following is true.

Vertical Angles
When two lines intersect, the pairs of vertical angles formed are equal.

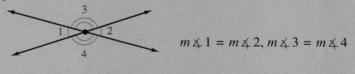

$m\angle 1 = m\angle 2, m\angle 3 = m\angle 4$

Try Problem 29.

5 Defining Parallel and Perpendicular Lines

Two lines, in a plane, that never meet are called **parallel lines.** We use the symbol // to mean *is parallel to*. Here we picture two parallel lines.

To help indicate that lines are parallel in a figure, we use special markings. The following figures illustrate how this is done.

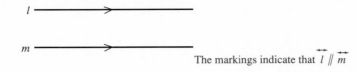

The markings indicate that $\overleftrightarrow{l} \parallel \overleftrightarrow{m}$

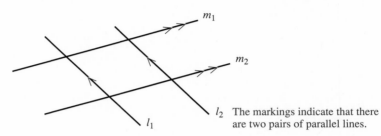

The markings indicate that there are two pairs of parallel lines.

When two lines intersect to form right angles, they are called **perpendicular lines.** The symbol ⊥ is used to mean *is perpendicular to*. Here are two figures showing pairs of perpendicular lines.

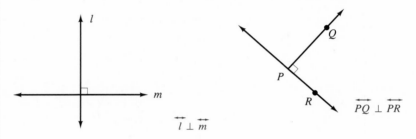

$\overleftrightarrow{l} \perp \overleftrightarrow{m}$ $\overleftrightarrow{PQ} \perp \overleftrightarrow{PR}$

6 Determining the Shortest Distance from a Point to a Line

Suppose point P is not on line l, $\overleftrightarrow{l}$, and you wish to travel from point P to $\overleftrightarrow{l}$ taking the shortest possible path. What path should you take?

Here we show several paths, the lengths of each, and the angle each makes with $\overleftrightarrow{l}$.

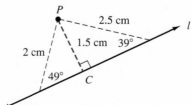

15.1 Points, Lines, and Angles   745

 Try These Problems

Draw and properly label each figure that is described.

30. Slanted lines *AB* and *CD* that are perpendicular.

31. Vertical lines $\vec{l}$ and $\vec{m}$ that are parallel.

32. Line segment $\overline{PQ}$ that is the shortest path from point *P* to line $\vec{l}$.

33. Line $\vec{m}$ that passes through point *Q* and is parallel to line $\vec{n}$.

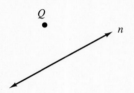

Observe that the shortest path from *P* to $\vec{l}$ is the line segment *PC*, $\overline{PC}$, which is perpendicular to $\vec{l}$. In general, the following is true.

> The shortest distance from a point to a line is the perpendicular distance from the point to the line.
>
>
>
> *PQ* is the shortest distance from point *P* to line *l*.

 Try Problems 30 through 33.

 Answers to Try These Problems

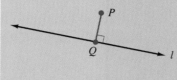

5. $AB = 2.5$ cm, $BC = 2$ cm, $AC = 1.5$ cm **6.** 11 cm **7.** $\overline{AB}$

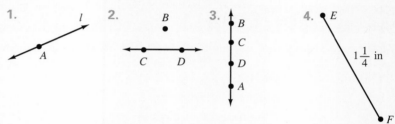

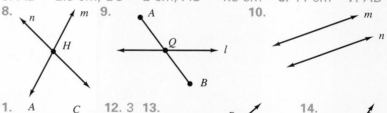

15. $\angle ADB$, $\angle DBC$, $\angle DAB$
16. 6 angles; $\angle DCA$, $\angle ACB$, $\angle ECB$, $\angle DCB$, $\angle ECA$, $\angle ECD$

 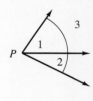

19. $m\angle 6 = 32°$ **20.** $m\angle 3 = 136°$ **21.** $m\angle 2 = 80°$

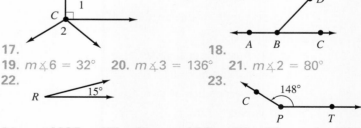

24. $m\angle MQP = 65°$ **25.** $m\angle ABC = 48°$ **26.** $m\angle 1 = 35°$
27. $m\angle FBE = 30°$, $m\angle FBA = 20°$ **28.** $60°$
29. $m\angle SRT = 50°$, $m\angle TRP = 130°$, $m\angle PRQ = 50°$

746   Chapter 15 Geometry

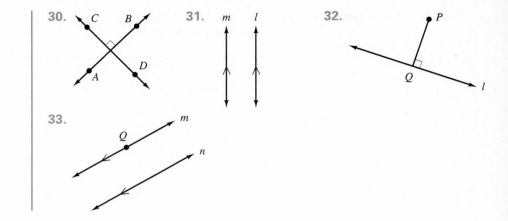

EXERCISES 15.1

Match each symbol with one appropriate phrase. Some of the phrases may not be used.

SYMBOL   PHRASE

1. A a. angle A
2. $\overleftrightarrow{AC}$ b. the measure of angle A
3. $\overline{AC}$ c. angle 9
4. $\overrightarrow{AC}$ d. an angle whose measure is 9°
5. AC e. the measure of $\angle 9$
6. $\overleftrightarrow{l}$ f. angle ABC, where points A, B, and C are any three points
7. $\angle A$ on the angle
8. $m\angle A$ g. angle ABC, where B is the vertex, and A and C are
9. $\angle ABC$ points on either side of the angle
10. $\angle 9$ h. points A and C
11. $\overleftrightarrow{l} // \overleftrightarrow{m}$ i. point A
12. $\overleftrightarrow{l} \perp \overleftrightarrow{m}$ j. point l
                 k. line l
                 l. a ray that begins at point C and passes through point A
                 m. a ray that begins at point A and passes through point C
                 n. line segment AC
                 o. the length of line segment AC
                 p. line AC
                 q. the measure of line AC
                 r. lines l and m
                 s. $\overleftrightarrow{l}$ is perpendicular to $\overleftrightarrow{m}$
                 t. l is parallel to $\overleftrightarrow{m}$

For Exercises 13 through 19, use the given figure and your ruler or protractor to find the following measurements. Measure lengths to the nearest $\frac{1}{8}$ inch and angles to the nearest degree.

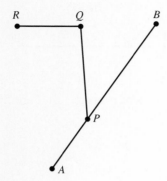

13. *PB*
14. The shortest distance from *P* to *R*
15. *PQ* + *QR*
16. $m \angle QPB$
17. $m \angle QPA$
18. $m \angle Q$
19. The shortest distance from point *Q* to $\overline{AB}$
20. Use the three-letter method to name eight different angles in the given figure.

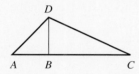

In Exercises 21 through 30, use a ruler or protractor to draw each figure. Label the figure appropriately.

21. Line segment $\overline{KL}$ with length 4.4 centimeters.
22. $\overrightarrow{CD}$ and $\overrightarrow{CF}$ with a common endpoint.
23. A vertical line and a horizontal line intersecting at point *R*.
24. $\angle PQR$ with measure 20°.
25. $\angle P$ with measure 115°.
26. Two acute angles, $\angle 1$ and $\angle 2$, with a common vertex and a common side.
27. Two obtuse angles, $\angle 3$ and $\angle 4$, with a common vertex but *not* a common side.
28. A vertical line and a slanted line intersecting to form a pair of vertical angles measuring 155°.
29. Two horizontal lines, $\overleftrightarrow{FG}$ and $\overleftrightarrow{RT}$, that are parallel.
30. Two slanted lines, $\vec{l}$ and $\vec{m}$, that are perpendicular.

In Exercises 31 through 34, do not use a ruler or a protractor.

31. In the figure *AB* = 8 meters, *BC* = 5 meters, and *AB* = *CD*. Find *AD*.

748 Chapter 15 Geometry

32. Find $m\angle FXG$.

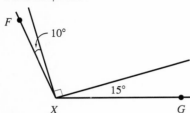

33. Find the value of x.

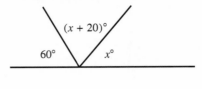

34. In the figure, $m\angle 1 = m\angle 4$, and $\angle 2$ and $\angle 3$ each have measures that are 5° more than $\angle 1$. Find $m\angle 4$.

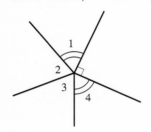

15.2	TRIANGLES AND QUADRILATERALS

OBJECTIVES
- Draw, recognize, and name triangles.
- Classify triangles as acute, obtuse, or right.
- Solve problems using the triangle inequality or the fact that the sum of the angles of a triangle is 180°.
- Draw, recognize, and name quadrilaterals.
- Solve problems using the fact that the sum of the angles of a quadrilateral is 360°.
- Classify a quadrilateral as a parallelogram, rectangle, square, rhombus, trapezoid, or none of these.
- Use a ruler or protractor to draw a geometric figure involving triangles or quadrilaterals.

1 Defining and Naming Triangles

A **triangle** is a geometric figure consisting of three points that do not lie in a straight line and the line segments that connect them. Here we picture triangle ABC, written $\triangle ABC$.

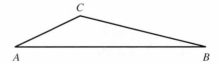

The points A, B, and C are called the **vertices of the triangle;** the segments $\overline{AB}$, $\overline{BC}$, and $\overline{CA}$ are called the **sides of the triangle;** the angles $\angle A$, $\angle B$, and $\angle C$ are called the **angles of the triangle.**

Try These Problems

1. How many triangles are in the figure? Name them.

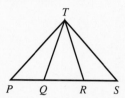

2. Can the measurements 7 feet, 9 feet, and 10 feet be the lengths of the sides of a triangle?

3. Can the measurements 12 centimeters, 35 centimeters, and 20 centimeters be the lengths of the sides of a triangle?

 Try Problem 1.

 Introducing the Triangle Inequality

Recall that the shortest path between two points is the line segment joining them. This gives us some important information about how the lengths of the sides of any triangle are related.

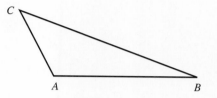

In general, the following is true. This result is known as the **triangle inequality.**

> ### The Triangle Inequality
> The length of one side of a triangle must be less than the sum of the lengths of the other two sides.

EXAMPLE 1 Can the measurements, 8 meters, 2 meters, and 5 meters, be the lengths of the sides of a triangle?

SOLUTION We must check to see that each measurement is less than the sum of the other two measurements.

$$8 + 2 = 10 \quad \text{and} \quad 5 < 10$$
$$8 + 5 = 13 \quad \text{and} \quad 2 < 13$$

but

$$2 + 5 = 7 \quad \text{and} \quad 8 > 7$$

Therefore, the given measurements cannot be the lengths of the sides of a triangle. ∎

 Try Problems 2 and 3.

3 Discovering the Sum of the Angles of a Triangle

There is an interesting result about the measures of the three angles of any triangle. To discover the result, try the following experiment.

Try These Problems

Use a protractor to measure the angles of these triangles. What is the sum of the measures of the three angles?

4.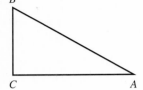

5.

EXPERIMENT

1. Draw any triangle on a piece of paper and cut out the triangle.

   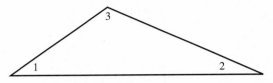

2. Cut or tear off the three angles.

   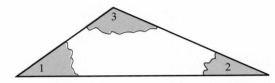

3. Place the three pieces together so that they have a common vertex.

      $m\angle 1 + m\angle 2 + m\angle 3 = 180°$

4. Observe that the three angles make a straight angle, that is, their measures total 180°.

This experiment leads us to the following general result.

Summing the Angles of a Triangle

The sum of the measures of the angles of any triangle is 180°.

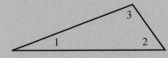 $m\angle 1 + m\angle 2 + m\angle 3 = 180°$

Here we picture some specific triangles. Observe in each case that the sum of the measures of the angles is 180°.

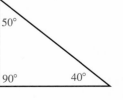 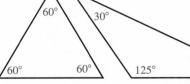

90° + 40° + 50° = 180°   60° + 60° + 60° = 180°   125° + 25° + 30° = 180°

Try Problems 4 and 5.

Try These Problems

Answer Problems 6 and 7 without using a protractor.

6. Find $m\angle 1$.

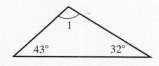

7. Find $m\angle KJL$ and $m\angle K$

EXAMPLE 2 Find the measure of $\angle 1$.

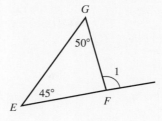

SOLUTION First find the measure of $\angle EFG$.

$$m\angle E + m\angle G + m\angle EFG = 180°$$
$$45° + 50° + m\angle EFG = 180°$$
$$95° + m\angle EFG = 180°$$
$$m\angle EFG = 180° - 95°$$
$$m\angle EFG = 85°$$

Now we find the measure of $\angle 1$ by using the fact that $\angle 1$ and $\angle EFG$ make up a straight angle.

$$m\angle 1 + m\angle EFG = 180°$$
$$m\angle 1 + 85° = 180°$$
$$m\angle 1 = 180° - 85$$
$$m\angle 1 = 95° \blacksquare$$

 Try Problems 6 and 7.

4 Classifying Triangles

Triangles with special characteristics are given special names. Here we look at five different types of triangles.

An **acute triangle** has three acute angles.

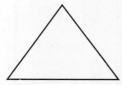

Acute triangle

An **obtuse triangle** has one obtuse angle and two acute angles.

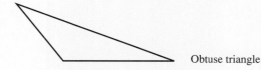

Obtuse triangle

Try These Problems

In Problems 8 through 12, use a ruler or protractor to draw each figure described.

8. An acute triangle with angles measuring 50°, 55°, and 75°.
9. An obtuse triangle with the obtuse angle measuring 150°.
10. A right triangle with the legs measuring 2 centimeters and 1.5 centimeters.
11. An isosceles triangle with the base angles measuring 40°.
12. An equilateral triangle with each side measuring $\frac{3}{4}$ inch.

A **right triangle** has one right angle and two acute angles. The sides of a right triangle are given special names. The two shorter sides that form the right angle are called the **legs.** The longest side that is opposite the right angle is called the **hypotenuse.**

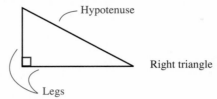

 Try Problems 8 through 10.

An **isosceles triangle** has two equal angles and two equal sides. The two equal angles are called the **base angles.**

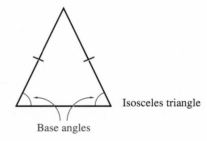

An **equilateral triangle** has three equal angles and three equal sides.

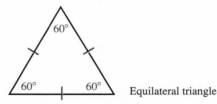

 Try Problems 11 and 12.

 Defining Quadrilaterals

A **quadrilateral** is a geometric figure consisting of four points in a plane, no three of which lie in a straight line, and the four line segments that connect these points that intersect only at their endpoints. Here we picture two quadrilaterals.

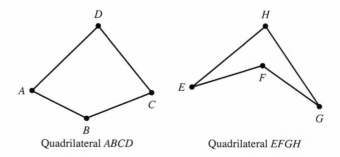

The quadrilateral ABCD is called a **convex quadrilateral** because when you draw a line segment between any two interior points, the entire line segment lies inside the quadrilateral. The quadrilateral *EFGH* is *not* convex because it is possible to connect two interior points with a line segment that extends outside the quadrilateral.

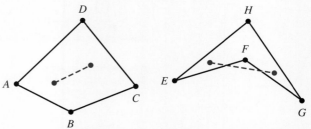

Convex quadrilateral *ABCD*   Nonconvex quadrilateral *EFGH*

In this text, we limit our study to convex quadrilaterals, therefore, when we use the term **quadrilateral,** we mean *convex* quadrilateral.

In the quadrilateral *ABCD* the points *A*, *B*, *C*, and *D* are called **vertices of the quadrilateral;** the segments *AB*, *BC*, *CD*, and *DA* are called the **sides of the quadrilateral;** the angles ∡*A*, ∡*B*, ∡*C*, and ∡*D* are called the **angles of the quadrilateral.**

6 Discovering the Sum of the Angles of a Quadrilateral

We have seen that the sum of the angles of any triangle is 180°. Now we can use that result to find the sum of the angles of a quadrilateral. Consider quadrilateral *ABCD* shown here. Draw a line from *B* to *D*, forming two triangles.

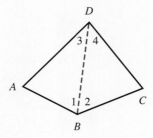

In △*ABD* and △*BCD* we have

$$m\angle A + m\angle 1 + m\angle 3 = 180°$$

and

$$m\angle 2 + m\angle C + m\angle 4 = 180°$$

Adding all of the angles together we obtain

$$m\angle A + \underbrace{m\angle 1 + m\angle 2} + m\angle C + \underbrace{m\angle 3 + m\angle 4} = 360°$$
$$m\angle A + m\angle B + m\angle C + m\angle D = 360°$$

We have shown that the sum of the angles of any quadrilateral is 360°.

Try These Problems

13. Use a protractor to measure each angle of this quadrilateral. What is the sum of the measures of the four angles?

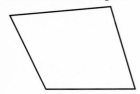

14. Find *x* without measuring.

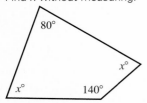

Summing the Angles of a Quadrilateral

The sum of the measures of the angles of any quadrilateral is 360°.

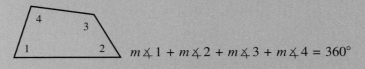

$m\angle 1 + m\angle 2 + m\angle 3 + m\angle 4 = 360°$

EXAMPLE 3 In quadrilateral *RSTU*, find the measure of $\angle T$ without measuring.

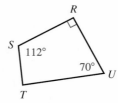

SOLUTION We use the fact that the sum of all four angles should be 360°.

$$m\angle S + m\angle R + m\angle U + m\angle T = 360°$$
$$112° + 90° + 70° + m\angle T = 360°$$
$$272° + m\angle T = 360°$$
$$m\angle T = 360° - 272°$$
$$m\angle T = 88°\ \blacksquare$$

 Try Problems 13 and 14.

7 Classifying Quadrilaterals

Quadrilaterals with special characteristics are given special names. Here we look at five different types of quadrilaterals.

A **parallelogram** is a quadrilateral with opposite sides both equal in measure and parallel. Also, the opposite angles of a parallelogram are equal. Here we picture parallelogram *ABCD*, written □*ABCD*.

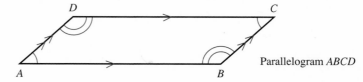

Parallelogram *ABCD*

A **rectangle** is a parallelogram with four right angles. Here we picture rectangle *WXYZ*.

Rectangle *WXYZ*

Try These Problems

For Problems 15 through 19, use a ruler or protractor to draw the figure described.

15. A parallelogram with one angle measuring 120°.
16. A rectangle with one side measuring $\frac{1}{2}$ inch and another side measuring $\frac{3}{4}$ inch.
17. A square with each side measuring 2 centimeters.
18. A rhombus with one side measuring 2.2 centimeters.
19. A trapezoid with the nonparallel sides of equal length.

A **square** is a rectangle with all four sides having equal measure. Here we picture square *RSTU*. Since a square is a rectangle, all four angles equal 90°.

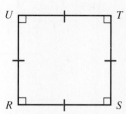

Square *RSTU*

 Try Problems 15 through 17.

A **rhombus** is a parallelogram with all four sides having equal measure. Since a rhombus is a parallelogram, opposite sides are parallel. Here we picture rhombus *JKLM*.

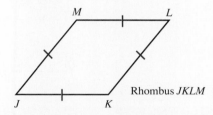

Rhombus *JKLM*

A **trapezoid** is a quadrilateral with only one pair of parallel sides. The parallel sides are not necessarily equal in length. Here we picture trapezoid *MNOP* with $\overline{PO}//\overline{MN}$.

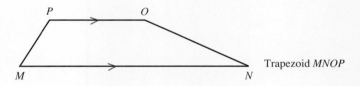

Trapezoid *MNOP*

 Try Problems 18 and 19.

 Answers to Try These Problems

1. 6 triangles; △*PTQ*, △*QTR*, △*RTS*, △*PTR*, △*QTS*, △*PTS*
2. yes   3. no, because 35 > 12 + 20
4. $m\angle A = 30°$, $m\angle B = 60°$, $m\angle C = 90°$, Sum = 180°
5. $m\angle F = 125°$, $m\angle G = 40°$, $m\angle E = 15°$, Sum = 180°
6. 105°   7. $m\angle KJL = 70°$, $m\angle K = 12°$

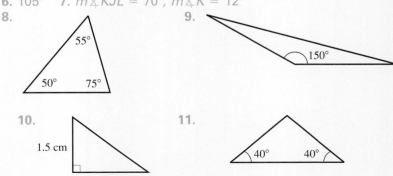

13. 75° + 100° + 115° + 70° = 360°   **14.** 70°

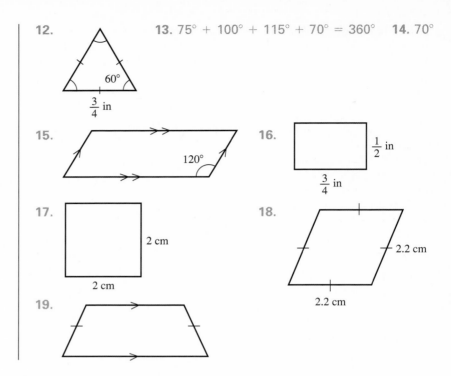

EXERCISES 15.2

1. How many triangles do you see in the figure? Name them.

Can these measurements be the lengths of the sides of a triangle?

2. 50 in, 23 in, 20 in **3.** 8.35 cm, 5.2 cm, 7 cm
4. $2\frac{1}{3}$ mi, $1\frac{3}{4}$ mi, $\frac{3}{4}$ mi **5.** 2 yd, 8 ft, 12 ft

In Exercises 6 through 13, find the value of x without measuring.

6.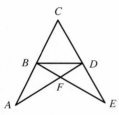

7.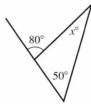

8.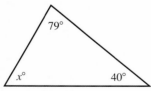

9.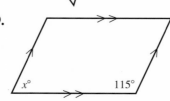

10. 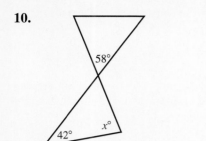   **11.**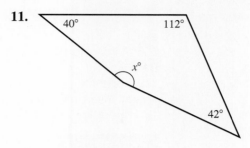

12. 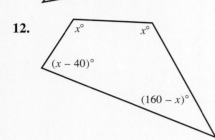   **13.**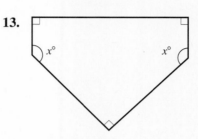

In Exercises 14 through 18, use a ruler or protractor to draw each figure if possible. If not possible, explain.

14. An acute triangle with one angle measuring 42°.

15. An equilateral triangle with one side measuring $\frac{3}{4}$ inch.

16. A triangle with angles measuring 50°, 40°, and 100°.

17. A rhombus with one angle measuring 60°.

18. A square with one side measuring 2 centimeters.

Solve.

19. A ladder is leaning against a vertical wall so that the angle the ladder makes with the floor is 70°. What is the measure of the angle the ladder makes with the wall? (Draw a picture.)

20. A ship sails away from a straight shore at an angle of 50°. After traveling in a straight line path for awhile, the ship makes a 90° turn and begins traveling in a straight line path to the shore. At what angle is the ship approaching the shore? (Draw a picture.)

15.3 PERIMETER AND AREA

OBJECTIVES
- Find the perimeter of a triangle, quadrilateral, or other geometric figure involving line segments.
- Draw the altitude that corresponds to a given base in a triangle, parallelogram, or trapezoid.
- Find the area of a triangle, parallelogram, or trapezoid.
- Find the area of composite geometric figures.
- Solve applications that involve finding the perimeter or area of a geometric figure.

Try These Problems

1. Find the perimeter of this parallelogram.

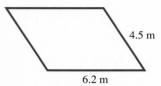

2. The perimeter of the triangle is 36 centimeters. Find the length of each side.

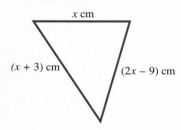

1 Working with Perimeter

The **perimeter** of a geometric figure means the distance all the way around it. Perimeter was also covered in Sections 1.4, 5.7, 6.7, 8.4, and 10.1. Here are more examples.

EXAMPLE 1 Find the perimeter of the given figure.

SOLUTION Add up the lengths of all five sides.

$$\text{Perimeter in meters} = 9 + 9 + 9 + 8.2 + 8.2$$
$$= 27 + 16.4$$
$$= 43.4$$

The perimeter is 43.4 meters. ∎

EXAMPLE 2 The perimeter of the given triangle is 320 feet. Find the length of the sides.

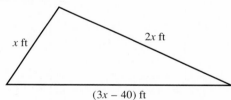

SOLUTION Because the perimeter is 320 feet, we have the following.

$$x + 2x + (3x - 40) = 320$$
$$6x - 40 = 320$$
$$6x = 360$$
$$x = 60 \quad \text{Length of one side in feet}$$

One side has length 60 feet. To find the lengths of the other two sides, evaluate the expressions $2x$ and $3x - 40$ for $x = 60$.

$$2x = 2(60) = 120 \quad \text{Length of a second side in feet}$$

$$3x - 40 = 3(60) - 40$$
$$= 180 - 40$$
$$= 140 \quad \text{Length of the third side in feet}$$

The sides measure 60 feet, 120 feet, and 140 feet. ∎

Try Problems 1 and 2.

2. Reviewing the Area of a Rectangle

Another measurement associated with plane figures is **area.** The area measures the extent of the region inside.

A rectangle that is 3 units wide and 5 units long has area 15 square units because 15 of the 1-unit squares fit inside. Here we picture such a rectangle.

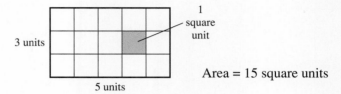

Observe that the area of this rectangle can be found by multiplying the width by the length. In fact, as you have seen in previous chapters, the area of any rectangle can be found by multiplying the width by the length.

Area of a Rectangle
The area of a rectangle is the length times the width.

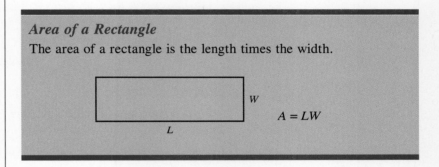

$A = LW$

The area of a rectangle was also covered in Sections 2.5, 4.5, 8.4, and 10.2. Here is another example.

EXAMPLE 3 Find the area of the unshaded region. Assume both quadrilaterals are squares.

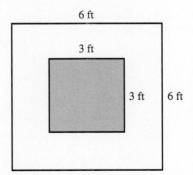

Try These Problems

3. Assume this figure is a rectangle within a square. Find the area of the shaded region.

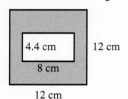

SOLUTION Because a square is a rectangle, we can find the areas of each of these squares by multiplying the length by the width. Then to find the area of the unshaded region, we need to subtract the area of the smaller square from the area of the larger square.

Area of smaller square in square feet = 3(3) = 9
Area of larger square in square feet = 6(6) = 36
Area of the unshaded region in square feet = 36 − 9 = 27

The area of the unshaded region is 27 square feet. ∎

 Try Problem 3.

3 Discovering a Formula for the Area of a Parallelogram

Next we look at how to find the area of a parallelogram. In parallelogram *ABCD*, draw a line from point *D* that is perpendicular to side *AB* at point *E*.

Cut off △*AED* and place it on the other end so that *AD* coincides with *BC*.

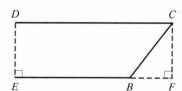

A rectangle is formed that has the same area as the original parallelogram.

$$\text{Area of parallelogram } ABCD = \text{area of rectangle } EFCD$$
$$= DE \cdot EF$$
$$= DE \cdot AB$$

The segment DE, $\overline{DE}$, or the length of $\overline{DE}$, DE, is called an **altitude** (or **height**) of the parallelogram, and side $\overline{AB}$ or the length of $\overline{AB}$, AB, is the corresponding **base**.

The above discussion leads us to a formula for finding the area of any parallelogram.

Try These Problems

4. In parallelogram *ABCD*, draw an altitude that corresponds to base *CD*.

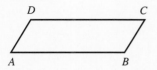

5. In parallelogram *ABCD*, draw an altitude that corresponds to base *AD*.

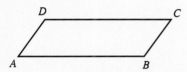

6. Find the area of this parallelogram.

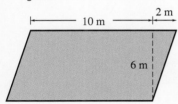

Area of a Parallelogram
The area of a parallelogram is the altitude times the base.

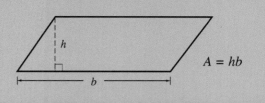

An **altitude** of a parallelogram is a segment drawn from one parallel side to the line containing the other and perpendicular to both of them. The corresponding **base** is either of those parallel sides. The length of the altitude is often called the altitude. Observe that we can find the area of a parallelogram using the altitude between either pair of parallel sides.

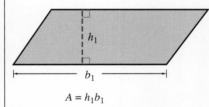

$A = h_1 b_1$

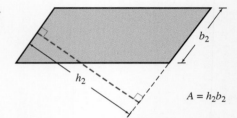

$A = h_2 b_2$

 Try Problems 4 and 5.

EXAMPLE 4 Find the area of this parallelogram.

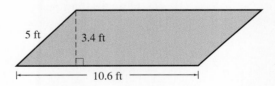

SOLUTION

$$\begin{aligned} \text{Area} &= (\text{altitude})(\text{base}) \\ \text{(sq ft)} &\quad\text{(ft)}\quad\text{(ft)} \\ &= (3.4)(10.6) \\ &= 36.04 \end{aligned}$$

The area is 36.04 square feet. ■

 Try Problem 6.

4 Discovering a Formula for the Area of a Triangle

Now we look at how to find the area of a triangle. We begin with a right triangle. Consider the right triangle ABC with $m\angle C = 90°$.

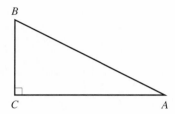

Make another copy of $\triangle ABC$ and place the two triangles so that the hypotenuses coincide and the two triangles form a rectangle.

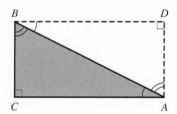

The area of the $\triangle ABC$ is half of the area of the entire rectangle $CADB$. We have the following.

$$\text{Area of the right } \triangle ABC = \frac{1}{2}(\text{area of rectangle } CADB)$$
$$= \frac{1}{2}(BC)(CA)$$

Observe that the area of right $\triangle ABC$ is half of the product of the legs.

The above discussion leads us to a formula for finding the area of any right triangle.

Area of a Right Triangle
The area of a right triangle is one half of the product of the legs.

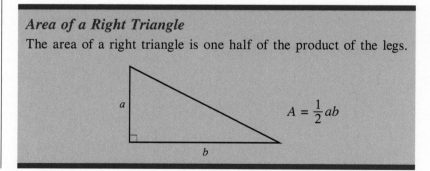

Try These Problems

7. Find the area of this triangle.

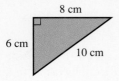

EXAMPLE 5 Find the area of this triangle.

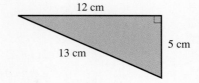

SOLUTION

$$\text{Area (sq cm)} = \frac{1}{2} \cdot \text{leg (cm)} \cdot \text{leg (cm)}$$

$$= \frac{1}{2}(12)(5)$$

The legs are the sides that form the right angle.

$$= 6(5)$$

$$= 30$$

The area of the triangle is 30 square centimeters. ■

Try Problem 7.

We have seen that the area of a right triangle is one half of the product of the legs. Now we look at how to find the area of other triangles that are not right triangles. Let $\triangle ABC$ represent any triangle.

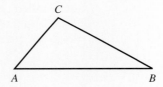

Make another copy of $\triangle ABC$ and place the two triangles together so that the sides with length BC coincide and the two triangles form a parallelogram.

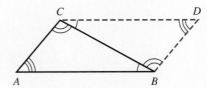

The area of $\triangle ABC$ is half of the area of the parallelogram and we know how to find the area of a parallelogram. Draw a line from point C that is perpendicular to $\overline{AB}$ at point E.

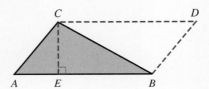

Try These Problems

In Problems 8 and 9, draw an altitude that corresponds to base AB.

8.
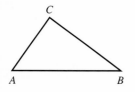

9.
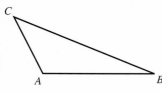

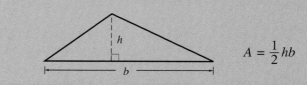

$$\text{Area of } \triangle ABC = \frac{1}{2}(\text{area of parallelogram } ABDC)$$

$$= \frac{1}{2} \cdot CE \cdot AB$$

Segment CE, $\overline{CE}$, or the length of $\overline{CE}$, CE, is called the **altitude** of $\triangle ABC$ and side $\overline{AB}$ or the length of $\overline{AB}$, AB, is called the corresponding **base**.

The above discussion leads us to the following general formula.

Area of any Triangle

The area of a triangle is one-half of the altitude times the base.

$$A = \frac{1}{2}hb$$

The **altitude** of a triangle is a segment drawn from one of the vertices perpendicular to the line containing the opposite side. The corresponding **base** is that opposite side. The length of the altitude is often called the altitude. We can use any one of the three altitudes of a triangle to find its area. Here we show the three ways to find the area of the given triangle, $\triangle ABC$.

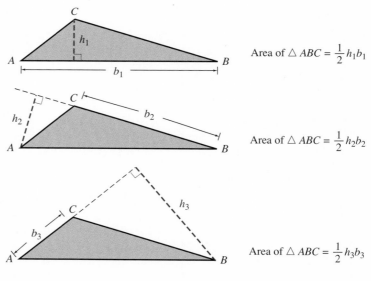

Area of $\triangle ABC = \frac{1}{2}h_1 b_1$

Area of $\triangle ABC = \frac{1}{2}h_2 b_2$

Area of $\triangle ABC = \frac{1}{2}h_3 b_3$

▲ **Try Problems 8 and 9.**

Try These Problems

In Problems 10 and 11, find the area of the given triangle.

10.

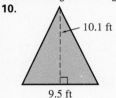

11.

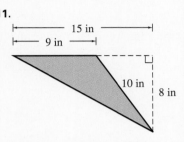

12. The figure shows the side of a house. At a cost of $0.25 per square foot, how much will it cost to paint the side of this house?

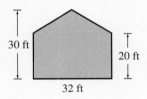

EXAMPLE 6 Find the area of triangle $\triangle GEF$.

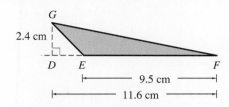

SOLUTION The altitude GD corresponds to the base EF. Therefore, the area of the triangle is as follows.

$$\begin{aligned}\text{Area} &= \frac{1}{2} \cdot GD \cdot EF \\ \text{(sq cm)} &\quad\quad \text{(cm)}\;\;\text{(cm)} \\ &= \frac{1}{2}(2.4)(9.5) \\ &= 1.2(9.5) \\ &= 11.4\end{aligned}$$

The area of $\triangle GEF$ is 11.4 square centimeters.

Try Problems 10 and 11.

EXAMPLE 7 A gardener charges $5 per square foot to design a flower garden. How much will it cost to have this triangular garden designed?

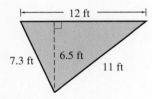

SOLUTION Because the gardener charges $5 per square foot, we need to find out how many square feet are in the triangular garden. That is, we need to find the area of the triangle.

$$\begin{aligned}\text{Area} &= \frac{1}{2}(\text{altitude})(\text{base}) \\ \text{(sq ft)} &\quad\quad\;\; \text{(ft)}\quad\;\;\text{(ft)} \\ &= \frac{1}{2}(6.5)(12) \\ &= 6(6.5) \\ &= 39\end{aligned}$$

The area of the triangular garden is 39 square feet. Now we can find the cost.

$$\begin{aligned}\text{Cost (\$)} &= \frac{\$}{\text{sq ft}} \cdot \text{sq ft} \\ &= 5 \cdot 39 \\ &= 195\end{aligned}$$

The cost for designing this garden is $195.

Try Problem 12.

5 Discovering a Formula for the Area of a Trapezoid

Next we look at how to find the area of a trapezoid. Consider trapezoid *ABCD*.

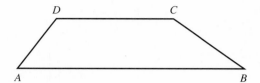

Draw a line from *D* to *B* and view the trapezoidal region as two triangular regions.

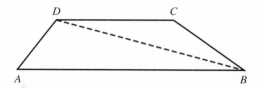

The area of the trapezoid equals the sum of the areas of the two triangles. Draw altitude *DE* of △*ABD* and altitude *BF* of △*BCD*.

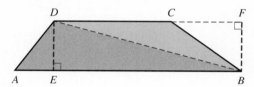

$$\text{Area of trapezoid } ABCD = \text{area of } \triangle ABD + \text{area of } \triangle BCD$$
$$= \frac{1}{2}(\text{alt})(\text{base}) + \frac{1}{2}(\text{alt})(\text{base})$$
$$= \frac{1}{2}(DE)(AB) + \frac{1}{2}(BF)(DC)$$

Because $DE = BF$, let h represent this length and substitute h for each of these lengths. Now we have the following.

$$\text{Area of trapezoid } ABCD = \frac{1}{2}h(AB) + \frac{1}{2}h(DC)$$
$$= \frac{1}{2}h(AB + DC) \qquad \text{Factor } \tfrac{1}{2}h \text{ from both terms.}$$

Segments $\overline{AB}$ and $\overline{DC}$ are called **bases** of the trapezoid. The bases are the two parallel sides. Segment $\overline{DE}$ or $\overline{BF}$ is called an **altitude** of the trapezoid. The altitude of a trapezoid is the segment drawn from one parallel side to the other and perpendicular to both. The lengths of the bases are often called bases and the lengths of the altitudes are often called altitudes.

The above discussion leads us to a formula for finding the area of a trapezoid.

Try These Problems

13. Find the area of this trapezoid.

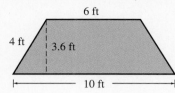

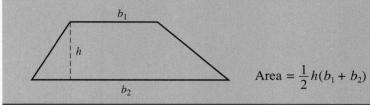

Area of a Trapezoid
The area of a trapezoid is one half of the altitude times the sum of the bases.

$$\text{Area} = \frac{1}{2}h(b_1 + b_2)$$

EXAMPLE 8 Find the area of this trapezoid.

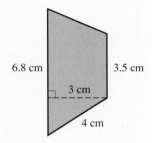

SOLUTION The altitude measures 3 centimeters and the bases measure 6.8 and 3.5 centimeters.

$$\begin{aligned}
\text{Area} &= \frac{1}{2}h(b_1 + b_2) \\
&= \frac{1}{2}(3)(6.8 + 3.5) \\
&= \frac{3}{2}(10.3) \\
&= \frac{30.9}{2} \\
&= 15.45 \text{ square centimeters} \quad \blacksquare
\end{aligned}$$

Try Problem 13.

EXAMPLE 9 The area of a trapezoid is 90 square inches. The bases measure 20 inches and 10 inches. Find the altitude.

SOLUTION First we view a picture of the situation.

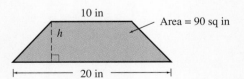

768 Chapter 15 Geometry

 Try These Problems

14. The longer base of a trapezoid is 6 feet more that the shorter base. The area of the trapezoid is 66 square feet and the altitude is 8 feet. Find the length of each base.

$$\text{Area} = \frac{1}{2}h(b_1 + b_2)$$

$$90 = \frac{1}{2}h(20 + 10)$$

$$90 = \frac{1}{2}h(30)$$

$$90 = 15h$$

$$\frac{90}{15} = \frac{15h}{15}$$

$$6 = h$$

The altitude measures 6 inches. ∎

 Try Problem 14.

 Answers to Try These Problems

1. 21.4 m **2.** 10.5 cm, 12 cm, 13.5 cm **3.** 108.8 sq cm

4. 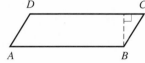 **5.**

6. 72 sq m **7.** 24 sq cm

8. 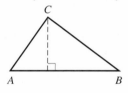 **9.**

10. 47.975 sq ft **11.** 36 sq in **12.** $200 **13.** 28.8 sq ft

14. $5\frac{1}{4}$ ft, $11\frac{1}{4}$ ft

EXERCISES 15.3

In Exercises 1 through 6, find the perimeter of each figure.

1.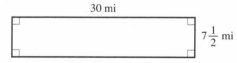

2.

3.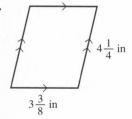

4.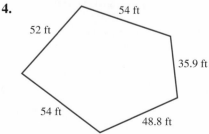

5.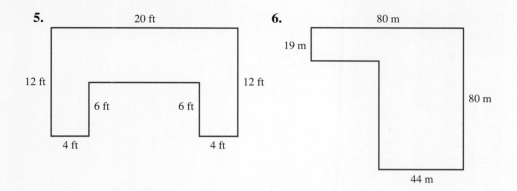

6.

7. In triangle *ABC*, draw the altitude that corresponds to base $\overline{AB}$.

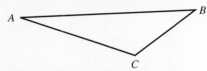

8. In triangle *GEF*, draw the altitude that corresponds to base $\overline{EF}$.

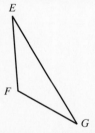

9. In trapezoid *RSTU*, draw an altitude.

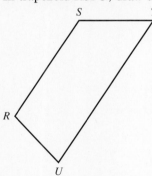

In Exercises 10 through 21, find the area of each figure.

10.   **11.**

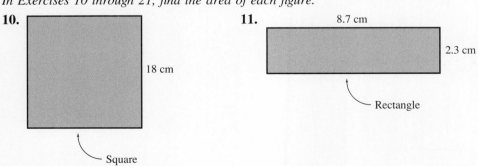

770 Chapter 15 Geometry

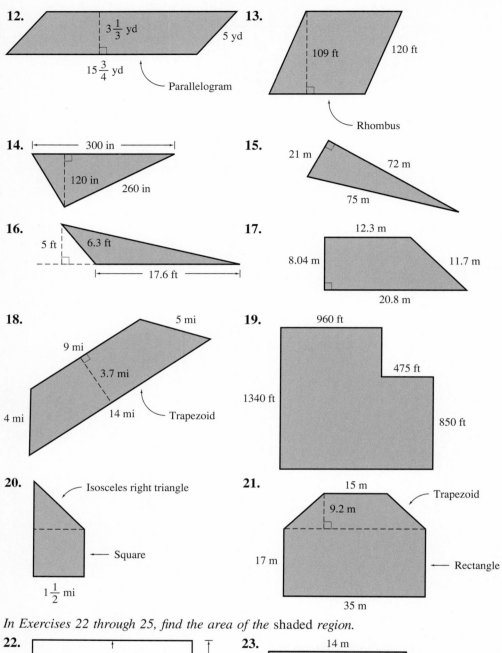

In Exercises 22 through 25, find the area of the shaded region.

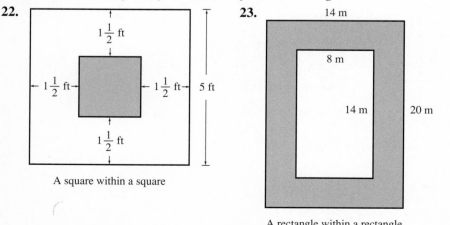

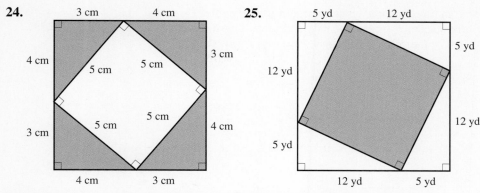

In Exercises 26 and 27, find the area of the unshaded regions.

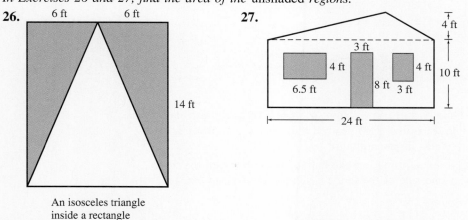

28. The area of a square is 169 square inches. Find the length of each side.

29. The length of a rectangle is 3 feet more than twice its width. The perimeter is 21 feet. Find the dimensions of the rectangle.

30. The area of a right triangle is 24 square meters. One leg measures 12 meters. Find the length of the other leg.

31. One side of a triangle measures 70 miles, the second side measures 120 miles less than the perimeter, and the third side measures 150 miles less than the perimeter. Find the length of the shortest side.

32. A plane flew from Miami to Memphis to Dallas and back to Miami. How far did the plane travel?

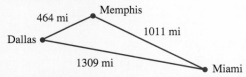

33. Metal stripping is to be attached around the edge of this metal plate. The stripping costs $1.20 per yard. What is the total cost of the stripping?

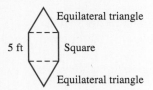

34. A farmer plans to fence in a rectangular grazing region for his cattle. He plans to take advantage of an adjacent river for one side and plans to use fencing that costs $4.50 per foot for the other three sides. How much will the fencing cost?

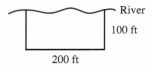

35. Carpeting costs $15.80 per square yard. How much does it cost to carpet this floor?

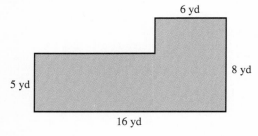

36. A metal plate in the shape of an isosceles right triangle weighs 8 grams per square centimeter. If one leg of the right triangle measures 7 centimeters, find the weight of the entire plate. (Draw a picture.)

37. A carpenter is building a fence, using trapezoidal slates as shown in the figure. What is the total area of one side of the fence if she places it along a distance of 36 feet?

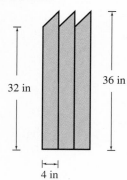

15.4 CIRCLES

OBJECTIVES
- Use a compass or ruler to draw a geometric figure that involves circles, radii, or diameters.
- Find the circumference and area of a circle.
- Find the circumference and area of a semicircular and quarter-circular region.
- Find the circumference and area of composite regions.
- Solve problems by using the formulas for the circumference and area of a circle.
- Solve applications that involve using results about circles.

Try These Problems

Use the figure to do Problems 1 through 5. Assume point A is the center of the circle.

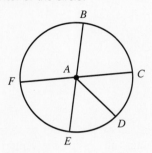

1. Name five segments that are radii.
2. Name two segments that are diameters.
3. Name five points that lie on the circle.
4. Use a ruler to measure the radius in centimeters.
5. Use a ruler to measure the diameter in centimeters.

Solve.
6. The radius of a circle is 4.5 meters. Find the diameter.
7. The diameter of a circle is 1018 inches. Find the radius.

Many objects in the world have circular shapes: wheels, compact discs, and our images of the sun and moon, to mention a few.

1 Defining a Circle and Its Basic Parts

A **circle** is the collection of all points in a plane that are the same distance from a fixed point. The fixed point is called the **center.** The distance from the center to each point on the circle is called the **radius.** We also refer to a line segment drawn from the center to a point on the circle as a radius. The plural for radius is *radii*. Here we picture a circle with center C and radius CP.

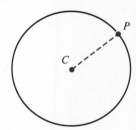

Circle with center C and radius CP

A line segment drawn from a point on the circle through the center to another point on the circle is called a **diameter.** The length of this line segment is also called a diameter. Here we picture a circle with center C and diameter PQ.

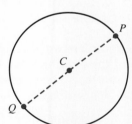

Circle with center C and diameter PQ

 Try Problems 1 through 3.

Study the picture below to see how the diameter compares to the radius.

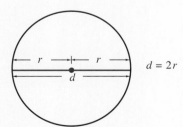

Observe that the *diameter of a circle is twice the radius.*

 Try Problems 4 through 7.

Try These Problems

For Problems 8 and 9, use a compass and ruler to draw and label the figure described.

8. A circle with center point *P* and radius *PQ* that measures $\frac{1}{2}$ inch.

9. A circle with center point *C* and diameter *AB* that measures 2 centimeters.

2 Using a Compass to Draw a Circle

An instrument used to draw a circle is called a **compass.** Here we show a compass with a pencil attached.

To draw a circle, open the compass to the desired radius, keep the sharp point at a fixed point on the paper, then rotate the compass around so that the pencil traces a circle.

 Try Problems 8 and 9.

3 Introducing the Number π and Working with a Formula for the Circumference of a Circle

The **circumference of a circle** is the distance all the way around. The ratio of the circumference to the diameter is the same for all circles. This ratio cannot be expressed as an ordinary decimal or fraction, but we can approximate it. One approximation commonly used is 3.14.

$$\frac{\text{Circumference}}{\text{Diameter}} \approx 3.14$$

Because the ratio cannot be expressed as an ordinary decimal or fraction, we introduce the symbol π to represent this ratio. That is, we have the following:

$$\frac{\text{Circumference}}{\text{Diameter}} = \pi \approx 3.14$$

Using a scientific calculator, we can get a much better approximation of π than 3.14. To obtain π, enter [INV] or [2ndF], then [π]. Here is a calculator approximation of π.

$$\pi \approx 3.1415927$$

Try These Problems

For Problems 10 and 11, express the answer in terms of π and approximate the answer to 3 decimal places.

10. Find the circumference of a circle with radius 16 feet.
11. Find the circumference of a circle with diameter 54.8 inches.

If we let C be the circumference of any circle, and d the diameter, we have the following relationship.

$$\frac{C}{d} = \pi$$

If we multiply both sides of the equation by d, we obtain a formula for the circumference of a circle in terms of the diameter.

$$\frac{C}{d} = \pi$$

$$\frac{\cancel{d}}{1} \cdot \frac{C}{\cancel{d}} = \pi d \quad \text{Multiply both sides by } d.$$

$$C = \pi d$$

Because a diameter is made up of two radii, this formula can be written in terms of the radius r by substituting $2r$ for d.

$$C = \pi d$$
$$C = \pi(2r) \quad \text{Substitute } 2r \text{ for } d.$$
$$C = 2\pi r \quad \text{The factors can be written in any order.}$$

Circumference of a Circle

The circumference of a circle is 2π times the radius, where $\pi \approx 3.1415927$.

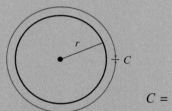

$$C = 2\pi r$$

EXAMPLE 1 Find the circumference of a circle with radius 1.8 centimeters. Express the answer in terms of π and approximate the answer to three decimal places.

SOLUTION We use the formula $C = 2\pi r$, with $r = 1.8$ centimeters.

$$C = 2\pi r$$
$$C = 2\pi(1.8)$$
$$C = 3.6\pi$$
$$C \approx 3.6(3.14159)$$
$$C \approx 11.310$$

The circumference is 3.6π centimeters or approximately 11.310 centimeters. ■

Try Problems 10 and 11.

USING THE CALCULATOR #17

WORKING WITH THE NUMBER π

In this feature we show how to do a calculation that involves π. Our discussion includes both the scientific and basic calculators.

A scientific calculator has a key $[\pi]$. It is often a second function key, so if this is the case, you will need to enter [2nd] or [INV] or [Shift] before entering $[\pi]$. Here we show how to calculate 5π on a scientific calculator.

ON A SCIENTIFIC CALCULATOR

To compute 5π
Enter   5 [×] [π] [=]
Result   15.707963

A basic calculator does not have the key $[\pi]$. You will have to enter the decimal approximation for π yourself. If you have several calculations to do with π, you could put the decimal approximation for π in the memory, then it would be available when you need it. Here we look at computing 5π on a basic calculator.

Compare the basic calculator result with the scientific calculator result. Notice that the results agree to only four decimal places. The scientific calculator result is more accurate because the scientific calculator used an approximation of π that was accurate to about 8 or 10 decimal places, while on the basic calculator, we entered 3.14159 for π which is accurate to only 5 decimal places.

On the basic calculator, you will have to enter an approximation for π. It is a good idea to enter at least two more decimal places than accuracy you are seeking. For example, if you want the answer accurate to 3 decimal places, then enter 3.14159 for π. If you want the answer accurate to only 2 decimal places, then you can get by with entering 3.1416 for π.

ON A BASIC CALCULATOR

To Compute 5π
Enter   5 [×] 3.14159 [=]
Result   15.70795

Calculator Problems

Evaluate. Approximate the answer to 3 decimal places. If you are working with a basic calculator, enter 3.14159 for π in the memory, then recall the memory as needed.

1. 9π

2. $36 - \pi$

3. $\dfrac{6\pi}{5}$

4. $8.4\pi - 6.7$

5. $\pi(7.2)^2$

6. $\sqrt{\dfrac{140}{\pi}}$

Try These Problems

For Problems 12 through 15, express the answer in terms of π and approximate the answer to 3 decimal places.

12. Find the radius of a circle with circumference 38 meters.
13. The distance around a turntable is 80 centimeters. Find the diameter of the turntable.
14. Find the perimeter of this quarter-circular region.

8 in

15. A flat piece of metal is in the shape of a semicircle. Find the perimeter of this piece of metal.

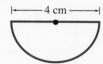
4 cm

EXAMPLE 2 The distance around a wheel is 176 inches. Find the diameter of the wheel. Express the answer in terms of π and approximate the answer to two decimal places.

SOLUTION We are given the distance around the wheel which is the circumference C of the wheel. We want to find the diameter d. The formula $C = 2\pi r$ can be written $C = \pi d$, because $2r = d$.

$$C = \pi d$$
$$176 = \pi d \quad \text{Substitute 176 for } C.$$
$$\frac{176}{\pi} = \frac{\pi d}{\pi} \quad \text{Divide both sides by } \pi.$$
$$\frac{176}{\pi} = d$$
$$d \approx \frac{176}{3.14159}$$
$$d \approx 56.02$$

The diameter of the wheel is $\dfrac{176}{\pi}$ inches or approximately 56.02 inches. ∎

Try Problems 12 and 13.

EXAMPLE 3 The figure shown here is a quarter-circle. Find its perimeter. Express the answer in terms of π and approximate the answer to 3 decimal places.

5 cm

SOLUTION To find the perimeter of this figure means to find the distance all the way around. This includes two straight line lengths of 5 centimeters each and the circular arc that is $\frac{1}{4}$ of the circumference of a circle with radius 5 centimeters. We have the following.

$$\begin{array}{c}\text{Perimeter} \\ \text{(cm)}\end{array} = \begin{array}{c}\text{radius} \\ \text{(cm)}\end{array} + \begin{array}{c}\text{radius} \\ \text{(cm)}\end{array} + \begin{array}{c}\text{quarter-circle} \\ \text{length} \\ \text{(cm)}\end{array}$$

$$= 5 + 5 + \frac{1}{4}(2\pi \cdot 5)$$
$$= 10 + \frac{1}{4}(10\pi)$$
$$= 10 + \frac{5\pi}{2}$$
$$= 10 + 2.5\pi$$
$$\approx 17.854$$

The perimeter is $10 + 2.5\pi$ centimeters, which is approximately 17.854 centimeters. ∎

Try Problems 14 and 15.

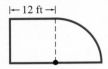 **Try These Problems**

For Problems 16 and 17, express the answer in terms of π and approximate the answer to 3 decimal places.

16. Find the perimeter of this figure, which is a square attached to a quarter-circle.

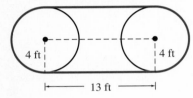

17. A belt is stretched around two wheels as shown in the diagram. How long is the belt?

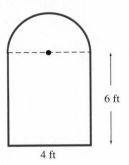

EXAMPLE 4 A Norman window is in the shape of a rectangle surmounted by a semicircle. Find the perimeter of this Norman window.

[figure: Norman window, 6 ft tall, 4 ft wide]

SOLUTION To find the distance around this window, we need to add the semicircular length to the three sides of the rectangle. First we find the length of the semicircle.

$$\text{Semicircle length} = \frac{1}{2} \cdot \text{circumference}$$
$$= \frac{1}{2} \cdot 2\pi \cdot \text{radius}$$
$$= \pi \cdot \text{radius}$$

Observe that the diameter of the circle is the same as the width across the bottom of the rectangle, 4 feet. Therefore, the radius is half of this or 2 feet. Now we substitute 2 feet for the radius.

$$\text{Semicircle length (ft)} = \pi \cdot \text{radius (ft)}$$
$$= \pi(2)$$
$$= 2\pi$$

Now we add the three sides of the rectangle to the semicircular length to find the distance all the way around the Norman window.

$$\text{Perimeter of window (ft)} = \text{semicircle length (ft)} + \text{three sides of rectangle (ft)}$$
$$= 2\pi + 6 + 6 + 4$$
$$= 2\pi + 16$$
$$\approx 22.283$$

The perimeter of the window is $2\pi + 16$ feet, which is approximately 22.283 feet. ■

 Try Problems 16 and 17.

4 Discovering and Working with a Formula for the Area of a Circle

We now perform an experiment that helps us to discover the formula for the *area of a circle*. In the experiment we cut the circular region into pieces and rearrange the pieces to form a region that is almost rectangular.

EXPERIMENT

1. Draw a circle on a piece of paper. Cut out the circle.

2. Cut the circle along a diameter to make two semicircles.

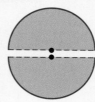

3. Draw several radii in each semicircle. Cut along the radii.

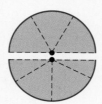

4. Fit the pieces together as shown.

The resulting shape resembles a rectangle. The more radii we cut the more like a rectangle the shape would be. The length of the rectangle is half the circumference and the width of the rectangle is the radius of the circle.

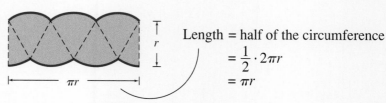

$$\text{Length} = \text{half of the circumference}$$
$$= \frac{1}{2} \cdot 2\pi r$$
$$= \pi r$$

Therefore, we have the following.

$$\begin{aligned}\text{Area of the circle} &= \text{area of the rectangle} \\ &= \text{length} \cdot \text{width} \\ &= \pi r \cdot r \\ &= \pi r^2\end{aligned}$$

This leads us to a formula for finding the area of a circle.

Try These Problems

For Problems 18 and 19, express the answer in terms of π and approximate the answer to 3 decimal places.

18. Find the area of the circle with radius 8 meters.

19. Find the area of the circle with diameter 10 inches.

Area of a Circle

The area of a circle is π times the radius squared, where $\pi \approx 3.1415927$.

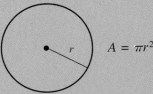

$$A = \pi r^2$$

EXAMPLE 5 Find the area of a circle with radius 5 feet. Express the answer in terms of π and approximate the answer to 2 decimal places.

SOLUTION

$A = \pi r^2$ with $r = 5$
$A = \pi(5)^2$
$A = 25\pi$
$A \approx 78.54$

The area of the circle is 25π square feet, which is approximately 78.54 square feet. ■

 Try Problems 18 and 19.

EXAMPLE 6 The area of a circle is 132 square centimeters. Find the diameter of the circle. Express the answer in terms of π and approximate the answer to 3 decimal places.

SOLUTION First we find the radius.

$A = \pi r^2$ with $A = 132$ square centimeters

$132 = \pi r^2$

$\dfrac{132}{\pi} = \dfrac{\cancel{\pi} r^2}{\cancel{\pi}}$ Divide both sides by π

$\dfrac{132}{\pi} = r^2$ To undo the square, take the positive square root of both sides.

$\sqrt{\dfrac{132}{\pi}} = r$ We want only the positive square root because r represents the length of a radius.

$r = \sqrt{\dfrac{132}{\pi}}$

Now we can find the diameter, which is 2 times the radius.

$\text{Diameter} = 2r$
$= 2\sqrt{\dfrac{132}{\pi}}$
≈ 12.964

Try These Problems

For Problems 20 through 22, express the answer in terms of π and approximate the answer to 3 decimal places.

20. Find the radius of a circle with area 5 square centimeters.
21. Find the diameter of a circle with area 60 square feet.
22. A region is bounded by a semicircle and three sides of a rectangle. Find the area of the region.

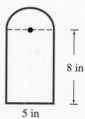

The diameter measures $2\sqrt{\dfrac{132}{\pi}}$ centimeters, which is approximately 12.964 centimeters. ■

 Try Problems 20 and 21.

EXAMPLE 7   Find the area of this hallway. It consists of a quarter-circle and two rectangles. Express the answer in terms of π and approximate the answer to the nearest tenth.

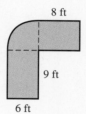

SOLUTION   First we find the area of the quarter-circle. Note that the radius is 6 feet, the same as the width of the rectangular portion of the hallway.

$$\begin{aligned}
\text{Area of quarter-circle (sq ft)} &= \tfrac{1}{4} \cdot \text{area of circle (sq ft)} \\
&= \tfrac{1}{4} \cdot \pi r^2 \\
&= \tfrac{1}{4} \cdot \pi(6)^2 \\
&= \tfrac{1}{4} \cdot 36\pi \\
&= 9\pi
\end{aligned}$$

The area of the quarter-circle is 9π square feet. Now we add the area of the quarter-circle to the areas of the two rectangular regions.

$$\begin{aligned}
\text{Area of hallway (sq ft)} &= \text{area of quarter-circle (sq ft)} + \text{areas of the two rectangles (sq ft)} \\
&= 9\pi + 6(9) + 6(8) \\
&= 9\pi + 54 + 48 \\
&= 9\pi + 102 \\
&\approx 130.3
\end{aligned}$$

The area of the hallway is $9\pi + 102$ square feet, which is approximately 130.3 square feet. ■

 Try Problem 22.

Try These Problems

Express the answer in terms of π and approximate the answer to 3 decimal places.

23. The figure shows two circles with the same center. Find the area of the shaded region.

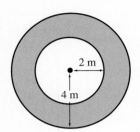

EXAMPLE 8 Find the area of the shaded region. Express the answer in terms of π and approximate the answer to the nearest hundredth.

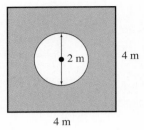

SOLUTION To obtain the area of the shaded region we need to subtract the area of the circle from the area of the entire square. Because the diameter of the circle is given to be 2 meters, the radius is 1 meter.

$$\begin{aligned}\text{Area of shaded region} &= \text{area of square} - \text{area of circle}\\ \text{(sq m)} &\quad\quad\text{(sq m)}\quad\quad\quad\text{(sq m)}\\ &= (4)(4) - \pi \cdot 1^2\\ &= 16 - \pi\\ &\approx 12.86\end{aligned}$$

The area of the shaded region is $16 - \pi$ square meters, which is approximately 12.86 square meters. ■

Try Problem 23.

Answers to Try These Problems

1. $\overline{AB}, \overline{AC}, \overline{AD}, \overline{AE}$, and $\overline{AF}$ 2. $\overline{FC}, \overline{EB}$ 3. B, C, D, E, and F
4. 1.5 cm   5. 3 cm   6. 9 cm   7. 509 in
8. 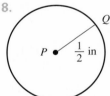   9.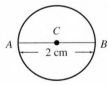

10. $32\pi \approx 100.531$ ft   11. $54.8\pi \approx 172.159$ in
12. $\dfrac{19}{\pi} \approx 6.048$ m   13. $\dfrac{80}{\pi} \approx 25.465$ cm
14. $16 + 4\pi \approx 28.566$ in   15. $2\pi + 4 \approx 10.283$ cm
16. $48 + 6\pi \approx 66.850$ ft   17. $26 + 8\pi \approx 51.133$ ft
18. $64\pi \approx 201.062$ sq m   19. $25\pi \approx 78.540$ sq in
20. $\sqrt{\dfrac{5}{\pi}} \approx 1.262$ cm   21. $2\sqrt{\dfrac{60}{\pi}} \approx 8.740$ ft
22. $3.125\pi + 40 \approx 49.817$ sq in   23. $12\pi \approx 37.699$ sq m

EXERCISES 15.4

1. Draw a circle with radii $\overline{AB}$, $\overline{AC}$, and $\overline{AD}$; and diameter $\overline{XY}$.
2. Draw a circle with center point C, diameter $\overline{DM}$, and $\vec{n}$ that intersects the circle in two points but is not a diameter.
3. The radius of a circle is $2\frac{1}{3}$ feet. Find the diameter.
4. The diameter of a circle is 5.3 centimeters. Find the radius.

In Exercises 5 through 18 express the answer in terms of π and approximate the answer to two decimal places.

5. Find the circumference and area of a circle with the given radius.
   a. 6 meters   b. $2\frac{1}{3}$ yards
6. Find the circumference and area of a circle with the given diameter.
   a. 300 feet   b. 2.8 meters
7. Find the perimeter and area of the semicircular region.

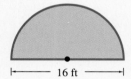

8. Find the perimeter and area of this quarter-circular region.

9. A region is bounded by a semicircle and three sides of a rectangle. Find the perimeter and area of the region.

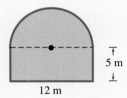

10. A region is bounded by a semicircle and two legs of a right triangle. Find the perimeter and area of the region.

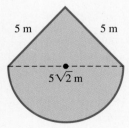

11. The shaded region is bounded by four quarter-circles. Find the perimeter and area of the shaded region.

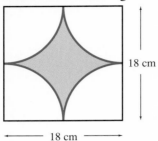

12. The figure shows a right triangle inscribed in a semicircle. Find the area of the shaded region.

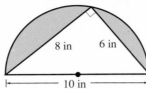

13. The circumference of an automobile tire is 200 centimeters. What is the diameter of this tire?
14. The area of a circular window is 30 square feet. Find the diameter of the window.
15. You wish to attach metal stripping around a circular wooden table top that is 2 yards in diameter. The stripping costs $4 per yard. If you buy 10% more than you need to allow for waste, how much does the stripping cost?
16. Betty plans to sew a lace border around a circular tablecloth that has a diameter of 90 inches. The lace border costs $5.60 per yard. What is the total cost of the lace?
17. A rectangular house is 50 feet by 30 feet. A dog is tied outside the house at one corner with a rope that is 20 feet long. How many square feet of ground can the dog roam?
18. A rectangular house is 50 feet by 30 feet. A dog is tied outside the house at one corner with a rope that is 40 feet long. How many square feet of ground can the dog roam?

CHAPTER 15 SUMMARY

KEY WORDS AND PHRASES

plane [15.1]
point [15.1]
geometric figure [15.1]
horizontal line [15.1]
vertical line [15.1]
oblique (slanted) line [15.1]
line segment [15.1]
endpoints [15.1]
intersecting lines [15.1]
ray [15.1]
endpoint of the ray [15.1]
angle [15.1]
vertex of an angle [15.1]
sides of an angle [15.1]
degree measurement [15.1]
right angle [15.1]
acute angle [15.1]

obtuse angle [15.1]
straight angle [15.1]
protractor [15.1]
vertical angles [15.1]
parallel lines [15.1]
perpendicular lines [15.1]
triangle [15.2]
triangle inequality [15.2]
acute triangle [15.2]
obtuse triangle [15.2]
right triangle [15.2]
legs of a right triangle [15.2]
hypotenuse of a right triangle [15.2]
isosceles triangle [15.2]
base angles [15.2]
equilateral triangle [15.2]

quadrilateral [15.2]
parallelogram [15.2]
rectangle [15.2]
square [15.2]
rhombus [15.2]
trapezoid [15.2]
perimeter [15.3]
area [15.3]
altitude (height) [15.3]
base [15.3]
circle [15.4]
center [15.4]
radius [15.4]
diameter [15.4]
compass [15.4]
circumference of a circle [15.4]
area of a circle [15.4]

SYMBOLS

$\overleftrightarrow{m}$	line m
$\overleftrightarrow{AB}$	line AB
$\overline{AB}$	line segment AB
AB	the length of the line segment, $\overline{AB}$.
$\overrightarrow{AB}$	ray AB
$\measuredangle A$	angle A
$m\measuredangle ABC$	the measure of angle ABC
$90°$	90 degrees
$\overleftrightarrow{l} \; // \; \overleftrightarrow{m}$	line l is parallel to line m.
$\overleftrightarrow{l} \perp \overleftrightarrow{m}$	line l is perpendicular to line m.
$\triangle ABC$	triangle ABC
$\square ABCD$	parallelogram $ABCD$
π	a real number that is approximately equal to 3.1415927

IMPORTANT RULES

The Sum of the Angles of a Triangle [15.2]
The sum of the measures of the angles of any triangle is 180°.

The Sum of the Angles of a Quadrilateral [15.2]
The sum of the measures of the angles of a quadrilateral is 360°.

Area Formulas [15.3]

Parallelogram   **Triangle**   **Trapezoid**

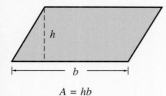   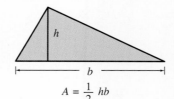   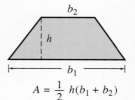

$A = hb$ $A = \frac{1}{2} hb$ $A = \frac{1}{2} h(b_1 + b_2)$

Circumference and Area of a Circle [15.4]

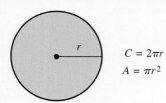

$C = 2\pi r$

$A = \pi r^2$

CHAPTER 15 REVIEW EXERCISES

In Exercises 1 through 5, use your ruler or protractor to draw each figure. Label the picture appropriately.

1. A horizontal line and a slanted line intersecting to form a pair of vertical angles measuring 50°.

2. Point P not on $\overleftrightarrow{l}$ such that the shortest distance from P to $\overleftrightarrow{l}$ is 1 centimeters.

786 Chapter 15 Geometry

3. A right triangle with one angle measuring 30°.
4. An obtuse triangle with one angle measuring 10°.
5. A rhombus with one side measuring $\frac{1}{2}$ inch.

In Exercises 6 through 13, find x without measuring.

6.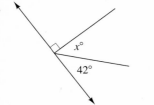

7.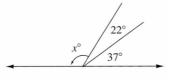

8.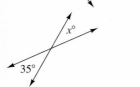

9.

10.

11.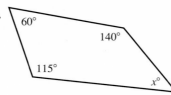

12.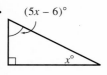

13.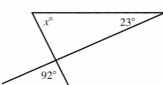

Which of these measurements can be the lengths of the sides of a triangle?

14. 18 cm, 33 cm, 17 cm
15. $5\frac{1}{2}$ yd, $3\frac{1}{2}$ yd, $2\frac{2}{3}$ yd

In Exercises 16 through 21, find the perimeter and area of each figure. Assume that angles that look like right angles are right angles. If the answer involves π, express the answer in terms of π and approximate the answer to two decimal places.

16.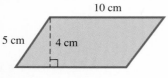

17.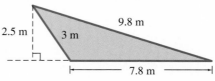

18.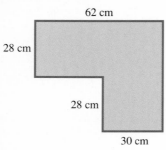

19.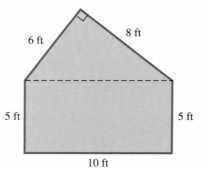

20. 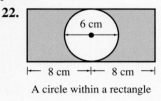 Semicircle, 8 in

21. Semicircle, 6 m, 6 m, 14 m

In Exercises 22 and 23, find the area of the shaded region. If the answer involves π, express the answer in terms of π and approximate the answer to two decimal places.

22. 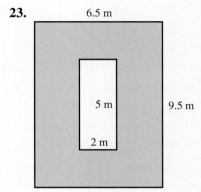 6 cm, 8 cm, 8 cm
A circle within a rectangle

23. 6.5 m, 5 m, 9.5 m, 2 m
A rectangle within a rectangle

In Exercises 24 through 30, solve. If the answer involves π, express the answer in terms of π and approximate the answer to two decimal places.

24. The perimeter of a parallelogram is 88 yards. The longest side is 3 times as long as the shortest side. Find the length of the longest side.
25. The bases of a trapezoid measure 24 feet and 28 feet. If the area of the trapezoid is 117 square feet, find the altitude.
26. Find the circumference and area of a circle with radius 19 centimeters.
27. The area of a quarter-circle is 200 square inches. Find the radius.
28. A metal plate in the shape of an isosceles right triangle weighs 7.5 ounces per square inch. If each leg of the triangle measures 16 inches, find the weight of the entire plate.
29. Jane rode in the inside lane of a merry-go-round that is 18 feet from the center. Her friend Herb rode in the outside lane that is 20 feet from the center. How much farther did Herb ride than Jane in 15 turns of the merry-go-round?
30. A water sprinkler sprays a lawn from the sprinkler to a distance of 5 meters all the way around. How many square meters of lawn does this sprinkler spray?

In Your Own Words

Write complete sentences to discuss each of the following. Support your comments with examples or pictures, if appropriate.

31. Discuss the difference between a line and a line segment.
32. Discuss the difference between parallel lines and perpendicular lines.
33. Discuss the difference between the circumference of a circle and the area of a circle.

34. A rectangle is a special parallelogram having some characteristics that differ from other parallelograms and some characteristics that are like other parallelograms. Discuss the characteristics that differ and those that are alike.

35. A right triangle has some characteristics that differ from other triangles and some characteristics that are like other triangles. Discuss the characteristics that differ and those that are alike.

CHAPTER 15 PRACTICE TEST

Use your ruler, protractor, or compass to draw and label each figure described. If it is not possible to draw the figure, briefly explain why it cannot be drawn.

1. $\angle PQR$ with measure 65°.
2. $\triangle ABC$ with $m\angle A = 20°$, $m\angle B = 60°$, and $m\angle C = 120°$.
3. A trapezoid with bases measuring 2 centimeters and 2.5 centimeters.
4. A circle with center P and diameter $\overline{AB}$ measuring 3 centimeters.

In Exercises 5 through 7, find x without measuring.

5.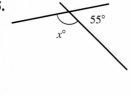
6.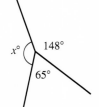
7.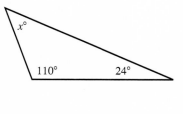

8. The figure shows an isosceles triangle inside a rectangle. Find the area of the shaded region.

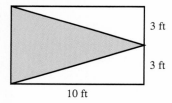

9. Find the perimeter of a parallelogram with one side measuring 8 inches and a second side measuring 12.4 inches.

10. The bases of a trapezoid measure 2.4 meters and 8.6 meters. The area is 132 square meters. Find the height.

In Exercises 11 and 12, if the answer contains π, express the answer in terms of π and approximate the answer to two decimal places.

11. The distance around a circular window is 6 feet. Find the diameter of the window.

12. The figure shows two circles with the same center. Find the area of the shaded region.

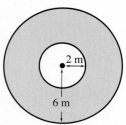

CUMULATIVE REVIEW EXERCISES: CHAPTERS 1–15

Factor each of these numbers as the product of primes.

1. 112
2. 650
3. 1449

Find the least common multiple of each group of numbers.

4. 54; 63; 14
5. 120; 400

In Exercises 6 through 12, translate each phrase to math symbols.

6. Forty-five thousand, eight hundred seven
7. Eleven twenty-thirds
8. Two and five-eighths
9. The opposite of w
10. Four less than twice y.
11. The sum of 6 and N.
12. The average of 5, w, and n.
13. Make a number line that shows the positions of the numbers $\frac{1}{5}, \frac{4}{5}, \frac{11}{5}$, and $\frac{18}{5}$.
14. Make a number line that shows the approximate positions of the numbers 2.5, 4.75, $\sqrt{3}$, $\sqrt{14}$, and π.

Evaluate for the given values of the variables.

15. $-3xy + 2y$ for $x = -6$ and $y = 25$
16. $x^2 + y^2$ for $x = 4$ and $y = 12$
17. $5\sqrt{c} - 4\sqrt{a}$ for $c = 36$ and $a = 81$
18. $400\,v^{-1} + 60$ for $v = 8$
19. $4\pi r^2$ for $r = 7.5$ (Express the answer in terms of π and approximate the answer to 2 decimal places.)
20. $\sqrt{a^2 + b^2}$ for $a = 6$ and $b = 7$ (Express the answer as a square root and approximate the answer to 2 decimal places.)

Remove parentheses and simplify.

21. $(H + 2H) + (15 - 4H)$
22. $16m - 6m(9 + 2m)$
23. $(-6y)(4xy)(3y^2)$
24. $6(9n - 7c + 6) + 5(8c - 2n + 7)$

Solve for the variable.

25. $3c + 1 = 5c - 13$
26. $\dfrac{5n}{2} + 3 = \dfrac{3}{4}$

Solve.

27. A Florida farmer harvests 51,600 pounds of oranges from one grove and 49,400 pounds of oranges from another grove. The oranges are packed in boxes with 25 pounds of oranges in each box. How many boxes are needed to pack these oranges?
28. Dr. Hoppe recommends that her patient take 4.2 grams of vitamin E each week. How many milligrams per day is this?
29. Patty worked 40 hours last week, which was $\frac{5}{6}$ times as long as Michael worked. Tom worked $\frac{5}{8}$ times as long as Michael.
    a. How long did Michael work?
    b. How long did Tom work?

30. Water is entering a reservoir at the rate of 4×10^5 gallons per minute.
    a. How much water has entered the reservoir after 30 minutes?
    b. How much water has entered the reservoir after 10^{-3} minute?
    c. How much water has entered the reservoir after 6 seconds?
    d. How long will it take for 3×10^7 gallons of water to enter the reservoir?

31. In the chemistry lab, Barnaby measured the temperature of a solution to be 25° Celsius. What is the temperature in Fahrenheit degrees? (Use the formula $F = \frac{9}{5}C + 32$.)

Set up an equation and solve.

32. The sum of two numbers is 5. Three times the larger number plus 6 times the smaller number equals 16. Find the two numbers.

33. The price of a calculator decreased by 15%. If the price after the decrease was $18.70, what was the original price of the calculator?

34. Mr. Conrad, a writer, invests his entire royalty check of $15,400. He buys some shares of transportation stock at $64 per share and some shares of utility stock at $36 per share. If the number of shares of ultility stock is 150 more than the number of shares of transportation stock, how many shares of each type of stock did he purchase?

The figure shows a right triangle inscribed in a semicircle. Use the figure to solve Exercises 35 through 39. If the answer contains π, express the answer in terms of π and approximate the answer to 2 decimal places.

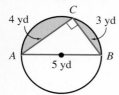

35. Find the perimeter of $\triangle ABC$.
36. Find the area of $\triangle ABC$.
37. Find the circumference of the circle.
38. Find the area of the circle.
39. Find the area of the shaded region.

APPENDIX TABLES

1. Multiplication Facts
2. Primes Less Than 100
3. Common Conversions: Fraction-Decimal-Percent
4. Perfect Squares
5. Choosing the Correct Operation
6. Percent Formulas
7. Units of Time
8. Units of Measurement: English System
9. Metric Prefixes
10. Units of Measurement: Metric System
11. English-Metric Conversions
12. Geometric Formulas

TABLE 1 MULTIPLICATION FACTS

×	0	1	2	3	4	5	6	7	8	9
0	0	0	0	0	0	0	0	0	0	0
1	0	1	2	3	4	5	6	7	8	9
2	0	2	4	6	8	10	12	14	16	18
3	0	3	6	9	12	15	18	21	24	27
4	0	4	8	12	16	20	24	28	32	36
5	0	5	10	15	20	25	30	35	40	45
6	0	6	12	18	24	30	36	42	48	54
7	0	7	14	21	28	35	42	49	56	63
8	0	8	16	24	32	40	48	56	64	72
9	0	9	18	27	36	45	54	63	72	81

TABLE 2 PRIMES LESS THAN 100

2	3	5	7	11
13	17	19	23	29
31	37	41	43	47
53	59	61	67	71
73	79	83	89	97

TABLE 3 COMMON CONVERSIONS: FRACTION-DECIMAL-PERCENT

Fraction	Decimal	Percent	Fraction	Decimal	Percent
$\frac{1}{100}$	0.01	1%	$\frac{3}{5}$	0.60	60%
$\frac{1}{20}$	0.05	5%	$\frac{5}{8}$	0.625	$62\frac{1}{2}$%
$\frac{1}{16}$	0.0625	$6\frac{1}{4}$%	$\frac{2}{3}$	$0.\overline{66}$	$66\frac{2}{3}$%
$\frac{1}{12}$	$0.08\overline{3}$	$8\frac{1}{3}$%	$\frac{7}{10}$	0.70	70%
$\frac{1}{10}$	0.10	10%	$\frac{3}{4}$	0.75	75%
$\frac{1}{8}$	0.125	$12\frac{1}{2}$%	$\frac{4}{5}$	0.80	80%
$\frac{1}{6}$	$0.1\overline{6}$	$16\frac{2}{3}$%	$\frac{5}{6}$	$0.8\overline{3}$	$83\frac{1}{3}$%
$\frac{1}{5}$	0.20	20%	$\frac{7}{8}$	0.875	$87\frac{1}{2}$%
$\frac{1}{4}$	0.25	25%	$\frac{9}{10}$	0.90	90%
$\frac{3}{10}$	0.30	30%	1	1.00	100%
$\frac{1}{3}$	$0.3\overline{3}$	$33\frac{1}{3}$%	$1\frac{1}{4}$	1.25	125%
$\frac{3}{8}$	0.375	$37\frac{1}{2}$%	$1\frac{1}{2}$	1.50	150%
$\frac{2}{5}$	0.40	40%	$1\frac{3}{4}$	1.75	175%
$\frac{1}{2}$	0.50	50%	2	2.00	200%

TABLE 4 PERFECT SQUARES

$1^2 = 1$	$6^2 = 36$	$11^2 = 121$	$16^2 = 256$	$21^2 = 441$
$2^2 = 4$	$7^2 = 49$	$12^2 = 144$	$17^2 = 289$	$22^2 = 484$
$3^2 = 9$	$8^2 = 64$	$13^2 = 169$	$18^2 = 324$	$23^2 = 529$
$4^2 = 16$	$9^2 = 81$	$14^2 = 196$	$19^2 = 361$	$24^2 = 576$
$5^2 = 25$	$10^2 = 100$	$15^2 = 225$	$20^2 = 400$	$25^2 = 625$

TABLE 5 CHOOSING THE CORRECT OPERATION

Operation	Situation
Addition +	1. You are looking for the total or whole. 2. You are looking for the result when a quantity has been increased. 3. You are looking for the perimeter of a figure.
Subtraction −	1. You are looking for one of the parts in a total or whole. 2. You are looking for the result when a quantity has been decreased. 3. You are looking for how much larger one quantity is than another. 4. You are looking for how much smaller one quantity is than another.
Multiplication ×	1. You are looking for a total quantity. $$\text{Total quantity} = \text{number of equal parts} \times \text{size of each part}$$ 2. You are looking for a total as in the following example. $$\text{Total miles} = \text{miles per hour} \times \text{number of hours}$$ $$= \frac{\text{miles}}{\text{hour}} \times \text{hours}$$ 3. You are looking for a number that is a certain amount times as large as another number. 4. You are looking for the area of a rectangle. $$\text{Area} = \text{length} \times \text{width}$$ 5. You are looking for a fraction of a number.
Division ÷	1. A total quantity is being separated into a number of equal parts. $$\frac{\text{Size of}}{\text{each part}} = \text{total quantity} \div \text{number of equal parts} = \frac{\text{total quantity}}{\text{number of equal parts}}$$ $$\frac{\text{Number of}}{\text{equal parts}} = \text{total quantity} \div \text{size of each part} = \frac{\text{total quantity}}{\text{size of each part}}$$ 2. You are looking for a missing multiplier in a multiplication statement. $$\frac{\text{Missing}}{\text{multiplier}} = \frac{\text{answer to the}}{\text{multiplication problem}} \div \frac{\text{given}}{\text{multiplier}}$$ 3. You are looking for the average of a collection of numbers. $$\frac{\text{Average of a}}{\text{collection of numbers}} = \frac{\text{sum of}}{\text{the numbers}} \div \frac{\text{how many}}{\text{numbers}}$$ 4. You are looking for an average rate as in the following example. $$\frac{\text{average rate}}{\text{in miles per hour}} = \frac{\text{number of miles}}{\text{number of hours}}$$ 5. You are looking for what ratio one number is to another number.

TABLE 6 PERCENT FORMULAS

Converting a Percent to a Decimal or Fraction	Drop the percent symbol and divide by 100.
Converting Decimals and Fractions to Percent Form	Multiply by 100 and attach the percent symbol.
Percent of a Number	Percent of a number = percent (%) × the number Amount = rate (%) × base
Percent as a Rate	Rate (%) = $\dfrac{\text{amount}}{\text{base}}$
What Percent of B is A?	Percent (%) = $\dfrac{A}{B}$
Sales Tax	Sales tax = sales tax rate (%) × marked price
Commission Based on Total Sales	Commission = commission rate (%) × total sales
Simple Interest Earned in One Year (Yield)	Yield = rate of return (%) × amount invested
Interest (Finance Charge) Paid on an Unpaid Balance	Interest = interest rate (%) × unpaid balance
Tip (Gratuity)	Tip = rate (%) × dinner bill
Percent Increase	Increase = percent increase (%) × original amount
Percent Decrease	Decrease = percent decrease (%) × original amount
Markup Based on Cost	Markup = markup rate (%) × cost
Markup Based on Selling Price	Markup = markup rate (%) × selling price

TABLE 7 UNITS OF TIME

1 calendar year (yr) = 365 days (da)
1 year = 12 months (mo)
1 year ≈ 52 weeks (wk)
1 week = 7 days
1 day = 24 hours (hr)
1 hour = 60 minutes (min)
1 minute = 60 seconds (sec)

TABLE 8 UNITS OF MEASUREMENT: ENGLISH SYSTEM

LENGTH
1 mile (mi) = 5280 feet (ft)
1 yard (yd) = 3 feet
1 yard = 36 inches (in)
1 foot (ft) = 12 inches

AREA
1 square mile (sq mi) = 640 acres
1 acre = 4840 square yards (sq yd)
1 square yard = 9 square feet (sq ft)
1 square yard = 1296 square inches (sq in)
1 square foot = 144 square inches

VOLUME
1 cubic yard (cu yd) = 27 cubic feet (cu ft)
1 cubic foot = 1728 cubic inches (cu in)
1 cubic foot = 7.48 gallons (gal)
1 gallon = 231 cubic inches
1 gallon = 4 quarts (qt)
1 quart = 2 pints (pt)
1 pint = 2 cups (c)
1 cup = 8 fluid ounces (fl oz)
1 fluid ounce = 2 tablespoons (T)
1 tablespoon = 3 teaspoons (t)

WEIGHT
1 ton = 2000 pounds (lb)
1 pound = 16 ounces (oz)
1 pound = 7000 grains (gr)
1 ounce = 437.5 grains

TABLE 9 METRIC PREFIXES

Prefix	Meaning
kilo-	1000 ×
hecto-	100 ×
deka-	10 ×
deci-	0.1 ×
centi-	0.01 ×
milli-	0.001 ×

TABLE 10 UNITS OF MEASUREMENT: METRIC SYSTEM

LENGTH
1 kilometer (km) = 1000 meters (m)
1 hectometer (hm) = 100 meters
1 dekameter (dam) = 10 meters
1 meter = 10 decimeters (dm)
1 meter = 100 centimeters (cm)
1 meter = 1000 millimeters (mm)

AREA
1 square kilometer (sq km) = 1,000,000 square meters (sq m)
1 hectare (ha) = 1 square hectometer (sq hm)
1 hectare = 10,000 square meters
1 are = 1 square dekameter (sq dam)
1 are = 100 square meters
1 square meter = 100 square decimeters (sq dm)
1 square meter = 10,000 square centimeters (sq cm)
1 square meter = 1,000,000 square millimeters (sq mm)

VOLUME
1 liter (ℓ) = 1 cubic decimeter (cu dm)
1 liter = 1000 cubic centimeters (cu cm)
1 liter = 1000 milliliters (mℓ)
1 milliliter = 1 cubic centimeter
1 milliliter = 1000 cubic millimeters (cu mm)

WEIGHT
1 kilogram (kg) = 1000 grams (g)
1 hectogram (hg) = 100 grams
1 dekagram (dag) = 10 grams
1 gram = 10 decigrams (dg)
1 gram = 100 centigrams (cg)
1 gram = 1000 milligrams (mg)

TABLE 11 ENGLISH-METRIC CONVERSIONS

LENGTH
1 mi ≈ 1.6 km
0.62 mi ≈ 1 km
39.37 in ≈ 1 m
1 in ≈ 2.54 cm

VOLUME
1.06 qt ≈ 1 ℓ
33.9 fl oz ≈ 1 ℓ

WEIGHT
2.2 lb ≈ 1 kg
1 lb ≈ 454 g

TABLE 12 GEOMETRIC FORMULAS

Rectangle
$P = 2\ell + 2w$
$A = \ell w$

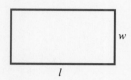

Rectangular Solid
$V = \ell w h$

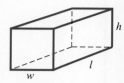

Triangle
$A = \dfrac{1}{2} bh$

Parallelogram
$A = bh$

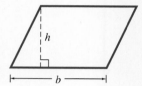

Trapezoid
$A = \dfrac{1}{2} h(b_1 + b_2)$

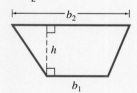

Circle
$C = 2\pi r$
$A = \pi r^2$

ANSWERS TO SELECTED EXERCISES

CHAPTER 1

Exercises 1.1 (page 6)
1. 5 **3.** 10 (ten) **5.** 5 **7.** 100,000,000 (one hundred million) **9.** 10,000 **11.** Aron **13.** Yes **15.** 14 **17.** 280 **19.** 2600 **21.** 15,500,000 **23.** 700,000 people **25.** forty-seven **27.** eight hundred six **29.** five thousand five hundred or fifty-five hundred **31.** one hundred three thousand, eighty-five **33.** five billion, seventy million **35.** thirty-five thousand dollars **37.** 900 **39.** 6830 **41.** 5000 **43.** 5,090,000 **45.** 45,000,000

Exercises 1.2 (page 12)
1. 44 **3.** 20 **5.** 21 **7.** 847 **9.** 31,035 **11.** 411 **13.** 1,195,598 **15.** $32 **17.** $1012 **19.** 38 **21.** 592 **23.** 2533 **25.** 2324 **27.** 59,805 **29.** 48,575 **31.** 233 **33.** 4962 **35.** 69,700 **37.** 448,458 **39.** 329,707 **41.** 31,638 **43.** 609,318 **45.** $38 **47.** 2063 mi **49.** 16 **51.** 19 **53.** 165 **55.** 697

Exercises 1.3 (page 17)
1. 18 + 17 = 35 **3.** 54 = 63 − 9 **5.** 50 + 360 = 410 **7.** 1023 + 978 = 2101 **9.** 3 = 7 − 4 **11.** 325 **13.** 3802 **15.** 2094 **17.** 6026 **19.** 6936

Exercises 1.4 (page 22)
1. $210 **3.** 6997 **5.** 32,000 pages **7.** decrease; $102,000 **9.** Burton; $340 **11.** 39 cm **13.** 154 ft **15.** $6500 **17.** $200,000 **19.** $80,000 **21.** increase; $40,000

Developing Number Sense #1 (page 13)
1. 1100 **2.** 800 **3.** 900 **4.** 1300 **5.** 16,000 **6.** 29,000 **7.** 4580 **8.** 856,000 **9.** 194,000

Using the Calculator #2 (page 14)
1. 12 **2.** 85 **3.** 9275 **4.** 591,968 **5.** 22 **6.** 258 **7.** 338 **8.** 5256 **9.** 54 **10.** 6463

Chapter 1 Review Exercises (page 25)
1. 3 **2.** 5 **3.** 1 (one) **4.** 10,000 (ten thousand) **5.** 0 **6.** 10,000,000 (ten million) **7.** 2099; 2200; 3000 **8.** 720 **9.** 1100 **10.** 18,000 **11.** 803,000 **12.** 12 **13.** 308 **14.** 57,024 **15.** 2,080,911 **16.** 65,000,000,000 **17.** eighteen **18.** four hundred eighty-nine **19.** two hundred six thousand, eight hundred one **20.** three billion, eight hundred thousand, five **21.** 665 mi **22.** 227 **23.** 750 **24.** 9239 **25.** 110,152 **26.** 177,289 **27.** 2577 **28.** 20,265 **29.** 12,021,626,700 **30.** 708 **31.** 7308 **32.** 6797 **33.** 70,068 **34.** 209,363 **35.** 319 **36.** 2906 **37.** 4935 **38.** 5297 **39.** 32,964 **40.** 149,000 **41.** 1,926,681 **42.** 9,015,513 **43.** 203,463 **44.** 428 **45.** 24,114 **46.** 683 **47.** 593,097 **48.** 92,800 **49.** 202,968,274 **50.** 174 **51.** 19,785 **52.** 328 **53.** 5090 **54.** 185,438 **55.** 10,503 **56.** 1000 − 72 = 928 **57.** 4300 + 486 = 4786 **58.** 26 + 193 = 219 **59.** 800 − 108 = 692 **60.** 13 **61.** 1702 **62.** 2048 **63.** 3804 **64.** 14 mi **65.** 245,222 **66.** $2020 **67.** 96 sq mi **68.** 190 ft **69.** $825 **70.** 39°F **71.** 46 ft **72.** 555 ft **73.** $192 **74.** $173 **75.** 363 mi **76.** Nancy; $4120

Chapter 1 Practice Test (page 28)

1. 6   **2.** 3   **3.** 100 (hundred)   **4.** 10,000,000 (ten million)   **5.** 746   **6.** ninety thousand
7. 98; 799; 801; 1002   **8.** 700   **9.** 98,000   **10.** 8123   **11.** 82,717   **12.** 175,469   **13.** 7087
14. 490   **15.** 415   **16.** 1596   **17.** 5826   **18.** 49,654   **19.** 39,180   **20.** 1171   **21.** 182
22. 1331   **23.** 452   **24.** 4908   **25.** $619   **26.** 21 ft   **27.** $27,800   **28.** 319 workers

CHAPTER 2

Exercises 2.1 (page 35)

1. 438   **3.** 924   **5.** 5400   **7.** 38,815   **9.** 2752   **11.** 8525   **13.** 201,344   **15.** 376,272
17. 280,232   **19.** 7,940,184   **21.** 1200   **23.** 72,000,000   **25.** 192,000   **27.** 170,000,000
29. 23,220   **31.** 12,587,146   **33.** 186,215,960   **35.** 195,156,000   **37.** 60   **39.** 432   **41.** 648
43. 14,250   **45.** $2844   **47.** $3990

Exercises 2.2 (page 45)

1. 9 R3   **3.** 65 R6   **5.** 874 R4   **7.** 1031 R1   **9.** 9 R20   **11.** 662 R60   **13.** 218 R15
15. 8893 R16   **17.** 32 R400   **19.** 62 R100   **21.** 235   **23.** 42,240   **25.** 600 R4   **27.** 6007 R68
29. 50,040   **31.** 8002 R20   **33.** $252   **35.** 40 chandeliers; 10 bulbs left over   **37.** 40 hr

Exercises 2.3 (page 56)

1. 12   **3.** 11   **5.** 300   **7.** 4,048,000   **9. a.** yes   **b.** yes   **c.** no   **d.** yes
11. a. yes   **b.** no   **c.** yes   **d.** yes   **13. a.** yes   **b.** yes   **c.** yes   **d.** yes
15. a. yes   **b.** yes   **c.** yes   **d.** no   **17. a.** yes   **b.** yes   **c.** yes   **d.** no
19. a. no   **b.** yes   **c.** yes   **d.** yes   **21. a.** no   **b.** yes   **c.** yes   **d.** yes
23. 3×3 **25.** $2 \times 2 \times 3$ **27.** $3 \times 3 \times 5$ **29.** 5×11 **31.** 2×13 **33.** 3×19
35. $2 \times 3 \times 19$ **37.** $5 \times 7 \times 7$ **39.** $2 \times 3 \times 5 \times 5$ **41.** $2 \times 2 \times 2 \times 2 \times 3 \times 5 \times 5$
43. $3 \times 3 \times 3 \times 11$ **45.** 3×29 **47.** $2 \times 2 \times 2 \times 3 \times 3 \times 7$

Exercises 2.4 (page 61)

1. $9\overline{)54}^{\,6}$ **3.** $30\overline{)1350}^{\,45}$ **5.** $15\overline{)75}^{\,5}$ **7.** $16 \div 8 = 2$ **9.** $600 \div 50 = 12$ **11.** $621 \div 3 = 207$
13. into   **15.** by   **17.** by   **19.** into   **21.** $17 \times 38 = 646$ **23.** $3 \times 11 = 33$ **25.** $2 \times 230 = 460$
27. 12   **29.** 162   **31.** 78,000   **33.** 19,000

Exercises 2.5 (page 75)

1. 120 min   **3.** 17 lb   **5.** 205 chandeliers; 10 bulbs   **7.** 32 mi   **9.** $530   **11.** 40 yd   **13.** $50,400
15. 55 mi   **17.** 32 hr   **19.** 105 hr   **21.** 1425 sq in   **23.** 45 in   **25.** 1620 sq in   **27.** $760
29. 220 sq ft   **31.** $1280   **33.** 110 mm   **35.** $37,500   **37.** $2550   **39.** 31 in   **41.** 73   **43.** 24 mi
45. $45,800   **47.** $28,627   **49.** 260 lb   **51.** 120 lb   **53.** 160 lb

Developing Number Sense #2 (page 46)

1. 560   **2.** 600   **3.** 1800   **4.** 48,000   **5.** 150,000   **6.** 21,000   **7.** 160,000   **8.** 14,000,000
9. 40   **10.** 20   **11.** 50   **12.** 225

Using the Calculator #3 (page 47)

1. 72   **2.** 3,900,000   **3.** 5278   **4.** 673,644   **5.** 2   **6.** 906   **7.** 887   **8.** 264   **9.** 4
10. 6080

Using the Calculator #4 (page 57)

4. 20   **5.** 6843   **6.** 135,000   **7.** 625   **8.** 23   **9.** 13×17 **10.** 7×23 **11.** 19×111
12. $13 \times 17 \times 19$

Using the Calculator #5 (page 78)

1. 75   **2.** 42,942   **3.** 20   **4.** 5225   **5.** $1181   **6.** 184 sq ft

Chapter 2 Review Exercises (page 80)

1. 1148 2. 12,623 3. 11,178 4. 217,674 5. 7,468,606 6. 2,625,104 7. 154,080
8. 6,717,512 9. 425,000 10. 5,760,000 11. 63,856 12. 15,460,242 13. 48 14. 126
15. 60 16. 980 17. 7 R3 18. 30 19. 839 20. 5415 R4 21. 2 R20 22. 600 R22
23. 555 24. 2003 R16 25. 1056 R8 26. 16 R346 27. 82 R220 28. 860 R400
29. 1751 R347 30. 5060 31. 25 32. 7 33. 384 34. 15 35. 436 36. 36,000
37. a. no b. yes c. yes d. no 38. a. no b. no c. yes d. yes
39. a. yes b. yes c. yes d. no 40. a. yes b. yes c. no d. yes
41. a. yes b. yes c. no d. yes 42. a. yes b. yes b. yes d. yes
43. a. no b. no c. yes d. yes 44. a. no b. no c. yes d. yes
45. $2 \times 3 \times 3$ 46. $2 \times 3 \times 3 \times 3$ 47. $2 \times 2 \times 2 \times 3 \times 7$ 48. $3 \times 3 \times 7 \times 7$
49. $2 \times 3 \times 5 \times 5 \times 5$ 50. $2 \times 2 \times 5 \times 7$ 51. $2 \times 2 \times 3 \times 3 \times 3 \times 5 \times 5$
52. $2 \times 2 \times 2 \times 2 \times 2 \times 2 \times 3 \times 5 \times 5 \times 5$ 53. $3 \times 3 \times 3 \times 13$ 54. $3 \times 5 \times 17$
55. $4\overline{)12}^{\,3}$ 56. $5\overline{)100}^{\,20}$ 57. $72 \div 8 = 9$ 58. $325 \div 25 = 13$ 59. into 60. by 61. by
62. into 63. $3 \times 16 = 48$ 64. $2 \times 400 = 800$ 65. $174 \div 6 = 29$ 66. $585 \div 13 = 45$
67. $4 \times 17 = 68$ 68. $60 \times 400 = 24,000$ 69. $9 \times 9 = 81$ 70. $8080 \div 16 = 505$ 71. 201
72. 6,000,000 73. 34 74. 3 75. 1900 76. 44 77. 8, 16, 24, 32, 40 78. yes; $17 \times 4 = 68$
79. $2250 80. $105 81. 122 82. 1080 assistants 83. 42 hr 84. $13,500 85. 154 sq ft
86. 7203 sq ft 87. 15 min 88. 194 m 89. $130,000 90. $11

Chapter 2 Practice Test (page 82)

1. 73,896 2. 538,479 3. 221,844 4. 4,800,000 5. 949,000 6. 63,856 7. 16,495,338
8. 42 9. 525 10. 839 11. 6082 R2 12. 555 13. 2003 R16 14. 860 R400 15. 5060
16. 1230 17. 405,000 18. yes 19. yes 20. $2 \times 3 \times 7$ 21. $2 \times 5 \times 5 \times 17$ 22. by
23. into 24. 454 25. 833 26. $325 27. 250 packages; no ears left over
28. 42 in 29. 29 oz 30. 25 cm 31. $3640 32. 1280 mi 33. 15 min

Cumulative Review Exercises: Chapters 1–2 (page 84)

1. sixty-four thousand, ninety-eight 2. 2,013,500 3. 150,000 4. 386,000 5. 7248
6. 238,000,000 7. 9060 8. 256 R 26 9. 4004 R 100 10. 98 R 228 11. $2 \times 2 \times 3 \times 3$
12. $5 \times 5 \times 11$ 13. $2 \times 2 \times 2 \times 3 \times 3 \times 5 \times 5 \times 5$ 14. $7 \times 7 \times 13$ 15. 6 16. 512
17. 4661 18. 1,410,640 19. 1461 20. 80 21. 135 22. 18
23. a. 98,000 sq ft b. 1260 ft 24. $165 25. $109 26. $275 27. 150 mi 28. 75 hr
29. a. 435 b. 87 c. 19 30. 142 customers 31. under, $34

CHAPTER 3

Exercises 3.1 (page 89)

1. $\dfrac{1}{4}$ 3. $\dfrac{1}{5}$ 5. $\dfrac{5}{6}$ 7. $\dfrac{3}{8}$ 9. $\dfrac{5}{5}$ or 1 11. $\dfrac{4}{3}, 1\dfrac{1}{3}$ 13. $\dfrac{13}{9}, 1\dfrac{4}{9}$ 15. $\dfrac{23}{6}, 3\dfrac{5}{6}$ 17. $\dfrac{8}{4}, 2$
19. $\dfrac{7}{5}, \dfrac{8}{3}, \dfrac{30}{13}$ 21. $\dfrac{8}{8}, \dfrac{4}{4}, \dfrac{13}{13}$ 23. $\dfrac{2}{5}, \dfrac{7}{12}$

Exercises 3.2 (page 93)

1. $\dfrac{1}{2}$ 3. $\dfrac{12}{5}$ 5. $\dfrac{2}{13}$ 7. $\dfrac{11}{30}$ 9. $\dfrac{5}{10}$ 11. $\dfrac{36}{1000}$ 13. seven-halves 15. six-elevenths
17. one hundredth 19. $6\dfrac{2}{5}$ 21. $60\dfrac{3}{100}$ 22. $5\dfrac{13}{1000}$ 23. nine and three-tenths
24. eighty and one-half 25. fourteen and seventeen-hundredths

Exercises 3.3 (page 98)

1. $\frac{2}{8}, \frac{1}{4}$ 3. $\frac{6}{10}, \frac{3}{5}$ 5. $\frac{6}{9}, \frac{2}{3}$ 7. $\frac{10}{12}, \frac{5}{6}$ 9. 15   11. 40   13. 143   15. 28   17. 8   19. 6
21. 35   23. 126   25. $\frac{8}{24}$ 27. $\frac{30}{42}$ 29. $\frac{60}{15}$

Exercises 3.4 (page 102)

1. 5   3. $7\frac{1}{5}$ 5. $40\frac{5}{6}$ 7. 25   9. $10\frac{11}{20}$ 11. 301   13. $\frac{13}{4}$ 15. $\frac{37}{5}$ 17. $\frac{55}{4}$ 19. $\frac{174}{5}$
21. $\frac{41}{11}$ 23. $\frac{159}{10}$ 25. not possible   27. 3   29. 2   31. 1   33. not possible   35. $2\frac{1}{13}$

Exercises 3.5 (page 108)

1. $\frac{2}{3}$ 3. $\frac{3}{2}$ or $1\frac{1}{2}$ 5. $\frac{3}{5}$ 7. $\frac{3}{4}$ 9. not possible   11. $\frac{3}{10}$ 13. $\frac{9}{2}$ or $4\frac{1}{2}$
15. $\frac{8}{7}$ or $1\frac{1}{7}$ 17. not possible   19. $\frac{3}{5}$ 21. $\frac{7}{11}$ 23. $\frac{3}{17}$ 25. $\frac{13}{2}$ or $6\frac{1}{2}$ 27. $\frac{11}{20}$
29. $\frac{11}{3}$ or $3\frac{2}{3}$ 31. $\frac{2}{7}$ 33. $\frac{5}{12}$ 35. $\frac{3}{8}$ 37. $\frac{3}{2}$ or $1\frac{1}{2}$ 39. $\frac{5}{4}$ or $1\frac{1}{4}$ 41. $\frac{1}{2}$ 43. $\frac{7}{9}$
45. $\frac{4}{5}$ 47. $\frac{35}{26}$ or $1\frac{9}{26}$ 49. $\frac{13}{15}$ 51. $\frac{7}{8}$ 53. $\frac{20}{7}$ or $2\frac{6}{7}$ 55. $\frac{9}{50}$ 57. $\frac{3}{20}$ 59. $\frac{23}{710}$

Exercises 3.6 (page 118)

1. 4 gal   3. $1\frac{1}{2}$ lb   5. $\frac{1}{5}$ ft   7. $\frac{1}{3}$ qt   9. $\frac{100}{3}$ or $33\frac{1}{3}$ km   11. $\frac{2}{7}$ 13. $\frac{2}{5}$ 15. $\frac{7}{3}$ or $2\frac{1}{3}$
17. $\frac{3}{5}$ 19. $1 \div 8$ 21. $19 \div 5$ 23. $3 \div 4$ 25. $4 \div 15$ 27. $12\overline{)1}$ 29. $5\overline{)12}$
31. $13\overline{)8}$ 33. $6\overline{)1}$ 35. $\frac{2}{3}$ 37. $\frac{4}{5}$ 39. $\frac{1}{5}$ 41. $\frac{1}{4}$ 43. $\frac{3}{11}$ 45. $3\frac{2}{5}$ 47. $\frac{7}{10}$ 49. $\frac{1}{200}$
51. $\frac{2}{3}$ 53. $\frac{3}{2}$ 55. $\frac{3}{4}$ 57. $\frac{8}{21}$ 59. $\frac{23}{50}$ 61. $\frac{7}{15}$ 63. $\frac{7}{20}$ 65. $\frac{34}{5}$ 67. $\frac{1}{9}$ 69. $\frac{1}{6}$ 71. $\frac{1}{5}$
73. $\frac{3}{8}$ 75. $\frac{1}{4}$

Using the Calculator #6 (page 109)

2. $\frac{7}{3}$ 3. $\frac{113}{6}$ 4. $\frac{971}{79}$ 5. $\frac{83}{4}$ 6. $3\frac{1}{9}$ 7. $23\frac{2}{15}$ 8. $28\frac{1}{4}$ 9. $3\frac{5}{6}$ 10. $\frac{2}{3}$
11. $\frac{50}{17}$ or $2\frac{16}{17}$ 12. $\frac{2}{5}$ 13. $\frac{69}{8}$ or $8\frac{5}{8}$ 15. $\frac{5{,}567{,}076}{6255}$ 17. $\frac{77}{221}$

Chapter 3 Review Exercises (page 122)

1. one-half   2. two-thirds   3. six-thirteenths   4. three and one-fourth
5. fourteen and two-fifths   6. twenty and thirteen one-hundredths
7. $\frac{2}{5}$ 8. $\frac{5}{9}$ 9. $\frac{9}{2}, 4\frac{1}{2}$ 10. $\frac{17}{6}, 2\frac{5}{6}$ 11. $\frac{8}{4}, 2$ 12. $\frac{15}{3}, 5$ 13. $\frac{21}{5}, \frac{11}{2}$ 14. $\frac{130}{21}, \frac{19}{4}$
15. $\frac{4}{4}, \frac{30}{30}$ 16. $\frac{15}{15}, \frac{1}{1}, \frac{7}{7}$ 17. $\frac{6}{9}, \frac{2}{3}$ 18. $\frac{8}{10}, \frac{4}{5}$ 19. $\frac{2}{6}, \frac{1}{3}$ 20. $\frac{6}{9}, \frac{2}{3}$ 21. 28   22. 156   23. 12
24. 180   25. $\frac{16}{56}$ 26. $\frac{32}{4}$ 27. $5\frac{7}{8}$ 28. 20   29. 19   30. $20\frac{3}{4}$ 31. $\frac{37}{7}$ 32. $\frac{141}{4}$
33. $\frac{77}{6}$ 34. $\frac{167}{23}$ 35. $5\frac{1}{3}$ 36. not possible   37. 1   38. not possible   39. 4   40. $20\frac{5}{6}$
41. $\frac{3}{10}$ 42. $\frac{3}{2}$ or $1\frac{1}{2}$ 43. not possible   44. $\frac{4}{7}$ 45. $\frac{9}{7}$ or $1\frac{2}{7}$ 46. not possible   47. $\frac{8}{15}$

48. $\dfrac{10}{27}$ 49. $\dfrac{9}{10}$ 50. $\dfrac{49}{6}$ or $8\dfrac{1}{6}$ 51. $\dfrac{11}{5}$ or $2\dfrac{1}{5}$ 52. $\dfrac{18}{35}$ 53. not possible 54. $\dfrac{2}{3}$ 55. 5
56. $\dfrac{3}{20}$ 57. $\dfrac{13}{3}$ or $4\dfrac{1}{3}$ 58. $\dfrac{3}{13}$ 59. $\dfrac{5}{6}$ 60. $\dfrac{6}{5} = 1\dfrac{1}{5}$ 61. $\dfrac{3}{5}$ 62. $\dfrac{5}{3} = 1\dfrac{2}{3}$ 63. $\dfrac{5}{3}$ or $1\dfrac{2}{3}$
64. $\dfrac{13}{2}$ or $6\dfrac{1}{2}$ 65. $\dfrac{7}{5}$ or $1\dfrac{2}{5}$ 66. $\dfrac{1}{3}$ 67. $\dfrac{1}{3}$ 68. $\dfrac{1}{20}$ 69. $\dfrac{1}{5}$ 70. 4 71. $\dfrac{3}{10}$ 72. $\dfrac{5}{12}$
73. $\dfrac{3}{4}$ 74. $\dfrac{4}{7}$ 75. $\dfrac{5}{3}$ or $1\dfrac{2}{3}$ lb 76. $\dfrac{1}{4}$ m 77. $\dfrac{7}{12}$ gal 78. $\dfrac{25}{8}$ or $3\dfrac{1}{8}$ gal 79. $\dfrac{4}{5}$ 80. $\dfrac{13}{50}$
81. $\dfrac{3}{7}$ 82. $\dfrac{1}{5}$

Chapter 3 Practice Test (page 125)

1. $\dfrac{4}{7}$ 2. $\dfrac{3}{8}$ 3. $\dfrac{13}{4}, 3\dfrac{1}{4}$ 4. $\dfrac{10}{12}, \dfrac{5}{6}$ 5. $\dfrac{5}{8}, \dfrac{6}{7}, \dfrac{1}{8}$ 6. 30 7. 18 8. 24 9. $\dfrac{2}{7}$
10. $\dfrac{5}{4}$ or $1\dfrac{1}{4}$ 11. $\dfrac{4}{5}$ 12. $\dfrac{2}{3}$ 13. $5\overline{)2}$ 14. seven-elevenths 15. four and three-fifths
16. $2\dfrac{3}{5}$ 17. $24\dfrac{17}{18}$ 18. 70 19. $\dfrac{99}{8}$ 20. $\dfrac{197}{20}$ 21. $\dfrac{5}{7}$ 22. $\dfrac{20}{7}$ or $2\dfrac{6}{7}$ 23. $\dfrac{6}{29}$ 24. $\dfrac{2}{7}$
25. $\dfrac{83}{150}$ 26. $\dfrac{2}{5}$ 27. $\dfrac{1}{5}$ 28. $\dfrac{27}{20}$ or $1\dfrac{7}{20}$ cm 29. $\dfrac{3}{8}\ell$ 30. $\dfrac{39}{80}$

CHAPTER 4

Exercises 4.1 (page 133)

1. $\dfrac{10}{3}$ or $3\dfrac{1}{3}$ 3. $\dfrac{6}{77}$ 5. $\dfrac{310}{243}$ or $1\dfrac{67}{243}$ 7. $\dfrac{130}{9}$ or $14\dfrac{4}{9}$ 9. $\dfrac{7}{19}$ 11. 32 13. $\dfrac{12}{7}$ or $1\dfrac{5}{7}$
15. $\dfrac{8}{15}$ 17. $\dfrac{9}{8}$ or $1\dfrac{1}{8}$ 19. $\dfrac{459}{5}$ or $91\dfrac{4}{5}$ 21. $\dfrac{1}{3}$ 23. $\dfrac{9}{4}$ or $2\dfrac{1}{4}$ 25. 200 27. $\dfrac{13}{6}$ or $2\dfrac{1}{6}$
29. $\dfrac{11}{36}$ 31. $\dfrac{21}{16}$ or $1\dfrac{5}{16}$ 33. $\dfrac{44}{49}$ 35. $\dfrac{57}{25}$ or $2\dfrac{7}{25}$ 37. $\dfrac{15}{7}$ or $2\dfrac{1}{7}$ 39. $\dfrac{39}{8}$ or $4\dfrac{7}{8}$ 41. $\dfrac{2}{5}$
43. $15 45. $42

Exercises 4.2 (page 138)

1. $\dfrac{6}{5}$ or $1\dfrac{1}{5}$ 3. $\dfrac{27}{20}$ or $1\dfrac{7}{20}$ 5. $\dfrac{9}{58}$ 7. 15 9. 2625 11. $\dfrac{1}{24}$ 13. $\dfrac{9}{10}$ 15. $\dfrac{3}{7}$ 17. $\dfrac{3}{5}$
19. 14 21. $\dfrac{35}{2}$ or $17\dfrac{1}{2}$ 23. 5 25. $\dfrac{125}{39}$ or $3\dfrac{8}{39}$ 27. 1 29. $\dfrac{3}{40}$
31. $\dfrac{27}{2}$ or $13\dfrac{1}{2}$ 33. $\dfrac{75}{38}$ or $1\dfrac{37}{38}$ 35. $\dfrac{1}{25}$ 37. $\dfrac{20{,}000}{7}$ or $2857\dfrac{1}{7}$ 39. $\dfrac{5}{3}$ or $1\dfrac{2}{3}$ ft
41. $\dfrac{7}{40}\ell$ 43. 50 bows

Exercises 4.3 (page 141)

1. 3 3. $\dfrac{42}{5}$ or $8\dfrac{2}{5}$ 5. $\dfrac{1}{4}$ 7. $\dfrac{1}{7}$ 9. $\dfrac{1}{6}$ 11. 72 13. 12 15. $\dfrac{800}{7}$ or $114\dfrac{2}{7}$ 17. $\dfrac{2}{7}$
19. $\dfrac{4}{5}$ 21. $\dfrac{3}{20}$ 23. 720 25. $\dfrac{500{,}000}{43}$ or $11{,}627\dfrac{39}{43}$

Exercises 4.4 (page 149)

1. $\dfrac{2}{5} \times 25 = 10$ 3. $\dfrac{3}{5} \times \dfrac{5}{6} = \dfrac{1}{2}$ 5. $\dfrac{3}{\frac{1}{4}} = 12$ 7. $\dfrac{40}{100} \times 200 = 80$ 9. $4 \div 2\dfrac{2}{3} = 1\dfrac{1}{2}$

11. $4\frac{1}{2} = \frac{1}{2} \times 9$ **13.** $\frac{7}{8} = 2 \times \frac{7}{16}$ **15.** 360   **17.** 105   **19.** 72   **21.** 4   **23.** $\frac{37}{10}$ or $3\frac{7}{10}$ **25.** 150   **27.** $\frac{17}{2}$ or $8\frac{1}{2}$ **29.** $\frac{203}{8}$ or $25\frac{3}{8}$ **31.** 90   **33.** 240   **35.** $\frac{1}{32}$ **37.** $\frac{1000}{7}$ or $142\frac{6}{7}$ **39.** $\frac{3}{100}$ **41.** 5400   **43.** 800   **45.** $\frac{5}{2}$ or $2\frac{1}{2}$

Exercises 4.5 (page 167)

1. $\frac{3}{2}$ or $1\frac{1}{2}$ gal   **3.** 28 slices   **5.** $\frac{1}{8}$ lb   **7.** 975 oz   **9.** $1512   **11.** $\frac{50}{7}$ or $7\frac{1}{7}$ min   **13.** $\frac{215}{32}$ or $6\frac{23}{32}$ gal   **15.** 6 ft per sec   **17.** $\frac{6}{25}$ gal per person   **19.** $\frac{40}{3}$ or $13\frac{1}{3}$ persons per sq mi   **21.** 70 words per min   **23.** 42 min   **25.** $\frac{13}{4}$ or $3\frac{1}{4}$ lb   **27.** $\frac{1}{30}$ lb   **29.** $\frac{40}{3}$ or $13\frac{1}{3}$ sq ft   **31.** 15 ft   **33.** 81 games   **35.** $\frac{3}{10}$ **37.** $\frac{1}{5}$ **39.** $18 per hr   **41.** 40¢ per ℓ **43.** $\frac{1}{3}$ **45.** 10,800 Asians

Developing Number Sense #3 (page 134)

1. more than 84   **2.** more than 84   **3.** less than 84   **4.** equal to 84   **5.** less than 84   **6.** more than 84   **7.** less than 682   **8.** more than 682   **9.** more than 682   **10.** more than 682   **11.** equal to 682   **12.** less than 682   **13.** less than $25\frac{1}{2}$ **14.** more than $25\frac{1}{2}$ **15.** more than $25\frac{1}{2}$ **16.** equal to $25\frac{1}{2}$

Chapter 4 Review Exercises (page 170)

1. 18   **2.** $\frac{61}{3}$ or $20\frac{1}{3}$ **3.** 18   **4.** $\frac{15}{2}$ or $7\frac{1}{2}$ **5.** $\frac{2}{15}$ **6.** $\frac{5}{14}$ **7.** $\frac{13}{14}$ **8.** 3   **9.** $\frac{21}{4}$ or $5\frac{1}{4}$ **10.** 10   **11.** $\frac{27}{10}$ or $2\frac{7}{10}$ **12.** $\frac{2}{7}$ **13.** 18   **14.** $\frac{3}{26}$ **15.** 208,000   **16.** $\frac{497}{12}$ or $41\frac{5}{12}$ **17.** $\frac{11}{2}$ or $5\frac{1}{2}$ **18.** 9   **19.** $69   **20.** $\frac{25}{8}$ or $3\frac{1}{8}$ **21.** $\frac{5}{9}$ **22.** $\frac{9}{5}$ or $1\frac{4}{5}$ **23.** $\frac{4}{15}$ **24.** $\frac{42}{25}$ or $1\frac{17}{25}$ **25.** $\frac{2}{9}$ **26.** $\frac{2}{5}$ **27.** $\frac{22}{3}$ or $7\frac{1}{3}$ **28.** $\frac{25}{24}$ or $1\frac{1}{24}$ **29.** 49   **30.** 480   **31.** 36   **32.** $\frac{125}{7}$ or $17\frac{6}{7}$ **33.** $\frac{9}{4}$ or $2\frac{1}{4}$ **34.** $\frac{9}{160}$ **35.** $\frac{625}{2}$ or $312\frac{1}{2}$ **36.** 30   **37.** $\frac{2}{3}$ ft   **38.** 9 doses   **39.** 410   **40.** 32   **41.** $\frac{3}{8}$ **42.** $\frac{14}{3}$ or $4\frac{2}{3}$ **43.** $\frac{9}{4}$ or $2\frac{1}{4}$ **44.** 440   **45.** 150   **46.** 320   **47.** $\frac{1}{4}$ **48.** $\frac{1}{4}$ **49.** $\frac{7}{16}$ **50.** 10   **51.** $\frac{125}{2}$ or $62\frac{1}{2}$ **52.** $\frac{1}{15}$ **53.** 36 workers   **54.** $30,000   **55.** $\frac{7}{4}$ or $1\frac{3}{4}$ lb   **56.** 24¢   **57.** $\frac{135}{4}$ or $33\frac{3}{4}$ sq yd   **58.** $\frac{3}{2}$ or $1\frac{1}{2}$ gal   **59.** $\frac{145}{14}$ or $10\frac{5}{14}$ lb   **60.** $\frac{169}{8}$ or $21\frac{1}{8}$ sq ft   **61.** $\frac{7}{2}$ or $3\frac{1}{2}$ hr   **62.** $189   **63.** $\frac{19}{4}$ or $4\frac{3}{4}$ ft   **64.** $\frac{1}{4}$

Chapter 4 Practice Test (page 172)

1. $\frac{6}{35}$ **2.** 2   **3.** 5100   **4.** $\frac{115}{2}$ or $57\frac{1}{2}$ **5.** $\frac{1}{12}$ **6.** $\frac{2}{5}$ **7.** $\frac{1}{108}$ **8.** $\frac{4}{9}$ **9.** $\frac{45}{22}$ or $2\frac{1}{22}$ **10.** $\frac{15}{7}$ or $2\frac{1}{7}$ **11.** 9   **12.** 9225   **13.** $\frac{8}{7}$ or $1\frac{1}{7}$ **14.** 27   **15.** 35   **16.** $\frac{7}{6}$ or $1\frac{1}{6}$

17. $37\frac{1}{2}$ servings **18.** $80 **19.** $\frac{22}{3}$ or $7\frac{1}{3}$ min **20.** $84 **21.** $\frac{2}{3}$ ft **22.** $\frac{9}{2}$ or $4\frac{1}{2}$ yd **23.** $\frac{4}{5}$ **24.** $3\frac{3}{5}$ hr

CHAPTER 5

Exercises 5.1 (page 177)

1. $\frac{5}{8}$ **3.** $\frac{2}{3}$ **5.** 1 **7.** $\frac{25}{18}$ or $1\frac{7}{18}$ **9.** $\frac{16}{3}$ or $5\frac{1}{3}$ **11.** $9\frac{3}{4}$ or $\frac{39}{4}$ **13.** 3 **15.** $8\frac{5}{6}$ **17.** $34\frac{5}{32}$ **19.** $22\frac{2}{5}$ **21.** 34 **23.** $157\frac{2}{3}$ **25.** 9 lb **27.** $28\frac{1}{4}$ in

Exercises 5.2 (page 182)

1. 2, 7 **3.** 31, 13, 17 **5.** 3 × 3 **7.** 2 × 2 × 3 **9.** 11 is prime **11.** 2 × 3 × 2 × 2 × 2 **13.** 2 × 2 × 2 × 2 × 2 × 2 **15.** 5 × 5 × 5 **17.** 97 is prime **19.** 7 × 13 **21.** 5 × 7 × 7 **23.** 18 **25.** 120 **27.** 84 **29.** 154 **31.** 36 **33.** 44 **35.** 225 **37.** 325 **39.** 884

Exercises 5.3 (page 189)

1. $\frac{31}{18}$ or $1\frac{13}{18}$ **3.** $\frac{11}{8}$ or $1\frac{3}{8}$ **5.** $\frac{4}{5}$ **7.** $\frac{81}{22}$ or $3\frac{15}{22}$ **9.** $\frac{5}{4}$ or $1\frac{1}{4}$ **11.** $\frac{131}{210}$ **13.** $35\frac{8}{21}$ **15.** $104\frac{49}{64}$ **17.** $64\frac{29}{102}$ **19.** $\frac{7}{5}$ or $1\frac{2}{5}$ **21.** $\frac{433}{210}$ or $2\frac{13}{210}$ **23.** $14\frac{10}{99}$ **25.** $113\frac{94}{105}$ **27.** $20\frac{1}{40}$ **29.** $32\frac{123}{260}$ **31.** $15\frac{11}{15}$ hr **33.** $22\frac{1}{12}$ ft

Exercises 5.4 (page 193)

1. $\frac{2}{3}$ **3.** $\frac{7}{9}$ **5.** $\frac{2}{5}$ **7.** $\frac{7}{20}$ **9.** $\frac{10}{12}$ **11.** $\frac{1}{4}$ **13.** $\frac{6}{6}$ **15.** $\frac{2}{5}$ **17.** $\frac{1}{4}, \frac{3}{4}, \frac{5}{4}, \frac{11}{4}$ **19.** $\frac{1}{2}, \frac{3}{5}, \frac{13}{20}$ **21.** $\frac{11}{24}, \frac{24}{11}, 3$ **23.** < **25.** > **27.** = **29.** < **31.** > **33.** $\frac{7}{8}$ in **35.** Paul

Exercises 5.5 (page 199)

1. $\frac{2}{3}$ **3.** $\frac{5}{8}$ **5.** $\frac{13}{36}$ **7.** $11\frac{1}{2}$ **9.** $5\frac{3}{8}$ **11.** $3\frac{5}{8}$ **13.** $23\frac{17}{36}$ **15.** $4\frac{1}{5}$ **17.** $9\frac{5}{12}$ **19.** $\frac{7}{18}$ **21.** $4\frac{2}{3}$ **23.** $5\frac{1}{2}$ **25.** $3\frac{29}{75}$ **27.** $49\frac{37}{39}$ **29.** $5\frac{13}{40}$ **31.** $7\frac{1}{4}$ ft **33.** $1\frac{7}{8}$ in **35.** $\frac{5}{6}$ of the job **37.** $\frac{13}{24}$ of the pizza **39.** $\frac{1}{3}$ **41.** $\frac{16}{15}$ or $1\frac{1}{15}$ **43.** $1\frac{7}{12}$ **45.** $\frac{3}{13}$

Exercises 5.6 (page 204)

1. $2\frac{1}{3} + \frac{2}{3} = 3$ **3.** $7 - 6\frac{1}{4} = \frac{3}{4}$ **5.** $\frac{7}{8} + \frac{1}{8} = 1$ **7.** $\frac{1}{2} + \frac{3}{2} = 2$ **9.** $\frac{3}{10} = \frac{1}{2} - \frac{1}{5}$ **11.** $5\frac{1}{5}$ **13.** $\frac{1}{27}$ **15.** 1 **17.** $2\frac{1}{4}$ **19.** $\frac{1}{40}$ **21.** $\frac{22}{15}$ or $1\frac{7}{15}$

Exercises 5.7 (page 214)

1. $\frac{5}{6}$ **3.** $\frac{7}{12}$ cup **5.** $\frac{3}{4}$ of her income **7.** $1\frac{3}{4}$ mi **9.** 3 c **11.** $3\frac{5}{16}$ lb **13.** $1\frac{3}{4}$ hr **15.** $6\frac{3}{16}$ in **17.** too large by $\frac{1}{16}$ in **19.** $5\frac{7}{12}$ hr **21.** $\frac{29}{48}$ **23.** $\frac{41}{12}$ or $3\frac{5}{12}$ mi per hr **25.** $\frac{41}{16}$ or $2\frac{9}{16}$ lb **27.** $300 **29.** Nancy, $\frac{1}{48}$ **31.** $41\frac{1}{4}$ ft **33.** $\frac{1}{3} + \frac{1}{4} + \frac{1}{8} + \frac{7}{24} = \frac{8}{24} + \frac{6}{24} + \frac{3}{24} + \frac{7}{24} = \frac{24}{24} = 1$ **35.** $9000 **37.** 11 in **39.** $\frac{11}{5}$ or $2\frac{1}{5}$ in

Developing Number Sense #4 (page 194)

1. b. $\frac{10}{20}$ c. $\frac{24}{48}$ g. $\frac{43}{86}$ 2. a. $\frac{8}{17}$ d. $\frac{99}{202}$ e. $\frac{87}{180}$ 3. a. $\frac{9}{17}$ b. $\frac{17}{30}$ f. $\frac{76}{150}$
4. $\frac{1}{4} < \frac{1}{2}$ 5. $\frac{5}{8} > \frac{1}{2}$ 6. $\frac{10}{18} > \frac{1}{2}$ 7. $\frac{18}{36} = \frac{1}{2}$ 8. $\frac{34}{70} < \frac{1}{2}$ 9. $\frac{42}{87} < \frac{1}{2}$

Developing Number Sense #5 (page 201)

1. 5   2. 8   3. 12   4. 46   5. 0   6. 1   7. 70   8. 150   9. 20   10. 10   11. 50
12. 180   13. 15   14. 6   15. 51   16. 16   17. 864   18. 280   19. 170   20. 140
21. 6600   22. 23

Using the Calculator #7 (page 217)

1. $\frac{7}{10}$ 2. $\frac{24}{203}$ 3. $3\frac{3}{4}$ 4. $\frac{11}{45}$ 5. $71\frac{13}{35}$ 6. $\frac{8}{19}$

Chapter 5 Review Exercises (page 218)

1. 13, 23   2. 2, 7, 17, 37   3. 5×7 4. $2 \times 2 \times 2 \times 3 \times 3$ 5. $2 \times 3 \times 13$ 6. $3 \times 3 \times 19$
7. 30   8. 42   9. 45   10. 132   11. 136   12. 180   13. $\frac{5}{4}$ or $1\frac{1}{4}$ 14. 12   15. $\frac{3}{5}$
16. $\frac{7}{9}$ 17. $16\frac{1}{6}$ 18. $\frac{17}{21}$ 19. $14\frac{13}{36}$ 20. $\frac{26}{33}$ 21. $35\frac{19}{42}$ 22. $\frac{67}{80}$ 23. $\frac{77}{52}$ or $1\frac{25}{52}$
24. $57\frac{8}{25}$ 25. $\frac{10}{9}$ 26. $\frac{3}{11}$ 27. $\frac{3}{4}$ 28. $\frac{8}{27}, \frac{1}{3}, \frac{4}{9}$ 29. $\frac{2}{7}, \frac{2}{5}, \frac{2}{3}$ 30. $\frac{47}{50}, \frac{19}{20}, 1$ 31. >
32. <   33. >   34. >   35. >   36. $\frac{1}{3}$ 37. $2\frac{3}{4}$ 38. $7\frac{2}{5}$ 39. $11\frac{4}{7}$ 40. $\frac{11}{24}$ 41. $9\frac{5}{9}$
42. $3\frac{17}{20}$ 43. $7\frac{1}{20}$ 44. $19\frac{87}{98}$ 45. $2\frac{1}{3} - 2 = \frac{1}{3}$ 46. $\frac{3}{5} + \frac{4}{5} = \frac{7}{5}$ 47. $\frac{2}{3} + \frac{5}{3} = \frac{7}{3}$
48. $5 - 2\frac{1}{3} = 2\frac{2}{3}$ 49. $\frac{3}{10} + \frac{2}{10} = \frac{1}{2}$ 50. $2\frac{4}{7}$ 51. $5\frac{1}{5}$ 52. $\frac{3}{8}$ 53. 6   54. Shirley
55. $8\frac{1}{8}$ ft   56. $66\frac{1}{6}$ ft   57. $5\frac{1}{2}$ ft   58. $12\frac{1}{4}$ hr   59. $3\frac{1}{4}$ ft   60. $1\frac{1}{2}$ lb   61. $9\frac{1}{2}$ lb   62. no
63. $\frac{23}{7}$ or $3\frac{2}{7}$ mi   64. $\frac{7}{20}$ of the voters   65. $\frac{1}{12}$ of the job   66. 1980 students   67. $18\frac{1}{6}$ mi per hr
68. $\frac{1545}{16}$ or $96\frac{9}{16}$ sq ft   69. 600 students   70. $\frac{1}{24}$

Chapter 5 Practice Test (page 221)

1. 17, 23   2. $2 \times 3 \times 7$ 3. $2 \times 5 \times 11$ 4. 60   5. 216   6. $\frac{4}{3}$ or $1\frac{1}{3}$ 7. $\frac{13}{18}$ 8. $15\frac{4}{75}$
9. $2\frac{5}{36}$ 10. $18\frac{13}{14}$ 11. $\frac{1}{2}$ 12. $\frac{7}{12}, \frac{2}{3}, \frac{3}{4}$ 13. $\frac{17}{24} < \frac{3}{4}$ 14. $\frac{1}{8}$ 15. $5\frac{37}{45}$ 16. $4\frac{1}{2}$ or $\frac{9}{2}$
17. $6\frac{7}{24}$ 18. $5\frac{1}{3}$ or $\frac{16}{3}$ 19. $\frac{1}{3} + \frac{2}{3} = 1$ 20. $9 - 1\frac{3}{4} = 7\frac{1}{4}$ 21. $\frac{1}{2}$ 22. $\frac{7}{5}$ or $1\frac{2}{5}$
23. $18\frac{1}{16}$ or $\frac{289}{16}$ 24. $1\frac{1}{8}$ in   25. $25\frac{1}{2}$ ft   26. $6\frac{1}{4}$ ft   27. $\frac{3}{4}$ cup   28. $7\frac{1}{4}$ or $\frac{29}{4}$ gal per min
29. $\frac{3}{5}$ of the soft drinks   30. $\frac{2349}{32}$ or $73\frac{13}{32}$ sq in

Cumulative Review Exercises: Chapters 1–5 (page 222)

1. thirteen hundredths   2. $\frac{67}{12}$ 3. $10\frac{2}{3}$ 4. $\frac{60}{7}$ or $8\frac{4}{7}$ 5. $\frac{17}{30}, \frac{3}{5}, \frac{2}{3}$ 6. $\frac{63}{9}$
7. $2 \times 3 \times 5 \times 23$ 8. $3 \times 7 \times 11$ 9. $2 \times 2 \times 2 \times 2 \times 3 \times 5 \times 5$ 10. $2 \times 2 \times 2 \times 5 \times 5 \times 17$

11. 315 **12.** 560 **13.** 480 **14.** $\frac{121}{15}$ or $8\frac{1}{15}$ **15.** $\frac{2}{5}$ **16.** $\frac{5}{2}$ or $2\frac{1}{2}$ **17.** $12\frac{1}{8}$ or $\frac{97}{8}$
18. $16\frac{5}{12}$ **19.** $\frac{10}{27}$ **20.** $\frac{1}{6}$ **21.** $12\frac{7}{8}$ **22.** $\frac{81}{112}$ **23.** $\frac{41}{36}$ or $1\frac{5}{36}$ **24.** 125 R 20 **25.** 66,038
26. 309,894 **27. a.** 4 in **b.** $\frac{21}{5}$ or $4\frac{1}{5}$ in **c.** $\frac{14}{15}$ in **28. a.** 6800 lb **b.** $\frac{3}{10}$ **c.** $\frac{25}{17}$
29. a. $\frac{75}{4}$ or $18\frac{3}{4}$ mi **b.** $\frac{195}{8}$ or $24\frac{3}{8}$ mi **c.** $\frac{4}{15}$ hr **d.** $\frac{16}{15}$ or $1\frac{1}{15}$ hr
30. a. 40 tons **b.** 460 tons **c.** $\frac{23}{25}$
31. a. $\frac{255}{2}$ or $127\frac{1}{2}$ ft **b.** 340 ft **c.** $\frac{21675}{4}$ or $5418\frac{3}{4}$ sq ft **d.** $21,675

CHAPTER 6

Exercises 6.1 (page 228)

1. 2 **3.** 7 **5.** 8 **7.** 1 **9.** $\frac{2}{10} + \frac{3}{100} + \frac{8}{1000}$ **11.** $70 + 6 + \frac{8}{1000}$ **13.** 0.57 **15.** 630.803
17. $\frac{3}{10}$ **19.** $\frac{58}{100}$ **21.** $\frac{6}{1000}$ **23.** 0.04 **25.** 14.032 **27.** 2.368 **29.** 0.9 **31.** 0.04
33. 12.3 **35.** 50.104 **37.** 50.3 **39.** thirty-four hundredths **41.** eight thousandths
43. twenty-four and two hundredths **45.** nine and one hundred five ten-thousandths
47. nine thousand

Exercises 6.2 (page 233)

1. a. 100¢ **b.** $1.00 **3. a.** 2000¢ **b.** $20.00 **5. a.** 25¢ **b.** $0.25 **7.** half dollar
9. dime **11.** $0.17 **13.** $0.46 **15.** $25.83 **17.** $800.09
19. a. fifty-five cents **b.** fifty-five hundredths dollars
21. a. eight dollars and three cents **b.** eight and three hundredths dollars
23. a. sixty-seven dollars **b.** sixty-seven dollars **25.** 348.92 **27.** six hundred seventy-five and $\frac{80}{100}$

Exercises 6.3 (page 236)

1. a. .43 **c.** .430 **d.** 0.430 **3. c.** $30.00 **e.** $30.0 **5.** 2.8 **7.** 0.03 **9.** 132.113
11. 8. **13.** .138 **15.** .5; .55; 5 **17.** .32; .327; .33 **19.** = **21.** <

Exercises 6.4 (page 240)

1. 0.32 **3.** 732.014 **5.** $18 **7.** $0.44 **9.** 9.0140 **11.** 2304.20 **13.** 23 **15.** 0.236
17. 790.3 **19.** 75

Exercises 6.5 (page 245)

1. 49.33 **3.** 83.915 **5.** 738.85 **7.** $599.91 **9.** 388.443 **11.** 14.53 **13.** 8.069 **15.** 145.68
17. 301.547 **19.** $93.21 **21.** 1972.095 **23.** 3769.33 **25.** 4.7 m **27.** 109.67 **29.** 44.834
31. 1198.18

Exercises 6.6 (page 248)

1. $0.8 + 0.4 = 1.2$ **3.** $18.7 - 7.3 = 11.4$ **5.** $72.8 + 0.6 = 73.4$ **7.** $8.2 = 8.6 - 0.4$ **9.** 7.63
11. 10.82 **13.** 20.17 **15.** 16.416 **17.** 6.6

Exercises 6.7 (page 254)

1. $16.75 **3.** 39.8 cm **5.** 38.54 m **7.** 44.1 gal **9.** 1.8 mi **11.** 2.9 in **13.** $0.59 **15.** 4.8 ft
17. 70 in **19.** 485 calories **21.** 0.4 yd **23.** $1.25 **25.** increased by $0.75

Chapter 6 Review Exercises (page 256)

1. 0   **2.** 7   **3.** 6   **4.** 1   **5.** 0.609   **6.** 804.037   **7.** $\frac{3}{1000}$ **8.** $7\frac{27}{100}$ or $\frac{727}{100}$ **9.** 0.0019
10. 14.15   **11.** 16,000   **12.** 0.016   **13.** 90.0203   **14.** 7.14
15. five hundredths   **16.** fifty and two ten-thousandths   **17.** three hundred four thousandths
18. three hundred four thousand   **19. a.** 5¢   **b.** $0.05   **20. a.** 25¢   **b.** $0.25
21. a. 100¢   **b.** $1.00   **22. a.** 500¢   **b.** $5.00   **23.** dime   **24.** penny   **25.** ten-dollar bill
26. half dollar   **27.** $0.47   **28.** $62.12   **29. a.** sixty-one cents   **b.** sixty-one hundredths dollar
30. a. two hundred dollars and seven cents   **b.** two hundred and seven hundredths dollars
31. a. .58   **d.** 0.58   **32. b.** $102   **c.** $102.
33. 0.39, 1, 3.9   **34.** .732323; 73.23222; 73.2323   **35.** >   **36.** =   **37.** >   **38.** >   **39.** 138.14
40. $43   **41.** 0.232   **42.** $204.84   **43.** 9.8176   **44.** 93.15   **45.** 86.382   **46.** 29.3407
47. 92.03   **48.** 1043.45   **49.** $726.11   **50.** $364.41   **51.** 71.42   **52.** $61.13   **53.** 427.46
54. 81.392   **55.** 5.4   **56.** 54.74   **57.** 2995.48   **58.** 79,748.6   **59.** 42.08   **60.** 112.4
61. 210.8 − 107 = 103.8   **62.** 0.7 + 0.6 = 1.3   **63.** 4.2 + 15.8 = 20   **64.** 19.7 − 2 = 17.7
65. 1.35   **66.** 2.99   **67.** 148.97   **68.** 11.02   **69.** 2.55 mi   **70.** 127.15 ft   **71.** $34.15
72. $1905.80   **73.** 81.5 in   **74.** 18.8 m

Chapter 6 Practice Test (page 259)

1. 93.4   **2.** 6.0019   **3.** 0.65   **4.** 48.007   **5.** thirteen ten-thousandths   **6.** seven and two-tenths
7. $0.25   **8.** 500¢   **9. b.** .900   **c.** 0.9   **d.** .9   **10.** 0.079; 0.701; 1.07   **11.** 0.007 > 0.00089
12. 26.084   **13.** $79   **14.** 0.0134   **15.** 136.11   **16.** 876.556   **17.** 11.397   **18.** $774.34
19. 224.56   **20.** 4921.4   **21.** 183.42   **22.** 7.7   **23.** 230.24 ft   **24.** 9.6 lb.   **25.** $52.50
26. 3.5 mi

CHAPTER 7

Exercises 7.1 (page 267)

1. .09 or 0.09   **3.** 32.76   **5.** 3517.4   **7.** 152.75262   **9.** 4512.95   **11.** 3   **13.** 0.555
15. 1.17072   **17.** 54   **19.** 23,000   **21.** 781.3   **23.** 1.08252   **25.** $13,870,000   **27.** 12.9792
29. 271,606.8   **31.** $20.30   **33.** $9.49

Exercises 7.2 (page 278)

1. 1.4   **3.** 0.16   **5.** 1.3   **7.** 42   **9.** 350   **11.** 1990.4   **13.** 67,000   **15.** 1900   **17.** 3006.2
19. 0.0305   **21.** 608.004   **23.** 0.0026   **25.** 400   **27.** $8.9\overline{3}$ **29.** $138.\overline{8}$ **31.** $0.0\overline{1}$ **33.** 0.875
35. 0.0175   **37.** $1.041\overline{6}$ **39.** $0.\overline{27}$ **41.** 13.817   **43.** 7.606   **45.** 0.009   **47.** 0.0004   **49.** 15.64
51. $2.61\frac{3}{7}$ **53.** $71.66\frac{2}{3}$ **55.** $2.85\frac{5}{7}$ **57.** 0.0456   **59.** 8.976   **61.** 0.000075   **63.** 3.85 in
65. 6.5 m

Exercises 7.3 (page 282)

1. 23.2   **3.** 0.0205   **5.** 3.81   **7.** 0.0125   **9.** 360   **11.** 9.763   **13.** 13   **15.** 8.05   **17.** 1.03
19. 500

Exercises 7.4 (page 287)

1. 0.3 × 0.2 = 0.06   **3.** 45.2 = 5 × 9.04   **5.** 41.6 = 2 × 20.8   **7.** 940   **9.** 0.025   **11.** 15.25
13. 135   **15.** 6.5   **17.** 500   **19.** 36.18   **21.** 20   **23.** 0.08

Exercises 7.5 (page 302)

1. $425.50   **3.** 486 m   **5.** 18 pieces   **7.** $16.60   **9.** 49.6 mi   **11.** 9.2 min   **13.** $27.23
15. $1.06   **17.** 0.337   **19.** 9.3 ft   **21.** $430.20   **23.** 15 doses   **25.** 537 mi   **27.** 14.6 gal
29. 121.6 g   **31.** 120   **33.** $2.18   **35.** 0.15 m   **37.** 4104.18 sq ft   **39.** $4.37   **41.** 128 games
43. $1.75   **45.** $1.40

Developing Number Sense #6 (page 279)

1. 34   **2.** 7.84   **3.** 0.0659   **4.** 0.92   **5.** 0.84   **6.** 0.0875   **7.** 0.00006   **8.** 0.1245   **9.** 0.003
10. 0.057   **11.** 0.0457   **12.** 0.0005   **13.** 0.0367   **14.** 0.0786   **15.** 0.8   **16.** 120

Developing Number Sense #7 (page 283)

1. more than 850   **2.** less than 850   **3.** less than 850   **4.** more than 850   **5.** equal to 850
6. less than 850   **7.** more than 850   **8.** more than 850   **9.** more than 47.2   **10.** less than 47.2
11. more than 47.2   **12.** less than 47.2

Chapter 7 Review Exercises (page 305)

1. 0.6   **2.** 0.0768   **3.** 220.5   **4.** 2578.8   **5.** 44.82   **6.** 64.68   **7.** 38.545   **8.** 98.4   **9.** 0.216
10. 63.45   **11.** 184,525   **12.** 6.2521   **13.** 2,375,000   **14.** 162   **15.** 100,800   **16.** 1080   **17.** 86
18. 78,000   **19.** 1575   **20.** 2374.152   **21.** 0.405   **22.** 0.0025   **23.** 3.4   **24.** 0.25   **25.** 0.056
26. 2.08   **27.** 20,780   **28.** 7300   **29.** 37,000   **30.** 28.125   **31.** 3.7   **32.** 0.0075   **33.** $0.\overline{5}$
34. $8.\overline{6}$ **35.** $1.1\overline{6}$ **36.** $573.\overline{3}$ **37.** $8.36\frac{1}{4}$ **38.** $103.33\frac{1}{3}$ **39.** $0.06\frac{2}{13}$ **40.** $104.16\frac{2}{3}$ **41.** 30.643
42. 0.57   **43.** 0.2692   **44.** 0.254   **45.** 5.834   **46.** 18.64   **47.** 0.75   **48.** 26.25   **49.** 2.48
50. 0.175   **51.** $8.8\overline{3}$ **52.** $0.041\overline{6}$ **53.** $0.\overline{703}$ **54.** $8.\overline{45}$ **55.** 76.5   **56.** 25.125   **57.** 68.75
58. 4600   **59.** 0.0125   **60.** 9450   **61.** 9.75   **62.** 1.3   **63.** 0.0372   **64.** 0.06   **65.** 17.2
66. 8.865   **67.** 0.375   **68.** 1.53   **69.** 0.725 lb   **70.** 13 pieces   **71.** 33.75 min   **72.** $23.62
73. 0.015 ton   **74.** 80 hr   **75.** 16 gal per min   **76.** $32.30   **77.** $408   **78.** 0.02   **79.** $2.84
80. $147   **81.** 2148 sq cm   **82.** 21.1 ft   **83.** $16.51   **84.** 0.21   **85.** 7.2 min   **86.** 32 mi

Chapter 7 Practice Test (page 307)

1. 61.92   **2.** 125.6049   **3.** 450,000   **4.** 60.8   **5.** 0.0117   **6.** 136   **7.** 2.875   **8.** 0.401
9. 351,000   **10.** $0.\overline{6}$ **11.** $1.78\overline{3}$ **12.** $481.\overline{481}$ **13.** $26.06\frac{2}{3}$ **14.** $0.01\frac{7}{8}$ **15.** $23.33\frac{1}{3}$
16. 40.033   **17.** 1.8823   **18.** 3.195   **19.** 0.625   **20.** 1.72   **21.** $0.2\overline{7}$ **22.** $6.\overline{30}$ **23.** 0.5
24. 5   **25.** 29.05   **26.** 1600   **27.** 400   **28.** 3100   **29.** 0.003125   **30.** 0.0656   **31.** $3007.20
32. $750   **33.** 15 slices   **34.** 1384 sq ft   **35.** 47.6 mi per hr   **36.** 4743.6 mi   **37.** 25.4 in

Cumulative Review Exercises: Chapters 1–7 (page 308)

1. $\frac{18}{5}$ or $3\frac{3}{5}$ **2.** $\frac{17}{20}$ **3.** $\frac{1}{400}$ **4.** $2\frac{1}{8}$ or $\frac{17}{8}$ **5.** 0.375   **6.** 0.055   **7.** 75.5   **8.** 23.6
9. $19\frac{1}{3}$ **10.** 2000   **11.** $\frac{3}{2}$ or $1\frac{1}{2}$ **12.** $1\frac{27}{56}$ or $\frac{83}{56}$ **13.** 14.34   **14.** 7.7   **15.** $\frac{15}{4}$ or $3\frac{3}{4}$
16. $\frac{1}{8}$ or 0.125   **17.** 397,800   **18.** 0.00625   **19.** 160   **20.** $58.67   **21.** $6\frac{1}{2}$ hr
22. a. 2600 sq ft   **b.** 210 ft   **c.** 166.4 fl oz   **d.** $5197.50   **23.** $1500

CHAPTER 8

Exercises 8.1 (page 319)

1. A. 3   B. 5   **3.** A. 22   B. 25   **5.** A. 5   B. 25   **7.** A. 8   B. 22
9. A. $\frac{2}{8}$ or $\frac{1}{4}$ B. $\frac{7}{8}$ **11.** A. $4\frac{1}{3}$ B. $5\frac{2}{3}$ **13.** A. $\frac{1}{5}$ B. $\frac{3}{5}$
15. A. $2\frac{2}{6}$ or $2\frac{1}{3}$ B. $3\frac{1}{6}$ **17.** A. 2.9   B. 3.0 or 3   **19.** A. 8.25   B. 9.75
21. A. 13.2   B. 13.6   **23.** A. 3.875   B. 4.375   **25.** $\frac{1}{8}$ **27.** $\frac{3}{4}$

Exercises 8.2 (page 329)

1. > **3.** < **5.** = **7.** > **9.** > **11. a.** $\frac{28}{100}$ **c.** $\frac{35}{125}$ **d.** 0.28
13. b. $\frac{3}{50}$ **c.** 0.0600 **d.** $\frac{60}{1000}$ **15. a.** $\frac{60}{5}$ **d.** 12.0 **17.** 8.1638109 **19.** $1\frac{13}{18}$ **21.** $8\frac{1}{12}$
23. 1.3 **25.** 2.1, 20.75, 21.5, 21.6 **27.** $\frac{3}{4}$, 1, $1\frac{1}{4}$, $\frac{4}{3}$ **29.** $\frac{23}{40}$, $\frac{3}{5}$, $0.63\overline{7}$, 0.64
31. 2.06, 20.6, $20\frac{11}{18}$, $20\frac{2}{3}$

Exercises 8.3 (page 334)

1. 1350 **3.** $1\frac{7}{12}$ **5.** 0.085 **7.** $13\frac{3}{8}$ **9.** 31.25 **11.** $4\frac{137}{180}$ **13.** 0.025 **15.** $\frac{1}{4}$ or 0.25
17. $8\frac{1}{6}$ **19.** $\frac{2}{125}$ or 0.016

Exercises 8.4 (page 345)

1. $\frac{3}{4}$ or 0.75 ft **3.** $\frac{1}{4}$ or 0.25 **5.** 550 boxes **7.** 30 servings
9. a. $93\frac{1}{2}$ or 93.5 sq in **b.** 39 in **11.** $17\frac{11}{12}$ in **13.** $2.54 **15.** 5.25 or $5\frac{1}{4}$ hr
17. $\frac{14}{15}$ or $0.9\overline{3}$ **19. a.** $\frac{1}{7}$ **b.** $\frac{8}{5}$ **21.** 52 words per min **23.** $7.12 **25.** 200 mi

Exercises 8.5 (page 355)

1. $1778.67 **3.** $3.61 **5.** $6\frac{1}{4}$ or 6.25 mi **7.** increased by $1.25 per share
9. a. 216.72 ft **b.** 2695.842 sq ft **11.** $\frac{9}{283}$ **13.** 375 acres **15.** $65\frac{1}{2}$ or 65.5 lb
17. 53.8 mi per hr **19.** $7.35 **21.** 3500 **23.** 2875

Exercises 8.6 (page 364)

1. no **3.** yes **5.** yes **7.** no **9.** yes **11.** no **13.** no **15.** yes **17.** 4 **19.** 60
21. 5.4 or $5\frac{2}{5}$ **23.** 3.75 **25.** 7.2 **27.** 35 **29.** $\frac{200}{3}$ or $66\frac{2}{3}$ **31.** yes **33.** no **35.** no
37. 75 defective bulbs **39.** 240 men **41.** 48 lb **43.** 5.1 oz **45.** 44 ft

Using the Calculator #8 (page 334)

1. 0.875 **2.** 1.143 **3.** 2.692 **4.** 0.371 **5.** 0.042 **6.** 24 **7.** 0.625 **8.** 0.722 **9.** 353.581
10. 12.75 **11.** 1.111 **12.** 0.216 **13.** 1300.083 **14.** 0.006 **15.** 33.943 **16.** 2.281
17. 2146.912 **18.** 2.837

Using the Calculator #9 (page 346)

1. 0.775 **2.** 0.025 **3.** 0.542 **4.** 0.201 **5.** 348.208 **6.** 347.792 **7.** 59.304 **8.** 17.304
9. 7.867 **10.** 0.533 **11.** 26.037 **12.** 10.732

Chapter 8 Review Exercises (page 366)

1. A. 35 B. 40 **2.** A. 25 B. 75 **3.** A. $\frac{1}{4}$ B. 2 **4.** A. $4\frac{2}{3}$ B. 6
5. A. 7.5 B. 7.8 **6.** A. 0.53 B. 0.59 **7. b.** 0.18 **c.** 0.1800 **d.** $\frac{27}{150}$
8. b. $\frac{39}{5}$ **d.** $7\frac{32}{40}$ **9. b.** $\frac{75}{3}$ **c.** 25.00 **10.** < **11.** > **12.** < **13.** =

14. $\frac{31}{36}, \frac{7}{8}, 0.8\overline{7}$ **15.** $0.044, \frac{4}{90}, 0.44, 4\frac{2}{5}$ **16.** $8\frac{2}{3}, 8\frac{3}{4}, 8\frac{5}{6}$ **17.** $0.5625, 0.58, \frac{5}{8}$ **18.** 27.2
19. $\frac{2}{5}$ or 0.4 **20.** $\frac{4}{27}$ **21.** $\frac{25}{4}$ or $6\frac{1}{4}$ or 6.25 **22.** 400 **23.** $27\frac{2}{3}$ **24.** 0.016
25. $315,000 **26.** $\frac{5}{12}$ **27.** $10\frac{1}{4}$ gal **28.** 20 slices **29.** $\frac{1}{3}$ **30. a.** 129.4 in **b.** 1029.3 sq in
31. 33.75 gal **32.** 4.2 yd per sec **33.** 9.2 ft **34.** $96.12 **35.** $\frac{1}{4}$ or 0.25
36. 62.4 sq ft **37.** $86.25 **38.** $771.85 **39.** $277.50 **40.** $5 **41.** 31.2 sec
42. Natalie at 55.5 words per min **43.** 12 hr **44.** $31\frac{7}{15}$ yd **45.** 0.8 in **46.** 2.4 in **47.** 8 in
48. 1.6 in **49.** 0.075 or $\frac{3}{40}$ **50.** May

Chapter 8 Practice Test (page 369)

1. A. 680 **B.** 720 **2. A.** $\frac{1}{8}$ **B.** $\frac{7}{8}$ **3. A.** 10 **B.** 10.6 **4. c.** $\frac{232}{25}$ **d.** $8\frac{32}{25}$ **5.** =
6. < **7.** $9.14, \frac{46}{5}, 9\frac{1}{4}$ **8.** $0.95, \frac{83}{10}, 7\frac{4}{3}, 9.2$ **9.** $7\frac{5}{6}$ **10.** $\frac{52}{125}$ or 0.416 **11.** $26.10
12. 76 servings **13.** 147.73 sq ft **14.** $\frac{3}{5}$ **15.** $1\frac{1}{2}$ hr **16.** $2176.50 **17.** 8.6 lb **18.** $6\frac{2}{3}$ sec
19. 57 m **20.** $\frac{1}{5}$ or 0.2 **21.** $17.18 **22.** $598.56

CHAPTER 9

Exercises 9.1 (page 378)

1. less than 1 **3.** more than 1 **5.** equal to 1 **7.** 0.06 **9.** 0.1 **11.** 1 **13.** 3.5 **15.** 0.042
17. 0.1532 **19.** 0.009 **21.** 0.00145 **23.** 1.005 **25.** 0.0025 **27.** 0.025 **29.** 0.3775
31. 0.00025 **33.** $0.008\overline{3}, 0.008\frac{1}{3}$ **35.** $0.\overline{3}, 0.33\frac{1}{3}$ **37.** $\frac{1}{100}$ **39.** $\frac{3}{4}$ **41.** $\frac{9}{25}$ **43.** $1\frac{3}{4}$ or $\frac{7}{4}$
45. $3\frac{1}{4}$ or $\frac{13}{4}$ **47.** $\frac{1}{250}$ **49.** $\frac{1}{3}$ **51.** $\frac{1}{30}$ **53.** $\frac{1}{200}$ **55.** $\frac{7}{1000}$ **57.** $\frac{7}{16}$ **59.** $\frac{29}{300}$
61. $\frac{3}{40,000}$ **63.** $\frac{1}{40}$ **65.** $\frac{9}{200}$ **67.** $\frac{3}{8}$ **69.** $\frac{1}{125}$ **71.** $\frac{9}{10,000}$ **73.** $\frac{1}{400}$ **75.** $\frac{17}{400}$

Exercises 9.2 (page 386)

1. 30 **3.** 18 **5.** 140 **7.** 12.5 or $12\frac{1}{2}$ **9.** 3 **11.** 14.094 **13.** 7.476 **15.** 12.3
17. 156.84 **19.** 12.96 **21.** 0.255 **23.** 3.584 **25.** 0.078 **27.** 150 **29.** $\frac{5}{3}$ or $1\frac{2}{3}$ **31.** $\frac{7}{12}$
33. 34 **35.** $\frac{1}{24}$ **37.** $\frac{1}{4}$ or 0.25 **39.** $19,600 **41.** 115 trucks

Exercises 9.3 (page 394)

1. 45% **3.** 90% **5.** 5.3% **7.** 100% **9.** 240% **11.** 108% **13.** 0.5% **15.** 0.038%
17. 67.2% **19.** 7.46% **21.** 50% **23.** 35% **25.** 180% **27.** 130% **29.** $37\frac{1}{2}$% or 37.5%
31. $31\frac{1}{4}$% or 31.25% **33.** $33\frac{1}{3}$% **35.** $26\frac{2}{3}$% **37.** $166\frac{2}{3}$% **39.** $316\frac{2}{3}$% **41.** 0.6% or $\frac{3}{5}$%
43. 0.1875% or $0.18\frac{3}{4}$% **45.** 0.03, 3% **47.** $\frac{4}{5}$, 0.8 **49.** $2\frac{3}{50}$ or $\frac{103}{50}$, 206%

51. 0.625, 62.5% or $62\frac{1}{2}$%   **53.** $\frac{3}{8}$, 0.375   **55.** $\frac{3}{500}$, 0.6% or $\frac{3}{5}$%   **57.** $0.1\overline{6}$ or $0.16\frac{2}{3}$, $16\frac{2}{3}$%
59. $\frac{2}{3}$, $0.\overline{6}$ or $0.66\frac{2}{3}$

Exercises 9.4 (page 403)

1. 120   **3.** 20%   **5.** 166,000   **7.** 0.1% or $\frac{1}{10}$%   **9.** 95   **11.** 0.8%   **13.** 5.3885   **15.** 408
17. 21   **19.** 3600   **21.** 100%   **23.** $66\frac{2}{3}$%   **25.** $\frac{8}{15}$ **27.** 155.5 or $155\frac{1}{2}$ **29.** 9000
31. $33\frac{1}{3}$%   **33.** $\frac{65}{3}$ or $21\frac{2}{3}$

Exercises 9.5 (page 409)

1. 45 students   **3.** 0.2% or $\frac{1}{5}$%   **5.** $151.20   **7.** 150%   **9.** 50%   **11.** 60%   **13.** 20 oz
15. 252 Toyotas   **17.** 40%   **19.** 25%   **21.** 600 persons

Exercises 9.6 (page 420)

1. $33.54   **3.** $23.50   **5.** 2.5%   **7.** $8840   **9.** $19,200   **11.** $65.36   **13.** 1.9%   **15.** $254.70
17. $7.60   **19.** $90   **21.** $28.74   **23.** $8704   **25.** $464.80   **27.** 40%   **29.** 75,000

Exercises 9.7 (page 431)

1. $1.74 per hr   **3.** 17.5%   **5.** 30%   **7.** 2.5 min   **9.** $60   **11.** $832   **13.** $0.91   **15.** 45%
17. 900%   **19. a.** $33\frac{1}{3}$%   **b.** 25%

Exercises 9.8 (page 447)

1. 155   **3.** 3700   **5.** 20%   **7.** 80   **9.** 125%   **11.** 640,000   **13.** 950   **15.** 1600   **17.** 27,200
19. 468.75   **21.** 104   **23.** $360   **25.** 266   **27.** 15 oz   **29.** $103   **31.** 9.5%   **33.** 18%
35. 2.1%   **37.** 952 sq ft   **39.** $70

Developing Number Sense #8 (page 423)

1. $0.90   **2.** $1.40   **3.** $2.50   **4.** $6.80   **5.** $0.45   **6.** $0.95   **7.** $1.95   **8.** $3.30
9. $1.50   **10.** $3.90   **11.** $9   **12.** $22.50   **13.** $1.10   **14.** $3.90   **15.** $7.20   **16.** $15.60
17. $18.40   **18.** $28.80   **19.** $32.20   **20.** $78.20

Using the Calculator #10 (page 433)

1. 360   **2.** 2.7   **3.** 476   **4.** $12.25   **5.** 660   **6.** 540   **7.** $80.08   **8.** $11.44   **9.** 2125
10. 849.49   **11.** 500 or 500.000   **12.** 540   **13.** 93.3 or 93.300   **14.** 40   **15.** 31.25%
16. 6400

Chapter 9 Review Exercises (page 450)

1. equal to 1   **2.** less than 1   **3.** less than 1   **4.** more than 1   **5.** 0.07   **6.** 0.68   **7.** 1.2
8. 0.067   **9.** 0.5208   **10.** 0.005   **11.** 0.00018   **12.** 0.0025   **13.** 0.075   **14.** 0.134   **15.** 3
16. 4.6   **17.** $\frac{1}{20}$ **18.** $\frac{4}{5}$ **19.** $\frac{16}{25}$ **20.** $\frac{3}{2}$ or $1\frac{1}{2}$ **21.** 2   **22.** $\frac{3}{1000}$ **23.** $\frac{1}{120}$ **24.** $\frac{7}{80}$
25. $\frac{2}{3}$ **26.** $\frac{1}{16}$ **27.** $\frac{1}{8}$ **28.** $\frac{4}{125}$ **29.** $\frac{9}{1000}$ **30.** $\frac{3}{5000}$ **31.** $\frac{201}{400}$ **32.** $\frac{1}{100,000}$ **33.** 8.4
34. 2.4064   **35.** 40   **36.** 33   **37.** 3.842   **38.** $\frac{75}{7}$ or $10\frac{5}{7}$ **39.** 16%   **40.** 4.6%   **41.** 870%
42. 0.1%   **43.** 40%   **44.** 418%   **45.** 50.2%   **46.** 0.37%   **47.** 67%   **48.** 60%   **49.** 35%
50. $12\frac{1}{2}$% or 12.5%   **51.** 800%   **52.** 250%   **53.** $83\frac{1}{3}$%   **54.** $533\frac{1}{3}$%

55. a. $\frac{5}{4}$ or $1\frac{1}{4}$ **b.** 1.25 **56. a.** $\frac{3}{4}$ **b.** 75%
57. a. $0.\overline{3}$ **b.** $33\frac{1}{3}\%$ or $33.\overline{3}\%$ **58. a.** $\frac{1}{40}$ **b.** 0.025 **59. a.** $\frac{1}{500}$ **b.** 0.2%
60. a. 0.875 **b.** 87.5% or $87\frac{1}{2}\%$ **61.** 1370 **62.** $83\frac{1}{3}\%$ **63.** 70 **64.** 17,000 **65.** 2.4%
66. 0.403 **67.** $\frac{1}{25}$ or 0.04 **68.** 300% **69.** $3\frac{1}{3}\%$ **70.** $16\frac{2}{3}$ **71.** 120 textbooks
72. 69 smokers **73.** 32% **74.** 227.5 g **75.** 0.8 oz **76.** 180 **77.** $148.12 **78.** 6.2%
79. $208,000 **80.** $5440 **81.** 1.75% **82.** $5.70 **83.** $305 **84.** 5% **85.** $12\frac{1}{2}\%$ or 12.5%
86. $2808 **87.** $20 million **88.** $5180 **89. a.** 125% **b.** $55\frac{5}{9}\%$ **90.** $1465.13

Chapter 9 Practice Test (page 452)

1. 0.264 **2.** 3.7 **3.** 0.005 **4.** 0.0625 **5.** $\frac{9}{20}$ **6.** $\frac{7}{2}$ or $3\frac{1}{2}$ **7.** $\frac{1}{125}$ **8.** $\frac{1}{15}$ **9.** 400%
10. 5.7% **11.** 90% **12.** $233\frac{1}{3}\%$ **13.** 2.1 **14.** 40% **15.** 3.75 **16.** 0.05% **17.** 2.345
18. 216 **19. a.** 280% **b.** 180% **20.** 42% **21.** $343.20 **22.** $360,000 **23.** $6.30
24. 2500 **25. a.** $12.15 **b.** $16.20

Cumulative Review Exercises: Chapters 1–9 (page 453)

1. $\frac{101}{252}$ **2.** $3\frac{65}{108}$ **3.** $\frac{129}{7}$ or $18\frac{3}{7}$ **4.** $\frac{15}{14}$ or $1\frac{1}{14}$ **5.** 26.11 **6.** 91.16 **7.** 3354 **8.** 320
9. 0.00364 **10.** 0.45 **11.** $\frac{77}{250}$ **12.** 0.018 **13.** $0.\overline{63}$ **14. a.** $\frac{1}{40}$ **b.** 0.025 **c.** 2.5%
15. $\frac{119}{18}$ or $6\frac{11}{18}$ **16.** 0.3 ℓ **17.** 19.7 cg **18. a.** $\frac{3}{8}$ **b.** $\frac{5}{8}$ **c.** 25 **d.** 25%
19. a. $1.33 **b.** 4% **c.** 104% **d.** $10.64 **20. a.** $99.11 **b.** $0.89

CHAPTER 10

Exercises 10.1 (page 465)

1. $1\frac{1}{2}$ in **3.** 6.9 cm **5.** $4\frac{3}{4}$ in **7.** 15 ft **9.** $\frac{1}{12}$ ft **11.** 8000 m **13.** 0.1 dam **15.** 4 ft
17. 9400 mm **19.** $\frac{2}{3}$ ft **21.** $\frac{5}{4}$ or $1\frac{1}{4}$ mi **23.** 12,672 in **25.** 6250 cm **27.** 9.5 m **29.** 0.8 cm
31. 750 mm **33.** $2\frac{1}{2}$ or 2.5 yd **35.** $34\frac{3}{4}$ ft **37.** $37.70 **39.** $2.80

Exercises 10.2 (page 479)

1. length **3.** area **5.** length **7.** volume **9.** area **11.** volume **13.** 40 sq yd **15.** 0.25 ha
17. $\frac{15}{16}$ or 0.9375 qt **19.** 2992 pt **21.** 0.328 dℓ **23.** 11,700 sq cm **25.** $382
27. 871.2 sq yd **29.** 78,000 cu cm **31.** 8160 cu cm **33.** 3630 mℓ **35.** 3500 cu cm **37.** $\frac{1}{20}$

Exercises 10.3 (page 487)

1. 600 lb **3.** 0.01 ton **5.** 23.4 g **7.** 20.9 hg **9.** 6 hr **11.** 8400 sec **13.** $1.53
15. 24 servings **17.** 300 g **19.** 48 min **21.** $5

Exercises 10.4 (page 492)

1. $16 per shirt   **3.** $4\frac{1}{2}$ or 4.5 lb per cu ft   **5.** 12,000 yd per min   **7.** 8200 g per m   **9.** $\frac{80}{7}$ or $11\frac{3}{7}$ oz per da   **11.** 2250 mi per hr   **13.** $283.20   **15.** 14 c   **17.** $0.018 per m$\ell$ **19.** 20 mi   **21.** 30 mi   **23.** $4\frac{1}{2}$ or 4.5 hr   **25.** $\frac{3}{4}$ or 0.75 lb per da   **27.** 7.5 mi per hr   **29.** 21 hr

Exercises 10.5 (page 502)

1. 131 oz   **3.** $5\frac{1}{5}$ hr   **5.** 17,120 sec   **7.** $12\frac{7}{8}$ gal   **9.** 26 yd 2 ft   **11.** 4 mi 2640 ft   **13.** 1 hr 1 min 40 sec   **15.** 2 wk 3 da 8 hr   **17.** 22 lb 2 oz   **19.** 15 hr 20 min 21 sec   **21.** 6 ft 4 in   **23.** 6 wk 5 da   **25.** 6 hr 52 min   **27.** 6 hr 54 min 9 sec   **29.** 34 lb 8 oz   **31.** 67 wk 5 da 18 hr   **33.** 8 gal 2 qt   **35.** 6 yd $1\frac{2}{3}$ ft   **37.** 27 min 31 sec   **39.** 18 yd 1 ft $11\frac{1}{4}$ in   **41.** 1 wk 5 da   **43.** 6 lb 7 oz   **45.** 10 gal 2 qt

Developing Number Sense #9 (page 467)

1. 3000 mm   **2.** 49,000,000 cm   **3.** 0.04 m   **4.** 0.03 dam   **5.** 6.75 ℓ **6.** 4500 mℓ **7.** 8 daℓ **8.** 24 hℓ **9.** 7400 cg   **10.** 98,000 dg   **11.** 0.006 g   **12.** 0.045 kg

Chapter 10 Review Exercises (page 504)

1. $2\frac{7}{8}$ in   **2.** 7 cm   **3.** $\frac{3}{4}$ in by $2\frac{1}{4}$ in, $\frac{27}{16}$ or $1\frac{11}{16}$ sq in   **4.** volume   **5.** length   **6.** area   **7.** volume   **8.** weight   **9.** volume   **10.** length   **11.** volume   **12.** length   **13.** area   **14.** weight   **15.** area   **16.** 5 yd   **17.** 0.5 m   **18.** 36 sq ft   **19.** 193.6 sq yd   **20.** 32 c   **21.** 0.082 ℓ **22.** 5670 cg   **23.** 64 oz   **24.** 6720 min   **25.** 1001 cu in   **26.** 95 qt   **27.** $8\frac{3}{7}$ wk   **28.** $496\frac{2}{3}$ min   **29.** 33 yd 1 ft   **30.** 8 tons 500 lb   **31.** 50 hr 7 min 35 sec   **32.** $\frac{5}{12}$ T per person   **33.** 1.22 g per m   **34.** 400 lb per load   **35.** 31.5 kg per wk   **36.** $4\frac{1}{5}$ or 4.2 ft per sec   **37.** $0.48 per lb   **38.** 3240 cℓ per hr   **39.** 9.9 ft per sec   **40.** 34 yd 1 ft   **41.** 18 lb 10 oz   **42.** 7 wk 5 da 20 hr   **43.** 47 lb 3 oz   **44.** 60 yd 3 in   **45.** 7 hr 31 min   **46.** 3 yd 1 ft 9 in   **47.** 3 wk $20\frac{1}{5}$ hr   **48.** 21.7 mi   **49.** $3.40   **50.** $\frac{145}{16}$ or $9\frac{1}{16}$ sq yd   **51.** $3790.80   **52.** 28 doses   **53.** 1.2 ℓ **54.** 1710 mℓ **55.** 520   **56.** $1.94   **57.** 40 min   **58.** 80%   **59.** 7.8 min   **60.** 89.6 oz   **61.** 7.4 sec   **62.** 1100 ft per sec   **63.** $\frac{1}{30}$ **64.** 5 hr 40 min   **65.** 1 lb 11 oz   **66.** 4 lb $12\frac{1}{3}$ oz   **67.** 80 segments

Chapter 10 Practice Test (page 507)

1. $1\frac{3}{4}$ in   **2.** 2 cm by 4.5 cm, 13.0 cm   **3.** 24 qt   **4.** 5500 lb   **5.** $20\frac{1}{3}$ yd   **6.** 0.26 g   **7.** 7080 mℓ **8.** 0.12 km   **9.** 9 lb 13 oz   **10.** $4\frac{1}{5}$ hr   **11.** 42 mi per hr   **12.** 320 cℓ per hr   **13.** 44 ft per sec   **14.** 0.09 cg per m   **15.** 8 wk 4 da   **16.** 12 gal 3 qt   **17.** 23 yd 6 in   **18.** 2 lb 14 oz   **19.** 18,480 sq cm   **20.** 38.25 mi   **21.** 30 pieces   **22.** 1 ft 6 in   **23.** 6 ¢ per oz   **24.** 59 min 40 sec   **25.** $2\frac{1}{12}$ c

CHAPTER 11

Exercises 11.1 (page 515)

1. $60,000 **3.** $25,000 **5.** 20% **7.** 28 mi per gal **9.** 640 mi **11.** 18.75 gal **13.** 14%
15. 1974; 1976; 1977; 1978; and 1979 **17.** 14%
19. a. U.S. Health Care Cost   **b.** U.S. Health Care Cost

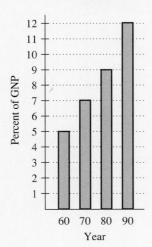
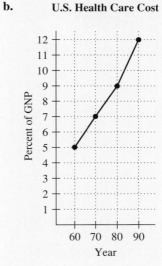

21. a.

Store	TURKEY PRICES Price (in dollars)
Thriftway Market	0.50
Savemore	0.75
Star Market	1.25
Miller's Foodmart	1.75

b. Turkey Prices **c.** Turkey Prices

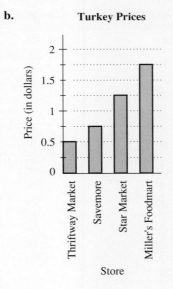
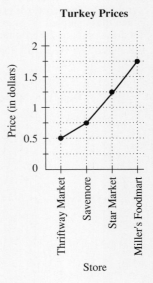

Exercises 11.2 (page 526)

1. 124 **3.** 90 **5.** $33\frac{1}{3}$% **7.** 20% **9.** 240 **11.** 126° **13.** A. 100.8° B. 43.2° C. 216°
15. A. 115° B. 85° C. 160° **17.** A. 70° B. 290°

19. a. 800 Adults Answer the Question "How many close friends do you have?"

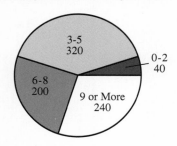

b. 800 Adults Answer the Question "How many close friends do you have?"

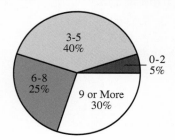

Exercises 11.3 (page 536)

1. a. 15.6 **b.** 16 **c.** 20   **3. a.** 305 **b.** 310 **c.** no mode
5. a. 12.4 **b.** 14.5 **c.** 18   **7. a.** 5.5 **b.** 4.0 **c.** 3.5
9. a. 10 **b.** $9\frac{5}{12}$ **c.** no mode   **11. a.** 9.4 min **b.** 9 min **c.** 9 min
13. a. $71.25 **b.** $50 **c.** $0 and $50   **15. a.** $34,500 **b.** $32,250 **c.** no mode
17. a. 3.38 **b.** 3.4 **c.** 3.2   **19. a.** 21\frac{19}{20}$ **b.** 22\frac{1}{4}$ **c.** no mode

Chapter 11 Review Exercises (page 538)

1. $6000   **2.** $180,000   **3.** $132,000   **4.** $\frac{2}{1}$

5. The professor's income is $5000 per month and the elementary teacher's income is $4000 per month. Therefore, the professor does not make twice as much as the elementary teacher. Since the squiggle ($\lessgtr$) is used on the vertical axis, the visual heights of the bars cannot be used to compare the incomes by ratios.

6. 0.8 million lb   **7.** 0.4 million lb   **8.** decreased by 0.2 million lb   **9.** 1.16 million lb per mo
10. $33\frac{1}{3}\%$

11. U.S. Exports to Mexico

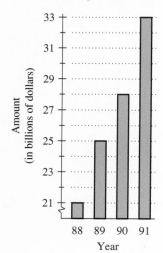

12. Housing Market Affordability Comparison

13. a.

UNMARRIED HOUSEHOLD COUPLES ON THE RISE	
Year	Number of Households (in millions)
1970	0.5
1975	1.25
1980	1.5
1985	2.25
1990	2.75

b.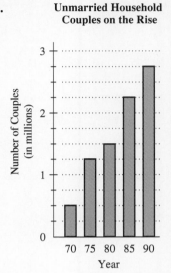

c.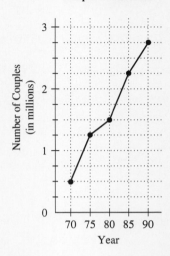

14. $15,000 **15.** $75,000 **16.** $15,000 **17.** Youngstown; $45,400 **18.** 20% **19.** 5% **20.** 2125
21. 2550 **22.** 54° **23.** 1.4 million lb **24.** 0.2 million lb **25.** 70% **26.** 10% **27.** 2.3 million lb
28. A. 144° B. 90° C. 126° **29.** A. 72° B. 54° C. 234°
30. A. 45° B. 110° C. 205°
31. 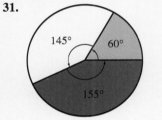 **32.** 25% **33.** 20%
34. 90° **35.** 72°

36.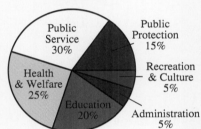

37. a. 1.8 glasses **b.** 2 glasses **c.** 0 glasses
38. a. 7.3 sec **b.** 7.2 sec **c.** no mode

Chapter 11 Practice Test (page 543)

1. 210 lb **2.** August, 5 lb **3.** 30 lb **4.** 207 lb **5.** $37\frac{1}{2}$% **6.** 160 **7.** 340 **8.** 153°

9. Taxi Mileage

10. a. 1.8 hr **b.** 1.5 hr **c.** 1 hr
11. a. 1.4 hr **b.** 1.5 hr **c.** no mode

Cumulative Review Exercises: Chapters 1–11 (page 544)

1. $\frac{146}{45}$ or $3\frac{11}{45}$ **2.** $\frac{3}{25}$ **3.** 1400 **4.** 675 **5.** $\frac{1}{30}$ **6.** $41\frac{61}{114}$

7. 14,058.5 **8.** 51,037.04 **9.** 18.225 **10.** 855 **11. a.** $2\frac{1}{3}$ ft **b.** 28 in

12. a. 105 min **b.** $1\frac{3}{4}$ hr **13. a.** 28 sq units **b.** 22 units **14.** 3 mi per hr

15. a. 4.8 lb **b.** $18.58 **c.** $15.20 **16.** 1200 aides **17.** 650 doses

18. a. 4 wk **b.** 28 days **c.** $\frac{3}{28}$ **d.** 10 days **19. a.** 1.25 **b.** 1 **c.** 0 and 1

20. a. Azcarate's Bicycling Time **b.** $2\frac{3}{5}$ or 2.6 hr per wk **c.** $\frac{3}{4}$ hr or 45 min **d.** 27%

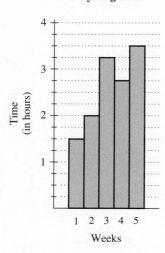

CHAPTER 12

Exercises 12.1 (page 557)

1. 48 **3.** 45 **5.** 16 **7.** 63 **9.** $\frac{2}{3}$ **11.** $\frac{12}{5}$ or $2\frac{2}{5}$ **13.** 2 **15.** 52 **17.** $\frac{13}{10}$ or $1\frac{3}{10}$

19. 64 **21.** 0 **23.** 32 **25.** 59 **27.** 5 **29.** 39 **31.** $\frac{12}{5}$ or $2\frac{2}{5}$ **33.** 18 **35.** 40

37. 96 **39.** 40 **41.** 38 **43.** 1120 **45.** 20 **47.** $\frac{14}{5}$ or $2\frac{4}{5}$ **49.** $\frac{2}{5}$ **51.** $\frac{11}{8}$ or $1\frac{3}{8}$

53. $\frac{13}{3}$ or $4\frac{1}{3}$ **55.** 13 **57.** 54 **59.** 60 **61.** 106 **63.** 220

Exercises 12.2 (page 565)

1. -120 **3.** 400 **5.** -25 **7.** 8 **9.** 6 **11.** -32 **13.** B. -2 C. -6

15. B. -10 C. -15 **17.** < **19.** > **21.** < **23.** 16 **25.** $-67, -36, -13, 0$ **27.** -17

29. 25 **31.** 14 **33.** 29 **35.** -62 **37.** 0 **39.** -17 **41.** -18 **43.** -5

Section 12.3 (page 570)

1. -7 **3.** -12 **5.** 5 **7.** -25 **9.** 0 **11.** -34 **13.** 8 **15.** -37 **17.** 55 **19.** -693

21. -3 **23.** 6 **25.** -49 **27.** 1 **29.** -71 **31.** -11 **33.** -52 **35.** $-31°F$ **37.** -18

Exercises 12.4 (page 581)

1. −5 **3.** 6 **5.** 2 **7.** −10 **9.** 75 **11.** −6 **13.** −90 **15.** −4 **17.** 70 **19.** 56
21. −4 **23.** −48 **25.** −24 **27.** −12 **29.** 3 − 4 = −1 **31.** −35 − 18 = −53
33. 80 − (−75) = 155 **35.** 11 **37.** −29 **39.** 34 **41.** 62 units **43.** 9500 ft **45.** 38°F
47. −13 **49.** −11 **51.** 10 **53.** −75 **55.** 29 **57.** −72 **59.** −28

Exercises 12.5 (page 592)

1. −32 **3.** 36 **5.** 84 **7.** −128 **9.** 0 **11.** 2600 **13.** −18,000 **15.** 424,512 **17.** −60
19. 280 **21.** −120 **23.** 0 **25.** 1500 **27.** 21 **29.** −22 **31.** −30 **33.** 20 **35.** 28
37. −5 **39.** 9 **41.** −15 **43.** 50 **45.** −1 **47.** 0 **49.** 6 **51.** −60 **53.** −87
55. no answer **57. a.** $\dfrac{-15}{2}$ **c.** −7.5 **d.** $\dfrac{15}{-2}$ **e.** $-\left(\dfrac{15}{2}\right)$ **59.** −3 **61.** $\dfrac{1}{9}$ **63.** $\dfrac{3}{5}$
65. $\dfrac{8}{5}$ or $1\dfrac{3}{5}$ **67.** 8 **69.** $\dfrac{-20}{3}$ or $-6\dfrac{2}{3}$ **71.** −40 **73.** 0.625 **75.** $-0.8\overline{3}$ **77.** $-1.\overline{3}$
79. −4.25

Exercises 12.6 (page 600)

1. A. $\dfrac{-2}{3}$ B. $\dfrac{-7}{3}$ or $-2\dfrac{1}{3}$ C. $\dfrac{-11}{3}$ or $-3\dfrac{2}{3}$

3. A. $-3\dfrac{1}{2}$ or $\dfrac{-7}{2}$ B. $-2\dfrac{1}{2}$ or $\dfrac{-5}{2}$ C. $\dfrac{1}{2}$

5. A. −0.2 B. −0.8 **7.** A. −0.25 B. 1.5 **9.** $\dfrac{-7}{10}$ **11.** $\dfrac{27}{20}$ or $1\dfrac{7}{20}$

13. $\dfrac{-25}{6}$ or $-4\dfrac{1}{6}$ **15.** $\dfrac{-5}{21}$ **17.** $\dfrac{-10}{3}$ or $-3\dfrac{1}{3}$ **19.** $\dfrac{10}{9}$ or $1\dfrac{1}{9}$ **21.** −25 **23.** 1.73
25. 9.4 **27.** −21.24 **29.** −8.55 **31.** −602 **33.** 127.68 **35.** 6 **37.** 800 **39.** 0.02
41. 6.106 **43.** $\dfrac{-13}{5}$ or $-2\dfrac{3}{5}$ **45.** $\dfrac{14}{3}$ or $4\dfrac{2}{3}$ **47.** 4.6 or $4\dfrac{3}{5}$ **49.** −0.612 or $\dfrac{-153}{250}$

Exercises 12.7 (page 606)

1. xy **3.** $\dfrac{y}{x}$ **5.** $x - y$ **7.** $y - 4$ **9.** $6 - 2y$ **11.** 31 **13.** −54 **15.** 175 **17.** −45
19. −17 **21.** 2 **23.** $\dfrac{5}{8}$ **25.** 5 **27.** −22.4 **29.** 1680 sq in **31.** 0.4 sq mi **33.** 54 ft
35. $7\dfrac{1}{2}$ or $\dfrac{15}{2}$ mi **37.** 212°F **39.** 5°F **41.** 10°C **43.** −20°C **45.** $D = -35$ **47.** $y = -5$

Using the Calculator #11 (page 558)

1. 165 **2.** 705 **3.** 487 **4.** 546 **5.** 5588 **6.** 0.033 **7.** 0.431 **8.** 411 **9.** 53.1
10. 19.882 **11.** 0.168 **12.** 3.3

Using the Calculator #12 (page 559)

1. 12 **2.** 78 **3.** 332.5 **4.** 9.15 **5.** 12 **6.** 0.5 **7.** 0.404 **8.** 3.931 **9.** $637.50
10. $23.46

Using the Calculator #13 (page 601)

1. −11 **2.** −46 **3.** −14 **4.** −2.5 **5.** 4455 **6.** 162 **7.** −54 **8.** 625 **9.** 22
10. −2250 **11.** −107.6 **12.** −0.675 or $\dfrac{-27}{40}$

Chapter 12 Review Exercises (page 609)

1. −2 **2.** −1 **3.** 504 **4.** −13 **5.** 32 **6.** 48 **7.** −17 **8.** −13 **9.** −56 **10.** 178
11. −84 **12.** 30 **13.** A. 30 B. −10 C. −20 **14.** A. 4 B. −2 C. −8 **15.** <
16. > **17.** −15, −8, 0, 12 **18.** −32 **19.** 48 **20.** −120 **21.** −11 **22.** −8 **23.** $\dfrac{-1}{8}$

24. 40 **25.** −$15, Monica is overdrawn by $15. **26.** −8°C **27.** 47 yd **28.** 3200 ft
29. a. $\frac{3}{8}$ **d.** 0.375 **30. b.** $\frac{-6}{5}$ **c.** $\frac{6}{-5}$ **e.** −1.2 **31.** −5.$\overline{6}$ **32.** 0.55
33. A. $\frac{2}{3}$ **B.** $\frac{-1}{3}$ **C.** $\frac{-4}{3}$ or $-1\frac{1}{3}$
34. A. $\frac{-7}{6}$ or $-1\frac{1}{6}$ **B.** $\frac{-3}{2}$ or $-1\frac{1}{2}$ **C.** $\frac{-8}{3}$ or $-2\frac{2}{3}$
35. A. −25.75 **B.** −26.5 **C.** −27.25 **36. A.** 2.5 **B.** −0.5 **C.** −1.5 **37.** $\frac{-3}{10}$
38. $\frac{-10}{81}$ **39.** −22.15 **40.** −51.25 **41.** 0.404 **42.** $\frac{9}{2}$ or $4\frac{1}{2}$ **43.** $5w$ **44.** $5 + w$
45. $w − 5$ **46.** $\frac{5}{w}$ **47.** 11 **48.** 2 **49.** −180 **50.** 36 **51.** $\frac{-19}{6}$ or $-3\frac{1}{6}$ **52.** −34.56
53. a. 2700 sq in **b.** 210 in **54. a.** 120.93 sq m **b.** 45.2 m **55.** −40°C **56.** $-11\frac{2}{3}$°C
57. 519 **58.** 8

Chapter 12 Practice Test (page 611)

1. 44 **2.** 0 **3.** 4 **4.** −15 **5.** −26 **6.** 35.1 **7.** −130 **8.** −42 **9.** −16 **10.** −122
11. 18 **12.** −1.24 **13. A.** −5 **B.** −9 **C.** −10 **14. A.** $\frac{2}{3}$ **B.** $\frac{-1}{3}$ **C.** $\frac{-5}{3}$ or $-1\frac{2}{3}$
15. A. −5.25 **B.** −3.75 **C.** −4.5 **16. b.** $\frac{-4}{5}$ **c.** $\frac{4}{-5}$ **e.** −0.8 **17.** −24, −15, 0, 8, 45
18. $n + 20$ **19.** $n − 20$ **20.** −180 **21.** −5 **22.** −51 **23.** 14,400 ft **24.** 14°F **25.** 12

CHAPTER 13

Exercises 13.1 (page 616)

1. 79^2 **3.** 19^6 **5.** 13^2 **7.** $(−9)^2$ **9.** 81 **11.** 64 **13.** 49 **15.** −64 **17.** −16 **19.** 41
21. 25 **23.** −71 **25.** −1 **27.** 1600 **29.** −36 **31.** −64 **33.** 32,010 **35.** 23 **37.** 80
39. 49 **41.** 375 **43.** −11 **45.** 16 **47.** 125 cu cm

Exercises 13.2 (page 620)

1. $\frac{1}{5}$ **3.** $\frac{1}{36}$ **5.** $\frac{1}{512}$ **7.** $\frac{1}{81}$ **9.** $\frac{1}{32}$ **11.** 1 **13.** 1 **15.** $\frac{5}{4}$ **17.** $\frac{2}{5}$ **19.** 4 **21.** $\frac{-1}{12}$
23. 1 **25.** $\frac{37}{25}$ **27.** $\frac{1}{6}$ **29.** $\frac{21}{8}$ **31.** $\frac{9}{2}$ **33.** $\frac{1}{225}$ **35.** $\frac{1}{20}$ **37.** −78
39. a. $2000 **b.** $500 **c.** $4000

Exercises 13.3 (page 626)

1. c. 2^5 **e.** 32 **3.** 5^8 **5.** 2^{23} **7.** 3^{-10} **9.** 12^4 **11.** n^8 **13.** y^5 **15. c.** $\frac{1}{9}$ **d.** 3^{-2}
17. 7^2 **19.** 10^{-2} or $\frac{1}{10^2}$ **21.** 3^{11} **23.** 4^5 **25.** x^4 **27.** n^{18} **29. c.** 256 **d.** 2^8 **31.** 4^{18}
33. 9^{14} **35.** 8^{-6} or $\frac{1}{8^6}$ **37.** 10^{32} **39.** a^{30} **41.** n^{24} **43. a.** 25 **b.** 49
45. a. 45 **b.** 9 **47.** In general, $x^2 + y^2$ does not equal $(x + y)^2$.
49. a. 35 **b.** 125 **51. a.** 124 **b.** 64 **53. a.** 216 **b.** 216 **55. a.** −27 **b.** −27
57. In general, $(xy)^3$ equals x^3y^3. **59. a.** 900 **b.** 900 **61. a.** 225 **b.** 225

Exercises 13.4 (page 633)

1. 100,000 **3.** 0.000 001 **5.** 700,000,000 **7.** 0.000 000 000 08 **9.** 3,300,000 m **11.** 4.98×10^9
13. 3.2×10^{10} **15.** 9×10^{-10} **17.** 4×10^{-6} **19.** 5.2×10^4 cu ft per sec
21. 8×10^{13} or 80,000,000,000,000 **23.** 1.3×10^7 or 13,000,000 **25.** 1.518×10 or 15.18
27. 4×10^3 or 4000 **29.** 2.5×10^{-6} or 0.000 0025 **31.** 7.5×10^2 or 750
33. 6×10^{12} or 6,000,000,000,000 cm **35.** 2×10^{-2} or 0.02 sec **37.** 1.32×10^{-1} or 0.132 sec

Exercises 13.5 (page 641)

1. 144　**3.** 7, −7　**5.** 8　**7.** 5　**9.** 20　**11.** −9　**13.** 2.449　**15.** 8.718　**17.** 6.731
19.

21.

23. $\sqrt{68}$, 8.25, 8.5, 9, $\sqrt{83}$　**25.** 5　**27.** 8　**29.** 50　**31.** 67　**33.** 6　**35.** 5　**37.** 60 ft

Using the Calculator #14 (page 617)

1. 81　**2.** 32　**3.** 11,390,625　**4.** 53,1441　**5.** 97.336　**6.** 0.269　**7.** 0.134　**8.** 2059.630

Using the Calculator #15 (page 634)

1. 4×10^{10}　**2.** 7.448×10^{13}　**3.** 1.2×10^{13}　**4.** 7×10^{10}　**5.** 6.25×10^{-12}　**6.** 2×10^{-10}
7. 5.264×10^{14}　**8.** 4×10^{9}

Developing Number Sense #10 (page 638)

1. 2, 3　**2.** 7, 8　**3.** 10, 11　**4.** 15, 16　**5.** 24, 25　**6.** 31, 32

Using the Calculator #16 (page 640)

1. 9　**2.** 53　**3.** 17.889　**4.** 9.088　**5.** 0.283　**6.** 0.866

Chapter 13 Review Exercises (page 642)

1. 1296　**2.** 6, −6　**3.** 125　**4.** −32　**5.** 25　**6.** −20　**7.** 64　**8.** 729
9. $\frac{1}{10000}$ or 0.0001　**10.** $\frac{1}{32}$　**11.** 9　**12.** 30　**13.** 225　**14.** $\frac{1}{16}$　**15.** −2　**16.** 280
17. $\frac{21}{80}$　**18.** 16　**19.** −106　**20.** 540　**21.** 43　**22.** −4　**23.** 152　**24.** 51　**25.** −8
26. 480　**27.** $\frac{5}{3}$　**28.** 95　**29.** $\frac{1}{49}$　**30.** $\frac{25}{144}$　**31.** 40　**32.** 14　**33.** 12^{11}　**34.** 10^{14}
35. 12^{-5} or $\frac{1}{12^5}$　**36.** 10^{-26} or $\frac{1}{10^{26}}$　**37.** 8^{20}　**38.** 4^{21}　**39.** x^4　**40.** x^{-14} or $\frac{1}{x^{14}}$
41. x^{-45} or $\frac{1}{x^{45}}$　**42.** a^4　**43.** 100,000　**44.** 0.000 000 001　**45.** 0.000 000 000 0059
46. 630,000,000　**47.** 1.75×10^6　**48.** 3×10^{14}　**49.** 7.3×10^{-3}　**50.** 4.72×10^{-10}
51. 5×10^5 or 500,000 gal per hr　**52.** 5×10^{-5} or 0.00005 cu m
53. 1.2×10^{10} or 12,000,000,000 m per min　**54.** 2.64×10^4 or 26,400 sec　**55.** $11,875
56. a. 400　**b.** 200　**c.** 50　**57.** 72 ft per sec
58.

59.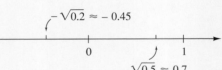

60. 6.5, $\sqrt{45}$, 7, $\sqrt{50}$, 7.5　**61.** 2, $\sqrt{5}$, 2.3, $\sqrt{6}$, 2.5　**62.** 7.616　**63.** 2.445　**64.** 0.866

Chapter 13 Practice Test (page 644)

1. 64　**2.** $\frac{1}{81}$　**3.** 11　**4.** −64　**5.** $\frac{7}{10}$　**6.** 13　**7.** 81　**8.** 600　**9.** $\frac{1}{12}$　**10.** $\frac{1}{14}$　**11.** 26
12. 11　**13.** 512 cu ft　**14.** 5^{11}　**15.** 9^{15}　**16.** y^{-9} or $\frac{1}{y^9}$　**17.** a^7　**18.** 700,000,000
19. 0.000 000 000 000 396　**20.** 4.5×10^{-11}　**21.** 9.4×10^5

22.

$\sqrt{20} \approx 4.47$; $\sqrt{23} \approx 4.8$ (on number line with 4.5, 5, 5.2, 5.5)

23. $\sqrt{62}$, 8, $\sqrt{67}$, 8.5   **24.** 9×10^4 or 90,000 dollars   **25.** 7×10^2 or 700 gal

Cumulative Review Exercises: Chapters 1–13 (page 645)

1. $\frac{71}{45}$ **2.** 123.425   **3.** 10   **4.** -17 **5.** 0.032   **6.** $5\frac{3}{4}$ **7.** 216   **8.** $\frac{1}{9}$ **9.** 50
10. 0.272   **11.** 20.588   **12.** 84.290   **13.** 0.511   **14.** 11   **15.** -60 **16.** 272   **17.** 40
18. 7.443   **19.** $\frac{1}{50}$ or 0.02   **20.** -4 **21.** 116   **22.** $\frac{1}{8}$
23. a. 3.7 mi   **b.** 3.5 mi   **c.** 3.0 mi   **24.** 19   **25.** $543   **26.** $10.80   **27.** 60%   **28.** $0.42
29. a. 8.75 gal   **b.** 280 mi   **c.** 420 mi   **30.** 35%, The fund earned 35%.
31. -10%, The fund lost 10%.   **32.** earned $129   **33.** lost $75

CHAPTER 14

Exercises 14.1 (page 651)

1. $a + 24$ **3.** $x - 5$ **5.** $23 + a + c$ **7.** $22 - r$ **9.** $w + 10$ **11.** $w - 19$ **13.** $15 + x - y$
15. $4x$ **17.** $96ab$ **19.** $-40x^3$ **21.** $5xw^3$ **23.** $3x^5$ **25.** $91x^4$ **27.** $-30x^3y^4$ **29.** $-105a^4b^6$

Exercises 14.2 (page 663)

1. $18w + 48$ **3.** $a^2 + ay$ **5.** $3ab + 3b^2 + 21b$ **7.** $-6x - 30y$ **9.** $15x - 3x^2 + 3xy$
11. $-15x + 45$ **13.** $-14x^2 + 35xy - 7x$ **15.** $6(x + y)$ **17.** $2(2m + 3n)$ **19.** $x(5 - 6y)$
21. $a(c - b + 8)$ **23.** $2a(5c + 4b)$ **25.** $7y(3x + 2y)$ **27.** $5(1 - a)$ **29.** $x(x^2 - 3x - 1)$
31. $7c(1 - 2a + x)$ **33.** $10y$ **35.** $11xy$ **37.** $5y^2$ **39.** $-8ac$ **41.** $6c$ **43.** $-5ac$
45. $12n - 11$ **47.** $23c^2 - 20c$ **49.** $10ac - 13c + 5a$ **51.** $-3x^2 + 6x + 8$ **53.** $-4x$ **55.** $-16c$
57. $8x + 6$ **59.** $11y - 14x$ **61.** $18x - 19$ **63.** $-24c + 66$ **65.** $7w^2 - 7w$ **67.** $-60cx$
69. $30c^2$ **71.** $30c + 5c^2$ **73.** $15x$ **75.** $15 + 2x$ **77.** $-56x^2$ **79.** $-4 + 9x$ **81.** $-26 - 3x$

Exercises 14.3 (page 674)

1. $x = 12$ **3.** $x = 21$ **5.** $y = -9$ **7.** $y = 6$ **9.** $w = -5$ **11.** $w = \frac{2}{5}$ **13.** $v = \frac{10}{7}$
15. $v = \frac{-25}{2}$ **17.** $v = 0$ **19.** $w = 56$ **21.** $c = -48$ **23.** $c = 7$ **25.** $a = -16$ **27.** $a = 7$
29. $m = 8$ **31.** $m = -2$ **33.** $n = -6$ **35.** $n = -7$ **37.** $x = \frac{5}{3}$ **39.** $x = \frac{1}{2}$ **41.** $x = \frac{-2}{3}$
43. $x = \frac{8}{7}$ **45.** $x = 3$ **47.** $x = 13$ **49.** $y = \frac{-3}{2}$ **51.** $m = \frac{-1}{2}$ **53.** $n = 0$ **55.** $x = 3$
57. $x = \frac{22}{5}$ **59.** $a = 9$ **61.** $a = \frac{7}{5}$

Exercises 14.4 (page 689)

1. 75 ft   **3.** -29 **5.** $270   **7.** 12 m   **9.** $50 - N = 18$ **11.** $\frac{N}{3} = 36$ **13.** $75 - 2N = 31$
15. $5(N + 13) = 75$ **17.** $L + (L - 15) = 155$ or $2L - 15 = 155$ **19.** $8x - x = 49$ or $7x = 49$
21. $2N - (50 - N) = 1$ or $3N - 50 = 1$ **23.** $\frac{86 + 78 + x}{3} = 80$ or $\frac{164 + x}{3} = 80$
25. $W(W + 5) = 204$ or $W^2 + 5W = 204$ **27.** $18(J + 10) + 38J = 1860$ or $56J + 180 = 1860$
29. -6 **31.** $\frac{3}{2}, \frac{7}{2}$ **33.** $\frac{-1}{2}, \frac{1}{2}, \frac{3}{2}$ **35.** $250   **37.** amount invested, $3600; amount spent, $1200
39. 37 ft, 74 ft   **41.** 210 $24 tickets; 340 $10 tickets

Exercises 14.5 (page 696)

1. $x = 4$ **3.** $y = -5$ **5.** $v = 5$ **7.** $a = \frac{-1}{3}$ **9.** $w = \frac{-4}{3}$ **11.** $w = -6$ **13.** $x = 6$
15. $y = \frac{-2}{3}$ **17.** $n = \frac{5}{3}$ **19.** $n = \frac{2}{3}$ **21.** $c = \frac{-3}{4}$ **23.** $c = \frac{-1}{2}$ **25.** $w = -20$ **27.** $w = \frac{15}{4}$
29. $c = \frac{-2}{3}$ **31.** -2 **33.** 21, 29 **35.** 12, 38 **37.** $-3, 6, 20$ **39.** 45 $12 tickets; 61 $8 tickets

Exercises 14.6 (page 710)

1. 5 **3.** $\frac{-1}{2}$ **5.** $\frac{7}{10}$ **7.** $\frac{17}{6}$ **9.** $\frac{7}{5}$ **11.** $\frac{2x}{3}$ **13.** $\frac{15c}{4}$ **15.** $\frac{12m}{13c}$ **17.** $\frac{x+1}{3}$ **19.** $\frac{m+8}{2}$
21. $\frac{25(3-c)}{3c}$ or $\frac{75-25c}{3c}$ **23.** $\frac{c-2}{2c}$ **25.** $\frac{2(c-4)}{3}$ or $\frac{2c-8}{3}$ **27.** $\frac{x+1}{8}$ **29.** $\frac{3a-1}{3a}$
31. $\frac{1}{4}$ **33.** $\frac{-7}{6}$ **35.** $\frac{5m}{4n}$ **37.** $\frac{21n}{10m}$ **39.** $\frac{m^2}{6}$ **41.** $\frac{3c}{50}$ **43.** $\frac{25}{22y}$ **45.** $\frac{4}{5y}$ **47.** $180m^2$
49. $\frac{8}{7b}$ **51.** $\frac{x+4}{4x}$ **53.** $\frac{3(w+4)}{20}$ or $\frac{3w+12}{20}$ **55.** $\frac{a-1}{2a}$ **57.** $\frac{7(m-1)}{6}$ or $\frac{7m-7}{6}$
59. $\frac{3(n-2)}{2n^2}$ or $\frac{3n-6}{2n^2}$ **61.** $\frac{5}{12}$ **63.** $\frac{17}{80}$ **65.** $\frac{3x}{5}$ **67.** $\frac{6-a}{9}$ **69.** $\frac{y}{30}$ **71.** $\frac{x+5}{15}$ **73.** $\frac{-2}{x}$
75. $\frac{12+2a}{3a}$ or $\frac{2(6+a)}{3a}$ **77.** $\frac{4-25y}{20y}$ **79.** $\frac{3y+2x}{xy}$ **81.** $\frac{2(m+x)}{3x}$ or $\frac{2m+2x}{3x}$ **83.** $\frac{3w-10}{5}$
85. $\frac{8+x}{x}$ **87.** $\frac{x-2}{2}$ **89.** $2x$ **91.** $\frac{3}{10}m$ or $\frac{3m}{10}$ **93.** $\frac{-17}{8}v$ or $\frac{-17v}{8}$ **95.** $\frac{11}{6}a$ or $\frac{11a}{6}$
97. $\frac{1}{5}y$ or $\frac{y}{5}$ **99.** $7.3n$ **101.** $2.55x$ **103.** $1.45y$ **105.** $0.35w$

Exercises 14.7 (page 727)

1. $x = 8$ **3.** $x = \frac{2}{3}$ **5.** $w = 2340$ **7.** $P = 60$ **9.** $A = \frac{35}{2}$ or $17\frac{1}{2}$ or 17.5 **11.** $w = 8.5$
13. $c = 4$ **15.** $c = 45$ **17.** $w = 7$ **19.** $\frac{15}{4}$ or $3\frac{3}{4}$ or 3.75 lb **21.** 130 acres of rice **23.** 30 ft
25. 0.625 or $\frac{5}{8}$ qt **27.** $3c$ **29.** $\frac{16}{15}$ or $1\frac{1}{15}$ yd **31.** $x = 72$ **33.** $x = \frac{49}{3}$ or $16\frac{1}{3}$
35. $w = 180$ **37.** $w = \frac{2}{15}$ **39.** $m = 80$ **41.** $m = 9.05$ **43.** $y = 250$ **45.** $y = 70$ **47.** $c = 36$
49. $c = 6$ **51.** $a = \frac{-14}{3}$ or $-4\frac{2}{3}$ **53.** $a = \frac{27}{4}$ or $6\frac{3}{4}$ or 6.75 **55.** $B = 14$ **57.** $B = 400$
59. $v = 640$ **61.** $v = 16$ **63.** $R = 20$ **65.** $R = 140$ **67.** $45,000 **69.** 6 hr **71.** 99
73. $8500 **75.** $750

Chapter 14 Review Exercises (page 730)

1. $7y + 28$ **2.** $20m - 30m^2$ **3.** $-12v^3 + 18v^2 - 6v$ **4.** $5(2x + 3)$ **5.** $a(3c - 1)$
6. $2y(y^2 + y - 1)$ **7.** $-1 - x$ **8.** $11 + 3x$ **9.** $60n^3$ **10.** $12n$ **11.** $15 - n$ **12.** $35w^5$
13. $12m^3y^5$ **14.** $13u$ **15.** $4c^2$ **16.** $10xy - 13y$ **17.** $15 - 11w$ **18.** $13x^2 - 2x + 7$
19. $10d - 1$ **20.** $14a - 29$ **21.** $c^2 - 11c$ **22.** $-6y^2$ **23.** $8v^2 - 10v + 10$ **24.** $x = 11$
25. $v = 5$ **26.** $c = 108$ **27.** $w = 30$ **28.** $x = 4$ **29.** $x = 3$ **30.** $y = \frac{3}{2}$ **31.** $w = \frac{5}{4}$
32. $v = \frac{-4}{3}$ **33.** $y = -46$ **34.** $x = 5$ **35.** $m = \frac{-16}{3}$ **36.** $y = \frac{10}{3}$ **37.** $x = 2$ **38.** $x = \frac{1}{20}$
39. $w = 1$ **40.** $v = \frac{-28}{5}$ or -5.6 **41.** $m = 24$ **42.** $m = \frac{-5}{48}$ **43.** $v = \frac{7}{4}$ or 1.75
44. $m = 0.75$ **45.** $n = 40$ **46.** $B = 0.004$ **47.** $x = 7$ **48.** $w = 56$ **49.** $w = 0.03$

824 Answers to Selected Exercises

50. $v = 12$ **51.** $c = 20$ **52.** $c = \frac{3}{10}$ or 0.3 **53.** $v = -1$ **54.** $w = \frac{44}{9}$ **55.** $A = 16$
56. $A = 160$ **57.** $v = 2.5$ or $\frac{5}{2}$ **58.** $w = 2$ **59.** $w = 15$ **60.** $m = 4250$ **61.** 12 **62.** $\frac{-35}{2}$
63. -10 **64.** $\frac{7}{5w}$ **65.** $\frac{8a}{5}$ **66.** $\frac{2(2n-1)}{3}$ or $\frac{4n-2}{3}$ **67.** $\frac{w+3}{2w}$ **68.** $\frac{3-x}{6}$ **69.** $\frac{y-7}{14}$
70. $\frac{8m}{5}$ **71.** $\frac{2}{3m}$ **72.** $\frac{1}{30x^2}$ **73.** $\frac{8m^2n^2}{3}$ **74.** $\frac{y(y+5)}{10}$ or $\frac{y^2+5y}{10}$
75. $2(c-3)$ or $2c-6$ **76.** $\frac{15(c+4)}{4}$ or $\frac{15c+60}{4}$ **77.** $\frac{5v}{3}$ **78.** $\frac{w+24}{6}$ **79.** $\frac{5x+3}{4}$
80. $\frac{11}{m}$ **81.** $\frac{3y-5x}{5y}$ **82.** $\frac{8}{15n}$ **83.** $\frac{24-v}{3}$ **84.** $\frac{4v+y}{y}$ **85.** $\frac{16m+9}{6m}$ **86.** $\frac{1}{5}B$ or $\frac{B}{5}$
87. $4B$ **88.** $\frac{6}{5}B$ or $\frac{6B}{5}$ **89.** $\frac{5}{3}B$ or $\frac{5B}{3}$ **90.** $0.5m$ **91.** $5.22m$ **92.** $7.5m$ **93.** $0.7m$
94. $2700 **95.** 1280 **96.** 22.5 lb **97.** 14 oz of ginger ale **98.** 180 calories **99.** $40.70
100. 0.125 gal of red paint **101.** $10\frac{5}{8}$ or 10.625 ft **102.** 18 hr **103.** $390,000 **104.** 3.6
105. $32.50 **106.** $295

Chapter 14 Practice Test (page 733)

1. $7a + 21$ **2.** $18x^2 - 12x$ **3.** $5(y+2)$ **4.** $8y(2y-1)$ **5.** v **6.** $9c^2$ **7.** $-1.1x$
8. $\frac{23}{20}W$ or $\frac{23W}{20}$ **9.** $13w - 2$ **10.** $20m^2$ **11.** $8m - 22$ **12.** $11x + 21$ **13.** $-48x^3$
14. $18x^2 + 48x - 10$ **15.** $y = 5$ **16.** $w = 7$ **17.** $x = \frac{5}{18}$ **18.** $c = 35$ **19.** $x = 10$
20. $B = -4$ **21.** $y = 750$ **22.** $\frac{-2}{3}$ **23.** $\frac{5a}{3}$ **24.** $\frac{3c+2}{c}$ **25.** $\frac{18xc}{7}$ **26.** $\frac{x-8}{28}$ **27.** $\frac{6w+5}{10}$
28. $\frac{21-2v}{3v}$ **29.** $\frac{2x+5}{x}$ **30.** $20,000 **31.** 225 mi **32.** 7.5 mi **33.** 120 gal per day

CHAPTER 15

Exercises 15.1 (page 746)

1. i **3.** n **5.** q **7.** a **9.** g **11.** t **13.** $1\frac{1}{4}$ in **15.** $1\frac{3}{4}$ in **17.** $141°$ **19.** $\frac{5}{8}$ in

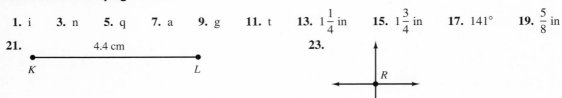

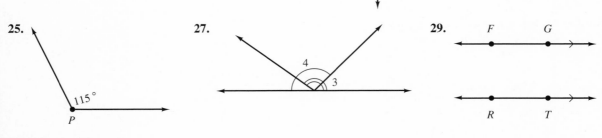

31. 21 m **33.** $50°$

Exercises 15.2 (page 756)

1. 8; $\triangle BCD$, $\triangle BDF$, $\triangle ABF$, $\triangle DEF$, $\triangle ABD$, $\triangle BDE$, $\triangle ACD$, $\triangle BCE$ **3.** yes **5.** yes **7.** $30°$ **9.** $65°$
11. $166°$ **13.** $135°$

15.   **17.** 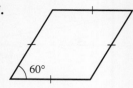   **19.** 20°

Exercises 15.3 (page 768)

1. 10.9 m   **3.** $15\frac{1}{4}$ in   **5.** 76 ft

7.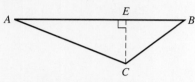

$\overline{CE}$ is the altitude corresponding to base $\overline{AB}$.

9.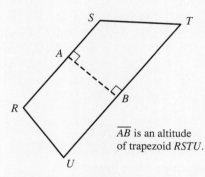

$\overline{AB}$ is an altitude of trapezoid $RSTU$.

11. 20.01 sq m   **13.** 13,080 sq ft   **15.** 756 sq m   **17.** 133.062 sq m   **19.** 1,690,150 sq ft
21. 825 sq m   **23.** 168 sq m   **25.** 169 sq yd   **27.** 226 sq ft   **29.** 2.5 ft by 8 ft   **31.** 50 mi
33. $12   **35.** $1548.40   **37.** 14,688 sq in or 102 sq ft

Exercises 15.4 (page 783)

1.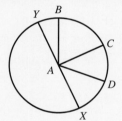

3. $\frac{14}{3}$ or $4\frac{2}{3}$ ft

5. a. $C = 12\pi \approx 37.70$ m, $A = 36\pi \approx 113.10$ sq m   **b.** $C = \frac{14\pi}{3} \approx 14.66$ yd, $A = \frac{49\pi}{9} \approx 17.10$ sq yd

7. $P = 8\pi + 16 \approx 41.13$ ft, $A = 32\pi \approx 100.53$ sq ft

9. $P = 6\pi + 22 \approx 40.85$ m, $A = 18\pi + 60 \approx 116.55$ sq m

11. $P = 18\pi \approx 56.55$ cm, $A = 324 - 81\pi \approx 69.53$ sq cm   **13.** $\frac{200}{\pi} \approx 63.66$ cm

15. 8.8π dollars $\approx \$27.65$ **17.** $300\pi \approx 942.48$ sq ft

Using the Calculator #17 (page 776)

1. 28.274   **2.** 32.858   **3.** 3.770   **4.** 19.689   **5.** 162.860   **6.** 6.676

Chapter 15 Review Exercises (page 785)

1. 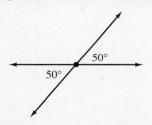   **2.**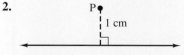

3. 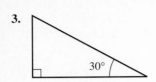   4. 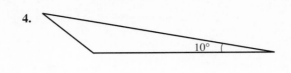   5.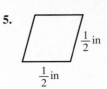

6. 48°   7. 121°   8. 35°   9. 167°   10. 15°   11. 45°   12. 16°
13. 65°   14. yes   15. yes   16. $P = 30$ cm, $A = 40$ sq cm   17. $P = 20.6$ m, $A = 9.75$ sq m
18. $P = 236$ cm, $A = 2576$ sq cm   19. $P = 34$ ft, $A = 74$ sq ft
20. $P = 4\pi + 8 \approx 20.57$ in, $A = 8\pi \approx 25.13$ sq in
21. $P = 7\pi + 26 \approx 47.99$ m, $A = \dfrac{49\pi}{2} + 84 \approx 160.97$ sq m
22. $96 - 9\pi \approx 67.73$ sq cm   23. 51.75 sq m   24. 33 yd   25. 4.5 ft
26. $C = 38\pi \approx 119.38$ cm, $A = 361\pi \approx 1134.11$ sq cm   27. $\sqrt{\dfrac{800}{\pi}} \approx 15.96$ in   28. 960 oz
29. $60\pi \approx 188.50$ ft   30. $25\pi \approx 78.54$ sq m

Chapter 15 Practice Test (page 788)

1. 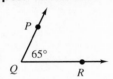   2. Impossible, because $20° + 60° + 120° = 200° > 180°$

3. 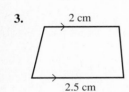   4. 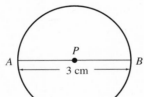   5. $125°$ 6. $147°$ 7. $46°$
8. 30 sq ft   9. 40.8 in   10. 24 m
11. $\dfrac{6}{\pi} \approx 1.91$ ft   12. $32\pi \approx 100.53$ sq m

Cumulative Review Exercises: Chapters 1–15 (page 789)

1. $2 \times 2 \times 2 \times 2 \times 7$ 2. $2 \times 5 \times 5 \times 13$ 3. $3 \times 3 \times 7 \times 23$ 4. 378 5. 1200 6. $45,807$
7. $\dfrac{11}{23}$ 8. $2\dfrac{5}{8}$ 9. $-w$ 10. $2y - 4$ 11. $6 + N$ 12. $\dfrac{5 + w + n}{3}$
13. 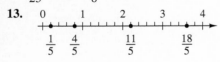   14.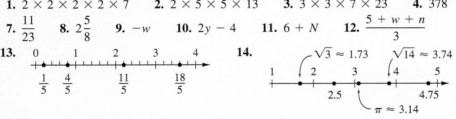

15. 500 16. 160 17. -6 18. 110 19. $225\pi \approx 706.86$ 20. $\sqrt{85} \approx 9.22$ 21. $15 - H$
22. $-38m - 12m^2$ 23. $-72xy^4$ 24. $44n - 2c + 71$ 25. $c = 7$ 26. $n = \dfrac{-9}{10}$ 27. 4040 boxes
28. 600 mg per day   29. a. 48 hr   b. 30 hr
30. a. 1.2×10^7 gal   b. 4×10^2 or 400 gal   c. 4×10^4 or $40,000$ gal   d. 75 min
31. $77°$F   32. $\dfrac{1}{3}, \dfrac{14}{3}$ 33. $\$22$ 34. 100 shares of transportation stock, 250 shares of utility stock
35. 12 yd   36. 6 sq yd   37. $5\pi \approx 15.71$ yd   38. $\dfrac{25\pi}{4} = 6.25\pi \approx 19.63$ sq yd
39. $\dfrac{25\pi}{8} - 6 = 3.125\pi - 6 \approx 3.82$ sq yd

INDEX

Absolute value, 563
Acute angle, 740
Acute triangle, 751
Addition
  of algebraic fractions, 705–710
  associative property of, see Addition, repeated
  on the calculator, 14, 217, 346–347
  with carrying, 9, 241–242
  commutative property of, see Addition, order of
  of decimals, 240–242, 597–598
  of fractions, 174–177, 183–189, 705–710
  of integers, 565–570
  language, 15–16, 202–203, 247, 331, 602
  of measures with more than one unit, 497
  of mixed numbers, 175–176, 185, 187–188
  order of, 15, 202, 247, 574, 648–649
  repeated, 8, 555, 568
  of signed fractions and decimals, 596–599
  of signed numbers, 565–570, 596–599
  of whole numbers, 8–9
Addition statement
  translating from English, 15–16, 202–203, 247, 331, 602
  missing number using box, 11, 199, 245
  missing number using variable, 666–667
Algebraic expressions
  coefficient, 653
  combining like terms, 659–663
  evaluating, 602–604, 699
  factor, 652, 654
  factored form, 654, 657
  factoring, 657–659
  fractions, 697–709
  multiplying, 649–651, 653–657, 703–704
  numerical coefficient, 653
  simplifying, 647–651, 659–663
  taking out factors common to all terms, 657–659
  term, 652, 654
Altitude
  of a parallelogram, 761
  of a trapezoid, 766
  of a triangle, 764
Angle(s)
  acute, 740
  around a point, 742
  central, 519–523
  degree measurement, 520–525, 739–740
  measure with protractor, 522–523, 740
  obtuse, 740
  of a quadrilateral, 753–754
  on one side of a line, 741–742
  right, 740
  of a triangle, 748–752

sides of an, 520, 737
straight, 740
vertex of an, 520, 737
vertical, 743
Application examples
  area of a circle, 780–782
  area of a rectangle, 68, 71–72, 161, 213, 297–298, 339–340, 352, 605
  area of a triangle, 677, 765
  averaging, 70, 211, 300, 353, 529–530, 533–535, 680, 726
  check writing, 232–233
  circumference of a circle, 777–778
  commission, 413–414, 444
  comparisons by addition and subtraction, 21, 209, 252, 683–685
  equal parts in a total quantity, 62–64, 111–112, 150–152, 288–289, 336–337
  equivalent rates, 64–66, 153–154, 290–291, 342–343, 349, 361–363, 442–446, 488–492, 716–718
  fraction of a number, 142, 162–163, 299, 341, 354
  interest earned, 445
  interest paid on an unpaid balance, 417
  more than one step, 71–73, 115–116, 212–213, 252, 348–355, 412–417, 424–430
  percent increase and decrease, 406, 423–427, 446, 726
  percent markup, 428–430
  perimeter, 18, 71, 206, 212, 252, 339, 348, 352, 463, 605, 675, 681, 686, 758
  rate as a multiplier, 67, 157–159, 293–296, 343–344, 349–351, 354, 464, 489–490
  ratio, 113–115, 148, 155, 165–166, 292, 363, 488–489
  restaurant tip, 418, 422
  sales tax, 18, 351, 411–412, 444
  a total as the sum of parts, 18–19, 206–207, 250, 335–336, 680–681, 684–688
  translating, 21, 69, 114–116, 162–166, 209, 212, 252–253, 298–299, 340–341, 404–409, 677–688, 694–696
  with unit conversion, 463–464, 469, 471–475, 477–478, 483–486, 488–492
Approximately equal to ($\approx$), 4, 237
Approximating
  a restaurant tip, 422
  by rounding off, 4–5, 13, 45–46, 200–201, 237–238
  by rounding up, 239
  a square root, 636, 638
  by truncating, 239
Area
  of a circle, 778–782, 797

of composite regions, 71–72, 352, 472, 766, 781–782
converting units of, 470–473
English units of, 470, 795
metric units of, 470, 796
of a parallelogram, 760–761, 797
of a quarter-circle, 781
of a rectangle, 67–68, 160–162, 164–166, 213–214, 297–298, 301, 338–340, 352, 468–469, 797
of a right triangle, 762–763
of a trapezoid, 766–768, 797
of a triangle, 762–765, 797
Associative property
  of addition, see Addition, repeated
  of multiplication, see Multiplication, repeated
Averaging
  mean, 69–70, 210–211, 300–301, 353, 529–530, 533–535, 726
  median, 530–535
  mode, 532–535

Bar, fraction, 86, 121
Bar graph
  making, 511–512, 516–517, 546
  reading, 24, 77, 217, 255, 304, 320, 357, 368, 422, 511–512, 515, 646
Base
  of an exponential expression, 612
  of a parallelogram, 760–761
  in a percent statement, 434–446
  of a trapezoid, 766
  of a triangle, 764
Base salary, 413
Basic calculator, 7
Borrowing in subtraction
  of decimals, 243–244
  of measures, 498
  of mixed numbers, 196–199
  of whole numbers, 10–11
Box, rectangular solid, 473–475
  volume of a, 474
Braces, 548
Brackets, 548

Canceling common factors
  in reducing fractions, 103–108, 322, 699–702
  in multiplying fractions, 128–133, 703–705
Carrying
  in addition, 9, 241–242
  in multiplication, 30–31
Celsius degrees (°C), 560, 606–607
Central angle, 519–523
Circle, 772–778
  area of a, 778–782
  center of a, 773
  central angle, 519–523
  circumference of a, 774–775, 777–778, 797
  compass, 774
  diameter of a, 773

quarter-, 777
radius of a, 773
sector of a, 518–521
semi-, 778
Circle graph,
  making, 519–525
  reading, 121, 169, 216, 386, 410, 518–519
Circumference of a circle, 774–775, 777–778, 797
Coefficient, 653
Combining like terms, 659–663
Comparing
  decimals, 234–236, 324–326
  fractions, 190–193, 326–328
  a fraction to $\frac{1}{2}$, 194
  integers, 562–563
  numbers by addition and subtraction, 20–21, 208–209, 251–253
  numbers by division, 113–117, 148, 292, 488
  percents to the number one, 372
  signed numbers, 562–563
  square roots to decimals, 636–638
  symbols, 191, 236, 324, 326, 562
  whole numbers, 3
Commission, 413–414, 419, 794
Commutative property
  of addition, see Addition, order of
  of multiplication, see Multiplication, order of
Converting number forms
  decimal to fraction, 226–228, 323, 393
  decimal to percent, 387–388, 393
  decimal to scientific notation, 629–630
  extra zero digits, 234, 263, 323
  fraction to decimal, 226–228, 272–273, 322, 393
  fraction to percent, 388–393
  improper fraction to mixed numeral, 99–100, 322
  mixed numeral to improper fraction, 100–101, 322
  percent to decimal, 372–374, 393
  percent to fraction, 375–378, 393
  raise to higher terms, 93–95, 322
  reduce to lowest terms, 102–108, 322
  scientific notation to decimal, 628–629
  simplify a fraction, 102–108, 322
  three ways to write a number, 393
  unnecessary zero digits, 234, 263, 323
  whole numerals to fractions, 96–97
Convex quadrilateral, 753
Cost
  to a manufacturer, 427–431
  markup based on, 427–429, 431
  to a retailer, 428–431
Cross product, 358, 435
Cube of a number, 612
Cubic units, 474, 475, 795–796

Index

Decimals
  addition, 240–242
  convert to fraction, 226–228, 323, 393
  convert to percent, 387–388, 393
  comparing, 234–236, 324–326
  dividing by 10, 100, 1000, and so on, 277
  division, 268–277
  English name of, 228
  multiplication, 260–267
  multiplying by 10, 100, 1000, and so on, 265
  place value, 224–225
  repeating, 271, 273, 276, 325–327, 374, 376
  subtraction, 242–244
  and whole numerals, 225, 263, 275, 323
Decimal fraction, 226
Decreased by, 15, 203, 247, 331
Degree
  Celsius (°C), 560, 606–607
  Fahrenheit (°F), 560, 606–607
  measure of an angle, 520–525, 739–740
Denominate numbers, *see* Measures, with more than one unit
Denominator
  of a fraction, 86
  least common, 178–182
Diameter of a circle, 773
Difference, 15, 21, 203, 208, 209, 213, 247, 251, 331, 348, 577–578
Digit, 1, 224
Discount, 20, 251, 426
Distance
  as length, 455–464
  between two numbers on the number line, 310–319, 561–563, 576–578, 593–595
  shortest between two points, 749
  shortest from point to line, 744–745
Distributive law of multiplication over addition, 653–659
Divided by, 59, 113, 148, 286, 333, 589, 602
Divided into, 59, 113, 148, 286, 333, 589, 602
Dividend, 37, 268
Division
  of algebraic fractions, 703–705
  on the calculator, 46, 334
  of decimals, 268–277, 305
  of exponential expressions, 623–624
  of fractions, 135–138
  of integers, 587–588
  long, 38–44, 269–277
  of measures with more than one unit, 500–502
  of mixed numbers, 135–138
  with remainder, 37–44
  with remainder in fractional form, 270–271
  repeated, 552–553, 586
  reversing the order of, 59, 113, 147, 286, 333, 589–590
  using scientific notation, 631–632
  of signed decimals, 598

of signed fractions, 597
of signed numbers, 587–588, 597–598
by 10, 100, 1000, and so on, 277
of whole numbers, 36–44, 99–101, 110–117
involving zero, 587–588
involving zeros in the quotient, 39, 40, 42–44, 269–270, 275–277
Divisibility rules, 53, 106
Divisible by, 48–53
Divisor, 37, 268
Double, 59, 146, 285

Endpoint(s)
  of a line segment, 736
  of a ray, 737
English measurement system, 456
  units of area, 470, 795
  units of length, 456, 795
  units of volume, 475, 795
  units of weight, 481, 795
English name for
  fractions, 91–92
  decimal numerals, 228
  mixed numerals, 92
  whole numerals, 3
Equations
  equivalent, 665
  solution, 665
  solving, 665–674, 691–694, 712–727
Equilateral triangle, 752
Equivalent equations, 665
Equality
  of decimals, 234, 263
  of fractions, 93–108
  of money values, 229–231
  of numbers, 93–108, 234, 263, 321–324, 393
Estimating, *see* Approximating
Evaluating
  algebraic expressions, 602–604, 699
  exponential expressions, 612–620
  formulas, 604–606
  numerical expressions, 547–557, 564, 579–580, 586, 596–599
  square-root expressions, 638–639
Even number, 50
Exponentiation
  as an operation, 613
  mixed with other operations, 613–615, 619–620
Exponent(s)
  laws of, 621–625
  negative integer, 617–620, 642
  positive integer, 614–615, 642
  as a power, 612, 627
  zero, 617–619, 642
Expanded form
  of a decimal numeral, 224
  of a whole numeral, 1–2
Expressions
  algebraic, 602, 647
  numerical, 548
  simplifying, 647–651, 659–663
  translating from English, 602

Factor
  as a multiplier, 29, 47, 139, 280

Factoring
  prime, 53–55
  by taking out factors common to all terms, 657–659
  by using the distributive law, 657–659
Fahrenheit degrees (°F), 560, 606–607
Finance charge, 416
Fluid ounce (fl oz), 475, 795
Formulas
  evaluating, 604–606
  geometric, 797
  percent, 419, 794
  with variables, 604, 675–677
Fraction(s)
  adding with like denominators, 174–177
  adding with unlike denominators, 183–189
  algebraic, 697–710
  convert to decimal, 226–228, 272–273, 322, 393
  convert to percent, 388–393
  comparing, 190–194, 326–328
  denominator, 86
  dividing, 135–138, 597, 704–705
  as division, 99, 110–113
  English name of, 90–92
  improper, 88, 99–101, 322
  mixed numerals, 88, 99–101, 322
  multiplying, 127–133
  numerator, 86
  picturing using shaded regions, 85–89, 93–96, 99
  raising to higher terms, 93–97
  as a ratio or rate, 113–115, 155–159, 292–296, 342–344
  reading and writing, 90–92
  reciprocal, 136
  reducing to lowest terms, 102–108, 322, 699–702
  simplifying, 102–108, 322, 699–702
  solving equations involving, 712–727
  subtracting, 195–199
Fraction bar
  meaning, 86, 121
  as a grouping symbol, 549–550

Geometric figure(s), 735
  angle, 737
  line, 735
  line segment, 735
  parallelogram, 754
  point, 735
  quadrilateral, 752
  ray, 737
  rectangle, 754
  rhombus, 755
  square, 755
  trapezoid, 755
  triangle, 748
Geometric formulas, 797
Gram (g), 481–482, 796
Graphs, 511–526
  bar, 511–512
  circle, 518–519, 523–525
  line, 512–513
  pictographs, 517

Grouping symbols
  braces, 548
  brackets, 548
  fraction bar, 549–550
  parentheses, 548
  square-root symbol, 638–639
  for holding a number, 547
  for defining the order to operate, 548–549

Half of, 146–147, 285
Height, *see also* altitude
  of a box, 473
  of a parallelogram, 760
  of a rectangular solid, 473
Horizontal axis, 511
Horizontal bars in a bar graph, 512
Horizontal display of metric units, 466–467
Horizontal line, 735
Hypotenuse of a right triangle, 752

Improper fractions
  and mixed numerals, 98–101
  and whole numerals, 89, 96–97, 99
  meaning, 88
Inequality symbols, *see* Comparing symbols
Integer(s)
  absolute value of, 563
  adding, 565–570
  comparing, 562
  dividing, 587–591
  exponent, 612–620
  multiplying, 582–585
  on the number line, 561–562
  opposite of, 564
  subtracting, 571–580
Intersect, 736
Interest
  earned on an investment, 414–415, 419, 794
  rate, 414, 419, 794
  rate of return, 414
  simple, 414–415
  compound, 415
  monthly rate, 416, 419
  paid on an unpaid balance, 416–417, 419, 794
Invert a fraction, 136
Isosceles triangle, 752

Language charts
  addition, 15, 202, 247, 331
  algebraic expressions, 602
  comparing by adding or subtracting, 21, 208, 251
  decrease or increase, 20, 208, 241
  division, 59, 148, 286, 333
  equality, 15, 58, 144, 202, 246, 284, 330
  exponential expressions, 612
  multiplication, 59, 144, 146, 284, 285, 332
  rates or ratios, 64, 117, 153, 155, 157, 290, 292
  reading money currency, 230, 231
  reading numbers, 3, 91–92, 113, 228
  subtraction, 15, 203, 247, 331
  unclear percent statements, 407

Index

Least common denominator (LCD), 178
Least common multiple (LCM)
  by listing multiples, 178
  meaning, 178
  by prime factoring, 179–182
Length
  of a box, 473
  English units of, 456, 795
  measuring, 455–457
  metric units of, 457, 796
  of a rectangle, 68, 160, 338–339, 468
  of a rectangular solid, 473
Like terms, 660
Line(s)
  horizontal, 735
  intersecting, 736
  number, 310–319, 561–562, 593–595
  oblique, 735
  parallel, 743–744
  perpendicular, 744
  segment, 735
  slanted, 735
  vertical, 735
Line graph, 512–513
Liter (ℓ), 475–476, 796
Long division
  of decimals, 269–277
  involving zeros in the quotient, 39, 40, 42–44, 269–270, 275–277
  of whole numbers, 38–44

Markup
  based on cost, 427–429, 431, 794
  based on the selling price, 429–431, 794
Markup rate, 428–429
Mass
  converting units of, 481–485
  metric units of, 482, 796
Mean, *see* Averaging
Measurement
  angle, 519–520, 739–741
  area, 67–68, 160–162, 164–166, 213–214, 297–298, 301, 338–340, 352, 468–469, 760–782
  capacity, 473–479
  English units of, 795
  length, 455–464
  mass, 481–485
  metric units of, 796
  perimeter, 18, 71, 206, 212, 252, 339, 348, 352, 463, 675, 758
  using a protractor, 522–523, 740
  rate, 64–66, 153–160, 289–296, 342–344, 349–350, 488–492
  using a ruler, 456–457
  temperature, 560, 606–607
  time, 485–487
  volume, 473–479
  weight, 481–485
Measures with more than one unit, 493–502
Measures of central tendency, *see* averaging
Median, 530–535
Meter, 456–457, 796
Metric prefixes, 458, 795

Metric system of measurement, 456
  units of area, 470, 796
  units of length, 457, 796
  units of mass, 482
  units of volume, 475, 796
Mixed numeral(s)
  converting to and from improper fractions, 98–101
  meaning, 88
  subtraction involving borrowing, 196–199
Mode, 532–535
Monthly interest rate, 416–417, 419
Multiples of a number, 48
Multiplication
  of algebraic expressions, 649–651, 653–657, 703–704
  of algebraic fractions, 703–704
  as repeated addition, 29, 127, 582
  associative property of, *see* Multiplication, repeated
  on the calculator, 46, 334
  commutative property of, *see* Multiplication, order of
  of decimals, 260–267
  by using the distributive law, 653–657
  of exponential expressions, 621–623
  of fractions, 127–133
  of integers, 582–585
  with measures involving more than one unit, 499, 502
  of mixed numbers, 128–129, 131–133
  of more than two quantities, 34, 132–133, 266–267, 553, 584–585, 647, 649–651
  order of, 59, 144, 284, 647, 649–651
  using scientific notation, 630–632
  of signed decimals, 598–599
  of signed fractions, 597
  of signed numbers, 582–586, 597–599
  by 10, 100, 1000, and so on, 265
  by 0.1, 0.01, and 0.001, 279
  repeated, 34, 132–133, 266–267, 553, 584–585, 647, 649–651
  short cuts involving zeros, 32–34, 265–266
  involving unnecessary zeros, 262–263
  of whole numbers, 29–35
Multiplication facts table, 30, 791
Multiplication statement
  missing number using box, 47–48, 139–141, 280–282, 395
  missing number using variable, 667–669
  translating from English, 59–60, 142–147, 283–285, 332, 395–403
Multiplier, 29, 47, 139, 280

Negative integer, 561
Negative integer exponents, 617–620, 642
Negative number, 561
Number line, 310–319, 561–562, 593–595

Numeral
  decimal, 224–225
  mixed, 88
  whole, 1–2
Numerator of a fraction, 86
Numerical coefficient, 653
Numerical expression, 548
  evaluating, 547–557, 564, 579–580, 586, 596–599

Oblique line, 735
Obtuse angle, 740
Obtuse triangle, 751
One
  as an exponent, 612, 617, 622
  as a numerical coefficient, 661
  as a ratio for unit conversion, 461, 471, 479, 491
Operation(s)
  absolute value of (| |), 563
  addition, (+), 8, 174, 241, 565
  division (÷), 36, 135, 268, 587
  exponentiation, 612–613
  multiplication (×), 29, 127, 260–261, 582–584
  opposite of (−), 564
  order of, *see* Order of operation
  square root ($\sqrt{\ }$), 635
  subtraction (−), 9, 195, 242, 571–572
Opposite of a number, 564
Order of operation
  and the calculator, 78, 558–559
  for positive integers, 548–557
  for signed fractions and decimals, 599
  for integers, 586

Parallel
  lines, 743–744
  markings, 744
  symbol (//), 743
Parallelogram, 754
  altitude of a, 760–761
  area of a, 761
  base of a, 760–761
  height of a, 760
  symbol (□), 754
Parentheses, 548
Perimeter, 18, 206, 253, 338–339, 758, 777–778, 797
Perpendicular
  lines, 744
  markings, 744
  symbol (⊥), 743
Percent(s)
  as a commission rate, 413–414
  comparing to the number one, 372
  conversions fraction-decimal-percent, 371, 393, 792
  convert decimal to, 387–388, 393
  convert to decimal, 372–374, 393
  convert fraction to, 388–393
  convert to fraction, 375–378, 393
  decrease, 426, 431, 726, 794
  increase, 423–425, 794
  as an interest rate, 414–417, 419, 794
  markup, 427–431, 794
  meaning, 370–371
  of a number, 379–385

  as a rate, 434–446, 794
  as a sales tax rate, 411–412, 419, 794
  symbol (%), 370–371
  as a tip rate, 418–419, 794
Percent statement
  amount in a, 434–446, 794
  base in a, 434–446, 794
  rate in a, 434–446, 794
  translate to a multiplication statement, 395–402
  translate to a proportion, 434–446
  containing unclear language, 407–409
Perfect squares, 636, 792
Pi (π), 774
Pictograph, 517
Pie chart, *see* Circle graph
Place value
  in a decimal, 224–225
  in a whole numeral, 1–3
Plane, 734
Point, 735
Positive
  integers, 561
  numbers, 561
  integer exponent, 612
Power(s)
  as an exponent, 612
  of ten, 627–628
Prime(s)
  factoring, 53–55, 179
  first sixteen, 105
  first ten, 53, 179
  less than a hundred, 792
  number, 53, 105, 179
Product(s)
  as the answer in a multiplication statement, 29, 47, 139, 280
  cross, 358, 435
Proportions
  applications of, 361–363, 441–447, 716–718
  missing number using box, 359–363, 436–447
  missing number using variable, 713–718
  use to solve a percent problem, 434–446
  solve by operating on both sides, 713–718
  solve by setting cross products equal, 359–363, 436–447
Protractor, 522–523, 740
Pure number, 499

Quotient, 37, *see also* Long division
Quadrilateral(s)
  angles of a, 753
  classifying, 754–755
  convex, 753
  nonconvex, 753
  sides of a, 753
  sum of the angles, 753–754
  vertices of a, 753

Radical ($\sqrt{\ }$), 635
Radius of a circle, 773
Raising
  a fraction to higher terms, 93–97
  a number to a power, 612

Rate(s)
  changing units of, 490–492
  commission, 413–414
  equivalent, 64–66, 153–156, 289–292, 342–344, 361–363, 716–718
  interest, 414–417, 419, 794
  markup, 427–431, 794
  as a multiplier, 66–67, 157–159, 293–296, 343, 350–351
  phrases that indicate, 64, 153, 155, 157, 290, 292, 342
  as a ratio, 155–156, 292–293, 342, 349, 482
  of return, see Rate, interest
  sales tax, 411–412, 419, 794
  testing equality of, 358–359
  tip, 418–419, 794
Ratio(s)
  as a comparison by division, 113–114, 292, 342, 349, 488
  language of, 114, 117, 148–149, 166–167, 286, 333
  as a percent, 371, 435
  as a rate, 155, 292, 358–359
  testing equality of, 357–358
Ray, 737
  endpoint of a, 737
Reading
  fractions, 91–92
  decimal numerals, 228
  mixed numerals, 92
  whole numerals, 3
Reciprocal of a number, 136
Rectangle
  area of a, 67–68, 160–162, 164–166, 213–214, 297–298, 301, 338–340, 352, 468–469, 797
  perimeter of a, 253, 338–339
  as a special parallelogram, 754
Rectangular solid, 473
  volume of a, 473–474
Remainder
  in a division problem, 37
  in fractional form, 270–271
Repeating decimal, 271, 273, 276, 325–327, 374, 376
Rhombus, 755
Right angle, 740
Right triangle, 752
  area of a, 762
  hypotenuse of a, 752
  legs of a, 752
Rounding
  to approximate, 13, 45, 200
  down, see Rounding by truncating
  off, 4–5, 237–238, 240
  by truncating, 239–240
  up, 239–240

Sales tax, 411–412, 419, 794
Sales tax rate, 411–412, 419, 794
Selling price
  for the manufacturer, 428, 431
  markup based on, 429–431
  for the retailer, 428–431
Service charge, 416
Scientific notation, 629
Signed number(s), 560
  absolute value of a, 563
  adding, 565–570
  comparing, 562
  distance between 576–577
  dividing, 587–588
  magnitude of a, 563
  multiplying, 582–585
  on the number line, 561
  opposite of a, 564
  order of operation, 586
  situations requiring, 560
  subtracting, 571–579
Sides
  of an angle, 520, 737
  of an equation, 665
  of a quadrilateral, 753
  of a triangle, 748
Simplifying
  fractions, 102–108, 322, 699–702
  algebraic expressions, 647–651, 659–663
  algebraic fractions, 699–702
  by combining like terms, 659–663
Slanted line, 735
Solution of an equation, 665
Solving equations, 665
  applications of, 682–688, 694–696, 716–718, 724–726
  with fractions or decimals, 712–724
  with the variable on both sides, 691–694
  with the variable on one side, 666–674
Solving proportions, see Proportions
Speed, 64
Square(s)
  as a special rectangle, 755
  of a number, 612, 635–636
  perfect, 636, 792
Square root ($\sqrt{}$)
  approximating, 637–638
  meaning of, 635
  mixing with other operations, 638–639
  on the number line, 637
Square units, 468–470, 795–796
Square yards (sq yd), 469, 470, 795
Statistics, 509
  table, 509–510
  bar graph, 511 see also Bar graph
  line graph, 512–513
  circle graph, 518, 523–524, see also Circle graph
  measures of central tendency, 529, see also Averaging
  pictograph, 517
Straight angle, 740
Subtraction
  with borrowing, 10–11, 196–199, 243–244, 498
  on the calculator, 14, 217, 346–347
  of decimals, 242–244
  of algebraic fractions, 706, 708
  of fractions, 195–199, 596, 706, 708
  language, 15–16, 202–204, 247–248, 331, 602
  of measures with more than one unit, 498

reversing the order of, 9, 15, 202, 247, 574–575
  of signed fractions and decimals, 596–597
  of signed numbers, 571–579, 596–597
  of whole numbers, 9–11
Subtraction statement
  translating from English, 15–16, 202–204, 247–248, 331
  missing number using box, 16, 204, 248
  missing number using variable, 666
Sum, 8, 15, 202, 247, 331, 602
Symbols
  absolute value (| |), 563
  addition (+), 1
  approximately equal to (≈), 4, 237
  division ($\overline{)}$, ÷), 36
  is equal to (=), 1
  fraction bar (—), 86
  fraction bar as division (—), 99, 110–111
  is less than (<), 191, 236
  magnitude (| |), 563
  is more than (>), 191, 236
  multiplication (×), 29
  multiplication dot (·), 547
  is not equal to (≠), 665
  parallelogram, (□), 754
  is parallel to (//), 743
  is perpendicular to (⊥), 744
  pi (π), 774
  opposite of (−), 564
  percent (%), 370
  square root ($\sqrt{}$), 635
  squiggle marking (≲), 511
  subtraction (−), 9
  triangle (△), 748
Table
  of multiplication facts, 30, 791
  for organizing data, 509
Temperature
  degrees Celsius (°C), 560, 606–607
  degrees Fahrenheit (°F), 560, 606–607
Terms
  in an addition statement, 8, 652
  combining like, 659–663
  like, 660
Time units, 485, 794
Translating, see Language charts and Application examples, translating
Trapezoid
  altitude of a, 766
  area of a, 766–767, 797
  bases of a, 766
  as a special quadrilateral, 755
Triangle, 748
  acute, 751
  angles of a, 748
  area, 677, 762–765, 797
  equilateral, 752
  isosceles, 752
  obtuse, 751
  perimeter, 18, 206, 758

right, 752
  sides of a, 748
  sum of the angles, 750
  symbol (△), 748
  vertices of a, 748
Triangle inequality, 749
Truncating, 239–240
Twice, 146, 285

Vertical angles, 743
Vertical axis, 511
Vertical bars in a bar graph, 511–512
Vertical line, 735
Vertex of an angle, 520, 737
Vertices
  of a quadrilateral, 753
  of a triangle, 748
Volume
  of a box, 473–474
  as capacity, 473–479
  converting units of, 476–478
  of a rectangular solid, 473–474
  units of, 475, 795, 796
Variable, 602
Variable expression, see Algebraic expression

Weight
  converting units of, 481–485
  English units of, 481, 795
  metric units of, 482, 796
Whole numbers, 3
  addition, 8–9
  comparing, 3
  divisibility rules for, 53
  division, 36–44, 99–101, 110–117
  expanded form, 1–2
  multiplication, 29–35
  on the number line, 310–312
  prime factoring of, 53–55
  rounding off, 4–5
  subtraction, 9–11
Whole numerals, 1
  convert to fraction, 96–97
  and decimals, 225, 263, 275, 323
  English name of, 3
  and improper fractions, 89, 96–97, 99
  place value, 2–3

Zero(s)
  absolute value of, 563
  added to a signed number, 566–567
  attaching extra when dividing decimals, 269–277
  digits in the quotient, 39, 40, 42–44, 269–270, 275–277
  division involving, 587–588
  as an exponent, 617–619
  multiplication shortcuts involving, 32–34
  as a necessary digit, 263
  opposite of, 564
  as the sum of opposites, 569
  as an unnecessary or extra digit, 263, 323